Moduli Spaces and Vector Bundles—New Trends

CONTEMPORARY MATHEMATICS

803

Moduli Spaces and Vector Bundles—New Trends

VBAC 2022 Conference
Moduli Spaces and Vector Bundles—New Trends
In honor of Peter Newstead's 80th Birthday
July 25–29, 2022
University of Warwick, Coventry, United Kingdom

Peter Gothen
Margarida Melo
Montserrat Teixidor i Bigas
Editors

AMS
AMERICAN MATHEMATICAL SOCIETY

2020 *Mathematics Subject Classification.* Primary 14H60, 14L24, 14D20, 14J60, 14L30, 53D30, 58D29, 14F45.

Library of Congress Cataloging-in-Publication Data

Names: Gothen, P. B. (Peter Beier), 1967– editor. | Melo, Margarida (Mathematician) editor. | Teixidor i Bigas, Montserrat, 1958– editor. | Newstead, P. E., honoree. | Moduli Spaces and Vector Bundles—New Trends (Conference) (2022 : University of Warwick)

Title: Moduli spaces and vector bundles—new trends / [edited by] Peter Gothen, Margarida Melo, Montserrat Teixidor i Bigas.

Description: Providence, Rhode Island : American Mathematical Society, 2024. | Series: Contemporary mathematics, 0271-4132 ; volume 803 | "VBAC 2022 Conference in honor of Peter Newstead's 80th Birthday, July 25–29 2022, University of Warwick, Coventry, United Kingdom." | Includes bibliographical references.

Identifiers: LCCN 2023053398 | ISBN 9781470472962 (paperback) | ISBN 9781470476465 (ebook)

Subjects: LCSH: Moduli theory. | Vector bundles. | Newstead, P. E. | AMS: Algebraic geometry – Curves – Vector bundles on curves and their moduli. | Algebraic geometry – Algebraic groups – Geometric invariant theory. | Algebraic geometry – Families, fibrations – Algebraic moduli problems, moduli of vector bundles. | Algebraic geometry – Surfaces and higher-dimensional varieties – Vector bundles on surfaces and higher-dimensional varieties, and their moduli. | Algebraic geometry – Algebraic groups – Group actions on varieties or schemes (quotients). | Differential geometry – Symplectic geometry, contact geometry – Symplectic structures of moduli spaces. | Global analysis, analysis on manifolds – Spaces and manifolds of mappings – Moduli problems for topological structures. | Algebraic geometry – (Co)homology theory – Topological properties.

Classification: LCC QA564 .M6435 2024 | DDC 516.3/5–dc23/eng/20240409
LC record available at https://lccn.loc.gov/2023053398

Contemporary Mathematics ISSN: 0271-4132 (print); ISSN: 1098-3627 (online)

DOI: https://doi.org/10.1090/conm/803

---

10 9 8 7 6 5 4 3 2 1   29 28 27 26 25 24

Contents

Preface

This volume is an outgrowth of the conference "Moduli Spaces and Vector Bundles—New Trends" that took place at the University of Warwick (U.K.), July 25–29, 2022. The conference used an hybrid format with over 50 people physically present and a larger number attending on-line. The theme of the conference was moduli theory in a broad sense. The main topics covered were the general theory of stability conditions in derived categories, non-reductive geometric invariant theory, Brill-Noether theory, and Higgs bundles and character varieties. There were twenty talks given by both established researchers and young mathematicians. The program was completed with a poster session both in person and on-line and a question and answer session with participation of S. Ramanan, Miles Reid, Peter Gothen and Peter Newstead. Titles, abstracts and in many cases videos from lectures are available at `https://vbac.wikidot.com/vbac2022`. The conference was organized in partnership with the Clay Mathematics Institute by Leticia Brambila-Paz, Gavin Brown, Antony Maciocia and the three editors of the present volume. We would like to take this opportunity to thank the sponsors of the conference: the Edinburgh Mathematical Society, the Foundation Compositio Mathematica, the London Mathematical Society, the National Science Foundation (USA), and the Warwick Mathematics Institute.

While the contents of this volume do not completely mirror conference presentations, its spirit and general themes are similar, and most authors were speakers at the conference. The volume includes both survey and original research articles. Most articles contain substantial background that we hope will help the novice get acquainted with the topics. The volume should therefore be a useful resource for both beginning and experienced researchers.

The conference, initially planned for 2021 but postponed due to the Covid epidemic, was dedicated to Peter Newstead on the occasion of his eightieth birthday. Peter Newstead has been a leading figure in algebraic geometry, especially moduli theory, since the 1960's. His book *Introduction to moduli problems and orbit spaces* provided access to a very abstract but crucial technique in the construction of moduli spaces. His early work on the homology of moduli spaces of vector bundles on curves has been generalized in many directions. His interest in Brill–Noether Theory and Coherent Systems for vector bundles on curves has grown into a field of its own.

Aside from his scientific contributions, Peter Newstead has been an extraordinary mentor, reliable organizer of conferences and indefatigable traveller and participant in all kinds of events. He has mentored large numbers of young researchers sometimes officially but most often unofficially. Peter Newstead is a generous mathematician willing to share his ideas and expertise with those around him. His efforts and leadership over the years have been instrumental for the careers of many young algebraic geometers.

Peter Newstead started the VBAC (Vector Bundles on Algebraic Curves) research group in 1994. Since then this group has been responsible for organizing annual workshops which have been highly influential in setting directions for future research and integrating young researchers in the community. During the covid lockdown, the conferences morphed into on-line bimonthly webinars and post covid both conferences and webinars are continuing.

Finally, no reference to the mathematical work of Peter Newstead would be complete without mentioning his wife Ann Newstead who supported not just Peter but also Peter's work and social connections talking to conference participants and collaborators, providing practical support and encouragement.

The editors of this volume are happy to dedicate it to Peter and Ann Newstead as a token both of their personal gratitude for the way they touched their lives and for Peter's general contribution to the progress of Mathematics.

Peter Gothen
Margarida Melo
Montserrat Teixidor i Bigas

Contemporary Mathematics
Volume **803**, 2024
https://doi.org/10.1090/conm/803/16092

Poincaré and Picard bundles on the moduli spaces of bundles on curves

Usha N. Bhosle

This paper is dedicated to Peter Newstead on his 80th birthday.

ABSTRACT. Poincare bundles and Picard bundles on the Jacobian of a smooth curve have been studied extensively classically. Peter Newstead and his collaborators have made major contributions to the theory of Poincaré and Picard bundles over the moduli spaces of vector bundles of higher ranks. We present old and new results on the Poincare bundles and Picard bundles on the moduli spaces of vector bundles on smooth curves and their generalisations to nodal curves. Finally we consider parabolic vector bundles. We prove the parabolic stability of Poincaré bundles over the moduli spaces of parabolic vector bundles, with parabolic structure of any type, on a smooth curve.

1. Introduction

The classical Picard bundles i.e. Picard bundles on the Jacobian of a smooth curve Y were first studied in Mattuck's paper (in the projective form) in 1961 [**M1**] in a very general setting. Picard sheaves on a Jacobian were studied by Mukai [**Mu**]. The (classical) Picard bundles were studied extensively by many authors including Mattuck, Gunning, Mukai, Kempf, Ein, Lazarasfeld (see [**M1**], [**M2**], [**G1**], [**G2**], [**EL**], [**FL**], [**EGK**], [**K1**], [**K2**]). Section 2 presents important results on the stability and ampleness of Picard bundles, Chern classes and cohomology of Picard bundles, and applications of Picard bundles. Deformations of Picard bundles were studied by Kempf [**K2**] and Mukai [**Mu**]. Some of these results were extended to nodal curves by myself and A. J. Parameswaran ([**BhP1**], [**BhP2**], [**Bh2**]). The Picard bundles on the Jacobian of a smooth curve are stable and ample. Stability is preserved for Picard bundles on the compactified Jacobian of a nodal curve, but the ampleness is lost.

In section 3, we move to the Poincaré and Picard bundles over the moduli spaces of vector bundles of higher ranks. For $n \geq 1$ and d integers, let $U(n,d)$ (respectively $U(n,L)$) denote the moduli space of semistable vector bundles of rank n and degree d (respectively with a fixed determinant isomorphic to a line bundle L of degree d) on a smooth curve Y. Let $U^s(n,d) \subset U(n,d)$ be the open subset corresponding stable bundles. If n and d are coprime, $U(n,d) = U^s(n,d)$ and there is a Poincaré (universal) bundle $\mathcal{U}$ on $U(n,d) \times Y$. For $d > n(2g-2)$, the direct

2020 *Mathematics Subject Classification.* Primary 14H60; Secondary 14D20.

This work was done during my tenure as an INSA Senior Scientist at the Indian Statistical Institute, Bangalore.

image of $\mathcal{U}$ on $U(n,d)$ is a vector bundle called a Picard bundle, denoted by $\mathbf{E}_{n,d}$. Let $\mathbf{E}_{n,L}$ be its restriction to $U(n,L)$. In general there is no universal family $\mathcal{U}$, but one can still get Projective Poincaré bundles on $U \times Y$, U an open subset of $U^s(n,d)$, and Projective Picard bundles on U.

In the coprime case, for $d > 2ng$, Y. Li proved the stability of $\mathbf{E}_{n,d}$ and results on its restriction to the Jacobian embedded in $U(n,d)$ [**Li**]. Peter Newstead and his collaborators (Vikraman Balaji, Indranil Biswas, Leticia Brambila-Paz, Tomas Gómez, $\cdots$) made major contributions to the study of Poincaré and Picard bundles, and Projective Poincaré and Picard bundles on higher rank moduli spaces (For example, [**BBrN**], [**BiBrGN**], [**BiBrN2**], [**BiBrN3**], [**BiR**]). They have studied the stability and deformations of these bundles. Some of their results were generalised to nodal curves ([**Bh4**], [**Bh5**], [**ABS**]). Projective Picard bundles on the moduli spaces of symplectic and orthogonal bundles were studied by Biswas I. and Gomez T. ([**BiG1**], [**BiG2**]). Picard bundles on Prym varieties were considered by L. Brambila-Paz, Gomez-González E. and Pioli F. [**BrGP**].

For a fixed $x \in Y$, the restriction $\mathcal{U}_x := \mathcal{U}|_{U(n,L)\times x}$ is (stable and) nef. It is semiample but not ample [**LM**, Theorem A]. If $g \geq 3$ or $g = n = 2$, then $\mathcal{U}_{L,x}$ is an ACM bundle for any ample line bundle on $U_Y(n,L)$, in general it is not Ulrich.

Fixing a vector bundle E_0 on Y and taking the direct image of $\mathcal{U} \otimes p_Y^* E_0$ on $U(n,d)$ one gets a torsionfree sheaf called a generalised Picard sheaf denoted by $W_{n,d}(E_0)$. Recently several results on the stability of these sheaves (dependent on the rank, degree and stability of E_0) and their deformations (in case they are locally free) were obtained by Indranil Biswas, Leticia Brambila-Paz and Peter Newstead (see [**BiBrN3**] for details). Many older results can be recovered from these recent results.

In the last section (Section 4), we shift our focus to parabolic vector bundles. Let $U_{par}(n,\xi)$ denote the moduli space of parabolic vector bundles (on a smooth curve) Y with parabolic structure over a divisor D. For positive integers k,ℓ and non negative integers $b_x, x \in D$ satisfying certain conditions, there is an ample line bundle Θ_{par} on $U_{par}(n,\xi)$. We assume that the greatest common divisor of n,d and the multiplicities in the parabolic structure $n_i(x), x \in D$, is 1. Then there is a universal family $\mathcal{U}_*$, of parabolic vector bundles, on $U_{par}(n,\xi) \times Y$.

THEOREM 1.1. *Let z be a closed point of Y distinct from the points in D. Let $\mathcal{U}_z$ be the restriction of the universal bundle to $U_{par}(n,\xi) \times z$. Then $\mathcal{U}_z$ is semistable with respect to the ample line bundle Θ_{par}.*

THEOREM 1.2. *The parabolic bundle $\mathcal{U}$ on $U_{par}(n,\xi) \times Y$ is parabolic stable with respect to any integral ample divisor of the form $a\Theta_{par} + bD_Y$ where $a, b > 0$ and D_Y is an ample divisor on Y.*

In case the parabolic structure is a Borel structure i.e., all the multiplicities are equal to 1, the theorems 1.1 and 1.2 were proved by Basu Suratno, Biswas Indranil and Dan Krishanu [**BaBiD**], we prove the general case. The proof uses (parabolic) spectral curves and the parabolic BNR correspondence.

2. The case of rank 1

2.1. Notation.

Let Y denote an irreducible projective curve with at most nodes as singularities and defined over an algebraically closed field K of characteristic 0. We recall that a point $y \in Y$ is a node if the completion of the local ring of Y at y is isomorphic to $K[[u,v]]/ < uv >$. Let $y_1, \cdots, y_k$ be the nodes of Y and Y_0 be the set of smooth points of Y. Let

$$p \colon X \to Y$$

be the normalisation map, $X_0 = p^{-1}Y_0$. Let $p^{-1}(y_j) := \{x_j, z_j\} \in X$ be the inverse image of the nodal point y_j in Y.

Let $g = g(Y) := h^1(Y, \mathcal{O}_Y)$ be the arithmetic genus of Y. Since Y is a Gorenstein curve, it has a locally free dualising sheaf ω_Y. Fix, a base point $t \in X$ different from all $x_j, z_j, j = 1, \cdots, k$ and sufficiently general. The image of t in Y will again be denoted by t.

2.2. The Compactified Jacobian.

For $d \in \mathbb{Z}$, let $J^d(Y)$ be the space of line bundles (locally free sheaves of rank 1) of degree d on Y. If Y is nonsingular, it is called the Jacobian of Y, it is an Abelian variety. If Y has nodes, $J^d(Y)$ is called the generalised Jacobian of Y, it is a non-singular quasi projective variety. Then the variety $J^d(Y)$ has a natural compactification $\overline{J}^d(Y)$ as the variety of torsion-free sheaves of rank 1 and degree d on Y. The variety $\overline{J}^d(Y)$ is not normal, it is a seminormal projective variety.

Let

$$h : \tilde{J}^d(Y) \to \overline{J}^d(Y)$$

be the normalisation. We denote by the same letter h the normalisation map for all d (if no confusion arises). The variety $\tilde{J}^d(Y)$ is non-singular, it is a bundle over $J(X)$ with each fibre a k-fold product of $\mathbb{P}^1$'s. The morphism h is an isomorphism over the smooth open subset $J^d(Y)$ of $\overline{J}^d(Y)$, hence $J^d(Y)$ can be identified with an open subset of $\tilde{J}^d(Y)$.

The cycles W_d : For $d \geq 1$, let $S^d(Y_0)$ denote the dth symmetric product of Y_0. We have the natural morphisms

$$f_d : S^d(Y_0) \to J^d(Y),$$

defined by

$$f_d(x_1, \cdots, x_d) := \mathcal{O}_Y(x_1 + \cdots + x_d)\,.$$

This morphism does not extend to $S^d(Y)$. We define the cycles W_d of dimension d on $\overline{J}^d(Y)$

$$W_d := \overline{f_d(S^d(Y))} \subseteq \overline{J}^d(Y) \tag{2.1}$$

as the closure of the image of the dth symmetric product (of Y) with the reduced scheme structure (see [**BhP1**] for details). We note that $W_{g-1} = \theta_Y$, the theta divisor.

By using the isomorphism from $\overline{J}^0(Y)$ with $\overline{J}^d(Y)$ defined by the translation by $\mathcal{O}_Y(dt)$, one can consider W_d as a cycle in $\overline{J}^0(Y)$ for all $d \geq 1$. One has a generalisation of the Poincaré formula

$$W_{g-d} = \frac{\theta_Y^d}{d!}\,,$$

as cycles modulo numerical equivalence [**BhP1**]. However, the two sides are not algebraically equivalent [**Ce**]. So the Poincaré formula is not valid in the Chow group.

Similarly we can define the cycles $\tilde{W}_d, d \geq 1$, on $\tilde{J}^0(Y)$, one has $h(\tilde{W}_d) = W_d\,; h^{-1}(W_d) = \tilde{W}_d$ and a similar generalised Poincaré formula [**BhP1**].

The Brill-Noether Schemes.

The Brill-Noether scheme $B_Y(1,d,r) \subset \overline{J}^d(Y)$ is the scheme whose underlying set is the set of torsion-free sheaves of rank 1 and degree d on Y with at least r independent sections. The "expected" dimension of $B_Y(1,d,r)$ is the Brill -Noether number $\beta_Y(1,d,r) = g(Y) - r(r-d-1+g(Y))$.

2.3. Picard sheaves on $\overline{J}(Y)$.

There is a Poincaré sheaf (universal sheaf) $\mathcal{L}$ on $\overline{J}^0(Y) \times Y$ such that for a point $L \in \overline{J}^0(Y)$, one has $\mathcal{L}|_{L\times Y}$ is isomorphic to the torsionfree sheaf L corresponding to the point L. This sheaf is unique up to tensoring by a line bundle on $\overline{J}^0(Y)$. Hence we normalise $\mathcal{L}$ by the condition that $\mathcal{L}|_{\overline{J}^0(Y)\times t} \cong \mathcal{O}_{\overline{J}^0(Y)}$, the trivial line bundle on $\overline{J}^0(Y)$. Let $p_J : \overline{J}^0(Y) \times Y \to \overline{J}^0(Y)$ and $p_Y : \overline{J}^0(Y) \times Y \to Y$ denote the projections. Let

$$\mathcal{L}(d) = \mathcal{L} \otimes p_Y^* \mathcal{O}_Y(dt)\,.$$

DEFINITION 2.1. We define the **Picard sheaves** on $\overline{J}^0(Y)$ as the direct image sheaves

$$\overline{E}_d := p_{J*}(\mathcal{L}(d)).$$

For $d \geq 2g-1$, the Picard sheaf is a vector bundle of rank $d+1-g$ on $\overline{J}^0(Y)$ called a **Picard bundle.**

For $d < 0, \overline{E}_d = 0$. For $0 \leq d \leq g-1$, the Picard sheaf E_d has support W_d. For $g \leq d \leq 2g-2$, it is a non locally free sheaf with generic rank $d+1-g$.

Henceforth we assume that $d \geq 2g-1$.

The Picard bundles on the Jacobian of a smooth curve have been studied extensively by several authors from 60's. The generalisations to nodal curves are done in last ten years or so.

Let E_d be the pullback of the Picard bundle $\overline{E}_d$ to $\tilde{J}^0(Y)$. The projective bundle $\mathbb{P}(E_d)$ has the following interesting interpretation which is well known in the case of nonsingular curves.

PROPOSITION 2.2. *The projective Picard bundle $\mathbb{P}(E_d)$ is isomorphic to a blowup of $S^d(Y)$, the dth symmetric power of Y.*

See [**ACGH**, Prop. 2.1, p. 309] for the smooth curve case, [**BhP2**, Proposition 5.4] for this generalisation to nodal curves. Over smooth curves, $\mathbb{P}(E_d)$ is isomorphic to $S^d(Y)$.

Now we describe $\tilde{J}(Y)$ and h explicitly. Let $\mathcal{L}' \to J(X) \times X$ be the Poincaré bundle. For all j, let $\mathcal{L}'_{x_j} := \mathcal{L}'|_{J(X)\times x_j}, \mathcal{L}'_{z_j} := \mathcal{L}'|_{J(X)\times z_j}$, both regarded as line bundles on $J(X)$. Define $P_j := \mathbf{P}(\mathcal{L}'_{x_j} \oplus \mathcal{L}'_{z_j})$. Then

$$\tilde{J}(Y) = P_1 \times_{J(X)} \cdots \times_{J(X)} P_k\,,$$

a fibre product over $J(X)$.

An element of $\tilde{J}(Y)$ is of the form $(L', F(L'))$ where $L' \in J(X), F(L') = (F_1(L'), \cdots, F_k(L'))$ where $F_j(L') \subset L'_{x_j} \oplus L'_{z_j}$ is a 1-dimensional subspace. Note that $p_*(L') \otimes \mathcal{O}_{y_j} \cong p_*(L'_{x_j} \oplus L'_{z_j})$. Then $h(L', F(L')) = L$ where L is given by

$$0 \to L \to p_* L' \to \oplus_j \, p_*(\frac{L'_{x_j} \oplus L'_{z_j}}{F(L')}) \to 0 \, .$$

If both $p_{x_j} : F_j(L') \to L'_{x_j}$ and $p_{z_j} : F_j(L') \to L'_{z_j}$ are isomorphisms for all j, then L is a line bundle and $L' = p^* L$, showing that h is an isomorphism over $J(Y)$.

For simplicity, let us first assume that Y has only 1 node so that $\tilde{J}(Y)$ is a $\mathbb{P}^1$-bundle over $J(X)$. It has the zero and infinity sections S_x and S_z with elements of the form (L', L'_x) and (L'', L''_z). One has $h(L', L'_x) = p_*(L'(-z)), h(L'', L''_z) = p_*(L''(-x))$. Hence for $N \in \overline{J}(Y) \setminus J(Y), N' = p^* N/torsion$, the fibre $h^{-1}N$ consists of exactly two points viz., $(N'(x), N'(x)_z)$ and $(N'(z), N'(z)_x)$. The morphism h does a twisted identification of S_x and S_z. If there are many nodes, $\overline{J}(Y)$ is obtained by twisted identification of the sections S_{x_j} and S_{z_j} (of P_j) for all j. It follows that h is quasi-finite. In fact h is finite.

2.4. Chern classes of the Picard bundle.

There is a naturally defined ample divisor on $J^d(X)$ (as well as $\overline{J}^d(Y)$), it is called a theta divisor. The associated line bundle is called the theta line bundle and is denoted by θ_X (respectively θ_Y). Let $\widetilde{\theta}_Y = h^*(\theta_Y)$.

THEOREM 2.3 ([**Bh2**, Theorem 2.3]). *One has the total Chern class (modulo numerical equivalence) $c(E_d^*) = e^{\widetilde{\theta}_Y}$ i.e.,*

$$c_i(E_d^*) = \frac{\widetilde{\theta}_Y^i}{i!} \, ,$$

for all i and hence $c(E_d) = e^{-\widetilde{\theta}_Y}$., i.e.,

$$c_i(E_d) = (-1)^i \frac{\widetilde{\theta}_Y^i}{i!} \, .$$

COROLLARY 2.4.

$$c_i(\overline{E_d}) = (-1)^i \frac{\theta_Y^i}{i!} \, ,$$

(modulo numerical equivalence).

PROOF. For a variety Z, let $N_i(Z)$ denote the Chow group of cycles of dimension i on Z modulo numerical equivalence.

The morphism $h : \tilde{J}(Y) \to \overline{J}(Y)$ is a finite surjective morphism of projective varieties of the same dimension. Hence the image, under h, of an irreducible closed subset in $\tilde{J}(Y)$ is an irreducible closed subset of the same dimension in $\overline{J}(Y)$ inducing the map $h_{*i} : N_i(\tilde{J}(Y)) \to N_i(\overline{J}(Y))$ for each i. Let $h_* = \sum_i h_{*i}$.

Given an irreducible closed subset S of dimension i in $\overline{J}(Y), h^{-1}S$ is a union of finitely many irreducible closed subsets of dimension i in $\tilde{J}(Y)$ giving a map h_i^* : $N_i(\overline{J}(Y)) \to N_i(\tilde{J}(Y)), h^* = \sum_i h_i^*$. One has $h_{*i} \circ h_i^* = Id_{\overline{J}}$. One has $E_d = h^* \overline{E}_d$ so that $c_i(E_d) = h_i^*(c_i(\overline{E}_d))$. Then $h_{*i} c_i(E_d) = h_{*i} \circ h_i^*(c_i(\overline{E}_d)) = c_i(\overline{E}_d)$. Thus

$$h_{*i} c_i(E_d) = c_i(\overline{E}_d) \, .$$

Note that $h_{i*} \widetilde{\theta}_Y = \theta_Y$. The Corollary now follows from Theorem 2.3 by taking h_*. $\square$

For $d = 2g-1$, the Picard bundle is of rank $g = \dim \overline{J}^0(Y)$. One can show that it has a section (unique upto a scalar) whose zero scheme is the reduced point W_0, thus proving that $\overline{J}^0(Y)$ has the week point property (see [**BhS2**] for details). In this special case, we have the following result on Chern classes in the Chow group of cycles on $\overline{J}^0(Y)$ (modulo rational equivalence).

THEOREM 2.5 ([**BhS2**]). *Let $d = 2g-1$.*

(1) *The g th Chern class $c_g(E_d^*) = \tilde{W}_0$ in the Chow group of $\tilde{J}^0(Y)$.*
(2) *The g th Chern class $c_g(\overline{E}_d^*) = W_0$ in the Chow group of $\overline{J}^0(Y)$.*
(3) *The first Chern class*

$$c_1(E_d^*) = c_1(det\ E_d^*) = \widetilde{\theta}_Y,\ c_1(\overline{E}_d^*) = c_1(det\overline{E}_d^*)$$

in the Chow group. Here $\widetilde{\theta}_Y$ is the pull back of the theta divisor on $\overline{J}^0(Y)$ to $\tilde{J}^0(Y)$. In fact,

$$det\ E_d^* \cong \mathcal{O}_{\tilde{J}^0(Y)}(\widetilde{\theta}_Y)\,,$$

the line bundle associated to the divisor $\widetilde{\theta}_Y$.

2.5. Cohomology of the Picard bundle.

If $Y = X_k$ denotes a nodal curve with k nodes with Picard bundle $E_{d,k}, \pi_k : X_{k-1} \to X_k$ is a partial normalisation obtained by blowing up one node and $\pi^k : X_0 \to X_k$, then one shows that

$$\pi_{k*}E_{d,k}^* \cong E_{d,k-1}^*,\ \pi_*^k E_{d,k}^* \cong E_{d,0}^*,\ (R^i\pi_k)_*E_{d,k}^* = 0$$

for $i > 0$. Using these, the cohomology of the Picard bundle is computed by induction on the number of nodes k [**BhS2**].

THEOREM 2.6. *Let $d \geq 2g-1$.*

(1) *For $g(X_0) \geq 1$, $h^0(\tilde{J}^0(Y), E_d^*) = 1$.*
(2) *Let $g(X_0) \geq 1$. Then*

$$h^i(\tilde{J}^0(Y), E_d^*) = {}_{(g(X_0)-1)}C_i \text{ for } 1 \leq i < g(X_0)\,,$$

$$h^i(\tilde{J}^0(Y), E_d^*) = 0 \text{ for } i \geq g(X_0)\,.$$

(3) *For $g(X_0) \geq 1$, $h^0(\overline{J}^0(Y), \overline{E}_d^*) = 1$.*
(4) *For $g(X_0) = 0$, $h^0(\tilde{J}^0(Y), E_d^*) = d+1$, $h^0(\tilde{J}^0(Y), E_d) = d-2k+1$.*
(5) *$h^g(\overline{J}^0(Y), \overline{E}_d^*) = 0$, $h^0(\overline{J}^0(Y), \overline{E}_d) = 0$.*
(6) *$h^0((det\ E_d^*) \otimes \omega_{\tilde{J}^0(Y)}) = 0$.*

The Poincaré sheaf $\mathcal{L}$ can be extended to a Poincaré sheaf $\overline{\mathcal{L}}$ on $\overline{J}^0(Y) \times \overline{J}^0(Y)$. Let $\overline{\mathcal{L}}_p$ be its restriction to $\overline{J}^0(Y) \times p$ for $p \in \overline{J}^0(Y)$. Then the cohomology of $E_d \otimes \overline{\mathcal{L}}_p$ can be computed using the Fourier-Mukai transform on the compactified Jacobian. Details can be found in [**Mu**] for Y smooth and [**BhS1**] for Y nodal.

2.6. Ampleness of the Picard bundles.

We recall that a vector bundle E is ample (respectively nef) if the line bundle $\mathcal{O}_{\mathbb{P}(E)}(1)$ is ample (respectively nef).

Ampleness in the smooth case.

For smooth curves, the ampleness of the Picard bundles was studied by Fulton W. and Lazarsfeld R. (1981).

THEOREM 2.7. **[FL]** *Let X be a nonsingular projective curve.*

(1) *The dual Picard bundle $\overline{E}_d^*$ on $J(X)$ is ample.*
(2) *If $\beta_X(1,d,r) = g(X) - r(r-d-1+g(X)) > 0$, then $B_X(1,d,r)$ are connected.*

NON-ampleness in nodal case.

Myself and Parameswaran A. J. showed that the result of Fulton and Lazarsfeld is not true over nodal curves [**BhP2**].

THEOREM 2.8. *Let $k \geq 1$ be the number of nodes of Y.*

(1) *The dual Picard bundles $\overline{E}_d^*$ and E_d^* are NOT ample if the curve Y has at least one node.*
(2) *However, E_d^* is ample on a general complete intersection subvariety of $\tilde{J}^0(Y)$ of codimension k.*

    In fact, let $Z \subset \tilde{J}^d(Y)$ be a subvariety. Then $E_d^* \mid Z$ is ample if and only if $h\,|_Z\colon Z \to J^d(X_0)$ is finite.
(3) *For any $d > 0$ and $\beta_Y(1,d,r) = g(Y) - r(r-d-1+g(Y)) > k$, the Brill-Noether loci $B_Y(1,d,r)$ are connected.*

We showed that the restriction of the Picard bundle $\tilde{E}_d^* = h^*(E_d^*)$ to the fibres of the morphism $\tilde{J}(Y) \to J(X)$ is not ample and hence $\tilde{E}_d^*$ is not ample. Since h is a finite morphism, it follows that E_d^* is not ample.

2.7. Stability of Picard bundle.

The following theorem was proved by Kempf G. (1990) for $d = 2g(X) - 1$ and by Ein L. and Lazarsfeld R. (1992) for $d \geq 2g(X)$.

THEOREM 2.9. *Let X be a nonsingular projective curve. The Picard bundle $\overline{E}_d$ on $J^d(X)$ is stable with respect to the theta line bundle on $J^d(X)$ for $g(X) \geq 1$.*

Nodal curve.

For $g(Y) = 1$, the stability of the Picard bundle $\overline{E}_d$ was proved using the machinery of Fourier-Mukai transforms by Hernandez Ruiperez D., Lopez Martin A.C., Sanchez Gomez D., Tejero Prieto C. [**HLST**].

For $g > 1$, the stability of the Picard bundle $\overline{E}_d$ was proved by myself and Parameswaran A.J.

THEOREM 2.10. *(see* [**BhP2**]*)*

(A) Let $d \geq 2g(Y), k =$ number of nodes.

(1) *The Picard bundle $\overline{E}_d$ on $\overline{J}^d(Y)$ is stable with respect to the theta divisor on $\overline{J}^d(Y)$ for $g(Y) > 0$.*
(2) *The Picard bundle E_d on $\tilde{J}^d(Y)$ is unstable (with respect to the pullback of the theta divisor) for $k > 0$.*
(3) *The Harder-Narasimhan filtration of E_d is (valid for $d \geq 2g(Y) - 1$)*

$$0 \subset F_d \subset E_d, \; ; \; F_d = \pi^* \pi_* E_d$$

where $\pi : \widetilde{J}(Y) \to J(X)$ is the projection.

For $g(X) \geq 1$, F_d is stable, with a nonsplit exact sequence

$$0 \to F_d \to E_d \to \bigoplus_{j=1}^{k} \mathcal{O}_{P_j}(-1) \to 0\,.$$

For $g(X) = 0$, $E_d \cong F_d \bigoplus (\bigoplus_{j=1}^{k} \mathcal{O}_{P_j}(-1))$
where F_d is a trivial bundle of rank $d - 2k + 1$.

(B) Let $d = 2g(Y) - 1$.
For $g(X) \geq 1$, both $\overline{E}_d$ and E_d are semistable.
For $g(X) = 0, d = 2k - 1, F_d = 0$ so that $E_d \cong \bigoplus_{j=1}^{k} \mathcal{O}_{P_j}(-1)$.

3. Poincaré and Picard bundles on Higher Rank Moduli spaces

3.1. Notations.

Let $n \geq 2$ and d be integers. Let $U_Y(n,d)$ denote the moduli variety of semistable torsionfree sheaves of rank n, degree d on the nodal curve Y. Then $U_Y(n,d) \supset U'_Y(n,d)$ where $U'_Y(n,d)$ is the open dense normal subvariety corresponding to vector bundles. The variety $U_Y(n,d)$ is seminormal but not normal, its open subvariety $U'^s_Y(n,d)$ corresponding to stable vector bundles is smooth.

Fix a line bundle L of degree d on Y. Denote by $U'_Y(n,L)$ the subvariety of $U'_Y(n,d)$ corresponding to vector bundles with determinant L. Let $U_Y(n,L)$ be the closure of $U'_Y(n,L)$ in $U_Y(n,d)$ with the reduced structure. We denote by $U'^s_Y(n,L)$ the open subset corresponding to stable vector bundles.

One has

$$\text{Pic } U'_Y(n,L) \cong \mathbb{Z}$$

and the restriction θ_L, of the theta line bundle to $U'_Y(n,L)$, is the ample generator of Pic $U'_Y(n,L)$ (see [**DN**] for a smooth curve, [**Bh1**] for a nodal curve).

3.2. Poincaré Bundles on $U_Y(n,L)$ and $U_Y(n,d)$.

We first assume that n and d are coprime. Then semistability is equivalent to stability. By a result of Narasimhan and Newstead [**N**, Theorem 5.12'], there is a Poincaré sheaf (a universal sheaf)

$$\mathcal{U} \longrightarrow Y \times U_Y(n,d)$$

such that for $E \in U_Y(n,d)$, the restriction of $\mathcal{U}$ to $Y \times \{E\}$ is isomorphic to the torsionfree sheaf corresponding to E. A universal sheaf is unique up to the pull back of a line bundle on $U_Y(n,d)$.

We first consider the stability of the Poincaré bundles $\mathcal{U}_L$ and $\mathcal{U}$.

Peter E. Newstead, V. Balaji and Leticia Brambila - Paz proved the following result over smooth curves [**BBrN**, Theorem 2.5].

THEOREM 3.1. *Let X be an integral projective smooth curve of genus $g \geq 2$ defined over an algebraically closed field. Let n and d be mutually coprime integers with $n \geq 2$ and $d > n(2g-2)$. Then:*

(1) The universal bundle $\mathcal{U}_L$ on $X \times U'_X(n,L)$ is stable with respect to any polarisation on $X \times U'_X(n,L)$.

(2) The universal bundle $\mathcal{U}$ on $X \times U_X(n,d)$ is stable with respect to any polarisation on $X \times U_X(n,d)$ of the form $a\alpha + b\theta$, a, b are positive integers, α is ample on X and θ is the generalised θ-divisor on X.

(3) The connected component of the moduli space of $\mathcal{U}_L$ containing $\mathcal{U}_L$ is isomorphic to the Jacobian of X.

The proof of this theorem uses spectral curves.

For Theorem 3.1, Part (3) one shows that any deformation of $\mathcal{U}_L$ is of the form $\mathcal{U}_L \otimes p_X^* N, N$ being a line bundle on X, so $N \mapsto \mathcal{U}_L \otimes p_X^* N$ gives the isomorphism. This argument is not extendable to the case of nodal curves.

Theorem 3.1 was partially generalised to nodal curves by me, my proof does not use spectral curves [**Bh4**].

THEOREM 3.2. *Let $x \in Y$ be a nonsingular closed point.*

1. *The restriction $\mathcal{U}_{L,x}$ of the Poincare sheaf to $x \times U_Y(n, L)$ is a vector bundle, it is stable with respect to the polarisation θ_L.*
2. *The restriction of $\mathcal{U}_{L,x}$ to $x \times U'_Y(n, L)$ is stable with respect to any polarisation.*

Theorem 3.2(1) generalises [**BBrN**, Proposition 2.4] to nodal curves. In fact, it improves [**BBrN**, Proposition 2.4] even in the case Y is smooth as [**BBrN**, Proposition 2.4] only proves semistability of $\mathcal{U}_{L,x}$ in the smooth case.

THEOREM 3.3. *Let Y be an integral nodal curve. Let $(n, d) = 1, d > n(2g - 2)$. Then the universal sheaf $\mathcal{U}$ (respectively $\mathcal{U}_L$) is stable with respect to any polarisation on $Y \times U_Y(n, d)$ (respectively $Y \times U_Y(n, L)$) of the form $a\alpha + b\theta$, a, b are positive integers, α is ample on Y and θ is the generalised θ-divisor on $U_Y(n, d)$ (respectively $U_Y(n, L)$).*

3.3. The restriction $\mathcal{U}_{L,x}$ of the Poincaré bundle is strictly Nef.

In this subsection, we assume that Y is an irreducible smooth projective curve of genus $g \geq 2$. Let $n \geq 2$ and d be integers such that $(n, d) = 1, 0 < d < n$. Assume that $\mathcal{U}_{L,x}$ is normalised in the sense that det $\mathcal{U}_{L,x} = \theta^\ell$ where ℓ is an integer such that $0 < \ell < n$ and $\ell d \equiv 1(\mod n)$.

THEOREM 3.4. *The vector bundles $\mathcal{U}_{L,x}$ and $\mathcal{U}^*_{L,x} \otimes \theta$ are strictly nef* [**LM**, Theorem A].

It is shown that $\mathcal{O}_{\mathcal{U}_{L,x}}(1)$ is semiample but it is not ample [**LM**, Lemma 4.6]. By a theorem of Lazarsfeld [**Laz**, Theorem 6.2.12. (iv)], $\mathcal{U}^*_{L,x} \otimes \theta$ is ample. Theorem 3.4 has some applications.

DEFINITION 3.5. Let Z be an n -dimensional variety with an ample line bundle A. A vector bundle F on Z is called an **ACM bundle** if

$$H^i(Z, F \otimes A^j) = 0\,,\ \forall\, 0 < i < n, j \in \mathbb{Z}\,.$$

An ACM bundle is Ulrich if (in addition)

$$H^0(Z, F \otimes A^{-1}) = 0\ ;\ H^0(Z, F) = r(F).deg(F)\,.$$

THEOREM 3.6. *Let n, d be two coprime positive integers, $n \geq 2, 0 < d < n$. If $g \geq 3$ or $g = n = 2$, then $\mathcal{U}_{L,x}$ is an ACM bundle for any ample line bundle on $U_Y(n, L)$.*

For $g = n = 2$ Theorem 3.6 is due to [**CKL**] and [**FK**]. The bundle $\mathcal{U}_{L,x}$ is not Ulrich in general. For example it is not Ulrich for $g = n = 2$ [**CKL**].

Application to Derived Categories.

Let $D^b(U_Y(n,L))$ be the bounded derived category of coherent sheaves on $U_Y(n,L)$. Let

$$\Phi_U : D^b(Y) \to D^b(U_Y(n,L))$$

be the Fourier-Mukai transform with kernel the normalised Poincaré bundle $\mathcal{U}_{L,x}$.

THEOREM 3.7. *Let n,d be two coprime integers, $n \geq 2, d < n$. If $g \geq n+3$, then the functor Φ_U is fully faithful , hence $D^b(Y)$ is embedded in $D^b(U_Y(n,L))$* [**LM**, Theorem B].

The embedding in Theorem 3.7 was proved for $n = 2, d = 1$ by M.S. Narasimhan [**Na1**], [**Na2**]. For $d = 1$ and $g \geq n+3$, it was proved by P. Belmans and S. Mukhopadhyay [**BM**]. For proving the theorem, Lee and Moon prove the following:

(1) $H^0(U_Y(n,L), \mathcal{U}_{L,x} \otimes \mathcal{U}^*_{L,x}) \cong \mathbb{C}$.
(2) $H^i(U_Y(n,L), \mathcal{U}_{L,x} \otimes \mathcal{U}^*_{L,x}) = 0$ for $i \geq 2$.
(3) $H^i(U_Y(n,L), \mathcal{U}_{L,x} \otimes \mathcal{U}^*_{L,y}) = 0$ for all $x \neq y$ and for all i.

Then the Bondal-Orlov criterion implies that Φ_U is fully faithful [**BO**].

3.4. The PICARD BUNDLES on $U_Y(n,d)$.

We define and study the Picard bundles on $U_Y(n,d)$ and $U_Y(n,L)$.

DEFINITION 3.8. Let π_U be the projection $Y \times U_Y(n,d) \to U_Y(n,d)$.

The direct image sheaves

$$\mathbf{E}_{n,d} := \pi_{U*}\mathcal{U} \text{ and } \mathbf{F}_{n,d} := R^1\pi_{U*}\mathcal{U}$$

are called the Picard sheaves on $U_Y(n,d)$.

For $d > (2g-2)n$ ($d \geq (2g-2)n$ if $n \geq 2$), the Picard sheaf $\mathbf{F}_{n,d} = 0$. Then the Picard sheaf $\mathbf{E}_{n,d}$ is a locally free sheaf of rank $d + n(1-g)$ and is called the **Picard bundle** on $U_Y(n,d)$.

For a fixed line bundle L, let $\mathbf{E}_L$ denote the restriction of $\mathbf{E}_{n,d}$ to $U_Y(n,L)$.

Stability of Picard bundle on $U_Y(n,L)$.

Yingchen Li proved the following theorem using the spectral curves and the BNR correspondence [**BNR**].

THEOREM 3.9. *Let Y be an irreducible smooth projective curve of genus $g \geq 2$ defined over an algebraically closed field. Let n and d be coprime integers with $n \geq 2$ and $d > 2ng$. Then the Picard bundle $\mathbf{E}_{n,d}$ over $U_Y(n,d)$ is stable with respect to the polarisation θ.*

For the fixed determinant variety over smooth curves, Biswas I., Brambila-Paz L., Gomez T. L. and Newstead P. E. proved the following theorem by a different method [**BiBrGN**].

THEOREM 3.10. *Let n and d be coprime integers with $n \geq 2$ and $d > n(2g-2)$. Fix a line bundle L of degree d on Y. Then:*

- *The Picard bundle $\mathbf{E}_L$ over $U_Y(n,L)$ is stable with respect to the polarisation θ_L.*
- *The restriction of the Picard bundle to $U'_Y(n,L)$ is stable with respect to any polarisation on $U'_Y(n,L)$.*

This theorem was generalised to integral projective curves of arithmetic genus $g \geq 2$, with at most nodes as singularities, defined over an algebraically closed field, by me [**Bh4**].

3.5. Restriction of the Picard bundle to the compactified Jacobian. We denote by $\overline{J}^{-1}(Y)$ the compactified Jacobian of Y, the space of torsion free sheaves of rank 1 and degree -1 on Y. Fix a general stable vector bundle E_0 on Y of rank n and degree $d+n$, where $-n < d \leq 0, (n,d) = 1$, such that p^*E_0 is stable. One can prove that there is an embedding

$$\alpha_Y : \overline{J}^{-1}(Y) \longrightarrow U_Y(n,d)\,;\; \alpha_Y : N \mapsto N \otimes E_0$$

(for details see [**Li**] in the smooth case, [**Bh3**] in the nodal case).

THEOREM 3.11.

(1) *The restriction of the Picard bundle* $\mathbf{E}_{n,d}$ *to* $\overline{J}^{-1}(Y)$ *is stable with respect to any theta divisor on* $\overline{J}^{-1}(Y)$ *if* $d > n(2g-1)$ *and it is semistable if* $d = n(2g-1)$ *(see* [**Li**, Theorem 2] *for a smooth* Y*,* [**Bh3**] *for a nodal curve* Y*).*

(2) *Let* Y *be a non-hyperelliptic nodal curve. Then the restriction of the Picard bundle* $\mathbf{E}_{n,n(2g-1)}$ *on* $U_Y(n, n(2g-1))$ *to* $\bar{J}(Y)$ *is stable with respect to any theta divisor on* $\bar{J}(Y)$ [**Bh4**].

3.6. Projective Poincaré and Picard bundles.

Now we move to the non-coprime case, in general there is no Poincaré bundle in this case. It is known that if Y has geometric genus $g(X) \geq 2$, then for n and d not coprime, there does not exist a Poincaré sheaf on $Y \times V$ for any Zariski open subset of $V \subset U_Y(n,L)$ (see [**R**] for Y a smooth curve, [**ABS**, Theorem 7.14, Corollary 2.11] for Y a nodal curve). However, there always exists a projective Poincaré bundle $\mathcal{PU}$ over $Y \times U_Y'^s(n,L)$ whose restriction to $Y \times \{E\}$ is isomorphic to $P(E)$ for all $E \in U_Y'^s(n,L)$ [**ABS**].

We start with some generalities. Let U be a big open subset of a projective variety W i.e., the codimension of the complement of U in W is at least 2. Let P be a projective bundle on U. Let P' be a projective subbundle of the restriction of P to a big open subset Z of U and $p' : P' \to Z$. Let $N' = N_{P'/P}$ be the normal bundle of P' in $P|_Z$ and $N = {p'}_*(N')$. Then N is a vector bundle on Z. We make the following definition for the stability of a projective bundle.

DEFINITION 3.12. The projective bundle P is stable(semistable) if for every subbundle P',

$$\deg N > 0 \; (\deg N \geq 0)$$

On smooth curves, this definition agrees with A. Ramanathan's definition of the stability of $PGL(n)$-bundles (see [**ABS**, Remark 2.4]).

PROPOSITION 3.13. *Let* Y *be an integral nodal curve of genus* $g \geq 2$ *with* k *nodes,* $k \geq 1, n \geq 2$.

(1) *The codimension of* $U_Y(n,L) - U_Y'(n,L)$ *in* $U_Y(n,L)$ *is at least* 2 [**Bh4**, Theorem 2.5].

(2) *Assume that if* $n = 2$ *then* $g \geq 3$. *Then the codimension of* $U_Y'(n,L) - U_Y'^s(n,L)$ *in* $U_Y'(n,L)$ *is at least* 3 [**ABS**, Theorem 6.1].

The proposition enables us to apply our definition of the stability of a projective bundle to projective bundles on $U_Y'^s(n,L)$ and $Y \times U_Y'^s(n,L)$.

THEOREM 3.14. *Let* Y *be an integral projective curve with at most nodes as singularities. Let* $x \in Y$ *be a nonsingular point. Define* $\mathcal{PU}_x = \mathcal{PU}|_{x \times U_Y'^s(n,L)}$.

(1) *The projective Poincaré bundle $\mathcal{PU}_x$ on $U_Y'^s(n,L)$ is stable for any polarisatiion* [**ABS**, Theorem 4.5].
(2) *Let η and θ_L be divisors defining the polarisation on Y and $U_Y'^s(n,L)$ respectively. Then the projective Poincaré bundle $\mathcal{PU}$ is stable with respect to $a\eta + b\theta_L$, $a,b > 0$* [**ABS**, Theorem 4.5].
(3) *Let $d > n(2g-1)$. Then the projective Picard bundle on $U_Y'^s(n,L)$ is stable for any polarisation* [**ABS**, Theorem 5.2].

For a smooth curve Y, the stability of the Projective Poincaré and Picard bundles was proved by I. Biswas, L. Brambila-Paz, Peter Newstead [**BiBrN2**], the generalisation to nodal curves is by Arusha C., myself and Sanjay Singh [**ABS**].

The basic idea of proofs in many of the results on the stability of Poincaré bundles, Picard bundles, projective Poincaré bundles and Picard bundles is similar. I would like to give a short sketch of the proof of the stability of projective Poincaré and Picard bundles.

We need the following results for the proofs.

DEFINITION 3.15. A torsionfree sheaf F on Y is $(0,1)$-stable if, for every proper subsheaf G of F (with a torsionfree quotient), one has

$$\frac{d(G)}{r(G)} < \frac{d(F)-1}{r(F)}.$$

PROPOSITION 3.16. *The $(0,1)$-stable vector bundles F, of rank n and determinant L' of degree d', form a non-empty open subset of the moduli space $U_Y^s(n,L')$ of stable torsionfree sheaves.*

The Common Idea of the Proof of Stability.

Take a nonsingular point x of Y and a $(0,1)$-stable vector bundle F of rank n and determinant $L(x), d(L) = d$. Let k_x denote the torsion sheaf of length 1 supported at x. Denote by $\mathbf{P} := \mathbf{P}(F_x^*)$ the projective space of lines in F_x^*. For every $\phi \in \mathbf{P}(F_x^*)$, we have a homomorphism $\phi : F \to k_x$ giving an exact sequence on Y

$$0 \longrightarrow E \longrightarrow F \xrightarrow{\phi} k_x \longrightarrow 0.$$

Since x is a nonsingular point of Y and F is locally free, it follows that E is a locally free sheaf of rank n and determinant L. Let $p_1 : Y \times \mathbf{P} \to Y, p_2 : Y \times \mathbf{P} \to \mathbf{P}$ be the projections and $i : \mathbf{P} \hookrightarrow Y \times \mathbf{P}, i(z) = (x,z)$ be the inclusion map. Varying ϕ, on $Y \times \mathbf{P}(F_x^*)$ we get the exact sequence

$$0 \longrightarrow \mathcal{E} \longrightarrow p_1^* F \longrightarrow i_* \mathcal{O}_{\mathbf{P}}(1) \longrightarrow 0,$$

with $\mathcal{E}$ a vector bundle.

The morphism $\psi_{F,x}$: If F is $(0,1)$-stable vector bundle, then E is a stable vector bundle. Hence by the universal property of $U_Y(n,L)$, we have a morphism

$$\psi_{F,x} \colon \mathbf{P}(F_x^*) \longrightarrow U_Y'^s(n,L),$$

where $U_Y'^s(n,L)$ the subvariety corresponding to stable vector bundles. One has

$$P(\mathcal{E}_x) \cong \psi_{F,x}^* \, \mathcal{PU}_x.$$

LEMMA 3.17. *The morphism $\psi_{F,x}$ is an isomorphism of $\mathbf{P}(F_x^*)$ onto its image in $U_Y'^s(n,L)$.*

Sketch of the proof of Stability of $P(\mathcal{U}_x)$.

Let $Z \subset U'^s_Y(n,L)$ be a big open subset and $P' \subset \mathcal{PU}_x|_Z$ be a projective subbundle. Let N' be the normal bundle of P' in $\mathcal{PU}_x|_Z$. Denote by N the direct image of N' on Z, it is a vector bundle. For proving the stability of $\mathcal{PU}_x$ we need to check that degree $N > 0$.

We note that both Pic $U'^s_Y(n,L)$ and Pic $\mathbf{P}(F^*_x)$ are isomorphic to $\mathbb{Z}$ and generated by the theta line bundle and $\mathcal{O}_{\mathbf{P}}(1)$ respectively. So the degree of a vector bundle W and the degree of $\psi^*_{F,x}(W)$ are a constant multiple of each other (independent of W). Hence to check that degree $N > 0$, it suffices to check that

$$deg(\psi^*_{F,x}(N)) > 0\,.$$

We check the latter in the final steps of the proof:

(1) For a general F, we show that $\psi^{-1}_{F,x}(Z) \subset \mathbf{P}(F^*_x)$ is a big open subset.

(2) We have $\mathcal{E}_x$ on $\mathbf{P}(F^*_x)$. Let $\mathcal{E}'_x = \mathcal{E}_x|_{\psi^{-1}_{F,x}(Z)}$. We prove that there exists a vector subbundle $V' \subset \mathcal{E}'_x$ such that

$$\psi^*_{F,x}P' = \mathbf{P}(V')\,;\quad \psi^*_{F,x}N = (V')^* \otimes (\mathcal{E}'_x/V')\,.$$

(3) We check that $\mu(V') < \mu(\mathcal{E}'_x)$ so that degree $\psi^*_{F,x}(N) > 0$.

Each step has a very technical proof.

The stability of $P\mathcal{U}$ follows from that of $P\mathcal{U}_x$ and of the restriction of $P\mathcal{U}$ to $Y \times \{E\}$.

The stability of Projective Picard bundles is proved using similar ideas, the proof differs only in the last two steps in the final steps.

4. Parabolic vector bundles

We now consider parabolic vector bundles and parabolic Hitchin bundles on smooth curves. Our aim is to generalise Theorem 3.1 to parabolic bundles. This was done in [**BaBiD**] under the assumption that the parabolic structure is a Borel structure i.e., all multiplicities are 1 (see the following Definition 4.1 for details).

We first start with necessary definitions and prove results needed for the proof of our theorem.

4.1. Preliminaries.

Let Y be an irreducible smooth projective curve and I a finite set of distinct points of Y. Let $D = \sum_{x\in I} x$ be the Cartier divisor in Y.

DEFINITION 4.1. A *quasi-parabolic structure* at a point $x \in I$ on a vector bundle E on Y is a flag of vector subspaces on the fibre E_x

$$E_x = F'_1(E_x) \supset \cdots \supset F'_{l_x}(E_x) \supset F'_{l_x+1}(E_x) = 0\,.$$

Let $c^i_x = \dim\, F'_i(E_x)$, then $\bar{n}_x = (c^1_x, \cdots, c^{l_x}_x)$ is called the type of the flag at x.

A *parabolic structure* on E over I is a quasiparabolic structure at $x \in I$ as above together with a rational number $\alpha_i(x)$ (called a parabolic weight) for each subspace $F'_i(E_x)$ such that

$$0 < \alpha_1(x) < \cdots < \alpha_{l_x}(x) < 1$$

Let $n_i(x) = \dim\, F'_i(E_x) - \dim F'_{i+1}(E_x), i = 1, \cdots, l_x$ and $r_i(x) = \sum_{j=1}^{i} n_j(x), i \geq 1$. The number $n_i(x)$ is called the multiplicity of $\alpha_i(x)$. Define

$$\bar{\alpha}(x) := (\alpha_1(x), \alpha_2(x), \cdots \alpha_{l_x}(x));\ \bar{n}(x) := (n_1(x), n_2(x), \cdots, n_{l_x}(x))\,.$$

A parabolic Hitchin bundle (E_*, ϕ) on Y with parabolic structure over I is a vector bundle E on Y together with a parabolic structure at each $x \in I$ and a homomorphism $\phi : E_* \to E_* \otimes L(D)$ of parabolic sheaves i.e., $\phi_x(F_i'(E_x)) \subset F_i'(E_x)$ for all i and $x \in I$. Here L denotes a fixed line bundle on Y.

A strongly parabolic Hitchin bundle (E_*, ϕ) on Y with parabolic structure over I is a vector bundle E on Y together with a parabolic structure at each $x \in I$ and a strongly parabolic homomorphism $\phi : E_* \to E_* \otimes L(D)$ of parabolic sheaves i.e., $\phi_x(F_i'(E_x)) \subset F_{i+1}'(E_x)$ for all i and $x \in I$.

For a parabolic Hitchin bundle (E_*, ϕ) as above, the parabolic degree is defined as

$$\text{par-deg}(E_*, \phi) := \text{degree}(E) + \sum_{x \in I} \sum_{i=1}^{l_x} \alpha_i(x) n_i(x) \,.$$

The parabolic slope is defined as

$$\text{par-}\mu(E_*) := \frac{\text{par-deg}(E_*)}{\text{rank}(E)} \in \mathbb{Q} \,.$$

DEFINITION 4.2. For a parabolic Hitchin bundle E_*, any nonzero subbundle $F \subset E$ has an induced parabolic structure. We denote by F_* the bundle F with the induced parabolic structure. A subbundle $F \subset E$ is called ϕ-invariant if $\phi(F) \subset F$. Then $(F_*, \phi|_F)$ is a ϕ-invariant subbundle of (E_*, ϕ).

A parabolic Hitchin bundle (E_*, ϕ) is called *stable* (respectively, *semistable*) if for every ϕ-invariant subbundle $F \subset E$, with $1 \le \text{rank}(F) < \text{rank}(E)$ one has

$$\text{par-}\mu(F_*) < \text{par-}\mu(E_*)$$

(respectively, $\text{par-}\mu(F_*) \le \text{par-}\mu(E_*)$) .

The (semi)stability of a strongly parabolic Hitchin bundle is defined similarly.

4.2. The moduli space of parabolic bundles.

Fix a line bundle ξ of degree d on Y. We assume that the parabolic weights are general so that semistability is same as stability for parabolic Hitchin bundles.

Let $U_{par}(n, \xi)$ denote the moduli space of parabolic vector bundles on Y. One has

$$\dim U_{par}(n, \xi) = (n^2 - 1)(g - 1) + \sum_{x \in D} \dim\, (SL(n)/P_x) \,, \tag{4.1}$$

where P_x is the parabolic subgroup of $SL(n)$ determined by the parabolic structure.

For positive integers k, ℓ and non negative integers $b_x, x \in D$ satisfying

$$\sum_{x \in d} \sum_{i=1}^{l_x} k(\alpha_{i+1}(x) - \alpha_i(x)) r_i(x) + n \sum_{x \in D} b_x + n\ell = k\chi \,, \tag{4.2}$$

where χ denotes the Euler characteristics of E i.e. $\chi = d + n(1 - g)$, there is an ample line bundle Θ_{par} on $U_{par}(n, \xi)$ having the following universal property. Let $\mathcal{E}_*$ be a family of parabolic vector bundles on Y parametrised by a scheme T and $\pi_T : T \times Y \to T$ the projection.

Define

$$\theta_T := (DetR_{\pi_T}\mathcal{E})^k \otimes \bigotimes_{x \in D} [det\mathcal{E}_{T \times x})^{b_x} \otimes \bigotimes_{i=1}^{l_x} (det(\mathcal{E}_{T \times x}/F_i(\mathcal{E}_{T \times x})))] \otimes det(\mathcal{E}_{T \times t})^{\ell} \,.$$

Then there is a morphism $f : T \to U_{par}(n, \xi)$ such that

$$\theta_T = f^*(\Theta_{par}).$$

4.3. The Hitchin map and the spectral curve.

For $x \in D$, we reorder the multiplicities $(n_1(x), \cdots, n_{l_x(x)})$ in descending order as $(n'_1(x), \cdots, n'_{l_x(x)}), n'_1(x) \geq \cdots \geq n'_{l_x}(x) > 0$. This gives a partition of n (see [**SWW**] for details). Let $(\mu_1(x) \geq \cdots \geq \mu_{l_x}(x) > 0)$ be the conjugate partition of n, given by

$$\mu_t(x) = num(\{c \mid n'_c(x) \geq t, 1 \leq c \leq l_x, \}) \tag{4.3}$$

where $num(S)$ denotes the number of elements in the set S. For $j = 1, \cdots, n$, define $\gamma_j(x)$ (level function) by

$$\gamma_j(x) = c \text{ if and only if } \sum_{t \leq c-1} \mu_t(x) < j \leq \sum_{t \leq c} \mu_t(x).$$

Define the strongly parabolic Hitchin base

$$\mathbf{A}'_p := \prod_{j=1}^{n} \mathbf{H}^0(Y, \omega_Y^j \otimes \mathcal{O}_Y(\sum_{x \in D}(j - \gamma_j(x)x)). \tag{4.4}$$

Here $\mathbf{H}^0(-)$ denotes the affine space underlying the vector space $H^0(-)$.

Let $\mathcal{H}^{str}_{par}(n, d)$ (respectively $\mathcal{H}^{str}_{par}(n, \xi)$) be the moduli space of strongly parabolic Hitchin bundles (E_*, ϕ) with E a vector bundle of rank n, degree d (respectively having $detE = \xi$), and $\phi : E \to E \otimes \omega_Y(D)$ a homomorphism. There is a Hitchin morphism

$$h_p : H^{str}_{par}(n, d) \longrightarrow \mathbf{A}'_p.$$

It is a proper morphism. Since $\mathcal{H}^{str}_{par}(n, \xi)$ is a closed subvariety, the restriction gives a proper Hitchin morphism

$$h^0_p : H^{str}_{par}(n, \xi) \longrightarrow \mathbf{A}^0_p,$$

where

$$\mathbf{A}^0_p := \prod_{j=2}^{n} \mathbf{H}^0(Y, \omega_Y^j \otimes \mathcal{O}_Y(\sum_{x \in D}(j - \gamma_j(x)x)); \ \dim \mathbf{A}^0_p = \ \dim U_{par}(n, \xi) \tag{4.5}$$

4.4. The Parabolic BNR correspondence.

For a general $a \in \mathbf{A}'_p$, there is a spectral curve Y_a which is smooth outside D [**BNR**] and a morphism $\pi : Y_a \to Y$ such that the fibre of the Hitchin map $h^{-1}(a) \cong \text{Pic}^{\delta'}(Y_a), \delta' = n^2(g-1) + 1 + \frac{n(n-1)}{2}deg(D)$. The BNR correspondence associates to a line bundle N' on Y_a the vector bundle $E = \pi_*(N)$ with a Hitchin field with values in $\omega_Y(D)$ induced by the $\pi_*(\mathcal{O}_{Y_a})$- structure on E. The covering π is totally ramified at points of D and Y_a is singular over points of D. Let $\tilde{Y}_a$ be a normalisation of Y and $\tilde{\pi} : \tilde{Y}_a \to Y_a \to Y$ the composite (see [**SWW**, subsection 4.2] for details). Then there are $n_1(x)$ (the length of conjugate partition) points of $\tilde{Y}_a$ over $x \in D$ with ramification degrees $(\mu_1(x), \cdots, \mu_{n_1(x)})$, where $\mu_t(x)$ are defined in (4.3) and the geometric genus of $\tilde{Y}_a$ is

$$g(\tilde{Y}_a) = \delta := n^2(g-1) + 1 + \sum_{x \in D} \dim\ (GL(n)/P'_x),$$

[**SWW**, Corollary 4.11].

Given a line bundle N on $\tilde{Y}_a$, the direct image $\tilde{\pi}_* N$ gets a $\tilde{\pi}_*\mathcal{O}_{\tilde{Y}_a}$-structure on it, making it an $\omega_Y(D)$-valued Hitchin bundle with a filtration with parabolic type P'_x over $x \in D$. Conversely, given a strongly parabolic bundle (E_*, ϕ) one shows that the parabolic data endow E with a $\tilde{\pi}_*\mathcal{O}_{\tilde{Y}_a}$-module structure and hence it is the direct image of a line bundle N on $\tilde{Y}_a$ (see the proof of [**SWW**, Theorem 4.2] for details). This gives the parabolic BNR correspondence between line bundles of degree δ on $\tilde{Y}_a$ and strongly parabolic Hitchin bundles of rank n, degree d on Y.

REMARK 4.3 (The Prym variety P_δ). For a general $a \in \mathbf{A}'_p$, the fibre of the strongly parabolic Hitchin map $h_p^{-1}(a) \cong \operatorname{Pic}^\delta(\tilde{Y}_a)$. For a general $a \in \mathbf{A}^0_p$, the fibre $(h^0_p)^{-1}(a)$ is isomorphic to the Prym variety

$$P_\delta := \{N \in \operatorname{Pic}^\delta(\tilde{Y}_a) | \tilde{\pi}_* N \text{ has determinant } \xi\}.$$

Given $z \in Y \setminus D$, we may assume that $\tilde{\pi}$ is unramified over z.

If Y is defined over an algebraically closed field of characteristic 0, then $\tilde{Y}_a$ and P_δ are smooth.

Define the nilpotent cone

$$Nil^0_p := (h^0_p)^{-1}(0) \subset H^{str}_{par}(n, \xi).$$

By corollary [**SWW**, Corollary 5.6], for any $a \in \mathbf{A}'_p$, dim $(h_p)^{-1}(0) \geq \frac{n(n-1)}{2} deg(D) + n^2(g-1)+1$ so that for any $a \in \mathbf{A}^0_p$, dim $(h^0_p)^{-1}(0) \geq \frac{n(n-1)}{2} deg(D) + (n^2-1)(g-1) =$ dim $U_{par}(n, \xi)$. We shall show that dim $(h^0_p)^{-1}(0) =$ dim $U_{par}(n, \xi)$.

The dimension of the reduced nilpotent cone $Nil^{0,red}_p$ can be determined algebraic geometrically using infinitesimal deformations as in [**SWW**, Section 6], we give a short sketch. The moduli space $H^{str}_{par}(n, \xi)$ is a geometric quotient by $PGL(V)$ of an algebraic scheme Q_ξ with a $PGL(V)$-action, let $q : Q_\xi \to H^{str}_{par}(n, \xi)$ be the projection. There is a universal family $(\mathcal{E}_*, \phi)$ over $Q_\xi \times Y$. One can show that there exists an irreducible open subset $W_0 \subset Nil^{0red}_p$ with generic point η such that the Higgs field ϕ_W on $q^{-1}(W^{red}_0)$ gives a filtration Ker $(\phi^i_{W_0})$ of $E_W := \mathcal{E}|_{q^{-1}(W^{red}_0)}$ consisting of vector bundles.

For a filtered vector bundle E, let $End^{0filt}(E)$ be the sheaf of local trace zero endomorphisms of E which preserve the filtration. For a parabolic vector bundle E_*, let $ParEnd^0(E)$ and $SParEnd^0(E)$ denote the sheaf of parabolic endomorphisms with trace zero and the sheaf of strongly parabolic endomorphisms with trace zero.

THEOREM 4.4. (1) *The space of infinitesimal deformations in W_0 of a strongly nilpotent Hitchin bundle (E_*, ϕ) with $detE = \xi$, is canonically isomorphic to the hypercohomology $\mathbb{H}^1(Y, \mathcal{A}^\bullet)$ of the following complex of sheaves on Y*

(4.6)

$$\mathcal{A}^\bullet : 0 \longrightarrow ParEnd^0(E) \cap End^{0filt} E \xrightarrow{ad\phi} SParEnd^0(E) \cap End^{0filt} E \otimes \omega_Y(D) \to 0$$

(2) *One has*

$$\dim_k (T_w W_0) = \dim_k \mathbb{H}^1(Y, \mathcal{A}^\bullet) = \dim U_{par}(n, \xi).$$

(3) *The Hitchin map h^0_p is flat and surjective.*

PROOF.

(1) This can be proved as in [**SWW**, Theorem 6.4], replacing endomorphisms by trace zero endomorphisms.
(2) Replacing $GL(n)$ by $SL(n)$, as in [**SWW**, Proposition 6.7], we can prove Part (2).
(3) This can be proved as in [**SWW**, Theorem 6.8] using Parts (1) and (2).

□

A stable parabolic bundle E_* is called *very stable* if it has no non-zero nilpotent Higgs field i.e., $(E_*, \phi) \notin Nil_p^0$ for any $\phi \neq 0$. One can show that the set of very stable parabolic bundles in $U_{par}(n,\xi)$ contains a nonempty open subset S of $U_{par}(n,\xi)$ [**BaBiD**].

Let $\tilde{f} : H_{par}^{str}(n,\xi) \dashrightarrow U_{par}(n,\xi)$ be the surjective forgetful map which forgets the parabolic structure.

COROLLARY 4.5. *For a general $a \in \mathbf{A}_p^0$, the restriction of the forgetful map $\tilde{f}$*

$$F_g : (h_p^0)^{-1}(a) \dashrightarrow U_{par}(n,\xi)$$

is a dominant rational map.

PROOF. Consider the composite map $\rho = \tilde{f} \times h_p^0$ i.e.,

$$\rho : H_{par}^{str}(n,\xi) \dashrightarrow U_{par}(n,\xi) \times \mathbf{A}_p^0 ; (E_*, \phi) \mapsto (E_*, h_p^0(E_*, \phi)) .$$

Note that ρ is a dominant map. Since dim $H_{par}^{str}(n,\xi) =$ dim $U_{par}(n,\xi)+$ dim $\mathbf{A}_p^0$, the morphism ρ is generically finite. There is an open subset $S \subset U_{par}(n,\xi)$ such that for $\rho^{-1}(E_*,0) = (E_*,0)$ for $(E_*,0) \in S \times \{0\}$. Hence $(h_p^0)^{-1}(a) \dashrightarrow U_{par}(n,\xi)$ is dominant for a general $a \in \mathbf{A}_p^0$. □

LEMMA 4.6. *Let Y be defined over an algebraically closed field of characteristic 0. The rational forgetful map F_g is defined on a big open subset of $(h_p^0)^{-1}(a)$ i.e. the open subset where it is defined has complement of codimension at least 2.*

PROOF. Note that under our assumptions $\tilde{Y}_a$ is smooth. Then there is a Poincaré family of line bundles $\mathcal{L}$ (on the Jacobian $J(\tilde{Y}_a) \times \tilde{Y}_a$ and hence) on $P_\delta \times \tilde{Y}_a$ which gives a family $(Id \times \tilde{\pi})_* \mathcal{L}$ of parabolic vector bundles of rank n, determinant ξ parametrised by P_δ. By [**WW**, Corollary 4.8], the subset of P_δ corresponding to stable parabolic bundles is an open subset whose complement is of codimension strictly greater than $(n-1)(g-1)$, which is at least 2 for $n \geq 2, g \geq 2$, proving the lemma. □

4.5. Semistability of the restriction of the Poincaré Bundle.

In this subsection we assume that the greatest common divisor of n, d and the multiplicities in the parabolic structure $n_i(x), x \in D$, is 1. Then there is a universal family $\mathcal{U}_*$, of parabolic vector bundles, on $U_{par}(n,\xi) \times Y$.

Let z be a closed point of Y distinct from the points in D. Let $\mathcal{U}_z$ be the restriction of the universal bundle to $U_{par}(n,\xi) \times z$. Note that $\mathcal{U}_z$ has no parabolic structure.

THEOREM 4.7 (Theorem 1.1). *$\mathcal{U}_z$ is semistable with respect to the ample line bundle Θ_{par}.*

PROOF. Our proof of this theorem is on the lines of the proofs of [**BBrN**, Theorem 1.5] and [**BaBiD**, Theorem 3.4].

Let $a \in \mathbf{A}_p^0$ be a general point and $\tilde{Y}_a$ the (parabolic) spectral curve chosen as in subsection 4.4.

Let $\mathcal{N}$ denote the restriction of the Poincare bundle on $J^\delta(\tilde{Y}_a) \times \tilde{Y}_a$ to $P_\delta \times Y_a$, where the Prym variety P_δ is defined in Remark 4.3. Let $Id \times \tilde{\pi} : P_\delta \times \tilde{Y}_a \to P_\delta \times Y$ be the product map. Then the direct image $(Id \times \tilde{\pi})_* \mathcal{N}$ is a Hitchin bundle of rank n with a Hitchin field

$$\Phi \in H^0(P_\delta \times Y, End((Id \times \tilde{\pi})_* \mathcal{N} \otimes p_Y^* \omega_Y(D))) .$$

and a parabolic structure over $P_\delta \times D$ induced by the $(Id \times \tilde{\pi})_* \mathcal{O}_{P_\delta \times \tilde{Y}_a}$-module structure on it (by Parabolic BNR correspondence in subsection 4.4). This family induces the (forgetful) rational map $F_g : P_\delta \dashrightarrow U_{par}(n, \xi)$ defined by mapping N to the parabolic vector bundle $\tilde{\pi}_*(N)$. Let $T_\delta \subset P_\delta$ be the subset where F_g is well defined. Then F_g is dominant by Corollary 4.5 and codim $(P_\delta \setminus T_\delta, P_\delta) \geq 2$ by Lemma 4.6. Define

$$\mathcal{E} := (Id \times \tilde{\pi})_* \mathcal{N})|_{T_\delta \times Y} .$$

By the universal property of the fine moduli space $U_{par}(n, \xi)$ we have the isomorphism

$$(F_g \times Id)^* \mathcal{U} \cong \mathcal{E} \otimes L_0 ,$$

for some line bundle L_0 on T_δ. Hence $F_g^* \, \mathcal{U}_z \cong \mathcal{E}|_{(\tilde{\pi})^{-1}(z)} \otimes L_0$ so that

$$F_g^* \, \mathcal{U}_z \ \cong \ (\oplus_{i=1}^n \mathcal{N}_{z_i}) \otimes L_0 ,$$

where $(\tilde{\pi})^{-1} z = \{z_1, \cdots, z_n\}$ is a set of distinct points of $\tilde{Y}_a$.

Since P_δ is smooth and codim $(P_\delta \setminus T_\delta, P_\delta) \geq 2$, the line bundle L_0 uniquely extends to a line bundle L_0' on P_δ. Since the line bundles $\mathcal{N}_{z_i}$ are all algebraically equivalent, the vector bundle $\oplus_i \mathcal{N}_{z_i} \otimes L_0'$ is semistable with respect to any ample line bundle on P_δ.

For the family $\mathcal{E}$ of parabolic vector bundles on T_δ, we define

$$\Theta_{T_\delta} := (DetR_{(\pi_{T_\delta})} \mathcal{E})^k \otimes M'$$

where

$$M' = \bigotimes_{x \in D} [(det \mathcal{E}_{T_\delta \times x})^{b_x} \otimes \bigotimes_{i=1}^{l_x} (det(\mathcal{E}_{T_\delta \times x} / F_i(\mathcal{E}_{T_\delta \times x})))] \otimes det(\mathcal{E}_{T_\delta \times t})^\ell .$$

Then by the universal property of Θ_{par}, we have

$$\Theta_{T_\delta} \ = \ F_g^*(\Theta_{par}) .$$

Since codim $(P_\delta \setminus T_\delta, P_\delta) \geq 2$, the line bundle M' on T_δ extends uniquely to a line bundle M on P_δ. Let θ_{P_δ} be the restriction of the theta line bundle, an ample line bundle, on $J^\delta(\tilde{Y}_a)$ to P_δ. By [**Li**, Theorem 4.3],

$$(DetR_{\pi_{T_\delta}} \mathcal{E})^k = m\theta_{P_\delta}^k |_{T_\delta}, \text{ for some } m \in \mathbb{Z}_{\geq 0} .$$

Then $F_g^*(\Theta_{par})$ is the restriction of the ample line bundle $H := m\theta_{P_\delta}^k \otimes M$ on P_δ. Since the extension to P_δ of $F_g^* \, \mathcal{U}_z$ is semistable with respect to H, the vector bundle $\mathcal{U}_z$ is semistable with respect to Θ_{par} by [**BBrN**, Lemma 2.1].

□

THEOREM 4.8 (Theorem 1.2). *The parabolic bundle $\mathcal{U}$ on $U_{par}(n, \xi) \times Y$ is parabolic stable with respect to any integral ample divisor of the form $a\Theta_{par} + bD_Y$ where $a, b > 0$ and D_Y is an ample divisor on Y.*

PROOF. This follows from Theorem 1.1 and [**BaBiD**, Lemma 3.1]. □

A generalisation of these results to the nodal curves is done in [**Bh6**].

References

[ABS] C. Arusha, U. N. Bhosle, and S. K. Singh, *Projective Poincaré and Picard bundles for moduli spaces of vector bundles over nodal curves*, Bull. Sci. Math. **166** (2021), Paper No. 102930, 34, DOI 10.1016/j.bulsci.2020.102930. MR4184304

[ACGH] E. Arbarello, M. Cornalba, P. A. Griffiths and J. Harris, *Geometry of algebraic curves*, Springer-Verlag (1985)

[AK] A. B. Altman and S. L. Kleiman, *Compactifying the Picard scheme*, Adv. in Math. **35** (1980), no. 1, 50–112, DOI 10.1016/0001-8708(80)90043-2. MR555258

[BaBiD] S. Basu, I. Biswas, and K. Dan, *Stability of the parabolic Poincaré bundle*, Int. J. Geom. Methods Mod. Phys. **15** (2018), no. 5, 1850081, 9, DOI 10.1142/S0219887818500810. MR3786513

[Bh1] U. N. Bhosle, *Picard groups of the moduli spaces of vector bundles*, Math. Ann. **314** (1999), no. 2, 245–263, DOI 10.1007/s002080050293. MR1697444

[Bh2] U. N. Bhosle, *Maps into projective spaces*, Proc. Indian Acad. Sci. Math. Sci. **123** (2013), no. 3, 331–344, DOI 10.1007/s12044-013-0135-6. MR3102377

[Bh3] U. N. Bhosle, *Embedding of a compactified Jacobian and theta divisors*, Manuscripta Math. **157** (2018), no. 3-4, 361–385, DOI 10.1007/s00229-018-1002-7. MR3858407

[Bh4] U. N. Bhosle, *Picard bundle on the moduli space of torsionfree sheaves*, Proc. Indian Acad. Sci. Math. Sci. **130** (2020), no. 1, Paper No. 34, 13, DOI 10.1007/s12044-020-00562-y. MR4110703

[Bh5] Usha N. Bhosle, *Poincaré bundle for the fixed determinant moduli space*, `arXiv:2012.068`, Jan 2020

[Bh6] U. N. Bhosle, *Hitchin maps and parabolic Hitchin maps on the moduli spaces of Hitchin sheaves on nodal curves*, Preprint 2023–2024.

[BhP1] U. N. Bhosle and A. J. Parameswaran, *On the Poincaré formula and the Riemann singularity theorem over nodal curves*, Math. Ann. **342** (2008), no. 4, 885–902, DOI 10.1007/s00208-008-0259-7. MR2443766

[BhP2] U. N. Bhosle and A. J. Parameswaran, *Picard bundles and Brill-Noether loci in the compactified Jacobian of a nodal curve*, Int. Math. Res. Not. IMRN **15** (2014), 4241–4290, DOI 10.1093/imrn/rnt069. MR3244926

[BhS1] U. N. Bhosle and S. K. Singh, *Fourier-Mukai transform on a compactified Jacobian*, Int. Math. Res. Not. IMRN **13** (2020), 3991–4015, DOI 10.1093/imrn/rny136. MR4120316

[BhS2] U. N. Bhosle and S. Singh, *Weak point property and sections of Picard bundles on a compactified Jacobian over a nodal curve*, Proc. Indian Acad. Sci. Math. Sci. **126** (2016), no. 3, 329–339, DOI 10.1007/s12044-016-0290-7. MR3531852

[BBrN] V. Balaji, L. Brambila-Paz, and P. E. Newstead, *Stability of the Poincaré bundle*, Math. Nachr. **188** (1997), 5–15, DOI 10.1002/mana.19971880102. MR1484665

[BiBrGN] I. Biswas, L. Brambila-Paz, T. L. Gómez, and P. E. Newstead, *Stability of the Picard bundle*, Bull. London Math. Soc. **34** (2002), no. 5, 561–568, DOI 10.1112/S0024609302001170. MR1912878

[BiBrN1] I. Biswas, L. Brambila-Paz, and P. E. Newstead, *Deformations of the generalised Picard bundle*, Topology **45** (2006), no. 2, 403–419, DOI 10.1016/j.top.2005.08.004. MR2193340

[BiBrN2] I. Biswas, L. Brambila-Paz, and P. E. Newstead, *Stability of projective Poincaré and Picard bundles*, Bull. Lond. Math. Soc. **41** (2009), no. 3, 458–472, DOI 10.1112/blms/bdp017. MR2506830

[BiBrN3] I. Biswas, L. Brambila-Paz, and P. E. Newstead, *Stability and deformations of generalised Picard sheaves*, `arXiv:1812.09732v3` [math. AG].

[BiG1] I. Biswas and T. L. Gómez, *Hecke correspondence for symplectic bundles with application to the Picard bundles*, Internat. J. Math. **17** (2006), no. 1, 45–63, DOI 10.1142/S0129167X06003357. MR2204839

[BiG2] I. Biswas and T. L. Gómez, *Hecke transformation for orthogonal bundles and stability of Picard bundles*, Comm. Anal. Geom. **18** (2010), no. 5, 857–890, DOI 10.4310/CAG.2010.v18.n5.a1. MR2805146

[BiR] I. Biswas and G. V. Ravindra, *On the Picard bundle*, Bull. Sci. Math. **133** (2009), no. 1, 51–55, DOI 10.1016/j.bulsci.2008.08.004. MR2483522

[BrGP] L. Brambila-Paz, E. Gómez-González, and F. Pioli, *On Picard bundles over Prym varieties*, Collect. Math. **52** (2001), no. 2, 157–168. MR1852035

[BM] P. Belmans and S. Mukhopadhyay, *Admissible subcategories in derived categories of moduli of vector bundles on curves*, Adv. Math. **351** (2019), 653–675, DOI 10.1016/j.aim.2019.05.019. MR3954042

[BNR] A. Beauville, M. S. Narasimhan, and S. Ramanan, *Spectral curves and the generalised theta divisor*, J. Reine Angew. Math. **398** (1989), 169–179, DOI 10.1515/crll.1989.398.169. MR998478

[BO] A. Bondal and D. Orlov, *Semiorthogonal decomposition for algebraic varieties*, Preprint, `arXiv:alg-geom/9506012`, 1995.

[Ce] G. Ceresa, *C is not algebraically equivalent to C^- in its Jacobian*, Ann. of Math. (2) **117** (1983), no. 2, 285–291, DOI 10.2307/2007078. MR690847

[CKL] Y. Cho, Y. Kim, and K.-S. Lee, *Ulrich bundles on intersections of two 4-dimensional quadrics*, Int. Math. Res. Not. IMRN **22** (2021), 17277–17303, DOI 10.1093/imrn/rnz320. MR4345828

[DN] J.-M. Drezet and M. S. Narasimhan, *Groupe de Picard des variétés de modules de fibrés semi-stables sur les courbes algébriques* (French), Invent. Math. **97** (1989), no. 1, 53–94, DOI 10.1007/BF01850655. MR999313

[EGK] E. Esteves, M. Gagné, and S. Kleiman, *Autoduality of the compactified Jacobian*, J. London Math. Soc. (2) **65** (2002), no. 3, 591–610, DOI 10.1112/S002461070100309X. MR1895735

[EL] L. Ein and R. Lazarsfeld, *Stability and restrictions of Picard bundles, with an application to the normal bundles of elliptic curves*, Complex projective geometry (Trieste, 1989/Bergen, 1989), London Math. Soc. Lecture Note Ser., vol. 179, Cambridge Univ. Press, Cambridge, 1992, pp. 149–156, DOI 10.1017/CBO9780511662652.011. MR1201380

[FL] W. Fulton and R. Lazarsfeld, *On the connectedness of degeneracy loci and special divisors*, Acta Math. **146** (1981), no. 3-4, 271–283, DOI 10.1007/BF02392466. MR611386

[FK] A. Fonarev and A. Kuznetsov, *Derived categories of curves as components of Fano manifolds*, J. Lond. Math. Soc. (2) **97** (2018), no. 1, 24–46, DOI 10.1112/jlms.12094. MR3764066

[G1] R. C. Gunning, *Some special complex vector bundles over Jacobi varieties*, Invent. Math. **22** (1973), 187–210, DOI 10.1007/BF01389673. MR335858

[G2] R. C. Gunning, *Riemann surfaces and generalized theta functions*, Ergebnisse der Mathematik und ihrer Grenzgebiete [Results in Mathematics and Related Areas], Band 91, Springer-Verlag, Berlin-New York, 1976. MR457787

[HLST] D. H. Ruipérez, A. C. López Martín, D. Sánchez Gómez, and C. Tejero Prieto, *Moduli spaces of semistable sheaves on singular genus 1 curves*, Int. Math. Res. Not. IMRN **23** (2009), 4428–4462, DOI 10.1093/imrn/rnp094. MR2558337

[K1] G. Kempf, *Toward the inversion of abelian integrals. I*, Ann. of Math. (2) **110** (1979), no. 2, 243–273, DOI 10.2307/1971261. MR549489

[K2] G. R. Kempf, *Rank g Picard bundles are stable*, Amer. J. Math. **112** (1990), no. 3, 397–401, DOI 10.2307/2374748. MR1055650

[Laz] R. Lazarsfeld, *Positivity in algebraic geometry. II: Positivity for vector bundles, and multiplier ideals*, Ergebnisse der Mathematik und ihrer Grenzgebiete. 3. Folge. A Series of Modern Surveys in Mathematics [Results in Mathematics and Related Areas. 3rd Series. A Series of Modern Surveys in Mathematics], vol. 49, Springer-Verlag, Berlin, 2004, DOI 10.1007/978-3-642-18808-4. MR2095472

[Li] Y. Li, *Spectral curves, theta divisors and Picard bundles*, Internat. J. Math. **2** (1991), no. 5, 525–550, DOI 10.1142/S0129167X91000296. MR1124282

[LM] K.-S. Lee and H.-B. Moon, *Positivity of the Poincaré bundle on the moduli space of Vector bundles and its applications*, arXiv:2106.04857v1 [Math.AG].

[M1] A. Mattuck, *Picard bundles*, Illinois J. Math. **5** (1961), 550–564. MR150141

[M2] A. Mattuck, *Symmetric products and Jacobians*, Amer. J. Math. **83** (1961), 189–206, DOI 10.2307/2372727. MR142553

[Mu] S. Mukai, *Duality between $D(X)$ and $D(\hat{X})$ with its application to Picard sheaves*, Nagoya Math. J. **81** (1981), 153–175. MR607081

[Na1] M. S. Narasimhan, *Derived categories of moduli spaces of vector bundles on curves*, J. Geom. Phys. **122** (2017), 53–58, DOI 10.1016/j.geomphys.2017.01.018. MR3713871

[Na2] M. S. Narasimhan, *Derived categories of moduli spaces of vector bundles on curves II*, Geometry, algebra, number theory, and their information technology applications, Springer Proc. Math. Stat., vol. 251, Springer, Cham, 2018, pp. 375–382, DOI 10.1007/978-3-319-97379-1_16. MR3880395

[N] P. E. Newstead, *Introduction to moduli problems and orbit spaces*, Tata Institute of Fundamental Research Lectures on Mathematics and Physics, vol. 51, Tata Institute of Fundamental Research, Bombay; Narosa Publishing House, New Delhi, 1978. MR546290

[R] S. Ramanan, *The moduli spaces of vector bundles over an algebraic curve*, Math. Ann. **200** (1973), 69–84, DOI 10.1007/BF01578292. MR325615

[SWW] X. Su, B. Wang, and X. Wen, *Parabolic Hitchin maps and their generic fibers*, Math. Z. **301** (2022), no. 1, 343–372, DOI 10.1007/s00209-021-02896-3. MR4405653

[WW] J. Wang and X. Wen, *On the moduli spaces of parabolic symplectic orthogonal bundles on curves*, arXiv:2101.02383v1 [Math.AG].

Mathematics and Statistics Unit, Indian Statistical Institute, Bangalore 560059, India

Email address: usnabh07@gmail.com

Contemporary Mathematics
Volume **803**, 2024
https://doi.org/10.1090/conm/803/16093

Topology of the moduli spaces of Higgs bundles over abelian varieties

Indranil Biswas, Carlos Florentino, and Azizeh Nozad

Dedicated to Peter E. Newstead

ABSTRACT. Let G be a complex reductive group and A be an Abelian variety of dimension d over $\mathbb{C}$. We determine the Poincaré polynomial and also the mixed Hodge polynomial of the moduli spaces $\mathcal{M}_A^H(G)$ of G-Higgs bundles over A. We show that these are normal varieties with symplectic singularities, when G is a classical semisimple group. For $G = \mathrm{GL}_n(\mathbb{C})$, we also compute Poincaré polynomials of natural desingularizations of $\mathcal{M}_A^H(G)$ and of G-character varieties of free abelian groups, in some cases. In particular, explicit formulas are obtained when $\dim A = d = 1$, and also for rank 2 and 3 Higgs bundles, for arbitrary $d > 1$.

1. Introduction

The study of the topology of moduli spaces of Higgs bundles over algebraic curves was initiated in the very first article on Higgs bundles, [**Hi**], where these geometric objects were defined by Hitchin. In particular, in that work and in the work of Gothen, [**Goth**], we find some of the first computations of their Poincaré polynomials. Most of the stronger results deal with the cases where these moduli spaces are smooth, and the curve has genus > 1.

More recently, the study of moduli spaces of G-Higgs bundles over elliptic curves, for a complex reductive group G, was started by Franco, Garcia-Prada and Newstead in [**FGPN**], and has been generalized to G-Higgs bundles over abelian varieties by Biswas [**BF**, **Bi**] and Franco-Tortella [**FT**].

In this context, the non-abelian Hodge correspondence – between the moduli spaces of G-Higgs bundles and the G-character varieties of surface groups – became a very important tool in understanding the geometry of both these classes of spaces. The study of character varieties of *free abelian* groups was initiated by Thaddeus [**Th**] and Sikora [**Sk**], from the point of view of the algebraic geometry and relation to mirror symmetry, and by Florentino, Lawton and Silva in [**FL**, **FS**] from the point of view of the underlying topology and mixed Hodge structures. These character varieties are generally singular but of orbifold type.

2020 *Mathematics Subject Classification.* Primary 14J60, 14E15; Secondary 14L30, 32S35.
Key words and phrases. G-Higgs bundles, character varieties, Abelian varieties, symplectic singularities.

Take a complex abelian variety A of dimension d and a reductive complex affine algebraic group G. For a semistable G-Higgs bundles over A, the underlying principal G-bundle is semistable. Let $\mathcal{M}_A^H(G)$ denote the moduli space of semistable G-Higgs bundles over A whose underlying principal G-bundle is in the connected component, containing the trivial principal G-bundle, of the moduli space of principal G-bundles. In this article we prove that, when G is $\mathrm{GL}_n(\mathbb{C})$ or a classical semisimple group, the moduli spaces $\mathcal{M}_A^H(G)$ have symplectic singularities as defined by Beauville in [**Be2**]. We also determine the Poincaré polynomials, as well as mixed Hodge polynomials, of $\mathcal{M}_A^H(G)$ and of the corresponding G-character varieties, namely the moduli spaces of equivalence classes of G-representations of $\pi_1(A) \cong \mathbb{Z}^{2d}$.

In the last section, we consider $G = \mathrm{GL}_n(\mathbb{C})$ and prove that $\mathcal{M}_A^H(G)$ (as well as the $\mathrm{GL}_n(\mathbb{C})$-character varieties of the free abelian groups) admit a symplectic (also called crepant) resolution only when $d = 1$, and that in this case the resolution is given by the Hilbert-Chow morphism.

We finish with an explicit calculation of the Poincaré polynomials of the resolutions of $\mathcal{M}_A^H(G)$ and of the G-character varieties of $\mathbb{Z}^{2d}$, in all cases when their resolutions are Hilbert schemes. These are precisely the crepant cases: $d = 1$ and arbitrary rank $n \in \mathbb{N}$; and the non-crepant cases: $n = 2$ or $n = 3$ and arbitrary $d > 1$.

2. Preliminaries on Character Varieties and Higgs Bundles

As before, G is a complex reductive affine algebraic group. Its Lie algebra will be denoted by $\mathfrak{g}$. Let X be an irreducible smooth projective variety over $\mathbb{C}$, with fundamental group $\pi_1(X)$; we will omit reference to the base point.

2.1. Character varieties. Since $\pi_1(X)$ is a finitely presented group, the space $\mathsf{Hom}(\pi_1(X), G)$ of group homomorphisms $\pi_1(X) \longrightarrow G$ has a natural structure of an affine complex algebraic variety. The adjoint action of G on itself produces an action of G on $\mathsf{Hom}(\pi_1(X), G)$. The (affine) geometric invariant theoretic quotient

$$\mathsf{Hom}(\pi_1(X), G) /\!\!/ G,$$

which is called the G-character variety of $\pi_1(X)$, is an affine variety whose algebra of regular functions consists of the G-invariant regular functions on $\mathsf{Hom}(\pi_1(X), G)$.

Let A be a complex abelian variety of dimension d. Denote by

$$\mathcal{M}_A^R(G) := \mathsf{Hom}(\pi_1(A), G)^0 /\!\!/ G \tag{2.1}$$

the connected component, containing the trivial homomorphism, of the G-character variety of $\pi_1(A) \cong \mathbb{Z}^{2d}$. It is an irreducible affine variety of dimension $2dr$, where r is the rank of G, as shown in [**Sk**, Thm. 2.1(1)] and [**FL**, Cor. 5.14].

The celebrated non-abelian Hodge correspondence produces a C^∞ identification between the G-character variety $\mathsf{Hom}(\pi_1(X), G) /\!\!/ G$ and a moduli space of G-Higgs bundles over X, whose definition we now recall.

2.2. Principal Higgs bundles over abelian varieties. For any principal G-bundle E on a smooth complex projective variety X, the adjoint bundle $E(\mathfrak{g})$ is the Lie algebra bundle on X associated to E for the G-module $\mathfrak{g} = \mathrm{Lie}(G)$ defined by the adjoint action. The vector bundle over X whose sheaf of sections is given by the holomorphic i-forms will be denoted by Ω_X^i.

Recall that a G-Higgs bundle over X is a pair $(E,\, \theta)$ where E is a holomorphic principal G-bundle over X and θ is a holomorphic section of $E(\mathfrak{g}) \otimes \Omega^1_X$ satisfying the equation $\eta(\theta \otimes \theta) = 0$, where

$$\eta : (E(\mathfrak{g}) \otimes \Omega^1_X)^{\otimes 2} \longrightarrow E(\mathfrak{g}) \otimes \Omega^2_X$$

is the homomorphism obtained by combining the Lie bracket operation on $E(\mathfrak{g})$ with the exterior product of 1-forms.

Fix an ample line bundle on X to define the degree of coherent sheaves on X. See [**BG**], [**Si1**], [**Sc**] for the definition of semistable G-Higgs bundles. Let $\widetilde{\mathcal{M}}^H_X(G)$ denote the moduli space of semistable G-Higgs bundles with vanishing rational characteristic classes [**Sc**], [**Si2**].

There are homeomorphisms

$$\mathsf{Hom}(\pi_1(X), G) /\!\!/ G \;\cong\; \widetilde{\mathcal{M}}^H_X(G), \tag{2.2}$$

given by the non-abelian Hodge theory (see [**Si1, Hi, BG**]). In particular, these two spaces have the same homotopy type as well as the rational cohomologies.

We will now consider G-Higgs bundles over abelian varieties.

As before, let A be a complex abelian variety of dimension d. For any semistable G-Higgs bundle $(E,\, \theta)$ on A, the underlying principal G-bundle E is semistable [**BG, BF**]. Denote by $\mathcal{M}_A(G)$ the connected component, containing the trivial principal G-bundle, of the moduli space of semistable topologically trivial principal G-bundles on A. Also, denote by $\mathcal{M}^H_A(G)$ the moduli space of semistable G-Higgs bundles over A such that the underlying principal G-bundle is in $\mathcal{M}_A(G)$.

3. Normality of the Moduli Spaces

Let G be a connected reductive group over $\mathbb{C}$. We start by recalling the descriptions of $\mathcal{M}^H_A(G)$ and of $\mathcal{M}^R_A(G)$ in [**Bi**].

Let $A^\vee := \mathrm{Pic}^0(A)$ be the dual abelian variety, parametrizing the isomorphism classes of topologically trivial holomorphic line bundles on A. Fix a polarization $\xi \in H^2(A; \mathbb{Q}) \bigcap H^{1,1}(A)$. So we have a nondegenerate pairing

$$H^0(A,\, \Omega^1_A) \otimes H^1(A,\, \mathcal{O}_A) \longrightarrow \mathbb{C}, \quad \alpha \otimes \beta \longmapsto \int_A \alpha \wedge \beta \wedge \xi^{d-1}.$$

This provides identifications

$$H^0(A,\, \Omega^1_A) \;=\; H^1(A,\, \mathcal{O}_A)^* \;=\; T^*_z A^\vee \tag{3.1}$$

for all $z \in A^\vee$.

Now, fix a maximal torus $T \subset G$ with the corresponding Weyl group W, and denote the multiplicative group $\mathbb{C}^*$ by $\mathbb{G}_m$. Then, the *cocharacter lattice* of T

$$\Lambda_T := \mathsf{Hom}(\mathbb{G}_m,\, T)$$

is naturally a W-module. Abbreviate by $\mathcal{M}^H := \mathcal{M}^H_A(\mathbb{G}_m)$ the moduli space of topologically trivial $\mathbb{G}_m$-Higgs bundles over A. From (3.1) we have $\mathcal{M}^H \cong T^*A^\vee$, the cotangent bundle of $A^\vee$. Recall that $\mathcal{M}^R_A(G)$ is the character variety $\mathsf{Hom}(\mathbb{Z}^{2d}, G)^0 /\!\!/ G$ (as in (2.1)), and let us denote $\mathcal{M}^R_A(\mathbb{G}_m) \cong \mathbb{G}_m^{2d}$ by $\mathcal{M}^R$.

THEOREM 3.1 ([**Bi**]). *There are natural algebraic bijections*

$$\varphi : (\mathcal{M}^H \otimes_{\mathbb{Z}} \Lambda_T)/W \longrightarrow \mathcal{M}^H_A(G),$$

and

$$\psi : (\mathcal{M}^R \otimes_{\mathbb{Z}} \Lambda_T)/W \longrightarrow \mathcal{M}^R_A(G),$$

where the Weyl group W acts naturally on Λ_T and trivially on $\mathcal{M}^H$ and $\mathcal{M}^R$.

In the case $G = \mathrm{GL}_n(\mathbb{C})$, the moduli spaces in Theorem 3.1 can be identified with symmetric products of smooth varieties. Given a quasi-projective complex variety X, we denote by $\mathsf{Sym}^n(X)$ its nth symmetric product, that is the quotient X^n/S_n where the symmetric group S_n acts on X^n by permuting the entries.

PROPOSITION 3.2. *Let $G = \mathrm{GL}_n(\mathbb{C})$ and A an abelian variety of complex dimension d. Then there are natural isomorphisms:*

$$\mathcal{M}^H_A(G) \cong \mathsf{Sym}^n(\mathcal{M}^H),$$

and

$$\mathcal{M}^R_A(G) \cong \mathsf{Sym}^n(\mathbb{G}^{2d}_m).$$

PROOF. This follows from [**FT**, Thm. 1.1]; we provide a sketch of the arguments for completeness. Let us start with the second isomorphism. In this case, the character variety is irreducible (see [**Sk**, Thm. 2.1(1)]), and we have

$$\mathcal{M}^R_A(G) = \mathsf{Hom}(\mathbb{Z}^{2d}, G)/G = T^{2d}/W, \tag{3.2}$$

where $T = \mathbb{G}^n_m$ is a maximal torus of $G = \mathrm{GL}_n(\mathbb{C})$ and W its Weyl group (see [**FS**]). Since $W = S_n$, the symmetric group on n elements, acts diagonally on T^{2d} using the conjugation action of W on T, we get:

$$T^{2d}/W = (\mathbb{G}^n_m)^{2d}/S_n \cong \mathsf{Sym}^n(\mathbb{G}^{2d}_m). \tag{3.3}$$

A similar construction works also for the moduli space of topologically trivial $\mathrm{GL}_n(\mathbb{C})$-Higgs bundles over A, which are rank n Higgs vector bundles. Indeed, a semistable rank n Higgs bundle over A, with topologically trivial underlying vector bundle, is S-equivalent to a direct sum of $\mathbb{C}^*$-Higgs bundles. The ordering in the direct sum is irrelevant, so we get:

$$\mathcal{M}^H_A(G) \cong (\mathcal{M}^H)^n/S_n = \mathsf{Sym}^n(\mathcal{M}^H)$$

as in the statement of the proposition. □

REMARK 3.3. For $G = \mathrm{GL}_n(\mathbb{C})$, the descriptions in Theorem 3.1 and in Proposition 3.2 are indeed compatible, since the Weyl group $W = S_n$ acts on the lattice $\Lambda_T = \mathsf{Hom}(\mathbb{G}_m, \mathbb{G}^n_m)$, which is isomorphic to $\mathbb{Z}^n$, by permuting the standard generators. Consequently, $(\mathcal{M}^H \otimes_{\mathbb{Z}} \Lambda_T)/S_n \cong \mathsf{Sym}^n(\mathcal{M}^H)$, and similarly for $\mathcal{M}^R_A(G)$.

Let us say that G is a classical semisimple group if it is one of $\mathrm{SL}_n(\mathbb{C})$, $\mathrm{Sp}_n(\mathbb{C})$ or $\mathrm{SO}_n(\mathbb{C})$, for some $n \in \mathbb{N}$. Combining the isosingularity theorem of Simpson [**Si2**, Thm. 10.6], with results of Sikora [**Sk**], we can show that the morphisms in Theorem 3.1 are actually isomorphisms in the classical semisimple case.

THEOREM 3.4. *Let G be a classical semisimple group. Then the maps φ and ψ in Theorem 3.1 are algebraic isomorphisms.*

PROOF. We start with the case of $\mathcal{M}^R_A(G)$. For a maximal torus $T \subset G$ with Weyl group W, we consider the composition of maps

$$T^{2d} = \mathsf{Hom}(\mathbb{Z}^{2d}, T) \subset \mathsf{Hom}(\mathbb{Z}^{2d}, G) \twoheadrightarrow \mathcal{M}^R_A(G) = \mathsf{Hom}(\mathbb{Z}^{2d}, G) /\!\!/ G.$$

This factors through a morphism $\chi : T^{2d}/W \longrightarrow \mathcal{M}^R_A(G)$, where W acts diagonally on T^{2d} using the natural action of W on T. Moreover, χ is an isomorphism when G is a classical semisimple group by [**Sk**, Thm 2.1(3)]. Since T^{2d}/W is normal, so is $\mathcal{M}^R_A(G)$. Being a bijective morphism to a normal target, by Zariski's main theorem we obtain that ψ is an isomorphism.

Now, the non-abelian Hodge correspondence between $\mathcal{M}^H(G)$ and $\mathcal{M}^R(G)$, and the same correspondence for $\mathbb{C}^*$, allows us to build the following diagram

$$\begin{array}{ccc} (\mathcal{M}^H \otimes_{\mathbb{Z}} \Lambda_T)/W & \longrightarrow & (\mathcal{M}^R \otimes_{\mathbb{Z}} \Lambda_T)/W \\ \varphi\downarrow & & \cong\downarrow\psi \\ \mathcal{M}^H_A(G) & \longrightarrow & \mathcal{M}^R_A(G), \end{array}$$

whose commutativity can be easily checked. By Simpson's isosingularity theorem [**Si2**, Thm. 10.6], both horizontal arrows are local isomorphisms for the étale topology (see also [**Ti**, Rem. 3.3]). Since a bijective algebraic morphism which is étale is an isomorphism, we conclude that φ is an algebraic isomorphism. □

The following is immediate from Proposition 3.2 and Theorem 3.4.

COROLLARY 3.5. *Assume that* $G = \mathrm{GL}_n(\mathbb{C})$ *or it is a classical semisimple group. Then* $\mathcal{M}^H_A(G)$ *and* $\mathcal{M}^R_A(G)$ *are irreducible and normal varieties.*

REMARK 3.6. This corollary extends [**FGPN**, Thm 3.7] to the context of general abelian varieties. The question of normality of $\mathcal{M}^H_A(G)$ and $\mathcal{M}^R_A(G)$ for general reductive G remains open, as far as we know. On the other hand, [**La**, Thm 4.16] showed that the moduli spaces of principal G-bundles $\mathcal{M}_A(G)$ are irreducible normal varieties, for general G, when $\dim A = 1$.

4. Mixed Hodge Polynomials

For background on mixed Hodge structures (abbreviated as MHS), we refer to [**PS**] and [**D**]. Given a quasi-projective algebraic variety X over $\mathbb{C}$, its mixed Hodge polynomial (MHP) μ_X is the polynomial in three variables whose coefficients are the Deligne-Hodge numbers

$$h^{k,p,q}(X) = \dim H^{k,p,q}(X),$$

where $H^{k,p,q}(X)$ is the (p,q) graded piece of $H^k(X;\mathbb{Q})$, the kth cohomology of X, with rational coefficients:

$$\mu_X(t,u,v) = \sum_{k,p,q\geq 0} h^{k,p,q}(X)\, t^k\, u^p\, v^q. \tag{4.1}$$

From μ_X we can obtain the Poincaré polynomial P_X and the Euler characteristic χ_X as specializations: $P_X(t) := \sum_{k\geq 0} b^k(X)\, t^k = \mu_X(t,1,1)$, where $b^k(X) = \dim H^k(X;\mathbb{Q})$ is the k-th Betti number. Note that $\chi_X = \mu_X(-1,1,1)$.

Let T be a maximal torus of G. Its Lie algebra will be denoted by $\mathfrak{t}$. The Weyl group W for T acts on both T and on $\mathfrak{t}$. The natural morphism from T^{2d}/W to $\mathcal{M}^R_A(G)$ [**Sk**, Thm. 2.1] induces an isomorphism of mixed Hodge structures; this yields the following result.

THEOREM 4.1. ([**FS**, Thm. 5.2], [**FLS**, Thm. 5.8]) *Let G be a connected reductive group. The mixed Hodge polynomial of $\mathcal{M}^R_A(G)$ is given by:*

$$\mu_{\mathcal{M}^R_A(G)}(t,\,u,\,v) = \frac{1}{|W|}\sum_{w\in W} (\det(I + t\,u\,v\,w))^{2d},$$

where $w \in W$ acts on $\mathfrak{t}$ and I is the identity automorphism of $\mathfrak{t}$.

In particular, the Poincaré polynomial of $\mathcal{M}_A^R(G)$ is (see also [**FLS**]):

$$P_{\mathcal{M}_A^R(G)}(t) = \frac{1}{|W|}\sum_{w\in W}(\det(I + t\,w))^{2d}. \tag{4.2}$$

Now, we can state the analogous result for the mixed Hodge polynomials of $\mathcal{M}_A^H(G)$ and of $\mathcal{M}_A(G)$, the connected component (of the trivial G-bundle) of the moduli space of semistable topologically trivial G-bundles on A.

THEOREM 4.2. *The mixed Hodge structures on $\mathcal{M}_A(G)$ and on $\mathcal{M}_A^H(G)$ coincide, and their mixed Hodge polynomial is given by:*

$$\mu_{\mathcal{M}_A^H(G)}(t,\, u,\, v) \;=\; \frac{1}{|W|}\sum_{w\in W}(\det(I + t\,u\,w)\det(I + t\,v\,w))^d,$$

where I is the identity automorphism on $\mathfrak{t}$.

PROOF. By [**D**, Thm. 2.3.5], any morphism of MHSs is strict with respect to both the weight and the Hodge filtration. This means that any algebraic map that induces an isomorphism in usual cohomology, induces also an isomorphism of MHSs. Now, the algebraic map

$$\mathcal{M}_A(G) \hookrightarrow \mathcal{M}_A^H(G) \tag{4.3}$$

given by the natural inclusion $[E] \longmapsto [(E,0)]$ for a G-bundle E, is a strong deformation retraction (see [**BF**]), so that these moduli spaces have isomorphic MHSs on their cohomologies.

In the same way, the algebraic bijection [**Bi**, Thm. 2.2]

$$\delta : (A^\vee \otimes_{\mathbb{Z}} \Lambda_T)/W \longrightarrow \mathcal{M}_A(G)$$

which induces isomorphisms in cohomology (see also [**BFM**]) gives also isomorphisms of MHSs:

$$H^*(\mathcal{M}_A(G)) \cong H^*\left((A^\vee \otimes_{\mathbb{Z}} \Lambda_T)/W\right) \cong H^*(A^\vee \otimes_{\mathbb{Z}} \Lambda_T)^W,$$

where the last space denotes the W-invariant subspaces of $H^*(A^\vee \otimes_{\mathbb{Z}} \Lambda_T)$, from which $\mu_{\mathcal{M}_A(G)}$ can now be obtained.

Observe now that $B := A^\vee \otimes_{\mathbb{Z}} \Lambda_T$ is also an abelian variety of dimension $n = dr$, where $r = \dim T$ is the rank of G. Moreover, it is well known that the mixed Hodge polynomial of any abelian variety B of dimension n is:

$$\mu_B(t,\, u,\, v) \;=\; ((1+tu)(1+tv))^n\,,$$

since B has a pure Hodge structure and it is generated in degree 1; moreover, in the decomposition

$$H^1(B) \;=\; H^1(A^\vee \otimes_{\mathbb{Z}} \Lambda_T) \;=\; H^{1,0}(B) \oplus H^{0,1}(B)$$

$H^{1,0}(B)$ can be naturally identified with the direct sum of $d = \dim A$ copies of $\mathfrak{t}$, the Lie algebra of T. Now, recall that if $V = \oplus_{k\geq 0} V^k$ is a (finite dimensional) graded $\mathbb{C}$-vector space and $g : V \longrightarrow V$ is a linear map preserving the grading, the graded-character of g, defined as the series in the formal variable t:

$$\chi_g(V) := \sum_{k\geq 0}\operatorname{tr}(g|_{V^k})\, t^k \in \mathbb{C}[t] \tag{4.4}$$

is additive and multiplicative, under direct sums and tensor products, respectively, and verifies:

$$\chi_g(\wedge^* V) = \det(I + t^\delta g),$$

whenever all elements of V have degree $\delta \in \mathbb{Z}$ (see, for example [**Se**]). In our case, we can write:

$$H^*(B) \cong \bigwedge^* H^1(B) \cong \bigwedge^* H^{1,0}(B) \otimes \bigwedge^* H^{0,1}(B) = \bigwedge^* \mathfrak{t}^d \otimes \bigwedge^* \bar{\mathfrak{t}}^d \cong (\bigwedge^* \mathfrak{t} \otimes \bigwedge^* \bar{\mathfrak{t}})^{\otimes d},$$

where, in terms of the triple grading, $\mathfrak{t}$ has degree $(1,1,0)$ and $\bar{\mathfrak{t}}$ has degree $(1,0,1)$ (corresponding, respectively, to the monomials tu and tv in (4.1)). Hence, applying (4.4) to the computation of the dimension of the W representation $H^*(B)$ (see, for example [**FLS**, Cor. 4.4]), we conclude that

$$\mu_{B/W}(t,u,v) = \frac{1}{|W|}\sum_{w\in W}\chi_w(H^*(B)) = \frac{1}{|W|}\sum_{w\in W}\det(I + tu\,w)^d\det(I + tv\,w)^d$$

as wanted. □

REMARK 4.3. Note that this formula specializes, for $u = v = 1$, to the same Poincaré polynomial as in Equation (4.2), as it should, given the homeomorphism $\mathcal{M}_A^H(G) \simeq \mathcal{M}_A^R(G)$.

On the other hand, the mixed Hodge structures on $\mathcal{M}_A^H(G)$ and on $\mathcal{M}_A^R(G)$ are very different: $\mathcal{M}_A^R(G)$ is a Stein analytic space (which implies that $\mathcal{M}_A^R(G)$ has no positive dimensional projective subvarieties), whereas $\mathcal{M}_A^H(G)$ has projective subvarieties, such as $\mathcal{M}_A(G)$. In particular, the MHS on $\mathcal{M}_A^H(G)$ is actually pure and, by contrast, $\mathcal{M}_A^R(G)$ has what is called a *balanced* or *Hodge-Tate* MHS.

In general, even though $\mathcal{M}^H$ (see Theorem 3.1) is a smooth quasi-projective variety, both $\mathcal{M}_A^R(G)$ and $\mathcal{M}_A^H(G)$ are generally singular algebraic varieties of orbifold type. In some cases, one can find natural desingularizations, and compute their MHP / Poincaré polynomials. We first consider their singularity types.

5. Symplectic singularities and resolutions

Let X be a normal and irreducible complex algebraic variety. We say that X is a *symplectic variety* if its smooth part $X^{sm} \subset X$ admits a holomorphic symplectic form ω (closed and non-degenerate) whose pull-back to any resolution $\pi : Y \longrightarrow X$ extends to a holomorphic 2-form $\widetilde{\omega}$ on Y (not necessarily non-degenerate). Equivalently, by Namikawa's theorem ([**Na1**, Thm. 6]), X is symplectic if and only if it has only rational Gorenstein singularities and its smooth part carries a holomorphic symplectic form.

If $\pi : Y \longrightarrow X$ is a resolution of singularities of a symplectic variety X, then π is called a *symplectic resolution* if the above extended form $\widetilde{\omega}$ is non-degenerate (so that it is a *de facto* symplectic form on Y). This notion is actually independent of the starting symplectic form ω on X^{sm}. Therefore, if $\pi : Y \longrightarrow X$ is a symplectic resolution, then the canonical divisor K_Y is trivial.

5.1. Symplectic singularities.

THEOREM 5.1. *Assume that $\mathcal{M}_A^H(G)$ is normal. Then, $\mathcal{M}_A^H(G)$ and $\mathcal{M}_A^R(G)$ are symplectic varieties, and have only rational Gorenstein singularities.*

PROOF. Let us first consider the G-Higgs bundle case. When $G = \mathrm{GL}_{nn}(\mathbb{C})$ the theorem follows from the fact that symmetric products of symplectic varieties are symplectic (see [**Be2**, Prop. 2.4]). Since $\mathcal{M}_A^H(G)$ is normal, we have the following isomorphism

$$\mathcal{M}_A^H(G) \cong (\mathcal{M}^H \otimes_{\mathbb{Z}} \Lambda_T)/W,$$

from Theorem 3.4. It is well known that $\mathcal{M}^H$, the moduli space of $\mathbb{C}^*$-Higgs bundles over A, is a hyperKähler manifold. In particular, it has a holomorphic symplectic form. Hence, both $\mathcal{M}^H$ and $X := \mathcal{M}^H \otimes_{\mathbb{Z}} \Lambda_T$ are smooth symplectic varieties. Now, the Weyl group W acts by automorphisms of X preserving the symplectic form, so that this form descends to a symplectic form ω on the smooth part

$$\mathcal{M}_A^H(G)^{sm} \cong (X/W)^{sm} = (X \setminus \bigcup_{g\in W\setminus\{1\}} X^g)/W,$$

where X^g is the fixed point locus for the action of the non-trivial element $g \in W \setminus \{1\}$. Since W acts by reflections on Λ_T, the codimension of each X^g coincides with the dimension of $\mathcal{M}^H$, which is $2d$. Hence, the complement of $(X/W)^{sm}$ in $X/W = \mathcal{M}_A^H(G)$ has also codimension $2d \geq 2$. Consequently, by [**Be2**, Prop 2.4], ω extends to a symplectic form on X/W. The case of $\mathcal{M}_A^R(G)$ is analogous. The rational Gorenstein property is immediate from [**Be2**, Prop. 1.3]. □

REMARK 5.2. In the case $G = \mathrm{SL}_n(\mathbb{C})$, the character variety $\mathcal{M}_A^R(G)$ has been previously shown to be symplectic, when A is an elliptic curve (see [**BS**]).

5.2. Hilbert-Chow morphism and symplectic resolutions. Given an algebraic variety X, the Hilbert-Chow morphism is:

$$\mathsf{Hilb}^n(X) \longrightarrow \mathsf{Sym}^n(X),$$

where $\mathsf{Hilb}^n(X)$ denotes the Hilbert scheme of zero dimensional subschemes of length n in the variety X, and the morphism sends a scheme to its support. When X is a smooth surface, the Hilbert-Chow morphism is a desingularization of the symmetric product. From now on, we only deal with the case $\mathrm{GL}_n = \mathrm{GL}_n(\mathbb{C})$.

COROLLARY 5.3. *When C is an elliptic curve,*

$$\mathsf{Hilb}^n(T^*C) \longrightarrow \mathsf{Sym}^n(T^*C)$$

is a symplectic (crepant) resolution of singularities of the moduli space $\mathcal{M}_C^H(\mathrm{GL}_n)$. In fact, this is the unique such resolution.

PROOF. This result appears in [**Ti**, Thm. 4.9]; it follows from Beauville ([**Be1**, Prop. 5]), since T^*C is an algebraic surface with the natural Liouville symplectic form, and has trivial canonical bundle (being a local statement, [**Be1**, Prop. 5] applies to non-projective surfaces). The uniqueness follows from Fu-Namikawa's theorem [**FN**, Cor. 2.3]. □

THEOREM 5.4. *Let A be an abelian variety of dimension $d \geq 1$, and $n = 2$ or 3. Then*

$$\mathsf{Hilb}^n(T^*A^\vee) \qquad (\textit{respectively,} \quad \mathsf{Hilb}^n(\mathbb{G}_m^{2d}))$$

are resolutions of singularities of $\mathcal{M}_A^H(\mathrm{GL}_n)$ (respectively, $\mathcal{M}_A^R(\mathrm{GL}_n)$). These are symplectic resolutions if and only if $d = 1$.

PROOF. In [**Ch2**, Thm. 3.2.2], Cheah showed that $\mathsf{Hilb}^2(X)$ and $\mathsf{Hilb}^3(X)$ are smooth varieties, when X is smooth. Since $\mathcal{M}^H \cong T^*A^\vee$ and $\mathbb{G}_m^{2d}$ are smooth, the result is clear. See also [**FT**, Rem. 5.11]. When $d = 1$ the resolutions are symplectic by Corollary 5.3.

When $d > 1$, we start by considering a general smooth symplectic variety X with an action by a finite group F that preserves the symplectic form. Then X/F is $\mathbb{Q}$-*factorial* and normal. Hence, any component E of the exceptional locus of a proper resolution $\pi : Z \longrightarrow X/F$ is of codimension 1. On the other hand, if π is a symplectic resolution, then by the semi-smallness property we have $2 = 2\mathsf{codim}(E) \geq \mathsf{codim}(\pi(E))$ (see [**Ka**, Lemma 2. 7]). As X/F is a normal variety, its singular locus is of codimension ≥ 2, and therefore $\mathsf{codim}(\pi(E)) = 2$. However, the singular locus of X/F is contained in $p(Y)$, where $p : X \longrightarrow X/F$ is the natural projection,

$$Y := \bigcup_{g \in G\setminus\{1\}} X^g,$$

and $X^g \subset X$ is the subvariety fixed by the action of the non-trivial element $g \in F\setminus\{1\}$. Hence $\mathsf{codim}(Y) \leq 2$ and since all components of Y are even dimensional (they are fixed loci under a symplectic action), Y has to be of pure codimension 2.

On the other hand, in view of the isomorphism

$$\mathcal{M}^H_A(\mathrm{GL}_n) \cong (\mathcal{M}^H \otimes_{\mathbb{Z}} \Lambda_T)/W$$

from Theorem 3.4, since W acts by reflections on Λ_T, the codimension of the fixed subvariety coincides with the dimension of $\mathcal{M}^H$, which is $2d > 2$. So, taking $X = \mathcal{M}^H \otimes_{\mathbb{Z}} \Lambda_T$ and $F = W$ in the previous paragraph we obtain a contradiction. This proves that $\mathsf{Hilb}^n(T^*A^\vee)$ is not a symplectic resolution of $\mathcal{M}^H_A(\mathrm{GL}_n)$ when $d > 1$.

Analogously, $\mathsf{Hilb}^n(\mathbb{G}_m^{2d}) \longrightarrow \mathcal{M}^R_A(\mathrm{GL}_n)$ is not a symplectic resolution, for $d > 1$. □

6. Topology of Resolutions via Hilbert Schemes

In this section, we consider $G = \mathrm{GL}_{nn}(\mathbb{C})$, and compute the Poincaré polynomials of the natural desingularizations of $\mathcal{M}^H_A(G)$ and of $\mathcal{M}^R_A(G)$ in all cases when these are given by the Hilbert-Chow morphism.

Our strategy is based on Cheah's identity for the generating series of Serre polynomials of Hilbert schemes. This is enough to get the Poincaré of the resolutions $\mathsf{Hilb}^n(T^*A^\vee)$ and $\mathsf{Hilb}^n(\mathbb{G}_m^{2d})$, given that the Hodge numbers on these moduli spaces have special symmetries.

6.1. The Serre polynomial and Hilbert E-series. The Serre polynomial (also known as E-polynomial) of a quasi-projective complex variety X is defined by:

$$E_X(u,v) := \mu^c_X(-1,u,v) = \sum_{k,p,q\geq 0} (-1)^k\, h^{k,p,q}_c(X)\, u^p\, v^q,$$

where $h^{k,p,q}_c(X) = \dim_{\mathbb{C}} H^{k,p,q}_c(X, \mathbb{Q})$ and we use the super/subscript c to refer to *compactly supported cohomology.*

Assuming that X is smooth (but not necessarily projective), the Poincaré polynomial $P_X(t) := \sum_{k=0}^{2d} \dim_{\mathbb{C}} H^k(X, \mathbb{Q})\, t^k$ (for usual cohomology) can be deduced from E_X in the pure cohomology case. Recall that X is said to have a *pure Hodge structure* if $h^{k,p,q}(X) = 0$ unless $k = p + q$ (which, for X smooth, is equivalent to requiring $h^{k,p,q}_c(X) = 0$ unless $k = p + q$).

LEMMA 6.1. *Let $E_X(u, v) \in \mathbb{Z}[u, v]$ be the E-polynomial of a smooth quasi-projective variety X, of complex dimension d. Then its Poincaré polynomial is given by*

$$P_X(t) = t^{2d} E_X(-t^{-1}, -t^{-1})$$

whenever the cohomology of X is pure. Moreover, $P_X(t) = E_X(-t, -t)$ if X is projective.

PROOF. This is a simple calculation. From the definition:

$$E_X(-t, -t) = \sum_{p,q=0}^{d} (-1)^{p+q} h_c^{p+q,p,q}(X)\,(-t)^p(-t)^q = \sum_{p,q=0}^{d} h_c^{p+q,p,q}(X)\,t^{p+q},$$

which is exactly the compactly supported Poincaré polynomial $P_X^c(t)$ since $\sum_{p+q=k} h_c^{p+q,p,q}(X) = \dim H_c^k(X;\mathbb{Q})$. Hence, the Lemma follows from Poincaré duality which implies $P_X^c(t) = t^{2d}P_X(t^{-1})$, in the quasi-projective case, and $P_X = P_X^c$ when X is projective. □

Let us now define the *Hilbert E-series of X*, which collects the Serre E-polynomials of $\mathsf{Hilb}^n X$, for all $n \in \mathbb{N}$.

DEFINITION 6.2. Let X be a complex quasi-projective variety. The *Hilbert E-series of* X is the formal power series (in the formal variable y) defined by:

$$\mathsf{HE}_X(u, v, y) := \sum_{n\geq 0} E_{\mathsf{Hilb}^n X}(u, v)\, y^n \in \mathbb{Z}[u, v][\![y]\!]. \tag{6.1}$$

In [**Ch1**], J. Cheah introduced a method for determining HE_X from the knowledge of E_X and of $\mathsf{HE}_{(\mathbb{C}^d,0)}$, where $\mathsf{Hilb}^n(\mathbb{C}^d, 0)$ are the *punctual Hilbert schemes* of subschemes of $\mathbb{C}^d$ which have length n, dimension 0, and are supported at the origin. To write this result in a convenient form, we introduce the so called *plethystic exponential* and *logarithm* functions. Given a power series of the form $f(u, v, y) = \sum_{n\geq 1}\sum_{p,q\geq 0} c_{p,q,n}u^pv^q\, y^n \in \mathbb{Z}[u, v][\![y]\!]$, we define:

$$\mathsf{PExp}[f(u, v, y)] := \exp\left(\sum_{n\geq 1}\frac{1}{n}f(u^n, v^n, y^n)\right) = \prod_{n\geq 1}\prod_{p,q\geq 1}(1 - u^pv^q\, y^n)^{-c_{p,q,n}}. \tag{6.2}$$

The equivalence of the two above expressions of PExp follows from the series expansion of the usual logarithm. The plethystic logarithm PLog is the inverse function of PExp (see for example [**FNZ1**]).

THEOREM 6.3 ([**Ch1**, Main Thm. sec. 2]). *Let X be smooth and quasi-projective of dimension d. Then, we have:*

$$\mathsf{HE}_X(u, v, y) = \mathsf{PExp}\left(E_X(u, v)\,\mathsf{PLog}(\mathsf{HE}_{(\mathbb{A}^d,0)}(u, v, y))\right).$$

PROOF. This formula is due to Cheah [**Ch1**, S 2], written in the form:

$$\mathsf{HE}_X(u, v, y) = \exp\left(\sum_{n\geq 1}\frac{1}{n}E_X(u^n, v^n)\, g_d(u^n, v^n, y^n)\right) \tag{6.3}$$

where $g_d(u, v, y) = \sum_{n\geq 1}\sum_{p,q\geq 0} a_d(p, q, n)\, u^pv^qy^n$, and the coefficients $a_d(p, q, n) \in \mathbb{Q}$ are defined by the infinite product expansion:

$$\mathsf{HE}_{(\mathbb{C}^d,0)}(u, v, y) = \prod_{n\geq 1}\prod_{p,q\geq 0}(1 - u^pv^q\, y^n)^{-a_d(p,q,n)}. \tag{6.4}$$

Using the product form of (6.2), Equation (6.4) is seen to be equivalent to

$$g_d(u, v, n) = \mathsf{PLog}(\mathsf{HE}_{(\mathbb{C}^d,0)}(u, v, y)),$$

and using the other form of (6.2) we get (6.3). □

To apply Theorem 6.3 to a surface S, since $\dim S = 2$, we now compute $\mathsf{PLog}(\mathsf{HE}_{(\mathbb{C}^2,0)})$.

COROLLARY 6.4. *Let S be a quasi-projective smooth surface. Then:*

$$\mathsf{HE}_S(u,\,v,\,y) = \mathsf{PExp}\left(\frac{E_S(u,\,v)\,y}{1-uvy}\right).$$

PROOF. In [**Ch2**, Cor. (4) of Thm. 3.3.3] the following expansion is obtained:

$$\sum_{n\geq 0} P_{\mathsf{Hilb}^n(\mathbb{C}^2,\,0)}(t)\,y^n = \prod_{n\geq 1}(1-t^{2n-2}y^n)^{-1},$$

so that $P_{\mathsf{Hilb}^n(\mathbb{C}^2,\,0)}(t)$ are polynomials in t^2. Since $\mathsf{Hilb}^n(\mathbb{C}^2,\,0)$ is smooth and projective, and has a cellular decomposition ([**Ch2**]), its Hodge structure is pure and balanced (i.e., $h_c^{k,p,q} = 0$ if $p \neq q$, or $k \neq p+q$) and Lemma 6.1 gives:

$$\mathsf{HE}_{(\mathbb{C}^2,0)}(u,v,y) = \sum_{n\geq 0} E_{\mathsf{Hilb}^n(\mathbb{C}^2,0)}(u,v)\,y^n = \prod_{n\geq 1}(1-(uv)^{n-1}y^n)^{-1}.$$

Hence, by definition of PLog:

$$\mathsf{PLog}\left(\mathsf{HE}_{(\mathbb{C}^2,0)}(u,v,y)\right) = \sum_{n\geq 1}(uv)^{n-1}y^n = \frac{y}{1-uvy}$$

and the Corollary follows from Theorem 6.3. □

6.2. The case of elliptic curves. Let C be an elliptic curve. We now use Corollary 6.4 to obtain the Poincaré polynomials of $\mathsf{Hilb}^n(T^*C)$, which are crepant resolutions of $\mathcal{M}_C^H(\mathrm{GL}_n)$, by Corollary 5.3, since $C = C^\vee$. For that, note that Equation (6.1), for a smooth surface S with pure cohomology, implies:

$$\mathsf{HE}_S(-t^{-1},-t^{-1},yt^4) := \sum_{n\geq 0} E_{\mathsf{Hilb}^n S}(-t^{-1},-t^{-1})\,t^{4n}y^n = \sum_{n\geq 0} P_{\mathsf{Hilb}^n S}(t)\,y^n.$$

Now, let $\mathcal{P}_n$ be the set of partitions of $n \in \mathbb{N}$, and denote a partition as:

$$\lambda = [1^{\lambda_1}2^{\lambda_2}\cdots n^{\lambda_n}] \in \mathcal{P}_n,$$

where $\lambda_j \geq 0$ is the number of parts of size $j = 1, \cdots, n$, so that $n = \sum_{j=1}^n j\lambda_j$.

THEOREM 6.5. *Let C be an elliptic curve and $n \in \mathbb{N}$. The Poincaré polynomials of $\mathsf{Hilb}^n(T^*C)$, the unique crepant resolution of $\mathcal{M}_C^H(GL_n\mathbb{C})$, are given by:*

$$P_{\mathsf{Hilb}^n(T^*C)}(t) = \frac{(t+1)^{2n}}{(t^2-1)^n}\sum_{\lambda\in\mathcal{P}_n}(t^{2\lambda_1}-1)\cdots(t^{2\lambda_n}-1)\,t^{2n-2|\lambda|},$$

where $|\lambda| = \sum_{j=1}^n \lambda_j$ is the number of parts of $\lambda \in \mathcal{P}_n$.

PROOF. The surface $S = T^*C$ is homotopic to C, which gives (note the factor uv from compactly supported cohomology) $E_{T^*C}(u,v) = uv(u-1)(v-1)$. So, from

Corollary 6.4:

$$\begin{aligned}\mathsf{HE}_{T^*C}(u,v,y) &= \mathsf{PExp}\left(\frac{(u-1)(v-1)\,uvy}{1-uvy}\right)\\ &= \frac{\mathsf{PExp}(\frac{uvy}{1-uvy})\,\mathsf{PExp}(\frac{(uv)^2y}{1-uvy})}{\mathsf{PExp}(\frac{uv^2y}{1-uvy})\,\mathsf{PExp}(\frac{u^2vy}{1-uvy})}\\ &= \prod_{n\geq 1}\frac{(1-y^nu^nv^{n+1})(1-y^nu^{n+1}v^n)}{(1-y^nu^{n+1}v^{n+1})(1-y^nu^nv^n)}\end{aligned}$$

and

(6.5)

$$\sum_{n\geq 0} P_{\mathsf{Hilb}^n T^*C}(t)\,y^n = \mathsf{HE}_{T^*C}(-t^{-1},-t^{-1},yt^4) = \prod_{n\geq 1}\frac{(1+y^nt^{2n-1})^2}{(1-y^nt^{2n-2})(1-y^nt^{2n})}.$$

To extract the coefficient of y^n we use the (partial fraction) expansion:

$$\frac{(1+\frac{z}{t})^2}{(1-\frac{z}{t^2})(1-z)} = 1+\sum_{k\geq 1} a_k(t)\,z^k,$$

where $a_k(t) = \frac{t+1}{t-1}(1-\frac{1}{t^{2k}})$, so that, replacing z successively by $(yt^2)^n$, (6.5) becomes:

$$\frac{(1+yt)^2}{(1-y)(1-yt^2)}\cdot\frac{(1+y^2t^3)^2}{(1-y^2t^2)(1-y^2t^4)}\cdots = \sum_{n\geq 0}\frac{(t+1)^n}{(t-1)^n}\sum_{\lambda\in\mathcal{P}_n}(t^{2\lambda_1}-1)\cdots(t^{2\lambda_n}-1)\,t^{2n-2|\lambda|}y^n,$$

as wanted. □

Corollary 6.6. *The Poincaré polynomial of* $\mathsf{Hilb}^n(\mathbb{G}_m^2)$ *coincides with the one of* $\mathsf{Hilb}^n(T^*C)$*, for all* $n\geq 1$.

Proof. The computations for $\mathsf{Hilb}^n(\mathbb{G}_m^2)$ follow the same procedure as for $\mathsf{Hilb}^n(T^*C)$, using now $E_{\mathbb{G}_m^2}(u,v) = (uv-1)^2$ and the fact that $\mathbb{G}_m^2 = (\mathbb{C}^*)^2$ has a *round* Hodge structure (its Hodge numbers verify $h^{k,p,q}\neq 0$ only when $k=p=q$, see [**FS**]). □

Remark 6.7. Our computations were done in the quasi-projective setting; in the projective context this agrees with results of Göttsche on Poincaré polynomials of Hilbert schemes of smooth projective surfaces, and also with the virtual motive computations in [**Gott**]. Note also that similar formulas for the case $G = \mathrm{SL}_n(\mathbb{C})$ can be obtained by the same method.

6.3. Higher dimensional abelian varieties. We now turn to the computation of Poincaré polynomials of $\mathsf{Hilb}^n(\mathcal{M}^H)$ (respectively, $\mathsf{Hilb}^n(\mathbb{G}_m^{2d})$) for abelian varieties A of higher dimension d, with $n = 2$ and 3 which, by Theorem 5.4 are natural resolutions of $\mathcal{M}_A^H(\mathrm{GL}_n)$ (respectively, $\mathcal{M}_A^R(\mathrm{GL}_n)$).

We use the following notation: $f_1(u,v) = E_{\mathcal{M}^H}(u,v)$, $f_1^*(u,v) := f_1(u^2,v^2)$ and $f_1^{**}(u,v) := f_1(u^3,v^3)$. Also, define, with $x = uv$,

$$f_2 := E_X\left(E_{\mathsf{Hilb}^2(\mathbb{A}^2,0)}-1\right) = E_X\left(x+\cdots+x^{d-1}\right)$$

$$f_3 := E_X\left(E_{\mathsf{Hilb}^3(\mathbb{A}^2,0)}-E_{\mathsf{Hilb}^2(\mathbb{A}^2,0)}\right) = E_X\left(\sum_{1\leq i\leq j\leq d-1} x^{i+j}\right) \tag{6.6}$$

(see [**Ch2**]).

PROPOSITION 6.8. *The E-polynomials of* $\mathsf{Hilb}^2(\mathcal{M}^H)$ *and* $\mathsf{Hilb}^3(\mathcal{M}^H)$ *are given by:*

$$E_{\mathsf{Hilb}^2(\mathcal{M}^H)} = f_2 + \frac{1}{2}(f_1^2 + f_1^*)$$
$$E_{\mathsf{Hilb}^3(\mathcal{M}^H)} = f_3 + f_1 f_2 + \frac{f_1^3}{6} + \frac{f_1 f_1^*}{2} + \frac{f_1^{**}}{3}.$$

PROOF. We will use Theorem 6.3. It is not difficult to show that

$$\mathsf{PLog}(\mathsf{HE}_{(\mathbb{C}^d,0)}(u,v,y)) = y + y^2(E_{\mathsf{Hilb}^2(\mathbb{C}^2,0)} - 1) + y^3(E_{\mathsf{Hilb}^3(\mathbb{C}^2,0)} - E_{\mathsf{Hilb}^2(\mathbb{C}^2,0)}) + \cdots$$

so that multiplying by $E_X(u,v)$, we can write:

$$E_X(u,v)\mathsf{PLog}(\mathsf{HE}_{(\mathbb{C}^d,0)}(u,v,y)) = yf_1(u,v) + y^2 f_2(u,v) + y^3 f_3(u,v) + \cdots,$$

where $f_1(u,v) = E_X(u,v)$. Hence we have

$$\begin{aligned}\mathsf{HE}_X(u,v,y) &= \mathsf{PExp}(E_X(u,v)\,\mathsf{PLog}(\mathsf{HE}_{(\mathbb{C}^d,0)}(u,v,y)))\\ &= \mathsf{PExp}(yf_1(u,v) + y^2 f_2(u,v) + y^3 f_3(u,v) + \cdots)\\ &= \exp(yf_1 + \frac{y^2}{2} f_1^* + \frac{y^3}{3} f_1^{**} + y^2 f_2 + y^3 f_3 + \dots)\\ &= 1 + f_1 y + (\frac{f_1^2}{2} + \frac{f_1^*}{2} + f_2)y^2 + (\frac{f_1^3}{6} + \frac{f_1 f_1^*}{2} + \frac{f_1^{**}}{3} + f_1 f_2 + f_3)y^3 + \cdots\end{aligned}$$

which proves the Proposition. □

Finally, the Poincaré polynomials of $\mathcal{M}_A^H(\mathrm{GL}_n)$ (respectively, $\mathcal{M}_{A^\vee}^R(\mathrm{GL}_n)$), are obtained in three steps: firstly, use $E_{T^*A}(u,v) = x^d(u-1)^d(v-1)^d$ and $E_{\mathbb{G}_m^{2d}}(u,v) = (x-1)^{2d}$ with $x = uv$, in the Equations (6.6). Secondly, get $E_{\mathsf{Hilb}^n(\mathcal{M}^H)}$ and $E_{\mathsf{Hilb}^n(\mathcal{M}^R)}$ for $n = 2,3$ from Proposition 6.8. Finally, since these are smooth varieties, use Lemma 6.1 (or the corresponding relation between E and P, in case of round cohomology) to get $P_{\mathsf{Hilb}^n(\mathcal{M}^H)}(t)$ and $P_{\mathsf{Hilb}^n(\mathcal{M}^R)}(t)$ for $n = 2,3$.

REMARK 6.9. The same method, using $E_{\mathbb{G}_m^r}(u,v) = (uv-1)^r$, provides also the Poincaré polynomials of the natural resolutions of the character varieties
$\mathsf{Hom}(\mathbb{Z}^r, \mathrm{GL}_n(\mathbb{C})) /\!\!/ \mathrm{GL}_n(\mathbb{C})$ for odd r, even though, in this case, $\mathbb{Z}^r$ is not a Kähler group, so it does not appear in the context of the non-abelian Hodge correspondence. The analogous results of Theorem 6.5 and Proposition 6.8, for the case $G = \mathrm{SL}_n(\mathbb{C})$ are also easy to obtain with the same methods.

Acknowledgements. We are grateful to the anonymous referee for detailed comments and for suggesting important corrections on a previous version. We thank Giosuè Muratore, Jaime Silva, Alfonso Zamora for useful discussions on these topics. IB is partially supported by a J. C. Bose Fellowship. CF is partially supported by FCT project UIDB/04561/2020. AN is supported by a grant from IPM, Iran.

References

[Be1] A. Beauville, *Variétés Kähleriennes dont la première classe de Chern est nulle* (French), J. Differential Geom. **18** (1983), no. 4, 755–782 (1984). MR730926

[Be2] A. Beauville, *Symplectic singularities*, Invent. Math. **139** (2000), no. 3, 541–549, DOI 10.1007/s002229900043. MR1738060

[Bi] I. Biswas, *Moduli spaces of bundles on an abelian variety*, Int. J. Geom. Methods Mod. Phys. **14** (2017), no. 2, 1750030, 10, DOI 10.1142/S021988781750030X. MR3599025

[BF] I. Biswas and C. Florentino, *Commuting elements in reductive groups and Higgs bundles on abelian varieties*, J. Algebra **388** (2013), 194–202, DOI 10.1016/j.jalgebra.2013.05.006. MR3061684

[BFM] A. Borel, R. Friedman, and J. W. Morgan, *Almost commuting elements in compact Lie groups*, Mem. Amer. Math. Soc. **157** (2002), no. 747, x+136, DOI 10.1090/memo/0747. MR1895253

[BG] I. Biswas and T. L. Gómez, *Connections and Higgs fields on a principal bundle*, Ann. Global Anal. Geom. **33** (2008), no. 1, 19–46, DOI 10.1007/s10455-007-9072-x. MR2369185

[BS] G. Bellamy and T. Schedler, *Symplectic resolutions of character varieties*, Geom. Topol. **27** (2023), no. 1, 51–86, DOI 10.2140/gt.2023.27.51. MR4584260

[Ch1] J. Cheah, *On the cohomology of Hilbert schemes of points*, J. Algebraic Geom. **5** (1996), no. 3, 479–511. MR1382733

[Ch2] J. Cheah, *Cellular decompositions for nested Hilbert schemes of points*, Pacific J. Math. **183** (1998), no. 1, 39–90, DOI 10.2140/pjm.1998.183.39. MR1616606

[D] P. Deligne, *Théorie de Hodge. II* (French), Inst. Hautes Études Sci. Publ. Math. **40** (1971), 5–57. MR498551

[FGPN] E. Franco, O. Garcia-Prada, and P. E. Newstead, *Higgs bundles over elliptic curves*, Illinois J. Math. **58** (2014), no. 1, 43–96. MR3331841

[FT] E. Franco and P. Tortella, *Moduli spaces of Λ-modules on abelian varieties*, Adv. Math. **318** (2017), 459–496, DOI 10.1016/j.aim.2017.08.006. MR3689747

[FL] C. Florentino and S. Lawton, *Topology of character varieties of Abelian groups*, Topology Appl. **173** (2014), 32–58, DOI 10.1016/j.topol.2014.05.009. MR3227204

[FLS] C. Florentino, S. Lawton, and J. Silva, *Mixed Hodge structures on character varieties of nilpotent groups*, Revista Matematica Complutense (2024). https://doi.org/10.1007/s13163-024-00490-9

[FNZ1] C. Florentino, A. Nozad, and A. Zamora, *Serre polynomials of SL_n- and PGL_n-character varieties of free groups*, J. Geom. Phys. **161** (2021), Paper No. 104008, 21, DOI 10.1016/j.geomphys.2020.104008. MR4180103

[FS] C. Florentino and J. Silva, *Hodge-Deligne polynomials of character varieties of free abelian groups*, Open Math. **19** (2021), no. 1, 338–362, DOI 10.1515/math-2021-0038. MR4261777

[FN] B. Fu and Y. Namikawa, *Uniqueness of crepant resolutions and symplectic singularities* (English, with English and French summaries), Ann. Inst. Fourier (Grenoble) **54** (2004), no. 1, 1–19. MR2069119

[Gott] L. Göttsche, *On the motive of the Hilbert scheme of points on a surface*, Math. Res. Lett. **8** (2001), no. 5-6, 613–627, DOI 10.4310/MRL.2001.v8.n5.a3. MR1879805

[Goth] P. B. Gothen, *The Betti numbers of the moduli space of stable rank 3 Higgs bundles on a Riemann surface*, Internat. J. Math. **5** (1994), no. 6, 861–875, DOI 10.1142/S0129167X94000449. MR1298999

[Hi] N. J. Hitchin, *The self-duality equations on a Riemann surface*, Proc. London Math. Soc. (3) **55** (1987), no. 1, 59–126, DOI 10.1112/plms/s3-55.1.59. MR887284

[Ka] D. Kaledin, *Symplectic singularities from the Poisson point of view*, J. Reine Angew. Math. **600** (2006), 135–156, DOI 10.1515/CRELLE.2006.089. MR2283801

[La] Y. Laszlo, *About G-bundles over elliptic curves*, Ann. Inst. Fourier (Grenoble) **48** (1998), no. 2, 413–424. MR1625614

[Na1] Y. Namikawa, *Extension of 2-forms and symplectic varieties*, J. Reine Angew. Math. **539** (2001), 123–147, DOI 10.1515/crll.2001.070. MR1863856

[PS] C. A. M. Peters and J. H. M. Steenbrink, *Mixed Hodge structures*, Ergebnisse der Mathematik und ihrer Grenzgebiete. 3. Folge. A Series of Modern Surveys in Mathematics [Results in Mathematics and Related Areas. 3rd Series. A Series of Modern Surveys in Mathematics], vol. 52, Springer-Verlag, Berlin, 2008. MR2393625

[Sc] A. H. W. Schmitt, *Geometric invariant theory and decorated principal bundles*, Zurich Lectures in Advanced Mathematics, European Mathematical Society (EMS), Zürich, 2008, DOI 10.4171/065. MR2437660

[Se] J.-P. Serre, *Linear representations of finite groups*, Graduate Texts in Mathematics, Vol. 42, Springer-Verlag, New York-Heidelberg, 1977. Translated from the second French edition by Leonard L. Scott. MR450380

[Sk] A. S. Sikora, *Character varieties of abelian groups*, Math. Z. **277** (2014), no. 1-2, 241–256, DOI 10.1007/s00209-013-1252-8. MR3205770

[Si1] C. T. Simpson, *Higgs bundles and local systems*, Inst. Hautes Études Sci. Publ. Math. **75** (1992), 5–95. MR1179076

[Si2] C. T. Simpson, *Moduli of representations of the fundamental group of a smooth projective variety. II*, Inst. Hautes Études Sci. Publ. Math. **80** (1994), 5–79 (1995). MR1320603

[Th] M. Thaddeus, *Mirror symmetry, Langlands duality, and commuting elements of Lie groups*, Internat. Math. Res. Notices **22** (2001), 1169–1193, DOI 10.1155/S1073792801000551. MR1862614

[Ti] A. Tirelli, *Symplectic resolutions for Higgs moduli spaces*, Proc. Amer. Math. Soc. **147** (2019), no. 4, 1399–1412, DOI 10.1090/proc/14339. MR3910407

Mathematics Department, Shiv Nadar University, NH91, Tehsil Dadri, Greater Noida, Uttar Pradesh 201314, India

Email address: indranil.biswas@snu.edu.in, indranil29@gmail.com

Departamento de Matemática, Faculdade de Ciências, Univ. de Lisboa, Edf. C6, Campo Grande 1749-016 Lisboa, Portugal

Email address: caflorentino@fc.ul.pt

School of Mathematics, Institute for Research in Fundamental Sciences (IPM), P.O. Box 19395-5746, Tehran, Iran

Email address: anozad@ipm.ir

Contemporary Mathematics
Volume **803**, 2024
https://doi.org/10.1090/conm/803/16094

Global properties of Higgs bundle moduli spaces

Steven Bradlow

Dedicated with gratitude and admiration to Peter Newstead on the occasion of his 80th birthday

ABSTRACT. The moduli spaces for Higgs bundles associated to real Lie groups and a closed Riemann surface have multiple connected components. This survey provides a compendium of results concerning the counting of these components in cases where the Lie group is a real forms of a complex simple Lie group. In some cases the components can be described quite explicitly.

CONTENTS

1. Introduction

The goals of this survey are quite modest: to collect in one place results on the global features of the moduli spaces of Higgs bundles on a Riemann surface, in particular

- results on the number of connected components, and
- descriptions of global structure

The moduli spaces in question, denoted here by $\mathcal{M}(\Sigma, G)$, are associated to a Riemann surface Σ and a Lie group G. We consider only closed surfaces, i.e. we do not allow the surface to have punctures or non-empty boundary. Many of the

2020 *Mathematics Subject Classification.* Primary 14D20, 14F45, 14H60.
Key words and phrases. Higgs bundles, moduli spaces, connected components.

basic results referenced in this survey apply if G is any real reductive Lie group[1], but we will assume that G is a non-compact real form of a connected complex semisimple Lie Group. The Lie algebras of such real forms thus encompass all the non-compact real forms of the complex simple Lie algebras.

One of the many points of interest concerning the moduli spaces $\mathcal{M}(\Sigma, G)$ is the so-called non-Abelian Hodge (NAH) correspondence. This correspondence identifies $\mathcal{M}(\Sigma, G)$ with a moduli space that depends only on the underlying real surface of Σ, say S. Denoted by $\mathrm{Rep}(\pi_1(S), G)$, this moduli space parameterizes conjugacy classes of homomorphisms from the fundamental group of S into G. Individually, the two moduli spaces have been viewed from very different perspectives, with each perspective revealing characteristics not easily seen from the other.

On the Higgs bundle side the moduli space depends on choices, specifically a choice of complex structure on the surface S. The reward for this arbitrary choice is that the Higgs moduli space is endowed with extra structures which serve as tools for exploring the moduli space. Perhaps surprisingly, these tools reveal properties which are independent of the complex structure and hence transfer to the representation variety side of the correspondence.

The correspondence relies on existence and uniqueness results for non-linear PDE's. As a result it is very difficult to pair individual Higgs bundles with specific representations or vice versa. The information gleaned on the representation varieties side of the NAH correspondence leans towards the properties of the individual representations in specific components, or the dynamics of the mapping class group action, with particularly striking results for so-called higher Teichmüller components. In contrast, the Higgs bundle tools are very effective at revealing topological information such as the number of connected components and, in some cases, other global features.

We will remain firmly on the Higgs bundle side of the NAH correspondence. There are by now several excellent surveys of G-Higgs bundles on Riemann surfaces and their moduli spaces (see Section 3). We will lean heavily on these and provide here a much abbreviated summary of the main points relevant for our purposes.

In keeping with our limited goals there are many key topics that are not covered in this survey. In particular we will say hardly anything about the many interesting aspects of the representation varieties $\mathrm{Rep}(\pi_1(S), G)$ and their components. On the Higgs bundle side of the NAH correspondence we will also be forced to give short shrift to many important topics, including

- Spectral data for G-Higgs bundles when G is a real form, and the use of monodromy in the Hitchin fibration to count components. The Hitchin fibration as described in Section 6 has singular fibers. Removing these leaves a true non-singular fibration over the regular locus in the base of the fibration. The monodromy around the singular fibers, in conjunction with spectral data for real Higgs bundles can be used to infer an upper bound on the number of connected components in $\mathcal{M}(\Sigma, G)$. If this agrees with a known lower bound then the method can provide an elegant way to count components. While spectral data for real Higgs bundles add significant extra facets to our understanding (see for example [**6**]), to date this method has provided new information about the number of connected components only in the cases of $\mathrm{GL}(2, \mathbb{R})$ and $\mathrm{SO}(2, 2)$ (see [**7**]).

[1] As defined in [**43**]

- Geometric structures detected by G-Higgs bundles. One of the many interesting features of so-called higher Teichmüller components of the representation varieties $\mathrm{Rep}(\pi_1(S), G)$ is their relation to geometric structures either on S itself or on related spaces (see [**55**] or most recently [**2**]). This may be viewed as a generalization of the way the Teichmüller space of S appears as a component of $\mathrm{Rep}(\pi_1(S), PSL(2,\mathbb{R}))$. Such generalizations have been explored mostly from the point of view of the representations, but G-Higgs bundles have increasingly been used in interesting new ways (see [**1**] for a recent survey).

Many of the results covered by this survey resulted initially from a case-by-case analysis, starting with Hitchin's groundbreaking study of $\mathrm{SL}(2,\mathbb{R})$-Higgs bundles in [**40**]. Indeed the techniques introduced there and in [**42**] permeate virtually all of the subsequent developments. This is certainly true for the many specific examples of non-compact real forms that have been individually studied (see the references in the Tables), but it is also true for the subsequent general frameworks that have followed. In addition to Hitchin's results for split real forms in [**42**], these include a general result for real forms of Hermitian type [**9**] and, most recently, the construction in [**10**] of so-called Cayley components of $\mathcal{M}(\Sigma, G)$ for real forms arising from distinguished ('magical') $\mathfrak{sl}(2,\mathbb{C})$-subalgebras.

Rather than starting with the most general results and then working out some noteworthy special cases, we will proceed in the opposite direction. The goal is to build intuition to make it easier to appreciate the mechanisms at work in the general unifying result. We hope to strike a balance between providing all the details and getting to the endpoint in a timely manner.

2. The real forms

Real forms of a complex simple Lie group $G^{\mathbb{C}}$ are by definition the fixed points of an anti-holomorphic Lie group involution on $G^{\mathbb{C}}$. These always include a compact real form, unique up to conjugation, for which we will denote the anti-holomorphic involution by

$$\kappa : G^{\mathbb{C}} \to G^{\mathbb{C}} \ . \tag{2.1}$$

If $\tau : G^{\mathbb{C}} \to G^{\mathbb{C}}$ is any other anti-holomorphic involution, then $\sigma = \tau\kappa$ defines a holomorphic involution. The holomorphic involution is called the *Cartan involution* for τ. If the compact real form is fixed, i.e. if κ is given, then the non-compact real forms are specified either by the other anti-holomorphic involutions or their holomorphic Cartan involutions.

The holomorphic and anti-holomorphic involutions on the Lie groups induce corresponding linear or conjugate-linear involutions on their Lie algebras. The real forms of the complex Lie algebra are the fixed point sets of the conjugate-linear involutions.

The real groups of interest in this survey are non-compact real forms of complex simple Lie groups. As such their Lie algebras are non-compact real forms of complex simple Lie algebras. In Table 1 we list all such real forms organized according to features which play a role in the rest of this survey, i.e. which are important for understanding the components of the Higgs bundle moduli spaces.

Some properties of the Higgs bundle moduli spaces depend only on Lie algebra data but others, for instance in the precise count of their connected components,

TABLE 1. Non-compact real forms of complex simple Lie algebras (with $\mathfrak{sp}(2n,\mathbb{R}), \mathfrak{so}(n,n+1), \mathfrak{f}_4$ listed twice because they have two magical structures)

$\mathfrak{g}^{\mathbb{C}}$	Split	Hermitian tube	Other Magical	Hermitian non-tube	Other non-Magical
$\mathfrak{sl}(n,\mathbb{C})$	$\mathfrak{sl}(n,\mathbb{R})$	$\mathfrak{su}(p,p)$ $(n=2p)$	–	$\mathfrak{su}(p,n-p)$ $1\leq p<\frac{n}{2}$	$\mathfrak{su}^*(2p)$ $(n=2p)$
$\mathfrak{so}(2n+1,\mathbb{C})$	$\mathfrak{so}(n,n+1)$	$\mathfrak{so}(2,2n-1)$	$\mathfrak{so}(p,2n+1-p)$ $2<p\leq n$	–	$\mathfrak{so}(1,2n)$
$\mathfrak{sp}(2n,\mathbb{C})$	$\mathfrak{sp}(2n,\mathbb{R})$	$\mathfrak{sp}(2n,\mathbb{R})$	–	–	$\mathfrak{sp}(2p,2n-2p)$ $(1\leq p\leq\frac{n}{2})$
$\mathfrak{so}(2n,\mathbb{C})$	$\mathfrak{so}(n,n)$	$\mathfrak{so}(2,2n-2)$ $\mathfrak{so}^*(4p)(n=2p)$	$\mathfrak{so}(p,2n-p)$ $2<p<n$	$\mathfrak{so}^*(4p+2)$ $(n=2p+1)$	$\mathfrak{so}(1,2n-1)$
$\mathfrak{g}_2$	$\mathfrak{g}_2^2$	–	–	–	–
$\mathfrak{f}_4$	$\mathfrak{f}_4^4$	–	$\mathfrak{f}_4^4$	–	$\mathfrak{f}_4^{-20}$
$\mathfrak{e}_6$	$\mathfrak{e}_6^6$	–	$\mathfrak{e}_6^2$	$\mathfrak{e}_6^{-14}$	$\mathfrak{e}_6^{-26}$
$\mathfrak{e}_7$	$\mathfrak{e}_7^7$	$\mathfrak{e}_7^{-25}$	$\mathfrak{e}_7^{-5}$	–	–
$\mathfrak{e}_8$	$\mathfrak{e}_8^8$	–	$\mathfrak{e}_8^{-24}$	–	–

are sensitive to the center of the group - a feature not detected by the Lie algebra. Among the connected Lie groups with a given Lie algebra there is always an adjoint form, i.e. a Lie group with trivial center. The adjoint form has a universal cover which is also a Lie group with the same Lie algebra. All the other groups are quotients of the universal cover and are finite covers of the adjoint group. Note that not all such covers are matrix groups. For example the finite covers of $\mathrm{SL}(2,\mathbb{R})$ cannot be realized as matrix groups. Though Higgs bundles can in principle be defined for all such finite covers we will consider only matrix groups.

Another feature of a group that is not determined by its Lie algebra is its number of connected components. Some real forms, notably $\mathrm{SO}(p,q)$, have more than one component. For these real forms the component of the identity (e.g. $\mathrm{SO}_0(p,q)$ in the case of $\mathrm{SO}(p,q)$) is still a real reductive Lie group as defined in [**43**]. Our survey includes such cases.

3. G-Higgs bundles

There are several recent surveys of Higgs bundles on Riemann surfaces and their moduli spaces (see [**23, 24, 33, 35, 45, 50, 53, 54**], and also Appendix A in [**5**]) so we will repeat here only the bare essentials.

Let S be a closed oriented surface of genus $g \geq 2$. We fix a complex structure on S and denote the resulting Riemann surface by Σ. We also fix a compatible Riemannian metric on Σ. In general terms, a G-Higgs bundle on Σ is a pair (E, Φ) where E is a holomorphic bundle on Σ and Φ, the Higgs field, is a holomorphic section of an associated vector vector bundle twisted by the canonical bundle on Σ. If G is a complex Lie group, then E is a holomorphic principal G-bundle and the associated vector bundle is the adjoint bundle $ad(E)$, i.e. a bundle with Lie algebra of G as fiber. The Higgs field is then a section of $ad(E) \otimes K_\Sigma$, where K_Σ is the canonical bundle. For our present purposes we need a more flexible framework in which G is a real Lie group. In this setting the description of both E and the bundle containing the Higgs field is less immediate than in the complex case. In all cases, the definition is motivated by properties described in the next sections.

3.1. Definition. Let G be a non-compact real form of a complexl semisimple[2] Lie group, $G^{\mathbb{C}}$. Let $H \subset G$ be a maximal compact subgroup of G and denote by $H^{\mathbb{C}}$ its complexification. The Cartan involution defining G induces a decomposition of the Lie algebra $\mathfrak{g}^{\mathbb{C}}$ of $G^{\mathbb{C}}$

$$\mathfrak{g}^{\mathbb{C}} = \mathfrak{h}^{\mathbb{C}} \oplus \mathfrak{m}^{\mathbb{C}}\ , \tag{3.1}$$

where the summands are the ± 1-eigenspaces. The $+1$-eigenspace $\mathfrak{h}^{\mathbb{C}}$ is the Lie algebra of $H^{\mathbb{C}} \subset G^{\mathbb{C}}$ where $H^{\mathbb{C}}$ is the complexification of a maximal compact subgroup $H \subset G$. The adjoint action of $H^{\mathbb{C}}$ preserves $\mathfrak{m}^{\mathbb{C}}$ and the resulting representation of $H^{\mathbb{C}}$ on $\mathfrak{m}^{\mathbb{C}}$ is the so-called *isotropy representation*

DEFINITION 3.1. A G-Higgs bundle on a Riemann surface Σ is a pair $(E_{H^{\mathbb{C}}}, \phi)$ where $E_{H^{\mathbb{C}}}$ is a holomorphic principal $H^{\mathbb{C}}$-bundle and Φ is a holomorphic section of the vector bundle $E_{H^{\mathbb{C}}}[\mathfrak{m}^{\mathbb{C}}] \otimes K_\Sigma$. Here $E_{H^{\mathbb{C}}}[\mathfrak{m}^{\mathbb{C}}]$ denotes the associated vector bundle (via the isotropy representation) with fiber $\mathfrak{m}^{\mathbb{C}}$, and K_Σ is the canonical bundle.

Two such Higgs bundles, say $(E_{H^{\mathbb{C}}}, \phi)$ and $(E'_{H^{\mathbb{C}}}, \phi')$, are isomorphic if there is a bundle isomorphism $u : E_{H^{\mathbb{C}}} \to E'_{H^{\mathbb{C}}}$ which pulls back Φ' to Φ, i.e. such that $u^*\Phi' = \Phi$.

The holomorphic bundle $E_{H^{\mathbb{C}}}$ may be viewed as an underlying smooth bundle, say $\mathbf{E}_{H^{\mathbb{C}}}$, together with a $\overline{\partial}$-operator $\overline{\partial}_E$ (also known as an anti-holomorphic partial connection because they arise as the anti-holomorphic part a connection on $\mathbf{E}_{H^{\mathbb{C}}}$). From this point of view a G-Higgs bundle is defined as follows.

DEFINITION 3.2. A G-Higgs bundle on $\mathbf{E}_{H^{\mathbb{C}}}$ is a pair $(\overline{\partial}_E, \Phi)$ where Φ is smooth holomorphic 1-forms with values in $\mathbf{E}_{H^{\mathbb{C}}}[\mathfrak{m}^{\mathbb{C}}]$, i.e. $\Phi \in \Omega^{1,0}(\mathbf{E}_{H^{\mathbb{C}}}[\mathfrak{m}^{\mathbb{C}}])$ with $\overline{\partial}_E \Phi = 0$.

The complex gauge group for $\mathbf{E}_{H^{\mathbb{C}}}$, i.e. $\mathcal{G}_{H^{\mathbb{C}}} = \Omega^0(Ad(\mathbf{E}_{H^{\mathbb{C}}}))$, acts on the space of pairs $(\overline{\partial}_E, \Phi)$ with the action preserving the holomorphicity condition, and such

[2]the framework may straightforwardly be enlarged to include reductive Lie groups but we will confine attention to the semisimple case

that the gauge orbits correspond to isomorphism classes of Higgs bundles as defined in Definition 3.1).

If $G = \mathrm{SL}(n, \mathbb{C})$ then the principal bundle can be viewed as a frame bundle and the definition is equivalent to the more familiar definition in terms of vector bundles, i.e. a rank n Higgs bundle is a pair (E, ϕ) in which E is a rank n holomorphic vector bundle with trivial determinant and the Higgs field is a holomorphic section of $End(E) \otimes K_\Sigma$.

The motivation for the definition comes in part from the following considerations

- the existence theorem (described in §3.2) for solutions to natural gauge-theoretic equations yields flat connections with holonomy in G, and hence a correspondence between G-Higgs bundles and surface group representations in G.
- good moduli spaces can be constructed parameterizing isomorphism classes of G-Higgs bundles.
- if σ is the involution which defines G as a real form of G^C then σ induces an involution on the moduli space of G^C-Higgs bundles, with the moduli space of G-Higgs bundles in the fixed point locus.

3.2. Equations.

The defining data for a G-Higgs bundle, including the Lie theoretic structures described above, facilitates the formulation of a system of PDE's which define a metric, h, on the principal bundle $E_{H^\mathbb{C}}$. Here a metric means a section of the associated bundle $E_{H^\mathbb{C}}(H^\mathbb{C}/H)$, where $H \subset H^\mathbb{C}$ denotes a maximal compact subgroup. Such a metric determines a unique connection on $E_{H^\mathbb{C}}$ – called the Chern connection – defined by compatibility with both the metric and the holomorphic structures. The metric reduces the structure group of $E_{H^\mathbb{C}}$ to H and thereby permits an extension of τ, the antiholomorphic involution which defines H, to a globally defined map. Combined with conjugation on forms, this yields a map

$$\tau_h : \Omega^{1,0}(E_{H^\mathbb{C}}[\mathfrak{m}^\mathbb{C}]) \to \Omega^{0,1}(E_{H^\mathbb{C}}[\mathfrak{m}^\mathbb{C}]) .$$

The Higgs bundle equations takes the form

$$F_h - [\Phi, \tau_h(\Phi)] = 0 \tag{3.2}$$

where F_h denotes the curvature of the Chern connection on $E_{H^\mathbb{C}}$. In the case where $G = \mathrm{SL}(n, \mathbb{C})$ so that h is just a Hermitian metric on the holomorphic vector bundle E, the curvature term is the curvature of the usual Chern connection, and τ_h is the map from $\Omega^{1,0}(End(E)) \to \Omega^{0,1}(End(E))$ defined by $\tau_h(\Phi) = -\Phi^{*_h}$ (where $*_h$ denotes the adjoint with respect to the metric h combined with conjugation on (1,0)-forms).

It is sometimes useful to fix a metric h on the smooth principal bundle $\mathbf{E}_{H^\mathbb{C}}$, i.e. fix a reduction to an H-bundle, say $\mathbf{E}_H$, and regard (3.2) as an equation for a pair (D_h, Φ) where D_h is a connection on $\mathbf{E}_H$, and Φ is a smooth section of the bundle $\mathbf{E}_H[\mathfrak{m}^\mathbb{C}] \otimes K_\Sigma$. Equivalently, with $\overline{\partial}_E$ given by the antiholorphic part of D_h, the equation may be viewed as an equation for a pair $(\overline{\partial}_E, \Phi)$ as in Definition 3.2. Equation (3.2) must then be complemented by the holomorphicity condition

$$\overline{\partial}_E \Phi = 0 \tag{3.3}$$

3.3. Stability and moduli spaces. The concept of stability comes from Geometric Invariant Theory and plays a pivotal role in the construction of moduli spaces. It is straightforward to define stability for Higgs vector bundles but more complicated for general G-Higgs bundles.

DEFINITION 3.3. A rank n trivial determinant Higgs vector bundle (E, ϕ) on Σ is *semi-stable* if

$$\frac{\deg(E')}{\operatorname{rank}(E')} \leq 0 \tag{3.4}$$

for all ϕ-invariant subbundles E', i.e. for all subbundles such that $\phi(E') \subset E' \otimes K_\Sigma$. It is *stable* if the inequality is strict for all non-trivial proper ϕ-invariant subbundles, and *polystable* if it decomposes as a direct sum of stable Higgs bundles of lower rank.

The definitions in the general setting of G-Higgs bundles is complicated by the more elaborate machinery needed to describe subobjects of principal bundles and to formulate the numerical replacement for the degree of a vector bundle. The details will not be needed in this survey. A good account can be found in [**24**]. The crucial consequence is that the set of isomorphism classes of polystable Higgs bundles can be endowed with the structure of a moduli space.

DEFINITION 3.4. The moduli space of G-Higgs bundles on Σ, denoted $\mathcal{M}(\Sigma, G)$, is the set of isomorphism classes of polystable G-Higgs bundles.

REMARK 3.5. In Definition 3.1 if the canonical bundle is replaced by any other line bundle on Σ, say L, we refer to the resulting pairs as *L-twisted G-Higgs pairs.* The stability concepts extend readily to these more general objects and permit the construction of moduli spaces. We will denote such moduli spaces by $\mathcal{M}_L(\Sigma, G)$. The moduli spaces of L-twisted G-Higgs pairs (with L being a power of K_Σ) will play a significant role in Sections 7 and 8.

The moduli spaces as defined above can be endowed with the structure of complex analytic varieties (see for example [**51**] for an account). This can also be seen analytically because of the relation between (poly)stability and the existence of a solution to the Higgs bundle equation.

THEOREM 3.6. *(Hitchin-Kobayashi correspondence) The G-Higgs bundle $(E_{H^\mathbb{C}}, \Phi)$ admits a metric satisfying the Higgs bundle equation if and only if $(E_{H^\mathbb{C}}, \Phi)$ is polystable.*

Both the existence of solutions to the Higgs bundle equations and also the property of polystability are properties of isomorphism classes, i.e. these properties are preserved by $H^\mathbb{C}$-gauge transformations. Thus we can idenitfy

$$\mathcal{M}(\Sigma, G) \simeq \left\{ (\overline{\partial}_E, \Phi) \,|\overline{\partial}_E \Phi = 0 \text{ and (3.2) has a solution} \right\} / \mathcal{G}_{H^\mathbb{C}} \tag{3.5}$$

3.4. The non-Abelian Hodge Correspondence. By extension of structure group $E_{H^{\mathbb{C}}}$ defines a holomorphic $G^{\mathbb{C}}$-bundle which we denote by $E_{H^{\mathbb{C}}}(G^{\mathbb{C}})$ or simply $E_{G^{\mathbb{C}}}$. The Chern connection on $E_{H^{\mathbb{C}}}$ induces a connection on $E_{G^{\mathbb{C}}}$, denoted by D_h. Using (3.1) we can now define a new connection

$$\nabla_h = D_h + \Phi + \Phi^{*_h} \tag{3.6}$$

This is, a priori, a $G^{\mathbb{C}}$-connection, i.e. an operator with values in the Lie algebra $\mathfrak{g}^{\mathbb{C}}$. However if h is a solution to (3.2) then ∇_h has two important features:

- ∇_h takes values in $\mathfrak{g}$ and
- ∇_h is a flat connection, i.e. it has zero curvature.

In particular, the holonomy of ∇_h defines a reduction of structure to G and defines a flat structure on the resulting G-bundle. The holonomy of this flat structure defines a representation of $\pi_1(\Sigma)$ in G. In this way, solutions to (3.2) define representations of $\pi_1(S)$ in G. Conversely, any representation $\rho : \pi_1(S) \to G$ defines a flat principal G-bundle on S via the construction

$$E_G = \tilde{S} \times_\rho G \, . \tag{3.7}$$

Here $\tilde{S}$ is the universal cover of S, viewed as a principal $\pi_1(S)$-bundle on S.

A theorem of Corlette (and also Donaldson) leads to the following fundamental relation

THEOREM 3.7. *There is a bijective correspondence between gauge equivalence classes of principal G-bundles on S with reductive flat structures and G-conjugacy classes of reductive representations in $Hom(\pi_1(S), G)$.*

The space of G-conjugacy classes of reductive representations in $Hom(\pi_1(S), G)$ is denoted by $\mathrm{Rep}(\pi_1(S), G)$. The combination of Corlette's theorem and the Hitchin-Kobayashi correspondence for Higgs bundles thus leads to an identification of moduli spaces:

THEOREM 3.8. *Let Σ be a closed smooth Riemann surface of genus $g \geq 2$ with underlying topological surface S, and let G be a real form of a complex semisimple Lie group. Then*

$$\mathcal{M}(\Sigma, G) \simeq \mathrm{Rep}(\pi_1(S), G) \tag{3.8}$$

This is known as the non-Abelian Hodge (NAH) correspondence. We emphasize that the moduli space $\mathrm{Rep}(\pi_1(S), G)$ depends only on the topology of S (i.e. its genus) and on the Lie group G, and parameterizes conjugacy classes of surface group representations. In contrast, $\mathcal{M}(\Sigma, G)$ depends not only on G but on the complex structure on the surface, and parameterizes G-Higgs bundles.

4. Component types and labels

On both sides of the NAH correspondence there are invariants coming directly from the topology of G. If G is connected then in $\mathrm{Rep}(\pi_1(S), G)$ such invariants arise from considering whether representations into G can be lifted into the universal cover of G. These obstruction can be interpreted as an element in $\pi_1(G)$. In $\mathcal{M}(\Sigma, G)$ the topological type of $E_{H^{\mathbb{C}}}$ is constant on connected components. The topological types of the principal bundles are classified by characteristic classes in $\pi_1(H) = \pi_1(G)$. If G is not connected the specification of topological types can be more subtle, but in all cases we get a partition of $\mathcal{M}(\Sigma, G)$ into sectors labelled by the topological types for principal $H^{\mathbb{C}}$-bundles, say

$$\mathcal{M}(\Sigma, G) = \coprod_{c \in \mathcal{I}} \mathcal{M}_c(\Sigma, G) \tag{4.1}$$

where $\mathcal{I}$ denotes the indexing set for the topological types and $\mathcal{M}_c(\Sigma, G)$ denotes the union of connected components in which the principal $H^{\mathbb{C}}$-bundles are of type $c \in \mathcal{I}$.

DEFINITION 4.1. The primary topological invariant of a connected component of $\mathcal{M}(\Sigma, G)$ is the topological type of the principal bundles in the Higgs bundles parameterized by that component.

If G is a complex simple group then (see [**26**]) there is a bijective correspondence between the topological types of G-bundles and the connected components of $\mathcal{M}(\Sigma, G)$. This is not true for the components of $\mathcal{M}(\Sigma, G)$ if G is a real group. The relation is complicated by factors including

- (Bounds) Not all topological types may occur. If G is a real form of Hermitian type (see Section 7) then the primary topological invariants determine a discrete real-valued invariant (the Toledo invariant). There is a (Milnor-Wood) bound on the Toledo invariant outside of which the components are empty.
- (Secondary invariants) The primary invariants may not distinguish connected components. If G admits magical structures (see section 8) then $\mathcal{M}(\Sigma, G)$ has components in which secondary topological structures emerge. The topological types of these secondary structures are needed to distinguish components with the same primary invariants.

In the complex case, the Higgs bundles can always be deformed to a Higgs bundle in which the Higgs field vanishes. On the other side of the NAH correspondence this corresponds to the fact that the representations can be deformed to representations which factor through the compact real form. In the next sections we see how the situation differs in the components of the moduli spaces for non-compact real forms, and how this affects the total number of connected components.

The rich structure of $\mathcal{M}(\Sigma, G)$ offers effective tools for detecting and counting its connected components. These include a real-valued function which provides Morse-theoretic information, and direct constructions (of various degrees of explicitness) of special components. Our main focus in the next sections is to describe such methods and the results they have produced.

5. Morse-theoretic approach

Let $(E_{H^{\mathbb{C}}}, \Phi)$ be a polystable G-Higgs bundle. The metric defined by the Higgs bundle equations (3.2) reduces the structure group of $E_{H^{\mathbb{C}}}$ to a maximal compact subgroup $H \subset H^{\mathbb{C}}$ and hence induces a unitary structure on the bundle $E_{H^{\mathbb{C}}}[\mathfrak{m}^{\mathbb{C}}] = E_H[\mathfrak{m}^{\mathbb{C}}]$. Combined with a fixed metric on the surface, and hence on K_Σ, this permits computation of a fiberwise norm $|\Phi(x)|_h$ at each point $x \in \Sigma$, and hence defines

$$f(E_{H^{\mathbb{C}}}, \Phi) = \|\Phi\|_h^2 = \int_\Sigma |\Phi|_h^2 dvol \ . \tag{5.1}$$

This is constant on isomorphism classes or, equivalently on $H^{\mathbb{C}}$-gauge orbits, and hence produces a well-defined map

$$\begin{aligned} f : \mathcal{M}(\Sigma, G) &\to \mathbb{R} \\ [E_{H^{\mathbb{C}}}, \Phi] &\mapsto \|\Phi\|_h^2 \end{aligned} \tag{5.2}$$

We will call this the Hitchin function as it was first defined by Hitchin who showed:

THEOREM 5.1. *This map is smooth on the smooth locus of the moduli space and is a proper map.*

The Hitchin function serves as a Morse-Bott function on the smooth locus in the moduli space. While the singularities in the moduli space complicate this relation, the properness of the function implies that f attains local minima on all connected components of the moduli space. Thus if $f_{min} \subset \mathcal{M}(\Sigma, G)$ denotes the locus of local minima, then the number of connected components of $\mathcal{M}(\Sigma, G)$ is bounded above by the number of connected components of f_{min}. Moreover the component counts are equal if the connected components of f_{min} all lie in distinct connected components of the moduli space.

Note that Higgs bundles of the form $(E_{H^{\mathbb{C}}}, 0)$, i.e with $\Phi = 0$, necessarily define local minima of f. We call these the *trivial local minima*. Crucially for the existence of many connected components, the trivial minima are not necessarily the only local minima.

One way to identify the non-trivial local minima of the Hitchin function is to use the relation between f and another key feature of the moduli spaces, namely a $\mathbb{C}^*$-action defined by

$$\lambda[E_{H^{\mathbb{C}}}, \Phi] = [E_{H^{\mathbb{C}}}, \lambda\Phi] \tag{5.3}$$

THEOREM 5.2. *The critical points of the Hitchin function coincide with the fixed points of the* $\mathbb{C}^*$ *action defined by* (5.3).

We now examine the two types of minima, namely minima with $\Phi = 0$ and minima with non-trivial Higgs field.

5.1. Minima with $\Phi = 0$. Suppose $(E_{H^{\mathbb{C}}}, 0)$ represents a point in $\mathcal{M}(\Sigma, G)$. Then $(E_{H^{\mathbb{C}}}, 0)$ must be polystable as a G-Higgs bundle. It follows that $E_{H^{\mathbb{C}}}$ must be polystable as a principal $H^{\mathbb{C}}$ bundle and must support a flat Hermitian Einstein metric, i.e. a reduction of structure to H with respect to which the Chern connection has zero curvature. This is not always possible. Depending on the topological type of $E_{H^{\mathbb{C}}}$ there may be obstructions to the existence of flat structures.

EXAMPLE 5.3. Let $G = \mathrm{SL}(2, \mathbb{R})$. In this case $H^{\mathbb{C}} = \mathbb{C}^*$ and the topological type of an $\mathrm{SL}(2, \mathbb{R})$-Higgs bundle is determined by an integer corresponding to the degree of a complex line bundle. All line bundles are stable, but only the zero degree line bundles admit flat structures.

Components which contain trivial minima are distinguished by the topological types of the principal $H^{\mathbb{C}}$-bundles at the minima. The G-Higgs bundles parameterized by these components can be deformed to G-Higgs bundle where the Higgs field is identically zero. On the other side of the NAH correspondence, the representations parameterized by the corresponding components of $\mathrm{Rep}(\pi_1(S), G)$ can be deformed to representations which factor through a maximal compact subgroup of G.

However, as the above example illustrates, not all such topological types occur as labels of components with trivial minima.

5.2. Minima with non-trivial Φ. It is important to note that the $\mathbb{C}^*$-action in (5.3) is on the moduli space, i.e. on isomorphism classes of G-Higgs bundles. The condition for an isomorphism class $[E_{H^{\mathbb{C}}}, \Phi] \in \mathcal{M}(\Sigma, , G)$ to be a fixed point is thus that for each $\lambda \in \mathbb{C}^*$ the Higgs bundle $(E_{H^{\mathbb{C}}}, \lambda\Phi)$ is isomorphic to $(E_{H^{\mathbb{C}}}, \Phi)$. Representing the Higgs bundles by pairs $(\overline{\partial}_E, \Phi)$ as in Definition 3.2 this means that for all $\lambda \in \mathbb{C}^*$ there is a gauge transformation $g(\lambda) \in \Omega^0(E_{H^{\mathbb{C}}}(H^{\mathbb{C}}))$ such that

$$(\overline{\partial}_E, \lambda\Phi) = g(\lambda)^*(\overline{\partial}_E, \Phi), \tag{5.4}$$

i.e. such that

$$g(\lambda)^{-1} \circ \overline{\partial}_E \circ g(\lambda) = \overline{\partial}_E \text{ and} \tag{5.5}$$

$$g(\lambda)^{-1} \circ \Phi \circ g(\lambda) = \lambda\Phi \ . \tag{5.6}$$

If the fixed point $[E_{H^{\mathbb{C}}}, \lambda\Phi]$ is stable then $g(\lambda)$ is uniquely determined[3]. The infinitesimal action at the fixed point induces weight-space decompositions

$$E_{H^{\mathbb{C}}}[\mathfrak{h}^{\mathbb{C}}] = \bigoplus_j E_{H^{\mathbb{C}}}[\mathfrak{h}^{\mathbb{C}}]_j \tag{5.7}$$

$$E_{H^{\mathbb{C}}}[\mathfrak{m}^{\mathbb{C}}] = \bigoplus_j E_{H^{\mathbb{C}}}[\mathfrak{m}^{\mathbb{C}}]_j \tag{5.8}$$

where the weights are integers or half integers. With respect to these decompositions Φ lies in $E_{H^{\mathbb{C}}}[\mathfrak{m}^{\mathbb{C}}]_1 \otimes K_\Sigma$ and its adjoint action increase weights by one, i.e.

$$ad_\Phi : E_{H^{\mathbb{C}}}[\mathfrak{h}^{\mathbb{C}}]_j \to E_{H^{\mathbb{C}}}[\mathfrak{m}^{\mathbb{C}}]_{j+1} \otimes K_\Sigma \tag{5.9}$$

We are primarily interested in the fixed points corresponding to local minima of the Hitchin function. Such fixed points may occur within the smooth locus of $\mathcal{M}(\Sigma, G)$ or at singular points. The latter occur at points represented by strictly polystable Higgs bundles, while the former are all represented by stable Higgs bundles. The following criterion (first introduced in [**42**]) is the main tool for identifying the smooth local minima.

THEOREM 5.4. *Let $[E_{H^{\mathbb{C}}}, \Phi]$ be a $\mathbb{C}^*$-fixed point in the smooth locus of $\mathcal{M}(\Sigma, G)$. Then $[E_{H^{\mathbb{C}}}, \Phi]$ is a local minimum of f if and only if either Φ is the zero section or the maps in* (5.9) *are sheaf isomorphisms for every $j > 0$.*

The criteria from Theorem 5.4 do not apply to fixed points outside the smooth locus of the moduli space. In all cases that have been examined so far, non-smooth

[3] For further details see the summary in Appendix A in [**5**] or for a more algebraic treatment see [**9**] for a discussion based on [**52**] of Hodge bundles in the framework of G-Higgs bundle.

critical points are analyzed by constructing explicit deformations which decrease the Hitchin function, thereby showing that they are not local minima.

The local minima of the Hitchin function can detect components but they provide no information about the global structure of the components. In the next sections we describe methods which reveal entire components. In some cases this is in the form of explicit descriptions of certain components, while in other cases the information is less complete.

6. Sections of Hitchin fibrations and Hitchin components

If G is a complex semisimple Lie group then the Higgs field in the pair (E_G, Φ) can be viewed as a holomorphic 1-form with values in the adjoint bundle $Ad(E_G) = E_G \times_{Ad} \mathfrak{g}$. If P is any homogeneous polynomial on $\mathfrak{g}$ invariant under the conjugation action of G, then $P(\Phi)$ is a well defined section of K_Σ^r where r is the degree of the polynomial. A basis for the ring of invariant polynomials thus defines a map, called the Hitchin fibration

$$\mathfrak{H} : \mathcal{M}(\Sigma, G) \to \bigoplus H^0(K^{r_i}) \tag{6.1}$$

where the r_i are the degrees of the invariant polynomials in the basis. Hitchin showed that the fibers of this fibration are generically tori and that it gives $\mathcal{M}(\Sigma, G)$ the structure of a algebraic completely integrable system. This fibration has many ramifications; we focus on just one here, namely the existence of sections.

First constructed in [**42**], Hitchin demonstrated the existence of sections

$$\Psi : \bigoplus H^0(K^{r_i}) \to \mathcal{M}(\Sigma, G) \tag{6.2}$$

with the property that their image lies in $\mathcal{M}(\Sigma, G_{split}) \subset \mathcal{M}(\Sigma, G)$, where G_{split} denotes the split real form of the complex group G. The construction is very explicit. The points in the image of Ψ can be represented by Higgs bundles (E, Φ) in which the bundle E is the same for all points of the section, and the Higgs field is described with reference to that bundle. For example, in the case of $G = \mathrm{SL}(2n, \mathbb{C})$, the bundle in the image of the section can be chosen to be

$$E = K_\Sigma^{-n+\frac{1}{2}} \oplus K_\Sigma^{-n+\frac{3}{2}} \oplus \cdots K_\Sigma^{n-\frac{1}{2}} \tag{6.3}$$

and the Higgs fields are given by

$$\Phi = \begin{bmatrix} 0 & 1 & 0 & 0 & \cdots & 0 \\ \alpha_2 & 0 & 1 & 0 & \cdots & 0 \\ \alpha_3 & \alpha_2 & 0 & 1 & \cdots & 0 \\ \vdots & \vdots & \vdots & \vdots & \cdots & \vdots \\ \alpha_{n-1} & \cdots & \alpha_3 & \alpha_2 & 0 & 1 \\ \alpha_n & \alpha_{n-1} & \cdots & \alpha_3 & \alpha_2 & 0 \end{bmatrix} \tag{6.4}$$

Here, for appropriate $i \in \{2, 3, \ldots, n\}$ and $j \in \{\frac{1-2n}{2}, \frac{3-2n}{2}, \ldots, \frac{2n-1}{2}\}$,

- 1 is the constant unit length section in $H^0(K_\Sigma^{-(j+1)} \otimes K_\Sigma^j \otimes K_\Sigma) = H^0(\mathcal{O}_\Sigma)$,
- $\alpha_i \in H^0(K_\Sigma^{-j} \otimes K_\Sigma^{j+i-1} \otimes K_\Sigma) = H^0(K_\Sigma^i)$

The Hitchin section in this case is defined by

$$\Psi(\alpha_1, \ldots, \alpha_n) = [(E, \Phi)] \tag{6.5}$$

where (E, Φ) are given by (6.3) and (6.4) respectively. Hitchin proved in [**42**] that the resulting Higgs bundles are stable, i.e. that they define points in $\mathcal{M}(\Sigma, \mathrm{SL}(2n, \mathbb{C}))$.

Moreover he showed that the image lies in $\mathcal{M}(\Sigma, \mathrm{SL}(2n,\mathbb{R}))$ and that the map defines a homeomorphism onto a connected component in $\mathcal{M}(\Sigma, \mathrm{SL}(2n,\mathbb{R}))$.

The construction generalizes so that for all complex semisimple G, the image of a Hitchin section is open and closed in $\mathcal{M}(\Sigma, G_{split})$ and thus defines a connected component. The construction of these components presents them as complex vector spaces of complex dimension $(g-1)\dim(G_{split})$ parameterized by global holomorphic differentials. The precise descriptions are given in Table 2. We will describe the construction in a more general context later (see the Cayley map in Section 8) but it is instructive to point out some key elements in the case of $\mathrm{SL}(2n,\mathbb{R})$. Foremost among those is the role of a principal $\mathfrak{sl}(2,\mathbb{C})$ subalgebra in $\mathfrak{sl}(2n,\mathbb{C})$, as described in the next section.

6.1. The underlying $\mathfrak{sl}$-triple. The structure group for the bundle in (6.3) is $T=\mathbb{C}^*$. Let $\mathcal{E}_T$ be the frame bundle for $K_\Sigma^{\frac{1}{2}}$. Define the group homomorphisms

$$T \to \mathrm{SL}(2,\mathbb{C}) \to \mathrm{SL}(2n,\mathbb{C}) \tag{6.6}$$

where the first arrow is the map $t \mapsto \begin{bmatrix} t & 0 \\ 0 & t^{-1}\end{bmatrix}$ and the second arrow is defined by the irreducible representation on $\mathbb{C}^{2n}$, i.e. defines a principally embedded $\mathfrak{sl}(2,\mathbb{C})$ subalgebra in $\mathfrak{sl}(2n,\mathbb{C})$. Using (6.6) to extend the structure group we can then construct the principal $\mathrm{SL}(2n,\mathbb{C})$-bundle $\mathcal{E}_T[\mathrm{SL}(2n,\mathbb{C})]$ thus realizing (6.3) as the associated $\mathbb{C}^{2n}$-bundle. Moreover, viewed from this perspective the bundle of trace-free endomorphsims, i.e. $End_0(E)$, is the associated bundle

$$End_0(E) = \mathcal{E}_T[\mathrm{SL}(2n,\mathbb{C})][\mathfrak{sl}(2n,\mathbb{C})] \tag{6.7}$$

Under the principal embedding, the image of the nilpotent element $f=\begin{bmatrix} 0 & 1 \\ 0 & 0\end{bmatrix}$ in $\mathfrak{sl}(2n,\mathbb{C})$ is

$$\tilde{f} = \begin{bmatrix} 0 & 1 & 0 & 0 & \cdots & 0 \\ 0 & 0 & 1 & 0 & \cdots & 0 \\ 0 & 0 & 0 & 1 & \cdots & 0 \\ \vdots & \vdots & \vdots & \vdots & \cdots & \vdots \\ 0 & 0 & 0 & 0 & \cdots & 1 \\ 0 & 0 & 0 & 0 & \cdots & 0 \end{bmatrix} \tag{6.8}$$

Via (6.6) the action of T on f and hence $\tilde{f}$ is given by $t\cdot\tilde{f} = t^{-2}\tilde{f}$. It follows that $\tilde{f}$ defines a global section of $End_0(E)\otimes\mathcal{E}_T^2 = End_0(E)\otimes K_\Sigma$.

The adjoint action of the principal $\mathfrak{sl}(2,\mathbb{C})$-subalgebra decomposes $\mathfrak{sl}(2n,\mathbb{C})$ into irreducible $\mathfrak{sl}(2,\mathbb{C})$-representations as

$$\mathfrak{sl}(2n,\mathbb{C}) = W_1 \oplus W_2 \oplus \cdots \oplus W_{n-1} \tag{6.9}$$

where $\dim(W_m) = 2m+1$ and W_1 is the principal $\mathfrak{sl}(2,\mathbb{C})$. An element in $\mathfrak{sl}(2n,\mathbb{C})$ of the form in (6.4) (where now the α_i are just numbers) can then be expressed as

$$\Phi = \tilde{f} + \alpha_1 e_1 + \ldots \alpha_l e_l \tag{6.10}$$

where e_i is a heighest weight vector for W_i. The subspaces spanned by the e_i (denoted by $<e_i>$) are preserved under conjugation by T, yielding subbundles

$$\mathcal{E}_T[\mathrm{SL}(2n,\mathbb{C})][<e_j>] \subset \mathcal{E}_T[\mathrm{SL}(2n,\mathbb{C})][\mathfrak{sl}(2n,\mathbb{C})] \tag{6.11}$$

TABLE 2. Hitchin/Teichmüller components in $\mathcal{M}(\Sigma, G_{split})$

G_{split}	Hitchin component
$\mathrm{SL}(n,\mathbb{R})$	$\bigoplus_{j=2}^{n} H^0(K_\Sigma^j)$
$\mathrm{SO}(n,n+1)$	$\bigoplus_{j=1}^{n} H^0(K_\Sigma^{2j})$
$\mathrm{Sp}(2n,\mathbb{R})$	$\bigoplus_{j=1}^{n} H^0(K_\Sigma^{2j})$
$\mathrm{SO}(n,n)$	$\bigoplus_{j=1}^{n-1} H^0(K_\Sigma^{2j}) \oplus H^0(K_\Sigma^{n})$
$G_{2,split}$	$H^0(K_\Sigma^2) \oplus H^0(K_\Sigma^6)$
$F_{4,split}$	$H^0(K_\Sigma^2) \oplus H^0(K_\Sigma^6) \oplus H^0(K_\Sigma^8) \oplus H^0(K_\Sigma^{12})$
$E_{6,split}$	$H^0(K_\Sigma^2) \oplus H^0(K_\Sigma^5) \oplus H^0(K_\Sigma^6) \oplus H^0(K_\Sigma^8) \oplus H^0(K_\Sigma^9) \oplus H^0(K_\Sigma^{12})$
$E_{7,split}$	$H^0(K_\Sigma^2) \oplus H^0(K_\Sigma^6) \oplus H^0(K_\Sigma^8) \oplus H^0(K_\Sigma^{10}) \oplus H^0(K_\Sigma^{12}) \oplus H^0(K_\Sigma^{14}) \oplus H^0(K_\Sigma^{18})$
$E_{8,split}$	$H^0(K_\Sigma^2) \oplus H^0(K_\Sigma^8) \oplus H^0(K_\Sigma^{12}) \oplus H^0(K_\Sigma^{14}) \oplus H^0(K_\Sigma^{18}) \oplus H^0(K_\Sigma^{20}) \oplus H^0(K_\Sigma^{24}) \oplus H^0(K_\Sigma^{30})$

Furthermore, if α_j is a section of K_Σ^j then $\alpha_j e_j$ is well defined as a section of

$$\mathcal{E}_T[\mathrm{SL}(2n,\mathbb{C})][< e_j >] \otimes K_\Sigma \subset \mathcal{E}_T[\mathrm{SL}(2n,\mathbb{C})][\mathfrak{sl}(2n,\mathbb{C})] \otimes K_\Sigma \ , \tag{6.12}$$

thus recasting (6.10) as an expression of the form (6.4).

REMARK 6.1. Hitchin points out (Section §4 in [**42**]) that Theorem 7 of [**44**] proves that there exists a basis $p_1, \dots p_l$ of invariant polynomials on $\mathfrak{g}$ such that

$$p_i(f + \alpha_1 e_1 + \dots \alpha_l e_l) = \alpha_i \tag{6.13}$$

where e_i are heighest weight vectors. Moreover, the degree of p_i is $2m_i + 1$, where $2m_i + 1$ is the dimension of V_i or equivalently m_i is the eigenvalue of ad_h on e_i. These fundamental invariants of the Lie algebra are called the exponents.

6.2. Local minima of the Hitchin function. The components of the moduli space $\mathcal{M}(\Sigma, G_{split})$ constructed via sections of the Hitchin fibration are called Hitchin or Teichmüller components. The Hitchin function can be evaluated on these components. It follows from the description of the Higgs fields as in (6.10) that the local minimum occurs at the Higgs bundle where the α_i are all zero, i.e. at the image of zero in the Hitchin section. The Higgs field at this minimum is $\Phi = \tilde{f}$. In particular it is not zero; indeed the Higgs field vanished nowhere on Hitchin components. The components constructed in this way are thus clearly distinct from components with trivial local minima.

The Hitchin components are examples of higher Teichmüller spaces (see for example [**55**] for more about this). Another source of higher Teichmüller spaces comes from real forms of Hermitian type, where distinguished components arise in a seemingly different way, as described in the next section. In Section 8 we will see that the underlying mechanism is actually the same as for the Hitchin components, with special $\mathfrak{sl}(2,\mathbb{C})$-subalgebras providing the key.

7. Maximal components and the Cayley correspondence

Suppose that G is a non-compact real form with maximal compact subgroup H such that the symmetric space G/H admits a G-invariant complex structure. Such real forms are said to be of Hermitian type. The following[4] is acomplete list

[4] Using Helgason's notation as in [**39**]

of the corresponding real Lie algebras if G is simple and connected:

$$\mathfrak{su}(p,q),\ \mathfrak{sp}(2n,\mathbb{R}),\ \mathfrak{so}(2,q),\ \mathfrak{so}^*(2n),\ \mathfrak{e}_6^{-14},\text{ and }\mathfrak{e}_7^{-25}.$$

These fall into two classes, referred to as *tube type* or *non-tube type*. The non-tube cases are $\mathfrak{su}(p,q)$ with $p \neq q$, $\mathfrak{so}^*(2n)$ with n odd, and $\mathfrak{e}_6^{-14}$; the rest are called tube type because their associated symmetric spaces G/H are biholomorphic to tubes, i.e. domains $V + i\Omega$ where V is a real vector space and $\Omega \subset V$ is a proper convex open cone.

Higgs bundles for real forms of Hermitian type have been studied on a case-by-case basis (see the references in the Tables) and more systematically in [**9**]. All the results are now subsumed under the umbrella of the Cayley components determined by magical $\mathfrak{sl}_2$-triples, as described in Section 8.

For all real form G of Hermitian type a discrete invariant called the Toledo number can be attached to a G-Higgs bundle or equivalently (via the NAH correspondence) to a surface group representation into G. The invariant is subject to a bound known as a Milnor-Wood bound. This allows one to distinguish *maximal* Higgs bundles or representations as those for which the Toledo invariant attains its maximal value. Since the invariant is constant on connected components, this leads to the following definition

DEFINITION 7.1. Let G be a real form of Hermitian type. A component of $\mathcal{M}(\Sigma, G)$ (respectively $\mathrm{Rep}(\pi_1(S), G)$) is called a *maximal component* if the Toledo invariant attains its maximal value on the Higgs bundles (respectively surface group representations) represented in that component.

If G is of tube type then the maximal components in $\mathrm{Rep}(\pi_1(S), G)$ have several special features, by virtue of which they are considered examples of higher Teichmüller spaces (see [**55**] for an excellent overview).

In the Higgs bundle moduli space, the special nature of the maximal components manifests as extra symmetries on the bundles and Higgs fields. As a result of these constraints the maximal components of $\mathcal{M}(\Sigma, G)$ acquire new descriptions as moduli spaces of Higgs-like pairs but for a new group, G', called the Cayley partner to G. We illustrate this phenomenon in the case of $G = \mathrm{Sp}(2n, \mathbb{R})$.

7.1. $\mathrm{Sp}(2n, \mathbb{R})$-Higgs bundles.

The maximal compact subgroup of $\mathrm{Sp}(2n, \mathbb{R})$ is $H = U(n)$ and the complexified Cartan decomposition can be given as

$$\mathfrak{sp}(2n,\mathbb{C}) = \mathfrak{h}^{\mathbb{C}} \oplus \mathfrak{m}^{\mathbb{C}},\text{ where} \tag{7.1}$$

$$\mathfrak{h}^{\mathbb{C}} = \left\{ \begin{bmatrix} A & 0 \\ 0 & -A^T \end{bmatrix} \;\middle|\; A \in \mathfrak{gl}(n) \right\}, \tag{7.2}$$

$$\mathfrak{m}^{\mathbb{C}} = \left\{ \begin{bmatrix} 0 & C \\ D & 0 \end{bmatrix} \;\middle|\; C, D \in \mathfrak{gl}(n),\ C = C^T, D = D^T \right\} \tag{7.3}$$

An $\mathrm{Sp}(2n, \mathbb{R})$-Higgs bundle can thus be described as a rank $2n$ Higgs vector bundle of the form $(V \oplus V^*, \Phi)$ where

- V is a rank n holomorphic vector bundles, and
- $\Phi = \begin{bmatrix} 0 & \beta \\ \gamma & 0 \end{bmatrix}$ with $\beta \in H^0(Sym^2(V) \otimes K_\Sigma)$ and $\gamma \in H^0(Sym^2(V^*) \otimes K_\Sigma)$.

The Toledo invariant is defined to be

$$\tau(V \oplus V^*, \Phi) = \deg(V) \tag{7.4}$$

and satisfies the bound

$$|\tau(V \oplus W, \Phi)| \leq n(g-1) \tag{7.5}$$

THEOREM 7.2. *If* $(V \oplus V^*, \Phi)$ *is a semistable* $\mathrm{Sp}(2n,\mathbb{R})$*-Higgs bundle with maximal Toledo invariant* $\tau = n(g-1)$ *then* γ *defines an isomorphism* $\gamma : V \simeq V^* \otimes K_\Sigma$ *with* $\gamma \in H^0(Sym^2(V^*) \otimes K_\Sigma)$.

Thus if we fix a square root of K_Σ, denoted by $K_\Sigma^{-1/2}$, and define $U = V \otimes K_\Sigma^{-\frac{1}{2}}$, then

- $q_U = \gamma \otimes 1_{K_\Sigma^{-\frac{1}{2}}} : U \to U^*$ defines an orthogonal structure, and
- $\theta = \beta \circ \gamma : U \to U \otimes K_\Sigma^2$ defines a K_Σ^2-twisted symmetric endomorphism.

Notice that the structure group for the orthogonal bundle is $O(n,\mathbb{C})$ and the K_Σ^2-twisted Higgs field takes values in $Sym^2(\mathfrak{gl}(n,\mathbb{C}))$. Moreover, the complexified Cartan decomposition for $\mathfrak{gl}(n,\mathbb{R})$ is

$$\mathfrak{gl}(n,\mathbb{C}) = \mathfrak{o}(n,\mathbb{C}) \oplus Sym^2(\mathfrak{gl}(n,\mathbb{C}) \tag{7.6}$$

Thus the defining data for maximal $\mathrm{Sp}(2n,\mathbb{R})$-Higgs bundles can be identified with the defining data for K_Σ^2-twisted G'-Higgs bundle where $G' = GL(n,\mathbb{R})$.

THEOREM 7.3. *The Cayley partner for* $\mathrm{Sp}(2n,\mathbb{R})$ *is* $GL(n,\mathbb{R})$ *and the map*

$$(U,\theta) \mapsto \left((U \otimes K_\Sigma^{\frac{1}{2}}) \oplus (U^* \otimes K_\Sigma^{-\frac{1}{2}}), \begin{bmatrix} 0 & \theta \circ q_U^{-1} \otimes 1 \\ q_u \otimes 1 & 0 \end{bmatrix} \right) \tag{7.7}$$

defines an isomorphism (the Cayley map)

$$\Psi : \mathcal{M}_{K^2}(\Sigma, GL(n,\mathbb{R})) \to \mathcal{M}_{max}(\Sigma, \mathrm{Sp}(2n,\mathbb{R})) \tag{7.8}$$

If G is any other simple group of Hermitian tube type, the maximal component $\mathcal{M}_{max}(\Sigma, G)$ similarly acquires a description as a moduli space of K^2-twisted G'-Higgs bundles where G' is the Cayley partner group. A list of the Cayley partners can be found in [**9**], where it is also shown that the Cayley partner of G can be characterized as the non-compact dual of H in $H^{\mathbb{C}}$, where $H \subset G$ is a maximal compact subgroup and $H^{\mathbb{C}}$ is its complexification.

The Cayley map gives a description of the maximal components which, while not as explicit as the parametrization of the Hitchin components, can nevertheless be useful in understanding global features of the component.

Like the special features of the Hitchin section, the appearance of these Cayley partners can be traced back to a distinguished $\mathfrak{sl}_2$-triple in the complexified Lie algebra. Indeed we will see a unifying construction in Section 8 but we briefly describe the specific case of $\mathrm{Sp}(2n,\mathbb{R})$ to illustrate the way it works in the case of real forms of Hermitian tube type.

7.2. The underlying $\mathfrak{sl}_2$-triple.

As above, if $(V \oplus V^*, \Phi)$ is a polystable $\mathrm{Sp}(2n,\mathbb{R})$-Higgs bundle with maximal Toledo invariant then

$$V \oplus V^* = (U \otimes K_\Sigma^{\frac{1}{2}}) \oplus (U^* \otimes K_\Sigma^{-\frac{1}{2}}) \simeq U \otimes (K_\Sigma^{\frac{1}{2}} \oplus K_\Sigma^{-\frac{1}{2}}) \tag{7.9}$$

The frame bundle for $V \oplus V^*$ is thus of the form $\mathcal{E}_{O(n,\mathbb{C})} \star \mathcal{E}_T$ where

- $\mathcal{E}_{O(n,\mathbb{C})}$ is the principal $O(n,\mathbb{C})$-bundle which defines U,
- $\mathcal{E}_T$ is the principal $\mathbb{C}^*$-bundle which defines $K_\Sigma^{\frac{1}{2}}$, and
- the $\star$ operation is a multiplication in $\mathrm{Sp}(2n,\mathbb{C})$, using the embeddings

$$A \mapsto \begin{bmatrix} A & 0 \\ 0 & A \end{bmatrix} \text{ for } A \in O(n,\mathbb{C}), \tag{7.10}$$

$$\lambda \mapsto \begin{bmatrix} \lambda I & 0 \\ 0 & \lambda^{-1} I \end{bmatrix} \text{ for } \lambda \in \mathbb{C}^* \tag{7.11}$$

Notice that

- the $\mathbb{C}^*$ in $\mathrm{Sp}(2n,\mathbb{C})$ lies in a copy of $\mathrm{SL}(2,\mathbb{C})$ embedded as $S = \mathrm{SL}(2,\mathbb{C}) \otimes I_n$, i.e. via

$$\begin{bmatrix} a & b \\ c & d \end{bmatrix} \mapsto \begin{bmatrix} aI_n & bI_n \\ cI_n & dI_n \end{bmatrix} \tag{7.12}$$

  Moreover, within $\mathrm{Sp}(2n,\mathbb{C})$ this $\mathrm{SL}(2,\mathbb{C})$-subgroup commutes with the $O(n,\mathbb{C})$-subgroup defined by (7.10), thereby allowing $\mathcal{E}_{O(n,\mathbb{C})} \star \mathcal{E}_T$ to be well defined as an $\mathrm{Sp}(2n,\mathbb{C})$-bundle.
- the summand $\mathfrak{m}^{\mathbb{C}}$ in (7.3) decomposes as $\mathfrak{m}^{\mathbb{C}}_+ \oplus \mathfrak{m}^{\mathbb{C}}{}_-$ where $D = 0$ in $\mathfrak{m}^{\mathbb{C}}_+$ and $C = 0$ in $\mathfrak{m}^{\mathbb{C}}_-$. Moreover if

$$T = \left\{ \begin{bmatrix} tI & 0 \\ 0 & t^{-1} I \end{bmatrix} \;\middle|\; t \in \mathbb{C} \right\} \tag{7.13}$$

  then the adjoint action of T has weight 0 on $\mathfrak{h}^{\mathbb{C}}$ and weights ± 2 on $\mathfrak{m}^{\mathbb{C}}_{\pm}$. The decomposition in (7.1) can thus be identified as

$$\mathfrak{sp}(2n,\mathbb{C}) = \mathfrak{m}^{\mathbb{C}}_- \oplus \mathfrak{h}^{\mathbb{C}} \oplus \mathfrak{m}^{\mathbb{C}}_+ \tag{7.14}$$

$$= \mathfrak{sp}(2n,\mathbb{C})_{-2} \oplus \mathfrak{sp}(2n,\mathbb{C})_0 \oplus \mathfrak{sp}(2n,\mathbb{C})_2 \tag{7.15}$$

  where the subscripts denote the weights of the T action
- The Lie algebra of S is an $\mathfrak{sl}(2,\mathbb{C})$-subalgebra spanned by the triple

$$\{f,h,e\} = \left\{ \begin{bmatrix} 0 & 0 \\ I_n & 0 \end{bmatrix}, \begin{bmatrix} I_n & 0 \\ 0 & -I_n \end{bmatrix}, \begin{bmatrix} 0 & I_n \\ 0 & 0 \end{bmatrix} \right\} \tag{7.16}$$

  Under the adjoint action the decomposition of $\mathrm{Sp}(2n,\mathbb{C})$ into irreducible $\mathfrak{sl}(2,\mathbb{C})$-representations takes the form

$$\mathfrak{sp}(2n,\mathbb{C}) = W_0 \oplus W_1 \tag{7.17}$$

  where W_m is a sum of irreducible representations of dimension $2m+1$.
- The decompositions in (7.17) and (7.15) are related by the fact that $\mathfrak{sp}(2n,\mathbb{C})_2$ is the highest weight space for W_1.

The Higgs field can thus be decomposed as

$$\Phi = \tilde{f} + \tilde{\theta} \tag{7.18}$$

where $\tilde{f}$ is defined by the nilpotent f in $\mathfrak{sl}(2,\mathbb{C})$ and $\tilde{\theta} = \begin{bmatrix} 0 & \theta \circ q_U^{-1} \otimes 1 \\ 0 & 0 \end{bmatrix}$ is determined by a highest weight element.

8. Magical triples and generalized Cayley maps

The underlying mechanism responsible for the Hitchin components when G is a split real form, or the maximal components when G is Hermitian of tube type, can be understood in a uniform way by focussing on the role of the distinguished $\mathfrak{sl}_2$-triples in the Lie algebras. From this perspective, the Hitchin components and the maximal components represent opposite extremes in a range which includes special components in the moduli spaces for G-Higgs bundles for G on a list which also includes $\mathrm{SO}(p,q)$ or certain exceptional real groups. The list of real forms which occur in this range is the same as the list which admit a positivity structure known as Θ-positivity. Introduced in [**37**], the Θ-positivity property plays a crucial role in higher Teichmüller theory, where it is responsible for many of the similarities between Teichmüller space and special components of $\mathrm{Rep}(\pi_1(S), G)$ known collectively as higher Teichmüller spaces (see [**38, 55**])

8.1. Magical triples.

Let $\mathfrak{g}^{\mathbb{C}}$ be a complex semisimple Lie algebra. By the Jacobson-Morozov theorem any non-zero nilpotent element $e \in \mathfrak{g}^{\mathbb{C}}$ can be completed to a triple (called an $\mathfrak{sl}_2$-triple) of nonzero elements $\{f, h, e\}$ satisfying

$$[h,e] = 2e \ , \ [h,f] = -2f \ , \ [e,f] = h \ , \tag{8.1}$$

i.e. which generate an $\mathfrak{sl}(2,\mathbb{C})$-subalgebra with h semisimple. Any $\mathfrak{sl}_2$-triple in $\mathfrak{g}^{\mathbb{C}}$ determines two decompositions of $\mathfrak{g}^{\mathbb{C}}$, namely

(1) into weight-spaces for the ad-action of h, denoted by

$$\mathfrak{g}^{\mathbb{C}} = \mathfrak{g}_0 \oplus \bigoplus_{j=1}^{M} \mathfrak{g}_j \tag{8.2}$$

where the subscript denotes the weight, and

(2) into irreducible $\mathfrak{sl}(2,\mathbb{C})$-representations, denoted

$$\mathfrak{g}^{\mathbb{C}} = W_0 \oplus \bigoplus_{j=1}^{M} W_{m_j} \tag{8.3}$$

where W_{m_j} is isomorphic to a direct sum of n_j copies (with $n_j > 0$) of the unique (m_j+1)- dimensional representation. The integer pairs $\{(m_j, n_j) \mid j = 1, 2, \cdots, M\}$ from (8.2) and (8.3) are called the *$\mathfrak{sl}_2$-data for the triple* $\{f, h, e\}$.

Note: The subspace of 1-dimensional representations, i.e. the summand W_0 in (8.3), is the centralizer of the the $\mathfrak{sl}(2,\mathbb{C})$-subalgebra generated by the $\mathfrak{sl}_2$-triple. We denote this by $\mathfrak{c}$.

A nilpotent e is called *even* if all weights in (8.2) are even. Since the highest ad_h-weight in W_k is k, for $\mathfrak{sl}(2,\mathbb{C})$-subalgebras determined by even nilpotents we get decompositions

$$W_{2m_j} = V_{2m_j} \oplus ad_f(V_{2m_j}) \oplus ad_f^2(V_{2m_j}) \oplus \cdots \oplus ad_j^{2m_j}(V_{2m_j}) \tag{8.4}$$

where $V_{2m_j} = W_{2m_j} \cap \mathfrak{g}_{2m_j}$ is the heighest weight space and $Z_{2m_j} = ad_f^{m_j}(V_{2m_j}) \subset \mathfrak{g}_0$.

DEFINITION 8.1. Define a vector space involution $\sigma : \mathfrak{g}^{\mathbb{C}} \to \mathfrak{g}^{\mathbb{C}}$ by the linear extension of

$$\sigma(x) = \begin{cases} x \text{ if } x \in V_0 \\ (-1)^{k+1} \text{if } x \in ad_f^k(V_j) \text{ for some } 0 \leq k \leq j \text{ and } j > 0 \end{cases} \tag{8.5}$$

In general σ does not preserve the Lie algebra structure on $\mathfrak{g}^{\mathbb{C}}$. In [**10**] we introduced the notion of a magical $\mathfrak{sl}_2$-*triple* defined as follows.

DEFINITION 8.2. The $\mathfrak{sl}_2$-triple $\{f, h, e\}$ is called *magical* if the involution σ preserves the Lie algebra structure on $\mathfrak{g}^{\mathbb{C}}$.

We have already seen two examples of magical $\mathfrak{sl}_2$-triples, namely

- principal $\mathfrak{sl}_2$-triples, and
- the triples associated to real forms of Hermitian tube type.

If the involution σ is magical, and thus a Lie algebra involution, then it has an associated anti-holomorphic involution[5] which defines a real form for $\mathfrak{g}^{\mathbb{C}}$.

DEFINITION 8.3. If $\{f, h, e\}$ is a magical $\mathfrak{sl}_2$-triple and τ_e is the antiholomorphic involution associated to the involution defined by (8.5), the real form of $\mathfrak{g}^{\mathbb{C}}$ defined by τ_e is called the *canonical real form associated to the magical* $\mathfrak{sl}_2$*-triple, or simply the magical real form.*

Let $G^{\mathbb{C}}$ be a connected complex Lie group with Lie algebra $\mathfrak{g}^{\mathbb{C}}$ and such that the involution σ integrates to an involution on $G^{\mathbb{C}}$. Denote also by $\tau_e : G^{\mathbb{C}} \to G^{\mathbb{C}}$ the involution which integrates τ_e. The fixed-point group defined by τ_e is called the *canonical real form of G defined by the magical* $\mathfrak{sl}_2$*-triple, or simply the magical real form of* $G^{\mathbb{C}}$.

THEOREM 8.4 ([**10**]). *A real form of a complex simple Lie algebra is the canonical real form associated to a magical* $\mathfrak{sl}_2$*-triple if and only if it is one of the following types:*

(1) *a split real form,*
(2) *a real form of Hermitian tube type,*
(3) $\mathfrak{so}(p, q)$ *with* $2 < p < q$,
(4) *the quaternionic real form of* $\mathfrak{f}_4, \mathfrak{e}_6, \mathfrak{e}_7$, *or* $\mathfrak{e}_8$ *(a.k.a.* $\mathfrak{f}_4^4, \mathfrak{e}_6^2, \mathfrak{e}_7^{-5}$, *or* $\mathfrak{e}_8^{-24}$*)*

REMARK 8.5. The four types in Theorem (8.4) are not mutually exclusive. The overlaps occur because the following real forms are split and hence also included in type (1):

- $\mathfrak{sp}(2n, \mathbb{R})$ (in type (2))
- $\mathfrak{so}(n, n+1)$ (in type (3))
- $\mathfrak{f}_4^4$ (in type (4))

Each of these real forms admits two distinct magical $\mathfrak{sl}_2$-triples, with the characteristics of each $\mathfrak{sl}_2$-triple determined by one of the two types to which the real form belongs.

[5]The antiholomorphic involution, say τ is unique up to conjugation and satisfies the condition that $\sigma\tau = \tau\sigma$ defines a compact real form.

8.2. Structures determined by magical $\mathfrak{sl}_2$-triples.

In the next three subsections we assume that $\{f, h, e\}$ is a magical subtriple in $\mathfrak{g}^{\mathbb{C}}$, and that $G^{\mathbb{C}}$ is as above, i.e. a connected complex Lie group with Lie algebra $\mathfrak{g}^{\mathbb{C}}$ and such that the involution σ integrates to an involution on $G^{\mathbb{C}}$. We note that in this case e is an even nilpotent so the subspaces $Z_{2j} = W_{2j} \cap \mathfrak{g}_0$ are non-empty. We denote the canonical real form associated to the magical $\mathfrak{sl}_2$-triple by G and its maximal compact subgroup by H. Then in addition to the decompositions (8.2) and (8.3) we have the Cartan decomposition

$$\mathfrak{g}^{\mathbb{C}} = \mathfrak{h}^{\mathbb{C}} \oplus \mathfrak{m}^{\mathbb{C}} \tag{8.6}$$

where $\mathfrak{h}^{\mathbb{C}}$ is the complexification of $\mathfrak{h} = Lie(H)$. We get the following structures (see [**10**] for details):

(1) **A real form of $\mathfrak{g}_0$:** The summand $\mathfrak{g}_0$ in (8.2), i.e. in the zero weight-space for ad_h, is a subalgebra of $\mathfrak{g}^{\mathbb{C}}$. If h is part of a magical $\mathfrak{sl}_2$-triple then $\mathfrak{g}_0$ admits a real form - called the Cayley real form - defined by the involution

$$\theta : \mathfrak{g}_0 \to \mathfrak{g}_0 \tag{8.7}$$

$$\theta = \begin{cases} +1 \text{ on } W_0 \\ -1 \text{ on } W_j \cap \mathfrak{g}_0 \text{ for } j > 0 \end{cases} \tag{8.8}$$

Denote this real form by $\mathfrak{g}_C$.

(2) **A decomposition of $\mathfrak{g}_C$:** If the magical $\mathfrak{sl}_2$-triple defined by the nilpotent e is not principal then there is a subalgebra $\mathfrak{g}(e) \subset \mathfrak{g}$ such that the $\mathfrak{sl}_2$-triple is principal in $\mathfrak{g}(e)$. In fact $\mathfrak{g}(e)$ is the centralizer of $\mathfrak{c} = W_0$ as defined in (8.3). The real form $\mathfrak{g}_C$ decomposes as

$$\mathfrak{g}_C = \mathfrak{g}_{0,ss} \oplus \mathbb{R}^{rank(\mathfrak{g}(e))} \tag{8.9}$$

where $\mathfrak{g}^{\mathbb{R}}_{0,ss}$ is either zero or a simple real Lie algebra.

(3) **The Cayley group:** The Lie algebra $\mathfrak{g}_C$ is does not uniquely determine a Cayley Lie group because it is insensitive to the center of the group. The requisite extra information is obtained as follows. Let $S \subset G^{\mathbb{C}}$ be the subgroup whose Lie algebra is the magical $\mathfrak{sl}_2$-triple. In the adjoint form of $\mathfrak{g}^{\mathbb{C}}$ this is isomorphic to $\mathrm{PSL}(2, \mathbb{C}) \simeq \mathrm{SO}(3, \mathbb{C})$; otherwise for the groups we consider, i.e. complex simple Lie groups, it isomorphic to $\mathrm{SL}(2, \mathbb{C})$. Let C be the centralizer of S in $G^{\mathbb{C}}$.

DEFINITION 8.6. The Cayley group of the magical $\mathfrak{sl}_2$-triple in $G^{\mathbb{C}}$ is the group

$$\mathrm{G}_C = \mathrm{G}_{0,ss} \times (\mathbb{R}^+)^{rank(\mathfrak{g}(e))}, \tag{8.10}$$

where $\mathrm{G}_{0,ss} \subset G^{\mathbb{C}}$ is a real Lie group with Lie algebra $\mathfrak{g}_{0,ss}$ and maximal compact subgroup $C \cap G$

8.3. Special G-Higgs bundles.

Recall that a G-Higgs bundle is a pair $(E_{H^{\mathbb{C}}}, \Phi)$ (see Section 3.1). If G is a magical real form then the structures described above can be used to construct G-Higgs bundles with special structure. These turn out to comprise a union of connected components in $\mathcal{M}(\Sigma, G)$.

By definition, the subgroups S and C commute in $G^{\mathbb{C}}$. It follows that principal S- and C-bundles, say $\mathcal{E}_S$ and $\mathcal{E}_C$, can be combined to produce a principal $G^{\mathbb{C}}$-bundle whose structure group factors through the image of $S \times C$ in $G^{\mathbb{C}}$. We denote this construction by $\mathcal{E}_S \star \mathcal{E}_C$. In particular, if $T \subset S$ is the $\mathbb{C}^*$-subgroup whose Lie algebra is spanned by h, and $\mathcal{E}_T$ is a principal T-bundle, then we can construct $\mathcal{E}_T \star \mathcal{E}_C$. We will take $\mathcal{E}_T$ to be the frame bundle for K_Σ if $S \simeq \mathrm{PSL}(2,\mathbb{C})$, or a square root $K_\Sigma^{\frac{1}{2}}$ if $S \simeq \mathrm{SL}(2,\mathbb{C})$.

A important subtlety is the difference between C and $C \cap H^{\mathbb{C}}$, where $H^{\mathbb{C}}$ is the complexification of a maximal compact subgroup of G. From (8.10) it is clear that $C \cap H^{\mathbb{C}}$ is the complexification of a maximal compact subgroup of the Cayley group G_C. Moreover, C and $C \cap H^{\mathbb{C}}$ have the same Lie algebra, namely W_0. They may however have different centers. For example, for $G_{\mathbb{R}} = \mathrm{SL}(n,\mathbb{R})$ we get $C = \mathbb{Z}_n$ but $H = \mathrm{SO}(n,\mathbb{C})$ and $C \cap \mathrm{SO}(n,\mathbb{C})$ is either $\mathbb{Z}_2$ (n even) or trivial (n odd) [See Remark 4.12 in [**10**]]

THEOREM 8.7. *Suppose that the Lie algebra of S is magical. Then $\mathcal{E}_T \star \mathcal{E}_C$ defines a principal $H^{\mathbb{C}}$-bundle if and only if the structure group of $\mathcal{E}_C$ reduces to $C \cap H^{\mathbb{C}}$.*

If $\mathcal{E}_T \star \mathcal{E}_C$ defines a principal $H^{\mathbb{C}}$-bundle

$$\mathcal{E}_{H^{\mathbb{C}}} = \mathcal{E}_T \star \mathcal{E}_C[H^{\mathbb{C}}] \tag{8.11}$$

then this can be used to construct G-Higgs bundles. In particular, recall that the maximal compact subgroup of $G_{0,ss}$ can be taken to be $H_{0,ss} = C \cap G$ (see (8.10)). Hence if $\mathcal{E}_C$ is a principal $H_{0,ss}^{\mathbb{C}}$-bundle (where $H_{0,ss}^{\mathbb{C}}$ denotes the complexification), then its structure group does reduce to $H^{\mathbb{C}}$. We assume this to be the case.

The Higgs fields must take values in summand $\mathfrak{m}^{\mathbb{C}}$ in (8.6), i.e. in the -1-eigenspace for the involution defined by (8.5). We get contributions from two sources:

(1) Under the ad-action S and C preserve the highest weights subspaces $V_{2m_j} \subset W_{2m_j}$ in (8.3), and by construction $V_{2m_j} \subset \mathfrak{m}^{\mathbb{C}}$. We thus get subbundles

$$\mathcal{E}_{H^{\mathbb{C}}}[V_{2j}] \subset \mathcal{E}_{H^{\mathbb{C}}}[\mathfrak{m}^{\mathbb{C}}] \tag{8.12}$$

(2) Since C acts trivially on f and the subspace spanned by f is preserved by T, we get a subbundle

$$\mathcal{E}_{H^{\mathbb{C}}}[< f >] = \mathcal{E}_T[< f >] \subset \mathcal{E}_{H^{\mathbb{C}}}[\mathfrak{m}^{\mathbb{C}}] \tag{8.13}$$

Recall that we take $\mathcal{E}_T$ to be the principal bundle for K_Σ or $K_\Sigma^{\frac{1}{2}}$, depending on whether S is $S = \mathrm{SL}(2,\mathbb{C})$ or $\mathrm{PSL}(2,\mathbb{C})$. Since $\mathbb{C}^* = T$ acts with weight -2 on f if $S = \mathrm{SL}(2,\mathbb{C})$ and weight -1 if $S = \mathrm{PSL}(2,\mathbb{C})$, in both cases we get

$$\mathcal{E}_T[< f >] \otimes K_\Sigma \simeq \mathcal{O} \tag{8.14}$$

and hence $f \in \mathfrak{m}^{\mathbb{C}}$ defines a constant section, which we denote also by f. In both cases we will refer to the pair $(\mathcal{E}_T, f)$ as the *uniformizing Higgs pair.*

Given any sections $\phi_{m_j} \in H^0(\mathcal{E}_{H^{\mathbb{C}}}[V_{2m_j}] \otimes K_\Sigma$ we can thus construct G-Higgs bundles of the special form

$$\left(\mathcal{E}_T \star \mathcal{E}_C[H^{\mathbb{C}}], f + \phi_{m_1} + \cdots + \phi_{m_M}\right) \tag{8.15}$$

8.4. Cayley map and exotic components.

The section f allows us to globalize the ad_f-action on $\mathfrak{g}^{\mathbb{C}}$ to define isomorphisms

$$ad_f^{m_j} : \mathcal{E}_{H^{\mathbb{C}}}[V_{2m_j}] \otimes K_\Sigma \to \mathcal{E}_C[Z_{2m_j}] \otimes K_\Sigma^{1+j} \tag{8.16}$$

where as above $Z_{2m_j} = W_{2m_j} \cap \mathfrak{g}_0$.

Two key facts are important:

(1) If $Z_{2m_j} \simeq \mathbb{C}$ (i.e. $\dim(Z_{2m_j}) = 1$) then $\mathcal{E}_{H^{\mathbb{C}}}[Z_{2m_j}] \simeq \mathcal{O}$ and we can identify $ad_f^j(\phi_{m_j})$ with a section of $K_\Sigma^{1+m_j}$, i.e.

$$ad_f^{m_j}(\phi_{m_j}) = q_{m_j} \in H^0(K_\Sigma^{1+m_j}) . \tag{8.17}$$

(2) If the $\mathfrak{sl}_2$-triple is magical then $\dim(Z_{2m_j}) > 1$ for at most one $j > 0$. Indeed, in the four cases listed in Theorem 8.4 this is trivially true in cases (1) and (2) ($\dim(Z_{2m_j}) = 1$ for all j in case (1), and in case (2) there is exactly one $j > 0$) and follows from calculations in the other cases (see theorem 4.3 in [**10**]).

Let m_c be the unique positive integer such that $\dim(Z_{2m_c}) > 1$. Recall that by definition, Z_{2m_j} lies in the -1-eigenspace for the involution that defines the Cayley real form. If ϕ_{m_c} is a section of $\mathcal{E}_C[V_{2m_c}] \otimes K_\Sigma$ and $\psi_{m_c} = ad_f^{m_c}(\phi_{m_c}) \in \mathcal{E}_{H^{\mathbb{C}}}[Z_{2j}] \otimes K_\Sigma^{2m_c+1}$ it then follows that $(\mathcal{E}_C, \psi_{m_c})$ defines a $K_\Sigma^{2m_c+1}$-twisted $G_{0,ss}$-Higgs bundle. The remaining indices m_j, i.e. those for which $\dim(Z_{2m_c}) = 1$ are determined by the exponents for the subalgebra $\mathfrak{g}(e)$ (see Lemma 5.7 in [**10**]).

The G-Higgs bundles of the form in (8.15) are thus determined by

- (Fixed data) $(\mathcal{E}_T, f)$, i.e. the uniformizing Higgs pair, and
- (Cayley parameters) $((\mathcal{E}_C, \psi), q_1, \cdots, q_N))$, i.e. a $K_\Sigma^{2m_c+1}$-twisted $G_{0,ss}$-Higgs bundle together with a collection of N holomorphic differentials where $N = rank(\mathfrak{g}(e))$.

The G-Higgs bundles constructed in this way are polystable if and only if the $K_\Sigma^{2m_c+1}$-twisted $G_{0,ss}$-Higgs bundle are polystable. We thus get a map

$$\Psi_e : \mathcal{M}_{K^{m_c+1}}(G_{0,ss}^{\mathbb{R}}) \times \bigoplus_{j=1}^{\operatorname{rank}(\mathfrak{g}(e))} H^0(K_\Sigma^{l_j+1}) \longrightarrow \mathcal{M}(G). \tag{8.18}$$

where the exponents l_j are the exponents of the Lie algebra $\mathfrak{g}(e)$.

THEOREM 8.8 (Theorem 7.1 and Corollary 7.6 in [**10**]). *The image of the map* (8.18), *which we call the Cayley map, is open and closed, and hence defines a union of connected component of $\mathcal{M}(G)$ homeomorphic to*

$$\mathcal{M}_{K^{m_c+1}}(G_{0,ss}^{\mathbb{R}}) \times \bigoplus_{j=1}^{\operatorname{rank}(\mathfrak{g}(e))} H^0(K_\Sigma^{l_j+1}) . \tag{8.19}$$

Each such component is locally irreducible and irreducible.

We can identify two extremes (see Table 3) corresponding to $\operatorname{rank}(\mathfrak{g}(e)) = 1$ and $\mathfrak{g}_{0,ss} = 0$. These two extremes correspond to cases in which the Cayley component comes entirely from the first factor in (8.19) (or more precisely from a factor of the form $\mathcal{M}_{K^{m_c+1}}(G')$, as described below) and, at the other extreme, to cases in which it comes entirely from the spaces of holomorphic differentials.

TABLE 3. Lie theoretic data for magical real forms, with the Lie algebras of the Cayley partner groups ($\mathfrak{g}'$) shown for the real forms of Hermitian tube type.

Real form	$\mathfrak{g}(e)$	$\mathfrak{g}_{0,ss}$	m_c	$\#\{dim(Z_{m_j}) = 1\}$
$\mathfrak{su}(n,n)$	$\mathfrak{sl}(2,\mathbb{C})$	$\mathfrak{sl}(n,\mathbb{C})$	1	0
$\mathfrak{so}(2,q)$	$\mathfrak{sl}(2,\mathbb{C})$	$\mathfrak{so}(1,q-1)$	1	0
$\mathfrak{sp}(2n,\mathbb{R})$	$\mathfrak{sl}(2,\mathbb{C})$	$\mathfrak{sl}(n,\mathbb{R})$	1	0
$\mathfrak{so}^*(4n)$	$\mathfrak{sl}(2,\mathbb{C})$	$\mathfrak{su}^*(2n)$	1	0
$\mathfrak{e}_7^{-25}$	$\mathfrak{sl}(2,\mathbb{C})$	$\mathfrak{e}_6^{-26}$	1	0
$\mathfrak{f}_4^4$	$\mathfrak{g}_2$	$\mathfrak{sl}(3,\mathbb{R})$	3	2
$\mathfrak{e}_6^2$	$\mathfrak{g}_2$	$\mathfrak{sl}(3,\mathbb{C})$	3	2
$\mathfrak{e}_7^{-5}$	$\mathfrak{g}_2$	$\mathfrak{sl}(3,\mathbb{H})$	3	2
$\mathfrak{e}_8^{-24}$	$\mathfrak{g}_2$	$\mathfrak{e}_6^{-26}$	3	2
$\mathfrak{so}(p,q)$ $2 < p < q$	$\mathfrak{so}(p-1,p)$	$\mathfrak{so}(1,q-p+1)$	$p-1$	$p-1$
split	$\mathfrak{g}$	0	-	$rank(\mathfrak{g})$

The first case, i.e. $\operatorname{rank}(\mathfrak{g}(e)) = 1$, corresponds to the real forms of Hermitian tube type. In these cases $\mathfrak{g}(e) = \mathfrak{sl}(2,\mathbb{C})$, $m_c = 1$, and there is just one summand in $\bigoplus_{j=1}^{\operatorname{rank}(\mathfrak{g}(e))} H^0(K_\Sigma^{l_j+1})$. The Cayley component has the form

$$\mathcal{M}_{K^2}(G_{0,ss}^{\mathbb{R}}) \times H^0(K_\Sigma^2) \tag{8.20}$$

Using the fact that $H^0(K_\Sigma^2)$ can be identified with $\mathcal{M}_{K^2}(\mathbb{R}^+)$ we see that the Cayley components given by (8.20) can be identified as moduli spaces of the form $\mathcal{M}_{K^2}(G')$, i.e. of K^2-twisted G' Higgs bundles where $G' = G_{0,ss}^{\mathbb{R}} \times \mathbb{R}^+$. The groups G' are precisely the Cayley partner groups identified in [**9**][6] and we thus recover the Cayley correspondence for maximal components in $\mathcal{M}(\Sigma, G)$, as described in Section 7.

The second case, i.e. $\mathfrak{g}_{0,ss} = 0$, corresponds to the split real forms. In this case the first factor in the Cayley component does not vanish entirely but is just a finite set and we recover the Hitchin components described in Section 6 as the images of Hitchin sections.

The other cases in Table 3 are a hybrid of the two extreme cases. The magical components in the associated moduli spaces are products of Cayley-like moduli spaces and Hitchin sections. For real forms with the exceptional Lie algebras in the table, the Cayley components are all of the form

$$\mathcal{M}_{K_\Sigma^4}(G_{0,ss}^{\mathbb{R}}) \times H^0(K_\Sigma^2) \oplus H^0(K_\Sigma^6) \tag{8.21}$$

where the Lie algebras of the groups $G_{0,ss}^{\mathbb{R}}$ are those shown in the table.

In the next section we describe the case of $\mathrm{SO}(p,q)$ with $2 < p < q$ in some detail because it is illustrative, and also because the details play a role in Section 10.1. For full details see [**5**].

[6]The identification of the Lie algebras is clear. In order to relate the Cayley partners to the groups $G_{0,ss}^{\mathbb{R}} \times \mathbb{R}^+$ one must take care to consider groups with the same centers, for example $U^*(2n) = (\mathrm{SU}^*(2n) \ltimes \mathbb{Z}_2) \times \mathbb{R}^+$.

8.5. SO(p,q)-Higgs bundles.

The magical $\mathfrak{sl}_2$-triples for which the Cayley real form is $\mathfrak{so}(p,q)$ (with If $2<p<q$) are principal in $\mathfrak{so}(2p-1,\mathbb{C})\subset\mathfrak{so}(p+q,\mathbb{C})$. They have

- $\mathfrak{g}(e)\simeq\mathfrak{so}(2p-1,\mathbb{C})$,
- $\mathfrak{g}_{0,ss}\simeq\mathfrak{so}(1,q-p+1)$, and
- $C=S(O(1,\mathbb{C})\times O(1,q-p+1))$ in $\mathrm{SO}(p,q)$

The integers $\{m_j\}$ in the decomposition (8.3) are $\{1,3,\dots,2p-3\}\cup\{p-1\}$ with $m_c=p-1$. This lead to Cayley maps of the form

$$\Psi:\mathcal{M}_{K^p}(\Sigma,\mathrm{SO}(1,q-p+1))\times\bigoplus_{j=1}^{p-1}H^0(K^{2j})\longrightarrow\mathcal{M}(\Sigma,\mathrm{SO}(p,q)),\tag{8.22}$$

We can explicitly describe these maps and the structure of the resulting Cayley components. We start with a convenient description of SO(p,q)-Higgs bundles. The maximal compact subgroup of SO(p,q) is $H=S(O(p)\times O(q)$. According to Definition 3.1 the bundle in an SO(p,q)-Higgs bundle should thus be a principal $S(O(p,\mathbb{C})\times O(q,\mathbb{C})$-bundle. We may use the standard representations of $O(p,\mathbb{C})$ and $O(q,\mathbb{C})$ to give a vector bundle description as follows.

Definition 8.9. An SO(p,q)-**Higgs bundle** on a closed surface Σ is a tuple $((V,Q_V),(W,Q_W),\eta)$, where (V,Q_V), (W,Q_W) are rank p and rank q holomorphic orthogonal vector bundles respectively such that $\det(V\oplus W)=\mathcal{O}$, and $\eta\in H^0(\mathrm{Hom}(W,V)\otimes K_\Sigma)$.

Remark 8.10. Unless it is essential we often suppress the orthogonal structures and denote the SO(p,q)-Higgs bundles by tuples (V,W,η).

Definition 8.9 applies also if $p=1$ and the Higgs fields are L-twisted, where L is any line bundle. In particular, a K_Σ^m**-twisted** SO$(1,n)$**-Higgs bundle** is a tuple $(\mathcal{I},W,\eta)$ where

- $\mathcal{I}^2=\mathcal{O}$, i.e. $\mathcal{I}$ denotes a square root of the trivial line bundle,
- W is an orthogonal bundle of rank n,
- η is a holomorphic map $\eta:W\to\mathcal{I}\otimes K_\Sigma^m$

The Cayley map (8.22) can then be explicitly described (see Section 4 in [**5**]) by

$$\Psi:\left([\mathcal{I},\widehat{W},\eta],q_2,q_4,\dots q_{2p-2}\right)\longmapsto[V,W,\eta]\tag{8.23}$$

with

$$(V,W,\eta)=\left(\mathcal{I}\otimes\mathcal{K}_{p-1},\ \mathcal{I}\otimes\mathcal{K}_{p-2}\oplus\widehat{W},\ \begin{bmatrix}q_2&q_4&\cdots&q_{2p-2}&\eta_{\widehat{W}}\\1&q_2&\cdots&q_{2p-4}&0\\\vdots&\ddots&\ddots&\vdots&\vdots\\0&\cdots&1&q_2&0\\0&\cdots&0&1&0\end{bmatrix}\right)\tag{8.24}$$

where

- $(\mathcal{I}^2,\widehat{W},\eta_{\widehat{W}})$ defines a K_Σ^p-twisted SO$(1,q-p+1)$ Higgs bunde,
- $q_j\in H^0(K_\Sigma^j)=H^0(Hom(K_\Sigma^i,K_\Sigma^{i+j-1})\otimes K_\Sigma)$ for $j=2,4,\cdots,2p-4$, and
- $\mathcal{K}_n=K_\Sigma^n\oplus K_\Sigma^{n-2}\oplus\cdots\oplus K_\Sigma^{2-n}\oplus K_\Sigma^{-n}$.

In [**5**] we showed directly that the $\mathrm{SO}(p,q)$-Higgs bundles of the form in (8.24) comprise a union of connected components in $\mathcal{M}(\Sigma, G)$. In order to relate this to the general results for Cayley components coming from magical $\mathfrak{sl}_2$-triples we proceed as follows.

Let $\mathrm{SO}(p,q) \subset \mathrm{SO}(p+q,\mathbb{C})$ be the group of unit determinant block matrices $\begin{bmatrix} A & B \\ C & D \end{bmatrix}$ such that

$$\begin{bmatrix} A^T & C^T \\ B^T & D^T \end{bmatrix} \begin{bmatrix} Q_p & 0 \\ 0 & -Q_q \end{bmatrix} \begin{bmatrix} A & B \\ C & D \end{bmatrix} = \begin{bmatrix} Q_p & 0 \\ 0 & -Q_q \end{bmatrix} \tag{8.25}$$

where Q_p and Q_q are positive definite symmetric bilinear forms on $\mathbb{C}^p$ and $\mathbb{C}^q$ respectively

Without loss of generality we assume that $p \leq q$ Then we can identify the subgroup $C = S(\mathrm{O}(1,\mathbb{C}) \times \mathrm{O}(q-p+1,\mathbb{C}))$ embedded via

$$(\det(A), A) \mapsto \begin{bmatrix} \det(A) I_p & 0 & 0 \\ 0 & \det(A) I_{p-1} & 0 \\ 0 & 0 & A \end{bmatrix} \tag{8.26}$$

and a copy of $T = \mathbb{C}^*$ embedded via

$$t \mapsto \begin{bmatrix} T_p(t) & 0 & 0 \\ 0 & T_{p-1}(t) & 0 \\ 0 & 0 & I_{q-p+1} \end{bmatrix} \tag{8.27}$$

where

$$T_n(t) = \begin{bmatrix} t^{n-1} & 0 & \cdots & 0 \\ 0 & t^{n-2} & \cdots & 0 \\ \vdots & & \ddots & 0 \\ 0 & \cdots & \cdots & t^{1-n} \end{bmatrix} \tag{8.28}$$

Let $\mathcal{E}_C$ be the frame bundle for $\mathcal{I} \oplus \widehat{W}$, and let $\mathcal{E}_T$ be the frame bundle for K_Σ. We can then write

$$(\mathcal{I} \otimes \mathcal{K}_{p-1}) \oplus \Big((\mathcal{I} \otimes \mathcal{K}_{p-2}) \oplus \widehat{W} \Big) \simeq \mathcal{E}_C \star \mathcal{E}_T [H^{\mathbb{C}}][\mathbb{C}^p \oplus \mathbb{C}^q] \tag{8.29}$$

where the $\star$ denotes the operation coming from the product of the subgroups C and T in $S(O(p,\mathbb{C}) \times O(q,\mathbb{C})) \subset \mathrm{SO}(p+q,\mathbb{C})$. This shows that the frame bundle for vector bundle $V \oplus W$ is precisely of the special form required in (8.15). We omit the details showing how to realize the Higgs fields in (8.24) as Higgs fields of the form in (8.15) except to note that

- the contribution from 1's corresponds to f,
- for $i = 2, 4, \ldots, 2p-2$ the contribution from the differentials q_i corresponds to $\phi_{m_{i-1}}$, and
- the contribution from $\eta_{\hat{W}}$ corresponds to $\phi_{m_{p-1}}$.

8.6. Local minima of the Hitchin function.

We note that (see Theorem 5.10 in [**5**])

THEOREM 8.11. *In* (8.24) *if the holomorphic differentials* q_i *are all zero and* $[\mathcal{I}, \widehat{W}, \eta]$ *represents a local minimum of the Hitchin function* (5.1) *on* $\mathcal{M}(\Sigma, \mathrm{SO}(1, q-p+1))$ *then the resulting* $\mathrm{SO}(p,q)$*-Higgs bundles define local minima of the Hitchin function* (5.1). *These are the only local minima with non-trivial Higgs fields.*

If $q-p+1>2$ then the local minima on $\mathcal{M}(\Sigma, \mathrm{SO}(1,q-p+1))$ are all trivial, i.e. have $\eta_{\widehat{W}}=0$. In contrast, $\mathcal{M}(\Sigma, \mathrm{SO}(1,2))$ admits non-trivial minima (see Section 11.3)

All possible topological types of $H^{\mathbb{C}}$-bundles occur in $\mathcal{M}(\Sigma, \mathrm{SO}(p,q))$ and trivial local minima occur with bundles of all such topological types. This, together with a count for $|\pi_0(\mathcal{M}(\Sigma, \mathrm{SO}(1,n))|$, leads to the count shown in Tables 6 and 7 of connected components in $\mathcal{M}(\Sigma, \mathrm{SO}(p,q))$.

The factors $\prod_{j=1}^{\mathrm{rank}(\mathfrak{g}(e))} H^0(K_\Sigma^{l_j+1})$ in (8.22) are contractible so the precise number of components in the image of the Cayley map (called *Cayley components*) is determined by $\pi_0(\mathcal{M}_{K^{m_c+1}}(G_{0,ss}))$. This in turn depends not only on Lie algebra data but also on the center of $G_{0,ss}$. The tables list cases where the component count has been done.

9. Remaining cases and loose ends

There are two types of real forms that have not yet been discussed: the non-tube Hermitian forms and those in the last column of Table 1. The non-tube Hermitian forms can be understood in terms of their tube-type cousins - see Section 9.1. For the real forms in the last column, the classical cases have been analyzed individually. There are no published results for the exceptional cases in the last column but it is reasonable to conjecture that for all of them the only local minima of the Hitchin function are at $\Phi=0$.

In addition to the above cases, we discuss the other remaining situations in which the tally of connected components is (potentially) incomplete.

9.1. Non-tube type. The Lie algebras $\mathfrak{su}(p,q)$ with $p\neq q$, $\mathfrak{so}^*(4m+2)$, and $\mathfrak{e}_6^{-14}$ are Hermitian real forms of non-tube type. They are not on the list in Theorem 8.4, but the results of section 8 apply in modified form. In each case there is a *maximal tube subalgebra*, denoted $\mathfrak{g}_T\subset\mathfrak{g}$, namely $\mathfrak{su}(p,p), \mathfrak{so}^*(4m)$, and $\mathfrak{so}(2,8)$ (see Section 6 in [**9**]).

If $G_T\subset G$ is the subgroup with Lie algebra $\mathfrak{g}_T$, and G_T^{Ad} is its adjoint form, then (see Theorem 6.2. [**9**]) the maximal components in the moduli spaces for G and G_T are related by a fibration

$$M(L_{\mathbb{C}})\to\mathcal{M}(\Sigma,G)_{max}\to\mathcal{M}(\Sigma,G_T^{Ad})_{max} \tag{9.1}$$

where $M(L_{\mathbb{C}})$ is the moduli space of polystable $L_{\mathbb{C}}$-bundles. If $G=\mathrm{SU}(p,q)$ then $L_{\mathbb{C}}=\mathrm{GL}(q-p,\mathbb{C})\rtimes\mathbb{Z}_p$ and $L_{\mathbb{C}}=\mathbb{C}^*$ in the other cases.

One consequence of (9.1) is that $\mathcal{M}(\Sigma,G)_{max}$ has lower than expected dimension. Moreover, since $M(L_{\mathbb{C}})$ is connected it also shows that $\mathcal{M}(\Sigma,G)_{max}$ has the same number of components as $\mathcal{M}(\Sigma,G_T^{Ad})_{max}$. For $G=\mathrm{SU}(p,q)$ or $\mathrm{SO}^*(4m+2)$ this does not yield new information. In the case of E_6^{-14}, where $L=U(1)$ and $G_T=Spin_0(2,8)$ this leads to the result that the $\mathcal{M}(\Sigma,E_6^{-14})_{max}$ has the same number of connected components as $\mathcal{M}(\Sigma,\mathrm{PSO}_0(2,8))_{max}$. As shown in [**29**] (Theorem 5.6), this number is 2.

9.2. Non-maximal components. For Hermitian real forms the Cayley components are maximal, i.e. the Toledo invariant is maximal on these components. A case by case analysis has shown that these are the only maximal components. It is also known that there is at least one non-empty connected component for every other value of the Toledo invariant strictly within its bounded range. The count depends on the number of connected components in the locus of local minima for the Hitchin function on these components. If $G = \mathrm{SU}(p,q)$ or $\mathrm{PU}(p,q)$ these non-maximal components are known to be connected, and the non-maximal components have been counted in the cases of $G = \mathrm{SO}(2,2)$, $\mathrm{SO}(2,3)$, $\mathrm{SO}_0(2,3) = \mathrm{PSp}(4,\mathbb{R})$, and $\mathrm{Sp}(4,\mathbb{R})$ (see the references in the tables). For the other classical matrix groups of Hermitian type the connectivity of the non-maximal components remains an open question.

9.3. Non-magical real forms. The real forms in the last column of Table 1 do not admit magical $\mathfrak{sl}$-triples and are not Hermitian of non-tube type.

The case $\mathrm{SO}(1,n)$ was analyzed in [**4**] where it was shown that the Hitchin function has only trivial local minima of on $\mathcal{M}(\Sigma, \mathrm{SO}(1,n))$ and hence that the moduli space is connected. In [**26**] they show that the Hitchin function has only trivial local minima of on $\mathcal{M}(\Sigma, U^*(2n))$. Since the maximal compact subgroup of $U^*(2n)$, i.e. $\mathrm{Sp}(2n)$, is connected, it follows that $\mathcal{M}(\Sigma, U^*(2n))$ is connected. The maximal compact subgroup for $U^*(2n)$ is the same as for $\mathrm{SU}^*(2n)$ and hence it follows that $\mathcal{M}(\Sigma, \mathrm{SU}^*(2n))$ is also connected.

There is reason to conjecture that if G is a real Lie group whose Lie algebra is one of the exceptional real forms in this category, namely $\mathfrak{e}_6^{-26}$ and $\mathfrak{f}_4^{-20}$, then all local minima on $\mathcal{M}(\Sigma, G)$ lie in $f^{-1}(0)$, i.e. have $\Phi = 0$ (see [**19**]).

Inspection of Table 1 shows that for all other cases there are local minima with $\Phi \neq 0$. For the classical matrix groups the case-by-case analyses show that the the non-trivial local minima (and hence the components on which they occur) are either Cayley components or components with non-maximal Toledo invariant. The latter occur only in the moduli spaces of G-Higgs bundles for G of Hermitian type.

9.4. Non-magical companions to Cayley components. The Cayley map detects connected components that are not accounted for simply by the topological types of the principal bundles. Via the NAH correspondence these components correspond to connected components in $\mathrm{Rep}(\pi_1(S), G)$ in which the representations cannot be deformed to representations which factor through a maximal compact subgroup of G. These features follow from the fact that the Higgs bundles in these components cannot be deformed to G-Higgs bundles with zero Higgs field.

This leaves open the possibility that in addition to the Cayley components there are other unaccounted for connected components in which the Higgs fields are never identically zero. This is known not to be the case if G is a split real form of a classical group or if $G = \mathrm{SO}(p,q)$. For the real forms of hemitian tube-type it is known that there are non-maximal components but, as discussed in Section 9.2, the precise number is not known in all cases.

For the exceptional real forms in Table 1 the issue could be settled if it were known that the Hitchin function has no local minima other than at zero or those which occur in the Cayley components. The analysis of local minima can be separated into two cases, namely smooth local minima and local minima outside the smooth locus of the moduli space. On the smooth locus the criterion in theorem 5.4 can be used to identify necessary conditions that must be satisfied at a smooth minimum. Showing that these cannot be satisfied except at known minima would rule out the possibility of undiscovered smooth local minima. The possibility of local minima at singular points has to be ruled out differently.

10. Beyond π_0

The local minima of the Hitchin function identify connected components on $\mathcal{M}(\Sigma, G)$, and the Cayley map reveals partial structural information about certain distinguished components. In some cases, which we describe below, the global structure can be described much more explicitly. The descriptions lead to global information which does not depend on the complex structure on Σ and thus applies to the representation varieties $\mathrm{Rep}(\pi_1(S), G)$ on the other side of the NAH correspondence.

10.1. Global descriptions. In [**40**] Hitchin described all components of the moduli space $\mathcal{M}(\Sigma, \mathrm{SL}(2,\mathbb{R}))$. Similar methods were used in [**15, 32**] to describe the maximal components in $\mathcal{M}(\Sigma, \mathrm{Sp}(4,\mathbb{R}))$, and in [**32**] to describe components in $\mathcal{M}(\Sigma, P\mathrm{Sp}(4,\mathbb{R}))$. All of these examples are essentially special cases of the description in [**20**] of distinguished components of $\mathcal{M}(\Sigma, \mathrm{SO}(n, n+1))$, as can be seen because of the isomorphisms $\mathrm{PSL}(2,\mathbb{R}) \simeq \mathrm{SO}_0(1,2)$ and $P\mathrm{Sp}(4,\mathbb{R}) \simeq \mathrm{SO}_0(2,3)$.

Though not obtained in this way, the special components described in [**20**] are in fact Cayley components. The $\mathrm{SO}(n, n+1)$-Higgs bundles in these components are all of the form in (8.24), but since $q - p + 1 = 2$, the orthogonal bundles $\widehat{W}$ in the K_Σ^p-twisted $\mathrm{SO}(1, q-p+1)$ Higgs bundles are $O(2,\mathbb{C})$-bundles. Such bundles are classified topologically by first and second Stiefel-Whitney classes $(sw_1, sw_2) \in H^1(\Sigma, \mathbb{Z}_2) \times H^2(\Sigma, \mathbb{Z}_2)$. If $sw_1 = 0$ then $\widehat{W}$ is an $\mathrm{SO}(2,\mathbb{C})$-bundle which may be taken to be of the form $L \oplus L^{-1}$, where L is a line bundle of non-negative degree. The degree, d, of L is bounded by $n(2g-2)$ and Collier shows in [**20**] that

THEOREM 10.1 (Theorem 4.11 in [**20**]). *For each degree in the range $0 < d \leq n(2g-2)$ the resulting component is diffeomorphic to a product*

$$\mathcal{F}_d \otimes \bigoplus_{j=1}^{n-1} H^0(K_\Sigma^{2j}) \tag{10.1}$$

where $\mathcal{F}_d$ is the total space of a rank $d + (2n-1)(g-1)$ vector bundle over the symmetric product $Sym^{n(2g-2)-d}(\Sigma)$. In particular

COROLLARY 10.2. *The components given by* (10.1) *are smooth and deformation retract onto the symmetric product $Sym^{n(2g-2)-d}(\Sigma)$.*

THEOREM 10.3. *If $d = 0$ then the resulting component is homeomorphic to a product*

$$\mathcal{F}_0 \otimes \bigoplus_{j=1}^{n-1} H^0(K_\Sigma^{2j}) \tag{10.2}$$

where $\mathcal{F}_0$ is a singular space which may be described as a GIT quotient $\tilde{\mathcal{F}}_0//O(2,\mathbb{C})$ of the space

$$\tilde{\mathcal{F}}_0 = \{(M, \mu, \nu) \mid M \in Pic^0(\Sigma),\ \mu \in H^0(M^{-1}K_\Sigma^n),\ \nu \in H^0(MK_\Sigma^n)\} \tag{10.3}$$

The components in which the first Stiefel-Whitney class of $\widehat{W}$ does not vanish can also be described (see Theorem 5.3 in [**20**]) and are of the form

$$\mathcal{F}_{sw_1}^{sw_2} \otimes \bigoplus_{j=1}^{n-1} H^0(K_\Sigma^{2j}) \tag{10.4}$$

with $\mathcal{F}^{sw_2}_{sw_1} = \tilde{\mathcal{F}}^{sw_2}_{sw_1}/(\mathbb{Z}_2 \oplus \mathbb{Z}_2)$ where $\tilde{\mathcal{F}}^{sw_2}_{sw_1}$ is a rank $(4n-2)(2g-2)$ vector bundle. The base of this vector bundle is one of the connected components of the Prym variety for the double cover of Σ determined by sw_1, with the component determined by sw_2.

THEOREM 10.4. *The components given by* (10.4) *deformation retract onto the quotient of the Prym variety by the* $\mathbb{Z}_2$*-action coming from the involution in the double cover.*

10.2. Other topological information. In [**14**] we exploit Morse-theoretic information at higher critical points of the Hitchin function to compute higher homotopy groups (i.e. π_i with $i > 0$) for components of the Higgs bundle moduli spaces $\mathcal{M}(\Sigma, \mathrm{GL}(n,\mathbb{C}))$ and $\mathcal{M}(\Sigma, U(p,q))$. In [**48**] the author proves, also by Morse theoretic methods, that the space consisting of two of the components in $\mathcal{M}(\Sigma, \mathrm{SL}(3,\mathbb{R}))$ is homotopically equivalent to the moduli space of polystable $O(3,\mathbb{C})$ bundles. The third component is the contractible Hitchin component.

11. Supplemental Notes

The tables in Section 12 summarize known information about the number of connected components in $\mathcal{M}(\Sigma, G)$ if G is a real form with Lie algebra in Table 1.

The number of connected components in $\mathcal{M}(\Sigma, G)$ is determined not just by the Lie algebra of G but also by the center of the group. This explains why the entries in the tables differ for groups with the same Lie algebra but different centers.

The entries in the table are obtained, for the most part, from the techniques outlined in this survey. The references contain the details but we record here some additional explanatory notes to give a flavor of the arguments or to explain some stand-out cases. The entries in [[brackets]] have been computed as lower bounds but are conjectured to be precise. The uncertainty comes from gaps in our knowledge concerning the non-trivial local minima of the Hitchin function, especially in components with non-maximal Toledo invariant, or outside the smooth locus of the components. It is reasonable to conjecture that the only connected components not labeled by topological invariants of the principal bundles, i.e. with non-trivial local minima of the Hitchin function, are Cayley components.

We organize this section differently to the way the tables are organized to emphasize the commonalities in the features which determine component counts in $\mathcal{M}(\Sigma, G)$ for the various categories of real forms.

11.1. Real forms of Hermitian symmetric type.

11.1.1. $\mathbf{SL(2,\mathbb{R})(\simeq Sp(4,\mathbb{R}) \simeq SU(1,1))}$*:* The SO(2)-representations (i.e. where $\Phi = 0$) correspond to stable *degree zero* line bundles, i.e. correspond to *zero Euler class* (aka Toledo invariant). The automorphism which switches $L \oplus L^{-1} \mapsto L^{-1} \oplus L$ lies in $O(2)$ but not in $SO(2)$. This explains why we get one component for each integer in $[1-g, g-1]$.

11.1.2. $\mathbf{SO^*(2n)}$ **and** $\mathbf{Sp(2n,\mathbb{R})}$*:* The lower bound includes the components at $\pm\tau_{max}$ and reflects the fact that

(1) there are no 'topological' components if the relevant Toledo invariant lies between zero and its extremal values. [The Higgs bundles are defined by triples (V, β, γ) where V is a $U(n)$-bundle of degree d; if $d \neq 0$ then V does not support flat connections], and

(2) there is at least one component for each non-zero, non-extremal value of the Toledo invariant (detected by the non-trivial minima of the Morse function),
(3) For $SO^*(2n)$ there is just one component with zero or maximal Toledo invariant ($\lfloor \frac{n}{2} \rfloor (2g-2)$). The total number of components is thus bounded below by the number of values for the Toledo invariant, with equality if and only if each of the non-maximal components is connected.

Thus the lower bound is exact unless there are undetected exotic components with non-zero non-extremal Toledo invariant. Note: the $SO^*(4n)$-moduli spaces have magical components but the $SO^*(4n+2)$ do not.

11.1.3. **$PSO^*(2n)$**. The maximal compact subgroup of $PSO^*(2n)$ is $U(n)/\mathbb{Z}_2$. Topological types of principal $U(n)/\mathbb{Z}_2$-bundles are classified by $(d, w) \in \mathbb{Z} \times \mathbb{Z}_2$ if n is even and $d \in \mathbb{Z}$ if n is odd. Not all $U(n)/\mathbb{Z}_2$-bundles lift to $U(n)$.

For $n = 2m$, it can be shown (see [**29**]) that $0 \leq |d| \leq 2m(g-1)$ and that $U(n)/\mathbb{Z}_2$-bundles lift to $U(n)$ iff $w = 0$. Using the Cayley correspondence one can show that there are two maximal components, one which lifts to $SO^*(2m)$, and one which does not. There are two components in which $\Phi = 0$, namely those labelled by $(d, w) = (0, w)$.

For $n = 2m + 1$ the invariant d must satisfy $0 \leq |d| \leq 2m(g-1)$. The $U(n)/\mathbb{Z}_2$-bundles lift to $U(n)$ iff d is even, in particular, if $|d|$ is maximal. There is no obstruction to lifting a principal bundle from $PSO^*(2m+1)$ to $SO^*(2n)$ (See also [**36**]). There is thus one maximal component since there is just one maximal component in $\mathcal{M}(\Sigma, , SO^*(2m+1)$. There is one component in which $\Phi = 0$, namely the one labelled by $d = 0$.

Note: $PSO^*(2)$ is the trivial group, while $PSO^*(4)$ and $PSO_0(2,2)$ are not simple.

11.1.4. **$SU(p,p)$**. The components detected by the non-trivial minima are all connected, except for the maximal $p = q$ case in which we get 2^{2g} components (because the fixed determinant condition is on $det(E)^2$). The number of non-maximal values taken by the Toledo invariant is

$$2(\tau_{max} - 1) + 1 = 2(p(g-1) - 1) + 1 = 2p(g-1) - 1 \tag{11.1}$$

Moreover if $p < q$ then the maximal components fiber over $M_0(U(q-p))$ with fiber $\mathcal{M}(\Sigma, SU(p,p))_{\tau_{max}}$, so the component count is the same as for $SU(p,p)$. Here $M_0(U(q-p))$ is the moduli space of degree zero, rank $q-p$) polystable vector bundles.

Note that there are no 'topological' components if the Toledo invariant is non-zero (since vector bundles with non-zero degree cannot support flat connections).

In [**32**] Gothen shows that for $SU(2,2)$ the components with Toledo invariant equal to zero or maximal are connected.

11.1.5. **$Sp(4,\mathbb{R})$**: The bound on d is $|d| \leq 2g-2$. There is one component ($d = 0$) where $\Phi = 0$, two each of the multiple components with $d = \pm(2g-2)$, and one component for each $0 < |d| < 2g-2$. The maximal components include the so called Gothen components (first identified in [**32**]) and the Teichmüller components.

11.1.6. **$Sp(2n,\mathbb{R})$**: The bound on d is $|d| \leq n(g-1)$. There is one component ($d = 0$) where $\Phi = 0$. There are two of each maximal component ($d = \pm n(g-2)$) and for each maximal value there are $2 \cdot 2^{2g}$ Cayley components with trivial minima and 2^{2g} Teichmüller components. There is at least one component for each $0 < |d| < n(g-1)$.

11.1.7. ***PSp(2n,ℝ)**: See* [**29**].

- n even: Topological types of principal bundle are determined by $(d,w) \in \mathbb{Z}\times\mathbb{Z}_2$ with $|d| \le n(g-1)$, for a total of $2\cdot(2n(g-1)+1) = 4n(g-1)+2$ types. Two types have $d = 0$ and two have $\pm$ maximal value. Each maximal value accounts for $2^{2g+1}+2$ components (if $n \ge 4$). 2^{2g} of the $\mathrm{PSp}(2n,\mathbb{R})$-components do not lift to $\mathrm{Sp}(2n,\mathbb{R})$.
- n odd: There is no obstruction to lifting from $P\mathrm{Sp}(2n,\mathbb{R})$ to $\mathrm{Sp}(2n,\mathbb{R})$ (See also [**36**]). Topological types of principal bundle are determined by $d \in \mathbb{Z}$ with $|d| \le 2n(g-1)$, for a total of $4n(g-1)+1$ types. One type has $d = 0$ and two have $\pm$ maximal value. Each maximal value accounts for 3 components (if $n \ge 3$). The 3.2^{2g} maximal $\mathrm{Sp}(2n,\mathbb{R})$-components map $2^{2g}:1$ onto the maximal $\mathrm{PSp}(2n,\mathbb{R})$-components

11.1.8. $\mathbf{SO(2,q)}$. The maximal compact subgroup of $\mathrm{SO}(2,q)$ is $H = S(O(2)\times O(q))$. Thus the primary topological invariants are given by $(a,b,c) \in H^1(S,\mathbb{Z}_2)\times H^2(S,\mathbb{Z}_2)\times H^2(S,\mathbb{Z}_2)$.

If the invariant $a \in H^1(S,\mathbb{Z}_2)$ is zero then structure group of the Higgs bundles reduces to $\mathrm{SO}_0(2,q)$. In this case the group is of Hermitian type and a new invariant appears (in this case, the new invariant comes from a lift of $b \in H^2(S,\mathbb{Z}_2)$ to $H^2(S,\mathbb{Z}) \simeq \mathbb{Z}$). The primary topological invariants are then $(l,c) \in H^2(S,\mathbb{Z})\times \times H^2(S,\mathbb{Z}_2)$ with $|l| \le 2g-2$. The integer l is, up to a normalizing factor, the Toledo invariant mentioned in Section 7. However in the group $SO(2,q)$ the sign of the Toledo invariant is not well defined, or said another way, positive and negative Toledo invariant for $\mathrm{SO}_0(2,q)$ can be conjugated to each other using $SO(2,q)$. When $|l|$ is maximal new invariants appear, with c being replaced by a pair in $\mathbb{Z}^{2g}\times\mathbb{Z}_2$.

- If $a \ne 0$ then all components admit only trivial minima, giving $(2^{2g}-1)\cdot 2\cdot 2$ such components.
- If $a = 0$, so that the invariants become $(|l|, c)$ as above then there are
  - two components with only trivial minima (with $l = 0$),
  - 2^{2g} Cayley components (with $|l| = 2g-2$), and
  - at least $2\cdot(2g-3)$ other components (with $0 < |l| < 2g-2$).

For $\mathbf{SO(2,2)}$ if $a \ne 0$ then the other two invariants remain $\mathbb{Z}_2$-valued. The non-zero minima in $\mathcal{M}(\Sigma, \mathrm{SO}(2,2)$ all lie in components with $a = 0$, i.e. where we have $\mathrm{SO}_0(2,2)$-Higgs bundles. In that case the maximal compact subgroup of $\mathrm{SO}_0(2,2)$ is $H = SO(2)\times SO(2)$ and the invariant $c \in H^2(S,\mathbb{Z}_2)$ also lifts to $H^2(S,\mathbb{Z})$. The primary topological invariants are $(l,m) \in H^2(S,\mathbb{Z})\times\times H^2(S,\mathbb{Z}) \simeq \mathbb{Z}\times\mathbb{Z}$, but subject to a Milnor-Wood bound which in this case is $l \ge 0$ and

$$l - 2g + 2 \le m \le -l + 2g - 2 \; . \tag{11.2}$$

Also, if $l = 0$, then only $|m|$ is an invariant.

For $\mathbf{SO(2,3)}$ the extra Cayley components come from the fact that if $q = 3$ then the $\mathrm{SO}(q)$-bundle reduces to an $S(O(1)\times O(2))$-bundle. Moreover, if the structure group reduces further to $\mathrm{SO}(1)\times\mathrm{SO}(2)$ then the second Stiefel-Whitney class is replaced by a new integer invariant.

For $\mathbf{SO_0(2,2)}$ all the minima in $\mathcal{M}(\Sigma, g)$ are connected subvarieties (see Section 6 of [**5**]) except when (l,m) equals $(0,2g-2)$ or $(2g-2,0)$; in each of those case there are 2^{2g} Hitchin components. There is one component (when $(l,m)=(0,0)$) with $\Phi=0$ at a local minimum.

Note: The 2^{2g} choices in the Hitchin components corresponds to an allowed twisting by a square root of $\mathcal{O}$ in both of the $\mathrm{SO}(2,\mathbb{C})$ bundles in the Higgs bundle. The constraint on the degree zero bundle is required to preserve the trivial determinant condition.

For $\mathbf{SO_0(2,3)}=\boldsymbol{PSp(4,\mathbb{R})}$, in [**3**] it is shown that there are $2(2^{2g}-1)+4g-3$ connected components for each of the maximal values of the integer topological invariant. In [**34**] it is proven that there is one connected components for each other values of the invariants. This gives $2(2(2^{2g}-1)+4g-3)+4(2g-3)+2=2^{2g+2}+16g-20$ connected components. The extra maximal components (compared to the number for $\mathrm{SO}_0(2,q), q>3$) arise in the same way as the extra Cayley components for $\mathrm{SO}(2,3)$, except now the components labelled by $\pm d$ must be counted separately. There are only two components in which the local minimum of the Hitchin function is at points with $\Phi=0$, namely those with $a=l=0$ as above.

11.1.9. $\mathbf{PSO_0(2,2n)}$.

The maximal compact subgroup is $(\mathrm{SO}(2)\times \mathrm{SO}(2n)/\mathbb{Z}_2$, leading to a classification of the principal bundles by $(d,w)\in\mathbb{Z}\times\mathbb{Z}_2$ with $0\le |d|\le 4g-4$ (see [**29**]). The bundles with even values of d lift to $\mathrm{SO}(2)\times\mathrm{SO}(2n)$-bundles. In particular, all bundles in the maximal components lift, so the maximal $\mathrm{SO}_0(2,2n)$-components surject onto $\mathcal{M}(\Sigma,\mathrm{PSO}(2,2n)$. There are 2^{2g} such $\mathrm{SO}_0(2,2n)$-components for each $w\in\mathbb{Z}_2$, but these get identified in $\mathcal{M}(\Sigma,\mathrm{PSO}(2,2n)$. The two components with $d=0$ are the only components where the Higgs field can vanish.

11.2. Split real forms.

11.2.1. $\mathbf{PSL(2,\mathbb{R})}$. In [**31**] there is a formula for the number of components of $Hom(\pi_1, G)$ where G is any n-fold cover of $\mathrm{PSL}(2,\mathbb{R})$, viz.:

$$2n^{2g}+\frac{4g-4}{n}-1 \text{ if } n \text{ divides } 2g-2 \tag{11.3}$$

$$2\left\lceil\frac{2g-2}{n}\right\rceil+1 \text{ otherwise} \tag{11.4}$$

$\mathrm{PSL}(2,\mathbb{R})$ can be identified with the identity component of $\mathrm{PGL}(2,\mathbb{R})$. The latter has two components distinguished by the sign of the determinant so the two components of $\mathrm{GL}(2,\mathbb{R})$ remain distinct in $\mathrm{PSL}(2,\mathbb{R})$).

11.2.2. $\mathbf{PSL(n,\mathbb{R})}$. In [**42**] The description of components for $\mathrm{PSL}(n,\mathbb{R})$ proceeds as follows:

- If n is odd, $\mathrm{PSL}(n,\mathbb{R})=\mathrm{SL}(n,\mathbb{R})$ and if $w_2(E)=0$ there are two components, if $w_2(E)\neq 0$ just one.
- If $n=2m$, there are four topological types. Two do not lift to $\mathrm{SL}(n,\mathbb{R})$ and so give connected spaces. Two do lift and if $w_2(E)\neq(g-l)m^2$ this

gives a connected space. If $w_2(E) = (g-l)m^2$ we have three components- the one containing minima with $\Phi = 0$ and two copies [7] of the Teichmüller component.

11.2.3. **PGL(2*n*, ℝ)**. Two factors complicate the counting of connected components of $\mathcal{M}(\Sigma, \mathrm{PGL}(2n,\mathbb{R}))$, namely (a) the maximal compact subgroup, $\mathrm{PO}(2n)$ has two components, and (b) $\mathrm{PO}(2n)$ is not a matrix group. The first factor affects the number of topological types of principal bundles, while the second means that vector bundle methods are not immediately available. Both are dealt with in [**48**].

11.3. The real forms SO(*p*, *q*).

The maximal compact subgroup of $\mathrm{SO}(p,q)$ is $H = S(O(p)\times O(q))$. Assuming $2 < p \leq q$, the primary topological invariants are thus

$$(a,b,c) \in H^1(S,\mathbb{Z}_2)\times H^2(S,\mathbb{Z}_2)\times H^2(S,\mathbb{Z}_2)$$

The invariant a distinguishes between $\mathrm{SO}(p,q)$- and $\mathrm{SO}_0(p,q)$-Higgs bundles, where $\mathrm{SO}_0(p,q)$-Higgs denotes the connected componet of the identity, with $a=0$ corresponding to the latter.

In Theorem 5.10 of [**5**] (see also theorem 8.24) we classify all the local minima on $\mathcal{M}(SO(p,q))$ for all $1\leq p\leq q$. This includes one $\Phi=0$ minima on each of the 2^{2g+2} non-empty component of the moduli space of polystable $S(O(p,\mathbb{C})\times O(q,\mathbb{C}))$-bundles. The other minima, i.e. those which detect the Cayley components, lie in sectors with topological invariants listed in the table in §6.1 of [**5**]. Except for $\mathrm{SO}(p,p+1)$, in all other cases this yields a total of 2^{2g+1} extra Cayley/exotic components.

If p is even then all Cayley components lie in sectors with $a=0$, while if p is odd then just two of the Cayley components have $a=0$. These considerations account for the dependence on the parity of p in the component counts for $\mathcal{M}(\Sigma, \mathrm{SO}_0(p,q))$.

11.3.1. **SO(*n*, *n* + 1)**. In the case of $\mathrm{SO}(p,q)=\mathrm{SO}(n,n+1)$ with $n\geq 3$ there are an additional $2n(g-1)-1$ so-called Collier components, first described in [**20**], due to the fact that $q-p+1=2$ if $q=p+1$ (see Section 10.1).

11.3.2. **SO(1, 2)**. For $\mathrm{SO}(1,2)$, $2(2^{2g}-1)$ of the topological components come from the classification of $O(2,\mathbb{C})$ bundles with $w_1\neq 0$, and one additional topological component comes from the component in which the bundles are of the form $(I,W) = (\mathcal{O}, L\oplus L^{-1})$ with $deg(L) = 0$. The other components have $0 < deg(L) \leq 2g-2$ and non-trivial Higgs field.

11.3.3. **SO(1, 1)**. In this special case $\mathrm{SO}(2,\mathbb{C}) \simeq \mathbb{C}^*$ and hence $\mathfrak{so}(2,\mathbb{C})$ is one-dimensional and thus admits no $\mathfrak{sl}(2,\mathbb{C})$-subalgebras! The maximal compact subgroup is $H = S(O(1)\times O(1)) \simeq \mathbb{Z}_2$, so an $\mathrm{SO}(1,1)$-Higgs bundle can be described as a pair $(\mathcal{I}, q)$ where $\mathcal{I}^2$ is a trivial line bundle and $q:\mathcal{I}\to\mathcal{I}K_\Sigma$, i.e. $q\in H^0(K_\Sigma)$. The primary topological invariant is $sw_1\in H^1(S,\mathbb{Z}_1)$. It follows that the moduli space is a union of 2^{2g} copies of $H^0(K_\Sigma)$, with the line bundle in each determined by

[7]The two copies get identified inside the $\mathrm{PSL}(2n,\mathbb{C})$-moduli space by an isomorphism in the non-identity component of $\mathrm{PGL}(2n,\mathbb{R})$ (see the last paragraph in [**42**]).

sw_1. Each component contains a point where $q = 0$. The only invariant polynomial on $\mathfrak{so}(2,\mathbb{C})$ is the (degree one) Pfaffian, defined by

$$P(\begin{bmatrix} 0 & b \\ -b & 0 \end{bmatrix}) = b\,. \tag{11.5}$$

The base of the Hitchin fibration is thus $H^0(K_\Sigma)$. There are 2^{2g} 'sections' given by the maps $q \mapsto [\mathcal{I}, q]$, i.e. by the identification of $H^0(K_\Sigma)$ with each component. These all contain points with $q = 0$, i.e. with vanishing Higgs field.

11.4. Exceptional real forms.

The information is least complete for real forms with exceptional Lie algebras, partly because the vector bundle versions of their G-Higgs bundles are difficult to work with, but also because less has been worked out for their maximal compact subgroups or (in cases where they admit magical triples) for their Cayley partners.

By [**42**] and also [**10**] we know that if G is a split real form of an exceptional group then $\mathcal{M}(\Sigma, G)$ contains at least one Cayley (in this case, Hitchin or Teichmüller) component. It also contains as many components with trivial local minima of the Hitchin function as there are topological types of principal H-bundles where H is the maximal compact subgroup of G.

For the quaternionic real forms of F_4, E_6, E_7 and E_8 the number of Cayley components is determined by the number of components in $\mathcal{M}_{K_\Sigma^4}(G_{0,ss}^{\mathbb{R}})$ in (8.21). This is computed in [**10**] for all cases except E_8. As stated there (see Remark 7.23) the expectation for E_8 is that there is just one Cayley component since the maximal compact of $G_{0,ss}^{\mathbb{R}}$ in that case (type F_4) is simply connected. It follows that the corresponding principal bundles have just one topological type. Moreover as stated at the beginning of this section, it is expected that in general the only connected components not labeled by topological invariants of the principal bundles are Cayley components.

By Theorem 8.4 the exceptional groups with Lie algebras $\mathfrak{e}_6^2, \mathfrak{e}_7^{-5}, \mathfrak{e}_8^{-24}, \mathfrak{f}_4^4$ are those for which $\mathcal{M}(\Sigma,,G)$ have Cayley components. The number of such components depends also on the center of the group or, equivalently on their fundamental groups. For groups of type F_4 or E_8, the adjoint forms are simply connected, so there is only one choice of group. For groups of type E_6 or E_7, the adjoint forms are not simply connected.

12. Tables

TABLE 4. $\mathfrak{g}^{\mathbb{C}} = \mathfrak{sl}(n, \mathbb{C})$

G	# with $\Phi = 0$	Cayley components	Total $\lvert\pi_0(\mathcal{M}(\Sigma, G))\rvert$	References
$\mathrm{SL}(2,\mathbb{R}) = \mathrm{Sp}(2,\mathbb{R}) = \mathrm{SU}(1,1)$	1	2^{2g+1}	$2^{2g+1} + 2g - 3$	**[31]**
$\mathrm{PSL}(2,\mathbb{R}) = \mathrm{PSp}(2,\mathbb{R}) = \mathrm{SO}_0(1,2)$	1	1	$4g - 3$	**[31]**
$\mathrm{PGL}(2,\mathbb{R}) = \mathrm{SO}(1,2)$	$2^{2g+1} - 1$	$(2g - 2)$	$2^{2g+1} + 2g - 3$	**[5, 6, 20, 48]**
$\mathrm{SL}(2m,\mathbb{R}), m > 1$	2	2^{2g}	$2^{2g} + 2$	**[42]**
$\mathrm{PSL}(2m+1,\mathbb{R})$	2	1	3	**[42]**
$\mathrm{PSL}(2m,\mathbb{R}), m > 1$	4	2	6	**[42]**
$\mathrm{PGL}(2m,\mathbb{R}), m > 1$	$2^{2g+1} + 1$	1	$2^{2g+1} + 2$	**[48]**
$\mathrm{SU}(p,q), p \leq q$	1	$2 \cdot 2^{2g}$	$2 \cdot 2^{2g} + 2p(g-1) - 1$	**[9, 13, 32]**
$\mathrm{PU}(p,q)$	$gcd(p,q)$	$2 \cdot gcd(p,q)$	$2(p+q)min\{p,q\}(g-1)$ $+gcd(p,q)$	**[11, 13]** **[16, 47]**
$U^*(2m)$	1	-	1	**[27]**

TABLE 5. $\mathfrak{g}^{\mathbb{C}} = \mathfrak{sp}(2n, \mathbb{C})$

G	# with $\Phi = 0$	Cayley components	Total $\lvert\pi_0(\mathcal{M}(\Sigma, G))\rvert$	References
$\mathrm{Sp}(4,\mathbb{R})$	1	$6 \cdot 2^{2g} + 4g - 8$	$6 \cdot 2^{2g} + 8g - 13$	**[30, 32]**
$\mathrm{PSp}(4,\mathbb{R}) = \mathrm{SO}_0(2,3)$	2	$4 \cdot 2^{2g} + 8g - 10$	$4 \cdot 2^{2g} + 16g - 20$	**[3, 34]**
$\mathrm{Sp}(2n,\mathbb{R}), n \geq 3$	1	$6 \cdot 2^{2g}$	$[[6 \cdot 2^{2g} + 2n(g-1) - 1]]$	**[25]**
$\mathrm{PSp}(4n+2,\mathbb{R}),\ n \geq 3$	1	6	$[[4(2n+1)(g-1) + 5]]$	**[29]**
$\mathrm{PSp}(4n,\mathbb{R}),\ n \geq 4$	2	$2^{2g+2} + 4$	$[[4(2n)(g-1) + 2 + 2^{2g+2}]]$	**[29]**
$\mathrm{Sp}(2p,2q)$	1	-	1	**[28, 49]**

TABLE 6. $\mathfrak{g}^{\mathbb{C}} = \mathfrak{so}(2n, \mathbb{C})$.

G	# with $\Phi = 0$	Cayley components	Total $\lvert\pi_0(\mathcal{M}(\Sigma, G))\rvert$	References
$\mathrm{SO}(1,1)$	2^{2g}	0	2^{2g}	**[5]**
$\mathrm{SO}(1,2n+1), n \geq 1$	2^{2g+1}	-	2^{2g+1}	**[5]**
$\mathrm{SO}_0(1,2n+1) n \geq 1$	2	-	2	**[4]**
$\mathrm{SO}(2,2)$	$2 \cdot 2^{2g+1} - 3$	2^{2g+1}	$6 \cdot 2^{2g} - 3 + 2g(2g-3)$	**[5, 7]**
$\mathrm{SO}(2,2n-2), n > 2$	$2 \cdot 2^{2g+1} - 2$	2^{2g+1}	$[[6 \cdot 2^{2g} + 4g - 8]]$	**[5]**
$\mathrm{SO}_0(2,2n-2), n > 2$	2	2^{2g+1}	$[[2^{2g+1} + 8g - 10]]$	**[9, 13, 22]**
$\mathrm{PSO}_0(2,2n-2), n \geq 2$	2	2	$[[8g - 6]]$	**[29]**
$\mathrm{SO}(p,2n-p), 2 < p \leq n$	$4 \cdot 2^{2g}$	$2 \cdot 2^{2g}$	$6 \cdot 2^{2g}$	**[5]**
$\mathrm{SO}_0(p,2n-p), 2 < p \leq n$	4	$2 \cdot 2^{2g}$ (p even) 2 (p odd)	$2 \cdot 2^{2g} + 4$ (p even) 6 (p odd)	**[5]** **[5]**
$\mathrm{SO}^*(4n), n \geq 2$	1	2	$[[4n(g-1)]]$	**[13, 18]**
$\mathrm{SO}^*(4n+2), n \geq 1$	1	-	$[[4n(g-1)]]$	**[13, 18]**
$\mathrm{PSO}^*(4n), n \geq 2$	1	2	$[[2 + 2n(g-1)]]$	**[29]**
$\mathrm{PSO}^*(4n+2), n \geq 1$	1	-	$[[1 + 2(n-1)(g-1)]]$	**[29]**

TABLE 7. $\mathfrak{g}^{\mathbb{C}} = \mathfrak{so}(2n+1, \mathbb{C})$

G	# with $\Phi = 0$	Cayley components	Total $\lvert\pi_0(\mathcal{M}(\Sigma, G))\rvert$	References
$\mathrm{SO}(1,2) = \mathrm{PGL}(2,\mathbb{R})$	$2^{2g+1} - 1$	$(2g-2)$	$2^{2g+1} + 2g - 3$	**[5, 6, 20, 48]**
$\mathrm{SO}_0(1,2) = \mathrm{PSL}(2,\mathbb{R})$	1	1	$4g - 3$	**[31]**
$\mathrm{SO}(1,2n), n > 1$	2^{2g+1}	-	2^{2g+1}	**[5]**
$\mathrm{SO}_0(1,2n) n \geq 1$	2	-	2	**[4]**
$\mathrm{SO}(2,3)$	$4 \cdot 2^{2g} - 2$	$2 \cdot 2^{2g} + 4g - 5$	$6 \cdot 2^{2g} + 8g - 13$	**[20, 34]**
$\mathrm{SO}_0(2,3) = P\mathrm{Sp}(4,\mathbb{R})$	2	$4 \cdot 2^{2g} + 8g - 10$	$4 \cdot 2^{2g} + 16g - 20$	**[3, 34]**
$\mathrm{SO}(2,2n-1), n > 2$	$2 \cdot 4^{2g} - 2$	$2 \cdot 2^{2g}$	$[[6 \cdot 2^{2g} + 4g - 8]]$	**[5]**
$\mathrm{SO}_0(2,2n-1) = \mathrm{PSO}_0(2,2n-1)$ $n > 2$	2	2^{2g+1}	$[[2^{2g+1} + 8g - 10]]$	**[9, 13]** **[22]**
$\mathrm{SO}(p,2n+1-p), 2 < p < n$	$4 \cdot 2^{2g}$	$2 \cdot 2^{2g}$	$6 \cdot 2^{2g}$	**[5]**
$\mathrm{SO}_0(p,2n+1-p), 2 < p < n$	4	$2 \cdot 2^{2g}$ (p even) 2 (p odd)	$2 \cdot 2^{2g} + 4$ (p even) 6 (p odd)	**[5]** **[5]**
$\mathrm{SO}(n,n+1), n > 2$	$4 \cdot 2^{2g}$	$2 \cdot 2^{2g}$ $+2n(g-1) - 1$	$6 \cdot 2^{2g}$ $+2n(g-1) - 1$	**[20]**

TABLE 8. Exceptional $\mathfrak{g}^{\mathbb{C}}$

G	# with $\Phi = 0$	Cayley components	Total $\lvert\pi_0(\mathcal{M}(\Sigma, G))\rvert$	References
E_6^6		[[1]]		**[10, 42]**
E_6^2		1 (simply connected) 3 (adjoint))		**[10]** **[10]**
E_6^{-14}		2		**[9, 29]**
E_6^{-26}	[[1]]	-	[[1]]	**[19]**
E_7^7		[[1]]		**[10, 42]**
E_7^{-25}		> 1		**[9]**
E_7^{-5}		1 (simply connected) 2 (adjoint))		**[10]** **[10]**
E_8^8		[[1]]		**[10, 42]**
E_8^{-24}		[[1]]		**[10]**
F_4^{-20}	[[1]]	-	[[1]]	**[19]**
F_4^4	1	3	4	**[10, 42]**
G_2^2		[[1]]		**[10, 42]**

Acknowledgments. There are many colleagues on both sides of the non-Abelian Hodge Correspondence, too many to list, who have immeasurably enriched my mathematical life in general and in particular my understanding of the work discussed here. Foremost among them are my main collaborators Oscar Garcia-Prada, Brian Collier, Peter Gothen, and Andre Oliveira, but others to whom I owe a great debt of gratitude include Bill Goldman, Nigel Hitchin, and Anna Wienhard. My sincere thanks to the editors of this volume for their truly Job-level of patience, and to the anonymous referee whose sharp insights forced clarifications of several poorly explained points and whose extraordinary diligence sanded down numerous rough edges in the text.

References

[1] D. Alessandrini, *Higgs bundles and geometric structures on manifolds*, SIGMA Symmetry Integrability Geom. Methods Appl. **15** (2019), Paper 039, 32 pp., DOI 10.3842/SIGMA.2019.039. MR3948927

[2] Daniele Alessandrini, Sara Maloni, Nicolas Tholozan, and Anna Wienhard, *Fiber bundles associated with Anosov representations*, arXiv:2303.10786, 2023.

[3] D. Alessandrini and B. Collier, *The geometry of maximal components of the* $\mathsf{PSp}(4,\mathbb{R})$ *character variety*, Geom. Topol. **23** (2019), no. 3, 1251–1337, DOI 10.2140/gt.2019.23.1251. MR3956893

[4] M. Aparicio Arroyo and O. García-Prada, *Higgs bundles for the Lorentz group*, Illinois J. Math. **55** (2011), no. 4, 1299–1326 (2013). MR3082870

[5] M. Aparicio-Arroyo, S. Bradlow, B. Collier, O. García-Prada, P. B. Gothen, and A. Oliveira, $\mathrm{SO}(p,q)$*-Higgs bundles and higher Teichmüller components*, Invent. Math. **218** (2019), no. 1, 197–299, DOI 10.1007/s00222-019-00885-2. MR3994589

[6] D. Baraglia and L. P. Schaposnik, *Cayley and Langlands type correspondences for orthogonal Higgs bundles*, Trans. Amer. Math. Soc. **371** (2019), no. 10, 7451–7492, DOI 10.1090/tran/7587. MR3939583

[7] D. Baraglia and L. P. Schaposnik, *Monodromy of rank 2 twisted Hitchin systems and real character varieties*, Trans. Amer. Math. Soc. **370** (2018), no. 8, 5491–5534, DOI 10.1090/tran/7144. MR3812111

[8] D. Baraglia and L. P. Schaposnik, *Cayley and Langlands type correspondences for orthogonal Higgs bundles*, Trans. Amer. Math. Soc. **371** (2019), no. 10, 7451–7492, DOI 10.1090/tran/7587. MR3939583

[9] O. Biquard, O. García-Prada, and R. Rubio, *Higgs bundles, the Toledo invariant and the Cayley correspondence*, J. Topol. **10** (2017), no. 3, 795–826, DOI 10.1112/topo.12023. MR3797597

[10] Steven Bradlow, Brian Collier, Oscar García-Prada, Peter B. Gothen, and André Oliveira, *A general Cayley correspondence and higher Teichmüller spaces*, arXiv:2101.09377, 2021.

[11] S. B. Bradlow, O. García-Prada, and P. B. Gothen, *Surface group representations and* $\mathrm{U}(p,q)$*-Higgs bundles*, J. Differential Geom. **64** (2003), no. 1, 111–170. MR2015045

[12] S. B. Bradlow, O. García-Prada, and P. B. Gothen, *Representations of surface groups in the general linear group*, Proceedings of the XII Fall Workshop on Geometry and Physics, Publ. R. Soc. Mat. Esp., vol. 7, R. Soc. Mat. Esp., Madrid, 2004, pp. 83–94. MR2123494

[13] S. B. Bradlow, O. García-Prada, and P. B. Gothen, *Maximal surface group representations in isometry groups of classical Hermitian symmetric spaces*, Geom. Dedicata **122** (2006), 185–213, DOI 10.1007/s10711-007-9127-y. MR2295550

[14] S. B. Bradlow, O. García-Prada, and P. B. Gothen, *Homotopy groups of moduli spaces of representations*, Topology **47** (2008), no. 4, 203–224, DOI 10.1016/j.top.2007.06.001. MR2416769

[15] S. B. Bradlow, O. García-Prada, and P. B. Gothen, *Deformations of maximal representations in* $\mathrm{Sp}(4,\mathbb{R})$, Q. J. Math. **63** (2012), no. 4, 795–843, DOI 10.1093/qmath/har010. MR2999985

[16] S. Bradlow, O. García-Prada, P. Gothen, and J. Heinloth, *Irreducibility of moduli of semi-stable chains and applications to* $\mathrm{U}(p,q)$*-Higgs bundles*, Geometry and physics. Vol. II, Oxford Univ. Press, Oxford, 2018, pp. 455–470. MR3931782

[17] S. B. Bradlow, O. García-Prada, and P. B. Gothen, *Maximal surface group representations in isometry groups of classical Hermitian symmetric spaces*, Geom. Dedicata **122** (2006), 185–213, DOI 10.1007/s10711-007-9127-y. MR2295550

[18] S. B. Bradlow, O. García-Prada, and P. B. Gothen, *Higgs bundles for the non-compact dual of the special orthogonal group*, Geom. Dedicata **175** (2015), 1–48, DOI 10.1007/s10711-014-0026-8. MR3323627

[19] Brian Collier, *Classification of character varieties with no higher Teichmüller components*, 2023, in preparation.

[20] B. Collier, $\mathsf{SO}(n, n+1)$*-surface group representations and Higgs bundles* (English, with English and French summaries), Ann. Sci. Éc. Norm. Supér. (4) **53** (2020), no. 6, 1561–1616, DOI 10.24033/asens.2454. MR4203035

[21] Brian Collier, *Various generalizations and deformations of* $\mathrm{PSL}(2,\mathbb{R})$ *surface group representations and their Higgs bundles*, Geometry and Physics: a Festschrift in honour of Nigel Hitchin, Oxford Press.

[22] B. Collier, N. Tholozan, and J. Toulisse, *The geometry of maximal representations of surface groups into* $\mathrm{SO}_0(2,n)$, Duke Math. J. **168** (2019), no. 15, 2873–2949, DOI 10.1215/00127094-2019-0052. MR4017517

[23] O. García-Prada, *Higgs bundles and surface group representations*, Moduli spaces and vector bundles, London Math. Soc. Lecture Note Ser., vol. 359, Cambridge Univ. Press, Cambridge, 2009, pp. 265–310. MR2537072

[24] O. García-Prada, *Higgs bundles and higher Teichmüller spaces*, Handbook of Teichmüller theory. Vol. VII, IRMA Lect. Math. Theor. Phys., vol. 30, Eur. Math. Soc., Zürich, [2020], pp. 239–285. MR4321175

[25] O. García-Prada, P. B. Gothen, and I. Mundet i Riera, *Higgs bundles and surface group representations in the real symplectic group*, J. Topol. **6** (2013), no. 1, 64–118, DOI 10.1112/jtopol/jts030. MR3029422

[26] O. García-Prada and A. Oliveira, *Connectedness of Higgs bundle moduli for complex reductive Lie groups*, Asian J. Math. **21** (2017), no. 5, 791–810, DOI 10.4310/AJM.2017.v21.n5.a1. MR3767265

[27] O. García-Prada and A. G. Oliveira, *Higgs bundles for the non-compact dual of the unitary group*, Illinois J. Math. **55** (2011), no. 3, 1155–1181 (2013). MR3069300

[28] O. García-Prada and A. G. Oliveira, *Connectedness of the moduli of* $\mathrm{Sp}(2p, 2q)$*-Higgs bundles*, Q. J. Math. **65** (2014), no. 3, 931–956, DOI 10.1093/qmath/hat045. MR3261975

[29] O. García-Prada and A. Oliveira, *Maximal Higgs bundles for adjoint forms via Cayley correspondence*, Geom. Dedicata **190** (2017), 1–22, DOI 10.1007/s10711-017-0224-2. MR3704810

[30] O. García-Prada and I. Mundet i Riera, *Representations of the fundamental group of a closed oriented surface in* $\mathrm{Sp}(4,\mathbb{R})$, Topology **43** (2004), no. 4, 831–855, DOI 10.1016/S0040-9383(03)00081-8. MR2061209

[31] W. M. Goldman, *Topological components of spaces of representations*, Invent. Math. **93** (1988), no. 3, 557–607, DOI 10.1007/BF01410200. MR952283

[32] P. B. Gothen, *Components of spaces of representations and stable triples*, Topology **40** (2001), no. 4, 823–850, DOI 10.1016/S0040-9383(99)00086-5. MR1851565

[33] P. B. Gothen, *Representations of surface groups and Higgs bundles*, Moduli spaces, London Math. Soc. Lecture Note Ser., vol. 411, Cambridge Univ. Press, Cambridge, 2014, pp. 151–178. MR3221295

[34] P. B. Gothen and A. G. Oliveira, *Rank two quadratic pairs and surface group representations*, Geom. Dedicata **161** (2012), 335–375, DOI 10.1007/s10711-012-9709-1. MR2994046

[35] O. Guichard, *An introduction to the differential geometry of flat bundles and of Higgs bundles*, The geometry, topology and physics of moduli spaces of Higgs bundles, Lect. Notes Ser. Inst. Math. Sci. Natl. Univ. Singap., vol. 36, World Sci. Publ., Hackensack, NJ, 2018, pp. 1–63. MR3837868

[36] O. Guichard and A. Wienhard, *Topological invariants of Anosov representations*, J. Topol. **3** (2010), no. 3, 578–642, DOI 10.1112/jtopol/jtq018. MR2684514

[37] O. Guichard and A. Wienhard, *Positivity and higher Teichmüller theory*, European Congress of Mathematics, Eur. Math. Soc., Zürich, 2018, pp. 289–310. MR3887772

[38] O. Guichard and A. Wienhard, *Generalizing Lusztig's total positivity*, `arXiv:2208.10114`, 2022.

[39] S. Helgason, *Differential geometry, Lie groups, and symmetric spaces*, Graduate Studies in Mathematics, vol. 34, American Mathematical Society, Providence, RI, 2001. Corrected reprint of the 1978 original, DOI 10.1090/gsm/034. MR1834454

[40] N. J. Hitchin, *The self-duality equations on a Riemann surface*, Proc. London Math. Soc. (3) **55** (1987), no. 1, 59–126, DOI 10.1112/plms/s3-55.1.59. MR887284

[41] N. Hitchin, *Stable bundles and integrable systems*, Duke Math. J. **54** (1987), no. 1, 91–114, DOI 10.1215/S0012-7094-87-05408-1. MR885778

[42] N. J. Hitchin, *Lie groups and Teichmüller space*, Topology **31** (1992), no. 3, 449–473, DOI 10.1016/0040-9383(92)90044-I. MR1174252

[43] A. W. Knapp, *Lie groups beyond an introduction*, Progress in Mathematics, vol. 140, Birkhäuser Boston, Inc., Boston, MA, 1996, DOI 10.1007/978-1-4757-2453-0. MR1399083

[44] B. Kostant, *Lie group representations on polynomial rings*, Amer. J. Math. **85** (1963), 327–404, DOI 10.2307/2373130. MR158024

[45] Q. Li, *An introduction to Higgs bundles via harmonic maps*, SIGMA Symmetry Integrability Geom. Methods Appl. **15** (2019), Paper No. 035, 30, DOI 10.3842/SIGMA.2019.035. MR3947036

[46] J. Li, *The space of surface group representations*, Manuscripta Math. **78** (1993), no. 3, 223–243, DOI 10.1007/BF02599310. MR1206154

[47] E. Markman and E. Z. Xia, *The moduli of flat* $\mathrm{PU}(p,p)$*-structures with large Toledo invariants*, Math. Z. **240** (2002), no. 1, 95–109, DOI 10.1007/s002090100364. MR1906709

[48] André Gama Oliveira, *Representations of surface groups in the projective general linear group*, Internat. J. Math. 22 (2):223–279, 2011.
Comments: v3 (`arXiv:0901.2314v3`) included an erratum (Section 12) which shows why Theorem 1.3 (stating that the Hitchin component in PSL(3,R) is homotopically equivalent to PSO(3)) is not correct, even though the original manuscript is left unchanged. This erratum has been published in Int. J. Math., 30, No. 2 (2019)

[49] L. P. Schaposnik, *Spectral data for G-Higgs bundles*, ProQuest LLC, Ann Arbor, MI, 2013. Thesis (D.Phil.)–University of Oxford (United Kingdom). MR3389247

[50] L. P. Schaposnik, *Higgs bundles—recent applications*, Notices Amer. Math. Soc. **67** (2020), no. 5, 625–634, DOI 10.1090/noti. MR4186330

[51] A. H. W. Schmitt, *Geometric invariant theory and decorated principal bundles*, Zurich Lectures in Advanced Mathematics, European Mathematical Society (EMS), Zürich, 2008, DOI 10.4171/065. MR2437660

[52] C. T. Simpson, *Moduli of representations of the fundamental group of a smooth projective variety. I*, Inst. Hautes Études Sci. Publ. Math. **79** (1994), 47–129. MR1307297

[53] J. Swoboda, *Moduli spaces of Higgs bundles—old and new*, Jahresber. Dtsch. Math.-Ver. **123** (2021), no. 2, 65–130, DOI 10.1365/s13291-021-00229-1. MR4254708

[54] R. A. Wentworth, *Higgs bundles and local systems on Riemann surfaces*, Geometry and quantization of moduli spaces, Adv. Courses Math. CRM Barcelona, Birkhäuser/Springer, Cham, 2016, pp. 165–219. MR3675465

[55] A. Wienhard, *An invitation to higher Teichmüller theory*, Proceedings of the International Congress of Mathematicians—Rio de Janeiro 2018. Vol. II. Invited lectures, World Sci. Publ., Hackensack, NJ, 2018, pp. 1013–1039. MR3966798

[56] E. Z. Xia, *The moduli of flat* $\mathrm{PGL}(2,\mathbf{R})$ *connections on Riemann surfaces*, Comm. Math. Phys. **203** (1999), no. 3, 531–549, DOI 10.1007/s002200050624. MR1700170

University of Illinois at Urbana-Champaign
Email address: `bradlow@illinois.edu`

Contemporary Mathematics
Volume **803**, 2024
https://doi.org/10.1090/conm/803/16095

Moduli of unstable bundles of HN-length two with fixed algebra of endomorphisms

L. Brambila-Paz and R. Ríos-Sierra

Dedicated to Peter E. Newstead on the occasion of his 80th birthday

ABSTRACT. Let X be a smooth, irreducible complex projective curve of genus $g \geq 2$ and $U_{\mu_1}(n,d)$ be the moduli scheme of indecomposable vector bundles over X with fixed Harder-Narasimhan type $\sigma = (\mu_1, \mu_2)$. In this paper, we give necessary and sufficient conditions for a vector bundle $E \in U\mu_1(n,d)$ to have $\mathbb{C}[x_1,\ldots,x_k]/(x_1,\ldots,x_k)^2$ as its algebra of endomorphisms. By fixing the dimension of the algebra of endomorphisms, we obtain a stratification of $U\mu_1(n,d)$, where each stratum $U\mu_1(n,d,k)$ is an algebraic variety, moreover, it is a coarse moduli space. A particular case of interest arises when the unstable bundles are simple. In such a case, the moduli space is fine. The topological properties of $U_{\mu_1}(n,d,k)$ will depend on the generality of the curve X. These results differ from the corresponding results for the moduli space of stable bundles, where non-emptiness, irreducibility, dimension are independent of the curve.

1. Introduction

Let X be a complex smooth projective variety. The moduli space of semistable bundles with fixed invariants on X is well-known to exist. In certain cases, this moduli space is fine, meaning that a universal family exists. However, constructing moduli spaces of unstable bundles requires additional information. For instance, when X has dimension 1 and the algebra of endomorphisms is fixed, the moduli spaces of unstable vector bundles of rank 2 and 3 have been constructed (references [**BM**] and [**R**], respectively). In [**St**], the moduli space of unstable bundles on the projective plane is considered, while in [**Ba**], it is explored for $X = \mathbb{P}^3(\mathbb{C})$, both taking into account the degree of instability. The case of very unstable bundles when $dimX > 2$ is studied by J. M. Drezet in [**D**]. Additionally, in [**HK**], the authors construct moduli spaces of pure sheaves with a fixed Harder-Narasimhan type, incorporating additional data called a m-rigidity. The moduli space of unstable sheaves using nonreductive GIT is studied by V. Hoskins, G. Berczi, J. Jackson, and F. Kirwan in [**BHJK**].

2020 *Mathematics Subject Classification.* Primary 14H60.

Key words and phrases. Unstable vector bundles, algebra of endomorphisms, moduli spaces.

Both authors acknowledge the support of CONACYT, in particular of CONACYT grant 251938. The first author is a member of the international research group VBAC (Vector Bundles on Algebraic Curves).

The approach used in this article to investigate unstable bundles over a complex projective curve X of genus $g \geq 2$ involves analyzing their algebra of endomorphisms. This strategy offers several advantages in terms of comprehending the behavior of these bundles. It not only facilitates the construction of coarse moduli spaces but also, in certain cases, allows for the construction of fine moduli spaces.

In order to present our results, it is important to recall that every vector bundle over X has a unique filtration known as the Harder-Narasimhan filtration:

$$0 = E_0 \subset E_1 \subset \cdots \subset E_m = E$$

This filtration satisfies the following properties: for $0 \leq i \leq m-1$, E_{i+1}/E_i is semistable and the sequence of slopes $\sigma = (\mu(E_1), \mu(E_2/E_1), \cdots, \mu(E_m/E_{m-1}))$ satisfies:

$$\mu(E_1) > \mu(E_2/E_1) > \cdots > \mu(E_m/E_{m-1}). \tag{1.1}$$

This sequence of slopes is referred to as the Harder-Narasimhan type (HN-type) of E. The moduli space of vector bundles with HN-type $\sigma = (\mu(E_1))$ is known as the moduli space of semistable bundles. Mumford [**Mu**] constructed this moduli space in the 1960s using Geometric Invariant Theory, while M.S. Narasimhan and S. Seshadri [**NS**] used representation theory. For a more general treatment, we refer the reader to [**Gi**], [**Ma**], [**Ma1**] and [**Si**]. Furthermore, the moduli space of certain non-simple semistable vector bundles with a fixed algebra of endomorphisms was constructed in [**Br1**], [**Br2**] and [**Br3**].

Let $U_{\mu_1}(n,d)$ denote the set of indecomposable vector bundles of rank n and degree d with coprime-type $\sigma = (\mu_1, \mu_2)$ (see Definition 2.1). When $\mu_1 - \mu_2 > 2g-1$ it follows that $U_{\mu_1}(n,d) = \emptyset$ (as shown in Proposition 2.6). However, for $0 < \mu_1 - \mu_2 \leq 2g-1$, $U_{\mu_1}(n,d)$ can be endowed with a projective scheme structure, that makes it a moduli scheme (see Theorem 4.6)[1]. Our purpose is to use one more numerical invariant to describe a stratification of $U_{\mu_1}(n,d)$ in such a way that each stratum becomes an algebraic variety and a coarse moduli space. This invariant is the dimension, as a $\mathbb{C}$-vector space, of the algebra of global endomorphisms.

First, we determine the structure of the algebra of endomorphisms of vector bundles in $U_{\mu_1}(n,d)$. If $\mu_1 = \frac{d_1}{n_1}$, write $d_2 = d - d_1$ and $n_2 = n - n_1$. The existence or absence of non-trivial endomorphisms depends on the construction of E (see Section 3). We prove that (see Corollary 3.2)

$$\text{if } E \in U_{\mu_1}(n,d) \text{ then } End(E) \cong \mathbb{C}[x_1, \ldots, x_k]/(x_1, \ldots, x_k)^2$$

where $0 \leq k \leq \frac{d_1 n_2 - d_2 n_1}{2} + n_1 n_2$.

We denote $A_k := \mathbb{C}[x_1, \ldots, x_k]/(x_1, \ldots, x_k)^2$ and $A_0 := \mathbb{C}$. For any integer $0 \leq k \leq \frac{d_1 n_2 - d_2 n_1}{2} + n_1 n_2$ we denote by $U_{\mu_1}(n,d,k)$ the set

$$U_{\mu_1}(n,d,k) := \{E \in U_{\mu_1}(n,d) : End(E) \cong A_k\}.$$

By fixing $\dim End(-)$ as an invariant, we establish the existence of a schematic stratification of a convenient variety Y with a universal property, based on the flattening stratification of an Ext^1-sheaf over Y (see to Theorem 4.1). To achieve this, we utilize sub-schemes and the concepts of twisted Brill-Noether theory (see [**HHN**]). This allows us to give an algebraic structure to $U_{\mu_1}(n,d,k)$ and prove the following theorems. Here, $B^k(\mathcal{U}_1, \mathcal{U}_2^*)$ denotes the twisted Brill-Noether locus of

[1] At the time of writing this article, some results on vector bundles of type $\sigma = (\mu_1, \mu_2)$ were obtained independently and using different methods in [**HJ**] and [**J**]

the product of two stable bundles with at least k sections, and $\mathcal{Y}_k$ represents those with exactly k sections (see (4.3) and Sections 4 for the precise definition of the twisted Brill-Noether locus $B^k(\mathcal{U}_1,\mathcal{U}_2^*)$ and the expected dimension).

THEOREM 1.1 (Theorem 4.6 and Corollary 4.7). *$U_{\mu_1}(n,d,k)$ is a coarse moduli space and $U_{\mu_1}(n,d,0)$ is a fine moduli space. Moreover, if $\mathcal{Y}_k$ is irreducible and smooth of dimension ρ, then $U_{\mu_1}(n,d,k)$ is also irreducible and smooth of dimension $\rho+h_1-1$. In particular, if $U_{\mu_1}(n,d,k)$ is non-empty, it implies that $B^k(\mathcal{U}_1,\mathcal{U}_2^*)$ is also non-empty.*

COROLLARY 1.2 (Corollary 4.8). *If $\mathcal{Y}_k$ is irreducible and smooth, then the cohomology groups $H^i(U_{\mu_1}(n,d,k),\mathbb{C})$ are isomorphic to $H^i(\mathcal{Y}_k,\mathbb{C})$ for $i\geq 0$.*

Of particular interest are vector bundles of HN-type $\sigma = (\frac{d-a}{n-1},a)$, where a is an integer. These bundles are indecomposable and unstable, representing extensions of a line bundle by a semistable bundle. The results obtained in this case are a reformulation of the existing results in Brill-Noether theory in terms of the moduli of unstable bundles. The Brill-Noether loci and the expected dimension $\beta(g,n_1,d_1,n_2,d_2,k)$ of $U_{\mu_1}(n,d,k)$ are discussed in Sections 2 and 4. Non-emptiness and topological properties of $U_{\mu_1}(n,d,k)$ are given in the following theorems.

THEOREM 1.3 (Theorem 5.1). *Under the conditions $0<d-an<2(n-1)$ and $(n-1,d-an,k)\neq(n-1,n-1,n-1)$, for $\mu_1=\frac{d-a}{n-1}$, $U_{\mu_1}(n,d,k)$ is non-empty if and only if $k\leq n-1+\frac{d-n(a+1)+1}{g}$. Moreover, if $U_{\mu_1}(n,d,k)$ is non-empty, it is irreducible, smooth, and of the expected dimension.*

There are special results for general and Petri curves;

THEOREM 1.4 (Theorem 5.4). *For integers $(g,n-1,d-na,k)$ satisfying the conditions given in Theorem 2.8,(5),(6) and (7), $U_{\mu_1}(n,d,k)$ is non-empty and has an irreducible component of the expected dimension. Moreover, if X is a Petri curve of genus $g\geq 2n-4$ with $n\geq 5$ then $U_{\mu_1}(n,d,k)$ is non-empty.*

As in the Brill-Noether theory for vector bundles, it is possible that the conclusions stated above do not hold for special curves. In these cases, the Brill-Noether locus and the corresponding moduli space may not be reduced. Therefore, the results presented here differ from the corresponding results for the moduli space of stable bundles, where non-emptiness, irreducibility, dimension, etc., are independent of the curve.

To the best of our knowledge, the following theorems also provide new results in the Brill-Noether and twisted Brill-Noether theory (see [**BN**]).

THEOREM 1.5 (Theorem 5.6). *Assume that $B(n_1,d_1,n_1+a)$ is non-empty where $a>0$. If $2n_1<d_1<a(g+1)$ and $d_2>2gm$ then for any $0\leq k\leq (d_2+m(1-g))(n_1+a)-(d_2n_1+d_1m+mn_1(1-g))$, $\mathcal{Y}_k\subset B^k(\mathcal{U}_1,\mathcal{U}_2^*)$ is non-empty. Moreover, if $\mu_1=\frac{d_1}{n_1}$ then $U_{\mu_1}(n,d,k)$ is non-empty, where $n=n_1+n_2$ and $d=d_1+d_2$.*

For Perti curves we prove

THEOREM 1.6. *(Theorem 5.7) Let X be a Petri curve of genus $g\geq 3$ and let $(\mathcal{O}(D),V)$ be a general generated linear system of degree $d_2\geq g+1$ with $\dim V = n_2+1$ where $n_2\leq 4$ or if $n_2\geq 5$, then $g\geq 2n_2-4$. Assume that $B(n_1,d_1,t)$ is non-empty and $\frac{d_2}{n_2}<\frac{d_1}{n_1}$. For any $0\leq k\leq n_2t-n_1d_2$, $\mathcal{Y}_k\subset B^k(\mathcal{U}_1,\mathcal{U}_2^*)$ is* *non-empty. Moreover, if* $n = n_2 + n_1$, $d = d_2 + d_1$ *and* $\mu_1 = \frac{d_1}{n_1}$, *then* $U_{\mu_1}(n, d, k)$, *is non-empty.*

In the remainder of the last section, necessary and sufficient conditions for $U_{\mu_1}(n, d, k)$ to be smooth on $E \in U_{\mu_1}(n, d, k)$ are given. Consider the HN-filtration $0 \subset E_1 \subset E$ of E in $U_{\mu_1}(n, d, k)$. Write $E/E_1 = F_1$. The diagram

$$\begin{array}{ccc} H^1(End(E_1)) \oplus H^1(End(F_1)) & \xrightarrow{d\Phi} & H^1(End(E_1 \otimes F_1^*)) \\ & \searrow^{\eta_E} & \downarrow \beta \\ & & H^0(E_1 \otimes F_1^*)^* \otimes H^1(E_1 \otimes F_1^*) \end{array}$$

is used to define η_E as $\eta_E := \beta \circ d\Phi$ and prove

THEOREM 1.7 (Theorem 5.9). $U_{\mu_1}(n, d, k)$ *is smooth at* E *and of the expected dimension if and only if* η_E *is surjective.*

In Section 2 we review some standard facts about unstable bundles. Section 3 focuses on the algebra of endomorphisms of unstable bundles. Sections 4 and 5 present the main results of the paper and provide their proofs.

Notation. The rank and degree of a vector bundle E are denoted by $rk(E)$ and $d(E)$, respectively. The slope is represented as a rational number $\mu(E) := \frac{d(E)}{rk(E)}$. We denote the space of homomorphisms from F to E as $Hom(F, E)$, which is defined as $Hom(F, E) := H^0(X, F^* \otimes E)$. Similarly, when $F = E$, the space of endomorphisms of E is denoted as $End(E)$, given by $End(E) = H^0(X, E \otimes E)$. When we refer to $< E \xrightarrow{f} F > \subset Hom(F, E)$, we mean the linear space generated by the homomorphism $f : E \to F$. The projection in the i-factor is denoted as p_i. The cohomology group $H^i(X, E)$ as $H^i(E)$ and its dimension as $h^i(E)$. Given an exact sequence

$$\rho : 0 \to G \to E \to F \to 0,$$

- by ρ^* we mean the dual sequence $\rho^* : 0 \to F^* \to E^* \to G^* \to 0$,
- by $(M \otimes \rho)$ we mean tensor the sequence ρ by the vector bundle M i.e.

$$(M \otimes \rho) : 0 \to M \otimes G \to M \otimes E \to M \otimes F \to 0.$$

- By $H^*(\rho)$ we mean the cohomology sequence of ρ

$$0 \to H^0(X, G) \to H^0(X, E) \to H^0(X, F) \xrightarrow{\delta} H^1(X, G) \to H^1(X, E) \to \dots.$$

To shorten notation, sometimes we write $E \in H^1(X, F^* \otimes G)$ instead of

$$(\rho : 0 \to G \to E \to F \to 0) \in H^1(X, F^* \otimes G)$$

2. Unstable bundles of HN-lenght 2

From now on, X will be a smooth irreducible complex projective curve of genus ≥ 2. Recall that a vector bundle E over X is semistable if for all proper subbundle $F \subset E$ the slopes satisfy the following inequality:

$$\mu(F) \leq \mu(E).$$

The vector bundle E is stable if the inequality is strict, and unstable if it is not semistable.

In the reference [**HN**], it was proven that any vector bundle over X has a unique filtration, called *Harder-Narasimhan filtration*:

$$0 = E_0 \subset E_1 \subset \cdots \subset E_m = E \tag{2.1}$$

such that for $1 \leq i \leq m$;

- E_i/E_{i-1} is semistable and
- $$\mu(E_1) > \mu(E_2/E_1) > \cdots > \mu(E_m/E_{m-1}). \tag{2.2}$$

To shorten notation, we write "HN-filtration" instead of Harder-Narasimhan filtration and F_i instead of the quotient E_{i+1}/E_i, when no confusion can arise. Note that $F_0 = E_1$. The HN-max and HN-min of E are defined as

$$\mu_{max}(E) := \mu(E_1) \quad \text{and} \quad \mu_{min}(E) := \mu(F_{m-1}).$$

The vector bundle E_1 is called *the maximal destabilizing subbundle.*

The following definitions will be used throughout the entire paper. Vector bundles G that satisfy $h^0(G) \cdot h^1(G) \neq 0$, are referred to as special bundles. These special bundles have $h^1(G)$ nonzero and have non-trivial sections, making their study fundamental in Brill-Noether Theory. On the other hand, general bundles are characterized by having either no non-trivial sections or $h^1(G) = 0$. i.e., $h^0(G) \cdot h^1(G) = 0$.

DEFINITION 2.1. Let $0 = E_0 \subset E_1 \subset \cdots \subset E_m = E$ be the Harder-Narasimhan filtration of E.

- The HN-type is the sequence of slopes $\sigma = (\mu(E_1), \ldots, \mu(E_m/E_{m-1}))$.
- The number m is called *the HN-length* of the HN-filtration.
- The HN-filtration is called of *simple type* if each E_i and F_i are simple for $i = 1, \ldots, m-1$.
- The HN-filtration is called of *coprime type* if the numbers in each slope $\mu(F_i)$ and $\mu(E_i)$ are coprime for $i = 1, \ldots, m-1$.
- The HN-filtration is called *HN-indecomposable* if each E_i and F_i are indecomposable $i = 1, \ldots, m-1$.
- E is called *HN-general* (respectively *HN-special*) if all the quotients F_i are general (respectively special).
- The extension $\rho_i : 0 \to E_i \to E_{i+1} \to F_i \to 0$ is called the HN(i)-extension and the sequence of extensions $\rho = (\rho_1, \cdots, \rho_{m-1})$ is called the *HN-sequence of* E.

REMARK 2.2. Let $0 = E_0 \subset E_1 \subset \cdots \subset E_m = E$ be the HN-filtration of E. It is important to note the following:

(1) The condition of being of simple/coprime type or HN-indecomposable apply to $i = 1, \ldots, m-1$. Thus, $E_m = E$ is not necessarily simple or indecomposable.
(2) Simple type implies HN-indecomposable type. However, HN-indecomposable type does not necessarily imply simple type.
(3) If $i \neq 1$, E_i is unstable and $0 = E_0 \subset E_1 \subset \cdots \subset E_i$ is the HN-filtration of E_i. Moreover, if the filtration of E is of coprime, simple, HN-indecomposable, or general type, the same holds for the HN-filtration of any E_i. This allows us to use induction on the HN-length.

(4) For each E_i we have an exact sequence

$$\rho_{i-1} : 0 \to E_{i-1} \to E_i \to F_{i-1} \to 0. \tag{2.3}$$

Thus, E is constructed by a successive sequence of extensions.

In this section, we will focus on vector bundles of HN-type $\sigma = (\mu_1, \mu_2)$. It should be noted that the value of μ_1 fixes the value of μ_2 and vice versa. Indeed, if μ_1 is determined by the quotient $\frac{d_1}{n_1}$ then $\mu_2 = \frac{d(E)-d_1}{rk(E)-n_1}$. Our goal is to describe some properties and parameterize vector bundles of HN-type $\sigma = (\mu_1, \mu_2)$.

Let

$$(\rho_1 : 0 \to E_1 \xrightarrow{i} E_2 \xrightarrow{p} F_1 \to 0) \in Ext^1(F_1, E_1) \cong H^1(X, F_1^* \otimes E_1). \tag{2.4}$$

be an extension of two semistable bundles with $\mu_1 := \mu(E_1) > \mu(F_1)$. It follows that E_2 is unstable and $0 \subset E_1 \subset E_2$ is its HN-filtration, with $E_2/E_1 = F_1$.

REMARK 2.3. Note that given two extensions $(\rho_1 : 0 \to E_1 \xrightarrow{i} E_2 \xrightarrow{p} F_1 \to 0)$ and $(\rho_1' : 0 \to E_1 \xrightarrow{i} E_2' \xrightarrow{p} F_1 \to 0)$ in $Ext^1(F_1, E_1)$, $E_2 \cong E_2'$ if and only if $\rho_1 \sim \rho_1'$ are in the same class in $Ext^1(F_1, E_1)/(Aut(E_1) \times Aut(F_1))$. Moreover, if E_1 and F_1 are simple, $E_2 \cong E_2'$ iff $\rho = \lambda\rho'$ with $\lambda \in \mathbb{C}^*$.

To prove our first results, we use the following lemmas.

LEMMA 2.4. *Let $\rho_1 : 0 \to E_1 \xrightarrow{i} E_2 \xrightarrow{p} F_1 \to 0$ be an extension of two semistable bundles with $\mu_1 := \mu(E_1) > \mu(F_1)$. If $\mu(E_1) - \mu(F_1) > 2g - 2$, $\rho_1 = 0$. Assume that $0 < \mu(E_1) - \mu(F_1) \leq 2g - 2$. If ρ_1 is non-trivial then*

(1) *$End(E_1) = Hom(E_1, E_2)$. Moreover, if E_1 is simple, $h^0(E_1^* \otimes E_2) = 1$ and $Hom(E_1, E_2) =< E_1 \xrightarrow{i} E_2 >$.*
(2) *$End(F_1) = Hom(E_2, F_1)$. Moreover, if F_1 is simple, $h^0(E_2^* \otimes F_1) = 1$ and $Hom(E_2, F_1) =< E_2 \xrightarrow{p} F_1 >$.*
(3) *If F_1 is simple then $Hom(F_1, E_1) = Hom(F_1, E_2)$.*

PROOF. The inequality $\mu(E_1) - \mu(F_1) > 2g-2$ implies that $\mu(F_1^* \otimes E_1) > 2g-2$. Hence, since $F_1^* \otimes E_1$ is semistable, $H^1(X, F_1^* \otimes E_1) = 0$.

If $0 < \mu(E_1) - \mu(F_1) \leq 2g - 2$ then $H^0(X, E_1^* \otimes F_1) = 0$ and (1) and (2) follow from the cohomology sequences

(2.5)

$$H^*(E_1^* \otimes (\rho_1)) : 0 \to H^0(X, E_1^* \otimes E_1) \to H^0(X, E_1^* \otimes E_2) \to H^0(X, E_1^* \otimes F_1) \xrightarrow{\delta_0} \cdots$$

and

(2.6)

$$H^*((\rho_1)^* \otimes F_1) : 0 \to H^0(X, F_1^* \otimes F_1) \to H^0(X, E_2^* \otimes F_1) \to H^0(X, E_1^* \otimes F_1) \xrightarrow{\delta_1} \cdots$$

(3).– Since F_1 is simple and $\rho_1 \neq 0$, the cohomology sequence $H^*(F_1^* \otimes (\rho_1))$

(2.7)

$$0 \to H^0(X, F_1^* \otimes E_1) \to H^0(X, F_1^* \otimes E_2) \to H^0(X, F_1^* \otimes F_1) \xrightarrow{\delta_2} H^1(X, F_1^* \otimes E_1) \cdots,$$

implies that $H^0(X, F_1^* \otimes E_1) = H^0(X, F_1^* \otimes E_2)$. □

LEMMA 2.5. *Assume E_1 and F_1 are semistable bundles with E_1 being simple and F_1 indecomposable. Then E_2 is indecomposable if and only if $\rho_1 \neq 0$.*

PROOF. One direction is trivial. Assume $\rho_1 \neq 0$. Suppose, contrary to our claim, that $E_2 = G_1 \oplus G_2$ and the extension

$$0 = \rho : 0 \to G_1 \xrightarrow{j} E_2 \xrightarrow{q} G_2 \to 0$$

splits. Denote by $\gamma : G_2 \to E_2$ the splitting morphism, i.e. $q \circ \gamma = id_{G_2}$. We can certainly assume that $\mu(G_2) \leq \mu(E_2)$, for if not, we replace G_2 by G_1.

The inclusion morphism $i : E_1 \to E_2$ in the extension ρ_1 induces the following diagram

$$\begin{array}{ccccccccc}
 & & 0 & & 0 & & 0 & & \\
 & & \downarrow & & \downarrow & & \downarrow & & \\
0 & \to & M & \to & E_1 & \overset{\sigma}{\to} & H & \to & 0 \\
 & & \downarrow & & \downarrow i & & \downarrow \nu & & \\
0 & \to & G_1 & \overset{j}{\to} & E_2 & \overset{q}{\rightleftarrows}_{\gamma} & G_2 & \to & 0
\end{array}$$

We have different cases.

CASE 1 ($H \neq 0$ and $M \neq 0$). In this case there is a contradiction to Lemma 2.4,(1) since the morphism $\gamma \circ \nu \circ \sigma : E_1 \to E_2$ is not a scalar multiple of the inclusion $i : E_1 \to E_2$. Note that $rk(M) < rk(E_1)$.

CASE 2 ($H = E_1$). The diagram

$$\begin{array}{ccccccccc}
 & & & & 0 & & 0 & & \\
 & & & & \downarrow & & \downarrow & & \\
 & & & & E_1 & = & E_1 & & \\
 & & & & \downarrow i & & \downarrow & & \\
0 & \to & G_1 & \overset{j}{\to} & E_2 & \overset{q}{\rightleftarrows}_{\gamma} & G_2 & \to & 0
\end{array}$$

induces the following diagram

$$\begin{array}{ccccccccc}
 & & & & 0 & & 0 & & \\
 & & & & \downarrow & & \downarrow & & \\
 & & & & E_1 & = & E_1 & & \\
 & & & & \downarrow i & & \downarrow & & \\
0 & \to & G_1 & \to & E_2 & \rightleftarrows_{\gamma} & G_2 & \to & 0 \\
 & & \| & & \downarrow & & \downarrow & & \\
0 & \to & G_1 & \to & F_1 & \overset{\sigma}{\to} & T & \to & 0 \\
 & & \downarrow & & \downarrow & & \downarrow & & \\
 & & 0 & & 0 & & 0 & & .
\end{array}$$

The semistability of F_1 implies that $\mu(G_1) \leq \mu(F_1) < \mu(E_2)$, hence that $\mu(E_2) < \mu(G_2)$, which is a contradiction to our assumption.

CASE 3 ($H = 0$). In this case we have the following diagram

$$
\begin{array}{ccccccccc}
 & & 0 & & 0 & & & & \\
 & & \downarrow & & \downarrow & & & & \\
 & & E_1 & = & E_1 & & & & \\
 & & \downarrow i & & \downarrow i & & & & \\
0 & \to & G_1 & \stackrel{j}{\to} & E_2 & \overset{q}{\underset{\gamma}{\rightleftarrows}} & G_2 & \to & 0 \\
 & & \downarrow & & \downarrow p & & \| & & \\
0 & \to & T_1 & \stackrel{\jmath}{\to} & F_1 & \stackrel{\ell}{\to} & G_2 & \to & 0 \\
 & & \downarrow & & \downarrow & & \downarrow & & \\
 & & 0 & & 0 & & 0 & & .
\end{array}
$$

If $0 \neq \eta := p \circ \gamma : G_2 \to F_1$ then

$$\ell \circ \eta = \ell \circ p \circ \gamma = q \circ \gamma = id_{G_2}$$

and hence F_1 is decomposable which is a contradiction, since F_1 is indecomposable.

If $\eta = 0$ then $G_2 \subset E_1 \subset G_1$, which is also a contradiction since $E_2 = G_1 \oplus G_2$. Therefore, E_2 is indecomposable, as claimed. □

Let $T = Ext^1(F_1, E_1) = H^1(X, F_1^* \otimes E_1)$. It is well known (see [**NR**]) that $\mathbb{P}(T)$ parameterize non-trivial extensions $\rho_1 : 0 \to E_1 \stackrel{i}{\to} E \stackrel{p}{\to} F_1 \to 0$ and that if $\mathbb{H}$ is the hyperplane bundle over $\mathbb{P}(T)$ the extension

$$0 \to p_1^*(E_1) \otimes p_2^*(\mathbb{H}) \to \mathcal{E} \to p_1^*(F_1) \to 0$$

that corresponds to the identity under the isomorphism

$$End(T) \cong H^1(X \times \mathbb{P}(T), p_1^*(F_1^* \otimes E_1) \otimes p_2^*(\mathbb{H}))$$

has universal properties (see [**NS**]). Thus, Lemma 2.5 leads us to the following result:

PROPOSITION 2.6. *Let $T = H^1(X, F_1^* \otimes E_1)$. If E_1 and F_1 are simple semistable vector bundles with $\mu(E_1) > \mu(F_1)$ then the pair $(\mathbb{P}(T), \mathcal{E})$ is a fine moduli space for indecomposable unstable bundles E with E_1 as the maximal destabilizing subbundle of E and $F_1 = E/E_1$. Furthermore, if $\mu(E_1) - \mu(F_1) > 2g - 2$, $\mathbb{P}(T)$ is empty.*

From now on, we assume that $0 < \mu(E_1) - \mu(F_1) \leq 2g - 2$. By Riemann-Roch, the dimension of $\mathbb{P}(T)$ is given by:

$$\dim \mathbb{P}(T) = h^0(F_1^* \otimes E_1) - \tilde{d} + \tilde{r}(g-1) - 1, \tag{2.8}$$

where $\tilde{d} := d(E_1)rk(F_1) - d(F_1)rk(E_1)$ and $\tilde{r} := rk(E_1)rk(F_1)$. The dimension of $\mathbb{P}(T)$ depends on $h^0(F_1^* \otimes E_1)$. Specifically, $\dim \mathbb{P}(T) > 0$ if and only if $h^0(F_1^* \otimes E_1) > \tilde{d} + \tilde{r}(1-g) + 1$. By analyzing the dimension of $h^0(F_1^* \otimes E_1)$ one can draw connections to the Brill-Noether loci and their expected dimensions. The behavior of $h^0(F_1^* \otimes E_1)$ within the framework of Brill-Noether theory provides insights into the existence and properties of special bundles. Furthermore, it contributes to our understanding of the moduli space of unstable vector bundles over the curve.

We will only briefly mention a few aspects of Brill-Noether Theory. The *Brill-Noether loci* are defined as follows:

$$B(n, d, k) := \{G \in M(n, d) : h^0(G) \geq k\},$$

where $M(n,d)$ is the moduli space of stable vector bundles of degree d and rank n over X. Moreover, these loci induce a filtration of the moduli space as follows:

$$M(n,d) \supseteq B(n,d,1) \supseteq \cdots \supseteq B(n,d,k) \supseteq B(n,d,k+1) \supseteq \cdots.$$

Let us introduce the notation Y_k for $B(n,d,k) - B(n,d,k+1)$. In other words,

$$Y_k = \{E \in B(n,d,k) : h^0(E) = k\}$$

is the subset of bundles in $B(n,d,k)$ that have exactly k global sections. The *Brill-Noether number* is denoted as

$$\rho(g,n,d,k) := n^2(g-1) + 1 - k(k - d + n(g-1)).$$

This number represents the expected dimension of the Brill-Noether loci $B(n,d,k)$. It's important to note that the Brill-Noether loci can also be defined on the moduli space of S-equivalent semistable bundles (see [**BGN**]).

REMARK 2.7. Denote the number χ as $\chi := d + n(1-g)$. It is worth noting that if $k \leq \chi$, then $B(n,d,k) = M(n,d)$. Therefore, to ensure proper Brill-Noether loci, we make the assumption that $k > \chi$.

For the convenience of the reader we provide a summary of the main results concerning these concepts and the relevant material without presenting the proofs, thus ensuring a self-contained exposition. For a comprehensive survey of the key findings in Brill-Noether Theory, refer to [**GT**]. Recall that X is a *Petri curve* if the multiplication map $\mu : H^0(X,L) \otimes H^0(X, K_X \otimes L^*) \to H^0(X, K_X)$ is injective for every line bundle L on X (see [**ACGH**]).

THEOREM 2.8. *Let G be a semistable vector bundle of rank n and degree d on X.*

(1) *If G is general,*

$$h^0(G) = \begin{cases} 0 & \text{if } 0 < \mu(G) < g-1, \\ d + r(1-g) & \text{if } g-1 \leq \mu(G) < 2(g-1). \end{cases}$$

(2) *If $B(n,d,k)$ is a proper locus, then every general point E belonging to a proper component of $B(n,d,k)$ has $h^0(E) = k$.*

(3) *(Clifford's Theorem for vector bundles* [**BGN**, Theorem 2.1]*) If G is a special bundle,*

$$h^0(G) \leq \frac{d}{2} + n.$$

(4) *(*[**BGN**] *and* [**Me1**]*) For any curve X, if $0 < \mu(G) < 2$,*

$$B(n,d,k) \neq \emptyset \text{ if and only if } n < d + (n-k)g \text{ and } (n,d,k) \neq (n,n,n).$$

Moreover, if $B(n,d,k)$ is not empty, then it is irreducible, of dimension $\rho(g,n,d,k)$ and $Sing(B(n,d,k)) = B(n,d,k+1)$. In particular, Y_k is irreducible, smooth and of dimension $\rho(g,n,d,k)$.

(5) *(*[**BMNO**]*) For general curve, $B(n,d,k)$ is non-empty if $d = nd' + d''$ and $n \leq d'' + (n-k)g$ and the following conditions are satisfied*

$$0 < d'' < 2n, \; d' \geq \frac{(s-1)(s+g)}{s}, \text{ with } 1 < s < g, \; (d'',k) \neq (n,n).$$

(6) *([**GT**] and [**Me**]) Let d, n, k be positive integers with $k > n$. Assume that $d_2 < n$, $k_2 < n$ and let $d = nd_1 + d_2$, $k = nk_1 + k_2$. For any curve, if one of the following conditions is satisfied:*

$$\begin{array}{ll} g - (k_1 + 1)(g - d_1 + k_1 - 1) \geq 1, & 0 = d_2 \geq k_2 \\ g - k_1(g - d_1 + k_1 + 1) > 1, & d_2 = k_2 = 0 \\ g - (k_1 + 1)(g - d_1 + k_1) \geq 1, & d_2 < k_2 \end{array}$$

$B(n, d, k)$ is non-empty. Moreover, if X is generic, $B(n, d, k)$ has one component of the expected dimension.

(7) *([**BBN1**]) Let X be a Petri curve of genus $g \geq 3$. If $n \geq 5$ and $g \geq 2n - 4$ then $B(n, d, n + 1)$ is non-empty.*

REMARK 2.9. The non-emptiness of the locus $B(n, d, n + 1)$ has been proven under various relations among n, d and g. We quote one of the results from Theorem 2.8,(7) for simplicity, but for more information, refer to [**Br**], [**BBN**] or [**BBN1**], as well as the bibliography provided there.

According to the above theorem, in certain cases, the non-emptiness and dimension of the moduli space $\mathbb{P}(T)$ differ from those of general or special curves. Theorem 2.8 will be relevant in Section 5 when considering families of extensions of semistable bundles.

The remaining part of this section will be devoted to the general properties of unstable bundles.

It is well known that $H^0(E) = 0$ for semistable vector bundles E of negative degree. However, unstable bundles of negative degree can have sections. Indeed, if F is a bundle of positive degree with $H^1(F) \neq 0$, any extension $\rho : 0 \to \mathcal{O} \to G \to F^* \to 0 \in H^1(F)$ defines an unstable bundle G of negative degree with at least one section. In the following proposition, we establish some well-known properties of semistable bundles for unstable bundles. These properties directly follow from the properties of the HN-filtration. While some of these properties may already be known, we include them in one proposition for completeness.

PROPOSITION 2.10. *Let E be an indecomposable unstable vector bundle of rank n and degree d, and let $0 = E_0 \subset E_1 \subset \cdots \subset E_m = E$ be its Harder-Narasimhan filtration.*

(1) *If $\mu_{min}(E) > 2g - 1$ then $H^1(X, E) = 0$ and E is (globally) generated. Moreover, indecomposable unstable bundles with a fixed HN-filtration are bounded.*
(2) *If E is HN-special, $h^0(E) \leq \frac{d}{2} + n$.*
(3) *If E is HN-general and $\mu_{min}(E) > g + 1$, $H^1(X, E) = 0$ and $h^0(E) = d(E) + rk(E)(1 - g)$. Moreover, if $0 < \mu_{max}(E) < g$, $h^0(E) = 0$.*

PROOF. Let $\rho_i : 0 \to E_i \to E_{i+1} \to F_i \to 0$ be the HN(i)-extension of E and consider the cohomology sequence:

$$0 \to H^0(X, E_i) \to H^0(X, E_{i+1}) \to H^0(X, F_i) \xrightarrow{\delta} \tag{2.9}$$

$$\xrightarrow{\delta} H^1(X, E_i) \to H^1(X, E_{i+1}) \to H^1(X, F_i) \to 0. \tag{2.10}$$

We recall that the inequalities hold:

$$\mu_{max}(E) = \mu(E_1) > \mu(F_1) > \cdots > \mu(F_{m-1}) = \mu_{min}(E) > 2g - 1.$$

We will prove the propositions by induction on the HN-length of E. Let's provide the proof for the case $m = 2$, using the HN(1)-extension.

(1).- Since $\mu(E_1) > \mu(F_1) = \mu_{min}(E_2) > 2g - 1$, from (2.10) we deduce that $H^1(E_2) = 0$, due to the semistability of E_1 and F_1. As both E_1 and F_1 are generated, we conclude that E_2 is also generated. The boundedness of indecomposable unstable bundles with a fixed HN-filtration follows from [**Ma**] (see also [**HK**]).

(2).- From (2.9), we have $h^0(E_2) \leq h^0(E_1) + h^0(F_1)$. Thus, combining with Theorem 2.8 (2) we obtain $h^0(E_2) \leq (d(E_1)/2) + rank(E_1) + (d(F_1)/2) + rank(F_1)$. Finally, using the fact that

$$h^0(E_1) + h^0(F_1) \leq \frac{d(E_1)}{2} + rk(E_1) + \frac{d(F_1)}{2} + rk(F_1)$$

we conclude:

$$h^0(E_2) \leq \frac{d(E_2)}{2} + rk(E_2).$$

(3).- Similarly to (2), we can deduce from (2.9) and (2.10), as well as applying Riemann-Roch and the assumption of HN-general filtration, that $H^1(X, E_2) = 0$ and $h^0(E_2) = d(E_2) + rank(E_2)(1 - g)$. Moreover, if $0 < \mu_{max}(E_2) < g$, then $h^0(E_2) = 0$.

From Remark 2.2, we can apply induction. By conducting a similar analysis as in the case $m = 2$, but now using the HN(i)-extension, we can show that (1), (2) and (3) hold for any $m > 2$. The details are left to the reader. □

It is known (see [**HL**, Lemma 1.3.3]) that if E and E' are two unstable vector bundles with $\mu_{min}(E) > \mu_{max}(E')$ then $Hom(E, E') = 0$. Our objective is to study $Hom(E, E')$ when $E = E'$, which corresponds to $End(E)$. For an unstable vector bundle E, $End(E)$ can be non-trivial and the presence or absence of non-trivial endomorphisms depends on the specific construction of E. These endomorphisms can arise from non-zero homomorphisms between its destabilizing subbundles. Therefore, a detailed analysis of the destabilizing subbundles and their interplay is crucial in determining whether $End(E)$ is trivial or non-trivial.

3. Algebra of endomorphisms

In this section, our goal is to analyze the algebra of endomorphisms for a HN-indecomposable vector bundle of HN-length 2. By utilizing the HN-filtration, we not only determine the structure of $End(-)$ for HN-length 2 but also provide an upper bound for its dimension, when the HN-length is greater than 2.

In the study of endomorphisms of vector bundles, it is known (see [**A**]) that for an indecomposable vector bundle E, the algebra of endomorphisms can be expressed as:

$$End(E) \cong (Id_E) \oplus Nil(E), \tag{3.1}$$

where $Nil(E)$ is the subset $Nil(E) \subset End(E)$ consisting of all nilpotent global endomorphisms of E. In [**Br3**], an upper bound for the dimension of $End(E)$ was provided for indecomposable semistable bundles of rank n. This bound is given by the formula:

$$\dim End(E) \leq 1 + \frac{n(n-1)}{2}. \tag{3.2}$$

The bound of $\dim End(E)$ for indecomposable bundles of HN-length 2 is established by our next proposition.

PROPOSITION 3.1. *Let E_2 be an indecomposable unstable bundle, and let $0 = E_0 \subset E_1 \subset E_2$ be its HN-filtration. If both E_1 and $F_1 = E_2/E_1$ are indecomposable, then the dimension of $End(E_2)$ is bounded by:*

$$\dim End(E_2) \leq 1 + \frac{n(n-1)}{2} + h^0(F_1^* \otimes E_1), \tag{3.3}$$

where $F_1 = E_2/E_1$.

PROOF. To prove the proposition, let's consider the HN(1)-sequence $\rho_1 : 0 \to E_1 \xrightarrow{\iota} E_2 \xrightarrow{p} F_1 \to 0$ of E_2. Assume $n_1 := rk(E_1)$ and $n_2 := rk(F_1)$. We will also use the following sequences:
(3.4)
$$0 \to H^0(X, E_1 \otimes E_2^*) \to H^0(X, E_2 \otimes E_2^*) \to H^0(X, F_1 \otimes E_2^*) \xrightarrow{\delta_0} H^1(X, E_1 \otimes E_2^*) \cdots$$
(3.5)
$$0 \to H^0(X, F_1^* \otimes F_1) \to H^0(X, E_2^* \otimes F_1) \to H^0(X, E_1^* \otimes F_1) \xrightarrow{\delta_1} H^1(X, F_1^* \otimes F_1) \cdots$$
(3.6)
$$0 \to H^0(X, F_1^* \otimes E_1) \to H^0(X, E_2^* \otimes E_1) \to H^0(X, E_1^* \otimes E_1) \xrightarrow{\delta_2} H^1(X, F_1^* \otimes E_1)$$
which are part of the cohomology sequences of $H^*((\rho_1) \otimes E_2^*)$, $H^*((\rho_1)^* \otimes F_1)$ and of $H^*((\rho_1)^* \otimes E_1)$, respectively.

From (3.4), we have the inequality:

$$\dim End(E_2) \leq h^0(E_1 \otimes E_2^*) + h^0(F_1 \otimes E_2^*). \tag{3.7}$$

Due to the semistability of F_1 and E_1 and the inequalities (2.2) we can conclude that $Hom(E_1, F_1) = H^0(X, E_1^* \otimes F_1) = 0$. Hence, from (3.5) we have

$$Hom(E_2, F_1) = H^0(X, F_1 \otimes E_2^*) = H^0(X, F_1^* \otimes F_1) = End(F_1). \tag{3.8}$$

From (3.6), we obtain the bound:

$$h^0(E_1 \otimes E_2^*) \leq h^0(F_1^* \otimes E_1) + h^0(E_1^* \otimes E_1).$$

Therefore, since E_1 and F_1 are semistable and indecomposables

$$\begin{aligned} \dim End(E_2) &\leq h^0(E_1 \otimes E_2^*) + h^0(F_1 \otimes E_2^*) \\ &= h^0(E_1 \otimes E_2^*) + \dim End(F_1) \\ &\leq h^0(F_1^* \otimes E_1) + \dim End(E_1) + \dim End(F_1) \\ &\leq h^0(F_1^* \otimes E_1) + \tfrac{n_1(n_1-1)}{2} + 1 + \tfrac{n_2(n_2-1)}{2} + 1 \quad \text{(see (3.2))} \\ &\leq h^0(F_1^* \otimes E_1) + 1 + \tfrac{(n_1^2+n_2^2)}{2} - \tfrac{n_1+n_2}{2} + 1 \\ &\leq h^0(F_1^* \otimes E_1) + 1 + \tfrac{n(n-1)}{2} + 1 - n_1 n_2 \\ &\leq h^0(F_1^* \otimes E_1) + 1 + \tfrac{n(n-1)}{2} \end{aligned}$$

as claimed. □

COROLLARY 3.2. *Let E_2 be an indecomposable bundle of rank n and let $0 = E_0 \subset E_1 \subset E_2$ be its HN-filtration. If both E_1 and $F_1 = E_2/E_1$ are simple, then $\dim End(E_2) = 1 + h^0(F_1^* \otimes E_1)$ and*

$$End(E_2) \cong \mathbb{C}[x_1, \ldots, x_k]/(x_1, \ldots, x_k)^2,$$

where $k = \dim Hom(F_1, E_1)$. Moreover, E_2 is simple if and only if $Hom(F_1, E_1) = 0$.

PROOF. Under the hypotheses of the corollary, Lemma 2.4, (2) shows that $h^0(F_1 \otimes E_2^*) = 1$. Using (3.4) we can conclude that for any scalar $\lambda \in \mathbb{C}$, $\delta_0(\lambda p) = 0$. Thus,

$$\dim End(E_2) = h^0(E_1 \otimes E_2^*) + h^0(F_1 \otimes F_1^*) = h^0(E_1 \otimes E_2^*) + 1.$$

It follows from (3.6) that $\delta_2(Id_{E_1}) \neq 0$, hence that $h^0(F_1^* \otimes E_1) = h^0(E_1 \otimes E_2^*)$, and finally that

$$\dim End(E_2) = 1 + \dim Hom(F_1, E_1).$$

From (3.1) we deduce that $\dim Nil(E) = h^0(F_1^* \otimes E_1)$. The map

$$\varphi : H^0(F_1^* \otimes E_1) \to Nil(E)$$

defined as $\sigma \mapsto \iota \circ \sigma \circ p$ is a well defined injective homomorphism. Therefore, $H^0(F_1^* \otimes E_1) \cong Nil(E)$. Moreover, $(\varphi(\sigma))^2 = 0$ and

$$(\varphi(\sigma_1)) \circ (\varphi(\sigma_2)) = (\varphi(\sigma_2)) \circ (\varphi(\sigma_1)) = 0.$$

Hence, we can conclude that $End(E) \cong \mathbb{C}[x_1, \dots, x_k]/(x_1, \dots, x_k)^2$ where $k = h^0(F_1^* \otimes E_1)$. Moreover, E_2 is simple if and only if $h^0(F_1^* \otimes E_1) = 0$, as desired. □

REMARK 3.3. Note that if the vector bundle $F_1^* \otimes E_1$ is special, then $h^0(F_1^* \otimes E_1) \cdot h^1(F_1^* \otimes E_1) \neq 0$. Thus, if $0 \neq \rho : 0 \to E_1 \to E_2 \to F_1 \to 0 \in H^1(X, F_1^* \otimes E_1)$, then E_2 is indecomposable and $\dim End(E_2) > 1$. On the other hand, if $F_1^* \otimes E_1$ is no special, then $h^0(F_1^* \otimes E_1) \cdot h^1(F_1^* \otimes E_1) = 0$. Consequently, either E_2 is decomposable or is simple.

The following proposition is a straightforward generalization for unstable bundles of HN-length $m > 2$.

PROPOSITION 3.4. *Let E be an indecomposable unstable vector bundle of rank n. If the HN-filtration $0 = E_0 \subset E_1 \subset \cdots \subset E_m = E$ is HN-indecomposable, then*

$$\dim End(E) \leq 1 + \frac{n(n-1)}{2} + \sum^{m-1} h^0(F_i^* \otimes E_i) \tag{3.9}$$

For the proof of Proposition 3.4 we will use the following lemma.

LEMMA 3.5. *For all $0 < i < j \leq m$*

(1) *$Hom(F_i, F_j) = 0$ and*
(2) *For all $0 < i \leq j \leq m$, $Hom(E_i, F_j) = 0$.*

PROOF. (1) Follows from the semistability of the quotients F_i's and the inequalities (2.2).

(2) The semistability of E_1 and F_1, along with the inequalities (2.2), imply that

$$H^0(X, F_1^* \otimes F_j) = H^0(X, E_1^* \otimes F_j) = 0,$$

Hence, from the cohomology sequence

$$H^*((\rho_1)^* \otimes F_j) : 0 \to H^0(X, F_1^* \otimes F_j) \to H^0(X, E_2^* \otimes F_j) \to H^0(X, E_1^* \otimes F_j) \to \cdots$$

we conclude that $H^0(X, E_2^* \otimes F_j) = 0$. Similarly, $H^0(X, E_i^* \otimes F_j) = 0$ for all $i \leq j \leq m$. The detailed verification is left to the reader. □

PROOF OF PROPOSITION 3.4. The proof is by induction on the HN-length. We begin with the base case, which is covered by Proposition 3.1. From the cohomology sequence $H^*(\rho_{m-1} \otimes E_m^*)$

$$\dim End(E_m) \leq h^0(E_{m-1} \otimes E_m^*) + h^0(F_{m-1} \otimes E_m^*). \tag{3.10}$$

We can proceed analogously to the proof of Proposition 3.1. From the cohomology sequence $H^*((\rho_{m-1})^* \otimes F_{m-1})$ and using Lemma 3.5,(2), we conclude:

$$H^0(F_{m-1} \otimes E_m^*) = H^0(F_{m-1}^* \otimes F_{m-1}).$$

From the cohomology sequence $H^*((\rho_{m-1})^* \otimes E_{m-1})$ we have the inequality

$$h^0(E_{m-1} \otimes E_m^*) \leq h^0(End(E_{m-1})) + h^0(F_{m-1}^* \otimes E_{m-1}),$$

recall that E_m is HN-indecomposable.

Now, the expression for $\dim End(E_m)$ becomes

$$\begin{aligned}
\dim End(E_m) &\leq h^0(End(E_{m-1})) + h^0(F_{m-1}^* \otimes E_{m-1}) + h^0(F_{m-1}^* \otimes F_{m-1}) \\
&\leq \tfrac{rk(E_{m-1})(rk(E_{m-1})-1)}{2} + \tfrac{rk(F_{m-1})(rk(F_{m-1})-1)}{2} + 2 + \\
&\quad \textstyle\sum^{m-2} h^0(F_i^* \otimes E_i) + h^0(F_{m-1}^* \otimes E_{m-1}) \\
&\leq 1 + \tfrac{rk(E_m)(rk(E_m)-1)}{2} + \textstyle\sum^{m-1} h^0(F_i^* \otimes E_i),
\end{aligned}$$

which is the desired conclusion. Therefore, by induction, Proposition 3.4 holds. □

The following result can be proven using similar techniques as Corollary 3.2.

COROLLARY 3.6. *Let E be an indecomposable unstable vector bundle of rank n of simple type. If $0 = E_0 \subset E_1 \subset \cdots \subset E_m = E$ is its HN-filtration, then*

$$\dim End(E) = 1 + h^0(F_{m-1}^* \otimes E_{m-1}). \tag{3.11}$$

Moreover, $h^0(F_{m-1}^* \otimes E_{m-1}) \leq \sum_0^{m-2} h^0(F_{m-1}^* \otimes F_i)$.

PROOF. We will provide a brief outline of the proof. We can proceed similarly to the proof of Corollary 3.2. The equality $\dim End(E) = 1 + h^0(F_{m-1}^* \otimes E_{m-1})$ follows from Lemma 3.5 and the cohomology of the exact sequences:

$$0 \to E_{m-1} \otimes E^* \to E \otimes E^* \to F_{m-1} \otimes E^* \to 0,$$

$$0 \to F_{m-1}^* \otimes F_{m-1} \to E^* \otimes F_{m-1} \to E_{m-1}^* \otimes F_{m-1} \to 0,$$

$$0 \to F_{m-1}^* \otimes E_{m-1} \to E^* \otimes E_{m-1} \to E_{m-1}^* \otimes E_{m-1} \to 0,$$

since E is of simple type, it is HN-indecomposable.

To establish the bound for $h^0(F_{m-1}^* \otimes E_{m-1})$ we consider the cohomology of the sequences:

$$0 \to E_{i-1} \otimes F_{m-1}^* \to E_i \otimes F_{m-1}^* \to F_{i-1} \otimes F_{m-1}^* \to 0,$$

and proceed by induction. The details of the proof are left to the reader. □

4. Moduli spaces

In this section Proposition 2.6 is presented in a more general context. Our aim is to consider families of semistable bundles. It is important to note that if the vector bundles E_1 and F_1 have automorphisms, the $\mathbb{C}^*$-action does not identify all the isomorphic classes of the vector bundles in $H^1(X, E_1 \otimes F_1^*)$. The simple vector bundles over X with fixed rank and degree have a coarse moduli space (see [**KO**, Corollary 6.5]). It is possible that the moduli space is non-separated, and according to the work of M. Artin it is an algebraic space. Therefore, to construct a moduli scheme of indecomposable vector bundles of HN-length 2 we will consider those of coprime type $\sigma = (\mu_1, \mu_2)$. It should be noted that if μ_1 is defined as $\mu_1 = \frac{d_1}{n_1}$, then $\mu_2 = \frac{d-d_1}{n-n_1} := \frac{d_2}{n_2}$. From now on, we will assume that $\mu_1 = \frac{d_1}{n_1} > \frac{d_2}{n_2} = \mu_2$, and we will denote by $U_{\mu_1}(n,d)$ the set of such bundles. Let

$$U_{\mu_1}(n,d,k) := \{E \in U_{\mu_1}(n,d) : \dim End(E) = 1 + k\}. \tag{4.1}$$

The moduli space $M(n_i, d_i)$ of semistable vector bundles of rank n_i and degree d_i for $1 = 1,2$ with $gcd(n_i, d_i) = 1$ carries a universal bundle $\mathcal{U}_i \to X \times M(n_i, d_i)$. In general, $\mathcal{U}_i$ is determined up to tensoring by a line bundle lifted from $M(n_i, d_i)$. In this paper $\mathcal{U}_i$ will remain fixed unless otherwise stated. Denote by p_{ij} the projection of $X \times M(n_1, d_1) \times M(n_2, d_2)$ onto the ij-factor. We will refer to $\mathcal{R}_{\mu_1}$ as the first direct image sheaf

$$\mathcal{R}_{\mu_1} := \mathcal{R}^1_{p_{23}}(p_{12}^*\mathcal{U}_1 \otimes p_{13}^*\mathcal{U}_2^*)$$

over $M(n_1, d_1) \times M(n_2, d_2)$.

With the notation $M_i := M(n_i, d_i)$, the following diagram summarizes the notation:

$$\begin{array}{ccccc} & p_{12}^*\mathcal{U}_1 \otimes p_{13}^*\mathcal{U}_2^* & & \mathcal{R}_{\mu_1} := \mathcal{R}^1_{p_{23}}(p_{12}^*\mathcal{U}_1 \otimes p_{13}^*\mathcal{U}_2^*) \\ & \downarrow & & \downarrow \\ \mathcal{U}_1 & X \times M_1 \times M_2 & \xrightarrow{\quad p_{23} \quad} & M_1 \times M_2 \\ \downarrow & \swarrow{\scriptstyle p_{12}} & {\scriptstyle p_{13}}\searrow & \\ X \times M_1 & & & X \times M_2. \end{array} \tag{4.2}$$

The following theorem follows from [**Gr**], [**L**], Lemma 2.5 and Lemma 3.5.

THEOREM 4.1. *If $\mu_1 - \frac{d_2}{n_2} \geq 2g-1$, then $U_{\mu_1}(n,d) = \emptyset$. If $0 < \mu_1 - \frac{d_2}{n_2} < 2g-1$, then $U_{\mu_1}(n,d)$ has a projective scheme structure that makes it a moduli scheme. Moreover, there is a natural isomorphism between $U_{\mu_1}(n,d)$ and $Proj(\mathcal{R}_{\mu_1})$.*

PROOF. Let $E \in U_{\mu_1}(n,d)$. Note that being of coprime type implies that E is an extension of two stable bundles, given by the exact sequence:

$$\rho_1 : 0 \to E_1 \to E \to F_1 \to 0.$$

Since $Ext^2(p_{13}^*\mathcal{U}_2, p_{12}^*\mathcal{U}_1) = 0$, it follows that $\mathcal{R}^1_{p_{23}}(p_{12}^*\mathcal{U}_1 \otimes p_{13}^*\mathcal{U}_2^*)_{|(E_1,F_1)} \to H^1(X, E_1 \otimes F_1^*)$ is an isomorphism. According to [**Gr**], [**L**], the coherent sheaf $\mathcal{R}_{\mu_1} := \mathcal{R}^1_{p_{23}}(p_{12}^*\mathcal{U}_1 \otimes p_{13}^*\mathcal{U}_2^*)$ parameterizes the classes of extensions of two stable bundles $(E_1, F_1) \in M_1 \times M_2$. From Remark 3.3,(1) E is indecomposable if and only if $\rho_1 \neq 0$. Therefore, the theorem follows from Remark 2.3. $\square$

REMARK 4.2. It is worth noting that the coherent sheaf $\mathcal{R}_{\mu_1}$ can be defined even when $\gcd(n_0, d_0) \neq 1$. Indeed, if $\gcd(n_i, d_i) \neq 1$, there exists an étale covering $\widetilde{M_i}$ of M_i where a universal bundle $\widetilde{\mathcal{U}_i}$ exists on $X \times \widetilde{M_i}$ (see [**NR**, Proposition 2.4]). Consequently, the coherent sheaf $\widetilde{\mathcal{R}_{\mu_1}} := \mathcal{R}^1_{p_{23}}(p_{12}^*\widetilde{\mathcal{U}_1} \otimes p_{13}^*\widetilde{\mathcal{U}_2^*})$ will parameterize extensions of two semistable bundles $(E_1, F_1) \in \widetilde{M_1} \times \widetilde{M_2}$. However, in this case Remark 3.3,(1) (see Remark 2.3) cannot be applied since semistable bundles may not be simple. Furthermore, it should be remembered that according to [**Br1**], two S-equivalent semistable (but not stable) vector bundles can have different endomorphism algebras.

In general, the coherent sheaf $\mathcal{R}_{\mu_1} = \mathcal{R}^1_{p_{23}}(p_{12}^*\mathcal{U}_1 \otimes p_{13}^*\mathcal{U}_2^*)$ is not locally free. We will provide a flattening stratification of $M_1 \times M_2$ for $\mathcal{R}_{\mu_1}$. To establish the schematic semi-continuity stratification of $M_1 \times M_2$ we will utilize the algebra of endomorphisms and twisted Brill-Noether theory. For a comprehensive overview of the theory, we refer the reader to [**BN**] and [**HHN**]. The construction of the determinantal varieties necessary for our purposes follows the approach outlined in [**ACGH**].

Let D be an effective divisor on X of degree $d_0 >> 0$ such that for any $(E_1, F_1) \in M_1 \times M_2$, $H^1(X, E_1 \otimes F_1^* \otimes \mathcal{O}(D)) = 0$. Consider the cohomology sequence:

$$0 \to H^0(X, E_1 \otimes F_1^*) \to H^0(X, E_1 \otimes F_1^* \otimes \mathcal{O}(D)) \xrightarrow{\phi} H^0(X, (E_1 \otimes F_1^*)_{|D})$$
$$\to H^1(X, E_1 \otimes F_1^*) \to 0,$$

which arises from the exact sequence:

$$0 \to E_1 \otimes F_1^* \to E_1 \otimes F_1^* \otimes \mathcal{O}(D) \to (E_1 \otimes F_1^*)_{|D} \to 0.$$

Thus, $h^0(E_1 \otimes F_1^*) \geq k$ if and only if $rk(\phi) < h^0(E_1 \otimes F_1^* \otimes \mathcal{O}(D)) - k$.

We aim to consider the variation of (E_1, F_1) in $M_1 \times M_2$. Let $\Gamma := D \times M_1 \times M_2$ be the divisor in $X \times M_1 \times M_2$. We have the following sequence

$$0 \to p_{12}^*\mathcal{U}_1 \otimes p_{13}^*\mathcal{U}_2^* \to p_{12}^*\mathcal{U}_1 \otimes p_{13}^*\mathcal{U}_2^* \otimes \mathcal{O}(\Gamma) \to \mathcal{N} \to 0,$$

over $X \times M_1 \times M_2$, where $\mathcal{N}$ is the quotient $p_{12}^*\mathcal{U}_1 \otimes p_{13}^*\mathcal{U}_2^* \otimes \mathcal{O}(\Gamma)/p_{12}^*\mathcal{U}_1 \otimes p_{13}^*\mathcal{U}_2^*$.

Since $H^1(E_1 \otimes F_1^* \otimes \mathcal{O}(D)) = 0$, the direct image induces the complex

$$\phi : \mathcal{K}_0 \to \mathcal{K}_1$$

of locally free sheaves over $M_1 \times M_2$ where $\mathcal{K}^0 := \mathcal{R}^0_{p_{23}}(p_{12}^*\mathcal{U}_1 \otimes p_{13}^*\mathcal{U}_2^* \otimes \mathcal{O}(\Gamma))$ is of rank $d(E_1 \otimes F_1^* \otimes \mathcal{O}(D)) + n_1 n_2(1-g)$ and $\mathcal{K}^1 := \mathcal{R}^0_{p_{23}}(\mathcal{N})$. Following the notation of [**HHN**], we denote $B^k(\mathcal{U}_1, \mathcal{U}_2^*)$ as the kth-determinantal variety of the complex $\phi : \mathcal{K}_0 \to \mathcal{K}_1$. That is,

$$(4.3) \qquad Supp(B^k(\mathcal{U}_1, \mathcal{U}_2^*)) := \{(E_1, F_1) \in M_1 \times M_2 : h^0(E_1 \otimes F_1^*) \geq k\}.$$

REMARK 4.3. To avoid confusion and simplify notation, we will adopt the following conventions from now on:

- $n = n_1 + n_2$,
- $d = d_1 + d_2$,
- $n_0 = n_1 n_2$,
- $d_0 = n_2 d_1 - n_1 d_2$,
- $h^0 = k$ and
- $h^1 = k - d_0 + n_0(g-1)$.

Hence, as kth-determinantal variety, the *expected dimension* of $B^k(\mathcal{U}_1,\mathcal{U}_2^*)$ is given by the number:

$$\begin{aligned}\rho(g,n_1,d_1,n_2,d_2,k) &:= \dim(M_1\times M_2)-h^0\cdot h^1\\ &= (n_1^2+n_2^2)(g-1)+2-k(k-d_0+n_0(g-1)).\end{aligned}$$

and $B^{k+1}(\mathcal{U}_1,\mathcal{U}_2^*)\subseteq SingB^k(\mathcal{U}_1,\mathcal{U}_2^*)$. Moreover, the filtration

$$M_1\times M_2\supset B^1(\mathcal{U}_1,\mathcal{U}_2^*)\supset\cdots\cdots\supset B^k(\mathcal{U}_1,\mathcal{U}_2^*)\supset B^{k+1}(\mathcal{U}_1,\mathcal{U}_2^*)\supset\cdots\cdots$$

defines a stratification by closed subsets. The difference $B^k(\mathcal{U}_1,\mathcal{U}_2^*)-B^{k+1}(\mathcal{U}_1,\mathcal{U}_2^*)$ will be denoted as $\mathcal{Y}_k$, i.e.

$$\mathcal{Y}_k:=B^k(\mathcal{U}_1,\mathcal{U}_2^*)-B^{k+1}(\mathcal{U}_1,\mathcal{U}_2^*).$$

That is,

$$\mathcal{Y}_k=\{(E_1,F_1)\in B^k(\mathcal{U}_1,\mathcal{U}_2^*):h^0(E_1\otimes F_1^*)=k\}.$$

For any $(E_1,F_1)\in\mathcal{Y}_k\subset B^k(\mathcal{U}_1,\mathcal{U}_2^*)\subset M_1\times M_2$,

$$\dim H^1(X,E_1\otimes F_1^*)=k-d(E_1\otimes F_1^*)+n_1n_2(g-1)=k-d_0+n_0(g-1).$$

Therefore, *the restriction of* $\mathcal{R}_{\mu_1}$ to $\mathcal{Y}_k$, denoted by $\mathcal{R}_{\mu_1}(\mathcal{Y}_k)\to\mathcal{Y}_k$, is locally free of rank $k-d_0+n_0(g-1)$. Let

$$\mathbb{P}(\mathcal{R}_{\mu_1}(\mathcal{Y}_k))\to\mathcal{Y}_k$$

be the projective bundle associated to $\mathcal{R}_{\mu_1}(\mathcal{Y}_k)\to\mathcal{Y}_k$.

The expected dimension of $\mathbb{P}(\mathcal{R}_{\mu_1}(\mathcal{Y}_k))$ is

$$\begin{aligned}\beta(g,n_1,d_1,n_2,d_2,k) &:= \dim(M_1\times M_2)-h^0h^1+h^1-1\\ &= \dim(M_1\times M_2)-h^1(h^0-1)-1\\ &= (n_1^2+n_2^2)(g-1)+1-(k-1)(k-d_0+n_0(g-1)).\end{aligned}$$

The remainder of this section will be devoted to establishing the relationship between the twisted Brill-Noether theory and the moduli spaces of unstable bundles.

REMARK 4.4. Note that $\mathcal{Y}_0$ is the locus where $Hom(F_1,E_1)=0$. Consequently, the indecomposable bundles $E\in H^1(X,E_1\otimes F_1^*)$ are simple (see Corollary 3.2). Assuming that $M_1\times M_2\neq B^1(\mathcal{U}_1,\mathcal{U}_2^*)$, the set

$$\mathcal{Y}_0=M_1\times M_2-B^1(\mathcal{U}_1,\mathcal{U}_2^*)$$

is open. Therefore, $\mathbb{P}(\mathcal{R}_{\mu_1}(\mathcal{Y}_0))$ is irreducible, and it has dimension

$$\dim\mathbb{P}(\mathcal{R}_{\mu_1}(\mathcal{Y}_0))=\dim M_1+\dim M_2+h^1-1.$$

REMARK 4.5.

(1) If $p_2:B^k(\mathcal{U}_1,\mathcal{U}_2^*)\subset M_1\times M_2\to M_2$ is the projection, then for $F_1\in p_2(B^k(\mathcal{U}_1,\mathcal{U}_2^*))$, the inverse image $p_2^{-1}(F_1)$ is given by the locus

$$B^k(\mathcal{U}_1,F_1^*):=\{E_1\in M_1:h^0(E_1\otimes F_1^*)\geq k\}.$$

(2) In particular, if $M_2=Pic^0(X)$ and $F_1=\mathcal{O}_X$ then $B^k(\mathcal{U}_1,\mathcal{O}_X)$ is the Brill-Noether locus $B(n_1,d_1,k)$. In this case, $\mathcal{Y}_k=Y_k$, where

$$Y_k=B(n_1,d_1,k)-B(n_1,d_1,k+1).$$

(3) In general, the product of two stable bundles is semistable. However, S. Ramanan proved that if $(n_0,d_0)=1$ or if one of the bundles is general, then $E\otimes F^*$ is stable (see[**CT**, Theorem 2.10.]). Thus, in this case, if $B^k(\mathcal{U}_1,\mathcal{U}_2^*)\neq\emptyset$ then $B(n_0,d_0,k)\neq\emptyset$.

(4) If $M_1 \times M_2 \neq B^k(\mathcal{U}_1,\mathcal{U}_2^*)$ and $B^k(\mathcal{U}_1,\mathcal{U}_2^*) \neq \emptyset$ then $\mathcal{Y}_k \neq \emptyset$.

Using the established relationship between the twisted Brill-Noether theory and the moduli spaces of unstable bundles, we can now state one of our main results. In doing so, we use the notation of Remark 4.3 and 4.1.

THEOREM 4.6. *$U_{\mu_1}(n,d,k) = \mathbb{P}(\mathcal{R}_{\mu_1}(\mathcal{Y}_k))$ is a coarse moduli space and $U_{\mu_1}(n,d,0)$ is a fine moduli space. Moreover, if $\mathcal{Y}_k$ is irreducible and smooth of dimension ρ, then $U_{\mu_1}(n,d,k)$ is also irreducible and smooth of dimension $\rho + h^1 - 1$.*

PROOF. Based on the results established so far and Corollary 3.2 it follows that

$$U_{\mu_1}(n,d,k) = \mathbb{P}(\mathcal{R}_{\mu_1}(\mathcal{Y}_k)).$$

If $E \in U_{\mu_1}(n,d,0)$, then E is simple, and from Corollary 3.2, $h^0(E_1 \otimes F_1^*) = 0$. Thus, $(\mathcal{R}^0_{p_{23}}(p_{12}^*\mathcal{U}_1 \otimes p_{13}^*\mathcal{U}_2^*))|_{\mathcal{Y}_0} = 0$. By [**L**, Corollary 4.5], we know that $\mathbb{P}(\mathcal{R}_{\mu_1}(\mathcal{Y}_0))$ parameterize a universal extension

$$0 \to q^*(p_{12}^*(\mathcal{U}_1)) \otimes p_2^*\mathbb{H} \to \mathcal{E} \to q^*p_{13}^*(\mathcal{U}_2) \to 0,$$

where $q : X \times \mathbb{P}(\mathcal{R}_{\mu_1}(\mathcal{Y}_0)) \to X \times \mathcal{Y}_0$ is the induced map, $p_2 : X \times \mathbb{P}(\mathcal{R}_{\mu_1}(\mathcal{Y}_0)) \to \mathbb{P}(\mathcal{R}_{\mu_1}(\mathcal{Y}_0))$ is the projection and $\mathbb{H}$ the hyperplane bundle over $\mathbb{P}(\mathcal{R}_{\mu_1}(\mathcal{Y}_0))$. The universal properties of the family $\mathcal{E}$ imply that the pair $(\mathbb{P}(\mathcal{R}_{\mu_1}(\mathcal{Y}_0)), \mathcal{E})$ is the fine moduli space for $U_{\mu_1}(n,d,0)$.

If $\mathcal{Y}_k$ is irreducible and smooth of dimension ρ, a straightforward computation shows that $U_{\mu_1}(n,d,k)$ is irreducible and smooth of dimension $\rho + h^1 - 1$, since $\mathbb{P}(\mathcal{R}_{\mu_1}(\mathcal{Y}_k))$ is a projective bundle with fibre $\mathbb{P}(H^1(X, E_1 \otimes F_1^*))$ at $(E_1, F_1) \in \mathcal{Y}_k$. □

COROLLARY 4.7. *If $U_{\mu_1}(n,d,k)$ is non-empty, then $B^k(\mathcal{U}_1,\mathcal{U}_2^*)$ is also non-empty.*

COROLLARY 4.8. *If $\mathcal{Y}_k$ is irreducible and smooth, then $H^i(U_{\mu_1}(n,d,k),\mathbb{C}) \cong H^i(\mathcal{Y}_k,\mathbb{C})$ for $i \geq 0$.*

REMARK 4.9. The moduli space of simple bundles of type

$$\sigma = (\mu(E_1), \ldots, \mu(E_m/E_{m-1})),$$

for $m > 2$, is considered in [**R1**].

REMARK 4.10. Theorem 4.6 expresses the equivalence between the existence and topology of $U_{\mu_1}(n,d,k)$ and that of twisted Brill-Noether loci. For certain values of (g,n,d,k), the non-emptiness, dimension, and irreducibility of $B^k(\mathcal{U}_1,\mathcal{U}_2^*)$, and of $\mathcal{Y}_k$, are known for general curve (see [**HHN**] and [**BN**]). Thus, similar to the Brill-Noether and twisted Brill-Noether theory for vector bundles, it is possible that for special curves, the moduli space $U_{\mu_1}(n,d,k)$ is even reduced. Thus, properties such as non-emptiness, irreducibility, and dimension will depend on the curve

5. Non-emptiness of $U_{\mu_1}(n,d,k)$

In this section, we will prove the non-emptiness of $U_{\mu_1}(n,d,k)$ for certain values of (g,n,d,k). Based on the results obtained so far, the non-emptiness of $U_{\mu_1}(n,d,k)$ is closely related to the non-emptiness of certain twisted Brill-Noether loci. We will establish the non-emptiness of specific twisted Brill-Noether loci that are of interest to us, and as far as we know, were not previously known.

For the non-emptiness of $U_{\mu_1}(n,d,k)$ first we consider the case of bundles of type $\sigma = (\mu_1, \mu_2)$ where $\mu_1 = \frac{d-a}{n-1}$ and $\mu_2 = a$ is an integer. In this case, the quotient of the maximal destabilizing subbundle is a line bundle of degree a. The non-emptiness of $U_{\mu_1}(n,d,k)$ in this case depends on the non-emptiness of $Y_k \subset B(n-1, d-an, k)$. The following theorems utilize Theorem 4.6 and the known results of the Brill-Noether loci $B(n-1, d-an, k)$ to establish the non-emptiness of $U_{\mu_1}(n,d,k)$.

THEOREM 5.1. *Let n, d, a and k be positive integers. Assume $0 < d - an < 2(n-1)$ and $(n-1, d-an, k) \neq (n-1, n-1, n-1)$. Then for $\mu_1 = \frac{d-a}{n-1}$, $U_{\mu_1}(n,d,k)$ is non-empty if and only if $k \leq n - 1 + \frac{d-n(a+1)+1}{g}$. Moreover, if $U_{\mu_1}(n,d,k)$ is non-empty, then it is irreducible and smooth of the expected dimension.*

PROOF. The theorem follows from Theorem 4.6, as it has been shown in [**BGN**, Theorem A] and [**Me1**, A-1 Théorème], $Y_k \subset B(n-1, d-an, k)$ has the required properties. □

REMARK 5.2. Note that under the assumptions of Theorem 5.1, the non-emptiness of $U_{\mu_1}(n,d,k)$ does not depend of the positivity of the expected dimension $\rho(n-1, d-an, 1, a, k)$.

For $g = 2$ we have a complete description.

COROLLARY 5.3. *Assume $g = 2$. If $0 < d - an < 2(n-1)$ and $0 \leq k \leq \frac{d-an}{2} + \frac{n-1}{2}$ then $U_{\mu_1}(n,d,k)$ is irreducible and smooth of the expected dimension. If $d - an \geq 2(n-1)$, $U_{\mu_1}(n,d,k) = \emptyset$.*

We now rephrase Theorem 2.8,(5),(6) and (7) as follows.

THEOREM 5.4. *Let $(g, n-1, d-na, k)$ be integers that satisfy the conditions given in Theorem 2.8,(5),(6) and (7). For general curve, $U_{\mu_1}(n,d,k)$ is non-empty and contains an irreducible component of the expected dimension. Moreover, if X is a Petri curve of genus $g \geq 3$ and $n \leq 4$, or if $n \geq 5$ and $g \geq 2n-4$, then $U_{\mu_1}(n,d,n)$ is also non-empty.*

According to our current knowledge, it is known that in the aforementioned case, there exists an irreducible component of the expected dimension. However, it is generally not known whether the set $B(n,d,k)$ is irreducible.

Now let's examine the general case using the notation provided in Remark 4.3. Recall that $B^k(\mathcal{U}_1, \mathcal{U}_2^*) \subset M_1 \times M_2$. Let M_{12} denote the moduli space of stable bundles $M(n_0, d_0)$. Based on the results obtained so far, we can deduce that if $B(n_0, d_0, k) = \emptyset$, then $U_{\mu_1}(n,d,k)$ is also empty. However, it does not imply that if $B(n_0, d_0, k) \neq \emptyset$, then $U_{\mu_1}(n,d,k) \neq \emptyset$. We require a $G_0 \in B(n_0, d_0, k)$ in the form $G_0 = F_1^* \otimes E_1$. Let $\Phi : M_1 \times M_2 \to M_{12}$ be the morphism defined as $(E, F) \mapsto E \otimes F^*$. This implies that

$$\Phi(B^k(\mathcal{U}_1, \mathcal{U}_2^*)) \subset B(n_0, d_0, k).$$

We aim to describe $\Phi(B^k(\mathcal{U}_1, \mathcal{U}_2^*))$ in certain cases.

REMARK 5.5. Consider a vector bundle G of rank m and degree d_2, generated by linear subspace $V \subset H^0(X, E)$ of dimension $k_2 = m + n_2$. Let $D_{E,V}^*$ be the kernel of the evaluation map (or the syzygy bundle). For simplicity, we introduce the temporary notation F_1^* for $D_{E,V}^*$. Thus, F_1^* has rank $rk(F_1) = n_2$, degree $d(F_1^*) = -d_2$ and fits into the following exact sequence:

$$0 \to F_1^* \to V \otimes \mathcal{O} \to G \to 0. \tag{5.1}$$

Now, tensoring (5.1) with a vector bundle E_1 of rank n_1 and degree d_1 with $h^0(E_1) \geq k_1$, we can analyze the injectivity of the multiplication map:

$$\mu_{V,E_1} : V \otimes H^0(E_1) \to H^0(G \otimes E_1)$$

by examining $h^0(F_1^* \otimes E_1)$. Indeed, if $H^0(F_1^* \otimes E_1) = 0$, then the multiplication map μ_{V,E_1} is injective. Using the cohomology sequence:

$$0 \to H^0(F_1^* \otimes E_1) \to V \otimes H^0(E_1) \to H^0(G \otimes E_1) \to$$

we obtain the inequality

$$\dim V \cdot h^0(E_1) - h^0(G \otimes E_1) \leq h^0(F_1^* \otimes E_1). \tag{5.2}$$

Therefore, for any positive number k satisfying:

$$k \leq k_1 k_2 - h^0(G \otimes E_1), \tag{5.3}$$

we have $k \leq h^0(F_1^* \otimes E_1) = \dim Hom(F_1, E_1)$.

With the given notation, we have $n_0 := rk(F_1^* \otimes E_1) = n_1 n_2$ and $d_0 := d(F_1^* \otimes E_1) = d_1 n_2 - d_2 n_1$. Additionally, let $\tilde{n} := rk(G \otimes E_1) = mn_1$ and $\tilde{d} := d(G \otimes E_1) = d_2 n_1 + d_1 m$. Our objective is to apply Remark 5.5 to situations where E_1 is a general element in $M(n_1, d_1) = M_1$, $F_1 \in M(n_2, d_2) = M_2$ and $k \geq 0$. Specifically, we are interested in cases where the multiplication map μ_{V,E_1} is not injective. In such instances, $(E_1, F_1^*) \in B^k(\mathcal{U}_1, \mathcal{U}_2^*)$ and $E_1 \otimes F_1^* \in B(n_0, d_0, k)$.

The following theorems establish the existence of certain Brill-Noether and twisted Brill-Noether loci, thus ensuring the non-emptiness of $U_{\mu_1}(n, d, k)$. The proofs differ from those presented in [**HHN**] and some values, to the best of our knowledge, have not been included in[**HHN**] or in Theorem 2.8 (see [**BN**]).

THEOREM 5.6. *Assume that $B(n_1, d_1, n_1 + a)$ is non-empty with $a > 0$. If $2n_1 < d_1 < a(g+1)$ and $d_2 > 2gm$, then for any $0 \leq k \leq (d_2 + m(1-g))(n_1 + a) - (d_2 n_1 + d_1 m + mn_1(1-g))$, the set $\mathcal{Y}_k \subset B^k(\mathcal{U}_1, \mathcal{U}_2^*)$ is non-empty. Moreover, if $\mu_1 = \frac{d_1}{n_1}$ then $U_{\mu_1}(n, d, k)$ is non-empty, where $n = n_1 + n_2$ and $d = d_1 + d_2$.*

PROOF. Let $G \in M(m, d_2)$ and $E_1 \in B(n_1, d_1, n_1 + a)$. Our goal is to prove that $0 \leq k_1 k_2 - h^0(G \otimes E_1)$, where $k_1 = n_1 + a$ and $k_2 = d_2 + m(1-g)$.

If $d_2 > 2gm$, G is generated and from [**Bu**, Theorem 1.2] we have the exact sequence

$$0 \to F_1^* \to H^0(G) \otimes \mathcal{O} \to G \to 0 \tag{5.4}$$

with $F_1 \in M(n_2, d_2)$ where $n_2 := h^0(G) - m = d_2 - mg$. From Remark (5.5) and (5.3)

$$h^0(G) \cdot (n_1 + a) \leq h^0(G) \cdot h^0(E_1) - h^0(G \otimes E_1) \leq h^0(E_1 \otimes F_1^*). \tag{5.5}$$

If $d_1 < a(g+1)$ and $d_2 > 2gm$ then $\frac{d_1}{a} + g - 1 < 2g < \frac{d_2}{m}$. Hence,

$$d_1 m + am(g-1) < ad_2. \tag{5.6}$$

Now, adding $d_2 n_1 + mn_1(g-1)$ to both sides of (5.6), we obtain:

$$d_2 n_1 + d_1 m + (n_1 + a)m(g-1) < d_2(n_1 + a) + n_1 m(g-1).$$

Hence,

(5.7) $0 < (n_1+a)(d_2+m(1-g))-(d_2n_1+d_1m+mn_1(1-g)) = k_1k_2-h^0(G\otimes E_1)$.

This shows that $0 \leq k \leq h^0(E_1\otimes F_1^*)$ for any $0 \leq k \leq (d_2+m(1-g))(n_1+a)-(d_2n_1+d_1m+mn_1(1-g))$. Therefore, the set $B^k(\mathcal{U}_1,\mathcal{U}_2^*)$ is non-empty, and furthermore, $U_{\mu_1}(n,d,k)$ is non-empty if $\mu_1 = \frac{d_1}{n_1}$ where $n = n_1+n_2$ and $d = d_1+d_2$. □

THEOREM 5.7. *Let X be a Petri curve of genus $g \geq 3$ and $(\mathcal{O}(D),V)$ be a general generated linear system of degree $d_2 \geq g+1$ and $\dim V = n_2+1$ with $n_2 \leq 4$ or if $n_2 \geq 5$ then $g \geq 2n_2-4$. Assume that $B(n_1,d_1,t)$ is non-empty and $\frac{d_2}{n_2} < \frac{d_1}{n_1}$. For any $0 \leq k \leq n_2t-n_1d_2$, $\mathcal{Y}_k \subset B^k(\mathcal{U}_1,\mathcal{U}_2^*)$ is non-empty. Moreover, if $n = n_2+n_1$, $d = d_2+d_1$ and $\mu_1 = \frac{d_1}{n_1}$, then $U_{\mu_1}(n,d,k)$, is non-empty.*

PROOF. Under the given conditions, there exists an exact sequence:

$$0 \to F_1^* \to V\otimes\mathcal{O} \to \mathcal{O}(D) \to 0 \tag{5.8}$$

where $F_1 \in M(n_2,d_2)$ (see [**Br**], [**BBN**] and [**BBN1**]).

Let $E_1 \in B(n_1,d_1,n_1+a)$ be such that $h^0(E_1) = t \geq n_1+a$. From Remark (5.5) and (5.3)

$$(n_2+1)\cdot t - h^0(\mathcal{O}(D)\otimes E_1) \leq h^0(F_1^*\otimes E_1). \tag{5.9}$$

By assuming $\frac{d_2}{n_2} < \frac{d_1}{n_1}$ and using the exact sequence:

$$0 \to E_1 \to \mathcal{O}(D)\otimes E_1 \to (E_1)_D \to 0 \tag{5.10}$$

we deduce that:

$$tn_2 - d_2n_1 < (n_2+1)\cdot t - h^0(\mathcal{O}(D)\otimes E_1) \leq h^0(F_1^*\otimes E_1). \tag{5.11}$$

Therefore, for any $0 \leq k \leq tn_2-d_2n_1$, $\mathcal{Y}_k \subset B^k(\mathcal{U}_1,\mathcal{U}_2^*)$ is non-empty. Moreover, if $n = n_2+n_1$, $d = d_2+d_1$ and $\mu_1 = \frac{d_1}{n_1}$, then $U_{\mu_1}(n,d,k)$, is non-empty, as claimed. □

REMARK 5.8. Under the assumptions of Theorems 5.6 and 5.7 it is possible to choose the values of d_2, n_2, n_1 and d_1 such that $(n_0,d_0) = 1$, implying that $B(n_0,d_0,k)$ is non-empty.

To conclude, we provide a description of a smooth point of $U_{\mu_1}(n,d,k)$, which is a projective bundle over $\mathcal{Y}_k$. Let $0 \subset E_1 \subset E$ be the HN-filtration of $E \in U_{\mu_1}(n,d,k)$, where we write $E/E_1 = F_1$. Consider a point (E_1,F_1^*) in $\mathcal{B}_{12}$, where $\mathcal{B}_{12}$ is an abbreviation for $B^k(\mathcal{U}_1,\mathcal{U}_2^*)$. We now describe the tangent bundle of $\mathcal{B}_{12}$ at a point (E_1,F_1^*). Since $\mathcal{B}_{12} \subset M_1\times M_2$, for any $z := (E_1,F_1) \in \mathcal{B}_{12}$ we have the following exact sequence:

$$0 \to T_z\mathcal{B}_{12} \to T_{E_1}M_1 \oplus T_{F_1}M_2 \to N \to 0. \tag{5.12}$$

In this sequence, N represents the normal bundle of $\mathcal{B}_{12}$. Furthermore, from the morphism $\Phi : M_1\times M_2 \to M_{12}$, we obtain:

$$T_{E_1}M_1 \oplus T_{F_1}M_2 \overset{d\Phi}{\to} T_zM_{1,2}. \tag{5.13}$$

In other words, there is a map from the tangent space of $E_1\oplus F_1$ to the tangent space of M_{12} at the point z. This map induces a map on cohomology:

$$H^1(X,End(E_1)) \oplus H^1(X,End(F_1)) \overset{d\Phi}{\to} H^1(X,End(E_1\otimes F_1^*)).$$

This sequence provides information about the tangent bundle of $\mathcal{B}_{12}$ at the point (E_1, F_1^*), giving us a better understanding of the geometry and smoothness properties of $U_{\mu_1}(n,d,k)$. The multiplication map is given by

$$H^0(X, E_1 \otimes F_1^*) \otimes H^0(X, (E_1 \otimes F_1^*)^* \otimes K) \to H^0(X, End(E_1 \otimes F_1^*) \otimes K)$$

Considering the dual sequence, we obtain the following sequence:

$$H^1(X, End(E_1 \otimes F_1^*)) \xrightarrow{\beta} H^0(X, E_1 \otimes F_1^*)^* \otimes H^1(X, E_1 \otimes F_1^*),$$

which gives us the morphism β. Thus,

$$\begin{array}{ccc} H^1(End(E_1)) \oplus H^1(End(F_1)) & \xrightarrow{d\Phi} & H^1(End(E_1 \otimes F_1^*)) \\ & \searrow^{\eta_E} & \downarrow \beta \\ & & H^0(E_1 \otimes F_1^*)^* \otimes H^1(E_1 \otimes F_1^*) \end{array}$$

where $\eta = \beta \circ d\Phi$ and the image of η is precisely the normal bundle. By applying [**HHN**, Proposition 3.5] and Theorem 4.6 we conclude the following result:

THEOREM 5.9. *$U_{\mu_1}(n,d,k)$ is smooth at the point E and has the expected dimension $\beta(g, n_1, d_1, n_2, d_2, k)$ if and only if η is surjective.*

Acknowledgments. The authors express their gratitude to the referee for providing valuable comments. They also acknowledge the helpful suggestions from Peter E. Newstead during the preparation of the paper. Special thanks are extended to Alfonso Zamora Saiz for bringing attention to certain unclear points.

References

[A] M. F. Atiyah, *Complex analytic connections in fibre bundles*, Trans. Amer. Math. Soc. **85** (1957), 181–207, DOI 10.2307/1992969. MR86359

[ACGH] E. Arbarello, M. Cornalba, P. A. Griffiths, and J. Harris, *Geometry of algebraic curves. Vol. I*, Grundlehren der mathematischen Wissenschaften [Fundamental Principles of Mathematical Sciences], vol. 267, Springer-Verlag, New York, 1985, DOI 10.1007/978-1-4757-5323-3. MR770932

[Ba] C. Bănică, *Sur les fibrés instables de rang* 2 *sur* $\mathbf{P}^3(\mathbf{C})$ (French), Arch. Math. (Basel) **43** (1984), no. 3, 250–257, DOI 10.1007/BF01247571. MR766432

[BPS] C. Bănică, M. Putinar, and G. Schumacher, *Variation der globalen Ext in Deformationen kompakter komplexer Räume* (German), Math. Ann. **250** (1980), no. 2, 135–155, DOI 10.1007/BF01364455. MR582514

[BHJK] G. Bérczi, V. Hoskins, J. Jackson and F. Kirwan, *Moduli of unstable sheaves*, In preparation.

[BN] L. Brambila-Paz and P. E. Newstead, *New examples of twisted Brill-Noether loci*, arXiv:2212.05573.

[Br] L. Brambila-Paz, *Non-emptiness of moduli spaces of coherent systems*, Internat. J. Math. **19** (2008), no. 7, 779–799, DOI 10.1142/S0129167X0800487X. MR2437070

[Br1] L. Brambila Paz, *Nonsimple semistable vector bundles over a curve*, Math. Z. **212** (1993), no. 3, 301–311, DOI 10.1007/BF02571660. MR1207295

[Br2] L. Brambila Paz, *Moduli of endomorphisms of semistable vector bundles over a compact Riemann surface*, Glasgow Math. J. **32** (1990), no. 1, 1–12, DOI 10.1017/S0017089500009010. MR1045081

[Br3] L. Brambila, *Algebras of endomorphisms of semistable vector bundles of rank* 3 *over a Riemann surface*, J. Algebra **123** (1989), no. 2, 414–425, DOI 10.1016/0021-8693(89)90054-9. MR1000495

[BGN] L. Brambila-Paz, I. Grzegorczyk, and P. E. Newstead, *Geography of Brill-Noether loci for small slopes*, J. Algebraic Geom. **6** (1997), no. 4, 645–669. MR1487229

[BM] L. Brambila-Paz and O. Mata-Gutiérrez, *Un teorema de Torelli para haces vectoriales inestables*, Aportaciones Matemáticas-Memorias de la Sociedad Matemática Mexicana 46 (2013), 03–16.

[BMNO] L. Brambila-Paz, V. Mercat, P. E. Newstead, and F. Ongay, *Nonemptiness of Brill-Noether loci*, Internat. J. Math. **11** (2000), no. 6, 737–760, DOI 10.1142/S0129167X00000350. MR1785516

[BBN] U. N. Bhosle, L. Brambila-Paz, and P. E. Newstead, *On coherent systems of type $(n, d, n + 1)$ on Petri curves*, Manuscripta Math. **126** (2008), no. 4, 409–441, DOI 10.1007/s00229-008-0190-y. MR2425434

[BBN1] U. N. Bhosle, L. Brambila-Paz, and P. E. Newstead, *On linear series and a conjecture of D. C. Butler*, Internat. J. Math. **26** (2015), no. 2, 1550007, 18, DOI 10.1142/S0129167X1550007X. MR3319666

[Bu] D. C. Butler, *Normal generation of vector bundles over a curve*, J. Differential Geom. **39** (1994), no. 1, 1–34. MR1258911

[CT] S. Casalaina-Martin and M. Teixidor i Bigas, *Singularities of Brill-Noether loci for vector bundles on a curve*, Math. Nachr. **284** (2011), no. 14-15, 1846–1871, DOI 10.1002/mana.200910093. MR2838286

[D] J.-M. Drézet, *Déformations des extensions larges de faisceaux* (French, with English summary), Pacific J. Math. **220** (2005), no. 2, 201–297, DOI 10.2140/pjm.2005.220.201. MR2200373

[Gi] D. Gieseker, *On the moduli of vector bundles on an algebraic surface*, Ann. of Math. (2) **106** (1977), no. 1, 45–60, DOI 10.2307/1971157. MR466475

[Gr] A. Grothendieck and J. A. Dieudonné, *Éléments de géométrie algébrique. I* (French), Grundlehren der mathematischen Wissenschaften [Fundamental Principles of Mathematical Sciences], vol. 166, Springer-Verlag, Berlin, 1971. MR3075000

[GT] I. Grzegorczyk and M. Teixidor i Bigas, *Brill-Noether theory for stable vector bundles*, Moduli spaces and vector bundles, London Math. Soc. Lecture Note Ser., vol. 359, Cambridge Univ. Press, Cambridge, 2009, pp. 29–50. MR2537065

[HN] G. Harder and M. S. Narasimhan, *On the cohomology groups of moduli spaces of vector bundles on curves*, Math. Ann. **212** (1974/75), 215–248, DOI 10.1007/BF01357141. MR364254

[HHN] G. H. Hitching, M. Hoff, and P. E. Newstead, *Nonemptiness and smoothness of twisted Brill-Noether loci*, Ann. Mat. Pura Appl. (4) **200** (2021), no. 2, 685–709, DOI 10.1007/s10231-020-01009-x. MR4229545

[HJ] V. Hoskins and J. Jackson, *Quotients by parabolic groups and moduli spaces of unstables objects*, `arXiv:2111.07429v1` [Math.AG], 2021.

[HK] V. Hoskins and F. Kirwan, *Quotients of unstable subvarieties and moduli spaces of sheaves of fixed Harder-Narasimhan type*, Proc. Lond. Math. Soc. (3) **105** (2012), no. 4, 852–890, DOI 10.1112/plms/pds022. MR2989806

[HL] D. Huybrechts and M. Lehn, *The geometry of moduli spaces of sheaves*, 2nd ed., Cambridge Mathematical Library, Cambridge University Press, Cambridge, 2010, DOI 10.1017/CBO9780511711985. MR2665168

[J] J. Jackson, *Moduli spaces of unstables objects: Sheaves of Harder-Narasimhan lenght 2*, `arXiv:2111.07428v1` [Math.AG], 2021.

[K] G. R. Kempf, *Instability in invariant theory*, Ann. of Math. (2) **108** (1978), no. 2, 299–316, DOI 10.2307/1971168. MR506989

[KO] S. Kosarew and C. Okonek, *Global moduli spaces and simple holomorphic bundles*, Publ. Res. Inst. Math. Sci. **25** (1989), no. 1, 1–19, DOI 10.2977/prims/1195173759. MR999347

[L] H. Lange, *Universal families of extensions*, J. Algebra **83** (1983), no. 1, 101–112, DOI 10.1016/0021-8693(83)90139-4. MR710589

[Ma] M. Maruyama, *Moduli of stable sheaves. I*, J. Math. Kyoto Univ. **17** (1977), no. 1, 91–126, DOI 10.1215/kjm/1250522815. MR450271

[Ma1] M. Maruyama, *Moduli of stable sheaves. II*, J. Math. Kyoto Univ. **18** (1978), no. 3, 557–614, DOI 10.1215/kjm/1250522511. MR509499

[Me] V. Mercat, *Le problème de Brill-Noether et le théorème de Teixidor* (French, with English summary), Manuscripta Math. **98** (1999), no. 1, 75–85, DOI 10.1007/s002290050126. MR1669595

[Me1] V. Mercat, *Le problème de Brill-Noether pour des fibrés stables de petite pente* (French), J. Reine Angew. Math. **506** (1999), 1–41, DOI 10.1515/crll.1999.005. MR1665673

[Mu] D. Mumford, *Projective invariants of projective structures and applications*, Proc. Internat. Congr. Mathematicians (Stockholm, 1962), Inst. Mittag-Leffler, Djursholm, 1963, pp. 526–530. MR175899

[NR] M. S. Narasimhan and S. Ramanan, *Deformations of the moduli space of vector bundles over an algebraic curve*, Ann. of Math. (2) **101** (1975), 391–417, DOI 10.2307/1970933. MR384797

[NS] M. S. Narasimhan and C. S. Seshadri, *Stable and unitary vector bundles on a compact Riemann surface*, Ann. of Math. (2) **82** (1965), 540–567, DOI 10.2307/1970710. MR184252

[R] R. Ríos-Sierra. *Moduli of unstable bundles of rank* 3, In Preparation

[R1] R. Ríos-Sierra. *On the moduli space of simple unstable bundles*, In preparation.

[Si] C. T. Simpson, *Moduli of representations of the fundamental group of a smooth projective variety. I*, Inst. Hautes Études Sci. Publ. Math. **79** (1994), 47–129. MR1307297

[St] S. A. Strømme, *Deforming vector bundles on the projective plane*, Math. Ann. **263** (1983), no. 3, 385–397, DOI 10.1007/BF01457140. MR704303

[T] M. Teixidor i Bigas, *Brill-Noether theory for stable vector bundles*, Duke Math. J. **62** (1991), no. 2, 385–400, DOI 10.1215/S0012-7094-91-06215-0. MR1104529

CIMAT, Apdo. Postal 402,C.P. 36240, Guanajuato, Mexico
Email address: lebp@cimat.mx

CIMAT, Apdo. Postal 402,C.P. 36240, Guanajuato, Mexico
Email address: rocio.rios@cimat.mx

Contemporary Mathematics
Volume **803**, 2024
https://doi.org/10.1090/conm/803/16096

The Brill-Noether Theory of the moduli spaces of sheaves on surfaces

Izzet Coskun, Jack Huizenga, and Howard Nuer

In honor of Peter Newstead's 80th birthday, with great admiration

ABSTRACT. In this paper, we survey recent developments in the Brill-Noether Theory of higher rank vector bundles on complex projective surfaces. We focus on weak Brill-Noether Theorems on rational and K-trivial surfaces and their applications.

1. Introduction

In this paper, we survey recent developments in the Brill-Noether Theory of higher rank vector bundles on complex projective surfaces. Throughout the paper we work over the field of complex numbers $\mathbb{C}$.

Brill-Noether Theory forms a cornerstone of classical curve theory. It describes representations of curves in $\mathbb{P}^n$ and is a main tool in the study of moduli spaces of curves. The cohomological behavior of a general line bundle $\mathcal{L}$ of degree d on a smooth curve C of genus g is determined by d and g (see Proposition 3.1). More generally, the cohomological behavior of a general stable vector bundle $\mathcal{V}$ of rank r and degree d on C is determined by r, d and g (see Proposition 3.8). Consequently, Brill-Noether Theory focuses on the geometry of loci of vector bundles with unexpected cohomological behavior.

In contrast, the cohomological behavior of a general stable bundle on a surface is not well-understood except for special surfaces or for special Chern characters. Crucially, the cohomology of a general stable bundle on a surface is not in general determined by the Euler characteristic and the slope. Consequently, before one can study the behavior of special vector bundles, one has to understand the behavior of the generic stable bundle. After some preliminaries §2 and reviewing the properties of the general stable bundles on a curve §3, in §4 we will contrast the behavior of bundles on curves and surfaces and give examples of new phenomena that occur.

2020 *Mathematics Subject Classification.* Primary 14L30, 14M15, 14M17; Secondary 14L35, 51N30.

Key words and phrases. Brill-Noether Theory, moduli spaces of sheaves, Bridgeland stability, Ulrich bundles.

During the preparation of this article the first author was partially supported by the NSF FRG grant DMS 1664296 and NSF grant DMS 2200684, the second author was partially supported by NSF FRG grant DMS 1664303.

For example, the stack of coherent sheaves of rank r and degree d on a curve is irreducible. On the other hand, the stack of coherent sheaves with fixed invariants on a surface is almost never irreducible. In many nice settings, focusing on stable sheaves results in an irreducible moduli space, although this is not always the case (see §4). Hence, stability plays a more central role in the case of surfaces. On a curve of genus $g \geq 2$, there exist stable bundles of every rank $r > 0$ and every degree. In contrast, there are restrictions on the possible Chern characters of stable sheaves on surfaces such as the Bogomolov inequality. Furthermore, we do not have a classification of the Chern characters of stable bundles on surfaces except in special cases. Finally, the cohomology of a general line bundle with fixed invariants on a surface may not be concentrated in a single degree. We often do not know the cohomology of certain line bundles even on relatively simple surfaces such as the blowup of $\mathbb{P}^2$ at 10 or more general points. This presents many challenges for studying the cohomology of higher rank bundles.

In §5, we will survey recent developments in computing the cohomology of the general stable sheaf on a surface. We will concentrate on two techniques. First, we will describe results on rational surfaces using the stack of prioritary sheaves. Second, we will describe results on K-trivial surfaces using Bridgeland stability conditions. We will survey cases when weak Brill-Noether Theorems hold, i.e., when the general stable sheaf has at most one nonzero cohomology group.

In §6, we will describe applications of weak Brill-Noether Theorems. We will concentrate on the following applications.

(1) The classification of Chern characters of stable bundles. After recalling the Drézet-Le Potier classification of stable bundles for $\mathbb{P}^2$, we will describe recent progress on Hirzebruch surfaces.
(2) The tensor product problem and the birational geometry of moduli spaces of sheaves. We will discuss the problem of computing the cohomology of the tensor product of two general stable sheaves and touch on applications to the construction of Brill-Noether divisors and the birational geometry of the moduli spaces.
(3) The classification of Ulrich bundles. We will discuss the relation between weak Brill-Noether Theorems and Ulrich bundles.

Finally, in §7, we will survey some recent work on the cohomology jumping loci on moduli spaces of sheaves on $\mathbb{P}^2$ following [**GLL22**]. Very little is known (or even conjectured) about Brill-Noether loci on moduli spaces of sheaves on surfaces. In general, the Brill-Noether loci are reducible with components of different dimensions even in rank 1. When $c_1(\mathbf{v})$ is a minimal positive class, however, then the Brill-Noether loci behave better as we will demonstrate.

2. Preliminaries

In this section, we collect basic facts and definitions concerning stable bundles. We refer the reader to [**CH15, HL10, LeP97**] for more details.

2.1. Stability.

We begin by recalling the definition of stability.

DEFINITION 2.1. Let X be a projective variety of dimension n equipped with an ample divisor H. The *H-slope* of a torsion free coherent sheaf $\mathcal{F}$ is defined by

$$\mu_H(\mathcal{F}) = \frac{c_1(\mathcal{F}) \cdot H^{n-1}}{\mathrm{rk}(\mathcal{F}) H^n}.$$

The sheaf $\mathcal{F}$ is called *μ_H-(semi)stable* if for every proper subsheaf $\mathcal{E} \subset \mathcal{F}$ we have

$$\mu_H(\mathcal{E}) \underset{(-)}{<} \mu_H(\mathcal{F}).$$

When H is fixed or irrelevant for the discussion, we will omit it from the notation.

For curves, the ample H does not play a role. Taking H to have degree 1, the slope becomes the degree over the rank of $\mathcal{F}$. For higher dimensional varieties, stability depends on H. Every torsion free sheaf $\mathcal{F}$ has a unique *Harder-Narasimhan filtration*

$$\mathcal{F}_1 \subset \cdots \subset \mathcal{F}_n = \mathcal{F}$$

such that $\mathcal{E}_i = \mathcal{F}_i/\mathcal{F}_{i-1}$ are μ_H-semistable with

$$\mu_H(\mathcal{E}_1) > \mu_H(\mathcal{E}_2) > \cdots > \mu_H(\mathcal{E}_n).$$

Furthermore, μ_H-semistable sheaves have a (not necessarily unique) Jordan-Hölder filtration into μ_H-stable sheaves. The reflexive hull of the associated graded sheaf is unique up to isomorphism [**HL10**, Corollary 1.6.10]. Two semistable sheaves are called *S-equivalent* if their associated graded sheaves are isomorphic. Consequently, stable vector bundles are the building blocks of all vector bundles.

If $\mathcal{V}$ and $\mathcal{W}$ are two μ_H-semistable sheaves and $\phi : \mathcal{V} \to \mathcal{W}$ is a nonzero homomorphism, then $\mu_H(\mathcal{W}) \geq \mu_H(\mathcal{V})$. Furthermore, if both are stable and equality holds, then $\mathcal{V} \cong \mathcal{W}$.

When X is a curve, Mumford constructed an irreducible, projective coarse moduli space $M_X(r,d)$ parameterizing S-equivalence classes of semistable sheaves of rank r and degree d. For higher dimensional varieties, a slightly different notion of stability is necessary.

DEFINITION 2.2. Let $\mathcal{F}$ be a pure dimensional coherent sheaf of dimension d. Then the Hilbert polynomial $P_{\mathcal{F}}(m)$ of $\mathcal{F}$ has the form

$$P_{\mathcal{F}}(m) = \chi(\mathcal{F}(mH)) = a_d \frac{m^d}{d!} + \text{l.o.t.}$$

The *reduced Hilbert polynomial* is defined by $p_{\mathcal{F}}(m) = \frac{P_{\mathcal{F}}(m)}{a_d}$. The sheaf $\mathcal{F}$ is *H-Gieseker (semi)stable* if for every proper subsheaf $\mathcal{E} \subset \mathcal{F}$, we have $p_{\mathcal{E}}(m) \underset{(-)}{<} p_{\mathcal{F}}(m)$ for $m \gg 0$.

For torsion free sheaves, $a_{\dim(X)} = \text{rk}(\mathcal{F})$. Consequently, by the Riemann-Roch Theorem, we conclude that μ_H-stability implies H-Gieseker stability and H-Gieseker semistability implies μ_H-semistability. When rH^n and $c_1(\mathcal{F}) \cdot H^{n-1}$ are relatively prime, then these four notions coincide. In general, all the implications are strict. Every torsion free sheaf has a Harder-Narasimhan filtration with respect to H-Gieseker semistability, and every H-Gieseker semistable sheaf has a Jordan-Hölder filtration with H-Gieseker stable quotients. Two H-Gieseker semistable sheaves are called *S-equivalent* if their associated graded objects are isomorphic.

Given a Chern character $\mathbf{v}$, Gieseker [**Gie77**] and Maruyama [**Mar78**] constructed a projective coarse moduli space $M_{X,H}(\mathbf{v})$ parameterizing S-equivalence classes of H-Gieseker semistable sheaves on X. Compared to the case of curves, the geometry of these moduli spaces are less well understood. The purpose of these notes is to explain aspects of the geometry of $M_{X,H}(\mathbf{v})$ when X is a surface and compare and contrast it with the case of curves.

PROPOSITION 2.3. *Let (X, H) be a polarized surface. If $\mathcal{V}$ and $\mathcal{W}$ are μ_H-semistable bundles such that $\mu_H(\mathcal{W}) > \mu_H(\mathcal{V}) + H \cdot K_X$, then* $\operatorname{Ext}^2(\mathcal{V}, \mathcal{W}) = 0$. *In particular, if $H \cdot K_X < 0$, then* $\operatorname{Ext}^2(\mathcal{V}, \mathcal{V}) = 0$.

PROOF. Let $\mathcal{V}$ and $\mathcal{W}$ be μ_H-semistable vector bundles. Then by Serre duality $\operatorname{ext}^2(\mathcal{V}, \mathcal{W}) = \hom(\mathcal{W}, \mathcal{V}(K_X))$. This is zero if $\mu_H(\mathcal{W}) > \mu_H(\mathcal{V}) + H \cdot K_X$ by stability. □

2.2. The Riemann-Roch Theorem. Let X be a smooth, projective curve of genus g and let $\mathcal{V}$ be a vector bundle of rank r and degree d. Then the Riemann-Roch Theorem computes the Euler characteristic $\chi(\mathcal{V})$

$$\chi(\mathcal{V}) = h^0(X, \mathcal{V}) - h^1(X, \mathcal{V}) = d + r(1 - g).$$

If X is a smooth projective surface and $\mathcal{F}$ is a sheaf of rank r, $\operatorname{ch}_1(\mathcal{F}) = c$ and $\operatorname{ch}_2(\mathcal{F}) = d$, then the Euler characteristic is given by

$$\chi(\mathcal{F}) = r\chi(\mathcal{O}_X) - \frac{K_X \cdot c}{2} + d,$$

where K_X is the class of the canonical bundle ω_X of X.

In these notes, we will be interested in $\chi(\mathcal{V} \otimes \mathcal{W})$. It is convenient to make a change of coordinates to express the Chern character with the logarithmic invariants. Given the Chern character $\mathbf{v}$ of a torsion free sheaf $\mathcal{F}$ on a smooth polarized surface (X, H), define the total slope ν and discriminant Δ of $\mathcal{F}$ by the following formulae

$$\nu(\mathbf{v}) = \frac{c_1(\mathbf{v})}{\operatorname{rk}(\mathbf{v})}, \qquad \Delta(\mathbf{v}) = \frac{\nu(\mathbf{v})^2}{2} - \frac{\operatorname{ch}_2(\mathbf{v})}{\operatorname{rk}(\mathbf{v})}.$$

Observe that these notions depend only on the Chern character of $\mathcal{F}$. The Chern character can be easily recovered from r, ν and Δ. The advantage of these invariants is that they are additive on tensor products

$$\nu(\mathcal{V} \otimes \mathcal{W}) = \nu(\mathcal{V}) + \nu(\mathcal{W}) \quad \text{and} \quad \Delta(\mathcal{V} \otimes \mathcal{W}) = \Delta(\mathcal{V}) + \Delta(\mathcal{W}).$$

In terms of these invariants, the Riemann-Roch formula reads

$$\chi(\mathcal{V}) = r(\mathcal{V})(P(\nu(\mathcal{V})) - \Delta(\mathcal{V})),$$

where

$$P(\nu) = \chi(\mathcal{O}_X) + \frac{1}{2}(\nu^2 - \nu \cdot K_X)$$

is the Hilbert polynomial of $\mathcal{O}_X$. More generally, if $\mathcal{V}$ and $\mathcal{W}$ are two torsion free sheaves,

$$\chi(\mathcal{V}, \mathcal{W}) = \sum_{i=0}^{2} (-1)^i \operatorname{ext}^i(\mathcal{V}, \mathcal{W}) = r(\mathcal{V})r(\mathcal{W})(P(\nu(\mathcal{W}) - \nu(\mathcal{V})) - \Delta(\mathcal{V}) - \Delta(\mathcal{W})),$$

where $\operatorname{ext}^i(\mathcal{V}, \mathcal{W}) = \dim(\operatorname{Ext}^i(\mathcal{V}, \mathcal{W}))$.

3. Brill-Noether for curves

In this section, a curve will always mean a smooth, irreducible, complex projective curve. We will review basic properties of the cohomology of stable vector bundles on curves. In the next section, we will contrast the cases of curves and surfaces.

3.1. Line bundles on curves. Let C be a curve of genus g. Line bundles of degree d on C are parameterized by $\mathrm{Pic}^d(C)$ which is isomorphic to a g-dimensional abelian variety. The Euler characteristic $\chi(\mathcal{L}) = d-g+1$ determines the cohomology of the *general* $\mathcal{L} \in \mathrm{Pic}^d(C)$.

PROPOSITION 3.1 (Weak Brill-Noether for line bundles). *Let C be a curve of genus g. Let $\mathcal{L}$ be a general line bundle of degree d on C. Then*

$$h^0(C, \mathcal{L}) = \max(0, d-g+1) \quad \textit{and} \quad h^1(C, \mathcal{L}) = \max(0, g-d-1).$$

PROOF. If $d < 0$, then $h^0(C, \mathcal{L}) = 0$ and $h^1(C, \mathcal{L}) = g-d-1$. Let $C^{(d)}$ denote the d-th symmetric product of C. If $0 \leq d \leq g-1$, since

$$\dim(C^{(d)}) = d < \dim(\mathrm{Pic}^d(C)) = g,$$

the natural map

$$\phi_d : \mathrm{Sym}^d(C) \to \mathrm{Pic}^d(C) \quad \text{sending} \quad D = \sum_{i=1}^{d} p_i \mapsto \mathcal{O}_C(D)$$

cannot be surjective. The fiber of ϕ_d is the linear system $|\mathcal{O}_C(D)|$. Hence, the general line bundle of degree d has no global sections.

Finally, if $d \geq g$, $h^0(C, \mathcal{L}) = d-g+1+h^1(C, \mathcal{L}) \geq 1$. Letting ω_C denote the canonical bundle of C, by Serre duality, $h^1(C, \mathcal{L}) = h^0(C, \omega_C \otimes \mathcal{L}^{-1})$. By Lemma 3.2, general points impose independent conditions on sections of a line bundle. Hence, if $\mathcal{L} = \mathcal{O}_C(D)$ for a general effective divisor D, then $h^0(C, \omega_C(-D)) = 0$ and $h^0(C, \mathcal{L}) = d-g+1$. □

LEMMA 3.2. *Let X be an irreducible projective variety and let $\mathcal{L}$ be a line bundle on X. If Z is a general set of m distinct points on X, then*

$$h^0(X, \mathcal{L} \otimes I_Z) = \max(h^0(X, \mathcal{L}) - m, 0).$$

PROOF. If W is a zero-dimensional scheme of length m, then $h^0(X, \mathcal{L} \otimes I_W) \geq \max(h^0(X, \mathcal{L}) - m, 0)$. We need to show that for a general set of points equality holds. This is trivially true if $m = 0$. By induction, suppose it holds for $m \leq m_0 - 1$. Since the symmetric product $X^{(m)}$ is irreducible, by the semi-continuity of cohomology, it suffices to exhibit one Z of length m_0 for which the lemma holds. Choose Z' of length $m_0 - 1$ for which

$$h^0(X, \mathcal{L} \otimes I_{Z'}) = \max(h^0(X, \mathcal{L}) - m_0 + 1, 0).$$

If $\mathcal{L} \otimes I_{Z'}$ has no global sections, then $h^0(X, \mathcal{L} \otimes I_Z) = 0$ and we are done. Hence, we may assume that $\mathcal{L} \otimes I_{Z'}$ has a nonzero section s. Take a point p where s does not vanish. Let $Z = Z' \cup p$. Then $h^0(X, \mathcal{L} \otimes I_Z) \leq h^0(X, \mathcal{L} \otimes I_{Z'}) - 1$ and the proof is complete. □

DEFINITION 3.3. A sheaf $\mathcal{F}$ on a projective variety X satisfies *weak Brill-Noether* if $\mathcal{F}$ has at most one nonzero cohomology group. The sheaf $\mathcal{F}$ is *nonspecial* if $H^i(X, \mathcal{F}) = 0$ for $i > 0$. Otherwise, $\mathcal{F}$ is called *special*.

On any curve, a general line bundle $\mathcal{L}$ satisfies weak Brill-Noether and, if $\chi(L) \geq 0$, $\mathcal{L}$ is nonspecial. Consequently, classical Brill-Noether theory focuses on loci of line bundles in $\mathrm{Pic}^d(C)$ with unexpected cohomology.

Recall that a g^r_d is a linear system of projective dimension r and degree d on the curve C of genus g. A base-point-free g^r_d corresponds to a nondegenerate morphism

$C \to \mathbb{P}^r$ of degree d. The curve C admits a nondegenerate map to $\mathbb{P}^r$ of degree at most d if and only if C has a g^r_d. The Brill-Noether number is defined by

$$\rho(g, r, d) = g - (r+1)(g - d + r).$$

Let $W^r_d(C)$ be the locus of line bundles $\mathcal{L} \in \mathrm{Pic}^d(C)$ such that $h^0(C, \mathcal{L}) \geq r+1$. The scheme $W^r_d(C)$ has a natural determinantal structure. The celebrated Brill-Noether Theorem determines the structure of $W^r_d(C)$ on a general curve.

THEOREM 3.4 (Brill-Noether). *Let C be a curve of genus g which is general in moduli.*

(1) *The curve has a g^d_r if and only if $\rho(g, r, d) \geq 0$* [**GH80**, Griffiths and Harris].
(2) *If $\rho \geq 0$, then $W^r_d(C)$ is normal, Cohen-Macaulay of dimension ρ and smooth away from $W^{r+1}_d(C)$* [**Gie82**, Gieseker].
(3) *$W^r_d(C)$ is irreducible if $\rho > 0$* [**FL81**, Fulton and Lazarsfeld].
(4) *When $\rho \geq 0$, the universal space $\mathcal{W}^r_d$ of g^r_d's has a unique component dominating the moduli space of curves* [**EH87**, Eisenbud and Harris].

Unlike Proposition 3.1, the Brill-Noether Theorem requires C to be general in moduli. For special curves, $W^r_d(C)$ may be reducible with components of larger than the expected dimension ρ. The structure of $W^r_d(C)$ on special curves is not fully understood and is an active area of research. For example, Larson, Larson and Vogt recently described $W^r_d(C)$ for a general curve C with fixed gonality [**LLV20**].

3.2. Higher rank vector bundles on curves. In this subsection, we will recall facts concerning stable bundles on curves and their Brill-Noether theory.

3.2.1. *Rational curves.* A vector bundle $\mathcal{V}$ on $\mathbb{P}^1$ is isomorphic to a direct sum of line bundles $\mathcal{V} \cong \bigoplus_{i=1}^r \mathcal{O}_{\mathbb{P}^1}(a_i)$ for a unique sequence of integers $a_1 \leq a_2 \leq \cdots \leq a_r$. In particular, there are no stable bundles of rank greater than one on $\mathbb{P}^1$. There exists a semistable vector bundle $\mathcal{V}$ of rank r and degree d on $\mathbb{P}^1$ if and only if r divides d, in which case $\mathcal{V} \cong \mathcal{O}_{\mathbb{P}^1}(\frac{d}{r})^{\oplus r}$.

The cohomology of the vector bundle $\mathcal{V} = \bigoplus_{i=1}^r \mathcal{O}_{\mathbb{P}^1}(a_i)$ is given by

$$h^0(\mathbb{P}^1, \mathcal{V}) = \sum_{a_i \geq 0} (a_i + 1) \quad \text{and} \quad h^1(\mathbb{P}^1, \mathcal{V}) = \sum_{a_i \leq -2} (-1 - a_i).$$

3.2.2. *Elliptic curves.* Let E be an elliptic curve. Vector bundles on E have been classified by Atiyah [**Ati57**]. There exists a stable vector bundle of rank r and degree d on E if and only if r and d are relatively prime. In this case, the determinant map gives an isomorphism between $M_E(r, d)$ and E. In particular, by taking direct sums, there exist semistable bundles of every rank r and degree d on E.

If $\mathcal{V}$ and $\mathcal{W}$ are two semistable bundles on E with $\mu(\mathcal{V}) > \mu(\mathcal{W})$, then

$$\mathrm{Ext}^1(\mathcal{W}, \mathcal{V}) \cong \mathrm{Hom}(\mathcal{V}, \mathcal{W})^* = 0$$

by semistability. By induction on the length of the Harder-Narasimhan filtration, we conclude that a vector bundle on E is a direct sum of semistable vector bundles. Hence, the cohomology of any bundle on E is determined by the cohomology of the bundles appearing in its Harder-Narasimhan filtration.

PROPOSITION 3.5. *Let $\mathcal{V}$ be a semistable bundle on an elliptic curve E. Then $\mathcal{V}$ satisfies weak Brill-Noether if and only if $\mathcal{O}_E$ is not a Jordan-Hölder factor of $\mathcal{V}$.*

PROOF. Let $\mathcal{V}$ be a stable vector bundle on E. If $\mu(\mathcal{V}) < 0$, then $H^0(E, \mathcal{V}) = 0$ by stability. Similarly, if $\mu(\mathcal{V}) > 0$, then $H^1(E, \mathcal{V}) = H^0(E, \mathcal{V}^*) = 0$. If $\mu(\mathcal{V}) = 0$ and $\mathcal{V}$ is stable, then $\mathcal{V}$ must be a line bundle of degree 0. The cohomology of $\mathcal{V}$ in that case vanishes except when $\mathcal{V} = \mathcal{O}_E$. We conclude that on an elliptic curve all stable bundles except $\mathcal{O}_E$ satisfy weak Brill-Noether. By considering the Jordan-Hölder filtration, we conclude that any semistable bundle $\mathcal{V}$ on E satisfies weak Brill-Noether provided that $\mathcal{O}_E$ is not a Jordan-Hölder factor of $\mathcal{V}$. It is easy to see that if $\mathcal{O}_E$ is a Jordan-Hölder factor, then $\mathcal{V}$ must have nonvanishing cohomology. □

EXAMPLE 3.6. The cohomology of a semistable bundle whose Jordan-Hölder factors contain $\mathcal{O}_E$ depends on the extension class. Let

$$0 \to \mathcal{O}_E \to \mathcal{V} \to \mathcal{O}_E \to 0$$

be the unique nontrivial extension of $\mathcal{O}_E$ by $\mathcal{O}_E$. Then $h^0(E, \mathcal{V}) = h^1(E, \mathcal{V}) = 1$, whereas $h^0(E, \mathcal{O}_E \oplus \mathcal{O}_E) = h^1(E, \mathcal{O}_E \oplus \mathcal{O}_E) = 2$.

3.2.3. *Curves of higher genus.* For the rest of this section, let C be a curve of genus $g \geq 2$. The following theorem describes the moduli spaces of semistable sheaves on C.

THEOREM 3.7. *Let C be a smooth curve of genus $g \geq 2$.*

(1) *Then the stack of coherent sheaves $\mathcal{C}oh_{d,r}$ of degree d and rank r on C is a smooth, irreducible Artin stack.*
(2) *The moduli space $M_C(r, d)$ is an irreducible, projective variety of dimension $r^2(g-1)+1$, which contains the locus of stable bundles as a dense open set. It is smooth at points corresponding to stable bundles. In particular, if r and d are relatively prime, then $M_C(r, d)$ is smooth.*

SKETCH OF PROOF. We refer the reader to [**Hof10**] for a detailed proof of (1). Briefly, using the Grothendieck Quot scheme, one can show that $\mathcal{C}oh_{d,r}$ is algebraic. Obstructions to deformations of a coherent sheaf $\mathcal{F}$ are contained in $\mathrm{Ext}^2(\mathcal{F}, \mathcal{F})$. Since $\mathrm{Ext}^2(\mathcal{F}, \mathcal{F}) = 0$ on C, $\mathcal{C}oh_{d,r}$ is smooth of dimension $-\chi(\mathcal{F}, \mathcal{F})$. Finally, by induction on rank and degree, one can show that $\mathcal{C}oh_{d,r}$ is connected to conclude (1). The irreducibility of $M_C(r, d)$ (when nonempty) follows from openness of semistability.

We now construct semistable vector bundles of rank r and degree d on curves of genus $g \geq 1$. First, assume r and d are coprime. Let $\frac{e}{s} < \frac{d}{r} < \frac{f}{t}$ be the three consecutive fractions in the Farey sequence for r. This implies that the middle fraction is the *mediant* of its two neighbors:

$$\frac{d}{r} = \frac{e+f}{s+t}.$$

Then s and t are strictly less than r and the e, s (similarly f, t) are coprime. By induction on the rank, there exist stable bundles $\mathcal{V}_1$ and $\mathcal{V}_2$ of rank and degree (s, e) and (t, f), respectively. Since

$$\chi(\mathcal{V}_2, \mathcal{V}_1) = et - sf + r(1-g) < 0,$$

there exists a nonsplit extension of the form

$$0 \to \mathcal{V}_1 \to \mathcal{V} \to \mathcal{V}_2 \to 0.$$

We claim that $\mathcal{V}$ is stable. Otherwise, let $\mathcal{W}$ be the maximal destabilizing subbundle. Since $\mu(\mathcal{W}) > \mu(\mathcal{V})$, $r(\mathcal{W}) < r(\mathcal{V})$ and $\frac{f}{t}$ is the next smallest fraction in the Farey series for r, we must have $\mu(\mathcal{W}) \geq \mu(\mathcal{V}_2)$. Hence, either $\mu(\mathcal{W}) > \mu(\mathcal{V}_2)$ and the natural map from $\mathcal{W}$ to $\mathcal{V}_2$ is zero or $\mu(\mathcal{W}) = \mu(\mathcal{V}_2)$. In the first case, there is an induced map from $\mathcal{W}$ to $\mathcal{V}_1$, which is a contradiction since $\mathcal{W}$ is stable and $\mu(\mathcal{V}_1) < \mu(\mathcal{W})$. In the second case, we must have $\mathcal{W} = \mathcal{V}_2$ and we get a splitting of the sequence, contrary to our assumptions.

Hence, when r and d are coprime, the moduli space $M_C(r,d)$ is nonempty. Furthermore, if $\mathcal{V}$ is a stable bundle, then

$$\hom(\mathcal{V},\mathcal{V}) = 1, \quad \text{and} \quad \operatorname{ext}^1(\mathcal{V},\mathcal{V}) = r^2(g-1)+1.$$

By basic deformation theory, $T_{[\mathcal{V}]}M_C(r,d) = \operatorname{Ext}^1(\mathcal{V},\mathcal{V})$ and $M_C(r,d)$ is smooth at $\mathcal{V}$ of dimension $r^2(g-1)+1$.

If r and d are not coprime, let $(r,d) = (kr', kd')$, where r' and d' are coprime. There exists a stable vector bundle $\mathcal{V}$ of rank and degree (r', d'). Taking the direct sum of k copies of $\mathcal{V}$, we obtain a semistable bundle of rank r and d. Hence, if $g \geq 1$, $M_C(r,d)$ is nonempty.

Now let $g \geq 2$. By induction on the gcd k, assume that there are stable bundles for all $k' < k$. The dimension of the stack $\mathcal{M}_C(r,d)$ is $r^2(g-1)$. Consider the first step of a Jordan-Hölder filtration

$$0 \to \mathcal{V}_1 \to \mathcal{V} \to \mathcal{V}_2 \to 0$$

with the rank and degree of $\mathcal{V}_i$ equal to (k_ir', k_id') and $k_1 + k_2 = k$. We have

$$\operatorname{ext}^1(\mathcal{V}_2,\mathcal{V}_1) = k_1k_2(r')^2(g-1) + \hom(\mathcal{V}_2,\mathcal{V}_1).$$

Since $\mathcal{V}_1$ is stable, either $\hom(\mathcal{V}_2,\mathcal{V}_1) = 0$ or $\mathcal{V}_1$ is a factor of the Jordan-Hölder filtration of $\mathcal{V}_2$. In the first case, the dimension of such extensions is bounded above by

$$(k_1r')^2(g-1) + k_1k_2(r')^2(g-1) + (k_2r')^2(g-1) < r^2(g-1).$$

In the latter case, $\mathcal{V}_1$ is determined up to finitely many choices and $\hom(\mathcal{V}_2,\mathcal{V}_1) \leq \frac{k_2}{k_1}$. Hence such loci is bounded by

$$\frac{k_2}{k_1} + k_1k_2(r')^2(g-1) + (k_2r')^2(g-1) < r^2(g-1).$$

We conclude that there must be stable bundles of rank r and degree d on C. □

PROPOSITION 3.8 (Higher rank weak Brill-Noether for curves). *Let $\mathcal{V}$ be a general semistable vector bundle of rank r and degree d on a curve C of genus g. Then $\mathcal{V}$ satisfies weak Brill-Noether:*

$$h^0(\mathcal{V}) = \max(0, d - r(g-1)), \quad \text{and} \quad h^1(\mathcal{V}) = \max(0, r(g-1) - d).$$

PROOF. The proposition is easy for $g = 0$ and follows from Proposition 3.5 for $g = 1$. Hence, we may assume that $g \geq 2$. By the semicontinuity of cohomology and Theorem 3.7, it suffices to exhibit one coherent sheaf of rank r and degree d with the expected cohomology. By Riemann-Roch, the Euler characteristic is given by $\chi(\mathcal{V}) = d + r(1-g)$. By the division algorithm, write $d = qr + s$ with $0 \leq s < r$. Consider a direct sum of line bundles

$$\mathcal{V} = \bigoplus_{i=1}^{r-s} \mathcal{L}_i \oplus \bigoplus_{j=1}^{s} \mathcal{L}'_j,$$

where $\mathcal{L}_i$ are general line bundles of degree q and $\mathcal{L}'_j$ are general line bundles of degree $q+1$. By Proposition 3.1, $h^0(\mathcal{L}_i) = \max(0, q-g+1)$ and $h^0(\mathcal{L}'_j) = \max(0, q+2-g)$. Since $h^0(\mathcal{V}) = \sum_i h^0(\mathcal{L}_i) + \sum_j h^0(\mathcal{L}'_j)$, we conclude that $h^0(\mathcal{V}) = 0$ if $q+2 \leq g$ or if $s=0$ and $q+1 \leq g$. Finally, we have $h^0(\mathcal{V}) = d + r(1-g)$ if $q+1 \geq g$. This proves the statement for h^0. The statement for h^1 follows by Serre duality. □

CAUTION 3.9. We warn the reader that points of $M_C(r,d)$ correspond to S-equivalence classes of bundles. Certain properties of bundles such as cohomology or global generation are not invariant under S-equivalence. The two bundles in Example 3.6 are S-equivalent, but they have different cohomology groups. Similarly, $\mathcal{O}_E \oplus \mathcal{O}_E$ is globally generated, whereas the nontrivial extension is not.

Since the cohomological behavior of the general stable bundle is as expected, Brill-Noether theory concentrates on the cohomology jumping loci. In higher rank, much less is known about cohomology jumping loci. We refer the reader to [**GM08**] for a survey. As in the case of line bundles, there are no interesting jumping loci as soon as the slope of the vector bundle is sufficiently large.

PROPOSITION 3.10. *Let C be a curve of genus g. If $\mu > 2g-2$, then every bundle $\mathcal{V}$ in $M_C(r,d)$ is nonspecial.*

PROOF. By Serre duality,

$$h^1(C, \mathcal{V}) = h^0(C, \mathcal{V}^* \otimes \omega_C) = \hom(\mathcal{V}, \omega_C).$$

Since $\mathcal{V}$ and ω_C are semistable and

$$\mu(\mathcal{V}) > 2g-2 = \mu(\omega_C),$$

we conclude that $\hom(\mathcal{V}, \omega_C) = 0$. Hence, every semistable bundle of slope greater than $2g-2$ has no higher cohomology. □

Consequently, the Brill-Noether problem is only interesting when $\mu(V) \leq 2g-2$. Since $h^1(C, \omega_C) = 1$, the bound in the proposition is sharp.

Several other properties of bundles such as global generation and ampleness play a central role in geometry. Recall that a vector bundle $\mathcal{V}$ is called *ample* if $\mathcal{O}_{\mathbb{P}\mathcal{V}}(1)$ is an ample line bundle on $\mathbb{P}\mathcal{V}$, or equivalently, if for every coherent sheaf $\mathcal{F}$ there exists an integer n_0 such that $\mathcal{F} \otimes \mathrm{Sym}^n(\mathcal{V})$ is globally generated for all $n \geq n_0$. For curves, it is easy to determine when the general stable bundle has these properties.

PROPOSITION 3.11. *Let C be a curve of genus $g \geq 2$.*

(1) [**La91, Su87**] *A general stable bundle of rank r and degree d on C is globally generated if $d - rg \geq 1$.*
(2) *If $\mu > 2g-1$, then every semistable vector bundle of rank r and degree d on C is globally generated.*
(3) [**Har71**, Theorem 2.4] *If $\mu > 0$, then every semistable bundle of rank r and degree d on C is ample.*

REMARK 3.12. Global generation is not in general an open property. For example, let $\mathcal{L}$ be a line bundle of degree 0 on a positive genus curve C. Then $\mathcal{L}$ is globally generated if and only if $\mathcal{L} \cong \mathcal{O}_C$. Hence, the locus of globally generated degree 0 line bundles is a single point in a g-dimensional abelian variety. However, the locus of globally generated bundles is open in the locus of nonspecial bundles.

PROOF. If $d-rg \geq 1$, then by Proposition 3.8, the general stable bundle has no higher cohomology and has at least $r+1$ sections. By the irreducibility of $\mathrm{Coh}_{d,r}$, it suffices to exhibit one globally generated nonspecial bundle of rank r and degree d.

A general line bundle $\mathcal{L}$ of degree $d \geq g+1$ has the property that $h^1(C, \mathcal{L}(-p)) = 0$ for every $p \in C$. Indeed, consider the locus

$$X_{d-1} := \{D \in C^{(d-1)} | h^0(C, D) \geq d-g+1\} \subset C^{(d)}.$$

By Proposition 3.8, the codimension of X_{d-1} is at least 1. The fibers of the natural map $\phi_d : C^{(d-1)} \to \mathrm{Pic}^{d-1}(C)$ of X_{d-1} have dimension at least $d-g$. Consequently, the image of X_{d-1} has dimension at most $g-2$. Hence, the codimension of the locus of line bundles of degree $d-1$ with unexpected cohomology is at least 2. Therefore, for the general $\mathcal{L} \in \mathrm{Pic}^d(C)$, $h^0(C, \mathcal{L}(-p)) = h^0(C, \mathcal{L}) - 1$ for every $p \in C$. By Serre duality, $h^1(C, \mathcal{L}(-p)) = 0$ for every $p \in C$.

By the Euclidean algorithm, write $d = rm + s$. If $m \geq g+1$, then we can take $\mathcal{V}$ to be an extension of the form

$$0 \to \mathcal{L}^{r-s} \to \mathcal{V} \to \mathcal{K}^s \to 0,$$

where $\mathcal{L}$ and $\mathcal{K}$ are general line bundles of degree m and $m+1$, respectively. We then have that $\mathcal{V}$ is nonspecial and $H^1(C, \mathcal{V}(-p)) = 0$ for every $p \in C$. Taking the long exact sequence associated to

$$0 \to \mathcal{V}(-p) \to \mathcal{V} \to \mathcal{V}|_p \to 0,$$

we conclude that $H^0(C, \mathcal{V})$ surjects onto $H^0(C, \mathcal{V}|_p)$ for every $p \in C$ and $\mathcal{V}$ is globally generated. If $m = g$, a more subtle argument is needed. We claim that the Brill-Noether locus in $M_C(r,d)$ parameterizing special vector bundles has codimension at least 2 if $d - r(g-1) \geq 1$ (see [**La91**, **Su87**]). Granting this claim, $\deg(\mathcal{V}(-p)) - r(g-1) \geq 1$. Consequently, $\mathcal{V}(-p)$ does not have h^1 for the general $\mathcal{V}$ and every $p \in C$. Hence, $\mathcal{V}$ is a globally generated, nonspecial bundle. To prove the claim, recall that the fibers of the determinant map $M_C(r,d) \to \mathrm{Pic}^d(C)$ are Fano varieties of Picard rank 1 [**DN89**]. Hence, it suffices to exhibit a complete curve in each fiber which is nonspecial. Furthermore, by taking direct sums, it suffices to do this when the rank and the degree are coprime. Now it is easy to construct such complete families by induction on the rank as in the proof of Theorem 3.7.

Now assume $\mathcal{V}$ is semistable and $\mu(\mathcal{V}) > 2g-1$. For any point $p \in C$, consider the exact sequence

$$0 \to \mathcal{V}(-p) \to \mathcal{V} \to \mathcal{V}_p \to 0.$$

We have that $H^1(C, \mathcal{V}(-p)) \cong H^0(C, \mathcal{V}^* \otimes \omega_C(p))$. Since $\mu(\mathcal{V}^* \otimes \omega_C(p)) < 0$ and is semistable, it cannot have any sections. Hence the evaluation map $H^0(C, \mathcal{V}) \to H^0(\mathcal{V}|_p)$ is surjective for every point p and $\mathcal{V}$ is globally generated.

Hartshorne [**Har71**] proves (3) by showing that a vector bundle $\mathcal{V}$ of positive degree on a curve all of whose symmetric powers $\mathrm{Sym}^n(\mathcal{V})$ are semistable is ample. In characteristic 0, by the Narasimhan-Seshadri Theorem, the symmetric powers of a semistable bundle are semistable. More generally, Hartshorne uses this fact to show that any vector bundle of positive degree on a smooth curve all of whose quotients have positive degree is ample. □

The locus of stable vector bundles that fail to be globally generated has been studied by Kopper and Mandal [**KM**]. Their main theorem is the following.

THEOREM 3.13 ([**KM**, Theorem 1.2]). *Let $N_C(r,d)$ be the locus of* stable *vector bundles of rank $r \geq 2$ and degree d on a smooth curve of genus $g \geq 2$ that fail to be globally generated. Assume $rg+1 \leq d \leq r(2g-1)-1$.*

(1) *$N_C(r,d)$ is nonempty and it has a component of expected codimension $d-rg$ and no component of smaller codimension.*
(2) *If $rg+g-1 \leq d \leq r(2g-1)-1$, then $N_C(r,d)$ is irreducible of the expected codimension $d-rg$.*

4. Pathologies for surfaces

In this section, we will contrast the behavior of stable bundles on surfaces with that of curves discussed in the last section. We will see that many of the nice properties that are valid for curves fail for surfaces. This makes studying Brill-Noether theory on surfaces much more challenging. Here are some of the new difficulties that arise.

(1) A general line bundle with fixed invariants on a surface can have more than one nonzero cohomology group. Hence, unlike on curves, the cohomology of a general line bundle on a surface is more delicate and can be challenging to compute.
(2) Unlike in the case of curves of genus $g \geq 2$, there are restrictions on the possible invariants of stable sheaves on surfaces. Determining the Chern characters of stable sheaves on a surface is a challenging problem.
(3) Unlike in the case of curves, the stack of coherent sheaves with fixed invariants on a surface is almost always reducible. Hence, stability plays a much greater role in determining the cohomology of a 'general' stable sheaf on a surface. Furthermore, on a surface X, the moduli space $M_{X,H}(\mathbf{v})$ itself may be reducible and even disconnected. Hence, it does not always even make sense to talk about a general stable sheaf with fixed invariants.

In this section, we will give some simple examples of these phenomena.

4.1. The cohomology of line bundles on surfaces. Proposition 3.1 showed that a general line bundle of degree d on *any* smooth curve has at most one nonzero cohomology group. This fails for surfaces. The general line bundle with a given class on a surface may have two or three nonzero cohomology groups. Moreover, whether weak Brill-Noether holds for a line bundle may depend on the isomorphism class of the surface.

EXAMPLE 4.1. We give several easy examples where weak Brill-Noether fails for line bundles on surfaces.

(1) Let X be a K3 surface. Then the Picard group of X is discrete and
$$h^0(X, \mathcal{O}_X) = h^2(X, \mathcal{O}_X) = 1.$$
(2) Let X be a very general surface of degree $d \geq 4$ in $\mathbb{P}^3$. By the Noether-Lefschetz Theorem, the Picard group of X is generated by $\mathcal{O}_X(1)$. Let $0 \leq k \leq d-4$. The long exact sequence associated to
$$0 \to \mathcal{O}_{\mathbb{P}^3}(k-d) \to \mathcal{O}_{\mathbb{P}^3}(k) \to \mathcal{O}_X(k) \to 0$$
implies that
$$h^0(X, \mathcal{O}_X(k)) = \binom{k+3}{3}, \quad h^1(X, \mathcal{O}_X(k)) = 0$$

$$h^2(X, \mathcal{O}_X(k)) = \binom{d-k-1}{3}.$$

Observe that both h^0 and h^2 are nonzero.

(3) Let $Y \subset \mathbb{P}^3$ be a very general surface of degree $d \geq 5$ and ℓ be a general line in $\mathbb{P}^3$. Let X denote the blowup of Y along $\ell \cap Y = \{p_1, \ldots, p_d\}$ and let E_i denote the exceptional divisor over p_i. Set $E = \sum_{i=1}^{d} E_i$. Let H be the pullback of the hyperplane class from Y. Then the Picard group of X is discrete. For an integer $1 \leq k \leq d-4$, the sections of $\mathcal{O}_X(kH - E)$ are given by hypersurfaces of degree k that vanish along $\ell \cap Y$. By Bézout's Theorem, these hypersurfaces of degree k must vanish along ℓ. By Serre duality,

$$h^2(X, kH - E) = h^0(X, (d-4-k)H + 2E).$$

Since $E_i \cdot E_j = -\delta_{i,j}$, $2E$ must be in the base locus of $\mathcal{O}_X((d-4-k)H+2E)$. Consequently, $h^2(X, kH - E) = h^0(X, (d-4-k)H)$. By an easy Euler characteristic computation, we conclude that

$$h^0(X, kH - E) = \binom{k+3}{3} - k - 1, \ h^1(X, kH - E) = d - k - 1,$$

$$h^2(X, kH - E) = \binom{d-k-1}{3}.$$

Hence, all three cohomology groups are nonzero. Observe that if X were the blowup of Y along d general points, the cohomology would instead be

$$h^0(X, kH - E) = \max\left(0, \binom{k+3}{3} - d\right),$$

$$h^1(X, kH - E) = \max\left(0, d - \binom{k+3}{3}\right),$$

$$h^2(X, kH - E) = \binom{d-k-1}{3}.$$

Hence, the generic cohomology of a line bundle depends on the isomorphism class of the surface.

(4) Let A be an abelian surface of Picard rank 1. Let X be the blowup of A at a point with exceptional divisor E. Let H denote the pullback of a very ample divisor on A. Let m be a nonnegative integer. Then any line bundle algebraically equivalent to $\mathcal{O}_X(H+mE)$ is of the form $\mathcal{L} \otimes \mathcal{O}_X(H+mE)$, where $\mathcal{L}$ is the pullback of a line bundle of degree 0 from A. The line bundle $\mathcal{L} \otimes \mathcal{O}_X(H)$ is the pullback of an ample line bundle from A, which has positive Euler characteristic and no higher cohomology by the Kodaira Vanishing Theorem. Consequently,

$$h^0(X, \mathcal{L} \otimes \mathcal{O}_X(H + mE)) > 0.$$

However,

$$\chi(\mathcal{L} \otimes \mathcal{O}_X(H + mE)) = \chi(\mathcal{O}_X) + \frac{1}{2}(H^2 - m^2 - m) < 0$$

if $m \gg 0$. We conclude that these line bundles have nontrivial h^0 and h^1.

In general, computing the cohomology of a line bundle on a surface can be a hard problem. For example, already for the blow-up of $\mathbb{P}^2$ in $m \geq 10$ very general points, we do not know the dimensions of the spaces of global sections of all line bundles. In this case there is a precise conjecture due Segre, Harbourne, Gimigliano and Hirschowitz.

CONJECTURE 4.2 (SHGH Conjecture [**Se60, Ha86, Gim87, Hi89**]). *Let X be a very general blow up of $\mathbb{P}^2$ at $m \geq 10$ points $p_1, \ldots, p_m$. Let H denote the pullback of the class of a line and let E_i denote the exceptional divisor over p_i. Let $D = dH - \sum_{i=1}^m n_i E_i$ be a divisor with $d > 0$ and $n_i \geq 0$. Then*

$$h^0(X, \mathcal{O}_X(D)) = \binom{d+2}{2} - \sum_{i=1}^{m} \binom{n_i+1}{2}$$

if and only if $\mathcal{O}_X(D)$ does not have a multiple (-1)-curve in its base locus.

The SHGH Conjecture, if true, provides an efficient algorithm for computing the cohomology of any line bundle on X. Despite steady progress, the conjecture remains open in general.

Given that weak Brill-Noether does not hold in general for line bundles on surfaces, we do not expect it to hold for higher rank bundles either. Unlike the case of curves, for surfaces already computing the cohomology of a general stable sheaf in a component of the moduli space is an interesting and challenging problem.

PROBLEM 4.3 (Weak Brill-Noether). Given an irreducible component of the moduli space $M_{X,H}(\mathbf{v})$, compute the cohomology of the general sheaf $\mathcal{F}$ in that component.

4.2. The cohomology of rank 1 sheaves. Once one knows the cohomology of line bundles on a surface X, understanding the general cohomology of rank 1 sheaves does not present new difficulties. If $\mathcal{V}$ is a rank 1 torsion-free coherent sheaf on a smooth projective surface, then $\mathcal{V} \cong I_Z \otimes \mathcal{L}$, where $\mathcal{L}$ is a line bundle and I_Z is the ideal sheaf of a zero-dimensional subscheme of X. The Hilbert scheme $X^{[n]}$ parameterizing length n zero-dimensional subschemes of X is a smooth, irreducible, projective variety of dimension $2n$. The long exact sequence associated to the standard exact sequence

$$0 \to I_Z \otimes \mathcal{L} \to \mathcal{L} \to \mathcal{O}_Z \to 0$$

implies that $H^2(X, I_Z \otimes \mathcal{L}) \cong H^2(X, \mathcal{L})$. If Z is a general set of n points on X, then the map $H^0(X, \mathcal{L}) \to H^0(X, \mathcal{O}_Z)$ has maximal rank. We conclude that $H^0(X, I_Z \otimes \mathcal{L}) = 0$ for a general set of points if and only if $n \geq H^0(X, \mathcal{L})$. Similarly, $h^1(X, I_Z \otimes \mathcal{L}) = h^1(X, \mathcal{L}) + n - h^0(X, \mathcal{L})$ for a general set of points Z. Hence, we can compute the cohomology of $I_Z \otimes \mathcal{L}$ for a general set of points Z purely based on the cohomology of $\mathcal{L}$.

REMARK 4.4. We caution the reader that rank 1 sheaves behave slightly differently from higher rank sheaves. When $n > 0$, the moduli space consists entirely of non-locally-free sheaves. Consequently, one cannot directly apply Serre duality. In fact, if $\mathcal{L}$ is a line bundle with $h^2(X, \mathcal{L}) \neq 0$, then $\mathcal{L} \otimes I_Z$ has nonvanishing h^1 and h^2 as soon as $n > h^0(X, \mathcal{L})$. In particular, if $\mathcal{L}$ is an ample line bundle on X with $h^0(X, \mathcal{L}) \neq 0$, then $I_Z \otimes \omega_X \otimes \mathcal{L}^*$ has nonvanishing h^1 and h^2 as soon as $n \geq 1$. In particular, one can only hope for weak Brill-Noether to hold for rank 1 sheaves when $h^2(X, \mathcal{L}) \neq 0$.

When the rank $r > 1$, there may be components of $M_{X,H}(\mathbf{v})$ that consist entirely of non-locally-free sheaves. However, such components do not exist for certain surfaces such as $\mathbb{P}^2$ and Hirzebruch surfaces. Moreover, when $\Delta(\mathbf{v}) \gg 0$, then the general member of the moduli space parameterizes a vector bundle. Consequently, one may apply Serre duality to compute the cohomology of the general sheaf.

4.3. The moduli space may be empty. Theorem 3.7 shows that on a curve of genus $g \geq 2$, the moduli space $M_C(r, d)$ is nonempty, irreducible and of the expected dimension for every $r \geq 1$ and every d. In contrast, there are restrictions on the possible Chern characters of stable sheaves on surfaces. For example, the Bogomolov inequality provides an important constraint.

THEOREM 4.5 (The Bogomolov inequality). *Let (X, H) be a polarized smooth surface and let $\mathcal{F}$ a μ_H-semistable sheaf. Then $\Delta(\mathcal{F}) \geq 0$.*

For abelian surfaces Bogomolov's inequality completely characterizes stable Chern characters. For some surfaces such as $\mathbb{P}^2$ or K3 surfaces one can easily strengthen the Bogomolov inequality.

EXAMPLE 4.6. Let $X = \mathbb{P}^2$ and let $\mathcal{V}$ be a torsion-free stable sheaf. Then

$$\mathrm{ext}^2(\mathcal{V}, \mathcal{V}) = \hom(\mathcal{V}, \mathcal{V}(-3)) = 0$$

by Serre duality and stability. Therefore,

$$\chi(\mathcal{V}, \mathcal{V}) = r^2(1 - 2\Delta(\mathcal{V})) = \hom(\mathcal{V}, \mathcal{V}) - \mathrm{ext}^1(\mathcal{V}, \mathcal{V}) \leq 1.$$

Consequently,

$$\Delta(\mathcal{V}) \geq \frac{1}{2}\left(1 - \frac{1}{r^2}\right).$$

We will later see that there are stronger restrictions on the Chern characters of stable sheaves on $\mathbb{P}^2$.

EXAMPLE 4.7. Let X be a K3 surface and let $\mathcal{V}$ be a torsion-free stable sheaf. Since $\hom(\mathcal{V}, \mathcal{V}) = \mathrm{ext}^2(\mathcal{V}, \mathcal{V}) = 1$, we have

$$\chi(\mathcal{V}, \mathcal{V}) = r^2(2 - 2\Delta(\mathcal{V})) \leq 2.$$

Hence, $\Delta(\mathcal{V}) \geq 1 - \frac{1}{r^2}$. On K3 surfaces this inequality characterizes the Chern characters of stable sheaves.

We say that a Chern character is μ_H-(semi)stable (respectively, H-Gieseker (semi)stable) if there exists a μ_H-(semi)stable (resp., H-Gieseker (semi)stable) sheaf with that Chern character. The following is a central open problem.

PROBLEM 4.8. Given a polarized surface (X, H) classify the Chern characters of μ_H or H-Gieseker semistable sheaves. Determine when there exist stable sheaves with the given Chern character.

The results concerning Problem 4.8 have two flavors. First, there are important asymptotic results for all surfaces guaranteeing the existence of stable bundles with sufficiently large discriminant.

THEOREM 4.9 (O'Grady). *Let (X, H) be a smooth polarized surface and let $\mathbf{v}$ be a Chern character with $r(\mathbf{v}) > 0$. If $\Delta_H(\mathbf{v}) \gg 0$ (where the necessary inequality depends on r, X and H), then the moduli space $M_H(\mathbf{v})$ is normal, generically smooth, irreducible and nonempty of the expected dimension. Furthermore, the slope stable sheaves are dense in $M_H(\mathbf{v})$.*

Second, Problem 4.8 has been studied and solved on certain surfaces including $\mathbb{P}^2$ [**DLP85**], K3 surfaces [**Yos99**], Abelian surfaces [**Yos01**], Hirzebruch surfaces [**CH21**], Enriques surfaces [**N16a, N16b, NY**], elliptic and bielliptic surfaces. There has been some progress for certain other classes of surfaces such as quintic and sextic surfaces in $\mathbb{P}^3$ at least in rank 2 [**MS11**]. However, the general problem remains wide open.

Not knowing the Chern characters of stable sheaves often presents difficulties in constructions. One is often forced to assume inequalities on the discriminant in constructions. There is however a general way to construct rank 2 bundles via the Serre construction, which we now recall.

4.4. The Serre Construction. Let X be a smooth projective surface. Let $Z \subset X$ be a zero-dimensional local complete intersection scheme of length n. Then Z satisfies the *Cayley-Bacharach property* with respect to a line bundle $\mathcal{L}$ on X if for any subscheme $W \subset Z$ of length $n-1$, any section of $\mathcal{L}$ vanishing on W vanishes on Z. The name is inspired by the classical Cayley-Bacharach Theorem which asserts that any cubic curve which contains 8 of the 9 intersection points of two cubic curves in $\mathbb{P}^2$ also contains the ninth.

THEOREM 4.10 (The Serre correspondence). *Let X be a smooth projective surface and $Z \subset X$ be a local complete intersection subscheme of dimension zero and length n. Let $\mathcal{L}$ be a line bundle on X. Then there exists an extension*

$$0 \to \mathcal{O}_X \to \mathcal{V} \to I_Z \otimes \mathcal{L} \to 0$$

with $\mathcal{V}$ locally free if and only if Z satisfies the Cayley-Bacharach property for $\omega_X \otimes \mathcal{L}$.

The Serre construction allows one to construct stable vector bundles of rank 2 on surfaces.

COROLLARY 4.11. *Let H be an ample divisor on X. Let $\mathcal{L}$ be a line bundle such that $c_1(\mathcal{L}) \cdot H > 0$. Let Z be a nonempty set of distinct points such that Z satisfies the Cayley-Bacharach property for $\omega_X \otimes \mathcal{L}$. Assume that $H^0(X, \mathcal{N} \otimes I_Z) = 0$ for any line bundle $\mathcal{N}$ with $c_1(\mathcal{N}) \cdot H \leq c_1(\mathcal{L}) \cdot H$. Then the general extension $\mathcal{V}$*

$$0 \to \mathcal{O}_X \to \mathcal{V} \to I_Z \otimes \mathcal{L} \to 0$$

is a μ_H-stable bundle.

PROOF. By Theorem 4.10, the general extension is a vector bundle. Suppose $\mathcal{N} \to \mathcal{V}$ is a destabilizing line subbundle. Then there must exist a nonzero map $\mathcal{N} \to I_Z \otimes \mathcal{L}$. Hence, $\mathcal{L} \otimes \mathcal{N}^{-1}$ is effective and nontrivial. In particular, $H^0(X, I_Z \otimes \mathcal{L} \otimes \mathcal{N}^{-1}) \neq 0$ and yet $0 < H \cdot c_1 \left(\mathcal{L} \otimes \mathcal{N}^{-1}\right) < H \cdot c_1(\mathcal{L})$, contrary to our assumptions. □

In general, it is hard to characterize loci of points on X that satisfy the Cayley-Bacharach property for $\mathcal{L} \otimes \omega_X$. However, there are easy conditions that ensure that it holds. For instance, we may choose Z general and $|Z|$ large, or we may choose $\mathcal{L}$ such that $H^0(X, \mathcal{L} \otimes \omega_X) = 0$.

4.5. The moduli space is not necessarily irreducible. Unlike in the case of curves, the stack of coherent sheaves with a fixed Chern character on a surface is in general reducible. Consequently, one cannot construct a coherent sheaf with the expected cohomology and deduce that the general stable sheaf will have the expected cohomology. Furthermore even the substack of semistable or stable sheaves can be reducible.

EXAMPLE 4.12 (Mestrano [**Me97**]). Let X be a very general surface of degree $d = 6$ in $\mathbb{P}^3$. By the Noether-Lefschetz Theorem, $\operatorname{Pic}(X) = \mathbb{Z}H$, where H is the hyperplane class. Let Z be a zero-dimensional scheme of length 11 of one of the following types:

I) Z is contained in the intersection of X with a twisted cubic curve C, or
II) Z is a general set of points in a hyperplane section of X.

Consider an extension of the form

$$0 \to \mathcal{O}_X \to \mathcal{V} \to I_Z(H) \to 0.$$

Since $\operatorname{ext}^1(I_Z(H), \mathcal{O}_X) = h^1(X, I_Z(3H)) = 1$, up to scaling there exist unique nonsplit extensions of this form. Since $K_X = 2H$, to apply Theorem 4.10 we need to check that Z satisfies the Cayley-Bacharach property with respect to $\mathcal{O}_X(3H)$. By Bézout's Theorem a hypersurface of degree 3 that contains at least 10 points of C contains all of C. Hence, Z satisfies the Cayley-Bacharach property in Case I. Similarly, in Case II, any hypersurface of degree 3 that contains 10 of the points contains the plane spanned by the points. Hence, Z satisfies the Cayley-Bacharach property in Case II. By Theorem 4.10, there are locally free sheaves $\mathcal{V}$ in both cases. Observe that $\mathcal{V}$ is necessarily μ_H-stable since $\mathcal{V}$ cannot admit a map from $\mathcal{O}_X(kH)$ for $k \geq 1$.

By a monodromy argument, one can check that the locus of bundles in Case I is irreducible of dimension 12, corresponding to a generically finite cover of the space of twisted cubics in $\mathbb{P}^3$. The tangent space to the moduli space at $\mathcal{V}$ is given by $\operatorname{Ext}^1(\mathcal{V}, \mathcal{V}) \cong H^1(X, \mathcal{V} \otimes \mathcal{V}^*)$. Since $\mathcal{V}$ is rank 2, we have $\mathcal{V}^* \cong \mathcal{V}(-H)$ and consequently

$$\mathcal{V} \otimes \mathcal{V}^* \cong (\mathcal{V} \otimes \mathcal{V})(-H) \cong (\operatorname{Sym}^2 \mathcal{V} \oplus \bigwedge^2 \mathcal{V})(-H) \cong \operatorname{Sym}^2(\mathcal{V})(-H) \oplus \mathcal{O}_X.$$

Since $H^1(X, \mathcal{O}_X) = 0$, we have $\operatorname{Ext}^1(\mathcal{V}, \mathcal{V}) = H^1\left(X, \operatorname{Sym}^2(\mathcal{V})(-H)\right)$. Using the standard exact sequence for symmetric powers

$$0 \to \mathcal{V}(-H) \to \operatorname{Sym}^2(\mathcal{V})(-H) \to I_{2Z \subset X}(H) \to 0,$$

where $I_{2Z \subset X}$ is the symbolic square of the ideal of Z in X, one shows that

$$\operatorname{ext}^1(\mathcal{V}, \mathcal{V}) = 12.$$

Hence, the bundles in Case I lie on a generically smooth component of the moduli space of dimension 12.

On the other hand, the locus of bundles in Case II has dimension at least 13. The choice of Z depends on 14 parameters, 3 for the choice of a hyperplane Λ and 11 for the choice of points in $\Lambda \cap X$. To go in the reverse direction, note that the bundles in Case II have $h^0(X, \mathcal{V}) = 2$, and from μ_H-stability it follows that a choice of a nonzero global section determines a quotient $\mathcal{V}/\mathcal{O}_X \cong I_Z(H)$ such that Z is in Case II from $h^0(X, \mathcal{V}) = 2$ and $h^1(X, \mathcal{O}_X) = 0$. Therefore, the locus in the moduli space coming from the bundles in Case I has dimension at least 13. We conclude that the moduli space has at least 2 irreducible components.

By using a similar construction, one can produce examples of moduli spaces with arbitrarily many components parameterizing sheaves on surfaces in $\mathbb{P}^3$.

THEOREM 4.13 ([**CH18c**]). *For any integer k, there exists an integer d_k such that if $d \geq d_k$, then a very general surface $X \subset \mathbb{P}^3$ of degree d has a moduli space of rank 2 sheaves with $c_1 = H$ that has at least k irreducible components.*

The moduli space does not even have to be connected. Okonek and Van de Ven [**OVdV86**] and Kotschick [**Ko89**] found examples of disconnected moduli spaces on elliptic surfaces with high Picard rank. These examples use ample classes that are very close to the fiber class. One can even find disconnected examples on complete intersection surfaces of Picard rank 1.

EXAMPLE 4.14. Let $d_1 \geq 5$ and $d_2 > d_1$ be two integers. Let D_1 be a smooth hypersurface in $\mathbb{P}^4$ that contains lines. Let D_2 be a very general hypersurface of degree d_2. Let X be the complete intersection of D_1 and D_2, which by Noether-Lefschetz Theory is a surface of Picard rank 1. Take a plane Λ containing a line ℓ of D_1. Then $\Lambda \cap D_1 = C \cup \ell$, where C is the residual curve of degree $d_1 - 1$. Let $Z = C \cap D_2$ be the zero dimensional subscheme of length $(d_1 - 1)d_2$. Then Z satisfies the Cayley-Bacharach property for $\omega_X(H)$ and up to scalars there is a unique nonsplit extension of the form

$$0 \to \mathcal{O}_X \to \mathcal{V} \to I_Z(H) \to 0.$$

One can compute that $h^0(X, \mathcal{V}) = 3$ and $h^1(X, \mathcal{V}) = 0$. Given a bundle $\mathcal{W}$ in the same irreducible component as $\mathcal{V}$, we have $h^0(X, \mathcal{W}) \geq 3$. The cokernel of the evaluation map of a section is $I_{Z'}(H)$, where Z' has length $(d_1-1)d_2$. Furthermore, since $h^0(X, I_{Z'}(H)) \geq 2$, Z' lies on a plane Λ'. By the Cayley-Bacharach Theorem, the residual d_2 points must be collinear and by Bézout's Theorem the line must be contained in D_1. With a slightly more careful analysis, one can deduce the following.

THEOREM 4.15 ([**CHK22**, Theorem 3.4]). *Let $\mathbf{v}$ be the Chern character of the sheaf $\mathcal{V}$ constructed in this example. For every connected (respectively, irreducible) component of the Fano scheme of lines $F_1(D_1)$ on D_1, $M_{X,H}(\mathbf{v})$ has a connected (respectively, irreducible) component of the same dimension.*

By letting the degree of D_1 tend to infinity, we can find threefolds D_1 in $\mathbb{P}^4$ that contain arbitrarily many isolated lines. We thus deduce the following.

COROLLARY 4.16 ([**CHK22**, Corollary 1.1]). *For any integer k, there exists a smooth complete intersection surface $X \subset \mathbb{P}^4$ with Picard rank 1 and a Chern character $\mathbf{v}$ on X such that $M_{X,H}(\mathbf{v})$ has at least k connected components.*

It is also interesting to note that the bundles $\mathcal{V}$ do not deform to bundles on complete intersections where D_1 does not contain a line.

Even on rational surfaces there are disconnected moduli spaces. Assuming the SHGH Conjecture 4.2, one can construct moduli spaces with arbitrarily many components for certain special polarizations (see [**CH23**]).

In this subsection, we have discussed the connectedness and the irreducibility of the moduli spaces. In general, there are many interesting questions about the topology of $M_{X,H}(\mathbf{v})$. A general expectation due to Donaldson, Gieseker and Li is that the moduli spaces become better behaved as Δ becomes large. O'Grady's Theorem 4.9 proves that the moduli space becomes irreducible for large Δ. There are several conjectures concerning the stabilization of other Betti numbers of the moduli spaces.

CONJECTURE 4.17 ([**CW22**, Conjecture 1.1]). *Let X be a smooth projective surface and let H be an ample line bundle. Fix a rank $r > 0$ and a first Chern class c. Then the ith Betti number of $M_{X,H}(r, c, \Delta)$ stabilizes to a constant $b_{i,\mathrm{Stab}}(X)$, which is independent of r, c and H, as Δ tends to ∞.*

The conjecture is known when X is a K3 or abelian surface and r and c are relatively prime. The conjecture is also known when X is a rational surface, $M_{X,H}(r, c, \Delta)$ do not contain strictly semistable sheaves and H is a polarization such that $H \cdot K_X < 0$ (see [**CW22**] for a discussion and references). The conjecture is wide open in general, especially for surfaces of general type.

5. Weak Brill-Noether

In this section, we will survey some results on the weak Brill-Noether Problem. Recall that an irreducible component of a moduli space of sheaves satisfies *weak Brill-Noether* if the general sheaf in that component has at most one nonzero cohomology group. In particular, if $\chi(\mathbf{v}) = 0$ and a component of the moduli space $M(\mathbf{v})$ satisfies weak Brill-Noether, then the general member of that component has no cohomology. The ideal situation occurs for $\mathbb{P}^2$. Let L be the class of a line on $\mathbb{P}^2$.

THEOREM 5.1 (Göttsche-Hirschowitz [**GH94**]). *A general stable bundle on $\mathbb{P}^2$ has at most one nonzero cohomology group.*

Consequently, the slope and the Euler characteristic determine the cohomology of the general stable bundle $\mathcal{V} \in M_{\mathbb{P}^2,L}(\mathbf{v})$.

- If $\chi(\mathcal{V}) < 0$, then $h^1(\mathbb{P}^2, \mathcal{V}) = -\chi(\mathcal{V})$ and $h^0(\mathbb{P}^2, \mathcal{V}) = h^2(\mathbb{P}^2, \mathcal{V}) = 0$.
- If $\chi(\mathcal{V}) \geq 0$ and $\mu_L(\mathcal{V}) \geq 0$, then $h^0(\mathbb{P}^2, \mathcal{V}) = \chi(\mathcal{V})$ and all other cohomology vanishes.
- If $\chi(\mathcal{V}) \geq 0$ and $\mu_L(\mathcal{V}) < 0$, then $h^2(\mathbb{P}^2, \mathcal{V}) = \chi(\mathcal{V})$ and all other cohomology vanishes.
- In particular, if $\chi(\mathcal{V}) = 0$, then the cohomology of $\mathcal{V}$ vanishes.

Observe that if $\mu_L(\mathcal{V}) < 0$, then $h^0(\mathbb{P}^2, \mathcal{V}) = 0$ by stability. By Serre duality and stability, if $\mu_L(\mathcal{V}) > -3$, then $h^2(\mathbb{P}^2, \mathcal{V}) = 0$.

REMARK 5.2. The sheaves $I_Z(-d)$ have nonvanishing h^1 and h^2 provided $|Z| \geq 1$ and $d \geq 3$. The general element of the corresponding moduli spaces are not locally free and Serre duality fails. The general stable sheaf of rank at least 2 on $\mathbb{P}^2$ is locally free, hence weak Brill-Noether holds for $M_{\mathbb{P}^2,L}(\mathbf{v})$ if the rank of $\mathbf{v}$ is at least 2.

We will discuss two general techniques for proving weak Brill-Noether Theorems on surfaces. The first uses prioritary sheaves and the second uses Bridgeland stability conditions. We begin by explaining the first.

5.1. Prioritary Sheaves. Let D be an effective divisor on a smooth surface X. A torsion free sheaf $\mathcal{V}$ is called *D-prioritary* if $\mathrm{Ext}^2(\mathcal{V}, \mathcal{V}(-D)) = 0$. We denote the stack of D-prioritary sheaves on X with Chern character $\mathbf{v}$ by $\mathcal{P}_{X,D}(\mathbf{v})$. The stack $\mathcal{P}_{X,D}(\mathbf{v})$ is an open substack of the stack of coherent sheaves on X with Chern character $\mathbf{v}$.

The condition $\mathrm{Ext}^2(\mathcal{V}, \mathcal{V}(-D)) = 0$ implies that the restriction map

$$\mathrm{Ext}^1_X(\mathcal{V}, \mathcal{V}) \to \mathrm{Ext}^1_D(\mathcal{V}|_D, \mathcal{V}|_D)$$

is surjective. Consequently, the general first order deformation of the sheaf $\mathcal{V}$ on X gives a general first order deformation of $\mathcal{V}|_D$ on D. For example, when D is a smooth rational curve and $\mathcal{V}$ is locally free on D, the restriction of a general deformation of $\mathcal{V}$ to D is balanced (i.e., the splitting type $\mathcal{V}|_D = \oplus_{i=1}^r \mathcal{O}_{\mathbb{P}^1}(a_i)$ satisfies $|a_i - a_j| \leq 1$ for all i, j).

Let H be an ample divisor on X. Suppose that $(K_X + D) \cdot H < 0$. Then a μ_H-semistable sheaf $\mathcal{V}$ is D-prioritary. By Serre duality,

$$\mathrm{ext}^2(\mathcal{V}, \mathcal{V}(-D)) = \mathrm{hom}(\mathcal{V}, \mathcal{V}(K_X + D)).$$

Since $(K_X + D) \cdot H < 0$, stability implies that the latter group vanishes. Hence, $\mathcal{V}$ is D-prioritary.

REMARK 5.3. The condition $(K_X + D) \cdot H < 0$ implies that $K_X \cdot H < 0$. Hence, mK_X cannot have any sections for $m > 0$. Therefore, this condition can only be satisfied for surfaces of Kodaira dimension $-\infty$. Furthermore, if $K_X + D$ is effective, there cannot be any D prioritary sheaves on X.

A projective surface X is *ruled* if it admits a morphism $X \to C$ onto a smooth curve where all the fibers are isomorphic to $\mathbb{P}^1$, equivalently if X is the projectivization of a vector bundle of rank 2 on C. A surface is *birationally ruled* if it admits a morphism $X \to C$ onto a smooth curve where the general fiber is $\mathbb{P}^1$. The following theorem of Walter makes prioritary sheaves a useful tool for studying moduli spaces of sheaves on birationally ruled surfaces.

THEOREM 5.4 (Walter [**Wal98**]). *Let X be a birationally ruled surface and let F be the fiber class. Then the stack of F-prioritary sheaves $\mathcal{P}_F(\mathbf{v})$ is irreducible whenever it is nonempty.*

This theorem generalizes an earlier theorem of Hirschowitz and Laszlo [**HL93**] which assets that $\mathcal{P}_{\mathbb{P}^2,L}(\mathbf{v})$ is irreducible when nonempty. The main advantage of working with prioritary sheaves is that they are easier than stable sheaves to construct. For example, $\mathcal{O}_{\mathbb{P}^2}(a) \oplus \mathcal{O}_{\mathbb{P}^2}(a+1)$ is L-prioritary on $\mathbb{P}^2$ but not stable.

REMARK 5.5. For rational surfaces, there are always polarizations H for which $(K_X + F) \cdot H < 0$. For minimal rational surfaces $K_X + F$ is anti-effective. Hence, $(K_X + F) \cdot H < 0$ for every ample class H. Every rational surface is obtained by blowing up a minimal rational surface at finitely many (possibly infinitely near) points. Let Y be the blowup of X at a point. If $(K_X + F) \cdot H < 0$, then $K_Y = \pi^* K_X + E$ and

$$(K_Y + F) \cdot \pi^* H = (K_X + F) \cdot H < 0.$$

The divisor $\pi^* H$ is not ample on Y, but nef. Any small perturbation H' of $\pi^* H$ by adding an ample divisor is ample. Hence, for an ample H' sufficiently close to $\pi^* H$, we still have $(K_Y + F) \cdot H' < 0$.

The utility of the notion of F-prioritary sheaves toward the weak Brill-Noether problem follows from openness of (semi)stability and semi-continuity of cohomology. Indeed, if $(K_X + F) \cdot H < 0$ and there exist (semi)stable sheaves with Chern character $\mathbf{v}$, then to show that $\mathbf{v}$ satisfies weak Brill-Noether it suffices to show that there exists an F-prioritary sheaf with at most one nonzero cohomology group. By semi-continuity of cohomology and irreducibility of the moduli stack, it follows that the general stable sheaf has at most one nonzero cohomology group. Hence, the moduli space satisfies weak Brill-Noether. This strategy can be used to show weak Brill-Noether on certain rational or birationally ruled surfaces.

5.2. Elementary modifications. Given a torsion free sheaf $\mathcal{V}$, a point $p \in X$ and a surjection $\phi : \mathcal{V} \twoheadrightarrow \mathcal{O}_p$, the kernel $\mathcal{V}'$ defined by the exact sequence

$$0 \to \mathcal{V}' \to \mathcal{V} \xrightarrow{\phi} \mathcal{O}_p \to 0 \tag{1}$$

is called the *elementary modification* of $\mathcal{V}$ with respect to ϕ. An elementary modification satisfies

$$r(\mathcal{V}') = r(\mathcal{V}), \quad \nu(\mathcal{V}') = \nu(\mathcal{V}), \quad \Delta(\mathcal{V}') = \Delta(\mathcal{V}) + \frac{1}{r}.$$

If $\mathcal{W}$ is a proper subsheaf of $\mathcal{V}'$ of smaller rank with $\mu_H(\mathcal{W}) \geq \mu_H(\mathcal{V}')$, then $\mathcal{W}$ is also a proper subsheaf of $\mathcal{V}$ with $\mu_H(\mathcal{W}) \geq \mu_H(\mathcal{V})$. Hence, if $\mathcal{V}$ is a μ_H-(semi)stable sheaf, then $\mathcal{V}'$ is also μ_H-(semi)stable. If $\mathcal{V}$ is a D-prioritary sheaf, then a general elementary modification is also D-prioritary.

CAUTION 5.6. Elementary modifications do not necessarily preserve Gieseker (semi)stability. For example, on $\mathbb{P}^2$ the sheaf $\mathcal{O}_{\mathbb{P}^2} \oplus \mathcal{O}_{\mathbb{P}^2}$ is Gieseker semistable, however any elementary modification is isomorphic to $I_p \oplus \mathcal{O}_{\mathbb{P}^2}$, which is not Gieseker semistable.

Since $h^i(\mathcal{O}_p) = 0$ for $i > 0$, we have that $H^2(X, \mathcal{V}') = H^2(X, \mathcal{V})$. The morphism ϕ corresponds to a choice of hyperplane Λ in the fiber of $\mathcal{V}$ over p. If $h^0(X, \mathcal{V}) > 0$, we can choose p and the hyperplane Λ so that one of the sections at p is not contained in Λ. In that case, the map $H^0(X, \mathcal{V}) \to H^0(X, \mathcal{O}_p)$ induced by ϕ is surjective. By the long exact sequence of cohomology associated to (1), we conclude that

$$h^0(X, \mathcal{V}') = h^0(X, \mathcal{V}) - 1, \quad h^1(X, \mathcal{V}') = h^1(X, \mathcal{V}).$$

The integrality of the Euler characteristic and the Riemann-Roch Theorem

$$\chi(\mathcal{V}) = r(\mathcal{V})(P(\nu(\mathcal{V})) - \Delta(\mathcal{V}))$$

imply that the difference of the discriminants of any two sheaves with the same rank and first Chern class is a multiple of $\frac{1}{r}$. Our discussion on elementary modifications yields the following theorem.

THEOREM 5.7. *Let $\mathcal{V}$ be a μ_H-(semi)stable (respectively, D-prioritary) sheaf of rank r, total slope ν and discriminant Δ_0 on a smooth projective surface X such that $h^2(X, \mathcal{V}) = 0$ and $\mathcal{V}$ has at most one nonzero cohomology group. Then there exists a μ_H-(semi)stable (respectively, D-prioritary) sheaf with rank r, total slope ν with at most one nonzero cohomology group for every $\Delta \geq \Delta_0$ for which (r, ν, Δ) is an integral Chern character.*

5.3. Weak Brill-Noether on $\mathbb{P}^2$. We are now ready to sketch the proof of Theorem 5.1] We will prove the following more general statement.

THEOREM 5.8. *Let $\mathcal{V}$ be a general prioritary sheaf on $\mathbb{P}^2$ with $\Delta(\mathcal{V}) \geq 0$ and rank at least 2. Then $\mathcal{V}$ has at most one nonzero cohomology group.*

PROOF. Since stable sheaves are prioritary and form a dense open substack of the stack of prioritary sheaves when nonempty, Theorem 5.1 follows by the semicontinuity of cohomology. By the irreducibility of the stack of prioritary sheaves and semicontinuity of cohomology, it suffices to exhibit one prioritary sheaf with at most one nonzero cohomology group. When the rank is at least 2, the general prioritary sheaf is locally free. Hence, by Serre duality, we may assume that $\mu(\mathcal{V}) \geq -\frac{3}{2}$. Given a rank r and a slope $\frac{c}{r} \geq -\frac{3}{2}$, we can find an L-prioritary sheaf $\mathcal{V} = \mathcal{O}_{\mathbb{P}^2}(a)^{\oplus r-s} \oplus \mathcal{O}_{\mathbb{P}^2}(a+1)^{\oplus s}$ with $a \geq -2$ of rank r and slope $\frac{c}{r}$. An easy computation shows that $\Delta(\mathcal{V}) \leq 0$. The sheaf $\mathcal{V}$ is nonspecial. Taking general elementary modifications, we obtain a prioritary sheaf that has at most one nonzero cohomology group for every integral Chern character. This concludes the proof of the theorem. □

5.3.1. *Gaeta resolutions.* On $\mathbb{P}^2$ one can write down a resolution for the general prioritary sheaf $\mathcal{V}$ with $\Delta \geq 0$. By Riemann-Roch, there exists a unique integer n such that $\chi(\mathcal{V}(-n)) \geq 0$ but $\chi(\mathcal{V}(-n-1)) < 0$. Set $n = a+2$. Then $\mathcal{V}$ has a resolution of the form

$$0 \to \mathcal{O}_{\mathbb{P}^2}(a)^{\alpha} \to \mathcal{O}_{\mathbb{P}^2}(a+1)^{\beta} \oplus \mathcal{O}_{\mathbb{P}^2}(a+2)^{\gamma} \to \mathcal{V} \to 0, \text{ or}$$
$$0 \to \mathcal{O}_{\mathbb{P}^2}(a)^{\alpha} \oplus \mathcal{O}_{\mathbb{P}^2}(a+1)^{\beta} \to \mathcal{O}_{\mathbb{P}^2}(a+2)^{\gamma} \to \mathcal{V} \to 0.$$

Here

$$\gamma = \chi(\mathcal{V}(-a-2)), \quad \alpha = -\chi(\mathcal{V}(-a-3)), \quad \beta = |\gamma - r(\mathcal{V}) - \alpha|.$$

The resolution is of the first type if $\gamma - r(\mathcal{V}) - \alpha \geq 0$ and otherwise of the second type. This generalizes Gaeta's resolution for the ideal sheaf of a general set of points on $\mathbb{P}^2$ [**Ga51**].

To see that such a resolution exists, given a Chern character $\mathbf{v}$ one can solve for a, α, β and γ and write a general Gaeta-type resolution. One then checks that $\mathcal{V}$ defined by such a resolution is prioritary and that the associated Kodaira-Spencer map is surjective (see [**CH18b**, §3] or [**CH20a**, §4] for more details). When $r(\mathcal{V}) \geq 2$, by a Bertini-type theorem, a sheaf defined by such a resolution is locally free. Consequently, we also deduce that the general prioritary sheaf of rank at least 2 is locally free. The existence of the Gaeta resolution also implies that the moduli spaces $M_{\mathbb{P}^2,L}(\mathbf{v})$ are unirational.

In general, we do not know a presentation for the general stable bundle on a surface. Having such presentations even for special families of surfaces would be very useful.

5.3.2. *Global generation.* The weak Brill-Noether Theorem allows one to classify Chern characters of nonspecial stable bundles that are globally generated.

THEOREM 5.9 ([**CH20a**, Corollary 5.3] (see also [**BGJ16**])). *The general member of $M_{\mathbb{P}^2,L}(\mathbf{v})$ is globally generated if and only if one of the following holds:*

- $\mathbf{v} = r(1,0,0)$ *and* $\mathcal{V} = \mathcal{O}_{\mathbb{P}^2}^{\oplus r}$
- $\mu(\mathbf{v}) > 0$ *and* $\chi(\mathbf{v}(-1)) \geq 0$
- $\mu(\mathbf{v}) > 0$, $\chi(\mathbf{v}(-1)) < 0$ *and* $\chi(\mathbf{v}) \geq r+2$

- $\mu(\mathbf{v}) > 0$, $\chi(\mathbf{v}(-1)) < 0$, $\chi(\mathbf{v}) \geq r+1$, *and* $\mathbf{v} = (r+1)\operatorname{ch}(\mathcal{O}_{\mathbb{P}^2}) - \operatorname{ch}(\mathcal{O}_{\mathbb{P}^2}(-2))$.

PROOF. If $\mathcal{V}$ is globally generated, then its determinant is globally generated, hence $\mu(\mathcal{V}) \geq 0$. If $\mu(\mathcal{V}) = 0$, then the Riemann-Roch Theorem implies that $\chi(\mathcal{V}) \leq \operatorname{rk}(\mathcal{V})$ with equality if and only if $\Delta(\mathcal{V}) = 0$. If $\mu(\mathcal{V}) = \Delta(\mathcal{V}) = 0$, then $\mathcal{V} = \mathcal{O}_{\mathbb{P}^2}^{\oplus r}$.

If $\chi(\mathbf{v}(-1)) \geq 0$, then the general sheaf in $\mathcal{P}_{\mathbb{P}^2,L}(\mathbf{v})$ has a Gaeta resolution with $a \geq 1$. Then the general sheaf is clearly a quotient of a globally generated bundle. If $\chi(\mathbf{v}(-1)) < 0$ and $\chi(\mathbf{v}) \geq \operatorname{rk}(\mathbf{v}) + 2$, then the general sheaf in $\mathcal{P}_{\mathbb{P}^2,L}(\mathbf{v})$ has a Gaeta resolution of the form

$$0 \to \mathcal{O}_{\mathbb{P}^2}(-2)^k \oplus \mathcal{O}_{\mathbb{P}^2}(-1)^l \to \mathcal{O}_{\mathbb{P}^2}^m \to \mathcal{V} \to 0, \text{ or}$$

$$0 \to \mathcal{O}_{\mathbb{P}^2}(-2)^k \to \mathcal{O}_{\mathbb{P}^2}(-1)^l \oplus \mathcal{O}_{\mathbb{P}^2}^m \to \mathcal{V} \to 0.$$

In the first case, $\mathcal{V}$ is the quotient of a globally generated vector bundle, hence globally generated. The most interesting case is the second one. By the assumption that $\chi(\mathcal{V}) \geq \operatorname{rk}(\mathcal{V}) + 2$, we have that $m \geq \operatorname{rk}(\mathcal{V}) + 2$. Therefore, $k \geq l+2$. To show that $\mathcal{V}$ is globally generated, it suffices to show that $H^1(\mathbb{P}^2, \mathcal{V} \otimes I_p) = 0$ for every point $p \in \mathbb{P}^2$. By the long exact sequence of cohomology, it suffices to show that the map

$$\phi : H^1(\mathbb{P}^2, I_p(-2))^k \to H^1(\mathbb{P}^2, I_p(-1))^l$$

is surjective. Consider the sequence

$$0 \to M \to \mathcal{O}_{\mathbb{P}^2}(-2)^k \to \mathcal{O}_{\mathbb{P}^2}(-1)^l \to 0.$$

Since the map is general, it is surjective and M is a vector bundle. Clearly M does not have any cohomology. Tensoring the standard exact sequence $0 \to I_p \to \mathcal{O}_{\mathbb{P}^2} \to \mathcal{O}_p \to 0$ with M, we see that $H^2(\mathbb{P}^2, I_p \otimes M) = 0$. Consequently, the map ϕ is surjective and $\mathcal{V}$ is globally generated. Finally, if $\chi(\mathcal{V}) = \operatorname{rk}(\mathcal{V}) + 1$ and $\mathcal{V}$ is globally generated, then there is a surjective map $\mathcal{O}_{\mathbb{P}^2}^{r+1} \to \mathcal{V}$. The kernel of this map is a line bundle $\mathcal{O}_{\mathbb{P}^2}(-d)$. If $d = 1$, then $\chi(\mathcal{V}(-1)) = 0$. If $d \geq 3$, then $\chi(\mathcal{V}) < r$ and it is not possible for the general prioritary sheaf with Chern character $\mathbf{v}$ to be globally generated. The only remaining possibility is for $d = 2$. In that case, $\chi(\mathcal{V}) = r+1$ and this is the Gaeta resolution of the general sheaf. This concludes the classification of globally generated Chern characters on $\mathbb{P}^2$. □

5.4. Weak Brill-Noether for Hirzebruch surfaces. Let $e \geq 0$ be an integer. Let $\mathbb{F}_e := \mathbb{P}(\mathcal{O}_{\mathbb{P}^1} \oplus \mathcal{O}_{\mathbb{P}^1}(e))$ denote the Hirzebruch surface. We refer the reader to [**Bea83, Cos06a**] for detailed information on the geometry of Hirzebruch surfaces. The surface $\mathbb{F}_e$ has a section E with self-intersection $-e$. Let F denote the class of a fiber. Then $\operatorname{Pic}(\mathbb{F}_e) = \mathbb{Z}F \oplus \mathbb{Z}E$. The intersection numbers are

$$E^2 = -e, \quad E \cdot F = 1, \quad \text{and} \quad F^2 = 0.$$

The effective cone of $\mathbb{F}_e$ is spanned by E and F. The nef cone of $\mathbb{F}_e$ is spanned by $E + eF$ and F. The minimal rational surfaces are $\mathbb{P}^2$ and $\mathbb{F}_e$ for $e \geq 0$, $e \neq 1$. The existence of curves with negative self-intersection provides an obstruction for the weak Brill-Noether property on Hirzebruch surfaces $\mathbb{F}_e$ with $e \geq 1$.

5.4.1. *Negative curves on a surface and the cohomology of vector bundles.* Suppose that a surface X contains a smooth curve C with $C^2 < 0$. Then by the Riemann-Roch Theorem,

$$\chi(\mathcal{O}_X(mC)) = \chi(\mathcal{O}_X) + \frac{m^2}{2}C^2 - \frac{m}{2}K \cdot C.$$

If $m \gg 0$, $\chi(\mathcal{O}_X(mC)) < 0$. Hence, both $H^0(X, \mathcal{O}_X(mC))$ and $H^1(X, \mathcal{O}_X(mC))$ are nonzero. In fact, if X is a surface such that

$$q = h^1(X, \mathcal{O}_X) = 0 \quad \text{and} \quad p_g = h^2(X, \mathcal{O}_X) = 0,$$

then $\mathcal{O}_X(mC)$ is an effective line bundle with nonvanishing h^1 provided that either

(1) $m \geq 1$ and either $C^2 < -1$ or the genus of C is at least 1, or
(2) $m \geq 2$ and C is an exceptional curve.

A similar phenomenon persists for higher rank vector bundles.

PROPOSITION 5.10. *Let X be a smooth projective surface that contains an irreducible curve C with $C^2 < 0$. Then for every rank r there exist infinitely many Chern characters $\mathbf{v}_i$ with rank r such that the moduli space $M_{X,H}(\mathbf{v}_i)$ is nonempty and does not satisfy weak Brill-Noether.*

PROOF. Let $M_{X,H}(\mathbf{v})$ be an irreducible moduli space of rank r sheaves where the general sheaf is μ_H-stable and has sections. Such moduli spaces exist by O'Grady's Theorem 4.9 and Serre vanishing. Let $\mathcal{V}$ be a general sheaf in $M_{X,H}(\mathbf{v})$. Let $\mathbf{v}_i$ be the Chern character of $\mathcal{V}(iC)$. Since $H^0(X, \mathcal{V}) \subset H^0(X, \mathcal{V}(iC))$, the sheaves $\mathcal{V}(iC)$ have global sections. On the other hand, by Riemann-Roch if $i \gg 0$, then $\chi(\mathbf{v}_i) < 0$, hence, they must also have h^1 and cannot satisfy weak Brill-Noether. □

Unlike $\mathbb{P}^2$, by Proposition 5.10, we cannot expect weak Brill-Noether to always hold for moduli spaces on $\mathbb{F}_e$ with $e \geq 1$. The next theorem shows that the existence of negative curves is the only obstruction for weak Brill-Noether on $\mathbb{F}_e$.

THEOREM 5.11 ([**CH20a**, Theorem 3.1]). *Let $\mathbf{v}$ be a Chern character with positive rank $r(\mathbf{v})$ and $\Delta(\mathbf{v}) \geq 0$. Then the stack of prioritary sheaves $\mathcal{P}_F(\mathbf{v})$ is nonempty and irreducible. Let V be a general sheaf in $\mathcal{P}_F(\mathbf{v})$.*

(1) *If $\nu(\mathbf{v}) \cdot F \geq -1$, then $h^2(\mathbb{F}_e, \mathcal{V}) = 0$.*
(2) *If $\nu(\mathbf{v}) \cdot F \leq -1$, then $h^0(\mathbb{F}_e, \mathcal{V}) = 0$.*
(3) *If $\nu(\mathbf{v}) \cdot F = -1$, then $h^1(\mathbb{F}_e, \mathcal{V}) = -\chi(\mathcal{V})$ and all other cohomology vanishes.*

Assume that $\nu(\mathbf{v}) \cdot F > -1$.

(4) *If $\nu(\mathbf{v}) \cdot E \geq -1$, then $\mathcal{V}$ has at most one nonzero cohomology group. If $\chi(\mathcal{V}) \geq 0$, then $h^0(\mathbb{F}_e, \mathcal{V}) = \chi(\mathcal{V})$, and if $\chi(\mathcal{V}) < 0$, then $h^1(\mathbb{F}_e, \mathcal{V}) = -\chi(\mathcal{V})$.*
(5) *If $\nu(\mathbf{v}) \cdot E < -1$, then $H^0(\mathbb{F}_e, \mathcal{V}) \cong H^0(\mathbb{F}_e, \mathcal{V}(-E))$ and the Betti numbers of $\mathcal{V}$ are inductively determined by (2) and (4).*

Observe that if $\nu(\mathbf{v}) \cdot F < -1$ and $\mathrm{rk}(\mathbf{v}) \geq 2$, then the general prioritary sheaf is locally free and Serre duality determines the cohomology.

SKETCH OF PROOF. The proof is similar to the proof of Theorem 5.1. A vector bundle $\mathcal{V}$ of the form

$$\mathcal{O}_{\mathbb{F}_e}(-E-(e+1)F)^{\oplus a} \oplus \mathcal{O}_{\mathbb{F}_e}(-F)^{\oplus b} \oplus \mathcal{O}_{\mathbb{F}_e}^{\oplus c} \text{ or}$$
$$\mathcal{O}_{\mathbb{F}_e}(-E-(e+1)F)^{\oplus a} \oplus \mathcal{O}_{\mathbb{F}_e}(-E-eF)^{\oplus b} \oplus \mathcal{O}_{\mathbb{F}_e}^{\oplus c}$$

is both E-prioritary and F-prioritary and has $\Delta(\mathcal{V}) \leq 0$. Moreover, $\mathcal{V}$ has no higher cohomology. If $\mathbf{v}$ satisfies $\nu(\mathbf{v}) \cdot F \geq -1$ and $\nu(\mathbf{v}) \cdot E \geq -1$, we can find a prioritary bundle of Chern character $\mathbf{v}$ by tensoring $\mathbf{v}$ by a nef line bundle $\mathcal{N}$ on $\mathbb{F}_e$ since tensoring with a nef line bundle preserves prioritariness and the discriminant. Furthermore, $\mathcal{V} \otimes \mathcal{N}$ has no higher cohomology on $\mathbb{F}_e$. We conclude that weak Brill-Noether holds for slopes satisfying $\nu(\mathbf{v}) \cdot F \geq -1$ and $\nu(\mathbf{v}) \cdot E \geq -1$. For slopes in the range $\nu(\mathbf{v}) \cdot F > -1$ and $\nu(\mathbf{v}) \cdot E < -1$, one can use the exact sequence

$$0 \to \mathcal{V}(-E) \to \mathcal{V} \to \mathcal{V}|_E \to 0$$

and the fact that $\mathcal{V}|_E$ is balanced on E to inductively compute the cohomology. □

As in the case of $\mathbb{P}^2$, the weak Brill-Noether Theorem allows one to find a Gaeta-type resolution for the general prioritary sheaf on Hirzebruch surfaces.

THEOREM 5.12 ([**CH20a**, Theorem 4.1]). *Let $\mathbf{v}$ be an integral Chern character on $\mathbb{F}_e$ of positive rank and assume that*

$$\Delta(\mathbf{v}) \geq \frac{1}{4} \text{ if } e=0, \quad \Delta(\mathbf{v}) \geq \frac{1}{8} \text{ if } e=1, \quad \Delta(\mathbf{v}) \geq 0 \text{ if } e \geq 2.$$

Then the general sheaf $\mathcal{V} \in \mathcal{P}_{\mathbb{F}_e,F}(\mathbf{v})$ admits a Gaeta-type resolution

$$0 \to L(-E-(e+1)F)^a \to L(-E-eF)^b \oplus L(-F)^c \oplus L^d \to \mathcal{V} \to 0, \tag{2}$$

for some line bundle L and nonnegative integers a, b, c, d.

Theorems 5.11 and 5.12 allow one to classify moduli spaces where the general bundle is a nonspecial, globally generated bundle.

THEOREM 5.13 ([**CH20a**, Theorem 5.1]). *Suppose $e \geq 1$. Let $\mathbf{v}$ be a Chern character on $\mathbb{F}_e$ such that $r(\mathbf{v}) \geq 2$, $\Delta(\mathbf{v}) \geq 0$ and $\nu(\mathbf{v})$ is nef. Then the general member of $\mathcal{P}_F(\mathbf{v})$ is globally generated if and only if one of the following holds.*

(1) *We have $\nu(\mathbf{v}) \cdot F = 0$ and*
$$\mathbf{v} = (r(\mathbf{v}) - m)\operatorname{ch}(\mathcal{O}_{\mathbb{F}_e}(aF)) + m\operatorname{ch}(\mathcal{O}_{\mathbb{F}_e}((a+1)F))$$
for integers $a, m \geq 0$.
(2) *We have $\nu(\mathbf{v}) \cdot F > 0$ and $\chi(\mathbf{v}(-F)) \geq 0$.*
(3) *We have $\nu(\mathbf{v}) \cdot F > 0$ and $\chi(\mathbf{v}(-F)) < 0$ and $\chi(\mathbf{v}) \geq r(\mathbf{v}) + 2$.*
(4) *We have $e = 1$, $\nu(\mathbf{v}) \cdot F > 0$, $\chi(\mathbf{v}(-F)) < 0$, $\chi(\mathbf{v}) \geq r(\mathbf{v}) + 1$ and*
$$\mathbf{v} = (r(\mathbf{v}+1)\operatorname{ch}(\mathcal{O}_{\mathbb{F}_1}) - \operatorname{ch}(\mathcal{O}_{\mathbb{F}_1}(-2E-2F)).$$

REMARK 5.14. For $\mathbb{F}_0 = \mathbb{P}^1 \times \mathbb{P}^1$, the statement needs to be slightly modified (see [**CH20a**, Theorem 5.2]). If F_1 and F_2 are the two rulings on $\mathbb{P}^1 \times \mathbb{P}^1$, then the general member of $\mathcal{P}_F(\mathbf{v})$ is globally generated if and only if one of the following holds.

(1) We have $\nu(\mathbf{v}) \cdot F_i = 0$ for some $1 \leq i \leq 2$ and
$$\mathbf{v} = (r(\mathbf{v}) - m)\operatorname{ch}(\mathcal{O}_{\mathbb{F}_e}(aF_i)) + m\operatorname{ch}(\mathcal{O}_{\mathbb{F}_e}((a+1)F_i))$$
for integers $a, m \geq 0$.

(2) We have $\nu(\mathbf{v})\cdot F_i > 0$ for $1 \leq i \leq 2$ and $\chi(\mathbf{v}(-F_j)) \geq 0$ for some $1 \leq j \leq 2$.
(3) We have $\nu(\mathbf{v}) \cdot F_i > 0$, $\chi(\mathbf{v}(-F_i)) < 0$ for $1 \leq i \leq 2$ and $\chi(\mathbf{v}) \geq r(\mathbf{v}) + 2$.

5.4.2. *Applications to ample bundles.* On curves stable vector bundles are ample if and only if their slope is positive. On higher dimensional varieties, it is easy to see that there can be no numerical characterization of ample vector bundles of higher rank. Ampleness is an open condition and we can ask when a general member of an irreducible component of a moduli space is ample.

PROBLEM 5.15. Let X be a smooth, projective variety. Classify Chern characters $\mathbf{v}$ on X for which there exists an ample vector bundle of Chern character $\mathbf{v}$.

This problem is wide open even for surfaces. Recently, Huizenga and Kopper [**HK22**] have made significant progress towards the solution of the problem for $\mathbb{P}^2$ and Hirzebruch surfaces. If $\mathcal{V}$ is an ample bundle on $\mathbb{P}^2$ or $\mathbb{F}_e$, then the restrictions $\mathcal{V}|_L$, $\mathcal{V}|_F$ and $\mathcal{V}|_E$ are ample, hence split as a direct sum of line bundles on $\mathbb{P}^1$ of positive degree. Huizenga and Kopper prove that on minimal rational surfaces, asymptotically this is the only obstruction to ampleness of the general stable bundle.

THEOREM 5.16 ([**HK22**, Theorem 4.1]). *Let $X = \mathbb{P}^2$ or $\mathbb{F}_e$. Let $\mathbf{v}$ be a stable Chern character on X such that*

(1) *if $X = \mathbb{P}^2$, $\nu(\mathbf{v}) \cdot L > 1$,*
(2) *if $X = \mathbb{P}^1 \times \mathbb{P}^1$, $\nu(\mathbf{v}) \cdot F > 1$ and $\nu(\mathbf{v}) \cdot E > 1$,*
(3) *if $X = \mathbb{F}_e$ with $e \geq 1$, $\nu(\mathbf{v}) \cdot F > 1$ and $\nu(\mathbf{v}) \cdot E \geq 1$.*

Then for $n \gg 0$, the general bundle $\mathcal{V} \in M_{X,H}(n\mathbf{v})$ is ample.

Huizenga and Kopper also classify globally generated ample bundles on minimal rational surfaces.

THEOREM 5.17 ([**HK22**, Theorem 5.1]). *Let $X = \mathbb{P}^2$ or $\mathbb{F}_e$. Let $\mathbf{v}$ be a Chern character such that the moduli space $M_{X,H}(\mathbf{v})$ is nonempty and the general sheaf is a globally generated vector bundle with no higher cohomology.*

(1) *if $X = \mathbb{P}^2$, $\nu(\mathbf{v}) \cdot L > 1 + \frac{1}{r}$,*
(2) *if $X = \mathbb{P}^1 \times \mathbb{P}^1$, $\nu(\mathbf{v}) \cdot F > 1$ and $\nu(\mathbf{v}) \cdot E > 1$,*
(3) *if $X = \mathbb{F}_e$ with $e \geq 1$, $\nu(\mathbf{v}) \cdot F > 1$ and $\nu(\mathbf{v}) \cdot E \geq 1$.*

Then the general sheaf $\mathcal{V} \in M_{X,H}(\mathbf{v})$ is ample.

5.5. Weak Brill-Noether for more general rational surfaces. The ideas and techniques used to study the weak Brill-Noether Problem on $\mathbb{P}^2$ and Hirzebruch surfaces can be extended to non-minimal rational surfaces. Here we will discuss general blowups of $\mathbb{P}^2$ and refer the reader to [**CH18b, CH18c**] for analogous statements for general blowups of Hirzebruch surfaces.

Let X_m denote the blowup of $\mathbb{P}^2$ at m general points $p_1, \ldots, p_m$. Let E_i denote the exceptional divisor lying over p_i. Let H denote the pullback of the class of a line. Then

$$\mathrm{Pic}(X_m) = \mathbb{Z}H \oplus \bigoplus_{i=1}^{m} E_i$$

and the intersection pairing satisfies

$$H^2 = 1, \quad H \cdot E_i = 0, \quad E_i \cdot E_j = -\delta_{i,j},$$

where $\delta_{i,j}$ is the Kronecker delta function. When $m \leq 8$, then the corresponding surface is a del Pezzo surface. Del Pezzo surfaces and $\mathbb{P}^1 \times \mathbb{P}^1$ are the Fano surfaces, that is the surfaces that have an ample anti-canonical bundle. We refer the reader to [**Bea83, Cos06b, Har77**] for more details on the geometry of del Pezzo surfaces.

Proposition 5.10 shows that we cannot expect weak Brill-Noether to always hold. However, the following higher rank generalization of the SHGH might hold.

CONJECTURE 5.18 ([**CH18b**, Conjecture 1.7]). *Let X be a blowup of $\mathbb{P}^2$ at m very general points. Let H be an ample class on X such that $H \cdot K_X < 0$. Let $\mathbf{v}$ be a Chern character such that $\nu(\mathbf{v})$ is nef. Then the general stable sheaf in $M_{X,H}(\mathbf{v})$ has at most one nonzero cohomology group.*

REMARK 5.19. One can make a bolder conjecture by weakening the condition that $\nu(\mathbf{v})$ is nef to $C \cdot \nu(\mathbf{v}) \geq -1$ for every (-1)-curve on X. This would mimic the SHGH Conjecture more closely. Even the weaker conjecture is open when $m \geq 5$.

When $\nu(\mathbf{v})$ is not too close to the boundary of the nef cone, one can find prioritary direct sums of line bundles with no higher cohomology. More precisely, let $\mathbf{v}$ be a Chern character of rank r and let the total slope be $\nu(\mathbf{v}) = \delta H - \sum_{i=1}^k \alpha_i E_i$. Then we have that

$$\delta = d + \frac{q}{r}, \quad \alpha_i = a_i + \frac{q_i}{r}$$

for some integers d, q, a_i and q_i with $0 \leq q < r$ and $0 \leq q_i < r$. Set

$$\gamma(\mathbf{v}) = \frac{q^2}{2r^2} - \frac{q}{2r} + \sum_{i=1}^k \left(\frac{q_i}{2r} - \frac{q_i^2}{2r^2} \right).$$

THEOREM 5.20 ([**CH18c**, Theorem 4.12]). *Let X be the blowup of $\mathbb{P}^2$ at m distinct points. Let $\mathbf{v}$ be a positive rank Chern character on X with total slope*

$$\nu(\mathbf{v}) = \delta H - \alpha_1 E_1 \cdots - \alpha_k E_k$$

with $\delta \geq 0$ and $\alpha_i \geq 0$. Suppose that the line bundle

$$\lfloor \delta \rfloor H - \lceil \alpha_1 \rceil E_1 \cdots - \lceil \alpha_k \rceil E_k$$

does not have higher cohomology. Assume that $\Delta(\mathbf{v}) \geq \gamma(\mathbf{v})$. Then the stack of prioritary sheaves $\mathcal{P}_{X,H-E_1}(\mathbf{v})$ is nonempty and the general sheaf in $\mathcal{P}_{X,H-E_1}(\mathbf{v})$ has at most one nonzero cohomology group.

Levine and Zhang in [**LZ19**] have studied the case of del Pezzo surfaces in greater detail. Let W_m denote the Weyl group acting on $\operatorname{Pic}(X_m)$. Let $\mathcal{P}_{H^W}(\mathbf{v})$ denote the stack parameterizing torsion free sheaves $\mathcal{V}$ such that $\operatorname{Ext}^2(\mathcal{V}, \mathcal{V}(-\sigma(H))) = 0$ for every $\sigma \in W_m$.

THEOREM 5.21 ([**LZ19**, Theorem 1.2]). *Let X_m be a del Pezzo surface with $m \leq 5$. Let $\mathbf{v}$ be a Chern character such that $H \cdot \nu(\mathbf{v}) \geq -2$ and $\mathcal{P}_{H^W}(\mathbf{v}) \neq \emptyset$.*

(1) *If $C \cdot \nu(\mathbf{v}) \geq -1$ for all (-1)-curves C on X_m, then $\mathbf{v}$ is nonspecial.*
(2) *If there exists $\sigma \in W_m$ such that $\nu(\mathbf{v}) \cdot \sigma(H) \leq -1$ or $\nu(\mathbf{v}) \cdot \sigma(H - E_i) \leq -1$ for some i, then $\mathbf{v}$ is nonspecial.*
(3) *Let $\nu(\mathbf{v}) \cdot \sigma(H) > -1$ and $\nu(\mathbf{v}) \cdot \sigma(H - E_i) > -1$ for all i and all $\sigma \in W_m$. Suppose C is a (-1)-curve such that $C \cdot \nu(\mathbf{v}) < -1$ and let π denote the map contracting C. Then $\mathbf{v}$ is nonspecial if and only if $\pi_*(\mathcal{V})$ is nonspecial for the general $\mathcal{V}$ and $\chi(\pi_*(\mathcal{V})) \leq 0$.*

As an application, when $m \leq 6$, Levine and Zhang classify $-K_{X_m}$-stable Chern characters in terms of a Drézet-Le Potier type condition. More generally, by studying blowups of $\mathbb{P}^2$ along collinear points and taking deformations, Zhao has studied the weak Brill-Noether Problem on general blowups of $\mathbb{P}^2$ [**Zha22**, Theorems 4.7 and 7.4] and has shown the nonemptiness of moduli spaces for certain Chern characters and polarizations [**Zha22**, Theorems 6.14 and 7.3].

5.6. Weak Brill-Noether for K-trivial surfaces. There has been recent progress for computing the cohomology of the general sheaf on K-trivial surfaces using Bridgeland stability conditions. In this subsection, we will introduce Bridgeland stability conditions and briefly explain the strategy for using them to prove weak Brill-Noether Theorems, concentrating on the case of K3 surfaces.

5.6.1. *Bridgeland stability.* Let $\mathcal{D}^b(X)$ denote the bounded derived category of coherent sheaves on X. Let $K(\mathcal{D}^b(X))$ denote the K-group of $\mathcal{D}^b(X)$ and let $Z : K(\mathcal{D}^b(X)) \to \mathbb{C}$ be a group homomorphism called *the central charge* which factors through the Chern character.

DEFINITION 5.22. A *Bridgeland stability condition* on $\mathcal{D}^b(X)$ is a pair $\sigma = (\mathcal{A}, Z)$, where $\mathcal{A}$ is an abelian category which is the heart of a bounded t-structure on $\mathcal{D}^b(X)$, and Z is a central charge satisfying the following properties:

(1) Positivity: If $0 \neq E \in \mathcal{A}$, then $Z(E) = re^{i\theta}$ with $r > 0$ and $0 < \theta \leq \pi$. Using Z, we can define $\mu_\sigma(E) = -\frac{\Re(Z(E))}{\Im(Z(E))}$. An object $E \in \mathcal{A}$ is σ-*(semi)stable* if for every subobject F in $\mathcal{A}$, we have $\mu_\sigma(F) \underset{(-)}{<} \mu_\sigma(E)$.
(2) Harder-Narasimhan property: Every object $E \in \mathcal{A}$ has a finite Harder-Narasimhan filtration with σ-semistable quotients of strictly decreasing slope.
(3) Support condition: For a fixed norm $|\cdot|$ on $H^*_{\text{alg}}(X, \mathbb{R})$, there exists a constant $C > 0$ such that for all σ-semistable $E \in \mathcal{A}$ we have $|Z(E)| \geq C|\operatorname{ch}(E)|$.

The set $\operatorname{Stab}(X)$ of Bridgeland stability conditions on X has the structure of a complex manifold [**Bri07**, Corollary 1.3]. Here we will be interested in very special stability conditions on surfaces constructed by Bridgeland [**Bri08**] and Arcara and Bertram [**AB13**].

Given an ample divisor H on a surface X, define two subcategories of the category of coherent sheaves $\operatorname{Coh}(X)$ by

$$\mathcal{T}_s = \{E \in \operatorname{Coh}(X) : \mu_H(G) > sH^2 \text{ for every quotient } G \text{ of } E\}$$
$$\mathcal{F}_s = \{E \in \operatorname{Coh}(X) : \mu_H(F) \leq sH^2 \text{ for every subsheaf } F \text{ of } E\}.$$

The pair $(\mathcal{T}_s, \mathcal{F}_s)$ forms a torsion pair in $\operatorname{Coh}(X)$, that is $\operatorname{Hom}(T, F) = 0$ if $T \in \mathcal{T}_s$ and $F \in \mathcal{F}_s$ and every coherent sheaf $\mathcal{V}$ fits in a unique exact sequence

$$0 \to T \to \mathcal{V} \to F \to 0,$$

where $T \in \mathcal{T}_s$ and $F \in \mathcal{F}_s$.

Given a torsion pair in the heart of a t-structure, one obtains the heart of a new t-structure by tilting. Explicitly, tilting $\operatorname{Coh}(X)$ with respect to $(\mathcal{T}_s, \mathcal{F}_s)$, we obtain the category $\mathcal{A}_s$ defined by

$$\mathcal{A}_s = \{E^\bullet \in \mathcal{D}^b(X) : H^{-1}(E^\bullet) \in \mathcal{F}_s, H^0(E^\bullet) \in \mathcal{T}_s, H^i(E^\bullet) = 0 \text{ for } i \neq -1, 0\}.$$

Define

$$Z_{s,t}(E) = -\int_X e^{-(s+it)H}\,\mathrm{ch}(E).$$

Then the pair $\sigma_{s,t} = (\mathcal{A}_s, Z_{s,t})$ is a Bridgeland stability condition for $s, t \in \mathbb{R}, t > 0$ [**AB13**]. Given a Chern character $\mathbf{v}$ and a stability condition $\sigma = (\mathcal{A}_s, Z_{s,t})$, let $M_\sigma(\mathbf{v})$ be the moduli space of σ-semistable objects in $\mathcal{A}_s$ with Chern character $\mathbf{v}$.

5.6.2. *Walls.* Given a Chern character $\mathbf{v}$, there is a locally finite wall-and-chamber decomposition of the upper (s,t)-half-plane such that within each chamber the σ-(semi)stable objects remain constant. If a semistable object E gets destabilized by a subobject F, then the σ-slopes of E and F must become equal for some stability conditions. Expressing the equality $\mu_\sigma(E) = \mu_\sigma(F)$ as a function of s and t, one obtains that the walls are either vertical lines or nested, disjoint semi-circles. Moreover, when t is sufficiently large (the large volume limit), the Bridgeland moduli space is isomorphic to $M_{X,H}(\mathbf{v})$.

5.6.3. *Weak Brill-Noether for K3 surfaces.* Let X be a K3 surface. Stable sheaves on K3 surfaces have been classified by Mukai [**Muk87**], O'Grady [**O'G99**] and Yoshioka [**Yos99**]. It is customary to phrase the classification in terms of the Mukai vector instead of the Chern character. Let

$$\mathbf{v}(E) := \mathrm{ch}(E)\sqrt{\mathrm{td}(X)} = (r(E), c_1(E), r(E) + \mathrm{ch}_2(E)),$$

where $\mathrm{td}(X)$ is the Todd class of X. There is a pairing between any two Mukai vectors given by

$$\langle \mathbf{v}, \mathbf{v}'\rangle = \langle (r,c,a), (r',c',a')\rangle := c\cdot c' - ra' - r'a = -\chi(\mathbf{v}, \mathbf{v}'),$$

where $c \cdot c'$ is the intersection product on $H^2(X,\mathbb{Z})$. A class is called *spherical* if $\mathbf{v}^2 = -2$ and *isotropic* if $\mathbf{v}^2 = 0$. A Mukai vector $\mathbf{v}$ is *primitive* if it is not divisible in $H^*_{\mathrm{alg}}(X,\mathbb{Z})$. We say a primitive Mukai vector $\mathbf{v} = (r,c,a)$ is *positive* if $\mathbf{v}^2 \geq -2$ and either

(1) $r > 0$; or
(2) $r = 0$, c is effective, and $a \neq 0$; or
(3) $r = c = 0$ and $a > 0$.

THEOREM 5.23. *Let X be a K3 surface and let $\mathbf{v} = m\mathbf{v}_0$ be a Mukai vector, where $\mathbf{v}_0$ is a primitive positive Mukai vector and $m > 0$. Then $M_{X,H}(\mathbf{v})$ is non-empty for any ample divisor H. If H is* generic *with respect to $\mathbf{v}$, then:*

(1) *The moduli space $M_{X,H}(\mathbf{v})$ is non-empty if and only if $\mathbf{v}_0^2 \geq -2$.*
(2) *If $m = 1$ or $\mathbf{v}_0^2 > 0$, then $\dim M_{X,H}(\mathbf{v}) = \mathbf{v}^2 + 2$.*
(3) *When $\mathbf{v}_0^2 = -2$, then $M_{X,H}(\mathbf{v})$ is a single point parameterizing the direct sum of m copies of a spherical bundle. When $\mathbf{v}_0^2 = 0$, then $\dim M_{X,H}(\mathbf{v}) = 2m$.*
(4) *When $\mathbf{v}_0^2 > 0$, $M_{X,H}(\mathbf{v})$ is a normal irreducible projective variety with $\mathbb{Q}$-factorial singularities.*

Now suppose that $\mathrm{Pic}(X) = \mathbb{Z}H$ with $H^2 = 2n$. If $\mathcal{V}$ is a stable sheaf with $\mathrm{ch}_1(\mathcal{V}) = dH$ with $d > 0$, then $H^2(X,\mathcal{V}) = 0$ by stability. Hence, we need to compute $H^0(X,\mathcal{V})$ and $H^1(X,\mathcal{V})$. The fact that $H^2(X,\mathcal{O}_X) = \mathbb{C}$ allows one to construct many counterexamples to weak Brill-Noether on K3 surfaces.

EXAMPLE 5.24 ([**CNY23**, Example 1.3]). The linear system $|H|$ defines a morphism $f : X \to \mathbb{P}^{n+1}$. The sheaf $\mathcal{V} = f^*T_{\mathbb{P}^{n+1}}$ is the unique stable sheaf in its moduli space. The pullback of the Euler sequence by f

$$0 \to \mathcal{O}_X \to \mathcal{O}_X(H)^{n+2} \to \mathcal{V} \to 0$$

implies that $h^0(X, \mathcal{V}) = n^2 + 4n + 3$ and $h^1(X, \mathcal{V}) = 1$.

EXAMPLE 5.25 ([**CNY23**, Example 6.3]). Let X be a double cover of $\mathbb{P}^2$ branched along a very general sextic curve. Then the pullback of an exceptional bundle on $\mathbb{P}^2$ is a spherical stable bundle on X, which is the unique member of its moduli space. For example, let f_k be the kth Fibonacci number with $f_1 = f_2 = 1$. For $k \geq 2$, the pullback of the corresponding exceptional bundle from $\mathbb{P}^2$ has resolution

$$0 \to \mathcal{O}_X^{\oplus f_{2k-2}} \to \mathcal{O}_X(H)^{\oplus f_{2k}} \to \mathcal{V}_k \to 0.$$

The bundle $\mathcal{V}_k$ has rank f_{2k-1} and

$$h^0(X, \mathcal{V}_k) = 2f_{2k} + f_{2k-1} \quad \text{and} \quad h^1(X, \mathcal{V}_k) = f_{2k-2}.$$

When $k = 2$, one recovers the $n = 1$ case of the previous example.

EXAMPLE 5.26 ([**CNY23**, Theorem 10.1]). Let C be a smooth member in $|H|$. Consider extensions of the form

$$0 \to \mathcal{O}_X(H)^r \to \mathcal{V} \to \mathcal{O}_C(L) \to 0,$$

where L is a general line bundle on C with Euler characteristic $2n - r$. Then $\mathbf{v}(\mathcal{V}) = (r, (r+1)H, n(r+2))$, and from the exact sequence we conclude that

$$h^0(X, \mathcal{V}) = r(n+2) + \max(0, 2n - r) \quad \text{and} \quad h^1(X, \mathcal{V}) = \max(0, r - 2n).$$

As the general element in $M_{X,H}(\mathbf{v})$ is given by such an extension, we get counterexamples to the weak Brill-Noether property when $r > 2n$.

These examples should convince the reader that the structure of the set of counterexamples to the weak Brill-Noether property on K3 surfaces is quite complicated and depends on arithmetic properties of the Mukai lattice. We now explain the strategy to use Bridgeland stability conditions to prove weak Brill-Noether.

5.6.4. *The strategy.* Let $\Delta \subset X \times X$ be the diagonal and let π_1 and π_2 denote the two projections from $X \times X$ to X. Let $\Phi^{I_\Delta}_{X \to X} : \mathcal{D}^b(X) \to \mathcal{D}^b(X)$ be the Fourier-Mukai transform acting by $\Phi^{I_\Delta}_{X \to X}(E) = \pi_{2*}(\pi_1^*(E) \otimes I_\Delta)$. For each $E \in \mathrm{Coh}(X)$, tensoring the exact sequence

$$0 \to I_\Delta \to \mathcal{O}_{X \times X} \to \mathcal{O}_\Delta \to 0$$

by $\pi_1^* E$ and pushing forward by π_2 gives the exact triangle

$$\Phi^{I_\Delta}_{X \to X}(E) \to R\Gamma(X, E) \otimes \mathcal{O}_X \to E$$

which induces the long exact sequence

$$0 \to \mathcal{H}^0(\Phi^{I_\Delta}_{X \to X}(E)) \to H^0(X, E) \otimes \mathcal{O}_X \xrightarrow{f} E \to \mathcal{H}^1(\Phi^{I_\Delta}_{X \to X}(E)) \to$$
$$H^1(X, E) \otimes \mathcal{O}_X \to 0 \to \mathcal{H}^2(\Phi^{I_\Delta}_{X \to X}(E)) \to H^2(X, E) \otimes \mathcal{O}_X \to 0.$$

Here f is the evaluation morphism. Let $F := \Phi^{I_\Delta}_{X \to X}(E))^\vee$ be the derived dual. Then

$$\mathcal{H}^i(\Phi^{I_\Delta}_{X \to X}(E))) = \mathcal{H}^i(F^\vee) = \mathcal{E}xt^i(F, \mathcal{O}_X).$$

LEMMA 5.27 ([**CNY23**, Lemma 3.1]). *Let E be a coherent sheaf with no zero-dimensional torsion and set $F := \Phi^{I_\Delta}_{X \to X}(E)^\vee$. Then*

(1) *F is a coherent sheaf if and only if E is nonspecial and generically globally generated.*
(2) *F is a torsion-free sheaf if and only if E is nonspecial and fails to be globally generated in at most finitely many points.*
(3) *F is a locally free sheaf if and only if E is nonspecial and globally generated.*

Lemma 5.27 transforms the weak Brill-Noether problem into showing that $\Phi^{I_\Delta}_{X\to X}(E)^\vee$ is a sheaf, and this is how Bridgeland stability becomes useful. Given the Mukai vector (r, dH, a) of E, there is a special chamber $\mathcal{C}$ in the (s,t)-plane right above the wall defined by $\mu_\sigma(I_x^\vee[1]) = \mu_\sigma(E)$, where I_x is the ideal sheaf of a point $x \in X$. Minamide, Yanagida and Yoshioka prove the following result.

THEOREM 5.28 ([**MYY18**, Theorem 4.9]). *If $\sigma \in \mathcal{C}$, then the moduli space of Bridgeland stable objects $M_\sigma(r, dH, a)$ is isomorphic to the Gieseker moduli space $M_{X,H}(a, dH, r)$ via $E \mapsto \Phi^{I_\Delta}_{X\to X}(E)^\vee$.*

Recall that for $t \gg 0$, the moduli space $M_\sigma(r, dH, a)$ is isomorphic to the Gieseker moduli space $M_{X,H}(r, dH, a)$. As we decrease t in order to reach the chamber $\mathcal{C}$, we may cross a number of Bridgeland walls. If not all of the sheaves in $M_{X,H}(r, dH, a)$ get destabilized along the way, then for $\sigma \in \mathcal{C}$, $M_{X,H}(r, dH, a)$ and $M_\sigma(r, dH, a)$ have a common open subset, so $M_{X,H}(r, dH, a)$ satisfies weak Brill-Noether by Lemma 5.27. Furthermore, if the general sheaf in $M_{X,H}(a, dH, r)$ is locally free, then the general sheaf in $M_{X,H}(r, dH, a)$ is globally generated.

However, along the way we may encounter totally semistable walls, where every sheaf gets destabilized. Bayer and Macrì [**BM14a, BM14b**] have classified the totally semistable walls. Using this classification, one can show that if $\mathbf{v} = (r, dH, a)$ is a Mukai vector with $r \geq 0$,$d > 0$, and $\mathbf{v}^2 \geq -2$ such that $M_{X,H}(\mathbf{v})$ does not satisfy weak Brill-Noether, then there must exists a spherical Mukai vector $\mathbf{v}_1 = (r_1, d_1 H, a_1)$ satisfying

(1) $0 < d_1 \leq d$
(2) $0 > \langle \mathbf{v}, \mathbf{v}_1 \rangle = 2ndd_1 - ar_1 - ra_1$
(3) $0 < dr_1 - d_1 r \leq da_1 - d_1 a.$

Determining whether there are totally semistable walls between the Gieseker chamber and $\mathcal{C}$ is a purely numerical problem, albeit a complicated one. Moreover, we obtain a dichotomy. Either there are no totally semistable walls above the chamber $\mathcal{C}$ and weak Brill-Noether holds, or the largest totally semistable wall provides a resolution of the general sheaf in $M_{X,H}(r, dH, a)$ which can then be used to compute the cohomology. This strategy yields the following sharp qualitative theorem.

THEOREM 5.29 ([**CNY23**]). *Let X be a K3 surface such that* $\mathrm{Pic}(X) \cong \mathbb{Z}H$ *with $H^2 = 2n$. Let $\mathbf{v} = (r, dH, a)$ be a Mukai vector with $\mathbf{v}^2 \geq -2$, $r \geq 2$ and $d > 0$.*

(1) *For each $r \geq 2$, there exists a finite set of tuples (n, r, d, a) for which $\mathbf{v}$ fails to satisfy weak Brill-Noether.*
(2) *If $n \geq r$, then $\mathbf{v}$ satisfies weak Brill-Noether.*
(3) *If $a \leq 1$, then $\mathbf{v}$ satisfies weak Brill-Noether.*
(4) *If $d \geq r\left\lfloor \frac{r}{n} \right\rfloor + 2$, then $\mathbf{v}$ satisfies weak Brill-Noether.*
(5) *Assume $a \geq 2$ and $n > 1$. If $n \geq 2r$ or $d \geq \left\lfloor \frac{2r}{n} \right\rfloor + 2$, then the general sheaf in $M_H(\mathbf{v})$ is globally generated.*

More importantly, the technique allows one to compute the cohomology of the general sheaf. For example, in [**CNY23**] the authors classify all the Mukai vectors of rank at most 20 on K3 surfaces of Picard rank 1 for which weak Brill-Noether fails and compute the cohomology of the general sheaf in these cases. Recently, Liu has developed these ideas further and obtained an algorithm for computing the cohomology of spherical bundles on K3 surfaces [**Liu23**].

Similar strategies can be applied to other K-trivial surfaces such as abelian surfaces and Enriques surfaces. For example, in the case of abelian surfaces of Picard rank 1, one obtains the following.

THEOREM 5.30 ([**CNY24**]). *Let X be an abelian surface with $\mathrm{NS}(X) = \mathbb{Z}H$ and let $\mathbf{v} = (r, dH, a)$ be a Mukai vector such that $r \geq 0$, $d > 0$ and $\mathbf{v}^2 \geq 0$. Then weak Brill-Noether holds for $M_{X,H}(\mathbf{v})$.*

REMARK 5.31. Yoshioka obtained a similar result by different methods in [**Yos09**, Lemma 3.13].

5.7. Asymptotic weak Brill-Noether. In general, the weak Brill-Noether problem is wide open. Given a rank $r \geq 2$ and a slope ν, if $\Delta \gg 0$, then $\chi(\mathbf{v}) < 0$. Moreover, by O'Grady's Theorem 4.9, we can assume that the moduli space $M_{X,H}(\mathbf{v})$ is irreducible and the general member is a slope-stable vector bundle. Potentially increasing Δ, after a series of elementary modifications, we can assume that the general sheaf has no global sections. By applying the same argument to the Serre dual, we can conclude that the general sheaf also has no h^2. We thus obtain an asymptotic weak Brill-Noether result.

PROPOSITION 5.32 ([**CH18c**, Proposition 7]). *Let $\mathbf{v}$ be a Chern character of rank at least 2. If $\Delta \gg 0$, depending on X, H, r and $\nu(\mathbf{v})$, then $H^i(X, \mathcal{V}) = 0$ unless $i = 1$.*

Similarly, by applying Serre Vanishing, one can show that the general sheaf in $M_{X,H}(\mathbf{v} \otimes mA)$ has no higher cohomology if A is ample and $m \gg 0$ (see [**CH18c**, Theorem 3.7]). It would be very useful to have good explicit bounds on m, especially for surfaces of general type.

6. Applications and generalizations of weak Brill-Noether

In this section, we discuss some applications and generalizations of weak Brill-Noether Theorems.

6.1. Classification of stable Chern characters. The weak Brill-Noether problem is closely tied to Problem 4.8, the problem of classifying Chern characters of stable sheaves.

6.1.1. *The classification of stable vector bundles on $\mathbb{P}^2$.* Drézet and Le Potier classified Chern characters of stable bundles on $\mathbb{P}^2$ [**DLP85, LeP97**]. We now briefly recall this classification.

Exceptional bundles. A coherent sheaf $\mathcal{V}$ is *exceptional* if $\mathrm{Hom}(\mathcal{V}, \mathcal{V}) = \mathbb{C}$ and $\mathrm{Ext}^i(\mathcal{V}, \mathcal{V}) = 0$ for $i > 0$. Exceptional sheaves on $\mathbb{P}^2$ have been classified by Drézet [**D87**]. Let $\mathcal{V}$ be an exceptional sheaf on $\mathbb{P}^2$ of rank r. First, since $\mathrm{Ext}^1(\mathcal{V}, \mathcal{V}) = 0$, every small deformation of $\mathcal{V}$ has to be trivial. Consequently,

$g^*(\mathcal{V}) \cong \mathcal{V}$ for every $g \in \mathbb{P}GL(3)$. We conclude that $\mathcal{V}$ cannot have any singularities and must be a vector bundle. By Riemann-Roch,

$$\chi(\mathcal{V},\mathcal{V}) = 1 = r^2(P(0) - 2\Delta) = r^2 - c_1^2 + 2r\,\mathrm{ch}_2\,.$$

Hence, r and c_1 are relatively prime and $\Delta(\mathcal{V}) = \frac{1}{2}\left(1 - \frac{1}{r^2}\right)$. Observe that the discriminant of an exceptional bundle is less than $\frac{1}{2}$ and that the slope determines the Chern character of an exceptional bundle.

If $E \subset \mathbb{P}^2$ is an elliptic curve embedded as a cubic, then applying $\mathrm{Hom}(\mathcal{V}, -)$ to the exact sequence

$$0 \to \mathcal{V}(-3) \to \mathcal{V} \to \mathcal{V}|_E \to 0$$

and using

$$\mathrm{Hom}(\mathcal{V},\mathcal{V}(-3)) = \mathrm{Ext}^2(\mathcal{V}(-3),\mathcal{V}(-3))^* = 0 \text{ and}$$

$$\mathrm{Ext}^1(\mathcal{V},\mathcal{V}(-3)) = \mathrm{Ext}^1(\mathcal{V}(-3),\mathcal{V}(-3))^* = 0,$$

we conclude that $\mathrm{Hom}(\mathcal{V}|_E,\mathcal{V}|_E) = \mathbb{C}$. Hence, the restriction of $\mathcal{V}$ to E is simple. Since a bundle on an elliptic curve is a direct sum of its Harder-Narasimhan factors, we conclude that $\mathcal{V}|_E$ is semistable. As E was arbitrary, it follows that $\mathcal{V}$ is semistable and thus stable since $\gcd(r, c_1) = 1$. Consequently, an exceptional bundle on $\mathbb{P}^2$ is stable. Conversely, a stable bundle $\mathcal{V}$ on $\mathbb{P}^2$ with $\Delta(\mathcal{V}) < \frac{1}{2}$ is exceptional by Riemann-Roch. Moreover, two exceptional bundles with the same slope are isomorphic since $\chi(\mathcal{V}_1,\mathcal{V}_2) = 1$ and hence there is a nontrivial homomorphism between $\mathcal{V}_1$ and $\mathcal{V}_2$ which must be an isomorphism by stability.

The line bundles $\mathcal{O}_{\mathbb{P}^2}(n)$ are exceptional. An *exceptional collection* on $\mathbb{P}^2$ is a triple of exceptional bundles $(\mathcal{V}_1,\mathcal{V}_2,\mathcal{V}_3)$ such that $\mathrm{Ext}^i(\mathcal{V}_k,\mathcal{V}_j) = 0$ for $1 \leq j < k \leq 3$ and all i. On $\mathbb{P}^2$, $(\mathcal{O}_{\mathbb{P}^2},\mathcal{O}_{\mathbb{P}^2}(1),\mathcal{O}_{\mathbb{P}^2}(2))$ is the *standard exceptional collection.* Given a pair of adjacent bundles in an exceptional collection $\mathcal{V},\mathcal{W}$, we can form the left and right mutations of the pair

$$0 \to \mathcal{V} \to \mathcal{W} \otimes \mathrm{Hom}(\mathcal{V},\mathcal{W})^* \to R_{\mathcal{W}}(\mathcal{V}) \to 0$$

$$0 \to L_{\mathcal{V}}(\mathcal{W}) \to \mathcal{V} \otimes \mathrm{Hom}(\mathcal{V},\mathcal{W}) \to \mathcal{W} \to 0.$$

For example, the Euler sequence exhibits $T_{\mathbb{P}^2}$ as $R_{\mathcal{O}_{\mathbb{P}^2}(1)}(\mathcal{O}_{\mathbb{P}^2})$. Drézet proves that on $\mathbb{P}^2$ the right $R_{\mathcal{W}}(\mathcal{V})$ and left $L_{\mathcal{V}}(\mathcal{W})$ mutations are again exceptional bundles. In fact, given an exceptional collection, $(\mathcal{V}_1,\mathcal{V}_2,\mathcal{V}_3)$, the collections $(\mathcal{V}_2, R_{\mathcal{V}_2}(\mathcal{V}_1),\mathcal{V}_3)$ and $(\mathcal{V}_1, L_{\mathcal{V}_2}(\mathcal{V}_3),\mathcal{V}_2)$ are again exceptional collections. This justifies the terminology right/left mutation. Furthermore, every exceptional bundle on $\mathbb{P}^2$ is obtained from line bundles by a sequence of mutations [**D87**]. One can systematically generate the slopes of all exceptional bundles on $\mathbb{P}^2$ by formalizing the process of mutations to obtain an explicit one-to-one correspondence $\varepsilon : \mathbb{Z}[\frac{1}{2}] \to \mathcal{E}$ between dyadic integers and the exceptional slopes, defined inductively by $\varepsilon(n) = n$ for an integer n and

$$\varepsilon\left(\frac{2p+1}{2^{q+1}}\right) = \varepsilon\left(\frac{p}{2^q}\right).\varepsilon\left(\frac{p+1}{2^q}\right),$$

where

$$\alpha.\beta = \frac{\alpha+\beta}{2} + \frac{\Delta_\beta - \Delta_\alpha}{3+\alpha-\beta}.$$

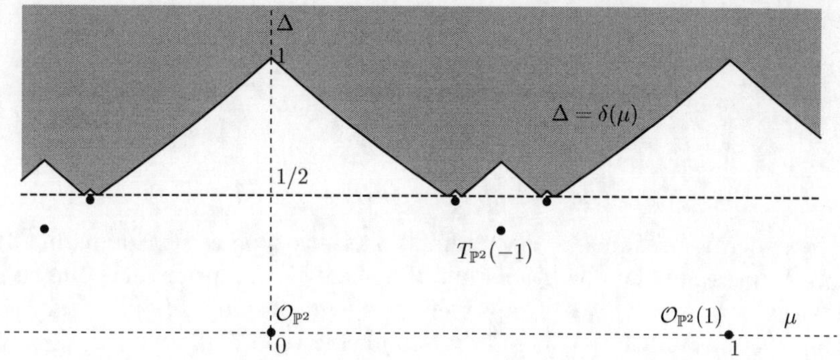

FIGURE 1. The Drézet–Le Potier curve. Chern characters of stable bundles with positive dimensional moduli spaces lie in the shaded region above the curve. The Chern characters of exceptional bundles are below the line $\Delta = \frac{1}{2}$.

Stable bundles in general. Given two stable bundles on $\mathbb{P}^2$ with $0 < \mu(\mathcal{V}) - \mu(\mathcal{W}) < 3$, we have that $\mathrm{Hom}(\mathcal{V}, \mathcal{W}) = 0$ and $\mathrm{Ext}^2(\mathcal{V}, \mathcal{W}) = \mathrm{Hom}(\mathcal{W}, \mathcal{V}(-3)) = 0$ by stability. Consequently,

$$\chi(\mathcal{V}, \mathcal{W}) = r(\mathcal{V})r(\mathcal{W})\left(P\left(\mu(\mathcal{W}) - \mu(\mathcal{V})\right) - \Delta(\mathcal{V}) - \Delta(\mathcal{W})\right) \leq 0.$$

Setting $\mathcal{V}$ to be an exceptional bundle $\mathcal{E}_\alpha$, gives an inequality for the discriminant $\Delta(\mathcal{W})$ of a stable bundle $\mathcal{W}$ in terms of its slope $\mu(\mathcal{W})$ provided $0 < \mu(\mathcal{E}_\alpha) - \mu(\mathcal{W}) < 3$. Similarly, setting $\mathcal{W}$ to be an exceptional bundle $\mathcal{E}_\alpha$, gives an inequality for the discriminant $\Delta(\mathcal{V})$ of a stable bundle $\mathcal{V}$ in terms of its slope $\mu(\mathcal{V})$ provided $0 < \mu(\mathcal{V}) - \mu(\mathcal{E}_\alpha) < 3$. Graphing the case of equality in the (μ, Δ)-plane for all exceptional bundles yields a fractal curve called the Drézet-Le Potier curve (see Figure 1). The Chern character of any stable bundle which is not exceptional must lie above the Drézet-Le Potier curve depicted by the shaded region. Conversely, using dimension estimates, one can show that if the Chern character lies above the Drézet-Le Potier curve, then the general bundle defined by the Gaeta resolution is stable. One thus obtains the main classification theorem on $\mathbb{P}^2$ due to Drézet and Le Potier.

THEOREM 6.1 (Drézet-Le Potier). *Let* $\mathbf{v}$ *be an integral Chern character of positive rank. There exists a Gieseker semistable sheaf with Chern character* $\mathbf{v}$ *if and only if* $\Delta(\mathbf{v}) \geq \delta(\mu(\mathbf{v}))$ *or* $\mathbf{v}$ *is a multiple of the Chern character of an exceptional bundle. When* $\Delta(\mathbf{v}) \geq \delta(\mu(\mathbf{v}))$*, the moduli space* $M_{\mathbb{P}^2,L}(\mathbf{v})$ *is an irreducible, normal projective variety of dimension* $r^2(2\Delta - 1) + 1$.

6.1.2. *Classification of stable Chern characters on Hirzebruch surfaces.* Using Theorems 5.11 and 5.12 as main tools, one can classify the Chern characters of stable sheaves on Hirzebruch surfaces $\mathbb{F}_e$ for generic polarizations (see [**CH21**]).

Let $H_m = E + (m+e)F$ on $\mathbb{F}_e$ with $m \in \mathbb{Q}_{>0}$. Then H_m is ample and every ample divisor is proportional to some H_m. To classify Chern characters, one first shows that an H_m-semistable sheaf is $H_{\lceil m \rceil + 1}$-prioritary and thus one needs to classify H_k-prioritary sheaves for positive integers k. Then one computes the Harder-Narasimhan filtration of the general H_k-prioritary sheaf. The Gaeta-type resolution resolves the first issue.

Let $\nu(\mathbf{v}) = \epsilon E + \varphi F$. Set

$$\psi := \varphi + \frac{1}{2}e(\lceil \epsilon \rceil - \epsilon) - \frac{\Delta}{1 - (\lceil \epsilon \rceil - \epsilon)},$$

and let

$$L_0 := L_{\lceil \epsilon \rceil, \lceil \psi \rceil} = \lceil \epsilon \rceil E + \lceil \psi \rceil F.$$

Then the general sheaf in $\mathcal{P}_{\mathbb{F}_e,F}(\mathbf{v})$ admits a Gaeta-type resolution as in (2) with $L = L_0$. Expressing the condition that the sheaf be H_k prioritary, one sees that we need $\chi(\mathbf{v}(-L_0 - H_k)) \leq 0$. Conversely, one can construct H_k-prioritary sheaves satisfying this inequality to conclude the following theorem.

THEOREM 6.2 ([**CH21**, Theorem 4.16]). *Let* $\mathbf{v}$ *be an integral Chern character of positive rank on* $\mathbb{F}_e$ *with* $\Delta(\mathbf{v}) \geq 0$ *and let* k *be a positive integer. Then the stack* $\mathcal{P}_{H_k}(\mathbf{v})$ *is nonempty if and only if*

$$\chi(\mathbf{v}(-L_0 - H_k)) \leq 0.$$

This theorem already provides stronger Bogomolov inequalities for the existence of semistable sheaves on $\mathbb{F}_e$.

Next, we will compute the Harder-Narasimhan filtration of the general H_m-prioritary sheaf. There exists an H_m-semistable sheaf if and only if the filtration is trivial. Suppose the general H_m-Harder-Narasimhan filtration has length ℓ and the graded pieces have Chern characters $\mathbf{v}_i = (r_i, \nu_i, \Delta_i)$. Then

$$\sum_{i=1}^{\ell} \mathbf{v}_i = \mathbf{v}.$$

Furthermore, the moduli spaces $M_{\mathbb{F}_e,H_m}(\mathbf{v}_i)$ are nonempty since the graded pieces are semistable sheaves. The fact that the sheaves are $H_{\lceil m \rceil + 1}$-prioritary and the Schatz-stratum corresponding to this Harder-Narasimhan filtration has codimension 0 lead to additional inequalities that place strong restrictions on $\mathbf{v}_i$.

First, the prioritary condition implies that the restriction of the general sheaf to a general rational curve in the class $H_{\lceil m \rceil}$ or $H_{\lfloor m \rfloor}$ has balanced splitting. This translates to the inequality that

$$|(\nu_i - \nu) \cdot H_m| < 1.$$

Next, a dimension computation shows that for the Schatz-stratum to have codimension 0, the following orthogonality relations hold

$$\chi(\mathbf{v}_i, \mathbf{v}_j) = 0 \quad \text{for} \quad i < j.$$

Hence, the slopes ν_i are restricted to a bounded region, and since the ranks r_i are bounded there are only finitely many possibilities for the ν_i. Furthermore, the orthogonality relations imply the discriminant Δ_i has to be the minimal possible discriminant of an H_m-semistable sheaf with rank r_i and total slope ν_i. Hence, there are finitely many possible $\mathbf{v}_i$ that can be the Chern characters of the graded pieces of the generic H_m-Harder-Narasimhan filtration.

Conversely, if one can find Chern characters $\mathbf{v}_i$, $1 \leq i \leq \ell$, that satisfy these constraints, then the general Harder-Narasimhan filtration has factors with these Chern characters.

THEOREM 6.3 ([**CH21**, Theorem 5.3]). *Let* $\mathbf{v}$ *be a Chern character such that* $\mathcal{P}_{H_{\lceil m \rceil}}(\mathbf{v})$ *is nonempty. Let* $\mathbf{v}_1, \ldots, \mathbf{v}_\ell \in K(\mathbb{F}_e)$ *be positive rank Chern characters satisfying the following properties:*

(1) $\sum_{i=1}^{\ell} \mathbf{v}_i = \mathbf{v}$,
(2) *The reduced Hilbert polynomials* q_i *of* $\mathbf{v}_i$ *are strictly decreasing* $q_1 > \cdots > q_\ell$,
(3) $\mu_{H_m}(\mathbf{v}_1) - \mu_{H_m}(\mathbf{v}_\ell) \leq 1$,
(4) $\chi(\mathbf{v}_i, \mathbf{v}_j) = 0$ *for* $i < j$,
(5) $M_{\mathbb{F}_e, H_m}(\mathbf{v}_i)$ *is nonempty,*

Then the Harder-Narasimhan filtration of the general sheaf in $\mathcal{P}_{H_{\lceil m \rceil}}(\mathbf{v})$ *has length* ℓ *and the factors have Chern characters* $\mathbf{v}_i$.

Therefore, determining the Harder-Narasimhan filtration of a general $H_{\lceil m \rceil}$-prioritary sheaf becomes a finite computational problem. In particular, one obtains an algorithm for classifying Chern characters of H_m-semistable sheaves. Using the fact that $K(\mathbb{F}_e) \cong \mathbb{Z}^4$, one can show $\ell \leq 4$ to further simplify the problem.

REMARK 6.4. On $\mathbb{P}^2$, either the subbundle or the quotient bundle in the general Harder-Narasimhan filtration of a general prioritary bundle is exceptional. Hence, one can give explicit inequalities for the discriminants of semistable bundles without computing the semistable bundles of lower rank. On $\mathbb{F}_e$, it may happen that neither the quotient nor the subbundle in a generic Harder-Narasimhan filtration of length 2 is an exceptional bundle (see [**CH21**] for explicit examples). In general, it is not enough to consider the constraints given by exceptional bundles.

6.2. The tensor product problem. A natural generalization of the Brill-Noether problem which plays a fundamental role in the birational geometry of moduli spaces of sheaves is the tensor product problem (see [**ABCH13, Hui16**]).

PROBLEM 6.5 (The tensor product problem). Let $\mathcal{V} \in M_{X,H}(\mathbf{v})$ and $\mathcal{W} \in M_{X,H}(\mathbf{w})$ be two stable sheaves on X. Compute the cohomology of $\mathcal{V} \otimes \mathcal{W}$.

More generally, one can ask for the cohomology of the tensor product of Bridgeland stable objects in the derived category of X. The problem is already interesting when $\mathcal{V}$ and $\mathcal{W}$ are general elements in their moduli spaces. When $\mathcal{W} = \mathcal{O}_X$, the problem reduces to computing the cohomology of $\mathcal{V}$. When $\mathcal{V}$ is general in its moduli, we recover the weak Brill-Noether problem.

When $\mathcal{W} = I_Z$ is an ideal sheaf of points and $\mathcal{V}$ is a line bundle, then the problem reduces to the classical interpolation problem asking for when points impose independent conditions on sections of a line bundle. More generally, when $\mathcal{V}$ is a higher rank sheaf, the problem is the higher rank interpolation problem asking for when points impose independent conditions on sections of a sheaf. In general, these problems are wide open, though some important special cases have been solved. For example, the higher rank interpolation problem has an explicit combinatorial solution for zero-dimensional monomial schemes on $\mathbb{P}^2$ [**CH14**, Theorem 1.4].

On $\mathbb{P}^2$ there is an almost complete answer to the tensor product problem for general sheaves. Let $\mathbf{v}$ be a stable Chern character on $\mathbb{P}^2$ that lies on or above the Drézet-Le Potier curve and let $\mathcal{V}$ be a general stable sheaf in $M_{\mathbb{P}^2,L}(\mathbf{v})$. Then the *associated exceptional bundle* E_+ is the exceptional bundle on $\mathbb{P}^2$ with the smallest slope such that if E' is any exceptional bundle with $\mu_L(E') > \mu_L(E_+)$, then $\chi(\mathcal{V} \otimes E') > 0$. It may happen that $\chi(\mathcal{V} \otimes E_+) < 0$, $\chi(\mathcal{V} \otimes E_+) = 0$ or $\chi(\mathcal{V} \otimes E_+) >$

0. By Theorem [**CHW17**, Theorem 4.1], the associated exceptional bundle exists. Briefly, by Riemann-Roch, the locus $\mathbf{v}^\perp := \{\mathbf{w} : \chi(\mathbf{v}\otimes\mathbf{w}) = 0\}$ is an upward parabola in the slope-discriminant plane. This parabola intersects the line $\Delta = \frac{1}{2}$ in two points which lie under peaks of the Drézet-Le Potier curve. The Drézet-Le Potier curve has infinitely many peaks (see Figure 1), each determined by an exceptional bundle. The associated exceptional bundle E_+ is the exceptional bundle lying under the rightmost peak determined in this way. (The points on the line $\Delta = \frac{1}{2}$ which do not lie under a peak form a generalized Cantor set $\mathfrak{C}$. The fact that the parabola $\mathbf{v}^\perp$ meets $\Delta = \frac{1}{2}$ at points that are not in $\mathfrak{C}$ relies on number-theoretic properties of $\mathfrak{C}$. For instance, the points of $\mathfrak{C}$ which are not endpoints of peaks have transcendental μ-coordinate.)

The answer to the problem is easiest to state in the case where $\chi(\mathbf{v}\otimes E_+) \le 0$. In the other case where $\chi(\mathcal{V}\otimes E_+) > 0$, we need to fix some additional notation. In this case we let

$$\mathbf{k} = \mathbf{v} - \chi(\mathcal{V}\otimes E_+)\,\mathrm{ch}(E_+)$$

be the Chern character of the mapping cone of the canonical evaluation map

$$E_+^* \otimes \mathrm{Hom}(E_+^*, \mathcal{V}) \to \mathcal{V}.$$

THEOREM 6.6 ([**CHK21**, Theorem 1.2]). *Let $\mathbf{v}$ and $\mathbf{w}$ be Chern characters of stable bundles on $\mathbb{P}^2$. Suppose $M_{\mathbb{P}^2,L}(\mathbf{v})$ is positive dimensional and $\mathbf{w}$ is sufficiently divisible (depending on $\mathbf{v}$). Let $\mathcal{V} \in M_{\mathbb{P}^2,L}(\mathbf{v})$ and $\mathcal{W} \in M_{\mathbb{P}^2,L}(\mathbf{w})$ be general bundles.*

1. *If $\chi(\mathbf{v}\otimes E_+) \le 0$, then either $H^0(\mathcal{V}\otimes\mathcal{W}) = 0$ or $H^1(\mathcal{V}\otimes\mathcal{W}) = 0$.*
2. *If $\chi(\mathbf{v}\otimes E_+) > 0$ and $\mathrm{rk}(\mathbf{k}) \le 0$, then either $H^0(\mathcal{V}\otimes\mathcal{W}) = 0$ or $H^1(\mathcal{V}\otimes\mathcal{W}) = 0$.*
3. *Suppose $\chi(\mathbf{v}\otimes E_+) > 0$ and $\mathrm{rk}(\mathbf{k}) > 0$.*
   (a) *If $\chi(\mathbf{k}\otimes\mathbf{w}) \ge 0$ or $\chi(\mathbf{w}\otimes E_+^*) \le 0$, then either $H^0(\mathcal{V}\otimes\mathcal{W}) = 0$ or $H^1(\mathcal{V}\otimes\mathcal{W}) = 0$.*
   (b) *Otherwise $\mathcal{V}\otimes\mathcal{W}$ is special and*

$$h^0(\mathcal{V}\otimes\mathcal{W}) = \chi(\mathbf{v}\otimes E_+)\chi(\mathbf{w}\otimes E_+^*) \quad \textit{and} \quad h^1(\mathcal{V}\otimes\mathcal{W}) = -\chi(\mathbf{k}\otimes\mathbf{w})$$

By applying Serre duality, the theorem computes the tensor product of two general stable bundles subject to the requirement that $\mathbf{w}$ is sufficiently divisible. One can give an explicit bound on how divisible $\mathbf{w}$ needs to be depending on $\mathbf{v}$. We do not know whether the divisibility is necessary on $\mathbb{P}^2$ or an artifact of the proof. The following example shows that on $\mathbb{P}^n$ for $n > 2$, some divisibility is necessary.

EXAMPLE 6.7 ([**CHS23**, Example 1.3]). Let $\mathcal{V}_m$ be the kernel of a general map

$$0 \to \mathcal{V}_m \to \mathcal{O}_{\mathbb{P}^3}(2)^{\oplus 4m} \to \mathcal{O}_{\mathbb{P}^3}(4)^{\oplus m} \to 0.$$

When $m = 1$, $h^0(\mathbb{P}^3, \mathcal{V}_1) = 6$ and $h^1(\mathbb{P}^3, \mathcal{V}_1) = 1$. On the other hand, when $m \ge 2$, then $h^1(\mathbb{P}^3, \mathcal{V}_m) = 0$.

6.3. Construction of Brill-Noether divisors and birational geometry. Let $\mathbf{v}$ and $\mathbf{w}$ be two stable Chern characters on X such that $\chi(\mathbf{v}\otimes\mathbf{w}) = 0$. Then given a general sheaf $\mathcal{W} \in M_{X,H}(\mathbf{w})$, we can define the virtual Brill-Noether divisor

$$D_{\mathcal{W}} := \{\mathcal{V} \in M_{X,H}(\mathbf{v}) | h^1(\mathcal{V}\otimes\mathcal{W}) \ne 0\}.$$

In general, $D_{\mathcal{W}}$ may not be a divisor. If $h^i(\mathcal{W}\otimes\mathcal{V}') = 0$ for all i and a general $\mathcal{V}'$, then $D_{\mathcal{W}}$ is an effective divisor. Hence, the tensor product problem in the special case of $\chi(\mathbf{v} \otimes \mathbf{w}) = 0$ is the key ingredient for constructing effective Brill-Noether divisors.

There has been significant progress in computing the ample and effective cones of $M_{X,H}(\mathbf{v})$ and running the minimal model program for these moduli spaces in recent years, especially when X is $\mathbb{P}^2$, a Hirzebruch, K3, abelian or Enriques surface. Surveying the developments in this direction would take us too far afield, so we refer the reader to [**ABCH13, BM14a, BM14b, BC13, CH16, CH15, CH18a, CHW17, Hui16, LZ19, MYY18, N16a, NY**] for some of these developments.

6.4. Constructions of Ulrich bundles on surfaces. A solution of the weak Brill-Noether problem allows one to construct Ulrich bundles on surfaces.

DEFINITION 6.8. Let $X \subset \mathbb{P}^n$ be a smooth, projective variety of dimension d. An *Ulrich bundle* $\mathcal{V}$ on X is a bundle that satisfies $H^i(X, V(-j)) = 0$ for $1 \leq j \leq d$ and all i.

Ulrich bundles play a central role in the study of Chow forms of a variety [**ESW03**], the minimal resolution conjecture (see [**AGO17**]) and Boij-Söderberg Theory (see [**ES11**]). For example, Eisenbud and Schreyer show that the cone of cohomology tables of X is the same as that of $\mathbb{P}^d$ if and only if X admits an Ulrich bundle [**ES11**]. Eisenbud and Schreyer raise the question whether every projective variety admits an Ulrich bundle. Existence is known in some cases including smooth curves [**ESW03**], complete intersections [**BaHU91**], del Pezzo surfaces [**CKM13**], certain rational surfaces [**ESW03, Kim16**], K3 surfaces [**AGO17, Fae18**], abelian surfaces [**Bea16**] and certain Enriques surfaces [**BN18**] among many others.

Let $\mathcal{V}$ be an Ulrich bundle for a polarized surface (X, H) with rank r, total slope ν and discriminant Δ. Since $\mathcal{V}(-H)$ and $\mathcal{V}(-2H)$ have no cohomology, the Euler characteristics of $\mathcal{V}(-H)$ and $\mathcal{V}(-2H)$ must vanish. By Riemann-Roch, we conclude that

$$\frac{1}{2}\nu^2 - \frac{1}{2}\nu \cdot K_X + \chi(\mathcal{O}_X) - \Delta = 0$$
$$\frac{1}{2}(\nu - H)^2 - \frac{1}{2}(\nu - H) \cdot K_X + \chi(\mathcal{O}_X) - \Delta = 0$$

Consequently, an Ulrich bundle on a surface satisfies

$$2\nu \cdot H = H^2 + H \cdot K_X \quad \text{and} \quad 2\Delta = \nu^2 - \nu \cdot K_X + 2\chi(\mathcal{O}_X). \tag{3}$$

Conversely, if we take $2\nu = H + K_X$, $M_{X,H}(r, \nu, \Delta)$ contains locally free sheaves, and it satisfies weak Brill-Noether, then the general locally free sheaf in $M_{X,H}(r, \nu, \Delta)$ is an Ulrich bundle.

Combining O'Grady's Theorem with Serre vanishing yields an asymptotic existence result on any smooth projective surface.

THEOREM 6.9 ([**CH20b**, Theorem 4.3]). *Let (X, H) be a smooth, polarized surface. There exists a positive integer m_0 such that for all $m \geq m_0$, the polarized surface (X, mH) admits an Ulrich bundle of every positive even rank. Moreover, if K_X (respectively, $K_X + H$) is divisible by 2 in* $\operatorname{Pic}(X)$ *and $2m \geq m_0$ (respectively, $2m + 1 \geq m_0$), then $(X, 2mH)$ (respectively, $(X, (2m + 1)H)$) admits an Ulrich bundle of every rank $r \geq 2$.*

A polarized variety (X, H) is of *Ulrich wild representation type* if one can find arbitrarily large dimensional families of Ulrich bundles on X with respect to H. An easy corollary Theorem 6.9 is the following:

COROLLARY 6.10 ([**CH20b**, Corollary 4.5]). *Let (X,H) be a smooth, polarized surface. Then there exists an integer m_0 such that for all $m \geq m_0$, (X,mH) is of Ulrich wild representation type.*

On specific surfaces, weak Brill-Noether gives detailed information on Chern classes of Ulrich bundles. For example, on Hirzebruch surfaces one obtains the following (see also V. Antonelli [**Ant18**]).

THEOREM 6.11 ([**CH20b**, Theorem 4.6]). *Let $H = aE + bF$ be an ample divisor on $\mathbb{F}_e$. Let $\mathbf{v} = (r,\nu,\Delta) = (r, \alpha E + \beta F, \Delta)$ be an integral Chern character with $r \geq 2$. There exists a locally free F-prioritary sheaf $\mathcal{V}$ with Chern character $\mathbf{v}$ satisfying*

$$H^i(\mathbb{F}_e, \mathcal{V}(-H)) = H^i(\mathbb{F}_e, \mathcal{V}(-2H)) = 0 \ \forall i$$

if and only if

$$a - 1 + \frac{ea(a-1)}{2b} \leq \alpha \leq 2a - 1 - \frac{ea(a-1)}{2b},$$

$$\beta = \left(e - \frac{b}{a}\right)(\alpha+1) + 3b - 1 - \frac{e}{2}(3a+1)$$

and

$$\Delta(\mathcal{V}) = \left(\frac{e}{2} - \frac{b}{a}\right)(\alpha^2 + (2-3a)\alpha + 2a^2 - 3a + 1).$$

In particular, there exists an Ulrich bundle of rank 2 for $H = aE + bF$ with invariants

$$\mathbf{v} = (r,\nu,\Delta) = \left(2, \left(\frac{3}{2}a - 1\right)E + \left(\frac{3}{2}b - \frac{e}{2} - 1\right)F, \frac{a}{8}(2b - ae)\right).$$

Similarly detailed theorems can be proven on certain blowups of $\mathbb{P}^2$ or $\mathbb{F}_e$ using the corresponding weak Brill-Noether statements. We refer the reader to [**CH20b**, Theorem 4.10 and 4.11] for precise statements. The weak Brill-Noether Theorems for K3 or abelian surfaces similarly yield classification results for Ulrich bundles. We state the result for K3 surfaces and refer the reader to [**CNY24**, Corollary 6.1] for abelian surfaces.

PROPOSITION 6.12 ([**CNY23**, Proposition 4.4]). *Let X be a K3 surface with $\mathrm{Pic}(X) = \mathbb{Z}H$. There exists an Ulrich bundle of rank r with respect to mH if and only if $2 \mid rm$. Moreover, when an Ulrich bundle of rank r exists, it has Mukai vector $\mathbf{v} = \left(r, \left(\frac{3rm}{2}\right)H, r(2m^2n - 1)\right)$. In particular, there exists an Ulrich bundle of any rank $r \geq 2$ with respect to $2H$.*

One can also prove results on variants of Ulrich bundles. For example, Eisenbud and Schreyer [**ES11**] conjectured that there exist bundles $\mathcal{V}$ on $\mathbb{P}^1 \times \mathbb{P}^1$ with natural cohomology, meaning that $\mathcal{V}(aF_1 + bF_2)$ has at most one nonzero cohomology group for any $a, b \in \mathbb{Z}$. The weak Brill-Noether theorem on $\mathbb{P}^1 \times \mathbb{P}^1$ immediately yields a positive solution (see also [**Sol18**]).

COROLLARY 6.13 ([**CH20b**, Corollary 4.13]). *Let F_1 and F_2 denote the two fiber classes on $\mathbb{P}^1 \times \mathbb{P}^1$. Let $\mathbf{v}$ be an integral Chern character such that $\mathrm{rk}(\mathbf{v}) \geq 2$ and $\Delta \geq 0$. Then $\mathcal{P}_{\mathbb{P}^1\times\mathbb{P}^1,F_1}(\mathbf{v})$ is nonempty, and the general $\mathcal{V} \in \mathcal{P}_{\mathbb{P}^1\times\mathbb{P}^1,F_1}(\mathbf{v})$ is locally free and has at most one nonzero cohomology group. In particular, the very general member of $\mathcal{P}_{\mathbb{P}^1\times\mathbb{P}^1,F_1}(\mathbf{v})$ is a bundle with natural cohomology.*

7. Cohomology jumping loci

When the weak Brill-Noether Theorem holds for $M_{X,H}(\mathbf{v})$, it is natural to try to describe the loci of bundles with unexpected cohomology. In this section, we will review recent developments on the Brill-Noether loci on $\mathbb{P}^2$ due to Gould, Lee and Liu [**GLL22**] and describe some generalizations to other surfaces.

Let $M_{X,H}(\mathbf{v})$ be an irreducible moduli space such that every sheaf $\mathcal{V}$ has $H^2(X,\mathcal{V}) = 0$. For example, this condition automatically holds on moduli spaces of sheaves on $\mathbb{P}^2$, a K3, an abelian or a Hirzebruch surface if $\nu(\mathbf{v})$ is effective. Assume that the locus of stable sheaves $M_{X,H}(\mathbf{v})^s$ is nonempty. Define the k-th Brill-Noether locus $B^k_{X,H}(\mathbf{v})$ as the closure of the locus of stable sheaves with at least k-independent sections:

$$B^k_{X,H}(\mathbf{v}) := \overline{\{\mathcal{V} \in M_{X,H}(\mathbf{v})^s | h^0(X,\mathcal{V}) \geq k\}} \subset M_{X,H}(\mathbf{v}).$$

The locus

$$B^k_{X,H}(\mathbf{v}) \cap M_{X,H}(\mathbf{v})^s$$

has a natural determinantal structure ([**CHW17**, Proposition 2.7] or [**CoMR10**, §2]), so every irreducible component of $B^k_{X,H}(\mathbf{v})$ has codimension at most the expected value $k(k - \chi(\mathbf{v}))$. By definition, we have $B^{k+1}_{X,H}(\mathbf{v}) \subset B^k_{X,H}(\mathbf{v})$ for $k \geq 0$.

PROBLEM 7.1. One can immediately raise the following natural problems.

(1) Determine when $B^k_{X,H}(\mathbf{v})$ is nonempty, equivalently determine

$$\max_{\mathcal{V} \in M_{X,H}(\mathbf{v})^s} (h^0(X,\mathcal{V})).$$

(2) Classify the irreducible components of $B^k_{X,H}(\mathbf{v})$ and determine their dimensions. In particular, determine when $B^k_{X,H}(\mathbf{v})$ is irreducible.
(3) Describe the singularities of $B^k_{X,H}(\mathbf{v})$.
(4) When do Torelli type theorems hold for $B^k_{X,H}(\mathbf{v})$, i.e., when can X be recovered from $B^k_{X,H}(\mathbf{v})$?

In general, very little is known about these problems. Even in rank 1, the loci $B^k_{X,H}(\mathbf{v})$ can have many components of different dimensions and are poorly understood.

EXAMPLE 7.2 ([**GLL22**, Proposition 5.3]). Let $\mathbf{v} = \mathrm{ch}(I_Z(3))$ on $\mathbb{P}^2$, where Z is a zero-dimensional scheme of length $|Z| = n \geq 10$. The Brill-Noether locus $B^2_{\mathbb{P}^2,L}(\mathbf{v})$ parameterizing $I_Z(3)$ with $h^0(\mathbb{P}^2, I_Z(3)) \geq 2$ has at least two irreducible components.

(1) Let $W_1 \subset \mathbb{P}^{2[n]}$ be the closure of the locus with $n-4$ points lying on a line and 4 general points. Then $\dim(W_1) = n + 6$.
(2) Let $W_2 \subset \mathbb{P}^{2[n]}$ be the closure of the locus with $n-1$ points lying on a smooth conic and 1 general point. Then $\dim(W_2) = n + 6$.

We claim that if $I_Z(3) \in B^2_{\mathbb{P}^2,L}(\mathbf{v})$ where Z has distinct points, then $Z \in W_1$ or $Z \in W_2$. Let C_1 and C_2 be two distinct curves of degree 3 vanishing on Z. By Bézout's Theorem, C_1 and C_2 must have a common curve D. Then D is either a conic (possibly reducible or nonreduced) or a line. If D is a conic, then the residual curves must be lines and they intersect in at most one point. Hence, $Z \in W_2$. If D is a line, then the residual curves must be conics. Since they do not have a common curve, their intersection must be a scheme of length 4. Hence, $Z \in W_1$. Since

W_1 and W_2 have the same dimension, they form distinct irreducible components of $B^2_{\mathbb{P}^2,L}(\mathbf{v})$.

A slightly more subtle argument allows Gould, Lee and Liu to deduce that when $\mathbf{v} = \mathrm{ch}(I_Z(k))$, where $|Z| > k^2$, the Brill-Noether locus $B^2_{\mathbb{P}^2,L}(\mathbf{v})$ has at least k components, typically not all of the same dimension.

EXAMPLE 7.3. The previous example is not special to $\mathbb{P}^2$. For simplicity, let X be a surface of Picard rank 1 generated by an effective ample divisor H and assume $H^1(X, \mathcal{O}_X) = 0$. Let m, n be two integers such that $m > 2$ is a sufficiently large odd integer and $n > m^2H^2$. Let $\mathbf{v} = \mathrm{ch}(I_Z(mH))$ with $|Z| = n$. For $0 < s < m$, let $W_s \subset X^{[n]}$ be the closure of the locus of points with $h^0(X, \mathcal{O}_X((m-s)H)) - 2$ general points on X and $n + 2 - h^0(X, \mathcal{O}_X((m-s)H))$ general points on a curve with class sH. Then

$$\dim(W_s) = h^0(X, \mathcal{O}_X(sH)) + h^0(X, \mathcal{O}_X((m-s)H)) + n - 2.$$

Note that $\dim(W_s) = \dim(W_{m-s})$ and if $s \gg 0$, then we can assume that the curve of degree s is irreducible. Hence, there are at least two components W_{s_0} and W_{m-s_0} with maximal dimension. The inclusion $I_Z(mH) \subset \mathcal{O}_X(mH)$ induces two independent curves C_1 and C_2 in $|mH|$ that contain Z. Let $I_Z(mH)$ be a general point of an irreducible component of $B^2_{X,H}(\mathbf{v})$. Since $|Z| > m^2H^2$, by Bézout's Theorem, C_1 and C_2 must have a common component of class kH. By the semicontinuity of the degree of the base curve, this component cannot contain W_s for $s > k$. Hence, if $B^2_{X,H}(\mathbf{v})$ is irreducible, we must have $k = 1$. There can be at most $h^0(X, \mathcal{O}_X((m-1)H)) - 2$ general points which do not lie on the common component. Hence, if this locus contains W_1, it must coincide with W_1. Then at least one of W_{s_0} or W_{m-s_0} is a component distinct from W_1 and we conclude that $B^2_{X,H}(\mathbf{v})$ is reducible.

In fact, by the semicontinuity of the degree of the base locus, observe that $W_s \not\subset W_t$ if $s > t$. Assuming that there are irreducible sections of $H^0(X, \mathcal{O}_X(kH))$ for every $k \geq 1$, then the loci W_s have no containments among each other and one can show that $B^2_{X,H}(\mathbf{v})$ has at least k components.

Once the Brill-Noether loci in rank 1 are reducible, one can make examples of higher rank reducible Brill-Noether loci.

EXAMPLE 7.4 ([**GLL22**, Example 5.5]). Consider extensions of the form

$$0 \to \mathcal{O}_{\mathbb{P}^2} \to \mathcal{V} \to I_Z(3) \to 0,$$

where $|Z| \geq 10$. Then $\chi(\mathcal{V}) = 11 - |Z|$. The Cayley-Bacharach condition is trivially satisfied, so there exists vector bundles of this form. Furthermore, if the points of Z are not collinear then there are no nonzero maps $\mathcal{O}_{\mathbb{P}^2}(2) \to I_Z(3)$, so $\mathcal{V}$ is stable. We have

$$\mathrm{ext}^1(I_Z(3), \mathcal{O}_{\mathbb{P}^2}) = |Z| - 1.$$

Set $\mathbf{v} = \mathrm{ch}(\mathcal{V})$ and $\mathbf{w} = \mathrm{ch}(I_Z(3))$. Now assume that $Z \in W_i$ for $i = 1$ or 2 as in Example 7.2. Then $\mathcal{V} \in B^3_{\mathbb{P}^2,L}(\mathbf{v})$. We thus obtain two loci in $B^3_{\mathbb{P}^2,L}(\mathbf{v})$ of dimension $2|Z| + 2$. The expected codimension of $B^3_{\mathbb{P}^2,L}(\mathbf{v})$ is $3(|Z| - 8)$. Since the moduli space has dimension $4|Z| - 12$, the expected dimension of $B^3_{\mathbb{P}^2,L}(\mathbf{v})$ is $|Z| + 12$. Observe that we obtain unexpectedly large dimensional loci in $B^3_{\mathbb{P}^2,L}(\mathbf{v})$ if $|Z| > 10$. One can also easily check that the two components of $B^2_{\mathbb{P}^2,L}(\mathbf{w})$ yield two distinct components of $B^3_{\mathbb{P}^2,L}(\mathbf{v})$.

Similar constructions using Example 7.3 produce reducible Brill-Noether loci on more general surfaces. Subbundles with a large space of sections is another source of components of Brill-Noether loci.

EXAMPLE 7.5 ([**GLL22**, Example 5.11]). Let $\mathbf{v}_k$ be a a Chern character on $\mathbb{P}^2$ with $\mu_L(\mathbf{v}_k) = \frac{2}{3}$ and $\Delta(\mathbf{v}_k) = \frac{5}{9} + \frac{k}{3}$. Then one can consider the following two types of bundles in $B^3_{\mathbb{P}^2,L}(\mathbf{v}_k)$.

(1) Let C be a smooth conic, D be a divisor of degree $1-k$ and $\mathcal{V}$ be a general extension of the form
$$0 \to \mathcal{O}_{\mathbb{P}^2}^{\oplus 3} \to \mathcal{V} \to \mathcal{O}_C(D) \to 0.$$
(2) Let $T_{\mathbb{P}^2}$ be the tangent bundle of $\mathbb{P}^2$, Z be a zero-dimensional scheme of length $k+1$ and $\mathcal{V}$ be defined by a nonsplit extension of the form
$$0 \to T_{\mathbb{P}^2}(-1) \to \mathcal{V} \to I_Z(1) \to 0.$$

One can check that in both cases $\mathcal{V}$ is stable and give distinct irreducible components of $B^3_{\mathbb{P}^2,L}(\mathbf{v}_k)$ when $k \gg 0$.

Using a systematic analysis of the ideas in these examples, Gould, Lee and Liu prove the following structure theorem on Brill-Noether loci in moduli spaces of sheaves on $\mathbb{P}^2$.

THEOREM 7.6 ([**GLL22**, Theorem 1.1]). *Let $M_{\mathbb{P}^2,L}(\mathbf{v})$ be a nonempty moduli space of sheaves on $\mathbb{P}^2$ with $\mu_L(\mathbf{v}) > 0$.*

(1) *For any $\mathcal{V} \in M_{\mathbb{P}^2,L}(\mathbf{v})$,*
$$h^0(\mathbb{P}^2, \mathcal{V}) \le \max\left(r(\mathcal{V}), P(c_1(\mathcal{V})) = \frac{1}{2}c_1(\mathcal{V})^2 + \frac{3}{2}c_1(\mathcal{V}) + 1\right).$$
In particular, $B^k(\mathbf{v}) = \emptyset$ if $k > \max(r(\mathbf{v}), P(c_1(\mathbf{v})))$.

(2) *The Brill-Noether locus $B^r(\mathbf{v})$ is nonempty.*

(3) *If $\mathrm{ch}_1(\mathbf{v}) = L$, then all the nonempty Brill-Noether loci are irreducible of the expected dimension.*

(4) *If $\mu_L(\mathbf{v}) > \frac{1}{2}$ and is not an integer and $\Delta(\mathbf{v}) \gg 0$, then $B^r(\mathbf{v})$ is reducible and contains components of both the expected and larger than expected dimension.*

The irreducibility in part (3) of the theorem also holds for more general surfaces. Let (X, H) be a polarized surface. Let $\mathbf{v} = (r, C, d)$ be a Chern character on X with $C \cdot H > 0$. Assume that there does not exist a class D on X with $0 < D \cdot H < C \cdot H$. On $\mathbb{P}^2$, the class of a line L satisfies this property. More generally, if X has Picard rank 1, then the ample generator of the Picard group has this property. In this case, the Brill-Noether loci are better behaved. The linear system $|C|$ cannot have any reducible or nonreduced elements (it can however be empty if C is not effective). Given a degree κ, let
$$\gamma_\kappa := \max\{h^0(B, \mathcal{L}) | B \in |C|, \deg(\mathcal{L}) = \kappa\}.$$
We have the easy bound $\gamma_\kappa \le \kappa + 1$, hence $\gamma_\kappa = 0$ when $\kappa < 0$. Combining the Riemann-Roch Theorem with Clifford's Theorem, one can obtain better bounds on γ_κ.

PROPOSITION 7.7. *Let (X, H) be a polarized surface. Let $\mathbf{v} = (r, C, d)$ be a Chern character on X with $C \cdot H > 0$. Assume that there does not exists a divisor class D with $0 < D \cdot H < C \cdot H$. Set $\kappa = d + \frac{C^2}{2}$. Then*

$$h^0(X, \mathcal{V}) \leq r + \gamma_\kappa.$$

In particular, if $\kappa < 0$, then $h^0(\mathcal{V}) \leq r$ for every $\mathcal{V} \in M_{X,H}(\mathbf{v})$ and $B^k_{X,H}(\mathbf{v}) = \emptyset$ for $k > r$. Conversely, if C is effective and $\Delta(\mathbf{v}) \gg 0$ (equivalently, $d \ll 0$), then $B^r_{X,H}(\mathbf{v})$ is nonempty.

PROOF. Let $\mathcal{V} \in M_{X,H}(\mathbf{v})$ be a stable sheaf such that $h^0(X, \mathcal{V}) \geq r$. Choosing r linearly independent sections induces a map $\phi : \mathcal{O}_X^{\oplus r} \to \mathcal{V}$. We claim that ϕ is an injective sheaf map. Suppose the image of ϕ is a sheaf $\mathcal{W}$ of rank $s < r$. Since $h^0(X, \mathcal{W}) \geq r$, we conclude that $\mu_H(\mathcal{W}) > 0$. Since C is the minimal class with $0 < C \cdot H$, we have that $c_1(\mathcal{W}) \cdot H \geq C \cdot H$. Since $s < r$, this contradicts the stability of $\mathcal{V}$. We conclude that the image of ϕ must have rank r, hence its kernel is zero. So we have a short exact sequence

$$0 \to \mathcal{O}_X^{\oplus r} \to \mathcal{V} \to \mathcal{O}_C(D) \to 0, \tag{4}$$

where $\mathcal{O}_C(D)$ is a pure rank one sheaf on a curve of class C since the quotient of a torsion-free sheaf by a locally free sheaf cannot have torsion supported in dimension zero. [**HL10**, Example 1.1.16]. By the long exact sequence on cohomology we conclude that

$$h^0(X, \mathcal{V}) \leq r + h^0(X, \mathcal{O}_C(D)) \leq r + \gamma_\kappa.$$

In particular, if $\kappa < 0$, then $h^0(X, \mathcal{V}) \leq r$.

If C is effective and $\Delta \gg 0$ (or equivalently, $d \ll 0$), then there exists nonsplit extensions of the form (4). It suffices to take Δ large enough so that

$$\chi(\mathcal{O}_C(D), \mathcal{O}_X) < 0.$$

Note that $\mathcal{V}$ is torsion-free (see [**GLL22**, Lemma 4.5]), stable and $h^0(X, \mathcal{V}) \geq r$. Hence, $B^r_{X,H}(\mathbf{v})$ is nonempty. □

REMARK 7.8. When $c_1(\mathcal{V})$ is not minimal, one can use the Grauert-Mülich Theorem and restriction to curves to give a bound on $\max_{\mathcal{V} \in M_{X,H}(\mathbf{v})^s}(h^0(X, \mathcal{V}))$ (see [**HL10**, §3.3]).

THEOREM 7.9. *Let (X, H) be a polarized surface such that $K_X \cdot H \leq 0$. Let $\mathbf{v} = (r, C, d)$ be a Chern character on X with $C \cdot H > 0$ and $\Delta(\mathbf{v}) \gg 0$. Assume that $h^0(X, \mathcal{L}) > 0$ and $H^i(X, \mathcal{L}) = 0$ for $i \geq 1$ for every $\mathcal{L}$ with $c_1(\mathcal{L}) = C$. Further assume that there does not exist a divisor class D with $0 < D \cdot H < C \cdot H$. Then $B^k_{X,H}(\mathbf{v})$ is empty if $k > r$ and nonempty, irreducible and of the expected dimension if $k \leq r$.*

PROOF. We prove the theorem by induction on the rank. We already know that $B^k_{X,H}(\mathbf{v}) = \emptyset$ if $k > r$ by Proposition 7.7. It is also easy to see this when $r = 1$. An element of $M_{X,H}(\mathbf{v})$ is of the form $I_Z(\mathcal{L})$, where $\mathcal{L}$ is a line bundle with class C. Since $\Delta(\mathbf{v}) \gg 0$, we have $|Z| \gg 0$. We only need that $|Z| > C^2$. By Bézout's Theorem, if there exists two independent curves C_1, C_2 in $|\mathcal{L}|$ containing Z, then they must have a common component. This is a contradiction since there are no reducible or nonreduced curves in the linear system $|\mathcal{L}|$ by the minimality of C.

On the other hand, if $k = 1$, then Z is a zero-dimensional scheme contained in an element of some linear system $|\mathcal{L}|$. Since every curve in $|\mathcal{L}|$ is reduced and irreducible and X is smooth, the Hilbert scheme on the curve is irreducible of dimension $|Z|$. We conclude that $B^1_{X,H}(\mathbf{v})$ is irreducible of dimension

$$|Z| + h^0(X,C) + \dim(\mathrm{Pic}^0(X)) - 1 = \dim(M_{X,H}(\mathbf{v}) - 1 + \chi(\mathbf{v})$$

as expected.

Suppose that the theorem holds for rank less than r. By Proposition 7.7, we may assume that $k \leq r$. For $\mathcal{V} \in B^k_{X,H}(\mathbf{v})$, consider the evaluation map $\phi : \mathcal{O}_X^{\oplus k} \to \mathcal{V}$. Then the saturation $\mathcal{W}$ of the image of ϕ has rank $s \leq k$. If $s < k$, then $\mathcal{W}$ is a subsheaf of $\mathcal{V}$ with $\mu_H(\mathcal{W}) > 0$. Since $C \cdot H$ is the minimal possible positive value and $r(\mathcal{W}) < r$, we get a contradiction to the stability of $\mathcal{V}$. We conclude that the evaluation map is injective. Hence, we obtain an exact sequence

$$0 \to \mathcal{O}_X^{\oplus k} \to \mathcal{V} \to \mathcal{V}' \to 0.$$

As above, $\mathcal{V}'$ cannot have torsion supported in dimension zero as the quotient of a torsion-free sheaf by a locally free one. In fact, when $k < r$, $\mathcal{V}'$ is a μ_H-stable sheaf. Indeed, if $\mathcal{V}'$ had either one dimensional torsion or a destabilizing subsheaf, then since the H-degree of $\mathcal{V}'$ is the minimal, taking the quotient by the torsion or destabilizing subsheaf would yield a torsion free sheaf Q with $\mu_H(Q) \leq 0$. Since Q would also be a quotient of $\mathcal{V}$, this would contradict the stability of $\mathcal{V}$. So we conclude that when $k < r$ any $\mathcal{V} \in B^k_{X,H}(\mathbf{v})$ is a nonsplit extension of a μ_H-stable $\mathcal{V}'$ of Chern character $\mathbf{v} - k\mathbf{v}(\mathcal{O}_X)$ by $\mathcal{O}_X^{\oplus k}$.

Conversely, consider nonsplit extensions of the form

$$0 \to \mathcal{O}_X^{\oplus k} \to \mathcal{V} \to \mathcal{V}' \to 0.$$

We have $\hom(\mathcal{V}', \mathcal{O}_X) = 0$ by stability. We split the proof into two cases, $K_X \cdot H < 0$ and $K_X \cdot H = 0$.

Suppose first that $K_X \cdot H < 0$. If $k < r - 1$, then

$$\mathrm{ext}^2(\mathcal{V}', \mathcal{O}_X) = \hom(\mathcal{O}_X, \mathcal{V}'(K_X)) = 0.$$

The latter follows from the fact that $\mu_H(\mathcal{V}'(K_X)) < 0$ since $r(\mathcal{V}') \geq 2$ and $c_1(\mathcal{V}')$ has minimal positive possible H-degree. If $k = r - 1$ and $K_X \cdot H < 0$, then $\mathrm{ext}^2(\mathcal{V}', \mathcal{O}_X) = \hom(\mathcal{O}_X, \mathcal{V}'(K_X)) = 0$ unless $\mathcal{V}'(K_X) = \mathcal{O}_X$. The latter cannot happen since $\Delta(\mathbf{v}) \gg 0$. We conclude that $\mathrm{ext}^1(\mathcal{V}', \mathcal{O}_X^{\oplus k})$ is constant for every $\mathcal{V}' \in M_{X,H}(\mathbf{v}')$ of dimension $-k\chi(\mathcal{V}', \mathcal{O}_X)$. Since $\Delta(\mathbf{v}) \gg 0$, we must also have $\Delta(\mathbf{v}') \gg 0$. By O'Grady's Theorem, the moduli space $M_{X,H}(\mathbf{v}')$ is irreducible of the expected dimension. We conclude that $B^k_{X,H}(\mathbf{v})$ is the image of a projective bundle over an irreducible space. Hence, $B^k_{X,H}(\mathbf{v})$ is irreducible and a dimension count shows that

$$\dim(B^k_{X,H}(\mathbf{v})) = \dim(M_{X,H}(\mathbf{v}')) - k\chi(\mathcal{V}', \mathcal{O}_X) - k^2.$$

We have $\mathbf{v}' = (r - k, c, d)$. Since $\Delta(\mathbf{v}), \Delta(\mathbf{v}') \gg 0$, we have that

$$\dim(M_{X,H}(\mathbf{v})) = 2r^2\Delta(\mathbf{v}) - (r^2 - 1)\chi(\mathcal{O}_X) \text{ and}$$

$$\dim(M_{X,H}(\mathbf{v}')) = 2r^2\Delta(\mathbf{v}') - ((r-k)^2 - 1)\chi(\mathcal{O}_X)$$

by O'Grady's Theorem. Hence, by Riemann-Roch

$$\dim(B^k_{X,H}(\mathbf{v})) = C^2 + (k - 2d)r + (kr - r^2 + 1)\chi(\mathcal{O}_X) - \frac{k}{2}C \cdot K_X - k^2.$$

On the other hand, the expected dimension of $B^k_{X,H}(\mathbf{v})$ is given by

$$\dim(M_{X,H}(\mathbf{v})) - k(k-\chi(\mathbf{v})) = 2r^2\Delta(\mathbf{v}) - (r^2-1)\chi(\mathcal{O}_X) - k(k-\chi(\mathbf{v}))$$

which by Riemann-Roch is equal to

$$C^2 + (k-2d)r + (kr - r^2 + 1)\chi(\mathcal{O}_X) - \frac{k}{2}C\cdot K_X - k^2.$$

Hence, $B^k_{X,H}(\mathbf{v})$ has the expected dimension.

When $k = r$, then we get an extension of the form (4) and the irreducibility follows by a similar argument using the irreducibility of the compactified Jacobian.

Finally, if $K_X \cdot H = 0$, then $\hom(\mathcal{O}_X, \mathcal{V}'(K_X)) = s$ along the Brill-Noether locus $B^s_{X,H}(\mathbf{v}')$ which by induction on the rank has codimension $s(s-\chi(\mathbf{v}'\otimes K_X))$. Since the latter is very large when $\Delta \gg 0$, these loci do not contribute new components. We conclude that the same estimate holds in this case as well. We conclude that the Brill-Noether locus $B^k_{X,H}(\mathbf{v})$ is irreducible, nonempty and of the expected dimension if $k \leq r$ and $\Delta(\mathbf{v}) \gg 0$. □

REMARK 7.10. Under further assumptions on X, one can give more explicit bounds on Δ in Theorem 7.9. For example, in addition to $\mathbb{P}^2$ [**GLL22**], the Brill-Noether loci where $c_1(\mathbf{v})$ is a minimal effective class have been studied for K3 surfaces by Leyenson [**Ley06, Ley12**] and Bayer [**Bay18**] and for abelian surfaces by Bayer and Li [**BL17**] (see also [**CNY24**]).

Acknowledgments. We would like to thank Arend Bayer, Aaron Bertram, Lawrence Ein, Ben Gould, Joe Harris, John Kopper, Daniel Levine, Yeqin Liu, Emanuele Macrì, Sayanta Mandal, Geoffrey Smith, Kōta Yoshioka and Matthew Woolf for many stimulating discussions on the geometry of moduli spaces of sheaves.

References

[Ant18] V. Antonelli, *Characterization of Ulrich bundles on Hirzebruch surfaces*, Rev. Mat. Complut. **34** (2021), no. 1, 43–74, DOI 10.1007/s13163-019-00346-7. MR4213051

[Ati57] M. F. Atiyah, *Vector bundles over an elliptic curve*, Proc. London Math. Soc. (3) **7** (1957), 414–452, DOI 10.1112/plms/s3-7.1.414. MR131423

[AGO17] M. Aprodu, G. Farkas, and A. Ortega, *Minimal resolutions, Chow forms and Ulrich bundles on K3 surfaces*, J. Reine Angew. Math. **730** (2017), 225–249, DOI 10.1515/crelle-2014-0124. MR3692019

[AB13] D. Arcara and A. Bertram, *Bridgeland-stable moduli spaces for K-trivial surfaces*, J. Eur. Math. Soc. (JEMS) **15** (2013), no. 1, 1–38, DOI 10.4171/JEMS/354. MR2998828

[ABCH13] D. Arcara, A. Bertram, I. Coskun, and J. Huizenga, *The minimal model program for the Hilbert scheme of points on $\mathbb{P}^2$ and Bridgeland stability*, Adv. Math. **235** (2013), 580–626, DOI 10.1016/j.aim.2012.11.018. MR3010070

[BaHU91] J. Herzog, B. Ulrich, and J. Backelin, *Linear maximal Cohen-Macaulay modules over strict complete intersections*, J. Pure Appl. Algebra **71** (1991), no. 2-3, 187–202, DOI 10.1016/0022-4049(91)90147-T. MR1117634

[Bay18] A. Bayer, *Wall-crossing implies Brill-Noether: applications of stability conditions on surfaces*, Algebraic geometry: Salt Lake City 2015, Proc. Sympos. Pure Math., vol. 97, Amer. Math. Soc., Providence, RI, 2018, pp. 3–27, DOI 10.1090/pspum/097.1/01668. MR3821144

[BL17] A. Bayer and C. Li, *Brill-Noether theory for curves on generic abelian surfaces*, Pure Appl. Math. Q. **13** (2017), no. 1, 49–76, DOI 10.4310/PAMQ.2017.v13.n1.a2. MR3858014

[BM14a] A. Bayer and E. Macrì, *Projectivity and birational geometry of Bridgeland moduli spaces*, J. Amer. Math. Soc. **27** (2014), no. 3, 707–752, DOI 10.1090/S0894-0347-2014-00790-6. MR3194493

[BM14b] A. Bayer and E. Macrì, *MMP for moduli of sheaves on K3s via wall-crossing: nef and movable cones, Lagrangian fibrations*, Invent. Math. **198** (2014), no. 3, 505–590, DOI 10.1007/s00222-014-0501-8. MR3279532

[Bea83] A. Beauville, *Complex algebraic surfaces*, London Mathematical Society Lecture Note Series, vol. 68, Cambridge University Press, Cambridge, 1983. Translated from the French by R. Barlow, N. I. Shepherd-Barron and M. Reid. MR732439

[Bea16] A. Beauville, *Ulrich bundles on abelian surfaces*, Proc. Amer. Math. Soc. **144** (2016), no. 11, 4609–4611, DOI 10.1090/proc/13091. MR3544513

[BC13] A. Bertram and I. Coskun, *The birational geometry of the Hilbert scheme of points on surfaces*, Birational geometry, rational curves, and arithmetic, Simons Symp., Springer, Cham, 2013, pp. 15–55, DOI 10.1007/978-1-4614-6482-2_2. MR3114922

[BGJ16] A. Bertram, T. Goller, and D. Johnson, *Le Potier's strange duality, quot schemes, and multiple point formulas for del Pezzo surfaces*, Selecta Math. (N.S.) **29** (2023), no. 5, Paper No. 75, DOI 10.1007/s00029-023-00880-9. MR4652967

[BN18] L. Borisov and H. Nuer, *Ulrich bundles on Enriques surfaces*, Int. Math. Res. Not. IMRN **13** (2018), 4171–4189, DOI 10.1093/imrn/rnx018. MR3829179

[BHU87] J. P. Brennan, J. Herzog, and B. Ulrich, *Maximally generated Cohen-Macaulay modules*, Math. Scand. **61** (1987), no. 2, 181–203, DOI 10.7146/math.scand.a-12198. MR947472

[Bri07] T. Bridgeland, *Stability conditions on triangulated categories*, Ann. of Math. (2) **166** (2007), no. 2, 317–345, DOI 10.4007/annals.2007.166.317. MR2373143

[Bri08] T. Bridgeland, *Stability conditions on $K3$ surfaces*, Duke Math. J. **141** (2008), no. 2, 241–291, DOI 10.1215/S0012-7094-08-14122-5. MR2376815

[CaG18] G. Casnati and F. Galluzzi, *Stability of rank two Ulrich bundles on projective $K3$ surfaces*, Math. Scand. **122** (2018), no. 2, 239–256, DOI 10.7146/math.scand.a-101999. MR3789442

[Che96] J. Cheah, *On the cohomology of Hilbert schemes of points*, J. Algebraic Geom. **5** (1996), no. 3, 479–511. MR1382733

[CKM13] E. Coskun, R. S. Kulkarni, and Y. Mustopa, *The geometry of Ulrich bundles on del Pezzo surfaces*, J. Algebra **375** (2013), 280–301, DOI 10.1016/j.jalgebra.2012.08.032. MR2998957

[Cos06a] I. Coskun, *Degenerations of surface scrolls and the Gromov-Witten invariants of Grassmannians*, J. Algebraic Geom. **15** (2006), no. 2, 223–284, DOI 10.1090/S1056-3911-06-00426-7. MR2199064

[Cos06b] I. Coskun, *The enumerative geometry of Del Pezzo surfaces via degenerations*, Amer. J. Math. **128** (2006), no. 3, 751–786. MR2230924

[CH14] I. Coskun and J. Huizenga, *Interpolation, Bridgeland stability and monomial schemes in the plane* (English, with English and French summaries), J. Math. Pures Appl. (9) **102** (2014), no. 5, 930–971, DOI 10.1016/j.matpur.2014.02.010. MR3271294

[CH15] I. Coskun and J. Huizenga, *The birational geometry of the moduli spaces of sheaves on $\mathbb{P}^2$*, Proceedings of the Gökova Geometry-Topology Conference 2014, Gökova Geometry/Topology Conference (GGT), Gökova, 2015, pp. 114–155. MR3381441

[CH16] I. Coskun and J. Huizenga, *The ample cone of moduli spaces of sheaves on the plane*, Algebr. Geom. **3** (2016), no. 1, 106–136, DOI 10.14231/AG-2016-005. MR3455422

[CH18a] I. Coskun and J. Huizenga, *The nef cone of the moduli space of sheaves and strong Bogomolov inequalities*, Israel J. Math. **226** (2018), no. 1, 205–236, DOI 10.1007/s11856-018-1687-z. MR3819692

[CH18b] I. Coskun and J. Huizenga, *Weak Brill-Noether for rational surfaces*, Local and global methods in algebraic geometry, Contemp. Math., vol. 712, Amer. Math. Soc., Providence, RI, 2018, pp. 81–104, DOI 10.1090/conm/712/14343. MR3832400

[CH18c] I. Coskun and J. Huizenga, *The moduli spaces of sheaves on surfaces, pathologies and Brill-Noether problems*, Geometry of moduli, Abel Symp., vol. 14, Springer, Cham, 2018, pp. 75–105. MR3968044

[CH20a] I. Coskun and J. Huizenga, *Brill-Noether theorems and globally generated vector bundles on Hirzebruch surfaces*, Nagoya Math. J. **238** (2020), 1–36, DOI 10.1017/nmj.2018.17. MR4092845

[CH20b] I. Coskun and J. Huizenga, *Brill-Noether problems, Ulrich bundles and the cohomology of moduli spaces of sheaves*, Mat. Contemp. **47** (2020), 21–72. MR4191133

[CH21] I. Coskun and J. Huizenga, *Existence of semistable sheaves on Hirzebruch surfaces*, Adv. Math. **381** (2021), Paper No. 107636, 96 pp., DOI 10.1016/j.aim.2021.107636. MR4215749

[CH23] I. Coskun and J. Huizenga, *Interpolation and moduli spaces of vector bundles on very general blowups of on very general blowups of* $\mathbb{P}^2$, to appear in Epijournal Geom. Algebr. (EPIGA).

[CHK21] I. Coskun, J. Huizenga, and J. Kopper, *The cohomology of general tensor products of vector bundles on* $\mathbb{P}^2$, Selecta Math. (N.S.) **27** (2021), no. 5, Paper No. 94, 46, DOI 10.1007/s00029-021-00707-5. MR4320843

[CHK22] I. Coskun, J. Huizenga, and J. Kopper, *Disconnected moduli spaces of stable bundles on surfaces*, Bull. Lond. Math. Soc. **54** (2022), no. 2, 812–824, DOI 10.1112/blms.12602. MR4453706

[CHS23] I. Coskun, J. Huizenga and G. Smith, Stability and cohomology of kernel bundles on $\mathbb{P}^n$, to appear in Michigan Math. J.

[CHW17] I. Coskun, J. Huizenga, and M. Woolf, *The effective cone of the moduli space of sheaves on the plane*, J. Eur. Math. Soc. (JEMS) **19** (2017), no. 5, 1421–1467, DOI 10.4171/JEMS/696. MR3635357

[CNY23] I. Coskun, H. Nuer, and K. Yoshioka, *The cohomology of the general stable sheaf on a K3 surface*, Adv. Math. **426** (2023), Paper No. 109102, 85 pp., DOI 10.1016/j.aim.2023.109102. MR4593904

[CNY24] I. Coskun and H. Nuer, *Weak Brill-Noether on abelian surfaces*, in preparation.

[CW22] I. Coskun and M. Woolf, *The stable cohomology of moduli spaces of sheaves on surfaces*, J. Differential Geom. **121** (2022), no. 2, 291–340, DOI 10.4310/jdg/1659987893. MR4466671

[CoMR10] L. Costa and R. M. Miró-Roig, *Brill-Noether theory for moduli spaces of sheaves on algebraic varieties*, Forum Math. **22** (2010), no. 3, 411–432, DOI 10.1515/FORUM.2010.023. MR2652705

[D86] J.-M. Drezet, *Fibrés exceptionnels et suite spectrale de Beilinson généralisée sur* $\mathbf{P}_2(\mathbf{C})$ (French), Math. Ann. **275** (1986), no. 1, 25–48, DOI 10.1007/BF01458581. MR849052

[D87] J.-M. Drezet, *Fibrés exceptionnels et variétés de modules de faisceaux semi-stables sur* $\mathbf{P}_2(\mathbf{C})$ (French), J. Reine Angew. Math. **380** (1987), 14–58, DOI 10.1515/crll.1987.380.14. MR916199

[DN89] J.-M. Drezet and M. S. Narasimhan, *Groupe de Picard des variétés de modules de fibrés semi-stables sur les courbes algébriques* (French), Invent. Math. **97** (1989), no. 1, 53–94, DOI 10.1007/BF01850655. MR999313

[DLP85] J.-M. Drezet and J. Le Potier, *Fibrés stables et fibrés exceptionnels sur* $\mathbf{P}_2$ (French, with English summary), Ann. Sci. École Norm. Sup. (4) **18** (1985), no. 2, 193–243. MR816365

[EH87] D. Eisenbud and J. Harris, *Irreducibility and monodromy of some families of linear series*, Ann. Sci. École Norm. Sup. (4) **20** (1987), no. 1, 65–87. MR892142

[ES11] D. Eisenbud and F.-O. Schreyer, *Boij-Söderberg theory*, Combinatorial aspects of commutative algebra and algebraic geometry, Abel Symp., vol. 6, Springer, Berlin, 2011, pp. 35–48, DOI 10.1007/978-3-642-19492-4_3. MR2810424

[ESW03] D. Eisenbud and F.-O. Schreyer, *Resultants and Chow forms via exterior syzygies*, J. Amer. Math. Soc. **16** (2003), no. 3, 537–579, DOI 10.1090/S0894-0347-03-00423-5. With an appendix by Jerzy Weyman. MR1969204

[Fae08] D. Faenzi, *Rank 2 arithmetically Cohen-Macaulay bundles on a nonsingular cubic surface*, J. Algebra **319** (2008), no. 1, 143–186, DOI 10.1016/j.jalgebra.2007.10.005. MR2378065

[Fae18] D. Faenzi, *Ulrich bundles on K3 surfaces*, Algebra Number Theory **13** (2019), no. 6, 1443–1454, DOI 10.2140/ant.2019.13.1443. MR3994571

[FL81] W. Fulton and R. Lazarsfeld, *On the connectedness of degeneracy loci and special divisors*, Acta Math. **146** (1981), no. 3-4, 271–283, DOI 10.1007/BF02392466. MR611386

[Ga51] F. Gaeta, *Sur la distribution des degrés des formes appartenant à la matrice de l'idéal homogène attaché à un groupe de N points génériques du plan* (French), C. R. Acad. Sci. Paris **233** (1951), 912–913. MR45083

[Gie77] D. Gieseker, *On the moduli of vector bundles on an algebraic surface*, Ann. of Math. (2) **106** (1977), no. 1, 45–60, DOI 10.2307/1971157. MR466475

[Gie82] D. Gieseker, *Stable curves and special divisors: Petri's conjecture*, Invent. Math. **66** (1982), no. 2, 251–275, DOI 10.1007/BF01389394. MR656623

[Gim87] A. Gimigliano, *On Linear Systems of Plane Curves*, ProQuest LLC, Ann Arbor, MI, 1987. Thesis (Ph.D.)–Queen's University (Canada). MR2635606

[GH94] L. Göttsche and A. Hirschowitz, *Weak Brill-Noether for vector bundles on the projective plane*, Algebraic geometry (Catania, 1993/Barcelona, 1994), Lecture Notes in Pure and Appl. Math., vol. 200, Dekker, New York, 1998, pp. 63–74. MR1651090

[GLL22] B. Gould, Y. Liu, and W. Lee, *Higher rank Brill-Noether theory on $\mathbb{P}^2$*, Int. Math. Res. Not. IMRN **24** (2023), 22096–22137, DOI 10.1093/imrn/rnac333. MR4681310

[GH80] P. Griffiths and J. Harris, *On the variety of special linear systems on a general algebraic curve*, Duke Math. J. **47** (1980), no. 1, 233–272. MR563378

[GM08] I. Grzegorczyk and M. Teixidor i Bigas, *Brill-Noether theory for stable vector bundles*, Moduli spaces and vector bundles, London Math. Soc. Lecture Note Ser., vol. 359, Cambridge Univ. Press, Cambridge, 2009, pp. 29–50. MR2537065

[Ha86] B. Harbourne, *The geometry of rational surfaces and Hilbert functions of points in the plane*, Proceedings of the 1984 Vancouver conference in algebraic geometry, CMS Conf. Proc., vol. 6, Amer. Math. Soc., Providence, RI, 1986, pp. 95–111. MR846019

[Har71] R. Hartshorne, *Ample vector bundles on curves*, Nagoya Math. J. **43** (1971), 73–89. MR292847

[Har77] R. Hartshorne, *Algebraic geometry*, Graduate Texts in Mathematics, No. 52, Springer-Verlag, New York-Heidelberg, 1977. MR463157

[Hi89] A. Hirschowitz, *Une conjecture pour la cohomologie des diviseurs sur les surfaces rationnelles génériques* (French), J. Reine Angew. Math. **397** (1989), 208–213, DOI 10.1515/crll.1989.397.208. MR993223

[HL93] A. Hirschowitz and Y. Laszlo, *Fibrés génériques sur le plan projectif* (French), Math. Ann. **297** (1993), no. 1, 85–102, DOI 10.1007/BF01459489. MR1238408

[Hof10] N. Hoffmann, *Moduli stacks of vector bundles on curves and the King-Schofield rationality proof*, Cohomological and geometric approaches to rationality problems, Progr. Math., vol. 282, Birkhäuser Boston, Boston, MA, 2010, pp. 133–148, DOI 10.1007/978-0-8176-4934-0_5. MR2605167

[Hui16] J. Huizenga, *Effective divisors on the Hilbert scheme of points in the plane and interpolation for stable bundles*, J. Algebraic Geom. **25** (2016), no. 1, 19–75, DOI 10.1090/jag/652. MR3419956

[HK22] J. Huizenga and J. Kopper, *Ample stable vector bundles on rational surfaces*, Comm. Algebra **50** (2022), no. 9, 3744–3760, DOI 10.1080/00927872.2022.2042548. MR4442469

[HL10] D. Huybrechts and M. Lehn, *The geometry of moduli spaces of sheaves*, 2nd ed., Cambridge Mathematical Library, Cambridge University Press, Cambridge, 2010, DOI 10.1017/CBO9780511711985. MR2665168

[Kim16] Y. Kim, *Ulrich bundles on rational surfaces with an anticanonical pencil*, Manuscripta Math. **150** (2016), no. 1-2, 99–110, DOI 10.1007/s00229-015-0802-2. MR3483171

[KM] J. Kopper and S. Mandal, *Non-globally generated bundles on curves*, Int. Math. Res. Not. IMRN **2** (2023), 1094–1118, DOI 10.1093/imrn/rnab276. MR4537321

[Ko89] D. Kotschick, *On manifolds homeomorphic to* $\mathbf{C}\mathrm{P}^2\#8\overline{\mathbf{C}\mathrm{P}}^2$, Invent. Math. **95** (1989), no. 3, 591–600, DOI 10.1007/BF01393892. MR979367

[LLV20] E. Larson, H. Larson and I. Vogt, Global Brill-Noether Theory over the Hurwitz Space, to appear in Geom. Topol.

[La91] G. Laumon, *Fibrés vectoriels spéciaux* (French, with English summary), Bull. Soc. Math. France **119** (1991), no. 1, 97–119. MR1101941

[LeP97] J. Le Potier, *Lectures on vector bundles*, Cambridge Studies in Advanced Mathematics, vol. 54, Cambridge University Press, Cambridge, 1997. MR1428426

[LZ19] D Levine and S Zhang, *Brill-Noether and existence of semistable sheaves for del Pezzo surfaces*, to appear in Ann. Inst. Fourier.

[Ley06] M. Leyenson, *On the Brill-Noether theory for K3 surfaces*, Cent. Eur. J. Math. **10** (2012), no. 4, 1486–1540, DOI 10.2478/s11533-012-0069-8. MR2925617

[Ley12] M. Leyenson, *On the Brill-Noether theory for K3 surfaces*, Cent. Eur. J. Math. **10** (2012), no. 4, 1486–1540, DOI 10.2478/s11533-012-0069-8. MR2925617

[LZ19] C. Li and X. Zhao, *Birational models of moduli spaces of coherent sheaves on the projective plane*, Geom. Topol. **23** (2019), no. 1, 347–426, DOI 10.2140/gt.2019.23.347. MR3921322

[Liu23] Y. Liu, *The cohomology of spherical bundles on K3 surfaces*, Preprint, arXiv:2210.11030v2.

[Mar78] M. Maruyama, *Moduli of stable sheaves. II*, J. Math. Kyoto Univ. **18** (1978), no. 3, 557–614, DOI 10.1215/kjm/1250522511. MR509499

[Me97] N. Mestrano, *Sur les espaces de modules des fibrés vectoriels de rang deux sur des hypersurfaces de* $\mathbf{P}^3$ (French), J. Reine Angew. Math. **490** (1997), 65–79, DOI 10.1515/crll.1997.490.65. MR1468925

[MS11] N. Mestrano and C. Simpson, *Moduli of sheaves*, Development of moduli theory—Kyoto 2013, Adv. Stud. Pure Math., vol. 69, Math. Soc. Japan, [Tokyo], 2016, pp. 77–172, DOI 10.2969/aspm/06910077. MR3586507

[MYY18] H. Minamide, S. Yanagida, and K. Yoshioka, *The wall-crossing behavior for Bridgeland's stability conditions on abelian and K3 surfaces*, J. Reine Angew. Math. **735** (2018), 1–107, DOI 10.1515/crelle-2015-0010. MR3757471

[Muk84] S. Mukai, *Symplectic structure of the moduli space of sheaves on an abelian or K3 surface*, Invent. Math. **77** (1984), no. 1, 101–116, DOI 10.1007/BF01389137. MR751133

[Muk87] S. Mukai, *On the moduli space of bundles on K3 surfaces. I*, Vector bundles on algebraic varieties (Bombay, 1984), Tata Inst. Fund. Res. Stud. Math., vol. 11, Tata Inst. Fund. Res., Bombay, 1987, pp. 341–413. MR893604

[N16a] H. Nuer, *Projectivity and birational geometry of Bridgeland moduli spaces on an Enriques surface*, Proc. Lond. Math. Soc. (3) **113** (2016), no. 3, 345–386, DOI 10.1112/plms/pdw033. MR3551850

[N16b] H. Nuer, *A note on the existence of stable vector bundles on Enriques surfaces*, Selecta Math. (N.S.) **22** (2016), no. 3, 1117–1156, DOI 10.1007/s00029-015-0218-6. MR3518547

[NY] H. Nuer and K. Yoshioka, *MMP via wall-crossing for moduli spaces of stable sheaves on an Enriques surface*, Adv. Math. **372** (2020), 107283, 119 pp., DOI 10.1016/j.aim.2020.107283. MR4125513

[O'G96] K. G. O'Grady, *Moduli of vector bundles on projective surfaces: some basic results*, Invent. Math. **123** (1996), no. 1, 141–207, DOI 10.1007/BF01232371. MR1376250

[O'G99] K. G. O'Grady, *Desingularized moduli spaces of sheaves on a K3*, J. Reine Angew. Math. **512** (1999), 49–117, DOI 10.1515/crll.1999.056. MR1703077

[OVdV86] C. Okonek and A. Van de Ven, *Stable bundles and differentiable structures on certain elliptic surfaces*, Invent. Math. **86** (1986), no. 2, 357–370, DOI 10.1007/BF01389075. MR856849

[Se60] B. Segre, *Alcune questioni su insiemi finiti di punti in geometria algebrica* (Italian), Univ. e Politec. Torino Rend. Sem. Mat. **20** (1960/61), 67–85. MR169119

[Sol18] P. Solis, *Natural cohomology on* $\mathbb{P}^1 \times \mathbb{P}^1$, Preprint.

[Su87] N. Sundaram, *Special divisors and vector bundles*, Tohoku Math. J. (2) **39** (1987), no. 2, 175–213, DOI 10.2748/tmj/1178228323. MR887936

[Wal98] C. Walter, *Irreducibility of moduli spaces of vector bundles on birationally ruled surfaces*, Algebraic geometry (Catania, 1993/Barcelona, 1994), Lecture Notes in Pure and Appl. Math., vol. 200, Dekker, New York, 1998, pp. 201–211. MR1651095

[Yos99] K. Yoshioka, Irreducibility of moduli spaces of vector bundles on K3 surfaces, arXiv:math/9907001

[Yos01] K. Yoshioka, *Moduli spaces of stable sheaves on abelian surfaces*, Math. Ann. **321** (2001), no. 4, 817–884, DOI 10.1007/s002080100255. MR1872531

[Zha22] J. Zhao, *Moduli spaces of sheaves on general blow-ups of* $\mathbb{P}^2$, Preprint, arXiv:2208.03619.

[Yos09] K. Yoshioka, *Stability and the Fourier-Mukai transform. II*, Compos. Math. **145** (2009), no. 1, 112–142, DOI 10.1112/S0010437X08003758. MR2480497

Department of Mathematics, Stat. and CS, University of Illinois at Chicago, Chicago, Illinois 60607
Email address: icoskun@uic.edu

Department of Mathematics, The Pennsylvania State University, University Park, Pennsylvania 16802
Email address: huizenga@psu.edu

Faculty of Mathematics, Technion, Israel Institute of Technology, Haifa, Israel
Email address: hnuer@technion.ac.il

Contemporary Mathematics
Volume **803**, 2024
https://doi.org/10.1090/conm/803/16097

Double covers of curves on Nikulin surfaces

Simona D'Evangelista and Margherita Lelli-Chiesa

Dedicated to Peter Newstead on the occasion of his 80th *birthday*

Abstract. We survey basic results concerning Prym varieties, the Prym-Brill-Noether theory initiated by Welters, and Brill-Noether theory of general étale double covers of curves of genus $g \geq 2$. We then specialize to curves on Nikulin surfaces and show that étale double covers of curves on Nikulin surfaces of standard type do not satisfy Welters' Theorem. On the other hand, by specialization to curves on Nikulin surfaces of non-standard type, we prove that general double covers of curves ramified at $b = 2, 4, 6$ points are Brill-Noether general; the case $b = 2$ was already obtained by Bud [The birational geometry of $\overline{\mathcal{R}}_{g,2}$ and Prym-canonical divisorial strata, math.AG/01718.v2] with different techniques.

1. Introduction

Double covers of complex curves are a classical and still very hot topic in algebraic geometry. Part of the interest in them stems from the fact that any étale double cover $\pi : \widetilde{C} \to C$ of a smooth irreducible curve C of genus $g \geq 2$ naturally defines a principally polarized abelian variety (P, Ξ) of dimension $g - 1$, known as the Prym variety of π. Prym varieties were introduced by Schottky and Jung [**SJ**] in relation to the Schottky problem and were named after the German mathematician Prym by Mumford ([**M1**]), who was the first to investigate them from an algebraic point of view. Despite a vast literature on the topic, many questions concerning the geometry of general double covers remain open, both in the étale and in the ramified case. We will here focus on Brill-Noether type questions.

We recall that the Brill-Noether theory of a general curve C of genus g is quite well understood since the 1980s. The Brill-Noether Thorem establishes that the Brill-Noether variety $W^r_d(C)$, parametrizing degree d line bundles on C with a space of global sections of dimension $\geq r+1$, is nonempty if and only if the so-called Brill-Noether number $\rho(g, r, d) := g - (r + 1)(g - d + r)$ is nonnegative. Furthermore, if nonempty, $W^r_d(C)$ is smooth of dimension $\rho(g, r, d)$ outside of $W^{r+1}_d(C)$ by the Gieseker-Petri Theorem. The existence part of the Brill-Noether Theorem is actually valid for any smooth curve of genus g and is due to Kleiman-Laksov

2020 *Mathematics Subject Classification.* Primary 14H51, 14H40, 14H30, 14J28.
Key words and phrases. Prym curves, Brill-Noether theory, Nikulin surfaces.
The second named author was supported in part by PRIN-2020 "Curves, Ricci flat Varieties and their Interactions" and by GNSAGA.

[KL] and Kempf **[Ke]**. The first proofs of both the non-existence part of the Brill-Noether Theorem and of the Gieseker-Petri Theorem, all based on degeneration techniques, were achieved by Griffiths-Harris **[Gr]**, Eisenbud-Harris **[EH1, EH2]**, and Gieseker **[Gi]**. A big breakthrough came with Lazarsfeld's alternative proof of the Gieseker-Petri Theorem **[La, Pa]**, that avoided any type of degeneration and proceeded instead by specialization to smooth curves lying on $K3$ surfaces.

If $\pi : \widetilde{C} \to C$ is a double cover of a general genus $g \geq 2$ curve C, the curve $\widetilde{C}$ is obviously non-general in moduli and it thus makes sense to investigate its Brill-Noether theory. In the étale case $\widetilde{C}$ has genus $2g-1$ and is never Petri general because $W^1_{2g-2}(\widetilde{C})$ turns out to be singular (cf. **[We]** and Section 3.2 below). Moreover, if g is even, $\widetilde{C}$ possesses a pencil of degree g which prevents it from being Brill-Noether general (cf. **[AF]** and Section 4 later in this paper). Hence, the Brill-Noether behaviour of $\widetilde{C}$ seems quite wild. The picture is much nicer if one considers instead the so-called Prym-Brill-Noether theory of $\widetilde{C}$. This is the study of the geometry of the Prym-Brill-Noether varieties V^r, that were introduced by Welters **[We]** and are related to the geometry of the Prym variety P; indeed, they are obtained by intersecting the Brill-Noether varieties $W^r_{2g-2}(\widetilde{C})$ with an appropriate translate of P living in $\mathrm{Pic}^{2g-2}(\widetilde{C})$. Using the degeneration techniques developed by Eisenbud and Harris, Welters proved (cf. Thm. 2.3 below) that for a general π the varieties V^r are smooth of dimension $\rho^-(g,r) := g-1-(r+1)r/2$. In particular, they are empty if $\rho^-(g,r) < 0$; conversely, V^r is nonempty if $\rho^-(g,r) \geq 0$ for any étale double cover π by a result of Bertram **[Ber]**. These statements are the analogues in Prym-Brill-Noether theory of the Brill-Noether Theorem and the Gieseker-Petri Theorem.

Perhaps discouraged by the intricate picture in the étale case, people ignored the Brill-Noether theory of $\widetilde{C}$ in the ramified case until recently, where Bud **[Bu]** proved that $\widetilde{C}$ is both Brill-Noether and Petri general (that is, it satisfies both the the Brill-Noether Theorem and the Gieseker-Petri Theorem) if $\pi : \widetilde{C} \to C$ is a double cover of a general genus g curve branched over 2 general points of C. This result is quite amazing as it highlights that ramified double covers behave better than étale ones from a Brill-Noether viewpoint.

This paper was born from the desire to make $K3$ surfaces and Lazarsfeld's techniques enter the picture. This is ensured by specialization to Nikulin surfaces, that is, primitively polarized $K3$ surfaces (S,H) endowed with a double cover $\pi_S : \widehat{S} \to S$ ramified along eight (-2)-curves, which are both pairwise disjoint and disjoint from smooth curves in the linear system $|H|$. The minimal model of $\widehat{S}$ is again a $K3$ surface $\widetilde{S}$ endowed with an involution having 8 fixed points. The Picard ranks of both S and $\widetilde{S}$ are ≥ 9 and, when equality holds, the Picard groups have been described in **[VGS, GS]**. It turns out that there are two types of Nikulin surfaces, depending on whether the embedding of the rank 9 sublattice generated by H and the (-2)-curves in $\mathrm{Pic}(S)$ is primitive or not: according to the terminology introduced in **[KLV1]**, (S,H) is called of standard type in the former case, and of non-standard type in the latter. Note that π induces an étale double cover $\pi|_C : \widetilde{C} := \pi^{-1}(C) \to C$ of any smooth curve $C \in |H|$, and the curve $\widetilde{C}$ can be identified with its image in $\widetilde{S}$. In the last decade Nikulin surfaces have been largely exploited in the study of étale double covers of curves (or equivalently, Prym curves) and their moduli space $\mathcal{R}_h$. In particular, Farkas and Verra **[FV2]** proved that for

$6 \neq h \leq 7$, general étale double covers of genus h curves live on Nikulin surfaces of standard type, and used this fact to describe the birational geometry of $\mathcal{R}_h$ in low genera. Furthermore, in the standard case the curve $\widetilde{C} \subset S$ has the gonality of a general étale double cover of a genus h curve **[AF]**. It is thus natural to wonder whether a different proof of Welters' Theorem can be obtained by specialization to étale double covers living on Nikulin surfaces of standard type. Unfortunately, we will answer negatively to this question proving the following theorem.

THEOREM. *Let $\pi : \widetilde{C} \to C$ be an étale double cover on a Nikulin surface S of standard type, with C in the primitive linear system $|H|$ of genus h. If $h > 7$ or $h = 6$, the curve $\widetilde{C}$ does not satisfy Welters' Theorem.*

Nikulin surfaces of non-standard type have not received as much attention, and their investigation essentially started in **[KLV1, KLV2]**. As emerged in **[KLV1]**, in the non-standard case the curve $\widetilde{C}$ is quite special as it possesses two theta characteristics with many sections cut out by two line bundles $R_1, R_2 \in \mathrm{Pic}(\widetilde{S})$. The main result of the paper suggests that Nikulin surfaces of non-standard type are instead the right environment to investigate double covers of curves ramified at $2, 4, 6$ points, which can be realized on a Nikulin surface of non-standard type by restricting π_S to the inverse image of smooth curves in the linear systems $|R_1|$ and $|R_2|$. By specialization to ramified double covers of curves on Nikulin surfaces of non-standard type, we obtain the following generalization of Bud's result.

THEOREM. *Let $\pi : \widetilde{C} \to C$ be a general double cover of a genus $g \geq 2$ curve ramified at 2, 4, or 6 points. Then the curve $\widetilde{C}$ is Brill-Noether general.*

The paper is organized as follows. Section 2 surveys the basic theory of Prym varieties and Prym-Brill-Noether theory, with particular attention to Welters' infinitesimal study of V^r in terms of the Prym-Petri map, which is the anti-invariant part $\mu_{0,L}^-$ of the Petri map $\mu_{0,L}$ of any line bundle $L \in V^r$. The sections ends with a discussion and interpretation of its invariant counterpart $\mu_{0,L}^+$, as well as with a conjectural picture concerning its injectivity. Section 3 recalls some prerequisites on theta characteristics in order to show, following an argument by Beauville **[Bea]**, that for any irreducible étale double cover $\pi : \widetilde{C} \to C$ the curve $\widetilde{C}$ possesses some invariant vanishing thetanulls: in particular, $\widetilde{C}$ is not Petri general. In Section 4 we concentrate on the Brill-Noether theory of $\widetilde{C}$ when π is general. More precisely, we recall Schwarz's non-existence result **[Sc]** concerning linear series whose Brill-Noether number is negative enough, and Aprodu-Farkas' theorem **[AF]** on the gonality of $\widetilde{C}$. The latter prevents $\widetilde{C}$ from being Brill-Noether general if the genus of C is even, while the Brill-Noether generality/speciality in the odd genus case is still unknown in full generality, up to our knowledge. We also recall Bud's result concerning the ramified case. Section 5 is focused on Nikulin surfaces of standard and non-standard type and on the proof of the above theorems.

2. Prym-Brill-Noether theory

2.1. Preliminaries on Prym varieties. Prym varieties are principally polarized abelian varieties arising from étale double covers of curves, and they are useful to link the geometry of curves to that of abelian varieties.

Let C be a complex smooth irreducible curve, and let $\pi : \widetilde{C} \to C$ be an irreducible étale double cover of C. Denoting by g and $\widetilde{g}$ the genus of C and $\widetilde{C}$, respectively, Hurwitz's formula yields $\widetilde{g} = 2g - 1$. The cover π induces the so-called *Norm map* between the Jacobians $J(C)$ and $J(\widetilde{C})$ of C and $\widetilde{C}$, respectively:

$$\begin{aligned} \mathrm{Nm}\colon J(\widetilde{C}) &\longrightarrow J(C) \\ \mathcal{O}_{\widetilde{C}}(D) &\longmapsto \mathcal{O}_C(\pi(D)). \end{aligned}$$

This fits in the following commutative diagram:

$$\begin{array}{ccc} \widetilde{C} & \xrightarrow{a_{y_0}} & J(\widetilde{C}) \\ {\scriptstyle\pi}\downarrow & & \downarrow{\scriptstyle\mathrm{Nm}} \\ C & \xrightarrow[a_{x_0}]{} & J(C), \end{array}$$

where a_{x_0} and a_{y_0} are the Abel-Jacobi maps associated with some fixed points $x_0 \in C$ and $y_0 \in \widetilde{C}$ such that $\pi(y_0) = x_0$.

In particular, the principally polarized Abelian varieties $(J(C), \Theta)$ and $(J(\widetilde{C}), \widetilde{\Theta})$ are related by two maps

$$\pi^* : J(C) \to J(\widetilde{C}), \quad \mathrm{Nm} : J(\widetilde{C}) \to J(C),$$

such that π^* and Nm are dual to each other and $\mathrm{Nm} \circ \pi^* : J(C) \to J(C)$ is multiplication by two ([**M1**]). Indeed, if $N = \mathcal{O}_C(D) \in J(C)$, then $\pi^* N = \mathcal{O}_{\widetilde{C}}(\pi^{-1}(D)) \in J(\widetilde{C})$ and thus $\mathrm{Nm}(\pi^* N) = \mathcal{O}_C(\pi(\pi^{-1}(D))) \in J(C)$; since π is a double cover, we get $\pi(\pi^{-1}(D)) = 2D$.

Denote by $\iota : \widetilde{C} \to \widetilde{C}$ the involution that interchanges the sheets of π. For every divisor $\widetilde{D}$ on $\widetilde{C}$, the following equality is straightforward:

$$\pi^{-1}(\pi(\widetilde{D})) = \widetilde{D} + \iota(\widetilde{D}).$$

It follows that

$$\pi^*(\mathrm{Nm}(M)) = M \otimes \iota^*(M), \quad \forall \quad M \in J(\widetilde{C}).$$

Since Nm is surjective, we also get that

$$\iota^*(\pi^*(N)) = \pi^* N, \quad \forall \quad N \in J(C),$$

that is, the involution ι^* acts as the identity on $\pi^*(J(C)) \subset J(\widetilde{C})$. On the other hand, ι^* acts as -1 on $\mathrm{Ker}(\mathrm{Nm}) \subset J(\widetilde{C})$. In [**M2**] Mumford decomposed the kernel of Nm into two irreducible components:

$$\begin{aligned} P_0 &:= \{M \otimes \iota^* M^\vee \mid M \in \mathrm{Pic}^0(\widetilde{C})\}, \\ P_1 &:= \{M \otimes \iota^* M^\vee \mid M \in \mathrm{Pic}^1(\widetilde{C})\}. \end{aligned}$$

More precisely, he proved the following:

LEMMA 2.1 ([**M2**] Lem. 1). *If L is a line bundle on $\widetilde{C}$ such that* $\mathrm{Nm}\, L \simeq \mathcal{O}_C$, *then $L \simeq M \otimes \iota^* M^\vee$ for some line bundle M on $\widetilde{C}$. Moreover, M can be chosen of degree* 0 *or* 1.

A classical theorem by Wirtinger ([**Wi**]) yields that the dimension of the space of global sections is constant mod 2 on P_0 and P_1 and it has opposite parity on the two components. The component $P := P_0$ of Ker(Nm) containing the origin is an abelian subvariety of $J(\widetilde{C})$ of dimension

$$\dim P = \dim J(\widetilde{C}) - \dim J(C) = g - 1;$$

furthermore, it turns out that the canonical polarization of $J(\widetilde{C})$ restricts to twice a principal polarization Ξ on P. The principally polarized abelian variety (P,Ξ) is called the *Prym variety* associated with π (cf. [**M1**, §3] for more details).

By the same argument [**M1**], the inverse image $\mathrm{Nm}^{-1}(\omega_C) \subset \mathrm{Pic}^{2g-2}(\widetilde{C})$ of the canonical line bundle $\omega_C \in \mathrm{Pic}^{2g-2}(C)$ breaks up in two components

$$P^+ := \{L \in \mathrm{Pic}^{2g-2}(\widetilde{C}) \mid \mathrm{Nm}(L) \simeq \omega_C,\ h^0(C,L) \equiv 0 \ (\mathrm{mod}\ 2)\},$$

$$P^- := \{L \in \mathrm{Pic}^{2g-2}(\widetilde{C}) \mid \mathrm{Nm}(L) \simeq \omega_C,\ h^0(C,L) \equiv 1 \ (\mathrm{mod}\ 2)\},$$

which are both translates of P.

We recall that the cover $\pi : \widetilde{C} \to C$ defines a class of order two $\eta \in J(C)[2]$ such that $\pi_*\mathcal{O}_{\widetilde{C}} = \mathcal{O}_C \oplus \eta$; viceversa, any non-trivial $\eta \in J(C)[2]$ determines an étale double cover of C by setting $\widetilde{C} := \mathrm{Spec}(\mathcal{O}_C \oplus \eta)$. Therefore, the datum of π is equivalent to that of the pair (C,η), which is called a *Prym curve* of genus g. The moduli space

$$\mathcal{R}_g := \{(C,\eta) \,|\, C \text{ smooth curve of genus } g,\ \eta \in J(C)[2]\,,\, \eta \not\simeq \mathcal{O}_C\}\,,$$

has attracted much attention in the last decades; we refer to [**F3**] for a very nice survey on its geometry.

2.2. Prym-Brill-Noether varieties and Welters' Theorem. In [**We**] Welters initiated a Brill-Noether study of the varieties P^- and P^+ and introduced the following closed subsets of $\mathrm{Nm}^{-1}(\omega_C)$, defined for any integer $r \geq -1$:

$$V^r(C,\eta) := \{L \in \mathrm{Pic}^{2g-2}(\widetilde{C}) \mid \mathrm{Nm}(L) \simeq \omega_C,\ h^0(C,L) \geq r+1,\ h^0(C,L) \equiv r+1 \ (\mathrm{mod}\ 2)\}.$$

Scheme-theoretically, these can be realized as the intersections

$$V^r(C,\eta) = W^r_{2g-2}(\widetilde{C}) \cap P^+ \quad \text{if } r \text{ is odd}, \tag{2.1}$$

$$V^r(C,\eta) = W^r_{2g-2}(\widetilde{C}) \cap P^- \quad \text{if } r \text{ is even},$$

where

$$W^r_{2g-2}(\widetilde{C}) = \left\{L \in \mathrm{Pic}^{2g-2}(\widetilde{C}) \mid h^0(C,L) \geq r+1\right\}$$

is the classical Brill-Noether variety; this justifies the name *Prym-Brill-Noether varieties* used for the loci $V^r(C,\eta)$. The following result proved in [**M2**] and [**Ha**] concerns their expected dimension.

PROPOSITION 2.2. *Let $L \in \mathrm{Nm}^{-1}(\omega_C)$ and set $h^0(L) = r+1$. Then, the dimension of $V^r(C,\eta)$ at L satisfies*

$$\dim_L(V^r(C,\eta)) \geq g-1-\binom{r+1}{2}.$$

We will refer to the number

$$\rho^-(g,r) := g-1-\binom{r+1}{2},$$

as the *Prym-Brill-Noether number.*

A priori one may hope that classical results of Brill-Noether theory carry over to étale double covers of smooth curves and to the varieties $V^r(C,\eta)$, but unfortunately this is not the case. Indeed, many statements of this theory, such as the Brill-Noether Theorem and the Gieseker-Petri Theorem, hold only for a general curve $\widetilde{C}$

of genus $\widetilde{g}$ and they fail in the case where $\widetilde{C}$ is the étale double cover of a genus g curve. Nevertheless, an analogue of the Gieseker-Petri Theorem for Prym varieties was obtained by Welters [**We**] by a degeneration argument similar to the one used by Eisenbud and Harris [**EH2**] in their proof of the classical statement.

Since $\omega_{\widetilde{C}} = \pi^*\omega_C$, the push-pull formula yields the decomposition

$$H^0(\widetilde{C}, \omega_{\widetilde{C}}) = H^0(C, \omega_C) \oplus H^0(C, \omega_C(\eta)). \tag{2.2}$$

into invariant and anti-invariant forms under the action of ι. Fix a line bundle $L \in \mathrm{Nm}^{-1}(\omega_C)$ and set $h^0(L) = r + 1$. Since the composition $\mathrm{Nm} \circ \pi^*$ is the multiplication by 2, the differential of Nm at L is $2({}^t\pi^*)$, where

$${}^t\pi^* : (H^0(\widetilde{C}, \omega_{\widetilde{C}}))^\vee \to (H^0(C, \omega_C))^\vee$$

is the transpose of the pull-back of differential forms. By taking kernels, we obtain the identification

$$T_L(\mathrm{Nm}^{-1}(\omega_C)) = (H^0(C, \omega_C(\eta)))^\vee \hookrightarrow (H^0(\widetilde{C}, \omega_{\widetilde{C}}))^\vee, \tag{2.3}$$

where the inclusion is the transpose of the projection map

$$\begin{aligned} p\colon H^0(\widetilde{C}, \omega_{\widetilde{C}}) &\longrightarrow H^0(C, \omega_C(\eta)) \\ \lambda &\longmapsto \frac{1}{2}(\lambda - \iota^*\lambda) \end{aligned}.$$

By classical Brill-Noether theory (cf., e.g., [**ACGH**]), we know that

$$T_L(W^r_{2g-2}(\widetilde{C})) = (\mathrm{Im}\mu_{0,L})^\perp \subset (H^0(\widetilde{C}, \omega_{\widetilde{C}}))^\vee, \tag{2.4}$$

where the so-called *Petri map*

$$\mu_{0,L} : H^0(\widetilde{C}, L) \otimes H^0(\widetilde{C}, \omega_{\widetilde{C}} \otimes L^\vee) \to H^0(\widetilde{C}, \omega_{\widetilde{C}})$$

is multiplication of global sections. In particular, the dimension of the tangent space at L to $W^r_{2g-2}(\widetilde{C})$ is:

$$\dim T_L(W^r_{2g-2}(\widetilde{C})) = \rho(2g-1, r, 2g-2) + \dim(\mathrm{Ker}\mu_{0,L}), \tag{2.5}$$

where the number $\rho(2g-1, r, 2g-2) := 2g - 1 - (r+1)^2$ equals the difference between the dimensions of the codomain and the domain of $\mu_{0,L}$. Recalling (2.1) and (2.3), we obtain:

$$T_L(V^r(C, \eta)) = (\mathrm{Im}\mu_{0,L})^\perp \cap (H^0(C, \omega_C(\eta))^\vee = (\mathrm{Im}(p \circ \mu_{0,L}))^\perp. \tag{2.6}$$

By applying π^* to the isomorphism $\mathrm{Nm}\, L \simeq \omega_C$, one gets $L \otimes \iota^* L \simeq \omega_{\widetilde{C}}$, or equivalently, $\omega_{\widetilde{C}} \otimes L^\vee \simeq \iota^* L$. Consider the following composition of maps:

$$\begin{array}{ccccccc} H^0(\widetilde{C}, L) \otimes H^0(\widetilde{C}, L) & \xrightarrow{1\otimes\iota^*} & H^0(\widetilde{C}, L) \otimes H^0(\widetilde{C}, \omega_{\widetilde{C}} \otimes L^\vee) & \xrightarrow{\mu_{0,L}} & H^0(C, \omega_{\widetilde{C}}) & \xrightarrow{p} & H^0(C, \omega_C(\eta)) \\ s \otimes t & \longmapsto & s \otimes \iota^* t & \longmapsto & s \cdot \iota^* t & \longmapsto & \frac{1}{2}(s \cdot (\iota^* t) - (\iota^* s) \cdot t), \end{array}$$

which is clearly skew-symmetric. By restriction to $\wedge^2 H^0(\widetilde{C}, L)$, we thus obtain the map

$$\begin{aligned} \mu^-_{0,L}\colon \wedge^2 H^0(\widetilde{C}, L) &\to H^0(C, \omega_C(\eta)) \\ s \wedge t &\mapsto \frac{1}{2}(s \cdot (\iota^* t) - (\iota^* s) \cdot t), \end{aligned}$$

which is called the *Prym-Petri map* of L. We can rewrite (2.6) in terms of $\mu_{0,L}^-$ as

$$T_L(V^r(C,\eta)) = (\mathrm{Im}\mu_{0,L}^-)^{\perp}.$$

Since $\rho^-(g,r)$ equals the difference between the dimensions of the codomain and the domain of $\mu_{0,L}^-$, we conclude that

$$\dim(T_L(V^r(C,\eta))) = \rho^-(g,r) + \dim(\mathrm{Ker}\mu_{0,L}^-), \tag{2.7}$$

and $V^r(C,\eta)$ is smooth of dimension $\rho^-(g,r)$ at L if and only if the Prym-Petri map is injective. Welters proved that this is the case for all $L \in \mathrm{Nm}^{-1}(\omega_C)$ provided that the cover π is general.

THEOREM 2.3 ([**We**] Thm. 1.11). *Let $(C,\eta) \in \mathcal{R}_g$ be general and let $\pi : \widetilde{C} \to C$ be the étale double cover defined by η. Then the Prym-Petri map $\mu_{0,L}^-$ is injective for all $L \in \mathrm{Nm}^{-1}(\omega_C)$.*

The result is analogous to the Gieseker-Petri Theorem, as it yields the smoothness of the Prym-Brill-Noether varieties $V^r(C,\eta)$ for a general (C,η) and their emptiness in the cases where $\rho^-(g,r) < 0$. The analogue of the existence part of the Brill-Noether Theorem for any Prym curve was instead established by Bertram:

THEOREM 2.4 ([**Ber**] Thm. 1.4). *Let (C,η) be a Prym curve of genus g and let $\pi : \widetilde{C} \to C$ be the étale double cover defined by η. If $\rho^-(g,r) \geq 0$, the Prym-Brill-Noether variety $V^r(C,\eta)$ is nonempty.*

Welters' proof is by degeneration to a reduced nodal curve C_0 consisting of a string of rational curves and g elliptic curves $E_1, \dots, E_g$, such that, if x_i and y_i are the two intersection points of any E_i with two adjacent components, the line bundle $\mathcal{O}_{E_i}(x_i - y_i)$ is not a torsion point in $\mathrm{Pic}^0(E_i)$. Consider a line bundle η_0 on C_0 which is trivial on every component except on E_g, where it restricts to a non-zero point of $\mathrm{Pic}^0(E_g)[2]$. The étale double cover $\widetilde{C}_0$ of C_0 defined by η_0 then looks as follows:

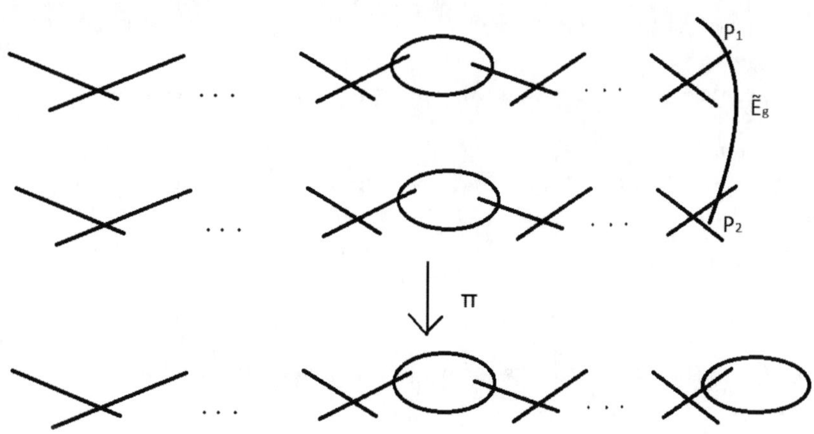

The points P_1 and P_2 on $\widetilde{E}_g$ are not $\mathbb{Z}$-independent; indeed, the line bundle $\mathcal{O}_{\widetilde{E}_g}(P_1 - P_2) \in \mathrm{Pic}^0(\widetilde{E}_g)$ is a 2-torsion point. This is why Eisenbud and Harris' proof [**EH2**] of the Gieseker-Petri Theorem fails for $\widetilde{C}_0$. However, using the theory of limit linear series and the Brill-Noether theory of a (non general) 2-pointed elliptic curve, Welters wrote down the explicit form of any element ρ in the kernel of the relevant Petri maps and, thanks to the skew-symmetry of $1 \otimes \iota^*$, concluded that such a ρ never lies in the kernel of the corresponding Prym-Petri map.

2.3. The map $\mu_{0,L}^+$. The decomposition (2.2) provides a splitting of the Petri map $\mu_{0,L}$ of any line bundle $L \in \mathrm{Nm}^{-1}(\omega_C)$ into a ι-anti-invariant part, namely, the Prym-Petri map $\mu_{0,L}^-$, and a ι-invariant part

$$\begin{aligned}\mu_{0,L}^+ \colon \mathrm{Sym}^2 H^0(\widetilde{C}, L) &\to H^0(C, \omega_C)\\ s\otimes t + t\otimes s &\mapsto s\cdot(\iota^* t) + (\iota^* s)\cdot t.\end{aligned}$$

As already remarked, the Prym-Petri map $\mu_{0,L}^-$ governs the smoothness of the varieties $V^r(C,\eta)$ and has thus been extensively investigated. By contrast, the map $\mu_{0,L}^+$ has been almost ignored so far. Up to our knowledge, it only appeared in the proof of the uniruledness of $\mathcal{R}_8$ due to Farkas and Verra [**FV2**].

We provide an interpretation of the map $\mu_{0,L}^+$ by considering the inclusion $V^r(C,\eta) \subset W^r_{2g-2}(\widetilde{C})$ and the induced short exact sequence of linear maps

$$0 \to T_L(V^r(C,\eta)) \to T_L W^r_{2g-2}(\widetilde{C}) \to N_{V^r(C,\eta)/W^r_{2g-2}(\widetilde{C}),L} \to 0,$$

where $N_{V^r(C,\eta)/W^r_{2g-2}(\widetilde{C}),L}$ is the normal space at the point L of $V^r(C,\eta)$ in $W^r_{2g-2}(\widetilde{C})$. By (2.5) and (2.7), we get:

$$\begin{aligned}&\dim_{\mathbb{C}}(N_{V^r(C,\eta)/W^r_{2g-2}(\widetilde{C}),L}) =\\ &\quad = \rho(2g-1, r, 2g-2) + \dim(\mathrm{Ker}\mu_{0,L}) - \rho^-(g,r) - \dim(\mathrm{Ker}\mu_{0,L}^-) =\\ &\quad = g - \frac{(r+1)(r+2)}{2} + \dim(\mathrm{Ker}\mu_{0,L}^+),\end{aligned}$$

where we have used that $\mathrm{Ker}\mu_{0,L} = \mathrm{Ker}\mu_{0,L}^+ \oplus \mathrm{Ker}\mu_{0,L}^-$ by construction. We set

$$\rho^+(g,r) := \rho(2g-1, r, 2g-2) - \rho^-(g,r) = g - \frac{(r+1)(r+2)}{2}$$

and notice that this number equals the difference between the dimensions of the codomain and the domain of $\mu_{0,L}^+$. We conclude that the dimension of the normal space $N_{V^r(\widetilde{X})/W^r_{2g-2}(\widetilde{X}),L}$ equals the dimension of the cokernel of $\mu_{0,L}^+$. More precisely, (2.4) and (2.6) imply that

$$N_{V^r(\widetilde{X})/W^r_{2g-2}(\widetilde{X}),L} = (\mathrm{Im}\mu_{0,L})^\perp \cap (H^0(\omega_C))^\vee = (\mathrm{Im}(q\circ\mu_{0,L}))^\perp,$$

where $q : H^0(\widetilde{C}, \omega_{\widetilde{C}}) \to H^0(C, \omega_C)$ is the projection mapping a form λ to $\frac{1}{2}(\lambda + \iota^*\lambda)$. The composition

$$q \circ \mu_{0,L} \circ (1 \otimes \iota^*) : H^0(\widetilde{C}, L) \otimes H^0(\widetilde{C}, L) \longrightarrow H^0(C, \omega_C)$$

maps a decomposable tensor $s \otimes t$ to the invariant form $s(\iota^* t) + (\iota^* s)t$ and is thus symmetric. Its restriction to $\mathrm{Sym}^2 H^0(\widetilde{C}, L)$ coincides with the map $\mu^+_{0,L}$ and this provides the identification

$$N_{V^r(\widetilde{X})/W^r_{2g-2}(\widetilde{X}),L} = (\mathrm{Im}\mu^+_{0,L})^\perp.$$

When $(C,\eta) \in \mathcal{R}_g$ is general, it is natural to wonder whether the map $\mu^+_{0,L}$ is injective for all $L \in \mathrm{Nm}^{-1}(\omega_C)$. Unfortunately, this trivially fails for any line bundle $L \in V^r(C,\eta)$ where r is any fixed integer r such that $\rho^-(g,r) \geq 0$ but $\rho(2g-1,r,2g-2) < 0$; indeed, the Prym-Petri map $\mu^-_{0,L}$ is injective by Theorem 2.3, but our assumptions prevent $\mu^+_{0,L}$ from being injective because $\dim \mathrm{Ker}\mu^+_{0,L} \geq -\rho^+(g,r) > 0$. More generally, the map $\mu^+_{0,L}$ fails to be injective for any line bundle $L \in V^r(C,\eta)$ as soon as

$$\rho^-(g,r) > \max\{-1, \rho(2g-1,r,2g-2)\}, \tag{2.8}$$

which implies $\rho^+(g,r) < 0$. The inequalities (2.8) can be rewritten only in terms of ρ as

$$-r \leq \rho(2g-1,r,2g-2) < r. \tag{2.9}$$

In this range, the equality $\dim \mathrm{Ker}\mu^+_{0,L} = -\rho^+(g,r)$ is the best one may hope for; since $V^r(C,\eta)$ is smooth at L by Theorem 2.3, this hope is equivalent to the expectation that $W^r_{2g-2}(\widetilde{C})$ and $V^r(C,\eta)$ coincide in a neighborhood of L.

EXPECTATION 2.5. *Let $(C,\eta) \in \mathcal{R}_g$ be general and let $\pi : \widetilde{C} \to C$ be the étale double cover defined by η. Then, the following hold:*

(i) *If $-r \leq \rho(2g-1,r,2g-2) < r$, then $W^r_{2g-2}(\widetilde{C}) = V^r(C,\eta)$. In particular, for all $L \in V^r(C,\eta)$ one has*

$$\dim(\mathrm{Ker}\mu_{0,L}) = \dim(\mathrm{Ker}\mu^+_{0,L}) = -\rho^+(g,r).$$

(ii) *If $\rho(2g-1,r,2g-2) \geq r$, then both $\mu_{0,L}$ and $\mu^+_{0,L}$ are injective for all $L \in V^r(C,\eta)$.*

The situation in the remaining cases $\rho(2g-1,r,2g-2) < -r$ is already clear, as $W^r_{2g-2}(\widetilde{C})$ turns out to be empty by a more general result of Schwarz [**Sc**] that we will recall in Section 4.

3. Theta-characteristics and vanishing thetanulls

3.1. Preliminaries on theta-characteristics. A *theta-characteristic* on a smooth irreducible curve C is a line bundle $\theta \in \mathrm{Pic}^{g-1}(C)$ such that $\theta^{\otimes 2} = \omega_C$. Theta-characteristics are of two types, called odd and even according to the parity of the dimension of the space of their global sections. The set of theta-characteristics on C, denoted by $\mathrm{Th}(C)$, is a principal homogeneous space for the space $J(C)[2]$ of two-torsion points in the Jacobian of C, that is, $J(C)[2]$ acts freely and transitively on $\mathrm{Th}(C)$. In particular, we have $|\mathrm{Th}(C)| = |J(C)[2]| = 2^{2g}$.

The $\mathbb{F}_2$-vector space $J(C)[2]$ is endowed with a nondegenerate symplectic form

$$\langle \cdot, \cdot \rangle : J(C)[2] \times J(C)[2] \to \mathbb{F}_2,$$

which is called *Weil pairing* and is defined as follows ([**M2**]). For any pair of points $\eta, \epsilon \in J(C)[2]$, one can write $\eta = \mathcal{O}_C(D)$ and $\epsilon = \mathcal{O}_C(E)$ for two divisors D and E on C with disjoint support, and choose rational functions f and g on C such that

$\operatorname{div}(f) = 2D$ and $\operatorname{div}(g) = 2E$. The condition on the supports of D and E ensures the validity of the so-called *Weil Reciprocity Law* (see **[Ha]**):

$$f((g)) = g((f)),$$

where the evaluation of a rational function h at a divisor Z whose support is disjoint from the set of zeros and poles of h is defined as $h(Z) := \prod_{p\in C} h(p)^{\mathrm{mult}_p Z}$. Hence, we get

$$\frac{f(2E)}{g(2D)} = \left(\frac{f(E)}{g(D)}\right)^2 = 1,$$

and $\frac{f(E)}{g(D)} = \pm 1$. The value of the Weil pairing at the pair (η, ϵ) is given by the following formula:

$$(-1)^{\langle \epsilon, \eta\rangle} = \frac{f(E)}{g(D)},$$

and one can check that this definition is independent of both the divisors D, E and the rational functions f, g.

Given a symplectic vector space $(V, \langle\cdot,\cdot\rangle)$ over $\mathbb{F}_2$, we denote by $Q(V)$ the set of quadratic forms on V with fixed polarity equal to the symplectic form $\langle\cdot,\cdot\rangle$, that is, all functions $q : V \to \mathbb{F}_2$ that satisfy the identity

$$q(x + y) = q(x) + q(y) + \langle x, y\rangle, \quad \forall\, x, y \in V.$$

Having fixed a symplectic basis $(e_1, \dots, e_g, f_1, \dots, f_g)$ of V, the so-called *Arf invariant* of a quadratic form $q \in Q(V)$ is defined as

$$\operatorname{arf}(q) := \sum_{i=1}^{g} q(e_i) \cdot q(f_i) \in \mathbb{F}_2,$$

this definition being independent of the basis. A quadratic form $q \in Q(V)$ is called *even* if $\operatorname{arf}(q) = 0$, and *odd* if $\operatorname{arf}(q) = 1$. The space of even (respectively, odd) quadratic forms is denoted by $Q(V)^+$ (resp., $Q(V)^-$). An easy computation gives $|Q(V)^+| = 2^{g-1}(2^g + 1)$ and $|Q(V)^-| = 2^{g-1}(2^g - 1)$.

Coming back to theta-characteristics, to any $\theta \in \operatorname{Th}(C)$ one associates the so-called *theta-form* $q_\theta : J(C)[2] \to \mathbb{F}_2$, by setting

$$q_\theta(\eta) := h^0(C, \eta \otimes \theta) + h^0(C, \theta) \bmod 2.$$

For any $\epsilon, \eta \in J(C)[2]$, the *Riemann-Mumford relation* (**[M2]**) yields:

$$h^0(C, \theta \otimes \epsilon \otimes \eta) + h^0(C, \theta \otimes \epsilon) + h^0(C, \theta \otimes \eta) + h^0(C, \theta) \equiv \langle \eta, \epsilon\rangle \bmod 2,$$

or equivalently, $q_\theta \in Q(J(C)[2])$. We thus identify the two spaces

$$\operatorname{Th}(C) = Q(J(C)[2]). \tag{3.1}$$

Since $\operatorname{arf}(q_\theta) = h^0(C, \theta) \bmod 2$ for any $\theta \in \operatorname{Th}(C)$, under this identification even (respectively, odd) theta-characteristics correspond to forms in $Q(J(C)[2])^+$ (resp., $Q(J(C)[2])^-$). Hence, on a genus g curve C there are precisely $2^{g-1}(2^g + 1)$ even theta-characteristics and $2^{g-1}(2^g - 1)$ odd ones.

Following Atiyah **[At]**, a *spin curve* of genus g is a pair (C, θ), where C is a smooth irreducible curve of genus g and $\theta \in \operatorname{Th}(C)$. Mumford (**[M2]**) and Atiyah (**[At]**) proved that the parity of a spin curve, that is, the value of its Arf invariant, is locally constant in families. As a consequence, the moduli space $\mathcal{S}_g$ parametrizing spin curves of genus g splits into two connected components $\mathcal{S}_g^+$ and $\mathcal{S}_g^-$: a pair (C, θ) lies in $\mathcal{S}_g^+$ if θ is an even theta-characteristic, and in $\mathcal{S}_g^-$ otherwise.

For each $g, r \geq 0$ one can define the locus

$$\mathcal{S}_g^r := \{(C,\theta) \in \mathcal{S}_g \mid h^0(C,\theta) \geq r+1 \text{ and } h^0(C,\theta) \equiv r+1 \bmod 2\}.$$

Harris ([**Ha**]) proved that the dimension of every component of $\mathcal{S}_g^r$ is bounded below by the number $3g-3-\binom{r+1}{2}$, to which we refer as the expected dimension of $\mathcal{S}_g^r$. It may well be the case that $\mathcal{S}_g^r$ is nonempty even when its expected dimension is negative. For instance, $\mathcal{S}_g^{[\frac{g-1}{2}]} \neq \emptyset$ contains hyperelliptic curves and is thus nonempty although in this case $3g-3-\binom{r+1}{2}$ is very negative. Existence of components of $\mathcal{S}_g^r$ having the expected dimension has been established by Farkas [**F2**] in the range $1 \leq r \leq 11, r \neq 10$ for all $g \geq g(r)$ where $g(r)$ is an explicit integer.

In the case where $r=1$, the locus $\mathcal{S}_g^1$ parametrizes curves having a so-called *vanishing thetanull*, that is, an effective even theta characteristic, and is a divisor in $\mathcal{S}_g^+$. Indeed, every theta characteristic θ on a general curve C of genus g has at most one section. This directly follows from the base-point-free pencil trick, which implies that, as soon as $H^0(C,\theta)$ contains a pencil V, the kernel of the Petri map

$$\mu_{0,V} : V \otimes H^0(C,\theta) \to H^0(C,\omega_C)$$

is nonzero; indeed, one has $\mathrm{Ker}\mu_{0,V} \simeq H^0(C,\mathcal{O}_C(B)) \neq 0$, where B is the base locus of V. The Gieseker-Petri Theorem thus excludes the existence of such a pencil if the curve C is general.

3.2. Invariant vanishing thetanulls on étale double covers. We denote by $\pi : \widetilde{C} \to C$ the irreducible étale double cover associated with a general Prym curve (C,η) of genus g. As explained at the end of Section 2, the Brill-Noether varieties $W^r_{2g-2}(\widetilde{C})$ governing the singularities of the theta divisor of $J(\widetilde{C})$ are nonempty as soon as $\rho(2g-1,r,2g-2) \geq -r$; in particular, for values of g for which there exists an r yielding $-r \leq \rho(2g-1,r,2g-2) < 0$, the curve $\widetilde{C}$ is Brill-Noether special.

We now fix g, r such that $\rho(2g-1,r,2g-2) \geq 0$; under this condition, one may still hope that $W^r_{2g-2}(\widetilde{C})$ is smooth and of the expected dimension. However, this expectation fails even for $r=1$, as observed by Welters himself [**We**, Rmk. 1.12]; in particular, $\widetilde{C}$ never satisfies the Gieseker-Petri Theorem. By the base-point-free pencil trick, $W^1_{2g-2}(\widetilde{C}) \setminus W^2_{2g-2}(\widetilde{C})$ is singular at any point defining a vanishing thetanull. On the other hand, a vanishing thetanull on $\widetilde{C}$ is easily obtained as pullback π^*M of a theta-characteristics M on C such that both M and $M \otimes \eta$ are odd. The following nice argument taken from [**Bea**, Prop. 4] counts the number of such Ms and thus in particular proves their existence.

For any $M \in \mathrm{Th}(C)$, the push-pull formula yields

$$h^0(\widetilde{C},\pi^*M) = h^0(C,\pi_*\pi^*M) = h^0(C,M) + h^0(C,M\otimes\eta) = q_M(\eta) \bmod 2.$$

We are looking for those $M \in \mathrm{Th}(C)$ such that $q_M(\eta)=0$ and $\mathrm{arf}(q_M)=1$. Pick $\epsilon \in J(C)[2]$ such that the Weil pairing $\langle\eta,\epsilon\rangle = 1$ and denote by $\Sigma \subset J(C)[2]$ the plane spanned by η and ϵ. Since $J(C)[2] = \Sigma \oplus \Sigma^\perp$, any $M \in \mathrm{Th}(C)$ is completely determined by the restrictions $q_M|_\Sigma$ and $q_M|_{\Sigma^\perp}$. The condition $q_M(\eta)=0$ yields $\mathrm{arf}(q_M|_\Sigma) = q_M(\eta)q_M(\epsilon) = 0$. Therefore, M is determined by the value $q_M(\epsilon) \in \mathbb{F}_2$ and by the restriction $q_M|_{\Sigma^\perp}$, which is a quadratic form on $\Sigma^\perp$ of arf invariant 1 (because $\mathrm{arf}(q_M) = \mathrm{arf}(q_M|_\Sigma) + \mathrm{arf}(q_M|_{\Sigma^\perp})$). Since $\dim_{\mathbb{F}_2}\Sigma^\perp = 2(g-1)$, we have $2^{g-2}(2^{g-1}-1)$ choices for $q_M|_{\Sigma^\perp}$ and, having the possibility of choosing $q_M(\epsilon)$, we obtain $2^{g-1}(2^{g-1}-1)$ theta-characteristics M on C as above. Since $\pi^*(M \otimes \eta) \simeq \pi^*M$, this construction provides precisely $2^{g-2}(2^{g-1}-1)$ vanishing thetanulls on $\widetilde{C}$, which are invariant under the covering involution.

4. Brill-Noether theory for double covers

Given the étale double cover $\pi : \widetilde{C} \to C$ of a general Prym curve $(C, \eta) \in \mathcal{R}_g$, it is natural to investigate not only the Prym-Brill-Noether varieties $V^r(C, \eta)$ but any Brill-Noether variety $W^r_d(\widetilde{C})$. However, not much is known about the Brill-Noether theory of $\widetilde{C}$. The following non-existence result follows from [**Sc**], where Schwarz proved a more general statement concerning étale cyclic covers of arbitrary degree.

THEOREM 4.1 ([**Sc**]). *Let $\pi : \tilde{C} \to C$ be the étale double cover associated with a general Prym curve $(C, \eta) \in \mathcal{R}_g$. Then the Brill-Noether variety $W^r_d(\widetilde{C})$ is empty if*

$$\rho(2g-1, r, d) < -r.$$

The above result can be easily proved using Welters' degeneration and applying [**F1**, Prop. 4.1] on the Brill-Noether theory of 2-pointed elliptic curves. For $r = 1$ Theorem 4.1 is known to be optimal and this implies that $\widetilde{C}$ is not Brill-Noether general if g is even. Indeed, the existence of a pencil of degree g on $\widetilde{C}$ in this case follows from the surjectivity (cf. [**ACGH**]) of the difference map

$$\begin{aligned} \phi_{\frac{g}{2}} : C^{\frac{g}{2}} \times C^{\frac{g}{2}} &\longrightarrow J(C) \\ (D, E) &\longmapsto \mathcal{O}_C(D-E), \end{aligned}$$

yielding $\eta = \mathcal{O}_C(D-E)$ for some effective divisors D, E both of degree $g/2$ on C. The pullback $\pi^*(\mathcal{O}_C(E)) \in \operatorname{Pic}^g(\widetilde{C})$ then satisfies

$$h^0(\widetilde{C}, \pi^*\mathcal{O}_C(E)) = h^0(C, \pi_*\pi^*\mathcal{O}_C(E)) = h^0(C, \mathcal{O}_C(E)) + h^0(C, \mathcal{O}_C(D)) \geq 2,$$

where we have again used the push-pull formula. On the other hand, in the odd genus case the gonality of $\widetilde{C}$ is maximal, as first proved by Aprodu and Farkas.

THEOREM 4.2 ([**AF**] Thm. 0.4). *Let $\pi : \tilde{C} \to C$ be the étale double cover associated with a general Prym curve $(C, \eta) \in \mathcal{R}_g$. Then the following hold:*

(i) *if $g \equiv 1$ mod 2, the curve $\tilde{C}$ has maximal gonality, that is,* $\operatorname{gon}(\tilde{C}) = g+1$*;*
(ii) *if $g \equiv 0$ mod 2, then* $\operatorname{gon}(\tilde{C}) = g$*.*

In both cases the Clifford index of $\widetilde{C}$ equals $\operatorname{gon}(\tilde{C}) - 2$.

Two different proofs are provided in [**AF**], one by degeneration and one by specialization to curves on Nikulin surfaces; the latter are particular $K3$ surfaces on which we will focus in the next section. Up to our knowledge, the following natural question remains open:

QUESTION 1. Let g be an odd positive integer such that the inequalities $-r \leq \rho(2g-1, r, 2g-2) < 0$ admit no integral solution $r \geq 1$. If $(C, \eta) \in \mathcal{R}_g$ is general, is the cover $\tilde{C}$ Brill-Noether general?

We stress that, quite surprisingly, ramified double covers of curves seem to behave better than étale ones from a Brill-Noether viewpoint. Following [**Bu**], let $\mathcal{R}_{g,2n}$ be the moduli space parametrizing irreducible double covers of smooth genus

g curves ramified at $2n$ points. The space $\mathcal{R}_{g,2n}$ can be alternatively defined as follows:

$$\mathcal{R}_{g,2n} := \{(C, x_1 + \ldots + x_{2n}, \eta) \mid [C] \in \mathcal{M}_g,\ x_i \in C\, \forall 1 \le i \le n$$
$$\eta \in \operatorname{Pic}^{-n}(C),\ \eta^2 = \mathcal{O}_C(-x_1 - \ldots - x_{2n})\}\,.$$

While studying its birational geometry, Bud proved the following astonishing result, which is in contrast to the étale case.

THEOREM 4.3 (**[Bu]** Thm. 4.1). *Let $\pi : \widetilde{C} \to C$ be the double cover associated with a general $(C, x+y, \eta) \in \mathcal{R}_{g,2}$. Then $\widetilde{C}$ is both Brill-Noether and Petri general.*

Bud's proof relies on the study of divisors on a compactification $\overline{\mathcal{R}_{g,2}}$ of the moduli space. In the next section we will provide a simpler proof of the Brill-Noether generality of $\widetilde{C}$ by specialization to ramified double covers of curves on Nikulin surfaces of non-standard type, and we will extend the result to covers ramified at 4 and 6 points.

5. Double covers of curves on Nikulin surfaces

5.1. Nikulin surfaces and their Picard group. Nikulin surfaces represent a rather special class of K3 surfaces, which has been studied in relation to various topics, including the theory of automorphisms **[VGS]**, the study of Prym curves **[FK]** and the birational geometry of their moduli spaces **[FV1, FV2, KLV1, KLV2]**. We recall their definition and some basic properties.

DEFINITION 1. A polarized *Nikulin* surface of genus $h \ge 2$ is a triple (S, M, H) consisting of a smooth $K3$ surface S and two line bundles $\mathcal{O}_S(M), H \in \operatorname{Pic} S$ that satisfy the following conditions:

- S contains 8 disjoint smooth rational curves $N_1, \ldots, N_8$ such that
$$N_1 + \cdots + N_8 \sim 2M.$$
- H is nef, $H^2 = 2(h-1)$ and $H \cdot M = 0$.

We say that (S, M, H) is *primitively polarized* if the class of H is primitive in $\operatorname{Pic} S$.

The line bundle M defines a non-trivial double cover $\pi : \widehat{S} \to S$ branched along $\sum_{i=1}^{8} N_i$. Denoting by $\sigma : \widehat{S} \to \widetilde{S}$ the blow-down of the eight (-1)-curves $E_i := \pi^{-1}(N_i) \subset \widehat{S}$, the surface $\widetilde{S}$ is a minimal $K3$ surface endowed with a so-called *Nikulin involution* $\iota \in \operatorname{Aut}(\widetilde{S})$ having eight fixed points corresponding to the images $\sigma(E_i)$ of the exceptional divisors. The quotient $\bar{S} := \widetilde{S}/\iota$ has eight ordinary double points and the quotient map $\bar{\pi} : \widetilde{S} \to \bar{S}$ fits in the following commutative diagram:

$$\begin{array}{ccc} \widehat{S} & \xrightarrow{\ \sigma\ } & \widetilde{S} \\ {\scriptstyle \pi}\downarrow & & \downarrow{\scriptstyle \bar{\pi}} \\ S & \xrightarrow[\ \bar{\sigma}\ ]{} & \bar{S} \end{array},$$

where $\bar{\sigma}$ is the contraction of the curves N_i to the eight nodes of $\bar{S}$.

DEFINITION 2. Let (S, M, H) be a polarized Nikulin surface of genus h. The rank 8 sublattice of $\operatorname{Pic} S$ generated by $N_1, \ldots, N_8$ and M is called *Nikulin lattice* and its denoted by $\mathbf{N} = \mathbf{N}(S, M)$.

Since the Picard group of a polarized Nikulin surface (S, M, H) contains both the Nikulin lattice and the polarization H, its rank is at least 9. Consider the rank 9 lattice

$$\Lambda_h = \Lambda(S, M, H) := \mathbb{Z}[H] \oplus_{\perp} \mathbf{N} \subset \operatorname{Pic} S;$$

the surface S is said to be a *Nikulin surface of standard type* if the embedding $\Lambda_h \subset \operatorname{Pic} S$ is primitive, and a *Nikulin surface of non-standard type* otherwise. Garbagnati and Sarti [**GS**] proved that in the latter case h is forced to be odd. The following result describes $\operatorname{Pic}(S)$ in the case of minimal Picard number.

PROPOSITION 5.1 ([**GS**], Prop. 2.1). *Let (S, M, H) be a genus h primitively polarized Nikulin surface of Picard number 9. Then either $\operatorname{Pic} S = \Lambda_h$ (standard case), or h is odd and $\Lambda_h \subset \operatorname{Pic} S$ has index two (non-standard case). In the latter situation, possibly after renumbering the curves N_i, one falls in one of these cases:*

- $h \equiv 3 \bmod 4$ *and there are $R_1, R_2 \in \operatorname{Pic} S$ such that*
$$R_1 \sim \frac{H - N_1 - N_2}{2}, \quad R_2 \sim \frac{H - N_3 - \cdots - N_8}{2};$$
in particular, one has $g(R_1) = (h+1)/4$, $g(R_2) = (h-3)/4$.
- $h \equiv 1 \bmod 4$ *and there are $R_1, R_2 \in \operatorname{Pic} S$ such that*
$$R_1 \sim \frac{H - N_1 - N_2 - N_3 - N_4}{2}, \quad R_2 \sim \frac{H - N_5 - N_6 - N_7 - N_8}{2};$$
in particular, one has $g(R_1) = g(R_2) = (h-1)/4$.

The growing interest in Nikulin surfaces is motivated from the fact that some Prym curves live on them. The condition $H \cdot M = 0$ in Definition 1 ensures that the double cover π restricts to an étale double cover $\pi|_C : \widetilde{C} := \pi^{-1}(C) \to C$ of any smooth curve $C \in |H|$; in other words, the pair $(C, M|_C)$ is a Prym curve of genus h. Since $\widetilde{C}$ is disjoint from the (-1)-curves E_i, the curve $\widetilde{C}$ can be identified with its image in $\widetilde{S}$ and we set $\widetilde{H} := \mathcal{O}_{\widetilde{S}}(\widetilde{C})$.

The Picard group of the $K3$ surface $\widetilde{S}$ was described by Van Geemen and Sarti [**VGS**] by investigating the action of the Nikulin involution ι on the cohomology group $H^2(\widetilde{S}, \mathbb{Z})$. It turns out that $\operatorname{Pic}(\widetilde{S})$ always contains the orthogonal complement $(H^2(\widetilde{S}, \mathbb{Z})^{\iota})^{\perp}$, which is isomorphic to the rank 8 lattice $E_8(-2)$. Since $\widetilde{H} \in E_8(-2)^{\perp} \subset \operatorname{Pic}(\widetilde{S})$, the Picard number of $\widetilde{S}$ is ≥ 9 and equality holds if and only if the same holds for S. The following proposition summarizes results by Van Geemen and Sarti [**VGS**, Prop. 2.2 and Prop. 2.7] and Aprodu and Farkas [**AF**, Prop. 4.3] in the standard case.

PROPOSITION 5.2. *Let (S, M, H) be a genus h polarized Nikulin surface of standard type and Picard number 9, and let $\widetilde{S}$ be the $K3$ surface with a Nikulin involution obtained from S. Then, the lattice*

$$\widetilde{\Lambda}_h := \mathbb{Z}\widetilde{H} \oplus E_8(-2)$$

has index 2 in $\operatorname{Pic}(\widetilde{S})$. The latter is generated by $\widetilde{\Lambda}_h$ and a class $\frac{\widetilde{H}+v}{2}$, where v is an element of $E_8(-2)$ satisfying $v^2 = -4$ if h is even and $v^2 = -8$ if h is odd.

We now turn to the non-standard case. By [**GS**, Prop. 3.5 (2)], for $i = 1, 2$ the linear system $|R_i|$ contains a smooth irreducible curve as soon as R_i has nonnegative self-intersection. We stress that $R_1 \cdot M = 2$ and $R_2 \cdot M = 6$ if $h \equiv 3 \bmod 4$, while $R_1 \cdot M = R_2 \cdot M = 4$ for $h \equiv 1 \bmod 4$. Therefore, for all smooth curves

$D_1 \in |R_1|$ and $D_2 \in |R_2|$, the cover π induces double covers $\pi_{D_1} : \pi^{-1}(D_1) \to D_1$ and $\pi_{D_2} : \pi^{-1}(D_2) \to D_2$ that are ramified at 2 and 6 points, respectively, if $h \equiv 3 \bmod 4$; on the other hand, both π_{D_1} and π_{D_2} are ramified at 4 points if $h \equiv 1 \bmod 4$. In any case, the curves $\pi^{-1}(D_1)$ and $\pi^{-1}(D_2)$ meet each exceptional curve E_i in at most 1 point and are thus isomorphic to their images in $\widetilde{S}$, that we denote by $\widetilde{D}_1$ and $\widetilde{D}_2$ respectively. It is trivial to check that

$$\sigma^*\widetilde{D}_1 - E_1 - E_2 \sim \pi^*D_1,\ \sigma^*\widetilde{D}_2 - E_3 - \cdots - E_8 \sim \pi^*D_2 \text{ if } h \equiv 3 \bmod 4, \tag{5.1}$$

$$\sigma^*\widetilde{D}_1 - E_1 - E_2 - E_3 - E_4 \sim \pi^*D_1,\ \sigma^*\widetilde{D}_2 - E_5 - E_6 - E_7 - E_8 \sim \pi^*D_2 \text{ if } h \equiv 1 \bmod 4.$$

Note that Hurwitz's formula always yields $g(\widetilde{D}_1) = g(\widetilde{D}_2) = (h+1)/2$.

The following result directly follows from [**VGS**, Prop. 2.2 and Prop. 2.7] in the non-standard case.

PROPOSITION 5.3. *Let (S, M, H) be a genus h polarized Nikulin surface of non-standard type and Picard number 9, and let $\widetilde{S}$ be the $K3$ surface with a Nikulin involution obtained from S. Then*

$$\mathrm{Pic}(\widetilde{S}) = \mathbb{Z}\widetilde{R} \oplus E_8(-2),$$

where $\widetilde{R}$ is a polarization of genus $(h+1)/2$ such that the curves $\widetilde{D}_1, \widetilde{D}_2$ as in (5.1) lie in $|\widetilde{R}|$.

5.2. Standard Nikulin surfaces. As highlighted in [**KLV1**, Prop. 2.3], a general curve $C \in |H|$ on a very general genus h primitively polarized Nikulin surface (S, M, H) of standard type is Brill-Noether general . Furthermore, it was proved by Aprodu and Farkas [**AF**, Thm. 1.5] that the étale double cover $\widetilde{C} \subset \widetilde{S}$ of C has the gonality of a general curve of genus $2h-1$ that covers a genus h curve, namely, $h+1$ if h is odd and h otherwise. It is thus natural to ask whether $\widetilde{C}$ is general from a Prym-Brill-Noether viewpoint, that is, if it satisfies Welters' Theorem. A positive answer would provide an alternative proof of Welters' result avoiding degeneration, in analogy with Lazarsfeld's proof of the Gieseker-Petri Theorem by specialization to curves on $K3$ surfaces. Unfortunately, the answer turns out to be negative as soon $h > 7$ and $h = 6$. This bound on the genus agrees with the following theorem by Farkas and Verra.

THEOREM 5.4 ([**FV2**], Thm. 0.2). *A general Prym curve $(C, \eta) \in \mathcal{R}_h$ lies on a Nikulin surface if and only if $h \leq 7$ and $h \neq 6$.*

We thus prove the following result.

THEOREM 5.5. *Let (S, M, H) be a genus h primitively polarized Nikulin surface of standard type with either $h > 7$, or $h = 6$. For any smooth curve $C \in |H|$, the double cover $\widetilde{C} \in |\widetilde{H}|$ of C defined by $M|_C$ does not satisfy Welter's Theorem.*

PROOF. Set $A := \frac{\widetilde{H}+v}{2} \in \mathrm{Pic}(\widetilde{S})$ with $v \in E_8(-2)$ as in Proposition 5.2. Since $\widetilde{C} \cdot v = 0$, the restriction $A|_{\widetilde{C}}$ is a line bundle on $\widetilde{C}$ of degree $2h-2$ such that:

$$r := h^0(\widetilde{C}, A|_{\widetilde{C}}) - 1 = h^0(\widetilde{S}, A) - 1 \geq \chi(A) - 1 = 1 + \frac{1}{2}\left(\frac{\widetilde{H}+v}{2}\right)^2 = \left\lfloor \frac{h}{2} \right\rfloor;$$

here, the first equality follows from the strong version of Bertini's Theorem due to Saint-Donat [**SD**] yielding $h^1(\widetilde{S}, A(-\widetilde{C})) = 0$, while in the last equality we have used that $v^2 = -4$ when h is even and $v^2 = -8$ otherwise.

Since $E_8(-2) = (H^2(\widetilde{S}, \mathbb{Z})^\iota)^\perp \subset \mathrm{Pic}(\widetilde{S})$, one has $\iota^*\widetilde{H} = \widetilde{H}$ and $\iota^* v = -v$. This implies $\iota^* A|_{\widetilde{C}} \simeq \omega_{\widetilde{C}} \otimes A|_{\widetilde{C}}^\vee$, or equivalently, $\mathrm{Nm}(A|_{\widetilde{C}}) = \omega_C$. Hence, $A|_{\widetilde{C}}$ defines an element of the Prym-Brill-Noether variety $V^r(C, M|_C)$. If h is odd, an easy computation gives

$$\rho^-(h,r) = -\frac{(h-1)(h-7)}{8},$$

which is negative when $h > 7$. Analogously, for even h one computes the Prym-Brill-Noether number

$$\rho^-(h,r) = -\frac{(h-2)(h-4)}{8},$$

which is negative for $h \geq 6$. Hence, the Prym-Petri map of $A|_{\widetilde{C}}$ cannot be injective if either $h > 7$ or $h = 6$, and this concludes the proof. □

5.3. Non-standard Nikulin surfaces. Let (S, M, H) be a very general genus h primitively polarized Nikulin surface of non-standard type. As remarked in [**KLV1**, Prop. 2.3 and Rmk. 2.4], in this case the line bundles $R_1, R_2 \in \mathrm{Pic}(S)$ prevent curves in $|H|$ from being Brill-Noether general; indeed, their restrictions to any smooth curve $C \in |H|$ define two theta-characteristics on C with negative Brill-Noether number. Proposition 3.5 (2) in [**GS**] yields the existence of a smooth irreducible curve in the linear systems $|R_i|$ for $i = 1, 2$ as soon as $c_1(R_i)^2 \geq 0$. Let us pick a smooth curve D lying in either $|R_1|, |R_2|$, and denote by $g \geq 1$ its genus. If $h \equiv 1 \bmod 4$, then $g = (h-1)/4$ and $(D, M|_C) \in \mathcal{R}_{g,4}$. If instead $h \equiv 3 \bmod 4$, then either $g = (h+1)/4$ and $(D, M|_C) \in \mathcal{R}_{g,2}$, or $g = (h-3)/4$ and $(D, M|_C) \in \mathcal{R}_{g,6}$, depending on whether D lies in $|R_1|$ or $|R_2|$. In any case the double cover $\widetilde{D} \subset \widetilde{S}$ of D defined by $M|_C$ has genus $(h+1)/2$.

PROPOSITION 5.6. *Let (S, M, H) be a general non-standard Nikulin surface of odd genus h and let D be a smooth curve in either $|R_1|$ or $|R_2|$. Then the following hold:*

(i) *D is always Brill-Noether general, and it is Petri general if it is general in its linear system;*
(ii) *the ramified double cover $\widetilde{D}$ of D defined by $M|_C$ is Brill-Noether general.*

PROOF. It is easy to verify that neither R_1 nor R_2 can be decomposed as the tensor product of two line bundles on S both satisfying $h^0 \geq 2$, and thus (i) follows proceeding as in Lazarsfeld's proof of Petri's Theorem [**La, Pa**]. Proposition 5.3 yields $\widetilde{D} \in |\widetilde{R}|$, where $\widetilde{R}$ is a generator of $\mathrm{Pic}(\widetilde{S})$. Again one can easily exclude the existence of $L_1, L_2 \in \mathrm{Pic}(\widetilde{S})$ with $\widetilde{R} \simeq L_1 \otimes L_2$ and $h^0(\widetilde{S}, L_i) \geq 2$. Hence, Lazarsfelds's Theorem implies that all smooth curves in the linear system $|\widetilde{R}|$ (thus, in particular, $\widetilde{D}$) are Brill-Noether general. □

A general curve $\widetilde{D}$ as in the above statement is highly expected to be Petri general, as well. However, this does not follow directly from Lazarsfeld's Theorem. Indeed, the latter only implies Petri generality for general curves in $|\widetilde{R}|$; however, a double cover $\widetilde{D}$ of a curve D in $|R_1|$ or $|R_2|$ is never general in the linear system $|\widetilde{R}|$, as it lies in either the ι-invariant or the ι-anti-invariant part of it.

By varying h, one obtains all possible values for the genus g of D. Recalling that the moduli space $\mathcal{R}_{g,2n}$ is irreducible for every $g \geq 2$ and $n \geq 0$ (cf. [**Bu**, Prop. 2.5]), Proposition 5.6 then implies the following generalization of Bud's result.

THEOREM 5.7. *Fix integers $g \geq 2$ and $n = 1, 2, 3$.*

Let $(D, x_1, \dots, x_{2n}, \eta) \in \mathcal{R}_{g,2n}$ be general and let $\widetilde{D}$ the double cover of D defined by η. Then the curve $\widetilde{D}$ is Brill-Noether general.

References

[AF] M. Aprodu and G. Farkas, *Green's conjecture for general covers*, Compact moduli spaces and vector bundles, Contemp. Math., vol. 564, Amer. Math. Soc., Providence, RI, 2012, pp. 211–226, DOI 10.1090/conm/564/11147. MR2894637

[ACGH] E. Arbarello, M. Cornalba, P. A. Griffiths, J. Harris, *Geometry of algebraic curves*, Grundl. Math. Wiss. 267, Springer Verlag, 1985.

[At] M. F. Atiyah, *Riemann surfaces and spin structures*, Ann. Sci. École Norm. Sup. (4) **4** (1971), 47–62. MR286136

[Bea] A. Beauville, *Vanishing thetanulls on curves with involutions*, Rend. Circ. Mat. Palermo (2) **62** (2013), no. 1, 61–66, DOI 10.1007/s12215-013-0109-7. MR3031568

[Ber] A. Bertram, *An existence theorem for Prym special divisors*, Invent. Math. **90** (1987), no. 3, 669–671, DOI 10.1007/BF01389185. MR914856

[Bu] A. Bud, *The birational geometry of $\overline{\mathcal{R}_{g,2}}$ and Prym-canonical divisorial strata*, math.AG/01718.v2.

[EH1] D. Eisenbud and J. Harris, *Divisors on general curves and cuspidal rational curves*, Invent. Math. **74** (1983), no. 3, 371–418, DOI 10.1007/BF01394242. MR724011

[EH2] D. Eisenbud and J. Harris, *A simpler proof of the Gieseker-Petri theorem on special divisors*, Invent. Math. **74** (1983), no. 2, 269–280, DOI 10.1007/BF01394316. MR723217

[F1] G. Farkas, *The geometry of the moduli space of curves of genus 23*, Math. Ann. **318** (2000), no. 1, 43–65, DOI 10.1007/s002080000108. MR1785575

[F2] G. Farkas, *Gaussian maps, Gieseker-Petri loci and large theta-characteristics*, J. Reine Angew. Math. **581** (2005), 151–173, DOI 10.1515/crll.2005.2005.581.151. MR2132674

[F3] G. Farkas, *Prym varieties and their moduli*, Contributions to algebraic geometry, EMS Ser. Congr. Rep., Eur. Math. Soc., Zürich, 2012, pp. 215–255, DOI 10.4171/114-1/8. MR2976944

[FK] G. Farkas and M. Kemeny, *The generic Green-Lazarsfeld secant conjecture*, Invent. Math. **203** (2016), no. 1, 265–301, DOI 10.1007/s00222-015-0595-7. MR3437872

[FV1] G. Farkas and A. Verra, *Moduli of theta-characteristics via Nikulin surfaces*, Math. Ann. **354** (2012), no. 2, 465–496, DOI 10.1007/s00208-011-0739-z. MR2965251

[FV2] G. Farkas and A. Verra, *Prym varieties and moduli of polarized Nikulin surfaces*, Adv. Math. **290** (2016), 314–328, DOI 10.1016/j.aim.2015.12.006. MR3451926

[GS] A. Garbagnati and A. Sarti, *Projective models of K3 surfaces with an even set*, Adv. Geom. **8** (2008), no. 3, 413–440, DOI 10.1515/ADVGEOM.2008.027. MR2427468

[Gi] D. Gieseker, *Stable curves and special divisors: Petri's conjecture*, Invent. Math. **66** (1982), no. 2, 251–275, DOI 10.1007/BF01389394. MR656623

[Gr] P. Griffiths and J. Harris, *On the variety of special linear systems on a general algebraic curve*, Duke Math. J. **47** (1980), no. 1, 233–272. MR563378

[Ha] J. Harris, *Theta-characteristics on algebraic curves*, Trans. Amer. Math. Soc. **271** (1982), no. 2, 611–638, DOI 10.2307/1998901. MR654853

[Ke] G. Kempf, *Schubert methods with an application to algebraic curves*, Publ. Math. Centrum, University of Amsterdam, 1971.

[KL] S. L. Kleiman and D. Laksov, *On the existence of special divisors*, Amer. J. Math. **94** (1972), 431–436, DOI 10.2307/2374630. MR323792

[KLV1] A. L. Knutsen, M. Lelli-Chiesa, and A. Verra, *Half Nikulin surfaces and moduli of Prym curves*, J. Inst. Math. Jussieu **20** (2021), no. 5, 1547–1584, DOI 10.1017/S1474748019000574. MR4311562

[KLV2] A. L. Knutsen, M. Lelli-Chiesa, and A. Verra, *Moduli of non-standard Nikulin surfaces in low genus*, Ann. Sc. Norm. Super. Pisa Cl. Sci. (5) **21** (2020), 361–384. MR4288606

[La] R. Lazarsfeld, *Brill-Noether-Petri without degenerations*, J. Differential Geom. **23** (1986), no. 3, 299–307. MR852158

[M1] D. Mumford, *Prym varieties. I*, Contributions to analysis (a collection of papers dedicated to Lipman Bers), Academic Press, New York-London, 1974, pp. 325–350. MR379510

[M2] D. Mumford, *Theta characteristics of an algebraic curve*, Ann. Sci. École Norm. Sup. (4) **4** (1971), 181–192. MR292836

[Pa] G. Pareschi, *A proof of Lazarsfeld's theorem on curves on $K3$ surfaces*, J. Algebraic Geom. **4** (1995), no. 1, 195–200. MR1299009

[SD] B. Saint-Donat, *Projective models of $K-3$ surfaces*, Amer. J. Math. **96** (1974), 602–639, DOI 10.2307/2373709. MR364263

[SJ] F. Schottky, H. Jung, *Neue Sätze über Symmetralfunktionen und die Abelschen Funktionen der Riemannschen Theorie*, S.-B. Preuss. Akad. Wiss. Berlin, Phys. Math. Kl. **1** (1909), 282–287.

[Sc] I. Schwarz, *Brill-Noether theory for cyclic covers*, J. Pure Appl. Algebra **221** (2017), no. 10, 2420–2430, DOI 10.1016/j.jpaa.2016.12.029. MR3646308

[VGS] B. van Geemen and A. Sarti, *Nikulin involutions on $K3$ surfaces*, Math. Z. **255** (2007), no. 4, 731–753, DOI 10.1007/s00209-006-0047-6. MR2274533

[We] G. E. Welters, *A theorem of Gieseker-Petri type for Prym varieties*, Ann. Sci. École Norm. Sup. (4) **18** (1985), no. 4, 671–683. MR839690

[Wi] W. Wirtinger, *Untersuchungen fiber thetafunktionen*, Teubner, Berlin, 1895.

DISIM, Università degli Studi dell'Aquila, Via Vetoio I, Loc. Coppito, 67100 L'Aquila, Italy

Email address: simona_devangelista@hotmail.it

Dipartimento di Matematica e Fisica, Università Roma Tre, Largo San Leonardo Murialdo 1, 00146 Roma, Italy

Email address: margherita.lellichiesa@uniroma3.it

Contemporary Mathematics
Volume **803**, 2024
https://doi.org/10.1090/conm/803/16098

O'Grady spaces and symplectic resolution of moduli spaces of Higgs bundles

Emilio Franco

Dedicated to Peter Newstead, in gratitude for all he taught me.

ABSTRACT. We describe here a degeneration of the symplectic desingularization of the moduli spaces of topologically trivial $\mathrm{GL}(2,\mathbb{C})$ and $\mathrm{SL}(2,\mathbb{C})$-Higgs bundles over a hyperelliptic curve, into O'Grady's ten and six dimensional exceptional examples of irreducible holomorphic symplectic manifolds. Most of this note is a survey of work on these degenerations by Donagi–Ein–Lazarsfeld, de Cataldo–Maulik–Shen and Felisetti–Mauri, although in certain cases we provide details that were missing the previous articles.

CONTENTS

1. Introduction

Peter's contribution to the theory of moduli spaces is invaluable. In this note we want to honor him by describing a beautiful relation between two important classes of moduli spaces.

2020 *Mathematics Subject Classification.* Primary 53D30.

The author has been supported by the Scientific Employment Stimulus program, funded by FCT (Portugal), by an FCT Investigator grant with fellowship reference CEECIND/04153/2017. He was previously supported by the project PTDC/MAT- GEO/2823/2014 also funded by FCT (Portugal) with national funds.

1.1. Background. A compact irreducible holomorphic symplectic manifold is a smooth compact simply connected Kähler manifold with a non-degenerate holomorphic 2-form which is unique up to scaling. Thanks to S.-T. Yau's proof of Calabi's conjecture [**Y1, Y2**], a compact irreducible holomorphic symplectic manifold is equivalent to a *compact hyperkähler manifold.* These manifolds play a central role in mathematical physics as they are one of three building blocks [**Be1**] for the construction of Ricci-flat manifolds, along with abelian varieties and Calabi–Yau manifolds. The list of known examples of compact irreducible holomorphic symplectic manifolds is surprisingly short [**Be4**] when compared with the other types of building blocks: only two series of examples in arbitrary dimension are known (Hilbert schemes of points on a K3 and generalized Kummer varieties as shown by Beauville in [**Be1**]), along with two exceptional examples in dimension ten and six obtained by O'Grady [**OG1, OG2**]. It is remarkable that all these examples are associated to moduli spaces of sheaves over symplectic surfaces, as the latter are equipped with Mukai's holomorphic symplectic form [**Mu2**]. In particular, O' Grady's exceptional examples appear after a symplectic resolution of singular moduli spaces of sheaves on symplectic surfaces. The existence of symplectic resolution of these singular moduli spaces has been studied in detail [**LS, KLS, Ki**], concluding that only the class of cases considered by O'Grady admit one, hence closing the door to the appearance of a new compact irreducible holomorphic symplectic manifold out of this method.

When we study moduli space of pure dimension 1 sheaves on a symplectic surface, it is possible to consider a morphism to a projective space using the fitting support. Beauville [**Be2**] showed that this provides an integrable system, known as the *Beauville–Mukai system.* In fact, a surpring result of Matsushita [**Ma**] implies that the only fibrations one can consider on compact holomorphic symplectic manifolds are integrable systems.

Higgs bundles were introduced by Hitchin [**Hi2**] and its moduli space has become an important object of study thanks to its rich geometry. Non-abelian Hodge theory [**Hi2, Si1, Si2, Do, Co**] shows that the moduli spaces of Higgs bundles and (projectively) flat connections are homeomorphic, and this equips the underlying manifold with a hyperKähler structure. Therefore, these moduli spaces constitute a great source of examples of *non-compact hyperKähler varieties*, hence holomorphic symplectic. These moduli spaces are smooth when the rank and the degree are chosen to be coprime and singular in the non-coprime case, as showed by Simpson [**Si2**]. It is then natural to ask whether these spaces admit a symplectic desingularization, a question that was first addressed by Kiem and Yoo [**KY**] via the study of the stringy E-functions. Using the work of Bellamy and Schedler [**BS**] on symplectic resolutions of character varieties, Tirelli [**Ti**] showed moduli spaces of Higgs bundles admit a symplectic resolution only when the rank and the genus of the base curve are both equal to 2. In this case, the desingularization was first provided by Yoo [**Yo**] applying the same techniques used by O'Grady [**OG1, OG2**], although after [**BS, Ti**] such symplectic resolution can be obtained by simply blowing-up the singular locus of the moduli space.

Another important feature of the moduli space of Higgs bundles is that it carries a morphism [**Hi1**] to a vector space, turning it into an integrable system named the *Hitchin system.* Thanks to the spectral correspondence [**Hi1, BNR, Si2**], the moduli space of Higgs bundles can be described as an open subset of the moduli space of pure dimension 1 sheaves on a certain ruled surface and one can interpret the Hitchin system in terms of the fibration obtained by considering the (fitting) support of these pure dimension 1 sheaves.

As we have seen, both the Beauville–Mukai and the Hitchin systems can be obtained by considering the fitting support on moduli spaces of pure dimension 1 sheaves on surfaces, the first one being compact and second one non-compact. These similarities become even more explicit with the construction of the Donagi–Ein–Lazarsfeld [**DEL**] degeneration of the Beauville–Mukai integrable system into the Hitchin integrable system by considering the normal cone degeneration of a curve inside a K3 surface. This has been generalized in several directions, de Cataldo, Maulik and Shen [**CMS2**] constructed a similar degeneration working on abelian surfaces in their proof of the P=W conjecture for SL_n, Sawon and Shen [**SS**] worked over a certain del Pezzo surface in order to construct a degeneration into the $\mathrm{Sp}(2m, \mathbb{C})$-Hitchin system, and the author studied in [**Fr**] the appearance of these degeneration over any smooth surface, obtaining a degeneration into the moduli space of L-twisted Higgs bundles. Sawon [**Sw**] wrote a beautiful survey on the subject.

1.2. Summary. In this note we provide a review of the previously mentioned degenerations and all the objects involved in these constructions: moduli spaces of sheaves in smooth surfaces, compact irreducible holomorphic symplectic manifolds, Higgs bundles, and the ruled surface associated to the canonical bundle of a curve.

We focus in a particular degeneration where the exceptional examples of O'Grady in dimensions ten and six are are taken to be our compact (irreducible) holomorphic symplectic varieties, and where we consider symplectic resolutions of the moduli space of $\mathrm{GL}(2, \mathbb{C})$ and $\mathrm{SL}(2, \mathbb{C})$-Higgs bundles over a curve of genus 2 as our non-compact holomorphic symplectic varieties. Felisetti and Mauri [**FM**], discussed the existence of such degenerations by considering, in the ten dimensional space, the degeneration provided by Donagi–Ein–Lazarsfeld [**DEL**], while relying on the work of de Cataldo–Maulik–Shen [**CMS2**] for the six dimensional example. Our main contribution consists on providing details that were missing in the construction of the degeneration of [**CMS2**]. Following [**CMS2**], the degeneration presented in the present paper could be used to study the occurence of P=W, by means of the associated specialisation morphism in cohomology. Another potential application of the degeneration we present here is the study of maximal variation of curves in the case of abelian varieties, generalizing recent work of Huybrechts and Dutta [**DH**] in the case of K3 surfaces.

1.3. Structure. In Section 2 we study the properties of moduli spaces of sheaves over smooth projective surfaces, which is a class of objects that are crucial both in the theory of compact irreducible holomorphic symplectic manifolds (when the base surface is either a K3 or an abelian surface) or in the theory of Higgs bundle (that are intimately related to certain sheaves over ruled surface).

In Section 3 we review some aspects of the theory of compact irreducible holomorphic symplectic manifolds, focusing on the construction of O'Grady's exceptional examples in dimensions ten and six.

Fix a smooth projective curve C and its canonical bundle K. We study in Section 4 some properties of the canonical ruled surface $\mathbb{K} = \mathbb{P}(\mathcal{O}_C \oplus K)$ as it plays an important role in the theory of Higgs bundles, which we address in Section 5. In the later, we study the moduli spaces of Higgs bundles for the groups GL_n and SL_n, recalling in which cases they admit a symplectic resolution.

In Section 6 we review the non-linear deformation described by Donagi, Ein and Lazarsfeld reproducing as well the degeneration of the holomorphic symplectic structure. These results provide a degeneration, as symplectic varieties, of the moduli space of sheaves over symplectic surfaces into the moduli space of GL_n-Higgs bundles. In Section 7 we describe the generalization to the SL_n-case in great detail.

Finally, in Section 8 we make use of all of the above to describe a degeneration of the O'Grady ten and six dimensional spaces into the symplectic resolution of, respectively, the Higgs moduli spaces for GL_2 and SL_2.

2. Moduli spaces of sheaves on surfaces

In this section we describe some properties of moduli spaces of sheaves in smooth projective surfaces, which will be the main objects in our study.

Consider a smooth projective surface S. We say that a coherent sheaf $\mathcal{F}$ on S is of pure dimension d when the dimension of its schematic support is d and all subsheaves are supported on dimension d subschemes. After equipping S with a polarization H, one can check that the corresponding Hilbert polynomial $P(\mathcal{F}, \mathrm{H})$ has degree d. Setting the coefficient multiplied by $d!$ to be its H-polarized rank $\mathrm{rk}(\mathcal{F}, \mathrm{H})$, we say that $\mathcal{F}$ is H-*stable* (*resp.* H-*semistable*) if it is of pure dimension d and every proper subsheaf $\mathcal{F}' \subset \mathcal{F}$ satisfies $P(\mathcal{F}', \mathrm{H})/\mathrm{rk}(\mathcal{F}', \mathrm{H}) < P(\mathcal{F}, \mathrm{H})/\mathrm{rk}(\mathcal{F}, \mathrm{H})$ (*resp.* $P(\mathcal{F}', \mathrm{H})/\mathrm{rk}(\mathcal{F}', \mathrm{H}) \leq P(\mathcal{F}, \mathrm{H})/\mathrm{rk}(\mathcal{F}, \mathrm{H})$) when $n \gg 0$. If $\mathcal{F}$ is semistable, one further say that it is polystable when it decomposes as a direct sum of stable sheaves.

Gieseker [**Gi**] constructed the moduli space $\mathrm{M}_S^{\mathrm{H}}(p)$ of pure dimension d H-semistable sheaves with Hilbert polynomial p, a construction that was extended by Simpson [**Si1**] to any projective varity. We denote by $\mathrm{M}_S^H(p)^{\mathrm{st}}$ the locus H-stable sheaves, and by $\mathrm{M}_S^H(p)^{\mathrm{str.ps}}$ the locus of strictly polystable sheaves.

In the case of a smooth surface S, the topological invariants of a sheaf $\mathcal{F}$ associated to the Hilbert polynomial are the H-polarized rank $\mathrm{rk}(\mathcal{F}, H)$, the first Chern class $c_1(\mathcal{F})$, and the Euler characteristic $\chi(\mathcal{F})$ (which contains information on the previous invariants together with the second Chern class $c_2(\mathcal{F})$). In order to simplify the computations, one introduces the *Mukai vector*, $v := \mathrm{ch}(\mathcal{F}) \cdot \sqrt{\mathrm{Todd}(S)}$. Observe that $v \in H^{2*}(S, \mathbb{Z})$ and the cup product in cohomology provides a pairing on the $\mathbb{Z}$-module $H^{2*}(S, \mathbb{Z})$, hence acquiring a lattice structure. The definition of the pairing is meant to satisfy

$$v^2 = \chi(\mathrm{End}(\mathcal{F})). \tag{2.1}$$

Since the Mukai vector contains all the information to reconstruct the Hilbert polynomial, we then abbreviate by $\mathrm{M}_S^{\mathrm{H}}(v)$ the moduli space of H-semistable sheaves S with Hilbert polynomial determined by the Mukai vector $v \in H^{2*}(S, \mathbb{Z})$. It is always possible write the Mukai vector as the multiple of a primitive class,

$$v = n \cdot v_0.$$

When $v = v_0$ is itself primitive, all coherent sheaves with such Mukai vector are forcely stable, so $\mathrm{M}^{\mathrm{H}}_S(v_0)^{\mathrm{st}} = \mathrm{M}^{\mathrm{H}}_S(v_0)$. A sheaf $\mathcal{F}$ is simple if its automorphism group is given by scalar multiplication, in particular $Hom_S(\mathcal{F},\mathcal{F}) \cong \mathbb{C}$ in that case. It can be shown that stable sheaves are simple by a standard argument.

Thanks to deformation theory, the tangent space to $\mathrm{M}^{\mathrm{H}}_S(v)$ at $\mathcal{F}$ corresponds with

$$\mathcal{T}_{\mathcal{F}}\mathrm{M}^{\mathrm{H}}_S(v) = \mathrm{Ext}^1_S(\mathcal{F},\mathcal{F}),$$

and, by Serre duality, the associated cotangent space is

$$\mathcal{T}^*_{\mathcal{F}}\mathrm{M}^{\mathrm{H}}_S(v) = \mathrm{Ext}^1_S(\mathcal{F},\mathcal{F}\otimes K_S).$$

A *symplectic surface* is a surface S endowed with a holomorphic symplectic form Ω_S. This is only the case when S is either a K3 or abelian surface. One can understand Ω_S as a non-vanishing section of the canonical bundle K_S which is forcely trivial. Due to Serre duality, one has the identification

$$Hom_S(\mathcal{F},\mathcal{F}) \cong \mathrm{Ext}^2_S(\mathcal{F},\mathcal{F})^*,$$

and both spaces are one dimensional when $\mathcal{F}$ is simple, implying

$$\dim_{\mathbb{C}} \mathrm{Ext}^1_S(\mathcal{F},\mathcal{F}) = v^2 + 2,$$

as a consequence of (2.1). This shows that the simple locus of $\mathrm{M}^{\mathrm{H}}_S(v)$ is smooth, and in particular $\mathrm{M}^{\mathrm{H}}_S(v_0)$ is itself smooth when v_0 is a primitive Mukai vector.

Thanks to Serre duality, $H^2(S,\mathcal{O}_S)$ is dual to $H^0(S,K_S) \cong \mathbb{C}$ in the case of symplectic surfaces. This was used by Mukai to construct a holomorphic symplectic form on $\mathrm{M}^{\mathrm{H}}_S(v)$.

THEOREM 2.1 ([**Mu2**]). *Let S be a symplectic surface. Then, the simple locus $\mathrm{M}^{\mathrm{H}}_S(v)$ is smooth and can be equipped with a holomorphic 2-form Ω_{M} defined by taking the trace of the Yoneda product composed with Ω_S,*

$$\mathrm{Ext}^1_S(\mathcal{F},\mathcal{F}) \wedge \mathrm{Ext}^1_S(\mathcal{F},\mathcal{F}) \xrightarrow{\circ} \mathrm{Ext}^2_S(\mathcal{F},\mathcal{F}) \xrightarrow{\mathrm{tr}} H^2(S,\mathcal{O}_S) \xrightarrow{\Omega_S(\cdot)} \mathbb{C}.$$

Furthermore, Ω_{M} is non-degenrate on the smooth locus of $\mathrm{M}^{\mathrm{H}}_S(v)$, defining a symplectic form there.

A *Poisson surface* is a surface S equipped with Poisson bi-vector Θ_S, a non-zero section of $\bigwedge^2 \mathcal{T}_S \cong K_S^{-1}$. The above construction was generalized independently by Bottacin [**Bo1**] and Markman [**Mr**] to the case of Poisson surfaces.

THEOREM 2.2 ([**Bo1**] and [**Mr**]). *Let S be is equipped with a Poisson bi-vector Θ_S. Then, one can define a (closed although possibly degenerate) Poisson structure Θ_{M} on the smooth locus of $\mathrm{M}^{\mathrm{H}}_S(v)$ by taking the trace of the Yoneda product composed with Θ_S,*

$$\begin{aligned}\mathrm{Ext}^1_S(\mathcal{F},\mathcal{F}\otimes K_S) \wedge \mathrm{Ext}^1_S(\mathcal{F},\mathcal{F}\otimes K_S) &\xrightarrow{\circ} \mathrm{Ext}^2_S(\mathcal{F},\mathcal{F}\otimes K^2_S) \\ &\xrightarrow{\mathrm{tr}} H^2(S,K^2_S) \xrightarrow{\langle\cdot,\Theta_S\rangle} H^2(S,K_S) \cong \mathbb{C}.\end{aligned}$$

On a smooth surface S, every pure dimension 1 sheaf $\mathcal{F}$ has a locally free resolution of length 2. This allow us to define the *fitting support* of a sheaf $\mathcal{F}$ as the determinant associated to this resolution, which we denote by $\mathrm{supp}(\mathcal{F})$. Observe that

$$c_1(\mathcal{F}) = [\mathrm{supp}(\mathcal{F})].$$

We then see that the Mukai vector of a pure dimension 1 sheaf $\mathcal{F}$ on a smooth surface S with canonical sheaf K_S takes the form

$$v = (v_1, v_2, v_3) = \left(0, [\mathrm{supp}(\mathcal{F})], \chi(\mathcal{F}) + \frac{1}{2}[K_S]\cdot[\mathrm{supp}(\mathcal{F})]\right),$$

and the pairing is simply given by the intersection of the supports,

$$\langle v, v'\rangle = v_2 \cdot v_2' = [\mathrm{supp}(\mathcal{F})]\cdot[\mathrm{supp}(\mathcal{F}')].$$

Hence, the square of these Mukai vectors are determined by the genus of the support

$$v^2 = 2\cdot g(\mathrm{supp}(\mathcal{F})) - 2.$$

For this class of Mukai vectors, the associated moduli spaces can be equipped with a fibration,

$$\begin{array}{ccc} \mathrm{M}_S^{\mathrm{H}}(v) & \longrightarrow & \mathrm{Hilb}_S([nC]) \\ \mathcal{F} & \longmapsto & \mathrm{supp}(\mathcal{F}), \end{array} \tag{2.2}$$

to the Hilbert scheme classifying dimension 1 subschemes of S with first Chern class equal to the second component of the Mukai vector. The fibre corresponding to a curve $A \in \mathrm{Hilb}_S([nC])$ amounts to the Simpson compactified Jacobian $\overline{Jac}_A^{\mathrm{H}}(a)$, classifying $\mathrm{H}|_A$-semistable pure dimension 1 sheaves on A of rank 1 and degree a.

REMARK 2.3. For a smooth projective K3 surface X one has that $H^1(X, \mathcal{O}_X)$ vanishes so $\mathrm{Pic}(X)$ is discrete and embeds into $H^2(X,\mathbb{Z})$. Then, the second component of the Mukai vector fixes the determinant and $\mathrm{Hilb}_X([nC])$ is the linear system $|nC|$. After the work of Beauville [**Be2**] the fibres of (2.2) are Lagrangian with respect to the Mukai form Ω_{M}. This gives rise to an algebraically completely integrable system

$$\mathrm{M}_X^{\mathrm{H}}(v) \longrightarrow |nC| \tag{2.3}$$

called the *Beauville–Mukai system*.

3. O'Grady's exceptional examples

It is a classical statement that K3 surfaces and abelian surfaces are the only smooth projective surfaces that can be endowed with a holomorphic symplectic structure. Abelian surfaces are not simply connected but K3 are, so the latter constitute the only example of compact irreducible holomorphic symplectic manifolds in dimension 2. Beauville [**Be1**] produced two series of irreducible holomorphic symplectic manifold in any dimension, namely Hilbert schemes of points on a K3 surface, $\mathrm{Hilb}^n(X)$, and generalized Kummer varieties of an abelian surface, $\mathrm{Kum}^n(A)$. As we have seen in Section 2, Mukai [**Mu2**] showed that the moduli space of torsion-free sheaves on a symplectic surface (*i.e.* K3 or abelian) can be equipped with a holomorphic symplectic form, so, whenever these moduli spaces are smooth they become examples of holomorphic symplectic manifolds. Nevertheless these spaces do not provide new examples of irreducible holomorphic symplectic manifolds since Yoshioka [**Ys1, Ys2**] showed that, when $\mathrm{M}_X^{\mathrm{H}}(v)$ is smooth, it is deformation equivalent to a Hilbert scheme $\mathrm{Hilb}^n(X)$, and similar results hold in the case of $\mathrm{M}_A^{\mathrm{H}}(v)$ and generalized Kummer varieties. Hence, $\mathrm{Hilb}^n(X)$ and $\mathrm{Kum}^n(A)$ were the only known examples of irreducible holomorphic symplectic manifolds until the appearance of O'Grady's exceptional examples in dimension ten [**OG1**] and six

[**OG2**], whose constructions rely on the existence of symplectic desingularization of singular moduli spaces.

Consider a singular variety M equipped with a holomorphic symplectic form Ω defined over the smooth locus M^{sm}. After Beauville [**Be3**] a symplectic desingularization of M is a resolution of singularities $\widetilde{M} \to M$ equipped with a holomorphic symplectic form $\widetilde{\Omega}$ that extends the pull-back of Ω over the preimage of the smooth locus. As proved in [**Be3**], when a symplectic desingularization exists, it is unique up to isomorphism.

For a particular choice of a primitive Mukai vector v_0 on a K3 surface X that satisfies $v_0^2 = 2$, O'Grady considered the moduli space $\mathrm{M}_X^{\mathrm{H}}(2v_0)$, which is a singular variety of dimension $(2v_0)^2 + 2 = 10$. Using Kirwan desingularization procedure [**Kw**], O'Grady [**OG1**] constructed a symplectic desingularization $\widetilde{\mathrm{M}}_X^{\mathrm{H}}(2v_0)$ and showed that it is an irreducible holomorphic symplectic variety.

Lehn and Sorger [**LS**] generalized O'Grady's symplectic desingularization of $\mathrm{M}_S^{\mathrm{H}}(2v_0)$ to any choice of primitive vector v_0 with $v_0^2 = 2$ on a symplectic surface S. Furthermore, they showed that the symplectic resolution $\widetilde{\mathrm{M}}_S^{\mathrm{H}}(2v_0)$ of $\mathrm{M}_S^{\mathrm{H}}(2v_0)$ can be achieved by a single blow-up at the reduced subscheme of the singular locus which coincides with the locus of strictly polystable sheaves $\mathrm{M}_S^{\mathrm{H}}(2v_0)^{\mathrm{str.ps}}$.

It is a natural question to ask if one can obtain more holomorphic symplectic varieties from symplectic desingularization of moduli spaces of the form $\mathrm{M}_S^{\mathrm{H}}(nv_0)$. This was answer negatively by Kaledin, Lehn and Sorger [**KLS**] and Kiem [**Ki**], who proved non-existence of symplectic resolutions of $\mathrm{M}_S^{\mathrm{H}}(nv_0)$ for choices of n and v_0 different than $n = 2$ and $v_0^2 = 2$. The work of Kaledin, Lehn and Sorger [**KLS**] is based in a local study of the singularities base on deformation theory, while the work of Kiem [**Ki**] relies on the computation of the stringy E-functions of the moduli spaces.

We summarize all this in the following theorem.

THEOREM 3.1 ([**OG1, OG2, Ki, KLS, LS, PR**]). *Suppose $v = nv_0$ with $v_0^2 \geq 2$ and $n \geq 2$. Let* H *be a v-generic ample divisor on the symplectic surface S. Then $\mathrm{M}_S^{\mathrm{H}}(v)$ is a variety with symplectic singularities.*

Furthermore, if and only if $n = 2$ and $v_0^2 = 2$ there exists a symplectic resolution of $\mathrm{M}_S^{\mathrm{H}}(v)$ which can be obtained by a single blow-up at the reduced subscheme of the singular locus, which amounts to $\mathrm{M}_S^{\mathrm{H}}(v)^{\mathrm{str.ps}}$, the locus of strictly polystable sheaves.

The symplectic resolutions appearing above allowed O'Grady to provide in [**OG1**] the first example of irreducible holomorphic symplectic manifold which is not deformation equivalent to $\mathrm{Hilb}^n(X)$ or $\mathrm{Kum}^n(A)$. Given a K3 surface X and a particular choice of the primitive Mukai vector v_0 satisfying $v_0^2 = 2$, O'Grady showed in [**OG1**] that the symplectic resolution $\widetilde{\mathrm{M}}_X^{\mathrm{H}}(2v_0)$ of the ten dimensional moduli space $\mathrm{M}_X^{\mathrm{H}}(2v_0)$ has second Betti number strictly bigger than 23 and therefore is not deformation equivalent to the previously known irreducible holomorphic symplectic manifolds $\mathrm{Hilb}^5(X)$ and $\mathrm{Kum}^n(A)$, whose Betti numbers are smaller or equal to 23. Rapagnetta [**Ra**] showed that the second Betti number of $\widetilde{\mathrm{M}}_X^{\mathrm{H}}(2v_0)$ is indeed 24. Perego and Rapagnetta [**PR**] proved that the symplectic resolutions $\widetilde{\mathrm{M}}_X^{\mathrm{H}}(2v_0)$, obtained by Kaledin–Lehn–Sorger [**KLS**] for any choice of primitive Mukai vector v_0 with $v_0^2 = 2$, are all deformation equivalent.

For future notation convenience, we pick X to be a K3 surface containing a smooth projective curve $C \subset X$ of genus 2. Note that C is an ample divisor of

X with self-intersection $C^2 = 2$, so the primitive Mukai vector $(0, C, \chi)$ satisfies $(0, C, -1)^2 = C^2 = 2$. We denote $\mathrm{M}_{\mathrm{ten}} := \mathrm{M}^C_X(0, 2C, -2)$ and we refer to its symplectic resolution $\widetilde{\mathrm{M}}_{\mathrm{ten}}$ as the *O'Grady's ten dimensional space* (although, our $\widetilde{\mathrm{M}}_{\mathrm{ten}}$ is only deformation equivalent to the variety studied in [**OG1**]). Let us summarize all the above in the following theorem.

THEOREM 3.2 ([**OG1**, **Ra**, **LS**, **PR**]). *O'Grady's ten dimensional space $\widetilde{\mathrm{M}}_{\mathrm{ten}}$ is an irreducible holomorphic symplectic manifold with $b_2(\widetilde{\mathrm{M}}_{\mathrm{ten}}) = 24$, and therefore it is not deformation equivalent to the Hilbert scheme of lenth 5 on a K3 surface, nor to the ten dimensional Kummer variety.*

For any primitive Mukai vector satisfying $v_0^2 = 2$, there exists a symplectic resolution of $\mathrm{M}^{\mathrm{H}}_S(2v_0)$ and it is deformation equivalent to O'Grady's ten dimensional space $\widetilde{\mathrm{M}}_{\mathrm{ten}}$.

Consider now an abelian surface A. Let us denote its dual variety by $\widehat{A} := \mathrm{M}^{\mathrm{H}}_A(1, 0, 0)$ (for any choice of polarization H), which is a fine moduli space for the classification of topologically trivial line bundles on A. Therefore, there exists a universal bundle $\mathrm{P} \to A \times \widehat{A}$ named the *Poincaré bundle.* Let us define the projections of the first factor $\mathrm{f} : A \times \widehat{A} \to A$ and to the second factor $\mathrm{g} : A \times \widehat{A} \to \widehat{A}$. Mukai considered in [**Mu1**] the corresponding integral functor,

$$\begin{array}{rccc} \Phi^{\mathrm{P}} : & D^b(A) & \longrightarrow & D^b(\widehat{A}) \\ & \mathcal{F}^\bullet & \longmapsto & R\mathrm{g}_!(\mathrm{P} \otimes \mathrm{f}^* \mathcal{F}^\bullet), \end{array} \tag{3.1}$$

showing that it gives a derived equivalence. Fixing $\mathcal{F}_0 \in \mathrm{M}^{\mathrm{H}}_S(v)$, on gets the associated *Albanese morphism*

(3.2)

$$\begin{array}{rccc} \mathfrak{a}_{\mathcal{F}_0} : & \mathrm{M}^{\mathrm{H}}_A(v) & \longrightarrow & A \times \widehat{A} \\ & \mathcal{F} & \longmapsto & (\det_{\widehat{A}}(\Phi^{\mathrm{P}}(\mathcal{F})) \otimes \det_{\widehat{A}}(\Phi^{\mathrm{P}}(\mathcal{F}_0))^{-1}, \det_A(\mathcal{F}) \otimes \det_A(\mathcal{F}_0)^{-1}). \end{array}$$

The pull-back through $\mathfrak{a}_{\mathcal{F}_0}$ of the symplectic form $\Omega_{A \times \widehat{A}}$, provides a non-trivial holomorphic symplectic form on $\mathrm{M}^{\mathrm{H}}_A(v)$. This implies that holomorphic symplectic forms on this space are not unique up to scaling, so we can not obtain an irreducible holomorphic symplectic manifold from it. In view of this, one considers the subvariety given by the kernel of the Albanese maorphism,

$$\mathrm{N}^{\mathrm{H}}_A(v) := \mathfrak{a}^{-1}_{\mathcal{F}_0}(\mathbf{1}_A, \mathcal{O}_A), \tag{3.3}$$

which was studied in detail by Yoshioka [**Ys1**, **Ys2**]. We denote by $\mathrm{N}^{\mathrm{H}}_A(v)^{\mathrm{st}}$ the locus of stable sheaves.

Recall the fibration (2.2) obtained by considering the fitting support of the sheaves. After the work of Beauville [**Be2**] the fibres of this map are Lagrangian. One can check that his picture restricts naturally to the reduced moduli space $\mathrm{N}^{\mathrm{H}}_A(0, nC, \chi)$ over abelian surfaces,

$$\begin{array}{ccc} \mathrm{N}^{\mathrm{H}}_A(0, nC, \chi) & \longrightarrow & |nC| \\ \mathcal{F} & \longmapsto & \mathrm{supp}(\mathcal{F}), \end{array} \tag{3.4}$$

giving rise to an algebraic completely integrable system that we call the *reduced Beauville-Mukai system.*

It can be shown that the Mukai form restricts to a symplectic form on $\mathrm{N}^{\mathrm{H}}_A(v)$ which is uniqe up to scaling. Furthermore, a symplectic resolution of $\mathrm{M}^{\mathrm{H}}_A(v)$ gives rise to a symplectic resolution of $\mathrm{N}^{\mathrm{H}}_A(v)$ in the cases studied in Theorem 3.1, and that such a resolution only exists when the corresponding resolution for $\mathrm{M}^{\mathrm{H}}_A(v)$ exists as well. We then have an analogous to Theorem 3.1.

THEOREM 3.3 ([**OG2, KLS, LS**]). *Suppose $v = nv_0$ with v_0 primitive, $v_0^2 \geq 2$ and $n \geq 2$. Let* H *be a v-generic ample divisor on the abelian surface A. Then $\mathrm{N}^{\mathrm{H}}_A(v)$ is a variety with symplectic singularities.*

Furthermore, if and only if $n = 2$ and $v_0^2 = 2$ there exists a symplectic resolution of $\mathrm{N}^{\mathrm{H}}_A(v)$ which can be which can be obtained by a single blow-up at the reduced subscheme of the singular locus, which amounts to $\mathrm{N}^{\mathrm{H}}_A(v)^{\mathrm{str.ps}}$, the locus of strictly polystable sheaves.

O'Grady [**OG2**] provided another example of irreducible holomorphic symplectic manifolds by desingularizing $\mathrm{N}^{\mathrm{H}}_A(2,0,2\chi)$, in this case of dimension 6. He also showed that its second Betti number is 8, different from $\mathrm{Hilb}^3(X)$ and the six dimensional Kummer varieties, and therefore not deformation equivalent to them. As before, the work of Lehn and Sorger [**LS**] provides a symplectic resolution of $\mathrm{M}^{\mathrm{H}}_A(2v_0)$ by blowing-up the singular locus. Also, the results of Perego and Rapagnetta on deformation equivalence extend to this case. Again, for future notation convenience, we pick a smooth projective curve C of genus $g = 2$ and we take A to be its Jacobian $A = Jac(C)$. The genus formula fixes the self-intersection of our genus 2 curve as $C^2 = 2$. Then, the primitive Mukai vector $(0, C, -1)$ squares to 2. We denote $\mathrm{M}_{\mathrm{six}} := \mathrm{M}^{\mathrm{H}}_A(0, 2C, -2)$ and $\mathrm{N}_{\mathrm{six}} := \mathrm{N}^{\mathrm{H}}_A(0, 2C, -2)$, where we have chosen $\mathcal{F}_0$ to be a sheaf supported on the double curve $2C$. We say that the symplectic resolution $\widetilde{\mathrm{N}}_{\mathrm{six}}$ is the *O'Grady's six dimensional space* (even if $\widetilde{\mathrm{N}}_{\mathrm{six}}$ is only deformation equivalent to the one described in [**OG2**]). We summarize the previous discussion in the following theorem.

THEOREM 3.4 ([**OG2, LS, PR**]). *O'Grady's six dimensional space $\widetilde{\mathrm{N}}_{\mathrm{six}}$ is an irreducible holomorphic symplectic manifold with $b_2(\widetilde{\mathrm{N}}_{\mathrm{six}}) = 8$, and therefore it is not deformation equivalent to the Hilbert scheme of length* 3 *on a K3 surface, nor to the six dimensional Kummer variety.*

For any primitive Mukai vector satisfying $v_0^2 = 2$, there exists a symplectic resolution of $\mathrm{N}^{\mathrm{H}}_A(2v_0)$ and it is deformation equivalent to O'Grady's six dimensional space $\widetilde{\mathrm{N}}_{\mathrm{six}}$.

4. The canonical ruled surface

Fixing a smooth projective curve C of genus g we denote its canonical bundle by K. We shall consider its total space $\mathrm{Tot}(K)$ and its projective compactification, the *canonical ruled surface*, namely the ruled surface with topological invariant $2g - 2$,

$$\mathbb{K} := \mathbb{P}(\mathcal{O}_C \oplus K),$$

The canonical ruled surface plays an important role in connection with Higgs bundles, so we present in this section some of its main properties.

The canonical ruled surface is naturally equipped with the projection $p : \mathbb{K} \to C$. One can easily construct two sections of p, namely $\sigma_0 \cong C$ that denotes the 0-section of K, and the infinity section $\sigma_\infty \cong C$, given by the 0-section of $\mathcal{O}_C$. Obviously, one has

$$\mathrm{Tot}(K) = \mathbb{K} - \{\sigma_\infty\}.$$

It is classical result, based on the properties of ruled surfaces (see [**Ha**, Sect. 2, Chap. V] for instance), that $H^2(\mathbb{K},\mathbb{Z})$ is generated by the classes of σ_0 and the fibres $F = p^{-1}(x_0)$, for some $x_0 \in C$, with

$$[\sigma_0]^2 = 2g-2,$$

$$[\sigma_0]\cdot[F] = 1,$$

and

$$[F]^2 = 0.$$

Furthermore, the class of the infinity section is

$$[\sigma_\infty] = [\sigma_0] - (2g-2)[F],$$

so it has negative self intersection

$$[\sigma_\infty]^2 = -(2g-2),$$

and,

$$[\sigma_0]\cdot[\sigma_\infty] = 0.$$

One can now derive the following technical result that will be used in Section 5.

LEMMA 4.1. *Let Σ be a projective curve inside* $\mathrm{Tot}(K) \subset \mathbb{K}$*, then*

$$\mathcal{O}_{\mathbb{K}}(\Sigma) \cong \mathcal{O}_{\mathbb{K}}(n\sigma_0)$$

for some $n \geq 1$. In particular,

$$\mathrm{Hilb}([n\sigma_0])|_{\mathrm{supp}\cap\sigma_\infty=\emptyset} \cong |n\sigma_0|_{\mathrm{supp}\cap\sigma_\infty=\emptyset}. \tag{4.1}$$

PROOF. First we compute the class $[\Sigma]$ in $H^2(\mathbb{K},\mathbb{Z})$. Since the latter is generated by $[\sigma_0]$ and $[F]$, one has

$$[\Sigma] = n[\sigma_0] + m[F]$$

for some $n, m \in \mathbb{Z}$. By hypothesis, Σ is contained in $\mathrm{Tot}(K)$, what implies that

$$[\Sigma]\cdot[\sigma_\infty] = 0.$$

Hence,

$$0 = (n[\sigma_0] + m[F])\cdot([\sigma_0] - (2g-2)[F]) = m,$$

so

$$[\Sigma] = [n\sigma_0].$$

Set $L := \mathcal{O}_{\mathbb{K}}(\Sigma) \otimes \mathcal{O}_{\mathbb{K}}(nC)^{-1}$. We shall next see that L is indeed the trivial bundle. Note that one has a $\mathbb{C}^*$ action on $\mathbb{K}$ given by scaling the fibres by the multiplication of $\tau \in \mathbb{C}^*$. We see that under this action Σ tends, when $\tau \to 0$, to a curve supported on σ_0, which must be the multiple curve $n\sigma_0$ as $[\Sigma] = [nC]$. Hence, one can construct a family $\mathcal{X}$ of curves inside $\mathbb{K}$ parametrized by $\mathbb{C}$, whose fibre over $\tau \in \mathbb{C}^*$ is $\tau(\Sigma)$ and its fibre over $t = 0$ is the multiple curve nC. Hence,

$$\mathcal{L} = \mathcal{O}_{\mathbb{K}\times\mathbb{C}}(\mathcal{X} - (n\mathbb{C}\times\mathbb{C}))$$

provides a family of line bundles, satisfying

$$\begin{aligned}\mathcal{L}|_{\tau\neq 0} \cong \mathcal{O}_{\mathbb{K}}(\tau(\Sigma)) \otimes \mathcal{O}_{\mathbb{K}}(nC)^{-1} &\cong (\tau^{-1})^*\mathcal{O}_{\mathbb{K}}(\Sigma) \otimes \mathcal{O}_{\mathbb{K}}(nC)^{-1} \\ &\cong \mathcal{O}_{\mathbb{K}}(\Sigma) \otimes \mathcal{O}_{\mathbb{K}}(nC)^{-1} \cong p^*L\end{aligned} \tag{4.2}$$

and

$$\mathcal{L}|_{\tau=0} \cong \mathcal{O}_{\mathbb{K}}(nC) \otimes \mathcal{O}_{\mathbb{K}}(nC)^{-1} \cong \mathcal{O}_{\mathbb{K}}.$$

It then follows that $L \cong \mathcal{O}_{\mathbb{K}}$, what concludes the proof of the first statement. The second statement follows immediately from the first. □

We will now study the open subset of the linear system $|n\sigma_0|_{\mathrm{supp}\cap\sigma_\infty=\emptyset}$. This open subset of the linear system $|n\sigma_0|$ contains curves that can be obtained by linear deformations of the multiple curve $n\sigma_0$. Linear deformations of a curve are classified by the sections of the normal bundle restricted to the curve. Hence, one has the identification

$$|n\sigma_0|_{\mathrm{supp}\cap\sigma_\infty=\emptyset} = H^0(n\sigma_0, \mathcal{O}_{\mathbb{K}}(n\sigma_0)|_{n\sigma_0}). \tag{4.3}$$

Let r denote the obvious projection of the multiple curve onto its reduced support and observe that $r_*\mathcal{O}_{\mathbb{K}}(n\sigma_0)|_{n\sigma_0}$ amounts to $\mathcal{O}_{\mathbb{K}}(n\sigma_0)|_{\sigma_0} \otimes r_*\mathcal{O}_{n\sigma_0}$ by the projection formula. The structural sheaf $\mathcal{O}_{n\sigma_0} \cong \mathcal{O}_\Sigma/\mathcal{I}_C^n$ decomposes, as an $\mathcal{O}_C$-module, into $\mathcal{O}_C \oplus (\mathcal{I}_C/\mathcal{I}_C^2) \oplus \cdots \oplus (\mathcal{I}_C^{n-1}/\mathcal{I}_C^n)$. The intersection of σ_0 with the canonical divisor is zero, so $\mathcal{I}_C/\mathcal{I}_C^2 \cong \mathcal{O}_{\mathbb{K}}(-\sigma_0)|_{\sigma_0} \cong K^{-1}$. Therefore,

$$H^0(n\sigma_0, \mathcal{O}_{\mathbb{K}}(n\sigma_0)|_{n\sigma_0}) \cong \bigoplus_{i=1}^n H^0(C, K^{\otimes i}),$$

so the open subset of the linear system $|n\sigma_0|_{\mathrm{supp}\cap\sigma_\infty=\emptyset}$ can be identified with

$$|n\sigma_0|_{\mathrm{supp}\cap\sigma_\infty=\emptyset} \cong \bigoplus_{i=1}^n H^0(C, K^{\otimes i}). \tag{4.4}$$

We finish this section with the study of the Poisson structure of $\mathbb{K}$. It is well known (see [**Ha**, Sect. 2, Chap. V]) that its canonical divisor is

$$K_{\mathbb{K}} = \mathcal{O}_{\mathbb{K}}(-2\sigma_\infty).$$

It follows that $\mathbb{K}$ is a Poisson surface, with $\Theta_{\mathbb{K}}$ being a section of $\bigwedge^2 \mathcal{T}_{\mathbb{K}} \cong K_{\mathbb{K}}^{-1} \cong \mathcal{O}_{\mathbb{K}}(2\sigma_\infty)$ vanishing with order 2 in σ_∞. Furthermore, $\mathrm{Tot}(K) = \mathbb{K} \setminus \{\sigma_\infty\}$ is equipped with a holomorphic symplectic form Ω_0 as it is a cotangent bundle, and the latter coincides with the one induced by the Poisson structure.

5. Moduli spaces of Higgs bundles and their symplectic resolutions

Let C be a smooth projective curve. A *Higgs bundle* is a pair (E, φ), where E is a holomorphic vector bundle on C, and $\varphi \in H^0(C, \mathrm{End}(E) \otimes K)$ is a holomorphic section of the endomorphisms bundle, twisted by the canonical bundle K. It is possible to construct [**Hi2, Si1, Si2, Ni**] the moduli space of rank n and degree d semistable (resp. stable) Higgs bundles on C which we denote by $\mathcal{M}_C(n,d)$ (resp. $\mathcal{M}_C(n,d)^{\mathrm{st}}$). This is a quasi-projective variety of dimension $2n^2(g-1)+2$. It is also connected, normal and irreducible after Simpson [**Si2**]. The moduli space of stable Higgs bundles $\mathcal{M}_C(n,d)^{\mathrm{st}}$ is a smooth open subset of $\mathcal{M}_C(n,d)$.

Taking a basis $(q_1, \dots, q_n)$ of the $\mathrm{GL}(n,\mathbb{C})$-invariant polynomials with $\deg(q_i) = i$, one can construct the *Hitchin fibration*,

$$\begin{array}{ccc} \mathcal{M}_C(n,d) & \longrightarrow & B_n := \bigoplus_{i=1}^n H^0(C, K^{\otimes i}) \\ (E,\varphi) & \longmapsto & (q_1(\varphi), \dots, q_n(\varphi)) . \end{array} \tag{5.1}$$

which is a surjective proper morphism [**Hi1**] endowing $\mathcal{M}_C(n,d)$ with the structure of an algebraically completely integrable system, called the GL_n-*Hitchin system.* In particular, the generic fibers of (5.1) are abelian varieties that admit a beautiful description [**Hi1, BNR, Si2, Sb**] in terms of (compactified) Jacobians of certain curves, called *spectral curves* in the literature, covering our base curve C. To describe these, consider the total space $\mathrm{Tot}(K)$ of the canonical bundle and the surjection $p : \mathrm{Tot}(K) \to C$ obtained by the structure morphism of the line bundle K over C. Observe that the pullback bundle $p^*K \to \mathrm{Tot}(K)$ admits a tautological section λ sending every point $k \in \mathrm{Tot}(K)$ to the point k on the fibre of p^*K over $p(k)$. For every $b = (b_1, \dots, b_n) \in B_n$, with $b_i \in H^0(C, K^i)$, we construct the *spectral curve* $C_b \subset \mathrm{Tot}(K)$ by considering the vanishing locus of the section of p^*K^n

$$\lambda^n + p^*b_1\lambda^{n-1} + \cdots + p^*b_{n-1}\lambda + p^*b_n.$$

The restriction of p to C_b is a ramified degree n cover over C that we denote by

$$p_b : C_b \longrightarrow C.$$

The genus formula and the triviality of the canonical bundle of the quasiprojective surface $\mathrm{Tot}(K)$, allow us to compute the genus $g(C_b) = n^2(g-1)+1$.

Let us now consider a torsion-free rank one sheaf $\mathcal{F}$ over C_b of Euler characteristic $d - n(g-1)$ and whose restriction to every irreducible component of C_b has rank 1. One has that the push-forward $E_{\mathcal{F}} := \pi_{b,*}\mathcal{F}$ is a vector bundle on C of rank n and degree d. Furthermore, pushing-forward the morphism associated to tensorization by the tautological section

$$\lambda : \mathcal{F} \to \mathcal{F} \otimes \pi^*K,$$

one obtains, thanks to the projection formula, a Higgs field for our vector bundle $E_{\mathcal{F}} = p_{b,*}\mathcal{F}$,

$$\varphi_{\mathcal{F}} := p_{b,*}\lambda : E_{\mathcal{F}} \longrightarrow E_{\mathcal{F}} \otimes K. \tag{5.2}$$

One can naturally check that the image of $(E_{\mathcal{F}}, \varphi_{\mathcal{F}})$ under (5.1) is b. The previous construction is indeed a 1-1 correspondence, called the *spectral correspondence*, between rank 1 torsion free sheaves on C_b and Higgs bundles whose image under the Hitchin morphism is $b \in B_n$ [**BNR**].

Furthermore, the (semi)stability of the rank 1 torsion-free sheaves $\mathcal{F}$ on C_b is equivalent to the (semi)stability of the corresponding Higgs bundle $(E_{\mathcal{F}}, \varphi_{\mathcal{F}})$. Therefore, the fibres of (5.1) are identified with the compactified Jacobians of the corresponding spectral curve, $\overline{Jac}^{\delta+d}(C_b)$, classifying torsion-free sheaves on C_b with rank 1 on each irreducible component of C_b.

Note that all the spectral curves C_b belong to the linear system $|n\sigma_0|$ in $\mathrm{Tot}(K)$. Over the ruled surface $\mathbb{K}$ we consider the moduli space of sheaves $\mathrm{M}^{\mathrm{H}}_{\mathbb{K}}(0, n\sigma_0, d - n(g-1))$ and inside it, take the open locus $\mathrm{M}^{\mathrm{H}_0}_{\mathbb{K}}(0, n\sigma_0, d - n(g-1))\big|_{\mathrm{supp}\cap\sigma_\infty=\emptyset}$ of those sheaves that do not intersect the infinity section σ_∞. After Lemma 4.1 and (4.4), the support morphism (2.2) restricts to

$$\mathrm{M}^{\mathrm{H}_0}_{\mathbb{K}}(0, n\sigma_0, d - n(g-1))\Big|_{\mathrm{supp}\cap\sigma_\infty=\emptyset} \longrightarrow B_n \cong |n\sigma_0|_{\mathrm{supp}\cap\sigma_\infty=\emptyset}.$$

In this context, the spectral correspondence leads to a second construction of the Higgs moduli space as the closed and dense subset of

$$\mathrm{M}^{\mathrm{H}_0}_{\mathbb{K}}(0, n\sigma_0, d - n(g-1))\Big|_{\mathrm{supp}\cap\sigma_\infty=\emptyset}$$

given by those sheaves whose restriction to its associated spectral curve has rank 1 on each irreducible component,

$$(5.3) \qquad \mathcal{M}_C(n,d) = \left\{ \begin{array}{c} \mathcal{F} \in \mathrm{M}_{\mathbb{K}}^{\mathrm{H}_0}(0, n\sigma_0, d - n(g-1)) \text{ s. t.} \\ \operatorname{supp}(\mathcal{F}) \subset \operatorname{Tot}(K) = \mathbb{K} \setminus \sigma_\infty \text{ and} \\ \operatorname{rk}(\mathcal{F}|_{Y_i}) = 1 \text{ for every irreducible} \\ \text{component } Y_i \text{ of } \operatorname{supp}(\mathcal{F}) \end{array} \right\}.$$

Following (4.4), under this identification, the Hitchin fibration can be interpreted as the support fibration (2.2),

$$(5.4) \qquad \mathcal{M}_C(n,d) \longrightarrow |n\sigma_0|_{\operatorname{supp}\cap\sigma_\infty = \emptyset}.$$

Fixing a point $x_0 \in C$, one can define the natural morphism

$$\begin{array}{rccc} a_{x_0}: & \mathcal{M}_C(n,d) & \longrightarrow & Jac^0(C) \times H^0(C,K) \\ & (E,\varphi) & \longmapsto & \left(\det(E) \otimes \mathcal{O}_C(dx_0)^{-1}, \operatorname{tr}(\varphi)\right). \end{array}$$

The kernel of this map is the moduli space of $\mathrm{SL}_n(n, \mathbf{C})$–Higgs bundles,

$$\mathcal{N}_C(n,d) := a_{x_0}^{-1}(\mathcal{O}_C, 0).$$

The locus of stable points $\mathcal{N}_C(n,d)$ is a smooth open subset of $\mathcal{M}_C(n,d)$. As before, $\mathcal{N}_C(n,d) = \mathcal{N}_C(n,d)^{\mathrm{st}}$ is smooth if and only if n and d are coprime. The Hitchin morphism (5.1) restricts to

$$(5.5) \qquad \begin{array}{ccc} \mathcal{N}_C(n,d) & \longrightarrow & V_n := \bigoplus_{i=2}^n H^0(C, K^{\otimes i}) \\ (E,\varphi) & \longmapsto & (q_2(\varphi), \dots, q_n(\varphi)), \end{array}$$

becoming an algebraically completely integrable system [**Hi1**] that we call the SL_n-*Hitchin system.*

One can define a holomorphic symplectic form on $\mathcal{M}_C(n,d)$ out of the construction of this moduli space as an infinite hyperKähler quotient. This holomorphic symplectic form coincides (up to scalling) with that obtained out of the Bottacin–Markman Poisson structure on the moduli space [**Bo2, Mr**]. Also, the holomorphic symplectic form is the extension of the standard holomorphic symplectic form defined on the cotangent of the moduli space of stable vector bundles, which is naturally included in $\mathcal{M}_C(n,d)$.

Using deformation theory [**Hi2, Si2, Ni**], one can compute the tangent space of the moduli spaces $\mathcal{M}_C(n,d)$ and $\mathcal{N}_C(n,d)$. Since stable implies simple also in this case, the stable loci $\mathcal{M}_C(n,d)^{\mathrm{st}}$ and $\mathcal{N}_C(n,d)^{\mathrm{st}}$ immediately smooth. A less straigh-forward analysis based on DGLA was used by Simpson [**Si2**] to show that the singular locus of $\mathcal{M}_C(n,d)$ and $\mathcal{N}_C(n,d)$ are indeed their strictly polystable loci $\mathcal{M}_C(n,d)^{\mathrm{str.ps}}$ and $\mathcal{N}_C(n,d)^{\mathrm{str.ps}}$.

The study of the existence of symplectic resolutions of the Higgs moduli spaces was first carried out by Kiem and Yoo [**KY**] for $\mathcal{N}_C(2,0)$ and different values of g, the genus of the curve. The method used here consists on the computation of the stringy E-function, which is a polynomial if and only if there exists a crepant resolution (recall that symplectic resolution implies crepant). Tirelli provided in [**Ti**] a complete description of the occurrence of symplectic resolutions of the Higgs moduli space for any rank and any genus. The results of [**Ti**] rely on Simpson's Isosingularity Theorem [**Si2**] and the work of Bellamy and Schedler [**BS**] on symplectic resolutions of character varieties associated to Riemann surfaces, generalizing the approach of Kaledin–Lehn–Sorger [**LS, KLS**] to the context of quiver and character varieties.

THEOREM 5.1 ([**KY**, **Ti**, **BS**]). *Let g denote the genus of the smooth projective curve C and assume $g \geq 2$. Then, $\mathcal{M}_C(n,0)$ and $\mathcal{N}_C(n,0)$ are varieties with a symplectic singularities and*

(1) *when $(n,g) = (2,2)$, there exist symplectic resolutions $\widetilde{\mathcal{M}}_C(2,0)$ and $\widetilde{\mathcal{N}}_C(2,0)$ of, respectively, $\mathcal{M}_C(2,0)$ and $\mathcal{N}_C(2,0)$ obtained after blowing-up the respective singular loci which amounts to their strictly polystable loci $\mathcal{M}_C(2,0)^{\mathrm{str.ps}}$ and $\mathcal{N}_C(2,0)^{\mathrm{str.ps}}$;*
(2) *when $(n,g) \neq (2,2)$, there does not exist a symplectic resolution of $\mathcal{M}_C(n,0)$ nor $\mathcal{N}_C(n,0)$.*

6. The non-linear deformation of Donagi, Ein and Lazarsfeld

In this section we review the non-linear deformation of Donagi, Ein and Lazarsfeld [**DEL**] for the case of K3 surfaces, later revisited by de Cataldo, Maulik and Shen in [**CMS2**] in the case of abelian surfaces (see the Appendix of [**FM**] as well). This degeneration was generalized by the author to the case of arbitrary smooth surfaces in [**Fr**], where, furthermore, it was described the behaviour of the holomorphic symlectic form.

Take a smooth symplectic surface S containing a genus $g > 1$ smooth curve

$$C \subset S.$$

Thanks to the triviality of the canonical bundle K_S, the adjunction formula implies that the normal bundle of C in S coincides with the canonical bundle of C,

$$N_{S/C} \cong K.$$

Consider the projective compactification $\mathbb{K}$. Recalling the properties of the canonical ruled surface from Section 4, one has $[\sigma_0]^2 = 2g-2 = [C]^2$ and $[\sigma_0] \cdot [K_{\mathbb{K}}] = 0 = [C] \cdot [K_S]$. Hence, the linear systems $|nC|$ and $|n\sigma_0|$ contain generic curves of equal genus,

$$g_{nC} = g_{n\sigma_0}, \tag{6.1}$$

and the topological invariants of the ideal sheaves describing these curves coincide.

Take the trivial family $S \times \mathbb{P}^1$ and consider the blow-up at $C \times \{0\}$,

$$\overline{\mathbf{S}} := Blow_{C\times\{0\}}\left(S \times \mathbb{P}^1\right), \tag{6.2}$$

which is non-singular as we are blowing-up a non-singular subvariety. Composing the structural morphism of the blow-up with the projection $S \times \mathbb{P}^1 \to \mathbb{P}^1$, one gets the structural morphism

$$\pi : \overline{\mathbf{S}} \to \mathbb{P}^1.$$

This map is flat and its fibres outside $0 \in \mathbb{P}^1$ are not affected by our construction, so for the generic fibre, at $t \neq 0$, we have

$$\overline{\mathbf{S}}_t \cong S.$$

Meanwhile, the central fibre, at $t = 0$, is the union

$$\overline{\mathbf{S}}_0 = \mathbb{K} \cup S, \tag{6.3}$$

where both irreducible components meet transversely at the image of infinity section $\sigma_\infty \subset \mathbb{L}$ identified with the curve $C \subset S$. Note that $\overline{\mathbf{S}}_0$ is a complete intersection subvariety. One also consider the open subset

$$\mathbf{S} := \overline{\mathbf{S}} - (S \times \{0\}), \tag{6.4}$$

and observe that the restriction there of the structural morphism gives, again, a flat family of surfaces over $\mathbb{P}^1$, but now the central fibre $\mathbf{S}|_{t=0} = \mathrm{Tot}(K)$ is not projective.

Denote by p_S the composition of the structural morphism of the blow-up with the projection of $S \times \mathbb{P}^1$ onto S. Being $\overline{\mathbf{S}}$ projective, one may choose a relative polarization

$$\mathbf{H} = p_S^*\mathrm{H} \otimes \mathcal{O}_{\overline{\mathbf{S}}}(\mathbb{K}). \tag{6.5}$$

For all $t \neq 0$ in $\mathbb{P}^1$, one trivially has that $\mathbf{H}|_{\overline{\mathbf{S}}_t} = \mathrm{H}$ while $H_0 := \mathbf{H}|_{\mathbb{K}}$ corresponds to $p^*(\mathrm{H}|_C) \otimes \mathcal{O}_{\mathbb{K}}(\sigma_\infty)$.

Since $C \times \{0\}$ is a Cartier divisor of $C \times \mathbb{P}^1$, one has that

$$\mathbf{C} := Blow_{C\times\{0\}}\left(C \times \mathbb{P}^1\right) \cong C \times \mathbb{P}^1 \tag{6.6}$$

embeds naturally into $\overline{\mathbf{S}}$. One can consider the fibration $\mathbf{C} \to \mathbb{P}^1$, whose generic fibres, at $t \neq 0$, are naturally identified with C,

$$\mathbf{C}_t \cong C \subset S = \overline{\mathbf{S}}_t.$$

Furthermore, the restriction of $\mathbf{C}$ to $t = 0$ coincides with the image of the zero section,

$$\mathbf{C}_0 \cong \sigma_0 \cong C,$$

inside $\mathbb{L} \subset \overline{\mathbf{S}}_0$. Note that the restriction to every point of $\mathbb{P}^1$ gives $\mathbf{C}_t \in H^2(\overline{\mathbf{S}}_t, \mathbb{Z})$. Using $\mathbf{C}$, consider the Mukai vector

$$\mathbf{v} = (0, [n\mathbf{C}], d - n(g-1)), \tag{6.7}$$

for some integer $d \in \mathbb{Z}$.

Following Simpson [**Si1**, Theorem 1.21] (see [**HL**, Theorem 4.3.7] for a specific treatment of the relative case), let us consider the moduli space $\mathrm{M}^{\mathbf{H}}_{\overline{\mathbf{S}}/\mathbb{P}^1}(\mathbf{v})$ of relatively pure dimension 1 $\mathbf{H}$-semistable relative sheaves over the $\mathbb{P}^1$ scheme $\overline{\mathbf{S}}$ having Mukai vector $\mathbf{v}$. By construction, $\mathrm{M}^{\mathbf{H}}_{\overline{\mathbf{S}}/\mathbb{P}^1}(\mathbf{v})$ comes equipped with the structural flat morphism to $\mathbb{P}^1$, which trivializes over $\mathbb{P}^1 - \{0\}$ having generic fibres equal to $\mathrm{M}_{S,\mathrm{H}}(\mathrm{v})$, and whose central fibre over $t = 0$ is $\mathrm{M}^{\mathbf{H}_0}_{\mathbb{K}\cup S}(\mathrm{v})$.

Let us consider the relative Hilbert scheme $\mathrm{Hilb}_{\overline{\mathbf{S}}/\mathbb{P}^1}([n\mathbf{C}])$ parametrizing ideal sheaves of curves in $\overline{\mathbf{S}}$ with first Chern class $[n\mathbf{C}]$ and let us denote by $\mathrm{Hilb}_{\mathbf{S}/\mathbb{P}^1}([n\mathbf{C}])$ the open subset given by those curves contained in $\mathbf{S}$. The determinant in cohomology is well defined and one can consider the determinant map to the relative Jacobian,

$$\det_{\mathbf{S}} : \mathrm{Hilb}_{\overline{\mathbf{S}}/\mathbb{P}^1}([n\mathbf{C}]) \to Jac_{\mathbb{P}^1}(\overline{\mathbf{S}}), \tag{6.8}$$

that restricts, of course, to $\mathrm{Hilb}_{\mathbf{S}/\mathbb{P}^1}([n\mathbf{C}])$. For $t \neq 0$, the fibre of (6.8) over $L \in Jac(S)$ corresponds to the linear system $|L|$ in S. After Lemma 4.1, one has that the whole slice of the Hilbert scheme is sent to a point,

$$\det_{\mathbf{S}}\left(\mathrm{Hilb}_{\mathbf{S}/\mathbb{P}^1}([n\mathbf{C}])|_{t=0}\right) = \mathcal{O}_{\overline{\mathbf{S}}_0}(n\sigma_0). \tag{6.9}$$

Furthermore, making use of (4.4), one can identify the whole central fibre of the Hilbert scheme with the Hitchin base

$$\mathrm{Hilb}_{\mathbf{S}/\mathbb{P}^1}([n\mathbf{C}])|_{t=0} \cong \bigoplus_{i=1}^{n} H^0(C, K^{\otimes i}) = B_n.$$

By specifying the relative (fitting) support of the sheaves in $\mathrm{M}^{\mathbf{H}}_{\overline{\mathbf{S}}/\mathbb{P}^1}(\mathbf{v})$, one has the following surjective morphism

$$\begin{array}{cccc} \mathbf{h}: & \mathrm{M}^{\mathbf{H}}_{\overline{\mathbf{S}}/\mathbb{P}^1}(\mathbf{v}) & \longrightarrow & \mathrm{Hilb}_{\overline{\mathbf{S}}/\mathbb{P}^1}([n\mathbf{C}]) \\ & \mathcal{F} & \longmapsto & \mathrm{supp}(\mathcal{F}). \end{array} \tag{6.10}$$

The morphism $\mathbf{h}$ commutes with the structural morphisms (to $\mathbb{P}^1$). Let us consider the restriction to those sheaves supported in $\mathbf{S} \subset \overline{\mathbf{S}}$,

$$\mathbf{h}^{-1}\left(\mathrm{Hilb}_{\mathbf{S}/\mathbb{P}^1}([n\mathbf{C}])\right) \longrightarrow \mathrm{Hilb}_{\mathbf{S}/\mathbb{P}^1}([n\mathbf{C}]).$$

Observe that $\mathbf{h}^{-1}\left(\mathrm{Hilb}_{\mathbf{S}/\mathbb{P}^1}([n\mathbf{C}])\right) \longrightarrow \mathbb{P}^1$ has generic fibres $\mathrm{M}^{\mathrm{H}}_{S}(v)$ over $t \neq 0$ and central fibre $\mathrm{M}^{H_0}_{\mathbb{K}}(v)|_{\mathrm{supp}\cap\sigma_\infty=\emptyset}$ at $t = 0$. The latter is a consequence of the equivalence between the H_0-semistability of a sheaf in $\mathrm{Tot}(K) \subset \mathbb{K}$ and its $\mathbf{H}_0$-semistability as a sheaf in $\mathrm{Tot}(K) \subset \mathbb{K} \cup S$. Having (5.3) in mind, we define the closed and dense subset of $\mathbf{h}^{-1}\left(\mathrm{Hilb}_{\mathbf{S}/\mathbb{P}^1}([n\mathbf{C}])\right) \subset \mathrm{M}^{\mathbf{H}}_{\overline{\mathbf{S}}/\mathbb{P}^1}(\mathbf{v})$, given by those sheaves that are either supported in a generic fibre, or those supported in the central fibre having rank 1 on each irreducible component of its support,

$$\mathbf{M}^{\mathbf{H}}_{\mathbf{S}/\mathbb{P}^1}(0, n\mathbf{C}, d-n(g-1)) := \left\{ \begin{array}{c} \mathcal{F} \in \mathrm{M}^{\mathbf{H}}_{\overline{\mathbf{S}}/\mathbb{P}^1}(\mathbf{v}) \text{ s. t. } \mathrm{supp}(\mathcal{F}) \subset \mathbf{S} \text{ and, whenever} \\ \mathrm{supp}(\mathcal{F}) \subset \mathrm{Tot}(K) \subset \overline{\mathbf{S}}_0, \text{ one has } \mathrm{rk}(\mathcal{F}|_{Y_i}) = 1 \\ \text{for every irreducible component } Y_i \text{ of } \mathrm{supp}(\mathcal{F}) \end{array} \right\}. \tag{6.11}$$

We can now see that the above construction provides the degeneration that we were looking for. This degeneration appeared first in [**DEL**] for K3 surfaces and in [**CMS1**] for abelian surfaces. A generalization to arbitrary smooth surfaces is provided in [**Fr**].

THEOREM 6.1 ([**DEL**]). *Consider a smooth symplectic surface surface S and a genus $g \geq 2$ smooth projective curve C inside S. One can construct $\mathbf{M}^{\mathbf{H}}_{\mathbf{S}}(0, n\mathbf{C}, d - n(g-1))$ and $\mathrm{Hilb}_{\overline{\mathbf{S}}/\mathbb{P}^1}([n\mathbf{C}])$ flat over $\mathbb{P}^1$, and a morphism between them commuting with the structural morphisms,*

$$\begin{array}{ccc} \mathbf{M}^{\mathbf{H}}_{\mathbf{S}/\mathbb{P}^1}(0, n\mathbf{C}, d - n(g-1)) & \longrightarrow & \mathrm{Hilb}_{\mathbf{S}/\mathbb{P}^1}([n\mathbf{C}]) \\ & \searrow \quad \swarrow & \\ & \mathbb{P}^1. & \end{array} \tag{6.12}$$

The horizontal arrow is a surjective fibration, trivial over $\mathbb{P}^1 - \{0\}$, and the generic fibre over $t \neq 0$ is the support morphism (2.2),

$$\mathrm{M}^{\mathrm{H}}_{S}(0, nC, d - n(g-1)) \to \mathrm{Hilb}_S([n\mathbf{C}]),$$

while the central fibre at $t = 0$ gives the Hitchin morphism (5.1),

$$\mathcal{M}_C(n, d) \to B_n = \bigoplus_{i=1}^{n} H^0(C, K^{\otimes i}).$$

Following [**Fr**], we now review the study of the behaviour of the symplectic structure along Donagi–Ein–Lazarsfeld degeneration, showing that it provides a deformation of symplectic varieties. We start by studying $K_{\overline{\mathbf{S}}/\mathbb{P}^1}$ and $K_{\overline{\mathbf{S}}_0}$ in the next proposition. For the shake of completion, we reproduce its proof that first appeared in [**Fr**].

PROPOSITION 6.2. *Let S be a symplectic surface and pick $C \subset S$ smooth curve of genus $g \geq 2$. One has the following*

(1) *$K_{\overline{\mathbf{S}}_0}$ is the line bundle given by $\mathcal{O}_{\mathbb{K}}(-\sigma_\infty)$ and $\mathcal{O}_S(C)$ identified along $\sigma_\infty \cong C$ by a natural isomorphism between $K \cong \mathcal{O}_{\mathbb{K}}(-\sigma_\infty)|_{\sigma_\infty}$ and $K \cong \mathcal{O}_S(C)|_C$;*
(2) *$H^0(\overline{\mathbf{S}}_0, K_{\overline{\mathbf{S}}_0}) = \mathbf{C}$ and every non-trivial section of $K_{\overline{\mathbf{S}}_0}$ vanishes on $\mathbb{K}$ and only on $\mathbb{K}$;*
(3) *$H^0(\overline{\mathbf{S}}_0, \omega^{-1}_{\overline{\mathbf{S}}_0}) = \mathbf{C}$ and every non-trivial section of $\omega^{-1}_{\overline{\mathbf{S}}_0}$ vanishes on S and only on S;*
(4) *$\pi_* K_{\overline{\mathbf{S}}/\mathbb{P}^1} \cong \mathcal{O}_{\mathbb{P}^1}(1)$;*
(5) *$\pi_* \omega^{-1}_{\overline{\mathbf{S}}/\mathbb{P}^1} \cong \mathcal{O}_{\mathbb{P}^1}(-1)$.*

PROOF. We first observe that $\overline{\mathbf{S}}$ is constructed by blowing-up a smooth subvariety of $S \times \mathbb{P}^1$, which is also smooth. Hence $\overline{\mathbf{S}}$ is smooth and has canonical bundle $K_{\overline{\mathbf{S}}} = r^* K_{S \times \mathbb{P}^1}(\mathbb{K})$, where we denote by r the structural morphism of the blow-up and we recall that $\mathbb{K}$ is the exceptional divisor. One can deduce that the singular central fibre $\overline{\mathbf{S}}_0 = \mathbb{K} \cup S$, where both components meet transversally, is a complete intersection variety, so it is Gorenstein and its dualizing sheaf $K_{\overline{\mathbf{S}}_0}$ is a line bundle. This implies that the relative dualizing sheaf $K_{\overline{\mathbf{S}}/\mathbb{P}^1}$ is a line bundle too, indeed it is $\mathcal{O}_{\overline{\mathbf{S}}}(\mathbb{K})$, as K_S is trivial.

Since $\overline{\mathbf{S}}_0$ is a complete intersection divisor in $\overline{\mathbf{S}}$, one has that $K_{\overline{\mathbf{S}}_0} = K_{\overline{\mathbf{S}}/\mathbb{P}^1}(\mathbb{K} + S)|_{\overline{\mathbf{S}}_0}$, where we abuse of notation by denoting the total transform of $S \times \{0\}$ simply by S. Then,

$$K_{\overline{\mathbf{S}}_0}|_{\mathbb{K}} \cong \mathcal{O}_{\overline{\mathbf{S}}}(2\mathbb{K})|_{\mathbb{K}} \otimes \mathcal{O}_{\overline{\mathbf{S}}}(S)|_{\mathbb{K}} \cong \mathcal{O}_{\mathbb{K}}(-2\sigma_\infty) \otimes \mathcal{O}_{\mathbb{K}}(\sigma_\infty) \cong \mathcal{O}_{\mathbb{K}}(-\sigma_\infty),$$

where we recall that $\mathcal{O}_{\overline{\mathbf{S}}}(-\mathbb{K})|_{\mathbb{K}}$ is the linearization $\mathcal{O}_{\mathbb{K}}(1) \cong \mathcal{O}_{\mathbb{K}}(\sigma_\infty)$. Similarly, the restriction of $K_{\overline{\mathbf{S}}_0}$ to S gives

$$K_{\overline{\mathbf{S}}_0}|_S \cong \mathcal{O}_{\overline{\mathbf{S}}}(2\mathbb{K})|_S \otimes \mathcal{O}_{\overline{\mathbf{S}}}(S)|_S \cong \mathcal{O}_S(2C) \otimes \mathcal{O}_S(-C) \cong \mathcal{O}_S(C),$$

where we recall that $K_S \cong \mathcal{O}_{\overline{\mathbf{S}}}(\mathbb{K} + S)|_S \cong \mathcal{O}_{\overline{\mathbf{S}}}(\mathbb{K})|_S \otimes \mathcal{O}_{\overline{\mathbf{S}}}(S)|_S$ is trivial, hence $\mathcal{O}_{\overline{\mathbf{S}}}(S)|_S \cong (\mathcal{O}_{\overline{\mathbf{S}}}(\mathbb{K})|_S)^{-1}$. Recalling that $K_{\mathbb{K}} = \mathcal{O}_{\mathbb{K}}(-2\sigma_\infty)$ and $K_S \cong \mathcal{O}_S$, applying adjunction, one gets that $\mathcal{O}_{\mathbb{K}}(-\sigma_\infty)|_{\sigma_\infty}$ and $\mathcal{O}_S(C)|_C$ both being isomorphic to K, the canonical bundle of $\sigma_\infty \cong C$, so the restriction of the line bundle $K_{\overline{\mathbf{S}}_0}$ provides a natural identification. This finishes the proof of (1).

Since $\mathcal{O}_{\mathbb{K}}(-\sigma_\infty)$ has no non-zero sections, every non-zero section of $K_{\overline{\mathbf{S}}_0}$ vanishes completely on $\mathbb{K}$, hence it is given by a section of $\mathcal{O}_S(C)$ vanishing identically at C. The set of those sections determines a 1-dimensional subspace of $H^0(S, \mathcal{O}_S(C))$ and (2) follows.

We proof (3) by observing that $h^0(\mathbb{K}, \mathcal{O}_{\mathbb{K}}(\sigma_\infty)) = 1$ and that $h^0(S, \mathcal{O}_S(-C)) = 0$. It follows that every non-zero section of $K^{-1}_{\overline{\mathbf{S}}_0}$ vanishes on S, so it is given by a section of $\mathcal{O}_{\mathbb{K}}(\sigma_\infty)$ which vanishes identically at σ_∞.

Recall that $\overline{\mathbf{S}}|_{\mathbb{P}^1 - \{0\}}$ is the trivial fibration $S \times (\mathbb{P}^1 - \{0\})$, so the restriction there of $K_{\overline{\mathbf{S}}/\mathbb{P}^1}$ is the trivial line bundle. After (2) and (3) and upper semicontinuity of cohomology, one has that $\pi_* K_{\overline{\mathbf{S}}/\mathbb{P}^1}$ and $\pi_* \omega^{-1}_{\overline{\mathbf{S}}/\mathbb{P}^1}$ are line bundles over $\mathbb{P}^1$, inverse to each other. Note that $K_{\overline{\mathbf{S}}/\mathbb{P}^1} \cong \mathcal{O}_{\overline{\mathbf{S}}}(\mathbb{K})$ comes naturally equipped with a section. Furthermore, $h^0(\mathcal{O}_{\overline{\mathbf{S}}}(\mathbb{K})) = 1$ by the properties of blow-up. It follows that $\pi_* K_{\overline{\mathbf{S}}/\mathbb{P}^1}$ has a single non-zero section (up to scaling), so (4) follows.

Finally, (5) follows from (4). □

In view of (2) and (3) of Proposition 6.2 we shall first construct a Poisson structure on $\mathrm{M}^{\mathbf{H}}_{\overline{\mathbf{S}}/\mathbb{A}^1}(\mathbf{v})$ and, then, derive the symplectic structure on $\mathbf{M}^{\mathbf{H}}_{\mathbf{S}/\mathbb{A}^1}(0, n\mathbf{C}, d - n(g-1))$. This is the context of the following theorem that first appeared in [**Fr**], and whose proof we reproduce for the shake of completion.

THEOREM 6.3. *Consider a smooth symplectic surface S and a genus $g \geq 2$ smooth projective curve C inside S. Then, there exists a relative Poisson structure Θ on $\mathrm{M}^{\mathbf{H}}_{\overline{\mathbf{S}}/\mathbb{A}^1}(\mathbf{v}) \to \mathbb{A}^1$ which defines a relative symplectic form $\mathit{\Omega}$ on $\mathbf{M}^{\mathbf{H}}_{\mathbf{S}/\mathbb{A}^1}(0, n\mathbf{C}, d - n(g-1)) \to \mathbb{A}^1$ coinciding (up to scaling) with the Mukai form Ω over the generic fibre $\mathrm{M}^{\mathrm{H}}_S(0, nC, d-n(g-1))$ over $t \neq 0$, and, on the central fibre $\mathcal{M}_C(n,d)$ at $t = 0$, with Ω_0 obtained by extending the canonical symplectic form on the cotangent of the moduli space of stable vector bundles.*

PROOF. Starting from (5) of Proposition 6.2, we pick $\mathbb{A}^1 \subset \mathbb{P}^1$ a trivialization of $\pi_* K^{-1}_{\overline{\mathbf{S}}/\mathbb{P}^1}$ containing $0 \in \mathbb{P}^1$ and choose a section $\vartheta \in H^0(\mathbb{A}^1, \pi_* K^{-1}_{\overline{\mathbf{S}}/\mathbb{A}^1})$. Following Theorem 2.2 one can construct a relative Bottacin–Markman Poisson structure Θ on $\mathrm{M}^{\mathbf{H}}_{\overline{\mathbf{S}}/\mathbb{A}^1}(\mathbf{v})$,

(6.13)

$$\mathrm{Ext}^1_{\mathcal{O}_{\overline{\mathbf{S}}_t}}(\mathcal{E}, \mathcal{E} \otimes K_{\overline{\mathbf{S}}_t}) \wedge \mathrm{Ext}^1_{\mathcal{O}_{\overline{\mathbf{S}}_t}}(\mathcal{E}, \mathcal{E} \otimes K_{\overline{\mathbf{S}}_t}) \xrightarrow{\circ} \mathrm{Ext}^2_{\mathcal{O}_{\overline{\mathbf{S}}_t}}(\mathcal{E}, \mathcal{E} \otimes K^2_{\overline{\mathbf{S}}_t}) \xrightarrow{\mathrm{tr}}$$
$$H^2(\overline{\mathbf{S}}_t, K^2_{\overline{\mathbf{S}}_t}) \xrightarrow{\langle \cdot, \vartheta \rangle} H^2(\overline{\mathbf{S}}_t, K_{\overline{\mathbf{S}}_t}) \cong \mathbf{C}.$$

Thanks to (3) of Proposition 6.2, we can check that the restriction of Θ to the subset $\mathcal{M}_C(n,d)$ of the central fibre is non-degenerate, hence defines a symplectic form there, as the tangent and the cotangent spaces to the moduli can be identified thanks to a trivialization of $K_{\mathbb{K}} = \mathcal{O}_{\mathbb{K}}(-2\sigma_\infty)$ over $\mathrm{Tot}(K) = \mathbb{K} - \{\sigma_\infty\}$. Furthermore, the canonical bundle is trivial over the generic fibres, $K_S \cong \mathcal{O}_S$, so in this case the cotangent space of $\mathrm{M}^{\mathbf{H}}_{\overline{\mathbf{S}}/\mathbb{A}^1}(\mathbf{v})|_t = \mathrm{M}^{\mathrm{H}}_S(v)$ is identified with the tangent space. Then, the Poisson form Θ_t over $t \neq 0$ is non-degenerate and defines a symplectic form which coincides with the Mukai form up to scaling. Hence, the restriction of Θ to $\mathbf{M}^{\mathbf{H}}_{\mathbf{S}/\mathbb{A}^1}(0, n\mathbf{C}, d - n(g-1))$ defines a relative symplectic form $\mathit{\Omega}$. □

Let us now consider the case of a K3 surface X containing the smooth curve C of genus 2. Since $H^1(X, \mathcal{O}_X) = 0$ by definition of a K3 surface, $\mathrm{Pic}(X)$ embeds into $H^2(X, \mathbb{Z})$ and one notes that $\mathrm{Hilb}_{\mathbf{X}/\mathbb{P}^1}([n\mathbf{C}])|_{t \neq 0}$ is the linear system $|nC|$ inside X. We observe as well, after Lemma 7.2 and (4.4), that the central fibre of $\mathrm{Hilb}_{\mathbf{X}/\mathbb{P}^1}([n\mathbf{C}])|_{t=0}$ is the open subset of curves described in the Hitchin base

$$B_n \cong |n\sigma_0|_{\mathrm{supp} \cap \sigma_\infty = \emptyset}.$$

Consider $\pi_* \mathcal{O}_{\overline{\mathbf{A}}}(n\mathbf{C})$ over $\mathbb{P}^1$, which is a trivial vector bundle over $\mathbb{P}^1 - \{0\}$ of rank n^2 after (7.1). The line bundle $\mathcal{O}_X(C)$ is big and nef, so one can compute $h^0(X, \mathcal{O}_X(C)) = 3$ and $h^0(X, \mathcal{O}_X(2C)) = 6$. In Section 3 we denoted $\mathrm{M}_{\mathrm{ten}} = \mathrm{M}^{\mathrm{H}}_X(0, 2C, -2)$ as its symplectic resolution provides the ten dimensional O'Grady space $\widetilde{\mathrm{M}}_{\mathrm{ten}}$. Denote accordingly

$$\mathbf{M}_{\mathrm{ten}} := \mathbf{M}^{\mathrm{H}}_{\mathbf{X}/\mathbb{A}^1}(0, nC, -2) \text{ and } \mathbf{B}_{\mathrm{ten}} := \mathrm{Hilb}_{\mathbf{X}/\mathbb{A}^1}([2\mathbf{C}]).$$

One has the following consequence of Theorems 6.1 and 6.3.

COROLLARY 6.4. *Given a smooth K3 surface S and a smooth projective genus 2 curve C inside S, one can construct $\mathbf{M}_{\text{ten}}$ and $\mathbf{B}_{\text{ten}}$ flat over $\mathbb{A}^1$, both fitting in the commuting diagram*

$$\begin{array}{ccc} \mathbf{M}_{\text{ten}} & \longrightarrow & \mathbf{B}_{\text{ten}} \\ & \searrow \quad \swarrow & \\ & \mathbb{A}^1. & \end{array}$$

The horizontal arrow is a surjective fibration, trivial over $\mathbb{A}^1 - \{0\}$, whose generic fibres coincide with the Beauville–Mukai system (2.3),

$$\mathrm{M}_{\text{ten}} \to |2C|,$$

and the central fibre at $t = 0$ gives the Hitchin system (5.1),

$$\mathcal{M}_C(2,0) \to H^0(C,K) \oplus H^0(C,K^2).$$

Furthermore, one can construct a relative symplectic form on $\mathbf{M}_{\text{ten}} \to \mathbb{A}^1$, which coincides (up to scaling) with Mukai's form on the generic fibres and, over the central fibre, with the standard symplectic form on the moduli space of Higgs bundles.

7. A non-linear deformation in the SL_n case

In this section we study the construction of the non-linear deformation of the reduced Beauville–Mukai system over an abelian variety into the SL_n-Hitchin system treated in [**CMS2**] and in [**FM**]. In these papers, the degeneration is described using the Donagi–Ein–Lazarsfeld degenerations and assuming that the kernel of the Albanese morphism of the generic fibres becomes, at the central fibre, the kernel of the trace and determinant map. The latter is stated in [**CMS2, FM**] with few details. The biggest contribution of this note is the rigorous proof of the existence of the mentioned degeneration.

For the rest of the article we shall consider A to be the Jacobian of a smooth genus 2 projective curve C,

$$A = Jac^0(C).$$

Taking $x_0 \in C$ such that $\mathcal{O}_C(2x_0) \cong K$, one gets a natural embedding of C into A as a theta divisor and a double cover $C \xrightarrow{2:1} \mathbb{P}^1$. The double cover induces an involution $i : C \to C$ which coincides with that given by inverting in A and restricting again to C. Hence $\mathcal{O}_C(x + i(x)) \cong K$. For every point $a \in A$, we denote by $t_a : A \to A$, the associated traslation in A, sending $a' \mapsto a' + a$. We know that $\mathcal{O}_A(C)$ is ample, and the associated principal polarization

$$\begin{array}{rccc} \lambda_C : & A & \longrightarrow & \widehat{A} \\ & a & \longmapsto & t_a^*\mathcal{O}_A(C) \otimes \mathcal{O}_A(C)^{-1}. \end{array}$$

is an isomorphism, so A is self-dual $A \cong \widehat{A}$. Furthermore, one can compute (see [**BL**] for instance)

$$h^0(A, \mathcal{O}_A(nC)) = n^2. \tag{7.1}$$

Since we recall that the dual determinant is defined in terms of a Fourier–Mukai transform, we focus now on the study of the associated integral functors. We have considered in Section 3 the Poincaré bundle $\mathrm{P} \to A \times \widehat{A}$. Let us recall as well the Fourier–Mukai transform (3.1) and the projections of the first factor

$\mathrm{f} : A \times \widehat{A} \to A$ and to the second factor $\mathrm{g} : A \times \widehat{A} \to \widehat{A}$. One has the canonical embedding of C as a theta divisor of $A = Jac(C)$. Consider $P_C \to C \times \widehat{A}$ to be the restriction of $\mathrm{P} \to A \times \widehat{A}$ to $C \times \widehat{A}$ and observe that this gives a universal bundle for the classification of (0-degree) line bundles on C. Taking $\mathrm{f}_C : C \times \widehat{A} \to C$ and $\mathrm{g}_C : C \times \widehat{A} \to \widehat{A}$, it is possible to construct an associated integral functor,

$$\begin{array}{rccc} \Phi^{P_C}: & D^b(C) & \longrightarrow & D^b(\widehat{A}) \\ & \mathcal{E}^\bullet & \longmapsto & R\mathrm{g}_{C,!}(P_C \otimes \mathrm{f}_C^* \mathcal{E}^\bullet). \end{array}$$

We are interested in the construction of an analog of Φ^{P} replacing A by $\mathrm{Tot}(K)$, where the sheaves classified by $\mathrm{M}_{\mathrm{six}}|_{t=0}$ are supported. Recall that we denote by $p : \mathbb{K} \to C$ the structural morphism and set

$$\mathrm{P}_0 := (p \times \mathbf{1}_{\widehat{A}})^* P_C \to \mathrm{Tot}(K) \times \widehat{A}.$$

We consider the obvious projections $\mathrm{f}_0 : \mathrm{Tot}(K) \times \widehat{A} \to \mathrm{Tot}(K)$ and $\mathrm{g}_0 : \mathrm{Tot}(K) \times \widehat{A} \to \widehat{A}$. We can consider the associated integral functor (not a derived equivalence)

$$\begin{array}{rccc} \Phi^{\mathrm{P}_0}: & D^b(\mathrm{Tot}(K)) & \longrightarrow & D^b(\widehat{A}) \\ & \mathcal{E}^\bullet & \longmapsto & R\mathrm{g}_{0,!}(\mathrm{P}_0 \otimes \mathrm{f}_0^* \mathcal{E}^\bullet). \end{array}$$

Recall that a coherent sheaf can naturally be seen as a complex in the derived category supported in 0 degree. One can show the following relation between the transforms of a Higgs bundle and its spectral data.

LEMMA 7.1. *Let (E, φ) be a Higgs bundle associated to the spectral data $\mathcal{E}$ supported on C, where $C \subset \mathrm{Tot}(K)$ is the spectral curve and $\mathcal{E}$ is a rank 1 torsion free sheaf supported on it. If we consider $\mathcal{E}$ as a sheaf on $\mathrm{Tot}(K)$, then*

$$\Phi^{\mathrm{P}_0}(\mathcal{E}) \cong \Phi^{P_C}(E).$$

PROOF. Consider the commuting diagram

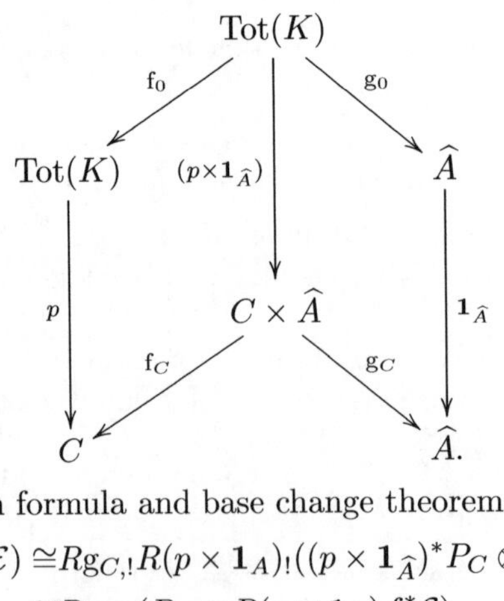

Applying the projection formula and base change theorems, one gets

$$\begin{aligned} \Phi^{\mathrm{P}_0}(\mathcal{E}) &\cong R\mathrm{g}_{C,!} R(p \times \mathbf{1}_A)_!((p \times \mathbf{1}_{\widehat{A}})^* P_C \otimes \mathrm{f}_0^* \mathcal{E}) \\ &\cong R\mathrm{g}_{C,!}(P_C \otimes R(p \times \mathbf{1}_{\widehat{A}})_! \mathrm{f}_0^* \mathcal{E}) \\ &\cong R\mathrm{g}_{C,!}(P_C \otimes \mathrm{f}_C^* R p_! \mathcal{E}). \end{aligned}$$

Finally, observe that, thanks to the spectral correspondence [**Hi1**, **Si2**, **BNR**], one has

$$p_! \mathcal{E} \cong (p|_C)_* \mathcal{E} \cong E,$$

and the proposition follows. $\square$

We are working with $A = Jac^0(C)$ which we recall once again that is self-dual. Note that $P_C \to C \times \widehat{A}$, constructed out of P and the embedding $C \subset A$, fixes canonically an isomorphism $\widehat{A} \cong A$. Denote by $\widehat{C} \subset \widehat{A}$ the image of $C \subset A$ under this isomorphism. Given a point in A associate to the line bundle $L \to C$, we denote by $\widehat{L} \to A$ the associated point in $\widehat{A}$, note that, by construction, the restriction of $\widehat{L}$ to $C \subset A$ amounts to L. The following technical lemma will be crucial for us.

LEMMA 7.2. *Let E be a vector bundle of rank n and degree d over a smooth projective curve C of genus 2, we have*

$$\det_{\widehat{A}}(\Phi^{P_C}(E)) \otimes \mathcal{O}_{\widehat{A}}\left(n\widehat{C}\right)^{-1} \cong \det_C(E) \otimes \mathcal{O}_C(dx_0)^{-1}.$$

PROOF. Combining [**Mu1**, Th 4.2] and [**Mu1**, Prop. 4.3], one gets, for $\ell \leq 1$, that $\mathcal{O}_C(\ell x_0)$ is WIT (recall that WIT stands for *Weak Index Theorem* and, essentialy, means that its Fourier–Mukai transform is a complex supported in a single degree, see [**Mu1**] for precise definition of this term) and $\Phi^{P_C}(\mathcal{O}_C(\ell x_0))$ is $\mathcal{O}_{\widehat{A}}(\widehat{C})$ in degree 1. In the case $\ell > 1$, one has that $\Phi^{P_C}(\mathcal{O}_C(\ell x_0)) \cong [-1]^* RHom_{\mathcal{O}_{\widehat{A}}}(F_{2-\ell}, \mathcal{O}_{\widehat{A}})$. In both cases, one has that

$$\det_{\widehat{A}}(\Phi^{P_C}(\mathcal{O}_C(\ell x_0))) \cong \mathcal{O}_{\widehat{A}}\left(\widehat{C}\right). \tag{7.2}$$

Given E, one can always provide a filtration of the form

$$0 \subsetneq E_1 \subsetneq \cdots \subsetneq E_{n-1} \subsetneq E_n = E, \tag{7.3}$$

where $E_i/E_{i-1} \cong L_i \otimes \mathcal{O}_C(\ell_i x_0)$, being L_i a line bundle of trivial degree. It follows naturally that

$$\det_C(E) \otimes \mathcal{O}_C(dx_0)^{-1} \cong L_1 \otimes \cdots \otimes L_n.$$

Since Φ^{P_C} preserves distinguished triangles, (7.3) implies

$$\det_{\widehat{A}}(\Phi^{P_C}(E)) \cong \det_{\widehat{A}}(\Phi^{P_C}(L_1 \otimes \mathcal{O}_C(\ell_1 x_0))) \otimes \cdots \otimes \det_{\widehat{A}}(\Phi^{P_C}(L_n \otimes \mathcal{O}_C(\ell_n x_0))).$$

After [**Mu1**, Proposition 3.13 (2)], the above equality becomes

$$\det_{\widehat{A}}\Phi^{P_C}(E) \cong t^*_{\widehat{L}_1}\left(\det_{\widehat{A}}\Phi^{P_C}(\mathcal{O}_C(\ell_1 x_0)\right) \otimes \cdots \otimes t^*_{\widehat{L}_n}\left(\det_{\widehat{A}}\Phi^{P_C}(\mathcal{O}_C(\ell_n x_0)\right),$$

and thanks to (7.2),

$$\det_{\widehat{A}}\Phi^{P_C}(E) \cong t^*_{\widehat{L}_1}\mathcal{O}_{\widehat{A}}\left(\widehat{C}\right) \otimes \cdots \otimes t^*_{\widehat{L}_n}\mathcal{O}_{\widehat{A}}\left(\widehat{C}\right).$$

Then, by means of the principal polarization $\lambda_{\widehat{C}} : \widehat{A} \overset{\cong}{\to} A$, the proof follows from the above. □

Consider the line bundle $P_{\mathbb{P}^1} \to (A \times \mathbb{P}^1) \times \widehat{A}$ denotes the pull-back of the Poincaré bundle $P \to A \times \widehat{A}$ introduced above. Recall that the normal cone degeneration $\mathbf{A}$ is given by $Blow_{C\times\{0\}}(A \times \mathbb{P}^1) - (A \times \{0\})$ and set

$$\mathcal{P} := (\beta \times_{\mathbb{P}^1} \mathbf{1}_A)^* P_{\mathbb{P}^1}|_{\mathbf{A}\times\widehat{A}},$$

where β is the structural morphism of the blow-up. Its generic fibre is

$$\mathcal{P}|_{t\neq 0} \cong P, \tag{7.4}$$

while at the central fibre,

$$\mathcal{P}|_{t=0} \cong P_0. \tag{7.5}$$

Considering the obvious projections

$$\begin{array}{ccc} & \mathbf{A}\times\widehat{A} & \\ {}^{f}\swarrow & & \searrow^{g} \\ \mathbf{A} & & \widehat{A}, \end{array}$$

we define the integral relative functor,

$$\begin{array}{rccc} \Phi^{\mathcal{P}}: & D^b_{\mathbb{P}^1}(\mathbf{A}) & \longrightarrow & D^b(\widehat{A}) \\ & \mathcal{E}^\bullet & \longmapsto & Rg_!(\mathcal{P}\otimes f^*\mathcal{E}^\bullet). \end{array} \tag{7.6}$$

Note that, for $t\in\mathbb{P}^1-\{0\}$, one has

$$\Phi^{\mathcal{P}}(\mathcal{E}^\bullet|_t)\cong\Phi^{\mathrm{P}}(\mathcal{E}|_t),$$

while for $t=0$,

$$\Phi^{\mathcal{P}}(\mathcal{E}^\bullet|_{t=0})\cong\Phi^{\mathrm{P}_0}(\mathcal{E}|_{t=0}).$$

Recall the family of divisors $\mathbf{C}\subset\mathbf{A}$ and take the associated family $2\mathbf{C}$ consisting on a family of double curves supported on $\mathbf{C}$. This comes naturally equipped with the closed immersion $j:\mathbf{C}\hookrightarrow 2\mathbf{C}$. Let E_0 be a vector bundle with $\det(E_0)\cong\mathcal{O}_C(dx_0)$. Recalling that $\mathbf{C}\cong C\times\mathbb{P}^1$, consider its pull-back q^*E_0 over $\mathbf{C}$ where $q:\mathbf{C}\to C$. One can consider the sheaf

$$\mathcal{E}_0:=j_*(q^*E_0),$$

which is a torsion free sheaf of rank 1 over $2\mathbf{C}$. Hence,

$$\det{}_{\mathbf{A}}(\mathcal{E}_0|_t)\cong\mathcal{O}_{\mathbf{A}_t}(2\mathbf{C}_t),$$

for every $t\in\mathbb{P}^1$. Furthermore, after Lemma (7.2), its dual determinant is

$$\det{}_{\widehat{A}}(\Phi^{\mathcal{P}}(\mathcal{E}_0|_t))\cong\det{}_{\widehat{A}}\Phi^{P_C}(E_0)\cong\mathcal{O}_{\widehat{A}}\left(n\widehat{C}\right), \tag{7.7}$$

for all $t\in\mathbb{P}^1$. Observe that at $t=0$ we have made use of Lemma 7.1.

Using the integral functor (7.6) and the determinant, we define,

$$\begin{array}{rccc} \alpha_{\mathcal{E}_0}: & \mathbf{M}^{\mathrm{H}}_{\mathbf{A}/\mathbb{A}^1}(0,n\mathbf{C},d-n) & \longrightarrow & A\times\widehat{A}\times\mathbb{P}^1 \\ & \mathcal{E} & \longmapsto & (\det_{\widehat{A}}(\Phi^{\mathcal{P}}(\mathcal{E}))\otimes\det_{\widehat{A}}(\Phi^{\mathcal{P}}(\mathcal{E}_0))^{-1},\det_{\mathbf{A}}(\mathcal{E})\otimes\det_{\mathbf{A}}(\mathcal{E}_0)^{-1}). \end{array}$$

One quickly observes that, at the generic fibres over $t\neq 0$, the restriction of the above morphism $\alpha_{\mathcal{E}_0}|_{t\neq 0}$ coincides with the Albanese morphism (3.2). Let us study the restriction to the central fibre over $t=0$.

PROPOSITION 7.3. *The restriction of the Albanese morphism, described in* (3.2)*, to the central fibre $\mathcal{M}_C(C,d)$ of the Donagi–Ein–Lazarsfeld degeneration $\mathbf{M}^{\mathbf{H}}_{\mathbf{A}/\mathbb{A}^1}(0,n\mathbf{C},d-n))\to\mathbb{A}^1$, becomes the determinant morphism of the underlying vector bundle,*

$$\begin{array}{rccc} \alpha_{\mathcal{E}_0}|_{t=0}: & \mathcal{M}_C(n,d) & \longrightarrow & A \\ & (E,\varphi) & \longmapsto & \det(E). \end{array}$$

PROOF. The proof follows from the construction of $\alpha_{\mathcal{E}_0}$, Lemma 7.2 and (6.9), which says that all curves in the Hitchin base B_n (understood as a subset of the linear system $|n\sigma_0|$) have determinant equal to $\mathcal{O}_{\mathbb{K}}(nC)$. □

REMARK 7.4. We observe from that the Albanese map at the central fibre do not impose any constraint in the set of spectral curves.

In view of Remark 7.4, in order to construct a degeneration of the reduced Beauville–Mukai system, we need to consider a closed subset of the kernel of the Albanese map $\alpha_{\mathcal{E}_0}^{-1}(\mathbf{1}_A, \mathcal{O}_{\mathbf{A}})$.

After (3.3) and the above discussion, we observe that $\alpha_{\mathcal{E}_0}^{-1}(\mathbf{1}_A, \mathcal{O}_{\mathbf{A}}) \to \mathbb{P}^1$ trivializes over $\mathbb{P}^1 - \{0\}$, and observe that our generic fibre is $\mathrm{N}_A^{\mathrm{H}}(0, nC, d-n)$. After Lemma 7.2 and (7.7), one has that the central fibre is

$$\alpha_{\mathcal{E}_0}^{-1}(\mathbf{1}_A, \mathcal{O}_{\mathbf{A}})|_{t=0} \cong \mathcal{M}_C(n,d)^{\det = \mathcal{O}_C(dx_0)}, \tag{7.8}$$

the locus of the Higgs moduli space of fixed determinant, but no restriction is imposed in the trace of the Higgs field.

Consider $\pi_* \mathcal{O}_{\overline{\mathbf{A}}}(n\mathbf{C})$ over $\mathbb{P}^1$, which is a trivial vector bundle over $\mathbb{P}^1 - \{0\}$ of rank n^2 after (7.1). By upper semicontinuity of cohomology, the coherent sheaf $\pi_* \mathcal{O}_{\overline{\mathbf{A}}}(2\mathbf{C})$ has torsion only at $t = 0$, killing the torsion, one may take the rank n^2 vector bundle $\mathbf{V} \subset \pi_* \mathcal{O}_{\overline{\mathbf{A}}}(n\mathbf{C})$ over $\mathbb{P}^1$. Projectivizing $\mathbf{V} \to \mathbb{P}^1$, one gets the $\mathbb{P}^{(n^2-1)}$-bundle $\overline{\mathcal{V}} \to \mathbb{P}^1$. The points of $\overline{\mathcal{V}}|_t$ represent $\mathbb{C}^*$-rays of sections of $\mathcal{O}_{\overline{\mathbf{A}}_t}(\mathbf{C}_t)$, which are completely determined by their vanishing locus. This provides the inclusion

$$\overline{\mathcal{V}} \subset \mathrm{Hilb}_{\overline{\mathbf{A}}/\mathbb{P}^1}([n\mathbf{C}]),$$

and note that generic fibres over $t \neq 0$ are identified with the linear system $|nC|$ in A. We may consider the restriction to those curves contained in $\mathbf{A} \subset \overline{\mathbf{A}}$,

$$\mathcal{V} := \overline{\mathcal{V}} \cap \mathrm{Hilb}_{\mathbf{A}/\mathbb{P}^1}([n\mathbf{C}]).$$

We now describe the central fibre of $\mathcal{V}$.

LEMMA 7.5. *Under the identification* (4.1) *and* (4.4), *one has that*

$$\mathcal{V}|_{t=0} \cong \bigoplus_{i=2}^{n} H^0(C, K^i), \tag{7.9}$$

coinciding with the base of the reduced Hitchin system (5.5).

PROOF. We recall from (4.1) and (4.3) that the central fibre of $\mathrm{Hilb}_{\mathbf{A}/\mathbb{P}^1}([n\mathbf{C}])$ corresponds to the space of linear deformations of the multiple curve nC inside $\mathbb{K}$,

$$\mathrm{Hilb}_{\mathbf{A}/\mathbb{P}^1}([n\mathbf{C}])|_{t=0} \cong |n\sigma_0|_{\mathrm{supp} \cap \sigma_\infty = \emptyset} = H^0(n\sigma_0, \mathcal{O}_{\mathbb{K}}(n\sigma_0)|_{n\sigma_0}).$$

Note that, since the normal bundle of C inside A coincides with the normal bundle of $\sigma_0 \cong C$ inside $\mathbb{K}$, both are described by the same ribbon that we shall denote by nC. Hence, we can identify the linear deformations for each of the fibres of $n\mathbf{C}$. The subspace $\mathcal{V}|_{t=0} \subset \mathrm{Hilb}_{\mathbf{A}/\mathbb{P}^1}([n\mathbf{C}])|_{t=0}$ is given by linear deformations of the curve induced by those that preserve the determinant within A, that is, those linear deformations describing an closed subset of the linear system $|nC|$ in A, the fibre of the map (6.8). By (7.1), one has that the dimension of $\mathcal{V}|_{t=0}$ is $n^2 - 1$.

The linear deformations of nC in A are classified by the sections of $\mathcal{O}_C(nC) \otimes (\mathcal{O}_C \oplus (\mathcal{I}_C/\mathcal{I}_C^2) \oplus \cdots \oplus (\mathcal{I}_C^{n-1}/\mathcal{I}_C^n))$, where $(\mathcal{I}_C/\mathcal{I}_C^2) \cong \mathcal{O}_C(-C) \cong K^{-1}$ as we lie on an abelian surface. Let us denote by W the subspace of linear deformations classified by sections of $\mathcal{O}_C(nC) \otimes (\mathcal{I}_C^{n-1}/\mathcal{I}_C^n) \cong K$. The elements of W vanish on the open subset $(n-1)C$ so they correspond to those deformations deforming the multiple curve nC into a multiple curve nC' with support C' different from C. All these linear deformations should contain those given by the differentials of $t_a : A \to A$, defined using the group structure. The former are classified by

$T_a A \cong H^1(A, \mathcal{O}_A) \cong H^0(C, K)$, whose dimension coincides with that of W so both are equal. We then know that no deformation classified by W preserves the determinant $\mathcal{O}_C(nC)$. Note that one can project the space of all deformations, $\mathrm{Hilb}_{\mathbf{A}/\mathbb{P}^1}([n\mathbf{C}])|_{t=0}$, into our subspace W. Under this projection, the subspace $\mathcal{V}|_{t=0}$ is sent to 0 as these deformations preserve the determinant. Hence, $\mathcal{V}|_{t=0}$ is a subspace of $\bigoplus_{i=2}^n H^0(C, K^i)$, and both coincide as they have the same dimension. □

According with (7.8) and Lemma 7.5, we define

$$\mathbf{N}^{\mathrm{H}}_{\mathbf{S}/\mathbb{P}^1}(0, n\mathbf{C}, d-n) := \alpha^{-1}_{\mathcal{E}_0}(\mathbf{1}_A, \mathcal{O}_{\mathbf{A}})|_{\mathcal{V}},$$

and, recalling the symplectic form Ω defined in Theorem 6.3, set

$$\bar{\Omega} := \Omega|_{\mathbf{N}}.$$

We observe that $\bar{\Omega}$ provides, fibrewise, a relative symplectic structure on $\mathbf{N}^{\mathrm{H}}_{\mathbf{S}/\mathbb{P}^1}(0, n\mathbf{C}, d-n)$. This leads to the main result of this section.

THEOREM 7.6. *Given a smooth projective curve C of genus $g = 2$ and consider its Jacobian $A = Jac^0(C)$. One can construct $\mathbf{N}^{\mathrm{H}}_{\mathbf{S}/\mathbb{P}^1}(0, n\mathbf{C}, d-n)$ and $\mathcal{V}$ flat over $\mathbb{P}^1$, both fitting in the commuting diagram*

$$\begin{array}{ccc} \mathbf{N}^{\mathrm{H}}_{\mathbf{S}/\mathbb{A}^1}(0, n\mathbf{C}, d-n) & \longrightarrow & \mathcal{V} \\ & \searrow \quad \swarrow & \\ & \mathbb{P}^1. & \end{array} \tag{7.10}$$

The horizontal arrow is a surjective fibration, trivial over $\mathbb{P}^1 - \{0\}$, and the generic fibres of $\mathbf{N}^{\mathrm{H}}_{\mathbf{S}/\mathbb{P}^1}(0, n\mathbf{C}, d-n)$ coincide with the reduced Beauville–Mukai system

$$\mathrm{N}^{\mathrm{H}}_A(0, nC, d-n) \to |nC|,$$

while the central fibre at $t = 0$ gives the SL_n-Hitchin system (5.5),

$$\mathcal{N}_C(n, d) \to \bigoplus_{i=2}^{n} H^0(C, K^i).$$

Furthermore, one can construct a relative symplectic form $\bar{\Omega}$ on $\mathbf{N}^{\mathrm{H}}_{\mathbf{S}/\mathbb{A}^1}(0, n\mathbf{C}, d-n) \to \mathbb{A}^1$, which coincides (up to scalling) with the reduced Mukai's form $\overline{\Omega}$ on the generic fibres and, over the central fibre, with the standard reduced symplectic form on the moduli space of Higgs bundles $\overline{\Omega}_0$.

PROOF. Everything follows easily from Theorems 6.1 and 6.3, and Lemmas 7.1, 7.2 and 7.5. □

Recall that, in Section 3, we denoted $\mathrm{N}_{\mathrm{six}} = \mathrm{N}^{\mathrm{H}}_A(0, 2C, -2)$, as the symplectic resolution of this moduli space is the six dimensional O'Grady space $\widetilde{\mathrm{N}}_{\mathrm{six}}$, what explains the subindex used in the notation. We shall use the same subindex to denote $\mathbf{N}_{\mathrm{six}} := \mathbf{N}^{\mathrm{H}}_{\mathbf{S}/\mathbb{A}^1}(0, 2\mathbf{C}, -2)$ and $\mathcal{V}_{\mathrm{six}}$.

COROLLARY 7.7. *Given a smooth projective genus* 2 *curve* C *and taking the abelian surface* $A = Jac^0(C)$, *one can construct* $\mathrm{N}_{\mathrm{six}}$ *and* $\mathcal{V}_{\mathrm{six}}$ *flat over* $\mathbb{P}^1$, *both fitting in the commuting diagram*

$$\begin{array}{ccc} \mathbf{N}_{\mathrm{six}} & \longrightarrow & \mathcal{V}_{\mathrm{six}} \\ & \searrow \quad \swarrow & \\ & \mathbb{A}^1. & \end{array}$$

The horizontal arrow is a surjective fibration, trivial over $\mathbb{A}^1 - \{0\}$, *whose generic fibres coincide with the reduced Beauville–Mukai system* (2.3),

$$\mathrm{N}_{\mathrm{six}} \to |2C|,$$

and the central fibre at $t = 0$ *gives the* SL_2*-Hitchin system* (5.1),

$$\mathcal{N}_C(2,0) \to H^0(C, K^2).$$

Furthermore, one can construct a relative symplectic form Ω_{six} *on* $\mathbf{N}_{\mathrm{six}} \to \mathbb{A}^1$, *which coincides (up to scalling) with the reduced Mukai's form* $\overline{\Omega}$ *on the generic fibres and, over the central fibre, with the standard reduced symplectic form on the moduli space of Higgs bundles* $\overline{\Omega}_0$.

8. Degenerating O'Grady spaces into symplectic resolutions of moduli spaces of Higgs bundles

In this last section we address the construction of a non-linear deformation of O'Grady ten and six dimensional spaces into the symplectic resolution of, respectively, the Higgs moduli spaces $\mathrm{M}_C(2,0)$ and $\mathrm{N}_C(2,0)$ where the genus of the curve is $g = 2$. As described in the Appendix of [**FM**], and as can be infered from Theorems 3.1, 3.3 and 5.1, such degeneration is obtained by blowing up the singular locus relative to the projection to $\mathbb{P}^1$.

Then, we shall first identify the singular locus of the fibres of our degenerating families, $\mathbf{M}_{\mathrm{ten}}$ and $\mathbf{N}_{\mathrm{six}}$, in order to perform a relative blow-up over these loci, giving rise to a relative symplectic resolution of singularities.

Pick a K3 surface X containing a genus 2 curve and consider the abelian surface A obtained as the Jacobian of the genus 2 curve C. We consider $\mathbf{M}^{\mathbf{H}}_{\mathbf{X}/\mathbb{P}^1}(0, \mathbf{C}, -1)$ and $\mathbf{M}^{\mathbf{H}}_{\mathbf{A}/\mathbb{P}^1}(0, \mathbf{C}, -1)$. Then, observe that the strictly semistable loci of the fibres of $\mathbf{M}_{\mathrm{ten}} \to \mathbb{P}^1$ and $\mathbf{N}_{\mathrm{six}} \to \mathbb{P}^1$ is given by

$$\mathbf{M}^{\mathrm{str.ps}}_{\mathrm{ten}} := \left\{ \mathcal{E}_1 \oplus \mathcal{E}_2 \text{ where } \mathcal{E}_i \in \mathbf{M}^{\mathbf{H}}_{\mathbf{X}/\mathbb{P}^1}(0, \mathbf{C}, -1) \right\}$$

and

$$\mathbf{N}^{\mathrm{str.ps}}_{\mathrm{six}} := \left\{ \mathcal{E}_1 \oplus \mathcal{E}_2 \text{ where } \mathcal{E}_i \in \mathbf{M}^{\mathbf{H}}_{\mathbf{A}/\mathbb{P}^1}(0, \mathbf{C}, -1) \right\} \cap \mathbf{N}_{\mathrm{six}}.$$

Finally, let us consider

$$\widetilde{\mathbf{M}}_{\mathrm{ten}} := Blow_{\mathbf{M}^{\mathrm{str.ps}}_{\mathrm{ten}}}(\mathbf{M}_{\mathrm{ten}})$$

and

$$\widetilde{\mathbf{N}}_{\mathrm{six}} := Blow_{\mathbf{N}^{\mathrm{str.ps}}_{\mathrm{six}}}(\mathbf{N}_{\mathrm{six}}),$$

and take the composition of the blow-up projections $\widetilde{\mathbf{M}}_{\mathrm{ten}} \to \mathbf{M}_{\mathrm{ten}}$ and $\widetilde{\mathbf{N}}_{\mathrm{six}} \to \mathbf{N}_{\mathrm{six}}$ with the structural morphisms $\mathbf{M}_{\mathrm{ten}} \to \mathbb{P}^1$ and $\mathbf{N}_{\mathrm{six}} \to \mathbb{P}^1$.

We finally arrive to the main result of these notes. Note that the construction of these degenerations has been described in the Appendix of [**FM**], although the degeneration of the holomorphic symplecic structures is not addressed there. This analysis appears in the theorem below, by means of Theorem 6.3 taken from [**Fr**].

THEOREM 8.1. *Let C be a genus 2 curve lying inside a K3 surface S. Then, the relative scheme $\widetilde{\mathbf{M}}_{\mathrm{ten}} \to \mathbb{P}^1$ is a non-linear deformation of symplectic varieties, from the ten dimensional O'Grady space $\widetilde{\mathrm{M}}_{\mathrm{ten}}$ into the symplectic resolution $\widetilde{\mathcal{M}}_C(2,0)$ of the Higgs moduli space $\mathcal{M}_C(2,0)$.*

Also, the relative scheme $\widetilde{\mathbf{N}}_{\mathrm{six}} \to \mathbb{P}^1$ is a non-linear degeneration of symplectic varieties, from the six dimensional O'Grady space $\widetilde{\mathrm{N}}_{\mathrm{six}}$ into the symplectic resolution $\widetilde{\mathcal{N}}_C(2,0)$ of the moduli space $\mathcal{N}_C(2,0)$ of Higgs bundles with 0 trace and trivial determinant.

PROOF. Since both $\widetilde{\mathbf{M}}_{\mathrm{ten}}$ and $\widetilde{\mathbf{N}}_{\mathrm{six}}$ are constructed by blowing-up the strictly polystable loci of, respectively, $\mathbf{M}_{\mathrm{ten}}$ and $\mathbf{N}_{\mathrm{six}}$, the proof follows after Theorems 3.1, 3.3 and 5.1, and Corollaries 6.4 and 7.7. We observe that the degeneration do preserve the singularities. □

Acknowledgments

The author wants to thank Peter Newstead for all his teachings and for sharing his passion about mathematics.

References

[Be1] A. Beauville, *Variétés Kähleriennes dont la première classe de Chern est nulle* (French), J. Differential Geom. **18** (1983), no. 4, 755–782 (1984). MR730926

[Be2] A. Beauville, *Systèmes hamiltoniens complètement intégrables associés aux surfaces K3* (French), Problems in the theory of surfaces and their classification (Cortona, 1988), Sympos. Math., XXXII, Academic Press, London, 1991, pp. 25–31. MR1273370

[Be3] A. Beauville, *Symplectic singularities*, Invent. Math. **139** (2000), no. 3, 541–549, DOI 10.1007/s002229900043. MR1738060

[Be4] A. Beauville, *Holomorphic symplectic geometry: a problem list*, Complex and differential geometry, Springer Proc. Math., vol. 8, Springer, Heidelberg, 2011, pp. 49–63, DOI 10.1007/978-3-642-20300-8_2. MR2964467

[BNR] A. Beauville, M. S. Narasimhan, and S. Ramanan, *Spectral curves and the generalised theta divisor*, J. Reine Angew. Math. **398** (1989), 169–179, DOI 10.1515/crll.1989.398.169. MR998478

[BS] G. Bellamy and T. Schedler, *On the (non)existence of symplectic resolutions of linear quotients*, Math. Res. Lett. **23** (2016), no. 6, 1537–1564, DOI 10.4310/MRL.2016.v23.n6.a1. MR3621098

[BL] C. Birkenhake and H. Lange, *Complex abelian varieties*, 2nd ed., Grundlehren der mathematischen Wissenschaften [Fundamental Principles of Mathematical Sciences], vol. 302, Springer-Verlag, Berlin, 2004, DOI 10.1007/978-3-662-06307-1. MR2062673

[Bo1] F. Bottacin, *Poisson structures on moduli spaces of sheaves over Poisson surfaces*, Invent. Math. **121** (1995), no. 2, 421–436, DOI 10.1007/BF01884307. MR1346215

[Bo2] F. Bottacin, *Symplectic geometry on moduli spaces of stable pairs*, Ann. Sci. École Norm. Sup. (4) **28** (1995), no. 4, 391–433. MR1334607

[CMS1] M. A. de Cataldo, D. Maulik, and J. Shen, *Hitchin fibrations, abelian surfaces, and the $P = W$ conjecture*, J. Amer. Math. Soc. **35** (2022), no. 3, 911–953, DOI 10.1090/jams/989. MR4433080

[CMS2] M. A. de Cataldo, D. Maulik, and J. Shen, *On the $P = W$ conjecture for SL_n*, Selecta Math. (N.S.) **28** (2022), no. 5, Paper No. 90, 21, DOI 10.1007/s00029-022-00803-0. MR4496523

[Co] K. Corlette, *Flat G-bundles with canonical metrics*, J. Differential Geom. **28** (1988), no. 3, 361–382. MR965220

[DEL] R. Donagi, L. Ein, and R. Lazarsfeld, *Nilpotent cones and sheaves on $K3$ surfaces*, Birational algebraic geometry (Baltimore, MD, 1996), Contemp. Math., vol. 207, Amer. Math. Soc., Providence, RI, 1997, pp. 51–61, DOI 10.1090/conm/207/02719. MR1462924

[Do] S. K. Donaldson, *Twisted harmonic maps and the self-duality equations*, Proc. London Math. Soc. (3) **55** (1987), no. 1, 127–131, DOI 10.1112/plms/s3-55.1.127. MR887285

[DH] Y. Dutta and D. Huybrechts, *Maximal variation of curves on K3 surfaces*, Tunis. J. Math. **4** (2022), no. 3, 443–464, DOI 10.2140/tunis.2022.4.443. MR4507328

[FM] C. Felisetti and M. Mauri, P = W *conjectures for character varieties with symplectic resolution* (English, with English and French summaries), J. Éc. polytech. Math. **9** (2022), 853–905, DOI 10.5802/jep.196. MR4436035

[Fr] E. Franco, *Degeneration of natural Lagrangians and Prymian integrable systems*, Math. Z. **302** (2022), no. 4, 2469–2511, DOI 10.1007/s00209-022-03135-z. MR4510173

[Gi] D. Gieseker, *On the moduli of vector bundles on an algebraic surface*, Ann. of Math. (2) **106** (1977), no. 1, 45–60, DOI 10.2307/1971157. MR466475

[Ha] R. Hartshorne, *Algebraic geometry*, Graduate Texts in Mathematics, No. 52, Springer-Verlag, New York-Heidelberg, 1977. MR463157

[Hi1] N. Hitchin, *Stable bundles and integrable systems*, Duke Math. J. **54** (1987), no. 1, 91–114, DOI 10.1215/S0012-7094-87-05408-1. MR885778

[Hi2] N. J. Hitchin, *The self-duality equations on a Riemann surface*, Proc. London Math. Soc. (3) **55** (1987), no. 1, 59–126, DOI 10.1112/plms/s3-55.1.59. MR887284

[HL] D. Huybrechts and M. Lehn, *The geometry of moduli spaces of sheaves*, 2nd ed., Cambridge Mathematical Library, Cambridge University Press, Cambridge, 2010, DOI 10.1017/CBO9780511711985. MR2665168

[KLS] D. Kaledin, M. Lehn, and Ch. Sorger, *Singular symplectic moduli spaces*, Invent. Math. **164** (2006), no. 3, 591–614, DOI 10.1007/s00222-005-0484-6. MR2221132

[KY] Y.-H. Kiem and S.-B. Yoo, *The stringy E-function of the moduli space of Higgs bundles with trivial determinant*, Math. Nachr. **281** (2008), no. 6, 817–838, DOI 10.1002/mana.200610643. MR2418849

[Ki] Y.-H. Kiem, *On the existence of a symplectic desingularization of some moduli spaces of sheaves on a $K3$ surface*, Compos. Math. **141** (2005), no. 4, 902–906, DOI 10.1112/S0010437X05001272. MR2148196

[Kw] F. C. Kirwan, *Partial desingularisations of quotients of nonsingular varieties and their Betti numbers*, Ann. of Math. (2) **122** (1985), no. 1, 41–85, DOI 10.2307/1971369. MR799252

[LS] M. Lehn and C. Sorger, *La singularité de O'Grady* (French, with English and French summaries), J. Algebraic Geom. **15** (2006), no. 4, 753–770, DOI 10.1090/S1056-3911-06-00437-1. MR2237269

[Ma] D. Matsushita, *Addendum: "On fibre space structures of a projective irreducible symplectic manifold" [Topology* **38** *(1999), no. 1, 79–83; MR1644091 (99f:14054)]*, Topology **40** (2001), no. 2, 431–432, DOI 10.1016/S0040-9383(99)00048-8. MR1808227

[Mr] E. Markman, *Spectral curves and integrable systems*, Compositio Math. **93** (1994), no. 3, 255–290. MR1300764

[Mu1] S. Mukai, *Duality between $D(X)$ and $D(\hat{X})$ with its application to Picard sheaves*, Nagoya Math. J. **81** (1981), 153–175. MR607081

[Mu2] S. Mukai, *Symplectic structure of the moduli space of sheaves on an abelian or $K3$ surface*, Invent. Math. **77** (1984), no. 1, 101–116, DOI 10.1007/BF01389137. MR751133

[Ni] N. Nitsure, *Moduli space of semistable pairs on a curve*, Proc. London Math. Soc. (3) **62** (1991), no. 2, 275–300, DOI 10.1112/plms/s3-62.2.275. MR1085642

[OG1] K. G. O'Grady, *Desingularized moduli spaces of sheaves on a $K3$*, J. Reine Angew. Math. **512** (1999), 49–117, DOI 10.1515/crll.1999.056. MR1703077

[OG2] K. G. O'Grady, *A new six-dimensional irreducible symplectic variety*, J. Algebraic Geom. **12** (2003), no. 3, 435–505, DOI 10.1090/S1056-3911-03-00323-0. MR1966024

[PR] A. Perego and A. Rapagnetta, *Deformation of the O'Grady moduli spaces*, J. Reine Angew. Math. **678** (2013), 1–34, DOI 10.1515/CRELLE.2011.191. MR3056101

[Ra] A. Rapagnetta, *Topological invariants of O'Grady's six dimensional irreducible symplectic variety*, Math. Z. **256** (2007), no. 1, 1–34, DOI 10.1007/s00209-006-0022-2. MR2282256

[Sb] D. Schaub, *Courbes spectrales et compactifications de jacobiennes* (French), Math. Z. **227** (1998), no. 2, 295–312, DOI 10.1007/PL00004377. MR1609069

[Sw] J. Sawon, *Lagrangian fibrations by Prym varieties*, Mat. Contemp. **47** (2020), 182–227. MR4191140

[SS] J. Sawon and C. Shen, *Deformations of compact Prym fibrations to Hitchin systems*, Bull. Lond. Math. Soc. **54** (2022), no. 5, 1568–1583. MR4499591

[Si1] C. T. Simpson, *Moduli of representations of the fundamental group of a smooth projective variety. I*, Inst. Hautes Études Sci. Publ. Math. **79** (1994), 47–129. MR1307297

[Si2] C. T. Simpson, *Moduli of representations of the fundamental group of a smooth projective variety. II*, Inst. Hautes Études Sci. Publ. Math. **80** (1994), 5–79 (1995). MR1320603

[Ti] A. Tirelli, *Symplectic resolutions for Higgs moduli spaces*, Proc. Amer. Math. Soc. **147** (2019), no. 4, 1399–1412, DOI 10.1090/proc/14339. MR3910407

[Y1] S. T. Yau, *Calabi's conjecture and some new results in algebraic geometry*, Proc. Nat. Acad. Sci. U.S.A. **74** (1977), no. 5, 1798–1799, DOI 10.1073/pnas.74.5.1798. MR451180

[Y2] S. T. Yau, *On the Ricci curvature of a compact Kähler manifold and the complex Monge-Ampère equation. I*, Comm. Pure Appl. Math. **31** (1978), no. 3, 339–411, DOI 10.1002/cpa.3160310304. MR480350

[Yo] S.-B. Yoo, *A desingularization of the moduli space of rank 2 Higgs bundles over a curve*, Taiwanese J. Math. **25** (2021), no. 2, 257–301, DOI 10.11650/tjm/200901. MR4255205

[Ys1] K. Yoshioka, *Some notes on the moduli of stable sheaves on elliptic surfaces*, Nagoya Math. J. **154** (1999), 73–102, DOI 10.1017/S0027763000025319. MR1689173

[Ys2] K. Yoshioka, *Moduli spaces of stable sheaves on abelian surfaces*, Math. Ann. **321** (2001), no. 4, 817–884, DOI 10.1007/s002080100255. MR1872531

Depto. Matemáticas, Facultad de Ciencias, Universidad Autónoma de Madrid, Campus de Cantoblanco 28049, Madrid, España.

Email address: emilio.franco@uam.es

Contemporary Mathematics
Volume **803**, 2024
https://doi.org/10.1090/conm/803/16099

Vinberg pairs and Higgs bundles

Oscar García-Prada

Dedicated to Peter Newstead on the occasion of his 80th birthday.

ABSTRACT. We explore the role of Vinberg pairs defined by cyclic gradings of a semisimple complex Lie algebra in the context of Higgs bundle theory.

1. Introduction

Let G be a semisimple complex Lie group with Lie algebra $\mathfrak{g}$ and X be a compact Riemann surface of genus $g \geqslant 2$. A G-Higgs bundle over X is a pair (E, φ) consisting of a holomorphic principal G-bundle and a holomorphic section φ of $E(\mathfrak{g}) \otimes K_X$, where $E(\mathfrak{g})$ is the vector bundle associated to the adjoint representation of G in $\mathfrak{g}$, and K_X is the canonical line bundle of X. There are suitable stability criteria for these objects and one can consider the moduli space $\mathcal{M}(G)$ of polystable G-Higgs bundles over X. Higgs bundles were introduced by Hitchin in [**39**, **40**], and their moduli spaces have been the object of intense research for more than 35 years.

The moduli space of G-Higgs bundles has an extremely rich geometry. Its gauge-theoretic interpretation as the moduli space of solutions to Hitchin's self-duality equations endows the smooth locus of the moduli space $\mathcal{M}(G)$ with a hyperkähler structure. The non-abelian Hodge correspondence, proved combining the work of Hitchin with work by Corlette [**16**], Donaldson [**18**] and Simpson [**72**], identifies topologically $\mathcal{M}(G)$ with the G-character variety of the fundamental group of X. Moreover, one has the Hitchin map from $\mathcal{M}(G)$ to an affine space obtained as a global version of the map $\mathfrak{g} \to \mathfrak{g} /\!\!/ G$ and the Chevalley restriction theorem $\mathfrak{g} /\!\!/ G \cong \mathfrak{t}/W(\mathfrak{t})$, where $\mathfrak{t} \subset \mathfrak{g}$ is a Cartan subalgebra, and $W(\mathfrak{t})$ is the Weyl group. The generic fibres of this map are abelian varieties, defining an algebraically completely integrable system, known as the Hitchin system [**4**, **19**, **40**]. The Hitchin fibration plays a very important role in mirror symmetry and Langlands duality [**44**], and has been a main ingredient in the proof of the Fundamental Lemma by Ngô [**55**].

Having introduced Higgs bundles, we introduce now Vinberg θ-pairs. To do this, let θ be an automorphism of G of order m for an integer $m > 0$. This defines an automorphism of order m of $\mathfrak{g}$, denoted also by θ, and hence a $\mathbb{Z}/m\mathbb{Z}$-grading

2020 *Mathematics Subject Classification.* Primary 14H60; Secondary 57R57, 58D29.

The author was partially supported by the Spanish Ministry of Science and Innovation, through the "Severo Ochoa Programme for Centres of Excellence in R&D (CEX2019-000904-S)" and grant PID2019-109339GB-C31.

of $\mathfrak{g}$, given by the decomposition

$$\mathfrak{g} = \bigoplus_{i \in \mathbb{Z}/m\mathbb{Z}} \mathfrak{g}_i \text{ with } \mathfrak{g}_i = \{x \in \mathfrak{g} \mid \theta(x) = \zeta^i x\} \text{ and } [\mathfrak{g}_i, \mathfrak{g}_j] \subset \mathfrak{g}_{i+j}, \tag{1.1}$$

where ζ is a primitive m-th root of unity. Let $G^\theta \subset G$ be the fixed-point subgroup. This is a reductive group with Lie algebra $\mathfrak{g}_0$, which acts via the adjoint representation on every $\mathfrak{g}_i$ in the decomposition given in (1.1). We refer to the pairs $(G^\theta, \mathfrak{g}_i)$ as Vinberg θ-pairs. Sometimes we also use this term for pairs in which G^θ is replaced by its connected component of the identity or its normalizer in G. These pairs are called θ-groups by Vinberg in [**76**], as well as Vinberg representations by other authors [**52–54, 58, 64, 75**]. In [**80**] Wallach calls them Vinberg pairs. Vinberg's theory is devoted to the geometric invariant theory and description of orbits for the action of G^θ on $\mathfrak{g}_i$. The theory of Vinberg θ-pairs has been extended to general fields of 0 characteristic and good positive characteristic not dividing the order of θ [**53, 54**].

Vinberg's theory is important of course in geometric invariant theory and representation theory, and has applications to many problems. These include the classification of trivectors of 9-dimensional space $\mathbb{C}^9$ by Elashvili–Vinberg [**78**] using a $\mathbb{Z}/3\mathbb{Z}$-grading of $\mathfrak{e}_8$ given by

$$\mathfrak{e}_8 = \mathfrak{sl}(9, \mathbb{C}) \oplus \Lambda^3(\mathbb{C}^9) \oplus \Lambda^6(\mathbb{C}^9).$$

Certain Vinberg θ-pairs have interesting connections to the arithmetic theory of elliptic curves and Jacobians (see Bhargava–Gross [**5**]). There is also a connection between the Vinberg θ-pair $(\mathrm{SL}(9,\mathbb{C})/\mu_3, \Lambda^3(\mathbb{C}^9))$ coming from the $\mathbb{Z}/3\mathbb{Z}$-grading of $\mathfrak{e}_8$ and the moduli space of genus 2 curves with some additional data (see Rains–Sam [**61**]). Vinberg's theory has also connections to physics related to del Pezzo surfaces: mysterious duality (see [**46**]).

The object of this paper is to explore the role of Vinberg θ-pairs in Higgs bundle theory. There are two main tasks at hand. First we want to generalize certain features of the theory of G-Higgs bundles to Vinberg θ-pairs —note that when $m = 1$, θ is the identity and the only Vinberg θ-pair that one has in this situation is simply $(G, \mathfrak{g})$. The first step for this is to observe that, in a similar way to which we have defined G-Higgs bundles over X, we can define a $(G^\theta, \mathfrak{g}_i)$-Higgs pair as a pair (E, φ) consisting of a principal G^θ-bundle E over X and a holomorphic section φ of $E(\mathfrak{g}_i) \otimes K_X$, where now $E(\mathfrak{g}_i)$ is the vector bundle associated to the adjoint representation of G^θ in $\mathfrak{g}_i$. As in the case of G-Higgs bundles, there are appropriate moduli spaces $\mathcal{M}(G^\theta, \mathfrak{g}_i)$ of polystable $(G^\theta, \mathfrak{g}_i)$-Higgs pairs. The second step in our programme is devoted to studying how these moduli spaces appear in connection with the moduli space of G-Higgs bundles.

This exploration is certainly not new. There has been extensive work in the case in which θ is an involution. In fact, this was already considered by Hitchin for $G = \mathrm{SL}(2, \mathbb{C})$ in his seminal paper [**39**]. A main motivation for this study has to do with the fact that the non-abelian Hodge correspondence can be extended to the involutive case, identifying topologically the moduli space $\mathcal{M}(G^\theta, \mathfrak{g}_1)$ with the G^σ-character variety of the fundamental group of X, where G^σ is a real form of G defined by an antiholomorphic involution σ of G obtained by composing θ with a commuting anti-holomorphic involution τ of G, corresponding to a compact real form of G. The existence of such τ is ensured by a result of Cartan [**13**]. The literature on this subject is extremely vast (look for example at the survey papers

[**22, 23**] for references). Of course $(G^\theta, \mathfrak{g}_0)$-Higgs pairs are just G^θ-Higgs bundles (note that, although we have defined G-Higgs bundles when G is semisimple, the theory can be easily extended to the case in which G is reductive). Both the moduli space of $(G^\theta, \mathfrak{g}_0)$-Higgs pairs and $(G^\theta, \mathfrak{g}_1)$-Higgs pairs appear inside the moduli space of G-Higgs bundles as fixed points of certain involutions of $\mathcal{M}(G)$ (see [**31**]). There are finite maps $\mathcal{M}(G^\theta, \mathfrak{g}_0) \to \mathcal{M}(G)$ and $\mathcal{M}(G^\theta, \mathfrak{g}_1) \to \mathcal{M}(G)$, whose images define a hyperkähler subvariety and a Lagrangian subvariety of $\mathcal{M}(G)$, respectively. Here the Lagrangian condition is with respect to the natural holomorphic symplectic structure of $\mathcal{M}(G)$ determined by the complex structure of X.

Another important feature that has also been pursued in the involutive case is the generalization of the Hitchin fibration [**28, 29, 59, 66, 67**]. Indeed, there is a generalization of the Hitchin map for the moduli space $\mathcal{M}(G^\theta, \mathfrak{g}_1)$. Here one exploits a version of the Chevalley restriction theorem for symmetric pairs proved by Kostant–Rallis [**49**], which establishes an isomorphism $\mathfrak{g}_1 /\!\!/ G^\theta \cong \mathfrak{a}/W(\mathfrak{a})$, where $\mathfrak{a}$ is a maximal abelian subalgebra of $\mathfrak{g}_1$ and $W(\mathfrak{a})$ is the little Weyl group. The work of Peón-Nieto [**59**] and [**28**] follows the cameral approach of Donagi–Gaitsgory [**19**] as well as that of Ngô [**55**] for the case of G-Higgs bundles, while the work of Schaposnik [**66, 67**] for the classical groups follows the spectral curve approach initiated by Hitchin [**40**]. From both points of view one can see that, in general, the generic fibre of the Hitchin fibration is no longer abelian, except when the real form G^σ corresponding to θ is quasi-split. This includes the split real forms (which exist for every semisimple G) and real forms whose Lie algebra is isomorphic to $\mathfrak{su}(n, n)$, $\mathfrak{su}(n, n+1)$, $\mathfrak{so}(n, n+2)$ and $\mathfrak{e}_6(2)$. From the spectral curve perspective this is well-illustrated in [**42**]. Further work on the Hitchin fibration in the symmetric pair case has been done by Leslie [**51**], where the motivation here is related to a version of the Fundamental Lemma for symmetric pairs, and also more recently by Hameister–Morrissey [**37**].

Our mission in this paper is two-fold. On the one hand we review several previous results put in the larger context of Vinberg's theory of θ-pairs, and on the other, we report on ongoing joint work with Miguel González, where we address some general issues regarding the geometry of moduli spaces of cyclic Higgs bundles associated to general Vinberg θ-pairs, and in particular the role of the Toledo invariant in this context [**24**].

In Section 2 we first recall the notion of $\mathbb{Z}/m\mathbb{Z}$-grading of a semisimple complex Lie algebra $\mathfrak{g}$, and give some examples. In addition to the case of involutions, we also describe for $G = \mathrm{SL}(n, \mathbb{C})$, the case defined by a cyclic quiver. We then define the notion of Vinberg θ-pairs, associated to a $\mathbb{Z}/m\mathbb{Z}$-grading determined by an order m automorphism of a semisimple complex Lie group G, and describe some of the main ingredientes of Vinberg's theory. A crucial one is a Chevalley type restriction theorem for the action of G^θ on any $\mathfrak{g}_i$ appearing in the $\mathbb{Z}/m\mathbb{Z}$-grading of $\mathfrak{g}$, for which there is a Cartan subspace and a little Weyl group, generalizing the original Chevalley theorem for the action of G on $\mathfrak{g}$ and the symmetric pair case proved by Kostant–Rallis. Of course this is the key for the definition below of a Hitchin map for the moduli spaces of $(G^\theta, \mathfrak{g}_i)$-Higgs pairs. We also review some facts studied in [**31**] about the parametrization of finite order automorphisms of G up to conjugation by inner automorphisms, and the relation to real forms of G in the involutive case. We finish the section recalling the notion of quasi-split real form and proposing a possible notion of quasi-splitness in the higher order case.

In Section 3 we define Higgs bundles over a compact Riemann surface associated to Vinberg θ-pairs $(G^\theta, \mathfrak{g}_i)$, defined by an order m automorphism θ of a semisimple complex Lie group G. A particular case are G-Higgs bundles obtained when $m = 1$, and hence θ is the identity automorphism. After recalling the appropriate stability criteria we define the moduli spaces of $(G^\theta, \mathfrak{g}_i)$-Higgs pairs. We then review work in [**31**] showing that there is a finite map from the moduli space $\mathcal{M}(G^\theta, \mathfrak{g}_i)$ of $(G^\theta, \mathfrak{g}_i)$-Higgs pairs to the moduli space $\mathcal{M}(G)$ of G-Higgs bundles, whose image is in the fixed point subvariety for an action on $\mathcal{M}(G)$ depending on θ and $\mathfrak{g}_i$ of a cyclic group of order m. The G-Higgs bundles in the image of this map are known as cyclic Higgs bundles, and have been extensively studied (see for example [**3, 15, 17, 21, 31, 50, 74**]). We finish the section recalling the non-abelian Hodge correspondence and explaining for which moduli spaces of $(G^\theta, \mathfrak{g}_i)$-Higgs pairs one may have such correspondence.

In Section 4 we start with a $\mathbb{Z}$-grading $\mathfrak{g} = \oplus_{i\in\mathbb{Z}}\mathfrak{g}_i$ of $\mathfrak{g}$, and consider pairs $(G_0, \mathfrak{g}_i)$, where G_0 is the connected group corresponding to $\mathfrak{g}_0$. These pairs, referred some times as Vinberg $\mathbb{C}^*$-pairs, play an essential role in the study of fixed points in $\mathcal{M}(G)$ under the $\mathbb{C}^*$-action — Hodge bundles. A crucial property proved by Vinberg [**77**] is that $\mathfrak{g}_i$ is a prehomogeneous vector space for the action of G_0 for every $i \neq 0$. We review the main ingredients of a theory recently introduced in [**6**], where a Toledo invariant is associated to a Hodge bundle and a bound, similar to the one given by the Milnor–Wood inequality for the Toledo invariant of a Higgs bundle for a real form of Hermitian type, is proved. We then consider a canonical $\mathbb{Z}/m\mathbb{Z}$-grading associated to a $\mathbb{Z}$-grading of $\mathfrak{g}$ and extend the inequality obtained for Hodge bundles in [**6**] to the cyclic Higgs bundles for this $\mathbb{Z}/m\mathbb{Z}$-grading.

Finally, in Section 5, we define the Hitchin map for the moduli space of $(G^\theta, \mathfrak{g}_1)$-Higgs pairs where θ is an automorphism of G of order $m > 0$. We briefly mention how one recovers for $m = 1$ the original Hitchin map for the moduli space $\mathcal{M}(G)$ of G-Higgs bundles and the Hitchin map in the symmetric pair case for $m = 2$. We also comment on some of the main difficulties to carry out the study for arbitrary m [**24**].

2. Graded Lie algebras and Vinberg θ-pairs

2.1. $\mathbb{Z}/m\mathbb{Z}$-gradings on Lie algebras. Let $m > 0$ be an integer, and let $\mathfrak{g}$ be a semisimple complex Lie algebra. A **$\mathbb{Z}/m\mathbb{Z}$-grading** on $\mathfrak{g}$ is a decomposition

$$\mathfrak{g} = \bigoplus_{i\in\mathbb{Z}/m\mathbb{Z}} \mathfrak{g}_i \text{ so that } [\mathfrak{g}_i, \mathfrak{g}_j] \subset \mathfrak{g}_{i+j}.$$

Let $\mu_m = \{z \in \mathbb{C}^* \text{ such that } z^m = 1\}$. Having a $\mathbb{Z}/m\mathbb{Z}$-grading on $\mathfrak{g}$ is equivalent to having a homomorphism $\tilde{\theta} : \mu_m \to \operatorname{Aut}(\mathfrak{g})$, where $\mathfrak{g}_i = \{x \in \mathfrak{g} \text{ such that } \tilde{\theta}(z)x = z^i x\}$ for every $z \in \mu_m$.

One can then choose a generator ζ of μ_m and define an element $\theta \in \operatorname{Aut}(\mathfrak{g})$ of order m given by

$$\theta(x) = \zeta^i x \text{ for every } x \in \mathfrak{g}_i.$$

Conversely, an element $\theta \in \operatorname{Aut}(\mathfrak{g})$ of order m, defines a $\mathbb{Z}/m\mathbb{Z}$-grading on $\mathfrak{g}$ with

$$\mathfrak{g}_i = \{x \in \mathfrak{g} \text{ such that } \theta(x) = \zeta^i x\}.$$

Let G be a semisimple complex algebraic group with Lie algebra $\mathfrak{g}$. We will very often assume that the $\mathbb{Z}/m\mathbb{Z}$-grading on $\mathfrak{g}$ is induced from an element of $\operatorname{Aut}(G)$

of order m, denoted also by θ. This is, of course, always the case if G is simply connected. Now, let $G_0 \subset G$ be the connected subgroup corresponding to $\mathfrak{g}_0$. The group G_0 is reductive and all the subspaces $\mathfrak{g}_i$ are stable under the adjoint action of G_0.

2.2. Examples.

2.2.1. *Symmetric pairs.* Let $m = 2$. Then a $\mathbb{Z}/2\mathbb{Z}$-grading $\mathfrak{g} = \mathfrak{g}_0 \oplus \mathfrak{g}_1$ defines a symmetric pair. These have been extensively studied in connection with the theory of symmetric spaces and real forms of $\mathfrak{g}$ and G.

2.2.2. *Cyclic quivers.* Let $m \geqslant 2$. Let V be a complex vector space equipped with a $\mathbb{Z}/m\mathbb{Z}$-grading $V = \oplus_{i \in \mathbb{Z}/m\mathbb{Z}} V_i$. Let $G = \mathrm{SL}(V)$. Define on $\mathfrak{g} = \mathfrak{sl}(V)$ the $\mathbb{Z}/m\mathbb{Z}$-grading given by

$$\mathfrak{g}_i = \{A \in \mathfrak{sl}(V) \text{ such that } A(V_j) \subset V_{j+i} \text{ for every } j \in \mathbb{Z}/m\mathbb{Z}\} \tag{2.1}$$

In this situation

$$G_0 = \mathrm{S}\Big(\prod_{i \in \mathbb{Z}/m\mathbb{Z}} \mathrm{GL}(V_i)\Big),$$

and the subspace $\mathfrak{g}_1$ is

$$\bigoplus_{i \in \mathbb{Z}/m\mathbb{Z}} \mathrm{Hom}(V_i, V_{i+1}).$$

We can define the **quiver** Q with m vertices indexed by $\mathbb{Z}/m\mathbb{Z}$ and arrows $i \mapsto i+1$ for each $i \in \mathbb{Z}/m\mathbb{Z}$. Then $\mathfrak{g}_1$ is the space of representations of Q where we put V_i at the vertex i, subject to the condition that the product of the determinants of the elements in $\mathrm{GL}(V_i)$ be 1. This can be represented by the diagram.

$$V_0 \xrightarrow{f_0} V_1 \xrightarrow{f_1} \cdots \xrightarrow{f_{m-2}} V_{m-1}, \qquad V_{m-1} \xrightarrow{f_{m-1}} V_0.$$

For other classical groups the action of G_0 on $\mathfrak{g}_1$ can be interpreted in terms of a cyclic quiver with some extra structure.

2.3. Vinberg θ-pairs. Let G be a semisimple complex algebraic group with Lie algebra $\mathfrak{g}$. Let $\theta \in \mathrm{Aut}(G)$ of order m. As explained above, θ defines a $\mathbb{Z}/m\mathbb{Z}$-grading $\mathfrak{g} = \oplus_{i \in \mathbb{Z}/m\mathbb{Z}} \mathfrak{g}_i$. Let G_0 be the connected subgroup of G corresponding to $\mathfrak{g}_0$. As mentioned above, the group G_0 is reductive and all the subspaces $\mathfrak{g}_i$ are stable under the adjoint action of G_0. In **[76]**, Vinberg studies the invariant theory for the action of G_0 on $\mathfrak{g}_i$. The pairs $(G_0, \mathfrak{g}_i)$ will be called **Vinberg θ-pairs**. These pairs are referred by Vinberg and other authors **[52–54, 58, 64, 75]** as θ-groups or Vinberg representations. In **[80]** Wallach calls them Vinberg pairs. Our terminology makes explicit the role of the automorphism θ.

In **[76]**, Vinberg observes that one can reduce such study to the action of G_0 on $\mathfrak{g}_1$. Indeed, for any k one may pass to a subalgebra $\mathfrak{g}'$ with a new grading modulo $m' = \frac{m}{(m,k)}$, by setting $\mathfrak{g}'_j = \mathfrak{g}_{jk}$ for $j \in \mathbb{Z}/m\mathbb{Z}$. We then have $\mathfrak{g}'_0 = \mathfrak{g}_0$ and $\mathfrak{g}'_1 = \mathfrak{g}_k$. We can thus focus mostly on the pair $(G_0, \mathfrak{g}_1)$. Recall that if $\mathfrak{t} \subset \mathfrak{g}$ is a Cartan subalgebra, and $W(\mathfrak{t})$ is the Weyl group, the **Chevalley restriction theorem** establishes an isomorphism

$$\mathfrak{g} /\!\!/ G \cong \mathfrak{t}/W(\mathfrak{t}).$$

Similarly, if θ is an involution of G and $\mathfrak{g} = \mathfrak{g}_0 \oplus \mathfrak{g}_1$ is the Cartan decomposition defined by θ, and $W(\mathfrak{a})$ is the **little Weyl group** defined by a maximal abelian subalgebra $\mathfrak{a} \subset \mathfrak{g}_1$, there is also a Chevalley restriction theorem stating

$$\mathfrak{g}_1 /\!\!/ G_0 \cong \mathfrak{a}/W(\mathfrak{a}),$$

where $G_0 \subset G$ is the connected subgroup corresponding to $\mathfrak{g}_0$.

One of the main results of Vinberg in **[76]** is a version of the Chevalley restriction theorem for a Vinberg θ-pair $(G_0, \mathfrak{g}_1)$ defined by a finite order element $\theta \in \mathrm{Aut}(G)$. A key concept is that of **Cartan subspace**, a linear subspace $\mathfrak{a} \subset \mathfrak{g}_1$ which is abelian as a Lie algebra, consists of semisimple elements, and is maximal with these two properties. Any two Cartan subspaces of $\mathfrak{g}_1$ are G_0-conjugate and any semisimple element of $\mathfrak{g}_1$ is contained in a Cartan subspace. Hence the dimension of $\mathfrak{a}$ is an invariant of the grading, called the **rank** of θ. Given a Cartan subspace $\mathfrak{a}$, the normalizer and centralizer of $\mathfrak{a}$ in G_0, as usual, are defined respectively by

$$N_{G_0}(\mathfrak{a}) = \{g \in G_0 \mid \mathrm{Ad}(g)\mathfrak{a} \subset \mathfrak{a}\},$$

and

$$C_{G_0}(\mathfrak{a}) = \{g \in G_0 \mid \mathrm{Ad}(g)x = x \quad \text{for every } x \in \mathfrak{a}\}.$$

Then the **little Weyl group** is defined by

$$W(\mathfrak{a}) = N_{G_0}(\mathfrak{a})/C_{G_0}(\mathfrak{a}).$$

The group $W(\mathfrak{a})$ is a finite linear group generated by semisimple transformations of $\mathfrak{a}$ fixing a hyperplane, and hence $\mathbb{C}[\mathfrak{a}]^{W(\mathfrak{a})}$ is a polynomial ring. As shown by Vinberg **[76]**, there is a version of the Chevalley restriction theorem, namely, the restriction of polynomial functions $\mathbb{C}[\mathfrak{g}_1] \to \mathbb{C}[\mathfrak{a}]$ induces an isomorphism of invariant polynomial rings

$$\mathbb{C}[\mathfrak{g}_1]^{G_0} \to \mathbb{C}[\mathfrak{a}]^{W(\mathfrak{a})}, \tag{2.2}$$

or equivalently,

$$\mathfrak{g}_1 /\!\!/ G_0 \cong \mathfrak{a}/W(\mathfrak{a}).$$

The fact that $W(\mathfrak{a})$ is a finite linear group generated by complex reflections implies, by results of Shephard–Todd **[69]**, and Chevalley **[14]**, that $\mathbb{C}[\mathfrak{a}]^{W(\mathfrak{a})} = \mathbb{C}[f_1, \cdots, f_r]$ is a polynomial algebra generated by r algebraically independent polynomials $f_1, \cdots, f_r$ whose degrees $d_1, \cdots, d_r$ are determined by the grading. In particular, the product of these degrees is the order of $W(\mathfrak{a})$.

In **[48]** Kostant showed that the quotient map $\mathfrak{g} \to \mathfrak{g} /\!\!/ G$ for the adjoint action of G on $\mathfrak{g}$ has a section, known as the **Kostant section**. This was extended in **[49]** to obtain the **Kostant–Rallis section** in the symmetric pair case for the quotient map $\mathfrak{g}_1 \to \mathfrak{g}_1 /\!\!/ G_0$ defined by the isotropy representation. The existence of a similar section for Vinberg θ-pairs for θ of higher order was conjectured by Popov **[60]**. In this context, such a section is referred as a **Kostant–Weierstrass section**, as we explain below. As outlined in **[79]**, for a particularly nice action of a reductive complex algebraic group G on a vector space V there exists an affine linear subvariety $W \subset V$, called a **Weierstrass section** such that the restricting to W induces an isomorphism $\mathbb{C}[V]^{\mathbb{C}} \to \mathbb{C}[W]$, that is, such that $W \hookrightarrow V$ is a section for he quotient morphism $V \to V /\!\!/ G$. This motivated the introduction of the term **Kostant–Weierstrass section** in **[58]** for the quotient $\mathfrak{g}_1 \to \mathfrak{g}_1 /\!\!/ G_0$ for a Vinberg θ-pair $(G_0, \mathfrak{g}_1)$.

The Popov conjecture was proved first by Panyushev in **[57]** when G_0 is semisimple, and in **[58]** when $\mathfrak{g}_1$ contains a regular nilpotent element of $\mathfrak{g}$. In **[52]** Levy studies Vinberg θ-pairs for fields of (good) positive characteristic and extends the Kostant-Rallis results when θ is an involution, and in **[53, 54]** he deals with the existence of the Kostant–Weierstrass section in the higher order case, including the case when the field is of zero characteristic. In particular, in **[53]** he shows the existence of a Kostant–Weierstrass section in zero and (good) positive characteristic for the classical groups, and in **[54]**, he shows the existence of such section for G of type G_2, F_4 and D_4. The general case is proved in **[64]** by Reeder–Levy–Yu–Gross. In **[64]**, these authors give also a complete classification of positive-rank gradings and their little Weyl groups for simple algebraic groups G.

2.4. Extended Vinberg θ-pairs. As above, let G be a semisimple complex algebraic group, and $\theta \in \operatorname{Aut}(G)$ of finite order. Let $(G_0, \mathfrak{g}_i)$ be a Vinberg θ-pair, where $\mathfrak{g}_i$ is a subspace appearing in the cyclic grading of $\mathfrak{g}$ defined by θ . Let $G^\theta \subset G$ be the subgroup of fixed points under the action of θ on G. The group G^θ is reductive and has G_0 as the identity component. We thus have an extension

$$1 \longrightarrow G_0 \longrightarrow G^\theta \longrightarrow \Gamma^\theta \longrightarrow 1,$$

where Γ^θ is the group of connected components, which is a finite group. Another relevant group is

$$G_\theta := \{g \in G \;:\; \theta(g) = c(g)g, \text{ with } c(g) \in Z(G)\},$$

which is the normalizer of G^θ in G (see **[31]**) for which there is an exact sequence

$$1 \longrightarrow G^\theta \longrightarrow G_\theta \longrightarrow \Gamma_\theta \longrightarrow 1,$$

where Γ_θ is a finite group. We can combine these two extensions to have the extension

$$1 \longrightarrow G_0 \longrightarrow G_\theta \longrightarrow \hat{\Gamma}_\theta \longrightarrow 1,$$

where $\hat{\Gamma}_\theta$ is the group of connected components of G_θ.

Now, the subspaces $\mathfrak{g}_i$ are stable under the adjoint action of both G^θ and G_θ. We can thus consider the pairs $(G^\theta, \mathfrak{g}_i)$ and $(G_\theta, \mathfrak{g}_i)$, to which we will refer also as Vinberg θ-pairs, using the more precise term **extended Vinberg θ-pairs** if needed. For these pairs, there are analogous result to (2.2), where the little Weyl group is defined replacing G_0 by G^θ and G_θ.

2.5. Finite order automorphisms of G and the clique map. Let G be a semisimple complex algebraic group. Let $\operatorname{Int}(G)$ be the normal subgroup of $\operatorname{Aut}(G)$ consisting of inner automorphisms of G and $\operatorname{Out}(G) = \operatorname{Aut}(G)/\operatorname{Int}(G)$. Given $\theta, \theta' \in \operatorname{Aut}(G)$, we say that $\theta \sim \theta'$ if there is an element $g \in G$ such that $\theta' = \operatorname{Int}_g \theta \operatorname{Int}_{g^{-1}}$. This defines an equivalence relation in $\operatorname{Aut}(G)$, so that that the quotient homomorphism $\pi : \operatorname{Aut}(G) \to \operatorname{Out}(G)$ descends to a map

$$\operatorname{Aut}(G)/\sim \longrightarrow \operatorname{Out}(G). \tag{2.3}$$

Let $m > 0$ be a integer and let $\operatorname{Aut}_m(G) \subset \operatorname{Aut}(G)$ be the set of elements of order m. If now $\theta, \theta' \in \operatorname{Aut}_m(G)$ and $\theta' = \operatorname{Int}_g \theta \operatorname{Int}_{g^{-1}}$, the two $\mathbb{Z}/m\mathbb{Z}$-gradings of $\mathfrak{g}$ given by $\mathfrak{g} = \oplus_{i \in \mathbb{Z}/m\mathbb{Z}} \mathfrak{g}_i$ and $\mathfrak{g} = \oplus_{i \in \mathbb{Z}/m\mathbb{Z}} \mathfrak{g}'_i$ are related by $\mathfrak{g}'_i = \operatorname{Ad}(g)\mathfrak{g}_i$. In particular, if G_0 and G'_0 are the connected subgroups of G corresponding to $\mathfrak{g}_0$ and $\mathfrak{g}'_0$, respectively, then $G'_0 = \operatorname{Int}_g G_0$ (similarly, $G^{\theta'} = \operatorname{Int}_g G^\theta$ and $G_{\theta'} = \operatorname{Int}_g G_\theta$).

The restriction of the map (2.3) to $\mathrm{Aut}_m(G)$ defines a map

$$cl : \mathrm{Aut}_m(G)/\sim \to \mathrm{Out}_m(G)$$

called the **clique map**, where here $\mathrm{Out}_m(G)$ is the image of $\mathrm{Aut}_m(G)$ under the map $\pi : \mathrm{Aut}(G) \to \mathrm{Out}(G)$. For an element $a \in \mathrm{Out}_m(G)$ we refer to the set $cl_m^{-1}(a)$ as the **clique** defined by a (or m-clique if we need to be more precise). To identify $cl_m^{-1}(a)$, let $\theta \in \mathrm{Aut}_m(G)$ so that $\pi(\theta) = a$ and consider the set

$$S_\theta := \{s \in G \ : \ s\theta(s)\cdots\theta^{m-1}(s) = z \in Z\}.$$

There is an action of Z, the centre of G, on S_θ by multiplication and of G given by

$$s \cdot g := g^{-1}s\theta(g) = \quad g \in G, s \in S_\theta.$$

In Section 2 of **[31]** it is proved that the map $S_\theta \to \mathrm{Aut}_m(G)$ given by $s \mapsto \mathrm{Int}_s\,\theta$ defines a bijection

$$S_\theta/(Z \times G) \longleftrightarrow cl^{-1}(a).$$

Moreove, there is an interpretation in terms of non abelian Galois cohomology since

$$S_\theta/(Z \times G) = H^1(\mathbb{Z}/m\mathbb{Z}, \mathrm{Ad}(G)),$$

where $\mathbb{Z}/m\mathbb{Z}$ acts on $\mathrm{Ad}(G)$ via θ.

2.6. Involutions and real forms of complex Lie groups. Let G be a complex Lie group. We say that a real Lie subgroup $G_\mathbb{R} \subset G$ (here we are considering the underlying real structure of G) is a **real form** of G if $G_\mathbb{R} = G^\sigma$, the fixed point set of a **conjugation** (antiholomorphic involution) σ of G.

Let $\mathrm{Conj}(G)$ be the set of conjugations of G. We can define in $\mathrm{Conj}(G)$ a similar equivalence relation to the one defined on $\mathrm{Aut}(G)$ in Section 2.5. Given $\sigma, \sigma' \in \mathrm{Conj}(G)$, we say that $\sigma \sim \sigma'$ if there is an element $g \in G$ such that $\sigma' = \mathrm{Int}_g\,\sigma\,\mathrm{Int}_{g^{-1}}$.

Let now G be semisimple. A basic result due to Cartan **[13]** establishes that given a compact conjugation $\tau \in \mathrm{Conj}(G)$, i.e. a conjugation τ so that G^τ is a maximal compact subgroup of G, and a class $\alpha \in \mathrm{Conj}(G)/\sim$, there exists a representative σ of α so that $\tau\sigma = \sigma\tau$. The same applies to a class in $\mathrm{Aut}_2(G)/\sim$, where $\mathrm{Aut}_2(G)$ is the set of holomorphic involutions of G. We thus have that the map $\sigma \mapsto \theta := \sigma\tau$ gives a bijection

$$\mathrm{Conj}(G)/\sim \longleftrightarrow \mathrm{Aut}_2(G)/\sim.$$

Combining this bijection with the map $cl : \mathrm{Aut}_2(G)/\sim \to \mathrm{Out}_2(G)$ (defined in Section 2.5), we obtain a map

$$\widehat{cl} : \mathrm{Conj}(G)/\sim \to \mathrm{Out}_2(G).$$

The set $\widehat{cl}^{-1}(1)$ consists of the equivalence classes of real forms of **Hodge type**. For example, for $G = \mathrm{SL}(n, \mathbb{C})$, these are the real forms $\mathrm{SU}(p, q)$, with $p + q = n$, including the compact form $\mathrm{SU}(n)$.

2.7. Quasi-split $\mathbb{Z}/m\mathbb{Z}$-gradings. In the symmetric pair situation of Example 2.2.1 where $m = 2$ and we have $\mathbb{Z}/2\mathbb{Z}$-grading $\mathfrak{g} = \mathfrak{g}_0 \oplus \mathfrak{g}_1$, there is the special class of quasi-split involutions θ, or equivalently quasi-split real forms of $\mathfrak{g}$ or G. Given an involution θ of G with $\mathbb{Z}/2\mathbb{Z}$-grading $\mathfrak{g} = \mathfrak{g}_0 \oplus \mathfrak{g}_1$, as above, let G_0 be the subgroup of G corresponding to $\mathfrak{g}_0$, and $\mathfrak{a} \subset \mathfrak{g}_1$ be a Cartan subspace. Let

$$\mathfrak{g}_1^{\mathrm{reg}} = \{x \in \mathfrak{g}_1 \mid \dim C_{G_0}(x) = \dim \mathfrak{a}\}$$

be the set of **regular** elements of $\mathfrak{g}_1$ for the adjoint action of G_0. Similarly one can define $\mathfrak{g}^{\mathrm{reg}}$ for the adjoint action of G.

$$\mathfrak{g}^{\mathrm{reg}} = \{x \in \mathfrak{g} \mid \dim C_G(x) = \dim \mathfrak{t}\},$$

where $\mathfrak{t}$ is a Cartan subalgebra of $\mathfrak{g}$.

The centralizer of $\mathfrak{a}$ in $\mathfrak{g}_0$ is defined as

$$C_{\mathfrak{g}_0}(\mathfrak{a}) = \{x \in \mathfrak{g}_0 \mid [x, y] = 0 \quad \text{for every} \quad y \in \mathfrak{a}\}.$$

One defines similarly $C_{\mathfrak{g}_1}(\mathfrak{a})$ and $C_{\mathfrak{g}}(\mathfrak{a})$.

There are several equivalent definitions for θ or the pair $(G_0, \mathfrak{g}_1)$ to be quasi-split (see for example [**28**]). The involution θ is **quasi-split** if any of the following conditions is satisfied:

(1) $C_{\mathfrak{g}_0}(\mathfrak{a})$ is abelian. In this case $C_g(\mathfrak{a}) = C_{\mathfrak{g}_0}(\mathfrak{a}) \oplus C_{\mathfrak{g}_1}(\mathfrak{a})$ is a Cartan subalgebra of $\mathfrak{g}$.

(2) There exists a Borel subgroup $B \subset G$ with a θ-invariant maximal torus $T \subset G$ such that $\theta(B) = B^{\mathrm{op}}$, where B^{op} is the opposite Borel subgroup.

(3) $\mathfrak{g}_1^{\mathrm{reg}} = \mathfrak{g}_1 \cap \mathfrak{g}^{\mathrm{reg}}$.

For $\mathfrak{g}$ simple the the real forms corresponding to quasi-split involutions are the split real forms of $\mathfrak{g}$ (this correspond to $C_{\mathfrak{g}_0}(\mathfrak{a}) = 0$) and the real forms (up to conjugation by an element of G) $\mathfrak{su}(n, n)$, $\mathfrak{su}(n, n+1)$, $\mathfrak{so}(n, n+2)$ and $\mathfrak{e}_6(2)$. In particular, there exist one and only one quasi-split class in the clique $\widehat{cl}^{-1}(a)$ of every element $a \in \mathrm{Out}_2(G)$.

For $m > 2$ we will make the following definition.

Definition 2.1. Let $(G_0, \mathfrak{g}_1)$ be a Vinberg θ-pair, and let $\mathfrak{a} \subset \mathfrak{g}_1$ be a Cartan subspace and $C_{\mathfrak{g}_0}(\mathfrak{a})$ be defined as above. Then $(G_0, \mathfrak{g}_1)$ is said to be quasi-split if $C_{\mathfrak{g}_0}(\mathfrak{a})$ is abelian. It is called split if $C_{\mathfrak{g}_0}(\mathfrak{a}) = 0$.

It would be interesting to investigate other characterizations of these conditions along the lines of the symmetric pair case.

Example 2.2. In the case of the cyclic quivers of Example 2.2.2, let k be the minimum of $\dim V_i$. Then the quasi-split condition for $(G_0, \mathfrak{g}_1)$ is equivalent to $\dim V_i = k$ or $\dim V_i = k + 1$ for every i. When $m = 2$, these correspond of course to the quasi-split real forms $\mathfrak{su}(k, k)$ and $\mathfrak{su}(k, k+1)$. The split condition for arbitrary m corresponds to $\dim V_i = 1$ for every i, which if $m = 2$ corresponds to the real form $\mathfrak{su}(1, 1)$.

3. Vinberg θ-pairs and Higgs bundles

For this this section, let X be a compact Riemann surface of genus $g \geqslant 2$ and let K_X be its canonical bundle.

3.1. Higgs pairs. Let G be a complex reductive Lie group with Lie algebra $\mathfrak{g}$ and nondegenerate G-invariant bilinear form $\langle\cdot,\cdot\rangle$. Let $\rho : G \to \mathrm{GL}(V)$ be a holomorphic representation. If $E \to X$ is a principal G-bundle, we will denote the V-bundle $E \times_G V$ associated to E via the representation ρ by $E(V)$.

A (G, V)-**Higgs pair** is a pair (E, φ) where E is a holomorphic principal G-bundle on X and φ is a holomorphic section of $E(V) \otimes K_X$.

When $V = \mathfrak{g}$ and the representation ρ is the adjoint representation, a $(G, \mathfrak{g})$-Higgs pair is called a G-**Higgs bundle**.

If E is a principal G-bundle and $\hat{G} < G$ is a subgroup, then a structure group reduction of E to $\hat{G}$ is a section σ of the bundle $E(G/\hat{G})$. Associated to such a reduction is a principal $\hat{G}$-subbundle $E_\sigma \subset E$ such that $E_\sigma(G)$ is canonically isomorphic to E. Let G be a complex reductive Lie group and $\rho : G \to \mathrm{GL}(V)$ be a holomorphic representation. Let $\hat{G} < G$ and $\hat{V} \subset V$ be a $\rho(\hat{G})$-invariant subspace. We say that a (G, V)-Higgs pair (E, φ) reduces to a $(\hat{G}, \hat{V})$-Higgs pair, if there is a holomorphic reduction $E_{\hat{G}}$ of E to $\hat{G}$ such that $\varphi \in H^0(E_{\hat{G}}(\hat{V}) \otimes K_X) \subset H^0(E_{\hat{G}}(V) \otimes K_X)$.

To form a moduli space of Higgs pairs, we need to define suitable notions of stability. We describe this below and refer to **[25]** for more details. Fix a maximal compact subgroup $K < G$ and let $\mathfrak{k}$ be its Lie algebra. An element $s \in i\mathfrak{k}$ defines subspaces of V via the representation ρ

$$V_s^0 = \{v \in V \mid \rho(e^{ts})v = v\} \text{ and } V_s = \{v \in V \mid \rho(e^{ts})(v) \text{ is bounded as } t \to \infty\}.$$

When $\rho : G \to \mathrm{GL}(\mathfrak{g})$ is the adjoint representation, $\mathfrak{g}_s = \mathfrak{p}_s \subset \mathfrak{g}$ is a parabolic subalgebra with Levi subalgebra $\mathfrak{g}_s^0 = \mathfrak{l}_s \subset \mathfrak{p}_s$. The associated subgroups $L_s < P_s$ are given by

$$L_s = \{g \in G \mid \mathrm{Ad}(g)s = s\} \text{ and } P_s = \{g \in G \mid \mathrm{Ad}(e^{ts})(g) \text{ is bounded as } t \to \infty\}.$$

Moreover, s defines a character $\chi_s : \mathfrak{p}_s \to \mathbb{C}$ by

$$\chi_s(x) = \langle s, x\rangle \text{ for } x \in \mathfrak{p}_s.$$

Let E be a G-bundle, $s \in i\mathfrak{k}$ and $P_s < G$ be the associated parabolic subgroup. A reduction of structure group of E to P_s is a P_s-subbundle $E_{P_s} \subset E$, this is equivalent to a section $\sigma \in \Gamma(E(G/P_s))$ of the associated bundle. We will denote the associated P_s-subbundle by E_σ. The degree of such a reduction will be defined using Chern–Weil theory. Since P_s is homotopy equivalent to the maximal compact $K_s = K \cap L_s$ of L_s, given a reduction of structure σ of E to P_s, there is a further reduction σ' of E to K_s which is unique up to homotopy. Let $E_{\sigma'} \subset E$ be the resulting K_s principal bundle. The curvature F_A of a connection A on $E_{\sigma'}$ satisfies $F_A \in \Omega^2(X, E_{\sigma'}(\mathfrak{k}_s))$. Thus, evaluating the character χ_s on the curvature we have $\chi_s(F_A) \in \Omega(X, i\mathbb{R})$, and we define the degree of σ as

$$\deg E(\sigma, s) = \frac{i}{2\pi} \int_X \chi_s(F_A).$$

If q is a rational number such that $q \cdot \chi_s$ exponentiates to a character $\tilde{\chi}_s : P_s \to \mathbb{C}^*$ and $\sigma \in \Gamma(E(G/P_s))$ is a reduction, then $E_\sigma \times_{\chi_s} \mathbb{C}^* = E_\sigma(\tilde{\chi}_s)$ is a line bundle and

$$\deg E(\sigma, s) = \frac{1}{q} \deg E_\sigma(\tilde{\chi}_s).$$

When s is in the centre of $\mathfrak{g}$ then $P_s = G$. In this case, the degree given in the equation above is simply the degree of E with respect to $\chi = \chi_s$, and will be denoted

$\deg_\chi(E)$. Again, if a multiple $q\cdot\chi$ exponentiates to a character $\tilde{\chi}: G\to\mathbb{C}^*$ we have

$$\deg_\chi(E) = \frac{1}{q}\deg E(\tilde{\chi}). \tag{3.1}$$

Let $d\rho : \mathfrak{g}\to\mathfrak{gl}(V)$ be the differential of ρ and let $\mathfrak{z}_\mathfrak{k}$ be the centre of $\mathfrak{k}$ and $\mathfrak{k} = \mathfrak{k}_{ss}+\mathfrak{z}_\mathfrak{k}$. Define

$$\mathfrak{k}_\rho = \mathfrak{k}_{ss}\oplus(\ker(d\rho|_{\mathfrak{z}_\mathfrak{k}}))^\perp.$$

We are now ready to define α-stability notions for $\alpha\in i\mathfrak{z}_\mathfrak{k}$.

Let $\alpha\in i\mathfrak{z}_\mathfrak{k}$. A (G,V)-Higgs pair (E,φ) is:

- **α-semistable** if for any $s\in i\mathfrak{k}$ and any holomorphic reduction $\sigma\in H^0(E(G/P_s))$ such that $\varphi\in H^0(E_\sigma(V_s)\otimes K_X)$, we have $\deg E(\sigma,s)\geqslant\langle\alpha,s\rangle$.
- **α-stable** if it is α-semistable and for any $s\in i\mathfrak{k}_\rho$ and any holomorphic reduction $\sigma\in H^0(E(G/P_s)$ such that $\varphi\in H^0(E_\sigma(V_s)\otimes K_X)$, we have $\deg E(\sigma,s)>\langle\alpha,s\rangle$.
- **α-polystable** if it is α-semistable and whenever $\sigma\in H^0(E(G/P_s))$ for $s\in i\mathfrak{k}$ satisfy $\varphi\in H^0(E_\sigma(V_s)\otimes K_X)$ and $\deg E(\sigma,s)=\langle\alpha,s\rangle$, there is a further holomorphic reduction $\sigma'\in H^0(E_\sigma(P_s/L_s))$ such that $\varphi\in H^0(E_{\sigma'}(V_s^0)\otimes K_X)$.

The **moduli space of α-polystable (G,V)-Higgs pairs** over X is defined as the set of isomorphism classes of α-polystable (G,V)-Higgs pairs and will be denoted by $\mathcal{M}^\alpha(G,V)$. A GIT construction of these spaces is given by Schmitt in **[68]** and by Simpson for the moduli space of 0-polystable G-Higgs bundles **[73]**.

When $\alpha=0$, we refer to 0-stability simply as stability (similarly for (semi, poly)stability), and denote the moduli space by $\mathcal{M}(G,V)$. The **moduli space of polystable G-Higgs bundles** will be denoted by $\mathcal{M}(G)$.

When G is a classical group, Higgs pairs can be studied in terms of vector bundles. For example, if $G=\mathrm{SL}(n,\mathbb{C})$, a G-Higgs bundle is equivalent to a pair (E,φ) where E is holomorphic vector bundle of rank n and trivial determinant, and φ is a homomorphism $E\to E\otimes K_X$ with vanishing trace. These are the original objects introduced by Hitchin in **[39]**. Here the stability condition is the familiar slope condition

$$\mu(F)<\mu(E),$$

where F is a nonzero proper subbundle of E preserved by φ, and $\mu(E)$ is the ratio between the degree of E and its rank.

3.2. Vinberg θ-pairs and cyclic Higgs bundles. Let G be a semisimple complex algebraic group with Lie algebra $\mathfrak{g}$ and centre $Z=Z(G)$. Let $\theta\in\mathrm{Aut}(G)$ be of order m. As explained in Section 2, θ defines a $\mathbb{Z}/m\mathbb{Z}$-grading $\mathfrak{g}=\oplus_{i\in\mathbb{Z}/m\mathbb{Z}}\mathfrak{g}_i$. Let $G_0\subset G$ be the connected subgroup corresponding to $\mathfrak{g}_0$. Recall also from Section 2 that G_0 is reductive and all the subspaces $\mathfrak{g}_i$ are stable under the adjoint action of G_0.

If now X is a compact Riemann surface, following the construction given in Section 3.1, we can consider $(G_0,\mathfrak{g}_i)$-Higgs pairs over X associated to the Vinberg θ-pairs $(G_0,\mathfrak{g}_i)$ and the moduli spaces $\mathcal{M}(G_0,\mathfrak{g}_i)$. In fact, we can also consider Higgs pairs associated to the extended Vinberg θ-pairs $(G^\theta,\mathfrak{g}_i)$ and $(G_\theta,\mathfrak{g}_i)$, for G^θ and G_θ as in Section 2.4 and the moduli spaces $\mathcal{M}(G^\theta,\mathfrak{g}_i)$ and $\mathcal{M}(G_\theta,\mathfrak{g}_i)$ for $i\in\mathbb{Z}/m\mathbb{Z}$.

The moduli spaces associated to Vinberg pairs do appear naturally inside the moduli space $\mathcal{M}(G)$ of G-Higgs bundles over X as fixed points under various actions of a cyclic group Γ of order m. To explain this, Let θ be a group automorphism of G. Let E be a principal G-bundle and $\theta(E) := E \times_\theta G$ be the principal G-bundle obtained by the extension of structure group defined by θ. Note that a Higgs field $\varphi \in H^0(X, E(\mathfrak{g}) \otimes K_X)$ produces a Higgs field $\theta(\varphi) \in H^0(X, \theta(E)(\mathfrak{g}) \otimes K_X)$, defining an action of $\operatorname{Aut}(G)$ on $\mathcal{M}(G)$. Because of gauge equivalence, this action descends to an action of the group $\operatorname{Out}(G) = \operatorname{Aut}(G)/\operatorname{Int}(G)$, where $\operatorname{Int}(G)$ is the normal subgroup of $\operatorname{Aut}(G)$ consisting of inner automorphisms of G. Now, the group $\mathbb{C}^*$ acts on $\mathcal{M}(G)$ by multiplication of the Higgs field.

Let $\mu_m = \{z \in \mathbb{C}^* \text{ such that } z^m = 1\}$ and let $\zeta \in \mu_m$ be a primitive m-th root of unity. Consider the homomorphism $\mu_m \to \operatorname{Aut}(G) \times \mathbb{C}^*$ defined by $\zeta \mapsto (\theta, \zeta)$. Let Γ be the image of this homomorphism. If now θ is of order m, the group Γ is isomorphic to μ_m and acts on $\mathcal{M}(G)$ by the rule

$$(E, \varphi) \mapsto (\theta(E), \zeta\theta(\varphi)).$$

The action of Γ depends only of the image $a = \pi(\theta)$ in $\operatorname{Out}(G)$, since $\operatorname{Int}(G)$ acts trivially on $\mathcal{M}(G)$. In **[31]** it is shown that the moduli space $\mathcal{M}(G^\theta, \mathfrak{g}_1)$ maps into $\mathcal{M}(G)$ by extending the structure group and that the image, denoted by $\widetilde{\mathcal{M}}(G, \mathfrak{g}_1)$, is in the fixed-point locus $\mathcal{M}(G)^\Gamma$. In fact, $\mathcal{M}(G)^\Gamma$ contains also $\widetilde{\mathcal{M}}(G_{\theta'}, \mathfrak{g}'_1)$ for $\theta' \in S_\theta$, where S_θ is defined in Section 2.5, and the subvariety $\widetilde{\mathcal{M}}(G_{\theta'}, \mathfrak{g}'_1) \subset \mathcal{M}(G)$ depends only on the class of $\theta' \in S_\theta/(Z \times G)$, i.e. on the element $[\theta']$ in the clique defined by $a = \pi(\theta)$ (see Section 2.5).

There is a partial converse to this result if we consider the smooth locus $\mathcal{M}_*(G) \subset \mathcal{M}(G)$ consisting of stable and simple G-Higgs bundles. This states (see Theorem 6.3 of **[31]**) that

$$\mathcal{M}_*(G)^\Gamma \subset \bigcup_{[\theta'] \in S_\theta/(Z \times G)} \widetilde{\mathcal{M}}(G^{\theta'}, \mathfrak{g}'_1).$$

The elements in $\widetilde{\mathcal{M}}(G^\theta, \mathfrak{g}_1)$ are called **cyclic** G-Higgs bundles and have been studied by several authors including **[3, 15, 17, 21, 31, 50, 74]**. In particular if $G = \operatorname{SL}(n, \mathbb{C})$, and $\theta \in \operatorname{Aut}(G)$ is an inner automorphism of order m, the Vinberg θ-pair $(G^\theta, \mathfrak{g}_1)$ is given by the cyclic quiver described in Example 2.2.2. Note that in this case, since $\operatorname{SL}(n, \mathbb{C})$ is simply connected, G^θ is connected and hence coincides with G_0.

In many of the papers referred above cyclic Higgs bundles are studied in the case in which θ is an inner automorphism, and these correspond to fixed points for the action of μ_m on $\mathcal{M}(G)$ given by multiplication of the Higgs field. In **[31]** this is studied also when θ is an outer automorphism, that is the cliques of non-trivial elements on $\operatorname{Out}_m(G)$ are considered. To be more precise, it seems appropriate to refer to the objects in $\widetilde{\mathcal{M}}(G^\theta, \mathfrak{g}_1)$ as **θ-cyclic G-Higgs bundles**.

The moduli spaces $\mathcal{M}(G^\theta, \mathfrak{g}_i)$ for general $\mathfrak{g}_i$ in the $\mathbb{Z}/m\mathbb{Z}$-grading of $\mathfrak{g}$ do also show up as fixed points in $\mathcal{M}(G)$. But now we have to consider the homomorphism $\mu_m \to \operatorname{Aut}(G) \times \mathbb{C}^*$ defined by $\zeta \mapsto (\theta, \zeta^i)$ and take as Γ the image of this homomorphism. Then, one has similar results to the ones described above for $\mathfrak{g}_1$ (see **[31]**).

Finally, the moduli spaces $\widetilde{\mathcal{M}}(G_\theta, \mathfrak{g}_i)$ for G_θ do also appear as fixed points in $\mathcal{M}(G)$, but now the action of Γ involves the action on $\mathcal{M}(G)$ of the group $H^1(X, Z)$

of isomorphism classes of Z-bundles on X given by the analogue of tensoring by a line bundle in the vector bundle case. This is studied in [**31**].

3.3. Non-abelian Hodge correspondence and cyclic gradings. Let G be a reductive Lie group (real or complex). A **representation** of $\pi_1(X)$ in G is a homomorphism $\rho\colon \pi_1(X) \to G$. The set of all such homomorphisms, $\mathrm{Hom}(\pi_1(X), G)$, is an analytic variety, which is algebraic if G is algebraic. The group G acts on $\mathrm{Hom}(\pi_1(X), G)$ by conjugation:

$$(g \cdot \rho)(\gamma) = g\rho(\gamma)g^{-1}$$

for $g \in G$, $\rho \in \mathrm{Hom}(\pi_1(X), G)$ and $\gamma \in \pi_1(X)$. If we restrict the action to the subspace $\mathrm{Hom}^+(\pi_1(X), g)$ consisting of reductive representations, the orbit space is Hausdorff. By a **reductive representation** we mean one that, composed with the adjoint representation in the Lie algebra of G, decomposes as a sum of irreducible representations. If G is algebraic this is equivalent to the Zariski closure of the image of $\pi_1(X)$ in G being a reductive group. The **moduli space of reductive representations** of $\pi_1(X)$ in G, or G-**character variety** of $\pi_1(X)$ is defined to be the orbit space

$$\mathcal{R}(G) = \mathrm{Hom}^+(\pi_1(X), G)/G.$$

If G is complex $\mathcal{R}(G)$ coincides with the GIT quotient

$$\mathcal{R}(G) = \mathrm{Hom}(\pi_1(X), G) /\!\!/ G.$$

It has the structure of an analytic variety which is algebraic if G is algebraic and is real if G is real or complex if G is complex.

Assume for the remaining part of this section that G is a semisimple complex Lie group. The **non-abelian Hodge correspondence** ([**16, 18, 39, 72**]) establishes a homeomorphism

$$\mathcal{M}(G) \cong \mathcal{R}(G). \tag{3.2}$$

Let $\theta \in \mathrm{Aut}(G)$ of order m, and $\mathfrak{g} = \oplus_{i\in\mathbb{Z}/m\mathbb{Z}}\mathfrak{g}_i$, the $\mathbb{Z}/m\mathbb{Z}$-grading defined by θ. Consider G_0, G^θ and G_θ as in Section 2.4. The correspondence (3.2) can be extended to a reductive group G by considering the subvariety $\mathcal{M}_0(G) \subset \mathcal{M}(G)$ consisting of G-Higgs bundles whose underlying G-bundle is topologically trivial. We thus have a homeomorphism

$$\mathcal{M}_0(G^\theta) \cong \mathcal{R}(G^\theta),$$

and similarly for G_0 and G_θ.

If now $m = 2$ and θ defines the $\mathbb{Z}/2\mathbb{Z}$-grading $\mathfrak{g} = \mathfrak{g}_0 \oplus \mathfrak{g}_1$, a generalization of the non-abelian Hodge correspondence (3.2) to real forms (see e.g. [**25, 31**]) gives a homeomorphism

$$\mathcal{M}(G^\theta, \mathfrak{g}_1) \cong \mathcal{R}(G^\sigma),$$

where G^σ is the real form of G defined the conjugation $\sigma = \theta\tau$, with τ a compact conjugation commuting with θ (see Section 2.6). The image of $\mathcal{M}(G^\theta, \mathfrak{g}_1)$ in $\mathcal{M}(G)$ defines a Lagrangian subvariety with respect to the natural holomorphic symplectic structure of $\mathcal{M}(G)$. If we replace G^θ by G_0 or G_θ we have to replace G^σ, respectively by the identity component of G^σ or G_σ, the normalizer of G^σ in G (see [**31**]).

4. $\mathbb{Z}$-gradings, Hodge bundles and cyclic Higgs bundles

In this section G is a semisimple complex Lie group with Lie algebra $\mathfrak{g}$ and Killing form B. Here we follow [6] to which we refer for details.

4.1. $\mathbb{Z}$-gradings and the Toledo character. A **$\mathbb{Z}$-grading** of a semisimple Lie algebra $\mathfrak{g}$ is a decomposition

$$\mathfrak{g} = \bigoplus_{j\in\mathbb{Z}} \mathfrak{g}_j \qquad \text{such that} \qquad [\mathfrak{g}_i, \mathfrak{g}_j] \subset \mathfrak{g}_{i+j}.$$

The subalgebra $\mathfrak{p} = \bigoplus_{j\geqslant 0} \mathfrak{g}_j$ is a parabolic subalgebra with Levi subalgebra $\mathfrak{g}_0 \subset \mathfrak{p}$. There is an element $\zeta \in \mathfrak{g}_0$ such that $\mathfrak{g}_j = \{X \in \mathfrak{g} \mid [\zeta, x] = jx\}$; the element ζ is called the **grading element** of the $\mathbb{Z}$-grading. Having a $\mathbb{Z}$-grading on $\mathfrak{g}$ is equivalent to having a homomorphism $\psi : \mathbb{C}^* \to \mathrm{Aut}(\mathfrak{g})$, defined by

$$\psi(z)|_{\mathfrak{g}_j} = z^j I.$$

Given a $\mathbb{Z}$-grading $\mathfrak{g} = \bigoplus_{j\in\mathbb{Z}} \mathfrak{g}_j$, let $G_0 < G$ be the centralizer of ζ; G_0 acts on each factor $\mathfrak{g}_j$. An important fact due to Vinberg [77] is that for each $j \neq 0$, $\mathfrak{g}_j$ is a **prehomogeneous vector space** for G_0. This means that $\mathfrak{g}_j$ (for $j \neq 0$) has an open G_0-orbit, which is then necessarily unique and dense (see [47, 65] for details on the theory of prehomogeneous vector spaces). The pairs $(G_0, \mathfrak{g}_i)$ are sometimes also referred as Vinberg pairs. To distinguish them from the Vinberg θ-pairs for θ an automorphism of finite order, to the prehomogeneous vector spaces $(G_0, \mathfrak{g}_i)$ obtained from a $\mathbb{Z}$-grading of $\mathfrak{g}$ we will call them **Vinberg $\mathbb{C}^*$-pairs.**

Without loss of generality, we can consider the Vinberg $\mathbb{C}^*$-pair $(G_0, \mathfrak{g}_1)$. Let $\Omega \subset \mathfrak{g}_1$ be the open G_0-orbit. Since $\mathfrak{g}_0$ is the centralizer of ζ, $B(\zeta, -) : \mathfrak{g}_0 \to \mathbb{C}$ defines a character. The **Toledo character** $\chi_T : \mathfrak{g}_0 \to \mathbb{C}$ is defined by

$$\chi_T(x) = B(\zeta, x)B^*(\gamma, \gamma)\ ,$$

where B^* is the bilinear form induced by B on the dual of $\mathfrak{g}$, and γ is the longest root such that $\mathfrak{g}_\gamma \subset \mathfrak{g}_1$. The normalization factor $B^*(\gamma, \gamma)$ guarantees that the Toledo character is independent of the choice of invariant bilinear form B. Let $e \in \mathfrak{g}_1$ and $\{f, h, e\}$ be an $\mathfrak{sl}_2$-triple with $h \in \mathfrak{g}_0$. We define the **Toledo rank** of e by

$$\mathrm{rk}_T(e) = \frac{1}{2}\chi_T(h),$$

and the **Toledo rank** of $(G_0, \mathfrak{g}_1)$ by

$$\mathrm{rk}_T(G_0, \mathfrak{g}_1) = \mathrm{rk}_T(e) \quad \text{for } e \in \Omega.$$

Given a $\mathbb{Z}$-grading $\mathfrak{g} = \bigoplus_{j\in\mathbb{Z}} \mathfrak{g}_j$, there is an involution $\theta : \mathfrak{g} \to \mathfrak{g}$ defined by

$$\theta|_{\mathfrak{g}_j} = (-1)^j \,\mathrm{Id}\,.$$

This corresponds to a real form of $\mathfrak{g}$ of Hodge type (and of G if there is a lift of θ to G).

4.2. Hodge bundles and the Toledo invariant. Let $\mathfrak{g} = \bigoplus_{j\in\mathbb{Z}} \mathfrak{g}_j$ be a $\mathbb{Z}$-grading with grading element $\zeta \in \mathfrak{g}_0$, and let $G_0 < G$ be the centralizer of ζ. Note that $\exp(\lambda\zeta)$ is in the centre of G_0 and $\mathrm{Ad}(\exp(\lambda\zeta))$ acts on each $\mathfrak{g}_j$ by $\exp(j\lambda) \cdot \mathrm{Id}$. Let X be a compact Riemann surface of genus $g \geqslant 2$. Let (E_{G_0}, φ) be a $(G_0, \mathfrak{g}_k)$-Higgs pair over X. Note that the central element $\exp(\frac{\lambda}{k}\zeta) \in G_0$ defines a holomorphic automorphism of E_{G_0} and acts on φ by multiplication by $\exp(\lambda)$. As a result we have an isomorphism of $(G_0, \mathfrak{g}_k)$-Higgs pairs

$$(E_{G_0}, \varphi) \cong (E_{G_0}, \lambda\varphi) \text{ for all } \lambda \in \mathbb{C}^*. \tag{4.1}$$

Extending the structure group defines a G-Higgs bundle (E_G, φ)

$$(E_G, \varphi) = (E_{G_0}(G), \varphi)$$

since $E_{G_0}(\mathfrak{g}_k) \subset E_{G_0}(\mathfrak{g}) \cong E_G(\mathfrak{g})$. Moreover, $(E_G, \varphi) \cong (E_G, \lambda\varphi)$ for all $\lambda \in \mathbb{C}^*$.

A G-Higgs bundle (E, φ) is called a **Hodge bundle of type** $(G_0, \mathfrak{g}_k)$ if it reduces to a $(G_0, \mathfrak{g}_k)$-Higgs pair. There is a natural $\mathbb{C}^*$-action on the moduli spaces of polystable G-Higgs bundles given by $\lambda \cdot (E, \varphi) = (E, \lambda\varphi)$. The $\mathbb{C}^*$-action is trivial on the moduli space of $(G_0, \mathfrak{g}_k)$-Higgs pairs by (4.1). As a result, polystable Higgs bundles which are Hodge bundles define fixed points of the $\mathbb{C}^*$-action on the moduli space of Higgs bundles. Simpson proved the converse **[72, 73]**. Namely, all $\mathbb{C}^*$-fixed points in the moduli space of Higgs bundles are Hodge bundles for some $\mathbb{Z}$-grading. In fact, if a Higgs bundle is a Hodge bundle, then polystability of the Higgs bundle is equivalent to polystability of the associated pair (see **[6]**). In particular, there is a well defined map of moduli spaces

$$\mathcal{M}(G_0, \mathfrak{g}_k) \to \mathcal{M}(G)$$

whose image consists of $\mathbb{C}^*$-fixed points, and every $\mathbb{C}^*$-fixed point in $\mathcal{M}(G)$ is in the image of such a map for some $(G_0, \mathfrak{g}_k)$ of some $\mathbb{Z}$-grading. Via the non-abelian Hodge correspondence, Hodge bundles correspond to **variations of Hodge structure** (see **[6]**).

Let (E, φ) be a $(G_0, \mathfrak{g}_1)$-Higgs pair and $\chi_T : \mathfrak{g}_0 \to \mathbb{C}$ be the Toledo character defined in Section 4.1. Then the **Toledo invariant** $\tau(E, \varphi)$ is defined to by

$$\tau(E, \varphi) = \deg_{\chi_T}(E), \tag{4.2}$$

where $\deg_{\chi_T}(E)$ is given by (3.1). Recall the definition of the Toledo rank $\mathrm{rk}_{\chi_T}(v)$ of a point $v \in \mathfrak{g}_1$ given in Section 4.1. We define the **Toledo rank** $\mathrm{rk}_T(\varphi)$ of a (G_0, g_1)-Higgs pair (E, φ) to be

$$\mathrm{rk}_T(\varphi) = \mathrm{rk}_T(\varphi(x)) \text{ for a generic } x \in X.$$

A main result proved in **[6]** is that if (E, φ) is a semistable, then one has the inequality

$$\tau(E, \varphi) \geqslant -\mathrm{rk}_T(\varphi)(2g-2), \tag{4.3}$$

referred in **[6]** as **Arakelov–Milnor inequality**, since it generalizes the Milnor–Wood inequality for the Toledo invariant of Higgs bundles for real forms of Hermitian type **[7]** (as we will see below), and is analogous to the Arakelov inequalities of classical variations of Hodge structure.

4.3. Hodge bundles and cyclic Higgs bundles.

Let

$$\mathfrak{g} = \bigoplus_{j \in \mathbb{Z}} \mathfrak{g}_j \tag{4.4}$$

be a $\mathbb{Z}$-grading with grading element $\zeta \in \mathfrak{g}_0$, and let $G_0 < G$ be the centralizer of ζ. The $\mathbb{Z}$-grading (4.4) defines a $\mathbb{Z}/k$-grading of $\mathfrak{g}$ for any integer $k > 0$, given by composing the inclusion homomorphism $\mu_k \subset \mathbb{C}^*$ of the group μ_k of k-th roots of unity with the homomorphism $\mathbb{C}^* \to \mathrm{Aut}(\mathfrak{g})$ corresponding to the $\mathbb{Z}$-grading. In particular, let $m \geqslant 2$ be the smallest integer in (4.4) for which $\mathfrak{g}_j = 0$ for every $|j| \geqslant m$. Then the $\mathbb{Z}/m\mathbb{Z}$-grading defined by (4.4) is given by

$$\mathfrak{g} = \bigoplus_{i \in \mathbb{Z}/m\mathbb{Z}} \overline{\mathfrak{g}}_i, \tag{4.5}$$

with

$$\overline{\mathfrak{g}}_i = \mathfrak{g}_i \oplus \mathfrak{g}_{i-m} \text{ for } i \in \{0, \cdots, m-1\}.$$

We will assume that the automorphism of $\mathfrak{g}$ of order m defining the $\mathbb{Z}/m\mathbb{Z}$-grading (4.5) lifts to an automorphism of G, and that $G^\theta = G_0$.

Let X be a compact Riemann surface. We want to study $(G_0, \overline{\mathfrak{g}}_i)$-Higgs pairs over X, which as explained in Section 3.2, correspond to θ-cyclic G-Higgs bundles. Without loss of generality we will consider $(G_0, \overline{\mathfrak{g}}_1)$-Higgs pairs. Now, let (E, φ) be a $(G_0, \overline{\mathfrak{g}}_1)$-Higgs pair over X. Notice that

$$E(\overline{\mathfrak{g}}_1) = E(\mathfrak{g}_1) \oplus E(\mathfrak{g}_{1-m}),$$

and hence, according to this, we can decompose $\varphi = \varphi^+ + \varphi^-$ with

$$\varphi^+ \in H^0(X, E(\mathfrak{g}_1) \otimes K_X) \text{ and } \varphi^- \in H^0(X, E(\mathfrak{g}_{1-m}) \otimes K_X).$$

We can now consider the Toledo invariant $\tau^+ = \tau(E, \varphi^+)$ as defined by (4.2) using the Toledo character χ_T^+ of $(G_0, \mathfrak{g}_1)$ determined by the grading element ζ. As mentioned above, if $\varphi^- = 0$ one has that the semistability of (E, φ) implies the inequality (4.3). But actually one has the following stronger result.

THEOREM 4.1. *Let $(E, \varphi^+ + \varphi^-)$ be a $(G_0, \overline{\mathfrak{g}}_1)$-Higgs pair over X with $\varphi^+ \neq 0 \neq \varphi^-$. Then if $(E, \varphi^+ + \varphi^-)$ is semistable one has*

$$\tau^+ \geqslant -\operatorname{rk}_T(\varphi^+)(2g-2).$$

The proof of Theorem 4.1 is given in **[24]** following **[6, 7]**.

We can also consider the Toledo invariant $\tau^- = \tau(E, \varphi^-)$ as defined by (4.2) using the Toledo character χ_T^- of $(G_0, \mathfrak{g}_{1-m})$. Notice that the grading element defining the Toledo character χ_T^- is $\zeta^- = \frac{\zeta}{1-m}$. As in the previous case, if $\varphi^+ = 0$ one has

$$\tau^- \geqslant -\operatorname{rk}_T(\varphi^-)(2g-2). \tag{4.6}$$

But now, in general we can not guarantee this inequality if $(E, \varphi^+ + \varphi^-)$ is semistable with $\varphi^+ \neq 0 \neq \varphi^-$. This happens however under some conditions, in particular when $m = 2$. In this case, the roles of $\mathfrak{g}_1$ and $\mathfrak{g}_{1-m} = \mathfrak{g}_{-1}$ are symmetric and we can apply Theorem 4.1, reversing the roles of φ^+ and φ^- to obtain (4.6) when $\varphi^+ \neq 0 \neq \varphi^-$.

In fact, when $m = 2$, $\tau := \tau^+ = -\tau^-$ and the semistability of $(E, \varphi^+, \varphi^-)$ implies

$$-\operatorname{rk}_T(\varphi^+)(2g-2) \leqslant \tau \leqslant \operatorname{rk}_T(\varphi^-)(2g-2). \tag{4.7}$$

When $\mathfrak{g}$ admits a a $\mathbb{Z}$-grading of the form $\mathfrak{g} = \mathfrak{g}_{-1} \oplus \mathfrak{g}_0 \oplus \mathfrak{g}_1$, the associated $\mathbb{Z}/2\mathbb{Z}$-grading $\mathfrak{g} = \overline{\mathfrak{g}}_0 \oplus \overline{\mathfrak{g}}_1$, with $\overline{\mathfrak{g}}_0 = \mathfrak{g}_0$ and $\overline{\mathfrak{g}}_1 = \mathfrak{g}_{-1} \oplus \mathfrak{g}_1$, corresponds to a real form of Hermitian type. If $\mathfrak{g}$ is simple a real form of Hermitian type is one of the classical real Lie algebras $\mathfrak{su}(p,q), \mathfrak{sp}(2n, \mathbb{R}), \mathfrak{so}^*(2n), \mathfrak{so}(2,n)$ or $\mathfrak{e}_6(-14), \mathfrak{e}_7(-25)$ in the exceptional case. In this situation, if the involution of $\mathfrak{g}$ can be lifted to an involution θ of G (which is necessarily inner) the moduli space $\mathcal{M}(G^\theta, \overline{\mathfrak{g}}_1)$ maps to the fixed point subvariety of the involution of $\mathcal{M}(G)$ given by $(E, \varphi) \mapsto (E, -\varphi)$ which is homeomorphic, via the non-abelian Hodge correspondence, to the character variety $\mathcal{R}(G^\sigma)$, where σ is an antiholomorphic involution of G corresponding to θ, and the symmetric space defined by the quotient G^σ by its maximal compact subgroup is a Hermitian symmetric space of the non-compact type. In this case, one has the Milnor–Wood inequality (4.7) for a semistable $(G^\theta, \overline{\mathfrak{g}}_1)$-Higgs pair $(E, \varphi^+, \varphi^-)$ as proved in [**7**]. This inequality leads to the classical Milnor–Wood inequality

$$|\tau| \leqslant 2r(g-1),$$

where r is the rank of the Hermitian symmetric space. Here τ coincides with the original invariant defined by Toledo for a representation of $\pi_1(X)$ in G^σ for which the above bound is proved in general by Burger–Iozzi–Wienhard [**12**].

4.4. Chains and cyclic SL$(n, \mathbb{C})$-Higgs bundles. A particular example of a $\mathbb{Z}/m\mathbb{Z}$-grading of a Lie algebra $\mathfrak{g}$ arising from a $\mathbb{Z}$-grading of $\mathfrak{g}$ in the way described in Section 4.3 is given by the cyclic quiver given in Example 2.2.2. In this example V is a complex vector space equipped with a $\mathbb{Z}/m\mathbb{Z}$-grading $V = \oplus_{i \in \mathbb{Z}/m\mathbb{Z}} V_i$ and $G = \mathrm{SL}(V)$. On $\mathfrak{g} = \mathfrak{sl}(V)$ there is a $\mathbb{Z}$-grading for which

$$G_0 = \mathrm{S}(\prod_{i \in \mathbb{Z}/m\mathbb{Z}} \mathrm{GL}(V_i)),$$

and

$$\mathfrak{g}_1 = \bigoplus_{i=1}^{m-1} \mathrm{Hom}(V_{i-1}, V_i) \text{ and } \mathfrak{g}_{1-m} = \mathrm{Hom}(V_{m-1}, V_0).$$

From here we obtain the $\mathbb{Z}/m\mathbb{Z}$-grading of $\mathfrak{g}$ described in Example 2.2.2, for which

$$\overline{\mathfrak{g}}_1 = \mathfrak{g}_1 \oplus \mathfrak{g}_{1-m}.$$

Let X be a compact Riemann surface of genus $g \geqslant 2$ with canonical bundle K_X. A $(G_0, \overline{\mathfrak{g}}_1)$-Higgs pair over X is equivalent to the K_X-twisted quiver bundle over X

$$E_0 \xrightarrow{\varphi_0} E_1 \xrightarrow{\varphi_1} \cdots \xrightarrow{\varphi_{m-2}} E_{m-1}, \qquad E_{m-1} \xrightarrow{\varphi_{m-1}} E_0 \tag{4.8}$$

with E_i holomorphic vector bundles over X of rank $n_i = \dim V_i$ and homomorphisms $\varphi_i : E_i \to E_{i+1} \otimes K_X$ for $i = 0, \cdots, m-1$. To such an object we can associate a G-Higgs bundle (E, φ) with

$$E = E_0 \oplus \cdots \oplus E_{m-1} \text{ and } \varphi = \begin{pmatrix} 0 & 0 & 0 & \cdots & \varphi_{m-1} \\ \varphi_0 & 0 & 0 & \cdots & 0 \\ 0 & \varphi_1 & 0 & \cdots & 0 \\ \vdots & \vdots & \vdots & \ddots & \vdots \\ 0 & 0 & \cdots & \varphi_{m-2} & 0 \end{pmatrix}.$$

This defines a morphism $\mathcal{M}(G_0, \overline{\mathfrak{g}}_1) \to \mathcal{M}(G)$, whose image is in the fixed point locus of the action of the group μ_m of m-th roots of unity on $\mathcal{M}(G)$ defined by sending (E, φ) to $(E, \zeta\varphi)$, where ζ is a primitive m-th root of unity. The Higgs bundle (E, φ) is thus a cyclic SL$(n, \mathbb{C})$-Higgs bundle as defined in Section 3.2.

The Hodge bundle defined by a quiver bundle (4.8) with $\varphi_{m-1} = 0$ is known in this case as a **chain** (see [**2, 26, 27**]) and is a fixed point in $\mathcal{M}(G)$ for the action of $\mathbb{C}^*$ given by rescaling the Higgs field. The study of the Toledo invariant and Arakelov–Milnor inequality for chains and corresponding cyclic Higgs bundles as described in Section 4.3 is treated in detail in [**24**]. We will just recall here the case $m = 2$, already studied in [**9, 10**].

The $m = 2$ case corresponds to $V = V_0 \oplus V_1$,

$$G_0 = \mathrm{S}(\mathrm{GL}(V_0) \times \mathrm{GL}(V_1)), \quad \text{and} \quad \bar{\mathfrak{g}}_1 = \mathrm{Hom}(V_0, V_1) \oplus \mathrm{Hom}(V_1, V_0),$$

and to the quiver representations described by the diagram

$$V_0 \underset{f_1}{\overset{f_0}{\rightleftarrows}} V_1.$$

A $(G^\theta, \mathfrak{g}_1)$-Higgs pair over X in the sense of Section 3.1 is equivalent to a 4-tuple $(E_0, E_1, \varphi_0, \varphi_1)$, consisting of holomorphic vector bundles E_0 and E_1 over X of ranks $n_0 = \dim V_0$ and $n_1 = \dim V_1$, respectively with $\det E_1 = (\det E_0)^{-1}$, and homomorphisms

$$\varphi_0 : E_0 \to E_1 \otimes K_X \quad \text{and} \quad \varphi_1 : E_1 \to E_0 \otimes K_X.$$

In the terminology of [**1**], this is equivalent to the K_X-twisted quiver bundle

$$E_0 \underset{\varphi_1}{\overset{\varphi_0}{\rightleftarrows}} E_1,$$

where the K_X-twisting means that the target of both homomorphisms are twisted by K_X. Let $(E_0, E_1, \varphi_0, \varphi_1)$ be an element in $\mathcal{M}(G_0, \bar{\mathfrak{g}}_1)$, then we can associate to it the $\mathrm{SL}(n, \mathbb{C})$-Higgs bundle (E, φ) with

$$E = E_0 \oplus E_1 \quad \text{and} \quad \varphi = \begin{pmatrix} 0 & \varphi_1 \\ \varphi_0 & 0 \end{pmatrix}.$$

This defines a morphism $\mathcal{M}(G^\theta, \mathfrak{g}_1) \to \mathcal{M}(G)$, whose image is in the fixed point locus of the involution of $\mathcal{M}(G) \to \mathcal{M}(G)$ defined by sending a G-Higgs bundle (E, φ) to $(E, -\varphi)$ (see Section 3.2, [**21, 31**]).

The objects in $\mathcal{M}(G_0, \bar{\mathfrak{g}}_1)$ are some times referred as $\mathrm{SU}(n_0, n_1)$-Higgs bundles since by the non-abelian Hodge correspondence explained in 3.3, the moduli space $\mathcal{M}(G_0, \bar{\mathfrak{g}}_1)$ is homemorphic to the $\mathrm{SU}(n_0, n_1)$-character variety of the fundamental group of X (see [**9, 10, 31**]). Indeed, for $G = \mathrm{SL}(n, \mathbb{C})$ the classes in $\mathrm{Conj}(G)/\sim$ corresponding to the trivial clique are represented (see [**38**] e.g.) by the conjugations σ_{n_0,n_1}, with $0 \leqslant n_0 \leqslant n_1$ and $n_0 + n_1 = n$ given by

$$\begin{aligned} \sigma_{n_0,n_1} : \mathrm{SL}(n, \mathbb{C}) &\to \mathrm{SL}(n, \mathbb{C}) \\ A &\mapsto I_{n_0,n_1} (\overline{A}^t)^{-1} I_{n_0,n_1}, \end{aligned}$$

where

$$I_{n_0,n_1} = \begin{pmatrix} I_{n_0} & 0 \\ 0 & -I_{n_1} \end{pmatrix}.$$

One has $G^{\sigma_{n_0,n_1}} = \mathrm{SU}(n_0,n_1)$. In particular $\tau := \sigma_{0,n}$ gives the compact real form $\mathrm{SU}(n)$. The elements in $\mathrm{Aut}_2(G)$ corresponding to σ_{n_0,n_1} are given by $\theta_{n_0,n_1} := \tau\sigma_{n_0,n_1}$, and hence $G^{\theta_{n_0,n_1}} = \mathrm{S}(\mathrm{GL}(n_0,\mathbb{C}) \times \mathrm{GL}(n_1,\mathbb{C}))$. Notice that the quasi-split real form in the trivial clique is $\mathrm{SU}(k,k)$ if $n = 2k$, and $\mathrm{SU}(k,k+1)$ if $n = 2k+1$.

A natural invariant associated to a $(G_0,\bar{\mathfrak{g}}_1)$-Higgs pair $(E_0,E_1,\varphi_0,\varphi_1)$ is given by

$$d = \deg E_0 = -\deg E_1.$$

The Toledo invariant in this case is $\tau = 2d$ and, if $(E_0,E_1,\varphi_0,\varphi_1)$ is semistable, one has as a particular case of (4.7)

$$-\operatorname{rank}(\varphi_1)(g-1) \leqslant d \leqslant \operatorname{rank}(\varphi_0)(g-1),$$

originally proved in **[9]**, which implies the classical Milnor–Wood inequality

$$|d| \leqslant \min\{n_0,n_1\}(g-1).$$

Let $\mathcal{M}_d \subset \mathcal{M}(G_0,\bar{\mathfrak{g}}_1)$ the subvariety consisting of elements with fixed Toledo invariant $2d$. In **[9–11]** it is proved that $\mathcal{M}_d$ is connected if $|d| < \min\{n_0,n_1\}(g-1)$ and has 2^{2g} connected components if $|d| = \min\{n_0,n_1\}(g-1)$. This result was proved by Hitchin for $\mathrm{SU}(1,1)$ in his original paper **[39]**.

A very special rigidity phenomenon takes place when $|d| = \min\{n_0,n_1\}(g-1)$. If $n_0 \neq n_1$ this manifests in the fact that the dimension of $\mathcal{M}_d$ is smaller than the expected dimension which is half of the dimension of $\mathcal{M}(G)$ (see **[9, 10]** for details). When $n_0 = n_1 = m$ and $|d| = m(g-1)$, then $\mathcal{M}_d$ is isomorphic as an algebraic variety to the moduli space of K_X^2-twisted $(\hat{G},\mathfrak{gl}(m,\mathbb{C}))$-Higgs pairs over X where

$$\hat{G} = \{A \in \mathrm{GL}(m,\mathbb{C}) \text{ such that } (\det A)^2 = 1\}.$$

This isomorphism is known as **Cayley correspondence** and takes place more generally for Hermitian real forms of **tube type** (see **[7, 9, 10]** for details). The 2^{2g} connected components of $\mathcal{M}_d$ in this case are known as **Cayley components** and via the non-abelian Hodge correspondence are homeomorphic to components of the corresponding character variety consisting entirely of discrete and faithful representations. These are examples of **higher Teichmüller spaces** (see **[23, 81]** for surveys on this subject and **[8, 33]** for more recent results). In the case of $\mathrm{SU}(1,1)$, as described by Hitchin **[39]**, these components are identified with the usual Teichmüller space of the underlying smooth surface to X. A similar rigidity phenomenon takes place also when the Toledo invariant is maximal for Hodge bundles **[6]** and cyclic Higgs bundles **[24]**.

5. Vinberg θ-pairs and the Hitchin fibration

Let G be a semisimple complex Lie group and let $\theta \in \mathrm{Aut}(G)$ be of order m. Consider the $\mathbb{Z}/m\mathbb{Z}$-grading defined by θ

$$\mathfrak{g} = \bigoplus_{i \in \mathbb{Z}/m\mathbb{Z}} \mathfrak{g}_i,$$

and the Vinberg θ-pair $(G^\theta,\mathfrak{g}_1)$. Let X be a compact Riemann surface and $\mathcal{M}(G^\theta,\mathfrak{g}_1)$ be the moduli space of $(G^\theta,\mathfrak{g}_1)$-Higgs pairs over X. Recall from Sections 2.3 and 2.4 (whose definitions and notions we follow) that by **[76]**

$$\mathbb{C}[\mathfrak{g}_1]^{G^\theta} \to \mathbb{C}[\mathfrak{a}]^{W(\mathfrak{a})} = \mathbb{C}[f_1,\cdots,f_r], \tag{5.1}$$

where $r = \dim \mathfrak{a} = \mathrm{rk}(G^\theta, \mathfrak{g}_1)$. Let $d_i = \deg f_i$.

Evaluating the polynomials f_i on the Higgs field we have the **Hitchin map**

$$h : \mathcal{M}(G^\theta, \mathfrak{g}_1) \to B(G^\theta, \mathfrak{g}_1) \cong \bigoplus_{i=1}^{r} H^0(X, K_X^{d_i}). \tag{5.2}$$

When $m = 1$, (5.1) is the classical Chevalley restriction Theorem for the adjoint action of G on $\mathfrak{g}$, and one has the original Hitchin map defined in **[40]** and the **Hitchin integrable system**

$$\mathcal{M}(G) \to B(G) \cong \bigoplus_{i=1}^{r} H^0(X, K^{d_i}).$$

Here $r = \mathrm{rk}\, G$ and $\{d_1, \cdots, d_r\}$ are the **exponents** of G. In **[40]** Hitchin gives a description of the generic fibres for the classical groups in terms of the Jacobian or appropriate Prym variety of a **spectral curve**. In **[19]** Donagi–Gaitsgory give a cameral curve description for general G as a **gerbe** with generic abelian fibres. This approach was reformulated by Ngô **[55]** in his proof of the **Fundamental Lemma**. In **[41]** Hitchin constructed a section of the Hitchin map which he identified with a connected component of the character variety of the fundamental group of X in a split real form of G. This is called a **Hitchin component** and is another instance of the higher Teichmüller spaces mentioned in Section 4.3.

The case $m = 2$ corresponds to symmetric pairs. Here, as explained in Section 3.3, the moduli space $\mathcal{M}(G^\theta, \mathfrak{g}_1)$ of $(G^\theta, \mathfrak{g}_1)$-Higgs pairs over X is homeomorphic to the character variety of $\pi_1(X)$ in the real form G^σ, where $\sigma = \tau\theta$, and τ is a compact antiholomorphic involution of G commuting with θ. The Hitchin map in this situation has been studied by Schaposnik for classical real forms using the spectral curve approach, and by Peón-Nieto **[59]** in terms of cameral covers for an arbitrary involution θ. From both points of view, one can see that the generic fibres are abelian if and only if the real form is quasi-split. This non-abelian phenomenon is very nicely illustrated for certain real forms in **[42]**. A construction of the gerbe in the quasi-split case, following the Donagi–Gaitsgory approach has been given in **[28]**. This has also been studied by Leslie **[51]** in relation to a symmetric pair version of the Fundamental Lemma. Further study of the gerbe structure is being carried out by Hameister–Morrissey **[37]**. A section of the Hitchin map in this case was constructed in **[29]** building upon the Kostant–Rallis section **[49]** and the Hitchin–Kostant section given by Hitchin in **[41]**

The study of the Hitchin map for general Vinberg θ-pairs with $m > 2$ is one of the main themes of **[24]**. There are several important questions to address here. One is regarding the quasi-split condition for the pair $(G^\theta, \mathfrak{g}_1)$ that was given in Section 2.7. While we have many examples of Vinberg θ-pairs satisfying this condition, we are not aware of a classification of quasi-split pairs in this generality. Similarly to the case of symmetric pairs, this condition ensures that the generic fibres are abelian. Another important issue regards the construction of a section of the Hitchin map. This is based on the existence of the Kostant–Weierstrass section for the quotient map $\mathfrak{g}_1 \to \mathfrak{g}_1 /\!\!/ G^\theta$ mentioned in Section 2.3.

Acknowledgments

The author wishes to thank Miguel González, Nigel Hitchin and Alastair King for very useful discussions.

References

[1] L. Álvarez-Cónsul and O. García-Prada, *Hitchin-Kobayashi correspondence, quivers, and vortices*, Comm. Math. Phys. **238** (2003), no. 1-2, 1–33, DOI 10.1007/s00220-003-0853-1. MR1989667

[2] L. Álvarez-Cónsul, O. García-Prada, and A. H. W. Schmitt, *On the geometry of moduli spaces of holomorphic chains over compact Riemann surfaces*, IMRP Int. Math. Res. Pap. (2006), Art. ID 73597, 82. MR2253535

[3] D. Baraglia, *Cyclic Higgs bundles and the affine Toda equations*, Geom. Dedicata **174** (2015), 25–42, DOI 10.1007/s10711-014-0003-2. MR3303039

[4] A. Beauville, M. S. Narasimhan, and S. Ramanan, *Spectral curves and the generalised theta divisor*, J. Reine Angew. Math. **398** (1989), 169–179, DOI 10.1515/crll.1989.398.169. MR998478

[5] M. Bhargava and B. H. Gross, *Arithmetic invariant theory*, Symmetry: representation theory and its applications, Progr. Math., vol. 257, Birkhäuser/Springer, New York, 2014, pp. 33–54, DOI 10.1007/978-1-4939-1590-3_3. MR3363006

[6] O. Biquard, B. Collier, O. García-Prada, and D. Toledo, *Arakelov-Milnor inequalities and maximal variations of Hodge structure*, Compos. Math. **159** (2023), no. 5, 1005–1041, DOI 10.1112/s0010437x23007157. MR4580300

[7] O. Biquard, O. García-Prada, and R. Rubio, *Higgs bundles, the Toledo invariant and the Cayley correspondence*, J. Topol. **10** (2017), no. 3, 795–826, DOI 10.1112/topo.12023. MR3797597

[8] S. B. Bradlow, B. Collier, O. García-Prada, P. B. Gothen, and A. Oliveira, *A general Cayley correspondence and higher Teichmüller spaces*, Preprint, arXiv:2101.09377, 2021.

[9] S. B. Bradlow, O. García-Prada, and P. B. Gothen, *Surface group representations and* $\mathrm{U}(p,q)$*-Higgs bundles*, J. Differential Geom. **64** (2003), no. 1, 111–170. MR2015045

[10] S. B. Bradlow, O. García-Prada, and P. B. Gothen, *Maximal surface group representations in isometry groups of classical Hermitian symmetric spaces*, Geom. Dedicata **122** (2006), 185–213, DOI 10.1007/s10711-007-9127-y. MR2295550

[11] S. Bradlow, O. García-Prada, P. Gothen, and J. Heinloth, *Irreducibility of moduli of semistable chains and applications to* $\mathrm{U}(p,q)$*-Higgs bundles*, Geometry and physics. Vol. II, Oxford Univ. Press, Oxford, 2018, pp. 455–470. MR3931782

[12] M. Burger, A. Iozzi, and A. Wienhard, *Surface group representations with maximal Toledo invariant*, Ann. of Math. (2) **172** (2010), no. 1, 517–566, DOI 10.4007/annals.2010.172.517. MR2680425

[13] E. Cartan, *Les groupes réels simples, finis et continus* (French), Ann. Sci. École Norm. Sup. (3) **31** (1914), 263–355. MR1509178

[14] C. Chevalley, *Invariants of finite groups generated by reflections*, Amer. J. Math. **77** (1955), 778–782, DOI 10.2307/2372597. MR72877

[15] B. Collier, *Finite order automorphisms of Higgs Bsundles: Theory and application*, ProQuest LLC, Ann Arbor, MI, 2016. Thesis (Ph.D.)–University of Illinois at Urbana-Champaign. MR3624308

[16] K. Corlette, *Flat G-bundles with canonical metrics*, J. Differential Geom. **28** (1988), no. 3, 361–382. MR965220

[17] S. Dai and Q. Li, *On cyclic Higgs bundles*, Math. Ann. **376** (2020), no. 3-4, 1225–1260, DOI 10.1007/s00208-018-1779-4. MR4081114

[18] S. K. Donaldson, *Twisted harmonic maps and the self-duality equations*, Proc. London Math. Soc. (3) **55** (1987), no. 1, 127–131, DOI 10.1112/plms/s3-55.1.127. MR887285

[19] R. Y. Donagi and D. Gaitsgory, *The gerbe of Higgs bundles*, Transform. Groups **7** (2002), no. 2, 109–153, DOI 10.1007/s00031-002-0008-z. MR1903115

[20] R. Donagi and T. Pantev, *Langlands duality for Hitchin systems*, Invent. Math. **189** (2012), no. 3, 653–735, DOI 10.1007/s00222-012-0373-8. MR2957305

[21] O. García-Prada, *Involutions of the moduli space of* $\mathrm{SL}(n,\mathbb{C})$*-Higgs bundles and real forms*, Vector bundles and low codimensional subvarieties: state of the art and recent developments, Quad. Mat., vol. 21, Dept. Math., Seconda Univ. Napoli, Caserta, 2007, pp. 219–238. MR2544088

[22] O. García-Prada, *Higgs bundles and surface group representations*, Moduli spaces and vector bundles, London Math. Soc. Lecture Note Ser., vol. 359, Cambridge Univ. Press, Cambridge, 2009, pp. 265–310. MR2537072

[23] O. García-Prada, *Higgs bundles and higher Teichmüller spaces*, Handbook of Teichmüller theory. Vol. VII, IRMA Lect. Math. Theor. Phys., vol. 30, Eur. Math. Soc., Zürich, 2020, pp. 239–285. MR4321175

[24] O. García-Prada and M. González, *Cyclic Higgs bundles and the Toledo invariant*, preprint 2024, arXiv:2403.00415.

[25] O. García-Prada, P. B. Gothen, and I. Mundet i Riera, *The Hitchin–Kobayashi correspondence, Higgs pairs and surface group representations*, arXiv:0909.4487 [math.AG], 2009.

[26] O. García-Prada and J. Heinloth, *The y-genus of the moduli space of* PGL_n*-Higgs bundles on a curve (for degree coprime to n)*, Duke Math. J. **162** (2013), no. 14, 2731–2749, DOI 10.1215/00127094-2381369. MR3127812

[27] O. García-Prada, J. Heinloth, and A. Schmitt, *On the motives of moduli of chains and Higgs bundles*, J. Eur. Math. Soc. (JEMS) **16** (2014), no. 12, 2617–2668, DOI 10.4171/JEMS/494. MR3293805

[28] O. García-Prada and A. Peón-Nieto, *Abelianization of Higgs bundles for quasi-split real groups*, Transform. Groups **28** (2023), no. 1, 285–325, DOI 10.1007/s00031-021-09658-9. MR4552177

[29] O. García-Prada, A. Peón-Nieto, and S. Ramanan, *Higgs bundles for real groups and the Hitchin-Kostant-Rallis section*, Trans. Amer. Math. Soc. **370** (2018), no. 4, 2907–2953, DOI 10.1090/tran/7363. MR3748589

[30] O. García-Prada and S. Ramanan, *Involutions of rank 2 Higgs bundle moduli spaces*, Geometry and physics. Vol. II, Oxford Univ. Press, Oxford, 2018, pp. 535–550. MR3931786

[31] O. García-Prada and S. Ramanan, *Involutions and higher order automorphisms of Higgs bundle moduli spaces*, Proc. Lond. Math. Soc. (3) **119** (2019), no. 3, 681–732, DOI 10.1112/plms.12242. MR3960666

[32] V. Gatti and E. Viniberghi, *Spinors of* 13*-dimensional space*, Adv. in Math. **30** (1978), no. 2, 137–155, DOI 10.1016/0001-8708(78)90034-8. MR513846

[33] O. Guichard, F. Labourie, and A. Wienhard, Positivity and representations of surface groups, preprint 2021, arXiv:2106.14584

[34] B. H. Gross, *On Bhargava's representation and Vinberg's invariant theory*, Frontiers of mathematical sciences, Int. Press, Somerville, MA, 2011, pp. 317–321. MR3050830

[35] L. Gruson and S. V. Sam, *Alternating trilinear forms on a nine-dimensional space and degenerations of* (3,3)*-polarized Abelian surfaces*, Proc. Lond. Math. Soc. (3) **110** (2015), no. 3, 755–785, DOI 10.1112/plms/pdu050. MR3342104

[36] L. Gruson, S. V. Sam, and J. Weyman, *Moduli of abelian varieties, Vinberg* θ*-groups, and free resolutions*, Commutative algebra, Springer, New York, 2013, pp. 419–469, DOI 10.1007/978-1-4614-5292-8_13. MR3051381

[37] T. Hameister and B. Morrissey, *The Hitchin fibration for symmetric pairs*, in preparation.

[38] S. Helgason, *Differential geometry, Lie groups, and symmetric spaces*, Pure and Applied Mathematics, vol. 80, Academic Press, Inc. [Harcourt Brace Jovanovich, Publishers], New York-London, 1978. MR514561

[39] N. J. Hitchin, *The self-duality equations on a Riemann surface*, Proc. London Math. Soc. (3) **55** (1987), no. 1, 59–126, DOI 10.1112/plms/s3-55.1.59. MR887284

[40] N. Hitchin, *Stable bundles and integrable systems*, Duke Math. J. **54** (1987), no. 1, 91–114, DOI 10.1215/S0012-7094-87-05408-1. MR885778

[41] N. J. Hitchin, *Lie groups and Teichmüller space*, Topology **31** (1992), no. 3, 449–473, DOI 10.1016/0040-9383(92)90044-I. MR1174252

[42] N. Hitchin and L. P. Schaposnik, *Nonabelianization of Higgs bundles*, J. Differential Geom. **97** (2014), no. 1, 79–89. MR3229050

[43] V. G. Kac, *Automorphisms of finite order of semisimple Lie algebras* (Russian), Funkcional. Anal. i Priložen. **3** (1969), no. 3, 94–96. MR251091

[44] A. Kapustin and E. Witten, *Electric-magnetic duality and the geometric Langlands program*, Commun. Number Theory Phys. **1** (2007), no. 1, 1–236, DOI 10.4310/CNTP.2007.v1.n1.a1. MR2306566

[45] A. King, private communication, 2022.

[46] A. King, Mysterious duality and exceptional bundles on del Pezzo surfaces, in preparation.

[47] A. W. Knapp, *Lie groups beyond an introduction*, 2nd ed., Progress in Mathematics, vol. 140, Birkhäuser Boston, Inc., Boston, MA, 2002. MR1920389

[48] B. Kostant, *Lie group representations on polynomial rings*, Amer. J. Math. **85** (1963), 327–404, DOI 10.2307/2373130. MR158024

[49] B. Kostant and S. Rallis, *Orbits and representations associated with symmetric spaces*, Amer. J. Math. **93** (1971), 753–809, DOI 10.2307/2373470. MR311837

[50] F. Labourie, *Cyclic surfaces and Hitchin components in rank 2*, Ann. of Math. (2) **185** (2017), no. 1, 1–58, DOI 10.4007/annals.2017.185.1.1. MR3583351

[51] S. Leslie, *An analogue of the Grothendieck-Springer resolution for symmetric spaces*, Algebra Number Theory **15** (2021), no. 1, 69–107, DOI 10.2140/ant.2021.15.69. MR4226983

[52] P. Levy, *Involutions of reductive Lie algebras in positive characteristic*, Adv. Math. **210** (2007), no. 2, 505–559, DOI 10.1016/j.aim.2006.07.002. MR2303231

[53] P. Levy, *Vinberg's θ-groups in positive characteristic and Kostant-Weierstrass slices*, Transform. Groups **14** (2009), no. 2, 417–461, DOI 10.1007/s00031-009-9056-y. MR2504929

[54] P. Levy, *KW-sections for Vinberg's θ-groups of exceptional type*, J. Algebra **389** (2013), 78–97, DOI 10.1016/j.jalgebra.2013.04.035. MR3065993

[55] B. C. Ngô, *Le lemme fondamental pour les algèbres de Lie* (French), Publ. Math. Inst. Hautes Études Sci. **111** (2010), 1–169, DOI 10.1007/s10240-010-0026-7. MR2653248

[56] A. L. Onishchik and E. B. Vinberg (Eds.), *Lie Groups and Lie Algebras III*, Encyclopedia of Mathematical Sciences, Vol. 41, Springer–Verlag 1994.

[57] D. I. Panyushev, *Regular elements in spaces of linear representations. II* (Russian), Izv. Akad. Nauk SSSR Ser. Mat. **49** (1985), no. 5, 979–985, 1120. MR810527

[58] D. I. Panyushev, *On invariant theory of θ-groups*, J. Algebra **283** (2005), no. 2, 655–670, DOI 10.1016/j.jalgebra.2004.03.032. MR2111215

[59] A. Peón-Nieto, *Higgs bundles, real forms and the Hitchin fibration*, Ph.D. Thesis, UAM-ICMAT, Madrid 2013.

[60] V. L. Popov, *Representations with a free module of covariants* (Russian), Funkcional. Anal. i Priložen. **10** (1976), no. 3, 91–92. MR417197

[61] E. M. Rains and S. V. Sam, *Invariant theory of* $\wedge^3(9)$ *and genus-2 curves*, Algebra Number Theory **12** (2018), no. 4, 935–957, DOI 10.2140/ant.2018.12.935. MR3830207

[62] A. Ramanathan, *Stable principal bundles on a compact Riemann surface*, Math. Ann. **213** (1975), 129–152, DOI 10.1007/BF01343949. MR369747

[63] A. Ramanathan, *Moduli for principal bundles over algebraic curves. II*, Proc. Indian Acad. Sci. Math. Sci. **106** (1996), no. 4, 421–449, DOI 10.1007/BF02837697. MR1425616

[64] M. Reeder, P. Levy, J.-K. Yu, and B. H. Gross, *Gradings of positive rank on simple Lie algebras*, Transform. Groups **17** (2012), no. 4, 1123–1190, DOI 10.1007/s00031-012-9196-3. MR3000483

[65] M. Sato and T. Kimura, *A classification of irreducible prehomogeneous vector spaces and their relative invariants*, Nagoya Math. J. **65** (1977), 1–155. MR430336

[66] L.P. Schaposnik, *Spectral data for G-Higgs bundles*, DPhil thesis, University of Oxford, 2013.

[67] L. P. Schaposnik, *Spectral data for* $U(m,m)$*-Higgs bundles*, Int. Math. Res. Not. IMRN **11** (2015), 3486–3498. MR3373057

[68] A. H. W. Schmitt, *Geometric invariant theory and decorated principal bundles*, Zurich Lectures in Advanced Mathematics, European Mathematical Society (EMS), Zürich, 2008, DOI 10.4171/065. MR2437660

[69] G. C. Shephard and J. A. Todd, *Finite unitary reflection groups*, Canad. J. Math. **6** (1954), 274–304, DOI 10.4153/cjm-1954-028-3. MR59914

[70] J.-P. Serre, *Cohomologie galoisienne* (French), 5th ed., Lecture Notes in Mathematics, vol. 5, Springer-Verlag, Berlin, 1994, DOI 10.1007/BFb0108758. MR1324577

[71] J. de Siebenthal, *Sur les groupes de Lie compacts non connexes* (French), Comment. Math. Helv. **31** (1956), 41–89, DOI 10.1007/BF02564352. MR94408

[72] C. T. Simpson, *Constructing variations of Hodge structure using Yang-Mills theory and applications to uniformization*, J. Amer. Math. Soc. **1** (1988), no. 4, 867–918, DOI 10.2307/1990994. MR944577

[73] C. T. Simpson, *Higgs bundles and local systems*, Inst. Hautes Études Sci. Publ. Math. **75** (1992), 5–95. MR1179076

[74] C. Simpson, *Katz's middle convolution algorithm*, Pure Appl. Math. Q. **5** (2009), no. 2, Special Issue: In honor of Friedrich Hirzebruch., 781–852, DOI 10.4310/PAMQ.2009.v5.n2.a8. MR2508903

[75] J. A. Thorne, *Vinberg's representations and arithmetic invariant theory*, Algebra Number Theory **7** (2013), no. 9, 2331–2368, DOI 10.2140/ant.2013.7.2331. MR3152016

[76] È. B. Vinberg, *The Weyl group of a graded Lie algebra* (Russian), Izv. Akad. Nauk SSSR Ser. Mat. **40** (1976), no. 3, 488–526, 709. MR430168

[77] È. B. Vinberg, *Classification of homogeneous nilpotent elements of a semi-simple graded Lie algebra*, Selecta Math. Soviet. **6** (1987) 15–35.

[78] E. B. Vinberg and A. G. Èlashvili, *Classification of trivectors of a 9-dimensional space*, Sel, Math. Sov. **7** (1988)

[79] È. B. Vinberg and V. L. Popov, *Invariant theory* (Russian), Algebraic geometry, 4 (Russian), Itogi Nauki i Tekhniki, Akad. Nauk SSSR, Vsesoyuz. Inst. Nauchn. i Tekhn. Inform., Moscow, 1989, pp. 137–314, 315. MR1100485

[80] N. R. Wallach, *Geometric invariant theory: Over the real and complex numbers*, Universitext, Springer, Cham, 2017, DOI 10.1007/978-3-319-65907-7. MR3700428

[81] A. Wienhard, *An invitation to higher Teichmüller theory*, Proceedings of the International Congress of Mathematicians—Rio de Janeiro 2018. Vol. II. Invited lectures, World Sci. Publ., Hackensack, NJ, 2018, pp. 1013–1039. MR3966798

[82] J. A. Wolf and A. Gray, *Homogeneous spaces defined by Lie group automorphisms. I*, J. Differential Geometry **2** (1968), 77–114. MR236328

Instituto de Ciencias Matemáticas, CSIC-UAM-UC3M-UCM, Nicolás Cabrera, 13–15, 28049 Madrid, Spain

Email address: `oscar.garcia-prada@icmat.es`

Contemporary Mathematics
Volume **803**, 2024
https://doi.org/10.1090/conm/803/16100

A guide to moduli theory beyond GIT

Tomás L. Gómez, Andres Fernández Herrero, and Alfonso Zamora

Dedicated to Peter E. Newstead, on his 80th birthday

ABSTRACT. In this survey we provide an overview of some recent developments in the construction of moduli spaces using stack-theoretic techniques. We will also explain the analogue of Harder-Narasimhan stratifications for general stacks, known as Θ-stratifications. As an application of the ideas exposed here, we address the moduli problem of principal bundles over higher dimensional projective varieties, as well as its different compactifications by the so-called principal ρ-sheaves. We construct a stratification by instability types whose lower strata admits a proper good moduli space of "Gieseker semistable" objects and a new Gieseker-type Harder-Narasimhan filtration for these objects. Detailed proofs of the latter results will appear elsewhere.

1. Introduction

The construction and study of moduli spaces has been one of the driving forces in algebraic geometry for decades. After the algebraic construction of the Picard scheme by Grothendieck, the study of the moduli of vector bundles came next. Not only has Peter Newstead been one of the pioneers in this study since his early papers in the 60's, but by founding the group VBAC (Vector Bundles on Algebraic Curves), he has been pivotal in building a large and successful community dedicated to this and related areas. It is an honour for us to dedicate this survey to him.

Since its inception by Mumford, Geometric Invariant Theory (GIT) [**MFK94**] has been a powerful way to construct moduli spaces for many moduli problems. When performing GIT, we do not attempt to classify all possible objects in a given moduli problem, but rather only those declared to be semistable. After adding the additional data of a "framing", these objects are parametrized by a scheme called the parameter space. Mumford's GIT provides a way to define the quotient of this scheme by the action of an algebraic group encoding the framing, revisiting Hilbert's 14th problem, to obtain a moduli scheme parametrizing semistable objects.

The intrinsic point of view in the Beyond GIT program developed in [**HL21**] does not focus on the subclass of semistable objects but on the whole moduli stack.

2020 *Mathematics Subject Classification.* Primary 14D20, 14D23, 14F06, 14J60, 14L24.

Key words and phrases. Moduli spaces, good moduli spaces, intrinsic GIT, Harder-Narasimhan filtration, Θ-stratifications, principal bundles, stacks, Gieseker stability, principal ρ-sheaves.

This work is supported by grants CEX2019-000904-S and PID2019-108936GB-C21 (funded by MCIN/AEI/ 10.13039/501100011033).

A generalization of the Hilbert-Mumford criterion leads to a new stacky numerical criterion through test stacks taking the role of former one-parameter subgroups. This makes the notion of semistability naturally appear in the stack, avoiding the tedious checking that semistable objects correspond to GIT semistable ones and vice-versa. Thanks to the recent paper [**AHLH19**], there are some intrinsic conditions that can be checked locally in an algebraic stack in order to show that it admits a separated good moduli space in the sense of Alper [**Alp13**].

Unstable objects usually come in a natural way with a maximal degeneration (a stacky Harder-Narasimhan filtration). The theory of Θ-stratifications in [**HL21**] can be used to find such maximal degenerations without having to consider non-canonically a GIT problem sufficiently large to contain the required unstable object as in [**GSZ15**]. This opens the door to understanding Harder-Narasimhan filtrations also in an intrinsic way, directly from the whole stack.

Harder-Narasimhan filtrations (also known as canonical filtrations) were first considered in [**HN75**] for vector bundles on curves. Shatz studied its behaviour in families in [**Sha77**]. In higher dimensions we have Gieseker semistablity, and we have Gieseker-Harder-Narasimahn filtrations for torsion free sheaves. Nitsure used them to produce schematic stratifications in [**Nit11**] (see also [**GN14**] for Λ-modules), which give rise to algebraic stacks corresponding to each Harder-Narasimhan type, thus defining a filtration of the stack of pure coherent sheaves by locally closed substacks.

As an application of this theory, we address the moduli problem of principal bundles over higher dimensional projective varieties, as well as its different compactifications by the so-called principal ρ-sheaves. We construct in [**GHZ24**] a stratification (the Θ-stratification of a given numerical invariant), whose lower open strata admit proper good moduli spaces (as in [**Alp13**]) of "Gieseker semistable" objects [**Gie77**]. This gives an intrinsic stack-theoretic construction to the already known moduli space of principal bundles on higher dimensional varieties [**Sch02, GS05, GLSS08, GLSS10**], as well as its compactifications.

For principal bundles, a Harder-Narasimhan filtration is usually called canonical reduction. They were announced by Ramanathan in [**Ram79**] and have been constructed over curves by Behrend [**Beh91**] and Biswas-Holla [**BH04**]. A reduction to a parabolic subgroup is also defined in [**AB83**] by a different method, and it is proved in [**AAB02**] to coincide with the canonical filtration. Gurjar and Nitsure shows that this reduction also gives a schematic stratification (i.e., a stratification of the stack) [**GN14, GN18**]. For principal bundles in positive characteristics, Gurjar and Nitsure [**GN16**] show that there are algebraic stacks corresponding to each HN-type, and these are radicial over the algebraic stack of all principal bundles. If the Behrend conjecture holds then these stacks are shown to define a stratification of the stack of all principal bundles (see also [**Hei08a**]). However, over higher dimension varieties, where a polynomial Gieseker stability condition is used to construct the compactification of the moduli space, canonical reductions for principal bundles in the literature only take into account the leading coefficients of the reduced Hilbert polynomial (analogous to slope stability as described in [**Mum63a**]). There has been partial attempts to construct a filtration that accounts for all of the terms in the Hilbert polynomial in lower rank [**Zam16**]. On the other hand, in [**BZ16**] it was shown that the notion of Gieseker Harder-Narasimhan parabolic reduction cannot be related to the Gieseker-Harder-Narasimhan filtration of the underlying vector bundles for faithful representations, which meant that the main strategies in the literature (e.g. [**AAB02**]) could not possibly apply if we try to account for all terms of the Hilbert polynomial.

An important outcome of the results in [**GHZ24**] is a new notion of a Gieseker-Harder-Narasimhan filtration for principal ρ-sheaves, which is constructed by iterating a filtration coming from the Θ-stratification (called the leading term HN filtration). The Gieseker Harder-Narasimhan filtration thus obtained consists of a parabolic reduction of the underlying rational principal bundle that measures instability by considering all degree terms of the numerical polynomials. Detailed proofs of the results contained in this survey will appear elsewhere.

NOTATION 1.1. For this survey we work over a fixed ground field k, which we assume to be of characteristic 0 unless explicitly stated. We denote by Sch_k the category of schemes over k. We will sometimes work instead over the smaller category Aff_k of affine schemes over k. For any k-schemes X and S, we will write X_S to denote the fiber product $X \times S$, which is a product over k unless otherwise stated.

2. Intrinsic approach to moduli problems

2.1. Algebraic stacks. Moduli theory rests on the insight that many classification problems in algebraic geometry can be parametrized by schemes. An example of this is the projective space $\mathbb{P}^n_k$, which classifies line bundles along with the choice of $n+1$ generating sections. In order to naturally deal with the algebraic structure of such classification problems, the Grothendieck school of algebraic geometry initiated the functorial point of view on such moduli problems.

We can view any scheme X as a functor h_X from $(\mathrm{Aff}_k)^{op}$ into sets via the Yoneda embedding. We say that functors of the form h_X are representable by schemes. We will equip Sch_k and Aff_k with the additional structure of the Grothendieck topology generated by fppf (finitely presented faithfully flat) covers. All representable functors satisfy descent with respect to this fppf topology.

Equipped with the functorial point of view, Grothendieck proved the representability of some of the central moduli problems in algebraic geometry: the Quot functor and the Picard functor. We refer the reader to the articles by Nitsure and Kleiman in [**FGI$^+$05**] for a nice exposition of these achievements. While the definition of the Quot functor intuitively captures the sought after moduli problem, surprises arise when one tries to parametrize line bundles on a given projective variety X. The natural functor classifying isomorphism classes of line bundles does not satisfy fppf descent, and therefore cannot be represented by a scheme. However, Grothendieck proved that the fppf sheafification $\mathrm{Pic}(X)$ of the natural functor is indeed represented by a scheme.

When the variety X is proper but not projective, Artin showed the fppf sheafification $\mathrm{Pic}(X)$ is represented by an algebraic space [**Art69**, Thm 7.3]. Algebraic spaces form a subcategory of all fppf sheaves. This subcategory provides a more robust setting to parametrize moduli problems, while remaining fundamentally geometric at the same time. We refer to [**Knu71**], [**Sta23**] for a detailed treatments of the theory of algebraic spaces. Artin's result on the representability of $\mathrm{Pic}(X)$ is one of the many achievements that hinted at the usefulness of enlarging the category of schemes in order to study moduli problems.

A more serious issue arises when one tries to generalize the Picard scheme to the setting of higher rank vector bundles on a given projective variety X. Even in the case when X is a curve, it becomes impossible to parametrize the moduli problem of all vector bundles in terms of a separated scheme. The main feature of this moduli problem is the existence of large nontrivial automorphisms of the objects, which makes the condition of descent for isomorphism classes a more delicate affair. One approach, pioneered by Mumford when X is a smooth projective curve, is to restrict to a special class of reasonably "rigid" vector bundles called stable vector bundles [**Mum63a**]. Alternatively, if one wants to be able to parametrize all vector bundles, then it becomes necessary to account for their automorphisms in order to naturally deal with the descent condition. This is elegantly encoded in the notion of an algebraic stack, which is another geometric enlargement of the category of schemes originally introduced by Deligne and Mumford [**DM69**].

For the rest of the survey we will assume that the reader has some familiarity with the theory of algebraic stacks. For an informal introduction we refer the reader to [**Gó1**] [**Hei10b**]; on the other hand we recommend [**CMW18**] [**Alp23**] for more detailed developments of the theory. Some standard references are [**LMB00**], [**Ols16**] and [**Sta23**].

We will favor the point of view of stacks as pseudofunctors valued in groupoids, which we believe is more intuitive for moduli problems. There should be no issue in translating things into the theoretically cleaner language of fibered categories thanks to [**FGI**$^+$**05**, Part 1, §3.1.3].

Recall that a stack $\mathcal{M}$ is a pseudofuctor from $(\mathrm{Sch}_k)^{op}$ into groupoids that satisfies fppf descent. We say that a stack is algebraic if it admits a smooth morphism $f : U \to \mathcal{M}$ representable by algebraic spaces such that the source U is a scheme. Such morphism f is called an atlas. Many of the properties of algebraic stacks (locally of finite type, quasi-compact, smooth, normal) can be encoded in terms of properties of some atlas.

Just as in the case of schemes, algebraic stacks are closed under the operation of fiber product. One can define the diagonal of a stack in a similar way as in the case of schemes. The main difference is that in the case of algebraic stacks the diagonal is not a monomorphism. Instead, the fibers of the diagonal encode automorphisms of the corresponding points of the stack.

Throughout this survey we will only work with stacks that are locally of finite type over k. Moreover we shall restrict our attention to stacks with affine diagonal.

2.2. Good moduli spaces. Many moduli problems in algebraic geometry, such as the moduli of vector bundles on a smooth projective curve, are parametrized by algebraic stacks [**Hei10b**]. As in the case of the Picard scheme or the substack of stable vector bundles, it is often possible to "rigidify" the functor to remove the automorphisms, thus obtaining a functor valued in sets that is representable by an algebraic space. Such algebraic space is called the moduli space associated to the stack.

When we are working with a stack $\mathcal{M}$ with finite inertia, the theorem of Keel and Mori [**KM97**] [**Sta23**, Tag 0DUT] provides a general procedure to obtain a "rigidification" in the form of an algebraic space M called the coarse moduli space as in the following.

DEFINITION 2.1. A morphism $f : \mathcal{M} \to M$ into an algebraic space M is called a coarse moduli space morphism (and M is called a coarse moduli space) if the following are satisfied

(1) f is initial among morphisms to algebraic spaces.
(2) For any algebraically closed field K, the natural transformation
$$f : \mathcal{M}(K) \to M(K)$$
induces a bijection on isomorphism classes of points.

There is a coarse moduli space morphism $\mathcal{M} \to M$ that relates both functors. Many geometric properties of the original algebraic stack $\mathcal{M}$ (and of the moduli problem parametrized by it) are reflected in the algebraic space M, which lives in a closer setting to the more geometric world of varieties.

When the objects parametrized by the algebraic stack $\mathcal{M}$ have infinite automorphisms (i.e. algebraic automorphism groups of positive dimension), then the power of the Keel-Mori theorem becomes unavailable. Unfortunately, many natural moduli problems arising in algebraic geometry have infinite symmetry groups, so it becomes important to develop technical tools to deal with this type of situation. Mumford's Geometric Invariant Theory (GIT) [**MFK94**] offers a powerful device to construct a moduli space in the case when the algebraic stack $\mathcal{M}$ is a quotient stack of the form $[X/G]$, where X is open inside a projective-over-affine variety and G is a reductive group. In this case, the properties of the corresponding morphism $[X/G] \to M$ are captured by the notion of a "good quotient".

Alper's notion of good moduli space is an elegant way to encode and generalize Mumford's theory to the setting of arbitrary algebraic stacks that do not necesarily arise as quotient stacks.

DEFINITION 2.2 (Good moduli space, [**Alp13**, Defn. 4.1]). A quasi-compact and quasiseparated morphism $\varphi : \mathcal{M} \to M$ from an algebraic stack $\mathcal{M}$ to an algebraic space M is called a good moduli space if both of the following properties are satisfied:

(a) The pushforward morphism $\varphi_* : \mathrm{QCoh}(\mathcal{M}) \to \mathrm{QCoh}(M)$ is exact.
(b) The induced morphism of structure sheaves $\mathcal{O}_M \to \varphi_*\mathcal{O}_{\mathcal{M}}$ is an isomorphism.

We call the algebraic space M the good moduli space of $\mathcal{M}$.

REMARK 2.3. The notion of good moduli space becomes quite restrictive if we work over a base of positive or mixed characteristic. Most of the classical moduli spaces that admit a good moduli space in characteristic 0 (such as the moduli of semistable vector bundles on a curve) do not admit a good moduli space over a field of positive characteristic. Instead, they admit morphisms $\mathcal{M} \to M$ to an algebraic space satisfying a weakening of condition (a) in Definition 2.2. A more appropriate concept, the notion of adequate moduli space, was introduced by Alper in [**Alp14**] to approach moduli problems in arbitrary characteristic.

The above Definition 2.2 is deceivingly concise; it has a lot of strong implications which link the properties of the moduli space M and the stack $\mathcal{M}$. The first main property justifies the name "the" moduli space.

THEOREM 2.4 ([**Alp13**, Thm. 6.6]). *Let $\mathcal{M}$ be a locally Noetherian algebraic stack with a good moduli space $\phi : \mathcal{M} \to M$. Then ϕ is initial among all maps from*

$\mathcal{M}$ into algebraic spaces. In particular the good moduli space $\phi : \mathcal{M} \to M$ is unique up to unique isomorphism.

Another important property of good moduli spaces is that they behave well with respect to base-change.

PROPOSITION 2.5 ([**Alp13**, Prop. 4.7(i)]). *Let $\mathcal{M}$ be an algebraic stack with a good moduli space $\phi : \mathcal{M} \to M$. For any morphism of algebraic spaces $X \to M$, the base-change $\phi_X : \mathcal{M} \times_M X \to X$ is a good moduli space for the fiber product $\mathcal{M} \times_M X$.*

We finally summarize some of the important properties of good moduli spaces in the following

PROPOSITION 2.6. *Let $\phi : \mathcal{M} \to M$ be a good moduli space. Then the following hold.*

(i) *ϕ is surjective and universally closed* [**Alp13**, Thm. 4.16(i)+(ii)].
(ii) *If ϕ has affine diagonal, then ϕ is of finite type* [**AHR20**, Thm. A.1].
(iii) *The fibers of ϕ are geometrically connected, each ϕ-fiber contains a unique closed point* [**Alp13**, Thm. 4.16(vii) + Prop. 9.1].
(iv) *If $\mathcal{M}$ is locally Noetherian, then so is M, and the push-forward ϕ_* preserves coherence of sheaves* [**Alp13**, Thm. 4.16(x)].
(v) *If $\mathcal{M}$ is of finite type over an excellent scheme S, then M is also of finite type over S* [**Alp13**, Thm. 4.16(xi)].

In addition, there are some results that relate the singularities of a stack and the singularities of its good moduli space. We would like to highlight the following recent generalization of a classical theorem of Boutot.

THEOREM 2.7 ([**BGLM21**, Thm. 5]). *Let $\mathcal{M}$ be a smooth algebraic stack with affine diagonal over a field k with good moduli space $\phi : \mathcal{M} \to M$. If M is quasiprojective, then M is of klt type.*

We end this section by providing some examples of good moduli spaces.

EXAMPLE 2.8 (Affine modulo linearly reductive). Let $X = \mathrm{Spec}(A)$ be an affine variety equipped with the action of a linearly reductive group G over k. We consider the quotient stack $[X/G]$. We denote by $A^G \subset A$ the subring of invariants. There is a morphism $\phi : [X/G] \to \mathrm{Spec}(A^G)$ which exhibits $\mathrm{Spec}(A^G)$ as a good moduli space for $[X/G]$. Indeed, let us check both properties of a good moduli space morphism.

(a) The abelian category of quasicoherent sheaves $\mathrm{QCoh}([X/G])$ on the stack $[X/G]$ is equivalent to the category of G-equivariant A-modules. The push-forward $\phi_*(-)$ takes a G-equivariant A-module to the module of $\phi_*(M) = M^G \subset M$ of G-invariants, viewed as a module over A^G. Since the algebraic group G is assumed to be linearly reductive, the functor of taking G-invariants $(-)^G$ is exact, and therefore $\varphi_*(-)$ is exact.
(b) From the description of $\phi_*(-)$ in (a) above, it follows that $\varphi_*(\mathcal{O}_{[X/G]}) = A^G = \mathcal{O}_{\mathrm{Spec}(A^G)}$.

Thanks to the étale slice theorem for algebraic stacks proved by Alper, Hall and Rydh [**AHR19**], algebraic stacks that admit good moduli spaces are étale locally of the form $[\mathrm{Spec}(A)/G]$ as in the example above. Therefore the example serves as an excellent local picture to keep in mind when dealing with good moduli spaces.

The other familiar example of good moduli space comes from GIT.

EXAMPLE 2.9 (GIT, [**Alp13**, §13.5]). Let G be a linearly reductive group over a field k. Let X be a projective-over-affine scheme over k equipped with an action of G. Let $\mathcal{L}$ be an ample G-equivariant line bundle on X. Let $X^{\mathcal{L}\text{-ss}} \subset X$ denote the open G-subscheme of $\mathcal{L}$-semistable points. Then the natural morphism $\phi : [X^{\mathcal{L}\text{-ss}}/G] \to \mathrm{Proj}\left(\bigoplus_{n\geq 0} H^0(X, \mathcal{L}^{\otimes n})^G\right)$ is a good moduli space for the stack $[X^{\mathcal{L}\text{-ss}}/G]$.

Note that Example 2.8 is a special case of Example 2.9, where we take the equivariant line bundle to be the structure sheaf $\mathcal{O}_X$, so that all points are semistable.

We should stress that not all stacks admit good moduli spaces. Having a good moduli space imposes some necessary conditions on the stack $\mathcal{M}$. For example, it forces every closed point of $\mathcal{M}$ to have linearly reductive stabilizers. This leads to an important question: what properties of an algebraic stack guarantee that it admits a good moduli space? In the next subsection we shall provide an answer to this question by explaining a set of intrinsic conditions [**AHLH19**] that are necessary and sufficient for a stack over k to admit a separated good moduli space.

2.3. Intrinsic construction of moduli spaces. One of the most useful foundational results in the theory of algebraic spaces and algebraic stacks is Artin's theorem [**Art74**] [**Sta23**, Tag 07SZ]. This theorem provides a set of intrinsic local criteria (now known as the Artin criteria) which guarantee that a given functor or pseudofunctor is represented by an algebraic space or an algebraic stack. Artin's criteria are one of the main tools in the construction of an algebraic stack that parametrizes a given moduli problem.

If we want to "rigidify" the moduli problem and look for the construction of good moduli spaces, then classically the main tool available is the theorem of Keel and Mori in the case when the moduli problem has finite inertia. One can first apply Artin's criteria to show representability by an algebraic stack, and then use the Keel-Mori theorem to construct the coarse space.

In the case when the moduli problem has infinite automorphisms, a recent breakthrough paper of Alper, Halpern-Leistner and Heinloth [**AHLH19**] provides a set of intrinsic local conditions for an algebraic stack to admit a separated good moduli space.

The main strategy for the construction of such a space is conceptually simple. First, the étale slice theorem for stacks due to Alper, Hall and Rydh [**AHR20**, **AHR19**] states that, under certain hypotheses, one can present an algebraic stack $\mathcal{M}$ étale locally as a quotient stack of the form $[\mathrm{Spec}(A)/G]$, where G is a linearly reductive group. We know how to construct moduli spaces for such stacks, by Example 2.8. Then one would hope to be able to glue such moduli spaces in order to obtain a good moduli space for the stack $\mathcal{M}$. The insight of [**AHLH19**] is that there are some intrinsic properties of the stack $\mathcal{M}$ that guarantee that: (1) the hypotheses of the slice theorem are satisfied; and (2) one can glue the good moduli spaces for the local models in order to obtain a separated good moduli space for $\mathcal{M}$.

This provides an intrinsic way of constructing moduli spaces associated to moduli problems in algebraic geometry. One can first apply the Artin criteria to prove that the natural groupoid-valued pseudofunctor is represented by an algebraic stack. Then, using the criteria for the existence of good moduli spaces in [**AHLH19**], one can construct the sought-after moduli space.

This strategy has been useful in cases where the techniques of GIT are not available. One such example is the construction of the moduli of Bridgeland semistable objects in the derived category of a smooth projective variety [**AHLH19**, §7], where the stack $\mathcal{M}$ is not known to be a quotient stack. Another example is the moduli of K-polystable Fano varieties [**ABHLX20**, **BLX22**], where the line bundle inducing semistability is not ample on the natural parameter space given by the Hilbert scheme. More recently, the same techniques have also been applied to obtain intrinsic constructions of moduli of quiver representations [**BDF⁺22**]. We would also like to mention the construction of moduli spaces of objects in abelian categories, treated in [**AHLH19**]. Another important question is the projectivity of the moduli spaces obtained via intrinsic methods [**Taj21**, **ABB⁺22**, **BDF⁺22**]; we will not treat this aspect of the theory in this survey.

To end this section, we would like to describe more carefully the criteria in [**AHLH19**]. They consist of filling conditions with respect to some morphisms from test stacks, reminiscent of the valuative criterion for properness in classical algebraic geometry.

We start by introducing the relevant test stacks. Let $\mathbb{G}_m$ denote the multiplicative group, whose group of S-points is the group of invertible functions $H^0(\mathcal{O}_S^\times)$ for any k-scheme S. We let $\mathbb{G}_m$ act on the affine space $\mathbb{A}^1_k = \operatorname{Spec}(k[t])$ linearly such that the weight of the coordinate t is -1.

Definition 2.10 (stack Θ). We denote by Θ the quotient stack $\left[\mathbb{A}^1_k/\mathbb{G}_m\right]$.

Notice that the stack Θ has two k-points: the closed point 0 with stabilizer $\mathbb{G}_m$ corresponding to the origin in $\mathbb{A}^1_k$, and the open point 1 with trivial stabilizer corresponding the open orbit in $\mathbb{A}^1_k$. If R is a discrete valuation ring over k with uniformizer ϖ, then the base-change $\Theta_R := \Theta \times_{\operatorname{Spec(k)}} \operatorname{Spec}(R)$ has a unique closed point $(0,0)$ corresponding to the simultaneous vanishing locus of t and ϖ inside $\mathbb{A}^1_R = \operatorname{Spec}(R[t])$.

Definition 2.11 (Θ-reductivity). An algebraic stack $\mathcal{M}$ over k is said to be Θ-reductive if for all discrete valuation rings R over k and all morphisms $f : \Theta_R \setminus (0,0) \to \mathcal{M}$, there is a unique dashed arrow as below making the diagram commutative

$$\begin{array}{ccc} \Theta_R \setminus (0,0) & & \\ \downarrow & \overset{f}{\searrow} & \\ \Theta_R & \dashrightarrow & \mathcal{M} \end{array}$$

We continue with a choice of discrete valuation ring R over k with uniformizer ϖ.

Definition 2.12 (stack $\overline{ST}_R$). We define the quotient stack

$$\overline{ST}_R := [\operatorname{Spec}(R[t,s]/(st-\varpi))/(\mathbb{G}_m)_R]\,,$$

where $(\mathbb{G}_m)_R = \mathbb{G}_m \times_{\operatorname{Spec}(k)} \operatorname{Spec}(R)$ acts with weight -1 on t and with weight 1 on s.

The stack $\overline{ST}_R$ has a unique closed point$(0,0)$ corresponding to the simultaneous vanishing locus of s, t, ϖ in $\operatorname{Spec}(R[t,s]/(st-\varpi))$.

DEFINITION 2.13 (S-completeness). An algebraic stack $\mathcal{M}$ over k is defined to be S-complete if for all discrete valuation rings R over k and all morphisms $f : \overline{ST}_R \setminus (0,0) \to \mathcal{M}$, there is a unique dashed arrow as below making the diagram commutative

$$\begin{array}{ccc} \overline{ST}_R \setminus (0,0) & & \\ \downarrow & \overset{f}{\searrow} & \\ \overline{ST}_R & \dashrightarrow & \mathcal{M} \end{array}$$

We refer the reader to [**AHLH19**, §3.5] for some intuition behind the filling condition in the definition above. We are now ready to state one of the main theorems in [**AHLH19**].

THEOREM 2.14 ([**AHLH19**, Thm. A]). *Let $\mathcal{M}$ be an algebraic stack with affine diagonal and of finite type over a field k of characteristic* 0. *Then $\mathcal{M}$ admits a separated good moduli space if and only if it is Θ-reductive and S-complete.*

REMARK 2.15. The pursue of the right analogue of Theorem 2.14 in positive or mixed characteristic is a question that is being actively investigated.

Our main concern now is to be able to identify moduli stacks that are Θ-reductive and S-complete. Unfortunately, many algebraic stacks arising in moduli theory (such as the moduli stack of vector bundles on a curve) are not Θ-reductive or S-complete. This is to be expected; we know from the work of Mumford that in order to construct a moduli space for vector bundles it is necessary to restrict to those vector bundles that are (semi)stable. This phenomenon is fairly common in moduli theory. It is often impossible to parametrize all objects in the moduli problem; instead one generally restricts their attention to a subset of objects satisfying a stability condition, which are "rigid" enough to admit a parametrization in terms of a moduli space.

Hence we put ourselves in the following situation: we start with a "large" stack $\mathcal{M}$ parametrizing all objects for a given moduli problem. The main question is: can we identify a reasonable open substack $\mathcal{M}^{\mathrm{ss}} \subset \mathcal{M}$ of "semistable objects" such that $\mathcal{M}^{\mathrm{ss}}$ admits a good moduli space? In the next subsection we provide an answer to this question via the notion of Θ-stability.

2.4. Θ-stability. In the setting of GIT, Mumford introduced a notion of semistability depending on the choice of an equivariant line bundle. Suppose that X is a projective variety over k equipped with an action of a reductive group G. Given a G-equivariant ample line bundle $\mathcal{L}$ on X, Mumford defined a G-stable open subset $X^{\mathrm{ss}} \subset X$ called the semistable locus. He showed that the stack $[X^{\mathrm{ss}}/G]$ admits a projective good moduli space. Hence GIT provides a powerful tool to identify appropriate open substacks $[X^{\mathrm{ss}}/G] \subset [X/G]$ depending on an equivariant line bundle $\mathcal{L} \in \operatorname{Pic}^G(X)$, which corresponds to a line bundle on the stack $[X/G]$.

The notion of Θ-stability, introduced by Halpern-Leistner [**HL21**] and Heinloth [**Hei17**], generalizes the notion of GIT stability to the setting of arbitrary algebraic stacks. It is inspired by the Hilbert-Mumford criterion [**MFK94**, Chpt. 2.1], which describes the semistable locus in GIT in terms of one-parameter subgroups of the reductive group G. In the case of a more general algebraic stack, the one-parameter subgroups are replaced by the notion of a Θ-filtration. This uses the test stack Θ introduced in Definition 2.10.

DEFINITION 2.16 (Θ-Filtration). Let $\mathcal{M}$ be an algebraic stack over k. Let $K \supset k$ be a field over k, and let $p : \mathrm{Spec}(K) \to \mathcal{M}$ be a K-point in $\mathcal{M}(K)$. A Θ-filtration of p consists of the data of a morphism $f : \Theta_K \to \mathcal{M}$ along with an isomorphism $\Theta|_{1_K} \simeq p$.

REMARK 2.17. In [**HL21**] a Θ-filtration is simply called a filtration. We stress the name Θ-filtration to clearly distinguish this intrinsic notion of filtration from the many other filtrations appearing in this survey.

Fix a line bundle $\mathcal{L}$ on the stack $\mathcal{M}$. For any filtration $f : \Theta_K \to \mathcal{M}$ of a point $p \in \mathcal{M}(K)$, the 0_K-fiber of the pullback $f^*(\mathcal{L})$ acquires a natural action of the stabilizer $(\mathbb{G}_m)_K$. We denote by $\mathrm{wt}\,(f^*(\mathcal{L})|_{0_K}) \in \mathbb{Z}$ the weight of the $\mathbb{G}_m$ action on the one dimensional 0_K-fiber.

DEFINITION 2.18 (Θ-stability). Let $\mathcal{M}$ be an algebraic stack over k, and let $\mathcal{L}$ be a line bundle on $\mathcal{M}$. Fix an algebraically closed field $K \supset k$ and a geometric point $p : \mathrm{Spec}(K) \to \mathcal{M}$. We say that p is Θ-semistable (with respect to $\mathcal{L}$) if for all Θ-filtrations $f : \Theta_K \to \mathcal{M}$ of the point p we have $\mathrm{wt}\,(f^*(\mathcal{L})|_{0_K}) \leq 0$.

The set of Θ-semistable geometric points with respect to a line bundle $\mathcal{L}$ yields a subset of the topological space of the stack $|\mathcal{M}|$ (as in [**Sta23**, Tag 0DQM]). We say that the Θ-semistable locus is open if this subset is open. In this case the Θ-semistable points consist exactly of the geometric points of an open substack, which we denote by $\mathcal{M}^{\mathcal{L}\text{-ss}} \subset \mathcal{M}$ (or just by $\mathcal{M}^{\mathrm{ss}}$, if the choice of the line bundle is implicit).

EXAMPLE 2.19 (Stability for vector bundles, [**Hei17**]). Let $\mathrm{Bun}_n(C)$ denote the stack of rank n vector bundles on a smooth projective curve C over k. There is a line bundle $L_{det} \in \mathrm{Pic}(\mathrm{Bun}_n(C))$ called the determinant line bundle, defined as follows.

By definition, there is a universal rank n vector bundle $\mathcal{E}_{univ}$ on the product $\mathrm{Bun}_n(C) \times C$. The derived push-forward $R(\mathrm{pr}_1)_*\mathcal{E}_{univ}$ under the first projection $\mathrm{pr}_1 : \mathrm{Bun}_n(C) \times C \to \mathrm{Bun}_n(C)$ is a perfect complex on $\mathrm{Bun}_n(C)$, since the morphism pr_1 is proper, schematic and flat. We set L_{det} to be the determinant $L_{det} = \det(R(\mathrm{pr}_1)_*\mathcal{E}_{univ})$ of the perfect complex $R(\mathrm{pr}_1)_*\mathcal{E}_{univ}$.

In [**Hei17**] it is proven that the semistable locus in $\mathrm{Bun}_n(C)$ with respect to the line bundle L_{det} is open and consists of the slope semistable vector bundles in the sense of Mumford. We already know by GIT that this substack of semistable vector bundles $\mathrm{Bun}_n(C)^{\mathrm{ss}}$ admits a good moduli space.

Now equipped with the notion of Θ-semistability, we have new tools to identify open substacks inside $\mathcal{M}$ that have a chance at admitting good moduli spaces. The question becomes: how can we check the sufficient existence criteria (Θ-reductivity and S-completeness) for the stack of Θ-semistable points $\mathcal{M}^{\mathrm{ss}} \subset \mathcal{M}$? We will answer this question later in this survey; this is Halpern-Leistner's Theorem 2.30 for intrinsic GIT. In order to state this theorem, we will need to gain some understanding of the complement of the Θ-semistable locus $\mathcal{M} \setminus \mathcal{M}^{\mathrm{ss}}$, which is called the Θ-unstable locus. And to acquire knowledge about the unstable locus and, as a consequence, gain insight on the Θ-semistable locus, Halpern-Leistner [**HL21**] introduced the concept of Θ-stratifications, to be explained in the next section.

2.5. Θ-stratifications. In this section we explain the notions of Θ-stratifications and numerical invariants introduced [**HL21**]. This aims to generalize to the setting of stacks the stratifications from classical GIT [**Sha77, Kem78, Hes81, Nes84, Kir84, Nit11**], which have also been translated to non-algebraic gauge theoretical frameworks [**BT05**]. For simplicity we work with an algebraic stack $\mathcal{M}$ with affine diagonal and locally of finite type over a field k of characteristic 0.

We denote by $\mathrm{Filt}(\mathcal{M}) = \mathrm{Map}(\Theta, \mathcal{M})$ the stack of Θ-filtrations. It is represented by an algebraic stack locally of finite type over k [**HL21**, Prop. 1.1.2]. There is a morphism $\mathrm{ev}_1 : \mathrm{Filt}(\mathcal{M}) \to \mathcal{M}$ given by evaluating at $1 \in \Theta$.

Definition 2.20 (Θ-stratum, [**HL21**, Defn. 2.1.1]). Let $\mathcal{X}$ be an open substack of $\mathcal{M}$. A Θ-stratum of $\mathcal{X}$ is a union of connected components of $\mathrm{Filt}(\mathcal{X})$ such that the restriction of ev_1 is a closed immersion.

We can think of a Θ-stratum as a closed substack of $\mathcal{X}$ that is identified with some connected components of $\mathrm{Filt}(\mathcal{X})$.

Definition 2.21 (Θ-stratification, [**HL21**, Defn. 2.1.2]). A Θ-stratification of $\mathcal{M}$ consists of a collection of open substacks $(\mathcal{M}_{\leq c})_{c\in\Gamma}$ indexed by a totally ordered set Γ. We require the following conditions to be satisfied

(1) $\mathcal{M}_{\leq c} \subset \mathcal{M}_{\leq c'}$ for all $c < c'$.
(2) $\mathcal{M} = \bigcup_{c\in\Gamma} \mathcal{M}_{\leq c}$.
(3) For all c, there exists a Θ-stratum $\mathfrak{S}_c \subset \mathrm{Filt}(\mathcal{M}_{\leq c})$ of $\mathcal{M}_{\leq c}$ such that
$$\mathcal{M}_{\leq c} \setminus \mathrm{ev}_1(\mathfrak{S}_c) = \bigcup_{c'<c} \mathcal{M}_{\leq c'}.$$
(4) For every point $p \in \mathcal{M}$, the set $\{c \in \Gamma \mid p \in \mathcal{M}_{\leq c}\}$ has a minimal element.

A useful way to define Θ-stratifications is via the use of numerical invariants in the sense of [**HL21**, Defn. 0.0.3]. Let us give in this paragraph a simplified account of how this works for intuition; a more precise mathematical discussion will follow afterwards. A numerical invariant yields a locally constant function ν on the topological space $|\mathrm{Filt}(\mathcal{M})|$ of the stack of Θ-filtrations. This function is taken to be valued in a totally ordered group Γ (usually $\Gamma = \mathbb{R}$ or more generally $\Gamma = \mathbb{R}[n]$ for polynomial numerical invariants). For any given geometric point p of the stack $\mathcal{M}$, we can consider the subspace $|\mathrm{Filt}(\mathcal{M}, p)| \subset |\mathrm{Filt}(\mathcal{M})|$ consisting of the images of all Θ-filtrations of p (by forgetting the identification $f(1) \simeq p$). It is often the case that the function ν attains a maximum on $|\mathrm{Filt}(\mathcal{M}, p)|$, which roughly measures the "instability" of the point p. For each $c \in \Gamma$, we can then define $|\mathcal{M}_{\leq c}| \subset |\mathcal{M}|$ to be the set of geometric points p such that the maximum of ν on $|\mathrm{Filt}(\mathcal{M}, p)|$ (i.e. the "degree of instability") is at most c. Under favorable circumstances, the subspace $|\mathcal{M}_{\leq c}| \subset |\mathcal{M}|$ is actually the set of geometric points of an open substack $\mathcal{M}_{\leq c} \subset \mathcal{M}$, and we obtain a Θ-stratification of $\mathcal{M}$.

For the rest of this section we shall give a more mathematically precise account of the previous paragraph. We shall restrict ourselves with numerical invariants valued in the group $\Gamma = \mathbb{R}[n]$ of polynomials in the variable n with coefficients in $\mathbb{R}$. This is sufficient for many applications, such as in the case of Gieseker stability discussed in later sections.

DEFINITION 2.22 (Total order on polynomials). For any two polynomials

$$p(n), q(n) \in \mathbb{R}[n],$$

we write $p(n) \leq q(n)$ if there exists some N such that $p(m) \leq q(m)$ for all $m \in \mathbb{R}$ with $m > N$.

The relation "$\leq$" in Definition 2.22 equips $\mathbb{R}[n]$ with the structure of a totally ordered abelian group.

DEFINITION 2.23 (Non-degenerate graded points). For $q \geq 1$, let $B\mathbb{G}_m^q := [\mathrm{Spec}(k)/\mathbb{G}_m^q]$. Let $K \supset k$ be a field. We say that a morphism $g : B(\mathbb{G}_m^q)_K \to \mathcal{M}$ is non-degenerate if the induced homomorphism $\gamma : (\mathbb{G}_m^q)_K \to \mathrm{Aut}(g(\mathrm{Spec}(K)))$ has finite kernel.

We note that the data of a morphism $B(\mathbb{G}_m^q)_K \to \mathcal{M}$ as in the definition above is equivalent to a pair (p, γ) of a point $p \in \mathcal{M}(K)$ and a homomorphism $\gamma : (\mathbb{G}_m^q)_K \to \mathrm{Aut}(p)$.

A numerical invariant is an assignment of certain scale-invariant functions to each non-degenerate graded point $g : B(\mathbb{G}_m^q)_K \to \mathcal{M}$. A more formal definition follows.

DEFINITION 2.24 (Polynomial numerical invariant). A polynomial numerical invariant ν on the stack $\mathcal{M}$ is an assignment defined as follows. Let $K \supset k$ be a field and let $p \in \mathcal{M}(K)$. Let $\gamma : (\mathbb{G}_m^q)_K \to \mathrm{Aut}(p)$ be a homomorphism with finite kernel. The polynomial numerical invariant ν assigns to this data a scale-invariant function $\nu_\gamma : \mathbb{R}^q \setminus \{0\} \to \mathbb{R}[n]$ such that:

(1) ν_γ is unchanged under field extensions $K \subset K'$.
(2) ν is locally constant in algebraic families. In other words, choose any scheme T, a morphism $\xi : T \to \mathcal{M}$, and a homomorphism $\gamma : (\mathbb{G}_m^q)_T \to \mathrm{Aut}(\xi)$ of T-group schemes with finite kernel. Then as we vary $t \in T$, we require that the function ν_{γ_t} is locally constant in T.
(3) Given a homomorphism $\phi : (\mathbb{G}_m^w)_K \to (\mathbb{G}_m^q)_K$ with finite kernel, the function $\nu_{\gamma \circ \phi}$ is the restriction of ν_γ along the inclusion $\mathbb{R}^w \hookrightarrow \mathbb{R}^q$ induced by ϕ.

A useful way to construct polynomial numerical invariants on a stack $\mathcal{M}$ is to use line bundles on $\mathcal{M}$, which are the analogues of equivariant polarizations in the classical setting of GIT. Fix a sequence of rational line bundles $(\mathcal{L}_n)_{n \in \mathbb{Z}}$, where each $\mathcal{L}_n$ is in the rational Picard group $\mathrm{Pic}(\mathcal{M}) \otimes_{\mathbb{Z}} \mathbb{Q}$ of the stack. For each morphism $g : (B\mathbb{G}_m^q)_K \to \mathcal{M}$, the pullback line bundle $g^*(\mathcal{L}_n)$ amounts to a rational character in $X^*(\mathbb{G}_m^q) \otimes_{\mathbb{Z}} \mathbb{Q}$. Under the natural identification $X^*(\mathbb{G}_m^q) \otimes_{\mathbb{Z}} \mathbb{Q} \cong \mathbb{Q}^q$, we can interpret this as a q-tuple of rational numbers $(w_n^{(i)})_{i=1}^q$, which we call the weight of $g^*(\mathcal{L}_n)$.

In practice we can often choose the line bundles $(\mathcal{L}_n)_{n \in \mathbb{Z}}$ in such a way that, for each fixed i, there is a polynomial $w^{(i)}(T) \in \mathbb{R}[T]$ such that the weight $w_n^{(i)}$ is $w^{(i)}(n)$. If this is the case, then for every g we can define an $\mathbb{R}$-linear function $L_g : \mathbb{R}^q \to \mathbb{R}[n]$ given by

$$L_g\left((r_i)_{i=1}^q\right) = \sum_{i=1}^q r_i \cdot w^{(i)}(n).$$

In order to obtain a scale invariant ν, we use a rational quadratic norm on graded points as in [**HL21**, Defn. 4.1.12], which is the analogue of a Weyl invariant quadratic form in the classical setting of GIT.

DEFINITION 2.25 (Rational quadratic norm on graded points). A rational quadratic norm b on graded points of the stack $\mathcal{M}$ is an assignment defined as follows. Let $K \supset k$ be a field and let $p \in \mathcal{M}(K)$. Let $\gamma : (\mathbb{G}_m^q)_K \to \mathrm{Aut}(p)$ be a homomorphism with finite kernel. Then b assigns to this data a positive definite quadratic norm $b_\gamma : \mathbb{R}^q \to \mathbb{R}$ with rational coefficients such that:

(1) b_γ is unchanged under field extensions $K \subset K'$.
(2) b is locally constant in algebraic families. In other words, choose any scheme T, a morphism $\xi : T \to \mathcal{M}$ and a homomorphism $\gamma : (\mathbb{G}_m^q)_T \to \mathrm{Aut}(\xi)$ of T-group schemes with finite kernel. Then as we vary $t \in T$, we require that the function b_{γ_t} is locally constant in T.
(3) Given a homomorphism $\phi : (\mathbb{G}_m^w)_K \to (\mathbb{G}_m^q)_K$ with finite kernel, the norm $b_{\gamma\circ\phi}$ is the restriction of b_γ along the inclusion $\mathbb{R}^w \hookrightarrow \mathbb{R}^q$ induced by ϕ.

CONSTRUCTION 2.26. Given a sequence of rational line bundles $(\mathcal{L}_n)_{n\in\mathbb{Z}}$ as above and a rational quadratic norm on graded points b, we define a polynomial numerical invariant ν as follows. For all non-degenerate $g : (B\mathbb{G}_m^q)_K \to \mathcal{M}$ with corresponding morphism $\gamma : (\mathbb{G}_m^q)_K \to \mathrm{Aut}(g(\mathrm{Spec}(K)))$, we set

$$\nu_\gamma(\vec{r}) = \frac{L_g(\vec{r})}{\sqrt{b_\gamma(\vec{r})}}$$

where $\vec{r} = (r_1, \ldots, r_q)$.

We next explain how a polynomial numerical invariant ν can be used top define a Θ-stratification on $\mathcal{M}$ (see [**HL21**, §4.1] for more details). We say that a Θ-filtration $f : \Theta_K \to \mathcal{M}$ is non-degenerate if the restriction $f|_0 : [0/(\mathbb{G}_m)_K] \to \mathcal{M}$ is non-degenerate. We regard ν as a function on the set of non-degenerate Θ-filtrations by defining $\nu(f) := \nu_{f|_0}(1) \in \mathbb{R}[n]$.

DEFINITION 2.27 (Semistability for numerical invariants). Given a polynomial numerical invariant ν on $\mathcal{M}$ and a point $p \in |\mathcal{M}|$, we say that p is ν-semistable if all non-degenerate Θ-filtrations f with $f(1) \simeq p$ satisfy $\nu(f) \leq 0$. Otherwise we say that p is ν-unstable.

For any ν-unstable point p, we set $M^\nu(p)$ to be the supremum of $\nu(f)$ over all Θ-filtrations f with $f(1) \simeq p$ (if such supremum exists). If p is ν-semistable, then by convention we set $M^\nu(p) = 0$. For any $c \in \mathbb{R}[n]_{\geq 0}$, let $\mathcal{M}_{\leq c}$ be the set of all points p satisfying $M^\nu(p) \leq c$, and we let $\mathcal{M}^{\nu\text{-ss}} := \mathcal{M}_{\leq 0}$ denote the set of ν-semistable points. If this is an open stratification of the stack coming from a Θ-stratification, then we say that ν defines a Θ-stratification.

REMARK 2.28. Generally one only considers non-degenerate Θ-filtrations, because these are the ones relevant for stability. Hence the adjective "non-degenerate" is usually omitted.

DEFINITION 2.29 (ν-Harder-Narasimhan filtration). For any ν-unstable point p, a Θ-filtration f of p is called a ν-Harder-Narasimhan filtration, abbreviated ν-HN filtration, if $\nu(f) = M^\nu(p)$.

If ν defines a Θ-stratification, then by [**HL21**, Lemma 2.1.4] every ν-unstable point admits a ν-Harder-Narasimhan filtration that is unique up to pre-composing with a ramified covering $\Theta \to \Theta$.

We should remark that not all numerical invariants ν coming from sequences of line bundles and rational quadratic norms define Θ-stratifications. In the next section we discuss some criteria that guarantee the existence of such a Θ-stratification and, in addition, imply that the open semistable locus $\mathcal{M}^{\nu\text{-ss}}$ admits a good moduli space.

2.6. Monotonicity and intrisic moduli theory. In this section we explain Halpern-Leistner's intrinsic GIT theory [**HL21**, §5]. We keep our assumption that $\mathcal{M}$ is an algebraic stack with affine diagonal and locally of finite type over a field k of characteristic 0.

A natural question to ask is: when does a polynomial numerical invariant ν define a Θ-stratification as described in the previous section? Another thing we would like to know is whether the locus $\mathcal{M}^{\nu\text{-ss}}$ of ν-semistable objects admits a good moduli space. A general strategy to address such questions proceeds in three steps:

(Step 1) Show that the polynomial numerical invariant ν satisfies some monotonicity conditions (as in Definition 2.33).
(Step 2) Prove that ν satisfies the HN boundedness condition (as in Definition 2.34).
(Step 3) Check that the semistable locus $\mathcal{M}^{\nu\text{-ss}}$ is a disjoint union of quasi-compact open substacks.

The following theorem describes this strategy more precisely.

THEOREM 2.30 (Intrinsic GIT, [**HL21**, Theorem B]). *Let $\mathcal{M}$ be an algebraic stack with affine diagonal and locally of finite type over a field k of characteristic* 0. *Let ν be a polynomial numerical invariant on $\mathcal{M}$ defined by a sequence of rational line bundles and a norm on graded points, as explained in the Construction* 2.26. *Then,*

(1) *If ν is strictly Θ-monotone (Definition* 2.33*) and satisfies the HN boundedness condition (Definition* 2.34*), then it defines a Θ-stratification of $\mathcal{M}$.*
(2) *Assume that ν satisfies the conditions in (1). In addition, suppose that ν is strictly S-monotone (Definition* 2.33*) and that the semistable locus $\mathcal{M}^{\nu\text{-ss}}$ can be written as a disjoint union of quasi-compact open substacks. Then $\mathcal{M}^{\nu\text{-ss}}$ has a separated good moduli space.*
(3) *If all the conditions in (2) are satisfied and if $\mathcal{M}$ satisfies the existence part of the valuative criterion for properness for complete discrete valuation k-algebras* [**Sta23**, Tag 0CLK], *then each quasi-compact open and closed substack of $\mathcal{M}^{\nu\text{-ss}}$ admits a good moduli space that is proper over k.*

Recall that, in Mumford's GIT, the quotient of a projective variety with an ample linearization is automatically projective. In many applications of GIT, this is the way in which properness of the moduli space is proved. But in Theorem 2.30 we see that, in order to prove properness, we have to show a valuative criterion. The prototype of such criteria is Langton's semistable reduction theorem for torsion free sheaves [**Lan75**]. There has been a lot of work extending this to principal bundles in different contexts (cf. [**BS02, BP03, Bal07, Hei08b, Hei10a**]).

We end this subsection by explaining each of the hypotheses that one needs to check in Theorem 2.30. We need some setup first.

NOTATION 2.31 ("Rigidified" Θ and $\overline{ST}_R$). Let R be a complete discrete valuation ring over k. Choose a uniformizer ϖ of R. We define $Y_{\Theta_R} := \mathbb{A}^1_R$ equipped with the $\mathbb{G}_m$-action that gives t weight -1. We have $\Theta_R = \left[\mathbb{A}^1_R / \mathbb{G}_m\right]$. Note that $\mathbb{A}^1_R$ contains a unique $\mathbb{G}_m$-invariant closed point cut out by the ideal (t, ϖ). We will denote this point by $(0,0)$.

We define $Y_{\overline{ST}_R} := \operatorname{Spec}\left(R[t,s]/(st-\varpi)\right)$. We equip $Y_{\overline{ST}_R}$ with the $\mathbb{G}_m$-action that gives s weight 1 and t weight -1. The isomorphism class of the $\mathbb{G}_m$-scheme $Y_{\overline{ST}_R}$ is independent of the choice of uniformizer ϖ. Define $\overline{ST}_R := \left[Y_{\overline{ST}_R} / \mathbb{G}_m\right]$. We will denote by $(0,0)$ the unique $\mathbb{G}_m$-fixed closed point in $Y_{\overline{ST}_R}$. We have that $(0,0)$ is cut out by the ideal (s,t).

NOTATION 2.32 (Weighted lines). Let $\kappa \supset k$ be a field and let $a \geq 1$ be an integer. We denote by $\mathbb{P}^1_\kappa[a]$ the $\mathbb{G}_m$-scheme $\mathbb{P}^1_\kappa$ equipped with the $\mathbb{G}_m$-action determined by the equation $t \cdot [x:y] = [t^{-a}x : y]$. We set $0 = [0:1]$ and $\infty = [1:0]$.

In the following definition we describe a simplified version of the monotonicity conditions in [**HL21**, §5].

DEFINITION 2.33 (Monotonicity). A polynomial numerical invariant ν on $\mathcal{M}$ is strictly Θ-monotone (resp. strictly S-monotone) if the following condition holds.

Let R be any complete discrete valuation ring and set $\mathfrak{X}$ to be Θ_R (resp. $\overline{ST}_R$), and $Y_\mathfrak{X}$ as in Notation 2.31. Choose a map $\varphi : \mathfrak{X} \setminus (0,0) \to \mathcal{M}$. Then, after maybe replacing R with a finite DVR extension, there exists a reduced and irreducible scheme Σ equipped with a $\mathbb{G}_m$-action along with $\mathbb{G}_m$-equivariant maps $f : \Sigma \to Y_\mathfrak{X}$ and $\widetilde{\varphi} : [\Sigma / \mathbb{G}_m] \to \mathcal{M}$ such that

(M1) The map f is proper, $\mathbb{G}_m$-equivariant, and its restriction induces an isomorphism $f : \Sigma_{Y_\mathfrak{X} \setminus (0,0)} \xrightarrow{\sim} Y_\mathfrak{X} \setminus (0,0)$.

(M2) The following diagram commutes

$$\begin{array}{ccc} \left[\left(\Sigma_{Y_\mathfrak{X}\setminus(0,0)}\right) / \mathbb{G}_m\right] & & \\ {\scriptstyle f}\downarrow & \searrow^{\widetilde{\varphi}} & \\ \mathfrak{X} \setminus (0,0) & \xrightarrow[\varphi]{} & \mathcal{M} \end{array}$$

(M3) Let κ denote a finite extension of the residue field of R. For any $a \geq 1$ and any finite $\mathbb{G}_m$-equivariant morphism $\mathbb{P}^1_\kappa[a] \to \Sigma_{(0,0)}$, we have $\nu\left(\widetilde{\varphi}|_{[\infty/\mathbb{G}_m]} \right) > \nu\left(\widetilde{\varphi}|_{[0/\mathbb{G}_m]} \right)$.

DEFINITION 2.34 (HN Boundedness). We say that a polynomial numerical invariant ν satisfies the HN boundedness condition if the following is always satisfied:

(HNB) Let T be an affine Noetherian scheme, equipped with morphism $g : T \to \mathcal{M}$. Then there exists a quasi-compact open substack $\mathcal{U}_T \subset \mathcal{M}$ such that the following holds. For all geometric points $t \in T$ with residue field $\kappa(t)$ and all Θ-filtrations $f : \Theta_{\kappa(t)} \to \mathcal{M}$ of the point $g(t)$ with $\nu(f) > 0$, there exists another filtration f' of $g(t)$ satisfying $\nu(f') \geq \nu(f)$ and $f'|_0 \in \mathcal{U}_T$.

This says that for the purposes of maximizing $\nu(f)$ among all Θ-filtrations of points in a bounded family (i.e. finding ν-HN filtrations as in Definition 2.29), it suffices to consider only those Θ-filtrations f such that the associated graded $f|_0$ lies in some other (possibly larger) bounded family.

3. Infinite dimensional GIT for torsion-free sheaves

In this section we illustrate the previous notions with one of the main examples in moduli theory: the moduli problem of torsion-free sheaves. This moduli space was constructed using GIT by Mumford [**Mum63b**] (stable vector bundles) and Seshadri (compactification with semistable vector bundles) [**Ses67**] over curves, Gieseker [**Gie77**] for surfaces, later Maruyama [**Mar78**] and Simpson [**Sim94**] in higher dimension.

3.1. Torsion-free sheaves. Let X be a smooth projective variety of dimension d over the field k, and fix an ample line bundle $\mathcal{O}(1)$ on X. Let T be a k-scheme and let $\mathcal{F}$ be a $\mathcal{O}_{X_T}$-module. We say that $\mathcal{F}$ is a relative torsion-free sheaf of rank r on X_T if it is T-flat, finitely presented, and for all points $t \in T$ the fiber $\mathcal{F}|_{X_t}$ is torsion-free of rank r (i.e. the dimension of the generic fiber is r).

Let us present the stack of coherent sheaves that is relevant for this moduli problem.

Definition 3.1 ([**GHZ24**, Defn. 2.9]). Denote by $\mathrm{Coh}_r^{tf}(X)$ the pseudofunctor from $(\mathrm{Aff}_k)^{op}$ into groupoids such that, for any affine scheme $T \in \mathrm{Aff}_k$, it gives

$$\mathrm{Coh}_r^{tf}(X)(T) = \left[\text{groupoid of } T\text{-flat relative torsion-free sheaves } \mathcal{F} \text{ of rank } r \text{ on } X_T \right]$$

The stack $\mathrm{Coh}_r^{tf}(X)$ is an open substack of $\mathrm{Coh}_r(X)$, the algebraic stack of coherent sheaves in X, and contains the open substack $\mathrm{Bun}_r(X)$ of rank r vector bundles on X.

Denote by $P_{\mathcal{F}}$ the Hilbert polynomial of a torsion-free sheaf $\mathcal{F}$ on X with respect to the line bundle $\mathcal{O}(1)$. This is a polynomial of degree d in the variable n

$$P_{\mathcal{F}}(n) = \sum_{i=0}^{d} \frac{a_i}{i!} n^i$$

with rational coefficients a_i (see [**HL97**, Lem. 1.2.1]). We define the reduced Hilbert polynomial as $\overline{p}_{\mathcal{F}} := \frac{1}{a_d} P_{\mathcal{F}}$. For each $0 \le i \le d-1$, define the ith slope of $\mathcal{F}$ as $\widehat{\mu}_i(\mathcal{F}) := \frac{a_i}{a_d}$, where $\widehat{\mu}_{d-1}(\mathcal{F}) = \frac{a_{d-1}}{a_d}$ is the usual slope of $\mathcal{F}$.

We define a family of line bundles on $\mathrm{Coh}_r^{tf}(X)$ as follows.

Definition 3.2 ([**GHZ24**, Defn. 4.1]). Given a k-scheme T, let $\pi_T : X_T \to T$ the structure morphism. Let $f : T \to \mathrm{Coh}_r^{tf}(X)$ be a morphism represented by a rank r torsion-free sheaf $\mathcal{F}$ of dimension d on X_T. For each integer n, define a line bundle $f^* M_n$ on T by

$$f^* M_n := \det R_{\pi_T *}(\mathcal{F}(n)).$$

The line bundles $f^* M_n$ for all $f : T \to \mathrm{Coh}_r^{tf}(X)$ are compatible with pullbacks and yield a line bundle M_n on $\mathrm{Coh}_r^{tf}(X)$. We set $b_d := \bigotimes_{i=0}^{d} M_i^{(-1)^{d-i}\binom{d}{i}}$, and we define

$$L_n = M_n \otimes b_d^{-\otimes \overline{p}_{\mathcal{F}}(n)}.$$

The following step in the construction is to define Θ-filtrations of the stack of torsion-free coherent sheaves, as in Definition 2.16. A Θ-filtration $f : \Theta_K \to \mathrm{Coh}_r^{tf}(X)$ of a K-point of the stack $\mathcal{F} \in \mathrm{Coh}_r^{tf}$ consists of

(a) A $\mathbb{G}_m$-equivariant family $\widetilde{\mathcal{F}}$ of torsion-free sheaves on $X \times \mathbb{A}^1_K$.
(b) An identification $\widetilde{\mathcal{F}}|_{X\times 1} \xrightarrow{\sim} \mathcal{F}$.

By using the Rees construction [**HL21**, Prop. 1.0.1] we can rewrite the definition of Θ-filtration in a much more explicit way.

PROPOSITION 3.3 ([**GHZ24**, Prop. 3.8]). *A Θ-filtration $f : \Theta_K \to \mathrm{Coh}_r^{tf}(X)$ of $\mathcal{F}$ corresponds to a $\mathbb{Z}$-filtration $(\mathcal{F}_m)_{m\in\mathbb{Z}}$ by $\mathcal{O}_{X_K}$-subsheaves of $\mathcal{F}$ satisfying*

(1) $\mathcal{F}_{m+1} \subset \mathcal{F}_m$.
(2) $\mathcal{F}_m = 0$ *for* $m \gg 0$ *and* $\mathcal{F}_m = \mathcal{F}$ *for* $m \ll 0$.
(3) $\mathcal{F}_m/\mathcal{F}_{m+1}$ *is torsion-free.*

The graded object at the origin $f|_{[0/\mathbb{G}_m]} : [0/\mathbb{G}_m] \to \mathrm{Coh}_r^{tf}(X)$ is the associated graded sheaf $\bigoplus_{m\in\mathbb{Z}} \mathcal{F}_m/\mathcal{F}_{m+1}$.

For each Θ-filtration the restrictions f^*L_n of the line bundles are given by $\mathbb{G}_m$-equivariant line bundles on $\mathbb{A}^1_{\mathbb{K}}$ and are determined by the fiber $L_n|_0$ considered as a 1-dimensional representation of $\mathbb{G}_m$. The integer classifying the character of $\mathbb{G}_m$ associated to this representation is called the weight $\mathrm{wt}(L_n|_0)$.

On the other hand, $M_n|_0$ is isomorphic to $\det\left(R_{\pi_K *}(\widetilde{\mathcal{F}}|_0)(n)\right)$, where the graded sheaf is $\widetilde{\mathcal{F}}|_0 = \bigoplus_{m\in\mathbb{Z}} \mathcal{F}_m/\mathcal{F}_{m+1}$. Given that $\mathbb{G}_m$ acts on $\mathcal{F}_m/\mathcal{F}_{m+1}$ with weight m, we get

$$\mathrm{wt}(M_n|_0) = \sum_{m\in\mathbb{Z}} \left(P_{\mathcal{F}_m}(n) - P_{\mathcal{F}_{m+1}}(n)\right)\cdot m$$

and

$$\mathrm{wt}(b_d|_0) = \sum_{m\in\mathbb{Z}} \left(\mathrm{rk}_{\mathcal{F}_m} - \mathrm{rk}_{\mathcal{F}_{m+1}}\right)\cdot m.$$

Therefore,

$$\mathrm{wt}(L_n|_0) = \mathrm{wt}(M_n|_0) - \overline{p}_{\mathcal{F}}(n)\cdot \mathrm{wt}(b_d|_0) = \sum_{m\in\mathbb{Z}} m\cdot\left(\overline{p}_{\mathcal{F}_m/\mathcal{F}_{m+1}}(n) - \overline{p}_{\mathcal{F}}(n)\right)\cdot \mathrm{rk}_{\mathcal{F}_m/\mathcal{F}_{m+1}}.$$

In order to set the polynomial numerical invariant as in Definition 2.24, we define a rational quadratic norm on the graded points of $\mathrm{Coh}_r^{tf}(X)$ (c.f. [**HLHJ21**, Section 4.1]). For simplicity we will focus only on $B\mathbb{G}_m^q$-points with $q = 1$.

If $g : (B\mathbb{G}_m)_K \to \mathrm{Coh}_r^{tf}(X)$ is a $\mathbb{Z}$-graded torsion-free sheaf $\overline{F} = \bigoplus_{m\in\mathbb{Z}} \overline{F}_m$ of rank r on X_K, define b_g to be the positive definite rational quadratic form on $\mathbb{R}$ given by

$$b_g(v) := \sum_{m\in\mathbb{Z}^q} \mathrm{rk}_{\overline{\mathcal{F}}_m}\cdot(m\cdot v)^2,$$

which is uniquely determined by its value in $v = 1$.

DEFINITION 3.4 ([**HLHJ21**, Defn. 4.3]). Let $f : \Theta_K \to \mathrm{Coh}_r^{tf}(X)$ be a Θ-filtration corresponding to the $\mathbb{Z}$-filtration $(\mathcal{F}_m)_{m\in\mathbb{Z}}$. We define the polynomial numerical invariant $\nu(f)$ to be

$$\nu(f) := \frac{\mathrm{wt}(L_n|_0)}{\sqrt{b_{f|_0}(1)}} = \frac{\sum_{m\in\mathbb{Z}} m\cdot(\overline{p}_{\mathcal{F}_m/\mathcal{F}_{m+1}} - \overline{p}_{\mathcal{F}})\cdot \mathrm{rk}(\mathcal{F}_m/\mathcal{F}_{m+1})}{\sqrt{\sum_{m\in\mathbb{Z}} m^2\cdot \mathrm{rk}(\mathcal{F}_m/\mathcal{F}_{m+1})}}. \tag{3.1}$$

3.2. Affine grassmannians, monotonicity and good moduli space. Affine grassmannians are the tool developed in [**HLHJ21**] to show the monotonicity properties (Definition 2.33) required to prove Theorem 2.30. The idea is to build an algebraic counterpart to infinite-dimensional Atiyah-Bott-Donaldson symplectic reduction.

We can define a fibered category $\mathrm{Coh}^{tf}_{r,rat}(X)$ whose objects are pairs $(D, \mathcal{F})$ of a divisor and torsion-free sheaf, and a morphism is a pair

$$(i : D_1 \hookrightarrow D_2, \psi : \mathcal{F}_2 \to \mathcal{F}_1)$$

where i is an inclusion and ψ is a homomorphism which is an isomorphism when restricted to the complement of D_2 (notice that this is not fibered in groupoids). This is called the stack of rational maps, see [**HLHJ21**, section 3.2] for more details. We have a natural morphism $\mathrm{Coh}^{tf}_r(X) \to \mathrm{Coh}^{tf}_{r,rat}(X)$, sending a torsion free sheaf $\mathcal{F}$ to the pair $(\emptyset, \mathcal{F})$. The fibers of this morphism are ind-projective ind-schemes. We call them affine Grassmannians, in analogy with the notion appearing in the moduli of vector bundles over curves. Changing the trivialization over the open subset, we can heuristically think of this morphism as being a local presentation of the stack $\mathrm{Coh}^{tf}_r(X)$ as the quotient of infinite-dimensional spaces, the ind-schemes, by the action of an infinite-dimensional group of rational gauge transformations.

This, together with the fact that the family of line bundles defined in the previous section is asymptotically ample on each projective stratum, will be the key to show monotonicity in Theorem 2.30. In the following, $\mathcal{F}$ will denote a family of torsion-free sheaves of rank r on X_S, where S is a quasi-compact k-scheme. Fix an effective Cartier divisor $D \hookrightarrow X_S$ flat over S and set $Q = X_S \setminus D$.

DEFINITION 3.5 (cf. [**HLHJ21**, Defn. 3.12]). We define $\mathrm{Gr}_{X_S,D,\mathcal{F}}$ to be the functor from Aff^{op}_S to sets such that for every affine scheme $T \in \mathrm{Aff}_S$, the set $\mathrm{Gr}_{X_S,D,\mathcal{F}}(T)$ consists of equivalence classes of pairs $(\mathcal{E}, \theta)$, where

(1) $\mathcal{E}$ is a family of rank r torsion-free sheaves on X_T.
(2) θ is morphism $\theta : \mathcal{F}|_{X_T} \to \mathcal{E}$ such that the restriction $\theta|_{Q_T}$ is an isomorphism.

The pairs $(\mathcal{E}_1, \theta_1)$ and $(\mathcal{E}_2, \theta_2)$ are equivalent if there is an isomorphism $\mathcal{E}_1 \xrightarrow{\sim} \mathcal{E}_2$ that identifies θ_1 and θ_2.

We remark that the fiber over $(D, \mathcal{F})$ of the natural morphism $\mathrm{Coh}^{tf}_r(X) \to \mathrm{Coh}^{tf}_{r,rat}(X)$ is $\mathrm{Gr}_{X_S,D,\mathcal{F}}$.

Now let j be the open immersion $j : Q \hookrightarrow X_S$ and choose $T \in \mathrm{Aff}_S$ and an element $(\mathcal{E}, \theta) \in \mathrm{Gr}_{X_S,D,\mathcal{F}}(T)$. Using that the unit $\mathcal{E} \to j_{T\,*}\, j_T^* \mathcal{E}$ is a monomorphism we can use the isomorphism $j_{T*}(j_T)^*\theta$ to understand $\mathcal{E}$ as a subsheaf of $j_{T\,*}\, j_T^* \mathcal{F}|_{X_T}$ and have $\mathcal{F}|_{X_T} \subset \mathcal{E} \subset j_{T\,*}\, j_T^* \mathcal{F}|_{X_T}$. We use the increasing sequence of sheaves

$$\mathcal{F}|_{X_T} \subset \mathcal{F}|_{X_T}(D_T) \subset \mathcal{F}_{X_T}(2D_T) \subset \cdots \subset \mathcal{F}|_{X_T}(nD_T) \subset \cdots$$

to define the truncated subfunctor $\mathrm{Gr}^{\leq N}_{X,D,\mathcal{F}} \subset \mathrm{Gr}_{X,D,\mathcal{F}}$, where

$$\mathrm{Gr}^{\leq N}_{X,D,\mathcal{F}}(T) := \left\{ \begin{array}{c} \text{pairs } (\mathcal{E}, \theta) \text{ in } \mathrm{Gr}_{X_S,D,\mathcal{F}}(T) \text{ such that} \\ \mathcal{E} \subset \mathcal{F}|_{X_T}(ND_T) \end{array} \right\}$$

For any rational polynomial $P \in \mathbb{Q}[n]$, define the open and closed subfunctor $\mathrm{Gr}^{\leq N,P}_{X_S,D,\mathcal{F}} \subset \mathrm{Gr}^{\leq N}_{X_S,D,\mathcal{F}}$ consisting of points $(\mathcal{E}, \theta)$ where $\mathcal{E}$ has Hilbert polynomial P on every fiber.

Recall that a strict ind-scheme is a colimit of closed immersions.

PROPOSITION 3.6 ([**HLHJ21**, Prop. 3.13]). *We have*

$$\mathrm{Gr}_{X_S,D,\mathcal{F}} = \varinjlim_{N>0} \mathrm{Gr}^{\leq N}_{X_S,D,\mathcal{F}}$$

(as presheaves on Aff_S*) and, for each* $N \geq 0$*,*

$$\mathrm{Gr}^{\leq N}_{X_S,D,\mathcal{F}} = \bigsqcup_{P\in\mathbb{Q}[n]} \mathrm{Gr}^{\leq N,P}_{X_S,D,\mathcal{F}},$$

where each $\mathrm{Gr}^{\leq N,P}_{X_S,D,\mathcal{F}}$ *is represented by a projective scheme of finite presentation over* S*. This induces a presentation of* $\mathrm{Gr}_{X_S,D,\mathcal{F}}$ *as an ind-projective strict ind-scheme over* S*.*

There is a natural forgetful morphism $\mathrm{Gr}_{X_S,D,\mathcal{F}} \to \mathrm{Coh}^{tf}_r(X)$ sending $(\mathcal{E},\theta)$ to $\mathcal{E}$. We denote again by L_n, b_d and M_n the line bundles on $\mathrm{Gr}_{X_S,D,\mathcal{F}}$ pulled back from those on $\mathrm{Coh}^{tf}_r(X)$ defined before, by abuse of notation.

PROPOSITION 3.7 ([**HLHJ21**, Prop. 3.19]). *There exists some integer* $m \gg 0$*, depending on* $N \in \mathbb{Z}_{\geq 0}$ *and* $P \in \mathbb{Q}[n]$*, such that, for all* $n \geq m$*, the restriction* $L_n^\vee|_{\mathrm{Gr}^{\leq N,P}_{X_S,D,\mathcal{F},\sigma}}$ *is* S*-ample.*

Using the affine grassmannians above and the rational filling condition proven in [**HLHJ21**, Lem. 4.7], one can show that the stack $\mathrm{Coh}^{tf}_r(X)$ satisfies the two monotonicity conditions in Definition 2.33. This type of argument was called "Infinite dimensional GIT" in [**HLHJ21**]. We refer the reader to the introduction in [**HLHJ21**] for a more detailed explanation of this argument, and how the affine grassmannian comes in.

THEOREM 3.8 ([**HLHJ21**, Thm. 4.8]). *The polynomial numerical invariant in Definition* 3.4 *is strictly* Θ*-monotone and strictly* S*-monotone on the stack* $\mathrm{Coh}^{tf}_r(X)$*.*

Recall the notion of HN boundedness in Definition 2.34. The following theorem was proven by a concrete optimization argument in [**HLHJ21**].

THEOREM 3.9 ([**HLHJ21**, Prop. 5.5]). *The polynomial numerical invariant in Definition* 3.4 *on the stack* $\mathrm{Coh}^{tf}_r(X)$ *satisfies the HN boundedness condition.*

Using the Θ-monotonicity condition satisfied by the stack of torsion-free coherent sheaves in Theorem 3.8, plus the HN boundedness condition of the polynomial numerical invariant ν in Theorem 3.9, it follows that there exists a Θ-stratification on $\mathrm{Coh}^{tf}_r(X)$ (Definition 2.16), as stated by the Intrinsic GIT Theorem 2.30(1).

THEOREM 3.10 ([**HLHJ21**, Thm. 5.11]). *The polynomial numerical invariant* ν *defines a* Θ*-stratification on the stack* $\mathrm{Coh}^{tf}_r(X)$*.*

From the polynomial numerical invariant in Definition 3.4 we can derive a definition of ν-stability by Definition 2.27 in terms of Θ-filtrations or, by Proposition 3.3, $\mathbb{Z}$-filtrations. Recall the total order on polynomials in Definition 2.22.

DEFINITION 3.11 (ν-semistability). Let $K \supset k$ be an algebraically closed field extension and let $\mathcal{F}$ be a torsion-free coherent sheaf on X_K. We say that $\mathcal{F}$ is

ν-semistable if for all $\mathbb{Z}$-filtrations $(\mathcal{F}_m)_{m\in\mathbb{Z}}$ we have

$$\sum_{m\in\mathbb{Z}} m \cdot (\overline{p}_{\mathcal{F}_m/\mathcal{F}_{m+1}} - \overline{p}_{\mathcal{F}}) \cdot \mathrm{rk}(\mathcal{F}_m/\mathcal{F}_{m+1}) \leq 0$$

Otherwise, $\mathcal{F}$ is called ν-unstable.

In [**HLHJ21**, Prop. 5.16] it is shown that a torsion-free coherent sheaf $\mathcal{F}$ is ν-semistable if and only if it is Gieseker semistable as in [**Gie77, Mar78, Sim94**]. Therefore the stack intrinsic definition of ν-stability for this polynomial numerical invariant happens to coincide with the usual polynomial Gieseker stability for torsion-free sheaves the in literature.

Using the definition of Θ-stratification, the substack of semistable torsion-free sheaves of rank r, denoted by $\mathrm{Coh}_r^{tf}(X)^{\nu\text{-ss}}$, is an open substack of $\mathrm{Coh}_r^{tf}(X)$. Given a rational polynomial $P \in \mathbb{Q}[n]$, we denote by $\mathrm{Coh}_r^{tf}(X)_P^{\nu\text{-ss}}$ the open and closed substack of $\mathrm{Coh}_r^{tf}(X)^{\nu\text{-ss}}$ parametrizing sheaves whose Hilbert polynomial is P.

PROPOSITION 3.12 ([**HLHJ21**, Prop. 5.18]). *The substack* $\mathrm{Coh}_r^{tf}(X)_P^{\nu\text{-ss}}$ *is quasi-compact.*

Finally, we gather Theorem 3.10, the S-monotonicity property shown in Theorem 3.8 and last Proposition 3.12 to prove the existence of a separated good moduli space by Theorem 2.30 (2). Together with the proof of the existence part of the valuative criterion for properness in [**HLHJ21**, Prop. 5.20] we obtain an intrinsic construction of a proper good moduli space of torsion-free sheaves as in Theorem 2.30 (3).

THEOREM 3.13 ([**HLHJ21**, Thm. 5.21]). *The stack* $\mathrm{Coh}_r^{tf}(X)_P^{\nu\text{-ss}}$ *admits a proper good moduli space.*

3.3. Leading term and Gieseker-Harder-Narasimhan filtrations. Let K be an algebraically closed field and let $\mathcal{F}$ be an unstable torsion-free coherent sheaf on X_K. Recall from Definition 2.29 that a Θ-filtration f, with associated underlying filtration $(\mathcal{F}_m)_{m\in\mathbb{Z}}$, is called a ν-Harder-Narasimhan filtration if the polynomial numerical invariant $\nu(f)$ achieves its maximal value. Having a Θ-stratification on $\mathrm{Coh}_r^{tf}(X)$ allows us to define a unique ν-HN filtration maximizing ν, as seen at the end of Section 2.5, for any field valued point of the stack.

PROPOSITION 3.14. *Let* $K \supset k$ *be an arbitrary field extension and let* $\mathcal{F}$ *be an unstable torsion-free coherent sheaf on* X_K. *Then,* $\mathcal{F}$ *admits a* ν*-Harder-Narasimhan filtration* $(\mathcal{F}_m)_{m\in\mathbb{Z}}$ *which is uniquely determined up to scaling all of the indexes by a constant rational number. We call this filtration* $(\mathcal{F}_m)_{m\in\mathbb{Z}}$ *the leading term HN filtration of* $\mathcal{F}$.

Depending on whether we use slope semistability by Mumford [**Mum63a**], or Gieseker semistability [**Gie77, Sim94**], we can consider two different filtrations, the slope Harder-Narasimhan filtration (slope HN) and the Gieseker Harder-Narasimhan filtration (Gieseker HN). Other intermediate polynomial stability conditions can be considered as in [**HL97**, Thm. 1.6.7].

The leading term HN filtration of an unstable torsion-free coherent sheaf $\mathcal{F}$ from Proposition 3.14 corresponds to an intermediate Harder-Narasimhan filtration where the sequence of Hilbert polynomials of the quotients find their differences not at the top-degree coefficient (as in the slope HN filtration), nor at any degree coefficient (as in the Gieseker HN filtration), but at an intermediate degree coefficient, from where the filtration takes its name. Let us explain this situation.

Start with the Gieseker HN filtration of $\mathcal{F}$ (c.f. [**HL97**, Thm. 1.3.4]):

$$0 = \mathcal{F}_0^{\mathrm{HN}} \subset \mathcal{F}_1^{\mathrm{HN}} \subset \mathcal{F}_2^{\mathrm{HN}} \subset \cdots \subset \mathcal{F}_n^{\mathrm{HN}} = \mathcal{F}$$

which verifies

$$\overline{p}_{\mathcal{F}_1^{\mathrm{HN}}} > \overline{p}_{\mathcal{F}_2^{\mathrm{HN}}/\mathcal{F}_1^{\mathrm{HN}}} > \cdots > \overline{p}_{\mathcal{F}_n^{\mathrm{HN}}/\mathcal{F}_{n-1}^{\mathrm{HN}}}$$

There exists a minimal index i such that for all $h > i$ we have that all h^{th} slopes are equal

$$\widehat{\mu}_h(\mathcal{F}_1^{\mathrm{HN}}) = \widehat{\mu}_h(\mathcal{F}_2^{\mathrm{HN}}/\mathcal{F}_1^{\mathrm{HN}}) = \cdots = \widehat{\mu}_h(\mathcal{F}_n^{\mathrm{HN}}/\mathcal{F}_{n-1}^{\mathrm{HN}}),$$

but i is the largest index for which we find at least one strict inequality in the sequence of ith slopes:

$$\widehat{\mu}_i(\mathcal{F}_1^{\mathrm{HN}}) \geq \widehat{\mu}_i(\mathcal{F}_2^{\mathrm{HN}}/\mathcal{F}_1^{\mathrm{HN}}) \geq \cdots \geq \widehat{\mu}_i(\mathcal{F}_n^{\mathrm{HN}}/\mathcal{F}_{n-1}^{\mathrm{HN}}).$$

Define indexes $h_0 < h_1 < h_2 < \cdots < h_j = n$ for the jumps of the $\widehat{\mu}_i$ sequence: let $h_0 = 0$ and

$$\{h_m\}_{m=1}^{j} = \left\{ 1 \leq l \leq n \mid \widehat{\mu}_i(\mathcal{F}_{l+1}^{\mathrm{HN}}/\mathcal{F}_l^{\mathrm{HN}}) < \widehat{\mu}_i(\mathcal{F}_l^{\mathrm{HN}}/\mathcal{F}_{l-1}^{\mathrm{HN}}) \right\}$$

and set $\mathcal{F}_k^{\text{l-term}} = \mathcal{F}_{h_k}^{\mathrm{HN}}$.

The filtration

$$0 = \mathcal{F}_0^{\text{l-term}} \subset \mathcal{F}_1^{\text{l-term}} \subset \mathcal{F}_2^{\text{l-term}} \subset \cdots \subset \mathcal{F}_j^{\text{l-term}} = \mathcal{F}$$

is called the unweighted leading term filtration of $\mathcal{F}$. We call i the leading term index, which coincides with the degree of the polynomial $\nu(f)$ has degree i, and call $\widehat{\mu}_i(\mathcal{F})$ the leading slope. Observe that, in the case that $\mathcal{F}$ is slope unstable, then $i = d - 1$ and hence the leading term HN filtration coincides with the slope HN filtration.

Definition 3.15 ([**HLHJ21**, Thm. 5.11] [**GHZ24**, Defn. 9.2]). Let $\mathcal{F}$ be a ν-unstable torsion-free coherent sheaf. If the unweighted leading term HN filtration for $\mathcal{F}$ is given by

$$0 = \mathcal{F}_0^{\text{l-term}} \subset \mathcal{F}_1^{\text{l-term}} \subset \mathcal{F}_2^{\text{l-term}} \subset \cdots \subset \mathcal{F}_j^{\text{l-term}} = \mathcal{F}$$

with leading slope $\widehat{\mu}_i(\mathcal{F})$, we define the (weighted) leading term HN filtration $(\mathcal{F}^{\mathrm{Can}})_{m\in\mathbb{Z}}$ as follows (up to scaling of the weights).

Set $\mu = \widehat{\mu}_i(\mathcal{F})$, $\mu_k = \widehat{\mu}_i\left(\mathcal{F}_k^{\text{l-term}}/\mathcal{F}_{k-1}^{\text{l-term}}\right)$. Choose a positive rational number L such that $L \cdot \mu \in \mathbb{Z}$ and $L \cdot \mu_k \in \mathbb{Z}$ for all k. Then,

(1) $\mathcal{F}_m^{\mathrm{Can}} = 0$ for $m > L(\mu_1 - \mu)$.
(2) Let $1 \leq k \leq j-1$. For each $L(\mu_k - \mu) \geq m > L(\mu_{k+1} - \mu)$, we set $\mathcal{F}_m^{\mathrm{Can}} = \mathcal{F}_k^{\text{l-term}}$.
(3) $\mathcal{F}_m^{\mathrm{Can}} = \mathcal{F}$ for $L(\mu_j - \mu) \geq m$.

This is precisely the leading term HN filtration where the kth jump happens at the integer $L(\mu_k - \mu)$.

The weighted leading term HN filtration recovers a destabilizing HN-type filtration at the maximal degree-coefficient of the Hilbert polynomial, the leading term, yielding a sequence of decreasing leading slopes. If we take the graded object of the filtration and iterate the process we will check all possible instability terms at all lower degrees in the polynomial. This means that we will arrive at the classical Gieseker HN filtration at the end. Let us explain this procedure (c.f. [**GHZ24**, Prop. 9.3]).

Denote the (weighted) leading term HN filtration in Definition 3.15 by $\left({}^{(1)}\mathcal{F}_m\right)_{m\in\mathbb{Z}}$ and the leading term index by i_1. Define the 1st associated graded Levi sheaf as

$$\mathrm{gr}_1(\mathcal{F}) = \left(\mathcal{F}_i^{l-term}/\mathcal{F}_{i-1}^{l-term}\right)_{i=1}^{j_1}$$

The recursion step consists of taking the leading term HN filtration of this graded object, which amounts to taking the leading term HN filtration for each $\mathcal{F}_i^{l-term}/\mathcal{F}_{i-1}^{l-term}$, with $i = 1, 2, \ldots, j_1$. Note that all leading term indices are strictly smaller than i_1.

This yields a lexicographic filtration indexed by $\mathbb{Z}^2$, denoted by $\left({}^{(2)}\mathcal{F}_{\vec{m}}\right)_{\vec{m}\in\mathbb{Z}^2}$, with jumps at the pairs

$$\vec{m}_1 > \vec{m}_2 > \cdots > \vec{m}_{j_2}$$

This way, the underlying unweighted filtration reads

$$0 \subset^{(2)} \mathcal{F}_{\vec{m}_1} \subset^{(2)} \mathcal{F}_{\vec{m}_2} \subset \cdots \subset^{(2)} \mathcal{F}_{\vec{m}_{j_2}} = \mathcal{F}$$

and satisfies that the Hilbert polynomials of the graded pieces ${}^{(2)}\mathcal{F}_{\vec{m}_i}/{}^{(2)}\mathcal{F}_{\vec{m}_{i-1}}$ are decreasing but their differences occur, at most, at a degree $i_2 < i_1$, i_2 being the leading term index of this second leading term HN filtration.

Continuing this process (i.e. taking the leading term HN filtration of the graded object while possible), we arrive to a qth filtration $\left({}^{(q)}\mathcal{F}_{\vec{m}}\right)_{\vec{m}\in\mathbb{Z}^q}$ with unweighted filtration

$$0 \subset^{(q)} \mathcal{F}_{\vec{m}_1} \subset^{(q)} \mathcal{F}_{\vec{m}_2} \subset \cdots \subset^{(q)} \mathcal{F}_{\vec{m}_{j_q}} = \mathcal{F}$$

whose graded pieces forming the qth associated Levi sheaf $\mathrm{gr}_q(\mathcal{F}) = \left({}^{(q)}\mathcal{F}_{\vec{m}_i}/{}^{(q)}\mathcal{F}_{\vec{m}_{i-1}}\right)$ do not admit any further filtering by any lower leading degree: they are Gieseker semistable. Therefore this final filtration is, indeed, the Gieseker HN filtration that we find in the literature [**Sim94, HL97**].

4. G-bundles on higher dimensional varieties

In this section we expose the theory of the "Beyond GIT" program for the moduli problem of G-bundles on higher dimensional varieties, providing a stack-intrinsic construction of the moduli space and a new notion of a Gieseker Harder-Narasimhan filtration. A more extensive treatment can be found in [**GHZ24**].

For the rest of this section X will denote a smooth projective variety of dimension d over an algebraically closed field k of characteristic 0.

4.1. Principal G-bundles. Let G be an algebraic reductive group. Let $\rho: G \hookrightarrow \mathrm{GL}(V)$ be a faithful representation. A principal G-bundle P on a scheme X induces a vector bundle $\mathcal{F} := P(V)$ of rank equal to $\dim V$. Using the converse construction, we want to describe a principal bundle as a pair consisting of a vector bundle and a reduction of structure group.

Let S be an affine k-scheme. Given a vector bundle $\mathcal{F}$ on X_S of rank $\dim V$, we define the associated principal $\mathrm{GL}(V)$-bundle

$$P_{\mathrm{GL}} = \underline{\mathrm{Iso}}(\mathcal{F}, V \otimes \mathcal{O}_{X_S}),$$

with the right action of $\mathrm{GL}(V)$ defined as the post-composition with the inverse of $h \in \mathrm{GL}(V)$:

$$(\beta, h) \mapsto \beta \cdot h := h^{-1} \circ \beta,$$

where we take the inverse to obtain an action on the right.

Define the scheme of reductions as the quotient

$$\mathrm{Red}_G(\mathcal{F}) = \underline{\mathrm{Iso}}(\mathcal{F}, V \otimes \mathcal{O}_{X_S})/G \longrightarrow X_S,$$

where G acts on the right via the representation ρ. A reduction of structure group is given by a section of $\mathrm{Red}_G(\mathcal{F})$ over X_S.

If $\dim X > 1$, then the moduli space of vector bundles does not satisfy the existence part of the valuative criterion for properness in general, because a family of vector bundles might degenerate to a torsion-free sheaf. Likewise, if we consider principal G-bundles we have to allow for more general objects to appear in the moduli problem. These will be called principal ρ-sheaves.

Before considering the general case, let us have a look at a simpler example. A subset $U \subset X_S$ is called big if for any point $s \in X$, the fiber $U_s \subset X_s = X$ is an open set with codimension in X at least two.

EXAMPLE 4.1. If $G = \mathrm{Sp}(r)$ is the symplectic group and ρ is the standard representation, then a reduction is the same thing as a homomorphism $\varphi : \mathcal{F} \otimes \mathcal{F} \to \mathcal{O}_{X_S}$ which is bilinear, skew-symmetric and non-degenerate. In this case, a ρ-sheaf on X will be a pair

$$(\mathcal{F}, \varphi : \mathcal{F} \otimes \mathcal{F} \to \mathcal{O}_X),$$

where $\mathcal{F}$ is a torsion-free sheaf and φ is a skew-symmetric bilinear form, which is non-degenerate when restricted to the open set $U_{\mathcal{F}}$ where $\mathcal{F}$ is locally free. Equivalently, a ρ-sheaf is a principal $\mathrm{Sp}(r)$-bundle P on a big open set $U \subset X$, a torsion-free sheaf $\mathcal{F}$ on X, and an isomorphism between the vector bundle associated to P and the restriction $\mathcal{F}|_U$.

REMARK 4.2. If we want our definition of stability to reduce to the definition of Ramanathan when $\dim X = 1$ [**Ram75**], then we have to require that the representation $\rho : G \to \mathrm{GL}(V)$ is central (meaning that the representation ρ sends the identity component of the center of G to the center of $\mathrm{GL}(V)$). Otherwise, for noncentral ρ our definition of semistability is stronger than Ramanathan's, see [**GHZ24**, Example 10.10] for an explanation. Notice that if G is semisimple, then ρ is automatically central. On the other hand, ρ cannot be central if the dimension of the center of G is larger than one; this is why we consider representations into products of general linear groups in subsection 4.4.

4.2. Principal ρ-sheaves. The representation $\rho : G \hookrightarrow \mathrm{GL}(V)$ gives a left action of G on V, hence a left action on $\det(V)$, and therefore $\det(V)$ acquires the structure of a right $k[G]$-comodule:

$$\det(V) \longrightarrow \det(V) \otimes k[G]. \tag{4.1}$$

Explicitly, $v \mapsto v \otimes \det(\rho(g))$.

Using the inverse $v \mapsto g^{-1}v$ we obtain a right action of G on V. This right action of G on V induces on V a left $k[G]$-comodule structure $V \to k[G] \otimes V$. We equip $V^\vee$ with the dual right $k[G]$-comodule structure:

$$c : V^\vee \longrightarrow V^\vee \otimes k[G]. \tag{4.2}$$

Explicitly, if $\{e_i\}$ is a basis of V and $\{e^i\}$ is the dual basis of $V^\vee$, then

$$c(e^i) = \sum_j e^j \otimes (\rho(g)^{-1})^i{}_j$$

(note that $A \mapsto (A^{-1})^i{}_j$ is an element of $k[\mathrm{GL}(V)]$). The value that it takes on an element $A \in \mathrm{GL}(V)$ is the (i,j)-entry of the inverse of A, expressed in the basis $\{e_i\}$.

These right $k[G]$-comodule structures (4.1) (4.2) induce a right $k[G]$-comodule structure on

$$E_V(\mathcal{F}) := \mathrm{Sym}^\bullet(\mathcal{F} \otimes_k V^\vee) \otimes_{\mathcal{O}_{X_S}} \mathrm{Sym}^\bullet(\det(\mathcal{F})^\vee \otimes_k \det(V)). \tag{4.3}$$

Therefore, the scheme $H(\mathcal{F}, V) := \underline{\mathrm{Spec}}_{X_S}(E_V(\mathcal{F}))$ (which is affine over X_S and of finite presentation) has a right action by G. Consider the GIT quotient

$$H(\mathcal{F}, V)/\!\!/G = \underline{\mathrm{Spec}}_{X_S}(E_V(\mathcal{F})^G),$$

where $E_V(\mathcal{F})^G$ is the subalgebra of G-invariants.

The restriction $H := H(\mathcal{F}, V)|_U$ to the open set $U \subset X_S$ where $\mathcal{F}$ is locally free is

$$H = \underline{\mathrm{Hom}}(\mathcal{F}|_U, V \otimes_k \mathcal{O}_U) \times \underline{\mathrm{Hom}}(\det(\mathcal{F}|_U)^\vee, \det(V^\vee) \otimes_k \mathcal{O}_U),$$

parametrizing pairs $(\alpha : \mathcal{F}|_U \to V \otimes_k \mathcal{O}_U, \beta : \det(\mathcal{F}|_U)^\vee \to \det(V^\vee) \otimes_k \mathcal{O}_U$. The equation $\det(\alpha)\beta = 1$ cuts out a closed subscheme which is isomorphic to

$$\underline{\mathrm{Iso}}(\mathcal{F}|_U, V \otimes_k \mathcal{O}_U).$$

Furthermore, the equation $\det(\alpha)\beta = 1$ is G-invariant, so it also cuts out a closed subscheme $(H/\!\!/G)_0 \subset H/\!\!/G$ which is isomorphic to

$$\underline{\mathrm{Iso}}(\mathcal{F}|_U, V \otimes_k \mathcal{O}_U)/G = \mathrm{Red}_G(\mathcal{F}|_U).$$

DEFINITION 4.3 ([**GHZ24**, Defn. 2.14]). Let $\mathcal{F}$ be a torsion-free sheaf on X. The scheme of reductions $\mathrm{Red}_G(\mathcal{F})$ is defined to be the scheme theoretic closure of $\mathrm{Red}_G(\mathcal{F}|_U)$ in $H(\mathcal{F}, V)/\!\!/G$.

The scheme $\mathrm{Red}_G(\mathcal{F}) \to X_S$ is affine of finite type and allows to give a definition of principal ρ-sheaf.

DEFINITION 4.4 ([**GHZ24**, Defn. 3.1]). A principal ρ-sheaf on X_S is a pair $(\mathcal{F}, \sigma)$ where

(1) $\mathcal{F}$ is a torsion-free sheaf on X_S.
(2) σ is a section of $\mathrm{Red}_G(\mathcal{F}) \to X_S$.

We have the following Hartogs-like result (see [**GHZ24**, Lem. 2.2]).

PROPOSITION 4.5 ([**GHZ24**, Prop. 2.18]). *Let $U \subset X_S$ be any big open set where $\mathcal{F}$ is locally free (note that such U always exists* [**GHZ24**, Lemma 2.5]*). There is a bijection between the sections of the morphism* $\mathrm{Red}_G(\mathcal{F}) \to X_S$ *and the sections on the restriction* $\mathrm{Red}_G(\mathcal{F}|_U) \to U$.

This allows us to give an equivalent definition of ρ-sheaf.

LEMMA 4.6. *Giving a principal ρ-sheaf on X_S is the same thing as giving a triple $(P, \mathcal{F}, \psi)$ where*

(1) $\mathcal{F}$ is a torsion-free sheaf on X_S, and U is a big open subset where $\mathcal{F}$ is locally free (we may take U to be the maximal such open).
(2) P is a principal G-bundle on U.
(3) ψ is an isomorphism between the vector bundle $P(\rho)$ associated to ρ, and the restriction $\mathcal{F}|_U$

$$\psi : P(\rho) \cong \mathcal{F}|_U.$$

Denote by $\mathrm{Bun}_\rho(X)$ the stack of principal ρ-sheaves on X.

DEFINITION 4.7. The algebraic stack of principal ρ-sheaves on X is the pseudofunctor defined by

$$T \mapsto \mathrm{Bun}_\rho(X)(T) = \big[\text{ groupoid of } \rho\text{-sheaves } (\mathcal{F},\sigma) \text{ on } X_T \big].$$

There is a forgetful morphism $\mathrm{Bun}_\rho(X) \to \mathrm{Coh}_r^{tf}(X)$ which is schematic, affine and of finite type ([**GHZ24**, Prop 3.6]). This allows to relate the constructions for ρ-sheaves to those of section 3 for torsion-free coherent sheaves.

4.3. Moduli space of ρ-sheaves. Here we introduce a polynomial numerical invariant on the stack $\mathrm{Bun}_\rho(X)$ of principal ρ-sheaves, which yields the notion of Gieseker stability. Once we prove that the requirements of Theorem 2.30 are met, it will be shown that the substack of semistable ρ-sheaves admits a good moduli space.

Let $\lambda : \mathbb{G}_m \to G$ be a one-parameter subgroup . We define a parabolic subgroup $P_\lambda \subset G$ associated to λ by

$$P_\lambda = \Big\{ g \in G \mid \lim_{x\to 0} \lambda(x)\, g\, \lambda(x)^{-1} \text{ exists} \Big\}$$

whose Levi subgroup is

$$L_\lambda = \Big\{ g \in G \mid \lim_{x\to 0} \lambda(x)\, g\, \lambda(x)^{-1} = g \Big\}.$$

Composing with the representation ρ we obtain a parabolic subgroup of $\mathrm{GL}(V)$:

$$P_\lambda^{\mathrm{GL}(V)} = \Big\{ g \in \mathrm{GL}(V) \mid \lim_{x\to 0} \rho\circ\lambda(x)\, g\, (\rho\circ\lambda(x))^{-1} \text{ exists} \Big\}$$

and it is clear that $P_\lambda = P_\lambda^{\mathrm{GL}(V)} \cap G$.

Let $(\mathcal{F},\sigma)$ be a principal ρ-sheaf on X. Let $P \to U \subset X$ be the associated principal G-bundle on the open set U where $\mathcal{F}$ is locally free, and let $\mathcal{G}_\lambda \to U' \subset U$ be a reduction of structure group of P to P_λ on a big open set $U' \subset X$. The inclusion $P_\lambda \subset P_\lambda^{\mathrm{GL}(V)}$ gives a principal $P_\lambda^{\mathrm{GL}(V)}$-bundle P' on U', which is a reduction of structure group to the parabolic $P_\lambda^{\mathrm{GL}(V)}$ of the principal $\mathrm{GL}(V)$-bundle associated to the vector bundle $\mathcal{F}|_{U'}$.

Giving a reduction to a parabolic of the linear group is the same thing as giving a filtration of the vector bundle $\mathcal{F}|_{U'}$, and such a filtration extends uniquely over X_S to a filtration of $\mathcal{F}$ by torsion-free sheaves (with the condition that successive quotients of the filtration are also torsion-free). Moreover, we use the weights of the one-parameter subgroup obtained by composition $\rho\circ\lambda : \mathbb{G}_m \to \mathrm{GL}(V)$ to obtain a $\mathbb{Z}$-weighted filtration $(\mathcal{F}_m)_{m\in\mathbb{Z}}$ of $\mathcal{F}$ as in Proposition 3.3.

Recall from Definition 2.16 that a Θ-filtration of a point p in a stack $\mathcal{M}$ is a morphism $f : \Theta = [\mathbb{A}^1/\mathbb{G}_m] \to \mathcal{M}$ together with an isomorphism $f(1) \simeq p$. Any Θ-filtrations $f : \Theta \to \mathrm{Bun}_\rho(X)$ of a ρ-sheaf $(\mathcal{F},\sigma)$ can be composed with the forgetful morphism $\mathrm{Bun}_\rho(X) \to \mathrm{Coh}_r^{tf}(X)$ to obtain a $\mathbb{Z}$-weighted filtration of $\mathcal{F}$ satisfying a series of compatibility conditions. This way, a Θ-filtration of a ρ-sheaf $(\mathcal{F},\sigma)$ can be understood as a $\mathbb{Z}$-weighted filtration $(\mathcal{F}_m)_{m\in\mathbb{Z}}$ of $\mathcal{F}$ (as in Proposition 3.3) which is associated to a reduction of structure group ([**GHZ24**, Prop. 3.12 and 3.16]). See [**GHZ24**, Sections 3.2 and 3.3] for details of this construction.

In a similar way to section 3.1, we can define a family of line bundles [**GHZ24**, Defn. 4.1] and a rational quadratic norm [**GHZ24**, Defn. 4.3] to define a polynomial numerical invariant ([**GHZ24**, Prop. 4.4]):

$$\nu(f) = \sqrt{A_d}\frac{\sum_{m\in\mathbb{Z}} m\cdot(\overline{p}_{\mathcal{F}_m/\mathcal{F}_{m+1}} - \overline{p}_{\mathcal{F}})\cdot \mathrm{rk}(\mathcal{F}_m/\mathcal{F}_{m+1})}{\sqrt{\sum_{m\in\mathbb{Z}} m^2\,\mathrm{rk}(\mathcal{F}_m/\mathcal{F}_{m+1})}}.$$

Note that this is the same expression as in Definition 3.4 for torsion-free sheaves, but for a ρ-sheaf $(\mathcal{F},\sigma)$ it accounts just for Θ-filtrations of $\mathrm{Bun}_\rho(X)$.

The polynomial numerical invariant ν defined on the stack $\mathrm{Bun}_\rho(X)$ can be shown to be strictly Θ-monotone and strictly S-monotone ([**GHZ24**, Prop. 4.16]) and to satisfy the HN boundedness condition ([**GHZ24**, Prop. 4.26]). Then it defines a Θ-stratification on $\mathrm{Bun}_\rho(X)$ by Theorem 2.30(1) ([**GHZ24**, Thm. 4.27]). This allows to define a notion of ν-stability which is an open condition.

Definition 4.8 ([**GHZ24**, Defn. 4.28]). A principal ρ-sheaf $(\mathcal{F},\sigma)$ is ν-semistable if for all one-parameter subgroups $\lambda : \mathbb{G}_m \to G$ and all reductions as above, we have

$$\sum_{m\in\mathbb{Z}} m\cdot(\overline{p}_{\mathcal{F}_m/\mathcal{F}_{m+1}} - \overline{p}_{\mathcal{F}})\cdot \mathrm{rk}(\mathcal{F}_m/\mathcal{F}_{m+1}) \le 0.$$

This definition coincides with the semistability notion in [**Sch02, GS05**]. The substack $\mathrm{Bun}_\rho(X)^{\nu\text{-ss}}$ of semistable ρ-sheaves satisfies the existence part of the valuative criterion for properness in [**GHZ24**, Thm. 5.8]. After fixing the degree ϑ of a ρ-sheaf $(\mathcal{F},\sigma)$ ([**GHZ24**, Defn. 6.3]) and the Hilbert polynomial P of $\mathcal{F}$, the substacks $\mathrm{Bun}_\rho(X)^{\nu\text{-ss},P}_\vartheta$ are quasi-compact ([**GHZ24**, Prop. 6.6], then by Theorem 2.30 (2,3) they do admit a proper good moduli space ([**GHZ24**, Thm. 6.8]). This is a stack intrinsic alternative to the constructions in [**Sch02, GS05**].

4.4. The scheme of G-reductions for products of representations. As we explained in Remark 4.2, if we want to consider reductive groups that are not semisimple, we need to deal with tuples of representations. In section 4.5 we will observe that when constructing the Gieseker HN filtration, even if we start with a semisimple group, the Levi factors we need to go through make necessary to consider non-semisimple groups.

Let $V^\bullet = (V^i)_{i=1}^b$ be a tuple of finite dimensional vector spaces. We consider a representation into the tuple

$$\rho = (\rho_1,\dots,\rho_b) : G \longrightarrow \mathrm{GL}(V^1)\times\cdots\times\mathrm{GL}(V^b).$$

Let $\mathcal{F}^\bullet = (\mathcal{F}^i)_{i=1}^b$ be a tuple of torsion-free sheaves on X_S. Using the construction (4.3), we consider the product scheme

$$H(\mathcal{F}^\bullet, V^\bullet) := \prod_{i=1}^b H(\mathcal{F}^i, V^i).$$

The group G has a right action on this product, induced by the right action on each factor defined by $\rho_i : G\to \mathrm{GL}(V^i)$. Let $H(\mathcal{F}^\bullet,V^\bullet)/\!\!/G$ be the GIT quotient. Let U be an open set where all sheaves $\mathcal{F}^i$ are locally free. On each factor $H(\mathcal{F}^i|_U, V^i)$ we have an equation $\alpha_i\det(\beta_i)=1$ as before, cutting out a divisor isomorphic to the scheme of isomorphisms $\mathrm{Iso}(\mathcal{F}^i|_U, V^i\otimes\mathcal{O}_U)$. Finally, we define $\mathrm{Red}_G(\mathcal{F}^\bullet)$ as the scheme theoretic closure of the subscheme

$$\prod \mathrm{Iso}(\mathcal{F}^i|_U, V^i\otimes\mathcal{O}_U) \cong \prod (H(\mathcal{F}^i|_U, V^i))_{\alpha^i\det(\beta^i)=1} \hookrightarrow H(\mathcal{F}^\bullet,V^\bullet)/\!\!/G.$$

DEFINITION 4.9 ([**GHZ24**, Defn. 7.1]). A principal ρ-sheaf on X_S is a pair $(\mathcal{F}^\bullet, \sigma)$ where

(1) $\mathcal{F}^\bullet$ is a tuple of torsion-free sheaves on X_S.
(2) σ is a section of $\mathrm{Red}_G(\mathcal{F}^\bullet) \to X_S$.

As in the case of a single representation, we denote by $\mathrm{Bun}_\rho(X)$ the stack of ρ-sheaves.

For any big open set U where all sheaves $\mathcal{F}^i$ are locally free, the section σ gives a reduction of structure group, from the principal $\mathrm{GL}(V^1) \times \cdots \times \mathrm{GL}(V^b)$-bundle associated to the sum of vector bundles $\mathcal{F}^1|_U \oplus \cdots \oplus \mathcal{F}^b|_U$, to a principal G-bundle $P \to U$ on U. Therefore, using the Hartogs-like Proposition 4.5, we have the following

LEMMA 4.10. *Giving a principal ρ-sheaf is the same thing as giving a triple $(\mathcal{F}^\bullet, P, \psi^\bullet)$ where*

(1) *P is a principal G-bundle on a big open set U.*
(2) *$\mathcal{F}^\bullet = (\mathcal{F}^i)_{i=1}^b$ is a tuple of b torsion-free sheaves, all of them locally free on U.*
(3) *$\psi^\bullet = (\psi^i)_{i=1}^b$ is a collection of isomorphisms*

$$\psi^i : P(\rho_i) \cong \mathcal{F}^i|_U .$$

Finally, we can extend to this case the notion of stability.

DEFINITION 4.11. A principal ρ-sheaf $(\mathcal{F}^\bullet, \sigma)$ on X is ν-semistable if for all one-parameter subgroups $\lambda : \mathbb{G}_m \to G$ and all reductions as described in subsection 4.3, we have

$$\sum_{i=1}^{b} \sum_{m \in \mathbb{Z}} m \cdot (\overline{p}_{\mathcal{F}^i_m/\mathcal{F}^i_{m+1}} - \overline{p}_{\mathcal{F}^i}) \cdot \mathrm{rk}(\mathcal{F}^i_m/\mathcal{F}^i_{m+1}) \le 0$$

4.5. Gieseker-Harder-Narasimhan filtration for ρ-sheaves. A unstable torsion-free sheaf has an associated canonical filtration, called the Harder-Narasimhan filtration, showing that any torsion-free sheaf can be constructed uniquely as a successive extension of semistable torsion-free sheaves [**HL97**, Thm. 1.3.4]. As we mentioned in subsection 3.3, this can be the slope HN or the Gieseker HN filtration, depending on the stability condition we are using.

In the case of principal bundles, there is already in [**Beh91**] a notion of Harder-Narasimhan reduction if we use Ramanathan's definition of semistability (which is the direct generalization of Mumford's slope condition in [**Mum63a**]). This is generalized to higher dimensional varieties in [**AAB02**]. However, if we use Gieseker's notion [**Gie77**] of semistability considering all terms of the Hilbert polynomial, the right notion of Harder-Narasimhan filtration had not been defined until the arrival of the "beyond GIT" theory [**HL21**]. The notion of semistability for ρ-sheaves (Definitions 4.8 and 4.11) can be obtained through the techniques of the "beyond GIT" program, in terms of Θ-filtrations directly from the stack picture. The advantage of this approach is that it provides a way to define a correct Gieseker Harder-Narasimhan filtration for moduli problems with a polynomial stability condition.

Let $(\mathcal{F}, \sigma)$ be a ρ-sheaf and let $f : \Theta \to \mathrm{Bun}_\rho(X)$ be a Θ-filtration of $(\mathcal{F}, \sigma)$. Recall from subsection 4.3 the definition of a polynomial numerical invariant which, when considering representations into tuples $V^\bullet$ as in subsection 4.4, is written as ([**GHZ24**, Section 7]):

$$\nu(f) = \sqrt{A_d}\frac{\sum_{i=1}^{b}\sum_{m\in\mathbb{Z}} m\cdot(\overline{p}_{\mathcal{F}^i_m/\mathcal{F}^i_{m+1}} - \overline{p}_{\mathcal{F}^i})\cdot \mathrm{rk}(\mathcal{F}^i_m/\mathcal{F}^i_{m+1})}{\sqrt{\sum_{i=1}^{b}\sum_{m\in\mathbb{Z}} m^2\,\mathrm{rk}(\mathcal{F}^i_m/\mathcal{F}^i_{m+1})}}.$$

We can rephrase (Definition 2.18) the condition of ν-semistability by saying that a ρ-sheaf $(\mathcal{F},\sigma)$ is ν-semistable if and only if for all Θ-filtrations $f:\Theta\to \mathrm{Bun}_\rho(X)$ with $f(1)\simeq(\mathcal{F},\sigma)$ we have $\nu(f)\le 0$.

Let $(\mathcal{F},\sigma)$ be a ν-unstable ρ-sheaf. Given that $\mathrm{Bun}_\rho(X)$ is endowed with a Θ-stratification it is shown in [**GHZ24**, Prop. 4.31] that there is a unique[1] Θ-filtration f giving the maximum value of $\nu(f)$. This is a ν-HN filtration of the point $(\mathcal{F},\sigma)$ in the stack $\mathrm{Bun}_\rho(X)$ as in Definition 2.29.

Associated to this maximal Θ-filtration f, we have a reduction of structure group to a parabolic subgroup $P_{\lambda_1}\subset G$ for some $\lambda_1:\mathbb{G}_m\to G$, and a $\mathbb{Z}$-filtration $(\mathcal{F}_m)_{m\in\mathbb{Z}}$ of $\mathcal{F}$ satisfying the compatibility conditions described in [**GHZ24**, Section 3.2]. By the same reasons explained in subsection 3.3 this is called the leading term HN filtration, which is the version for ρ-sheaves of the filtration described there for torsion-free coherent sheaves. In order to obtain a filtration of Gieseker type, meaning a filtration being able to compare all terms of Hilbert polynomials and whose graded object is semistable, we need to reproduce the algorithm described in that subsection 3.3 translated to the stack $\mathrm{Bun}_\rho(X)$.

Using the projection $P_{\lambda_1}\to L_{\lambda_1}$ from the parabolic group to its Levi factor, this reduction induces a principal ρ_{λ_1}-sheaf with structure group L_{λ_1}, the Levi factor of P_{λ_1}. The representation of this Levi is of the form

$$\rho_{\lambda_1}: L_{\lambda_1}\to \mathrm{GL}(V^1)\times\cdots\times\mathrm{GL}(V^b),$$

and the tuple of sheaves $\mathcal{F}'^\bullet$ is given by the successive quotients of the filtration $(\mathcal{F}_m)_{m\in\mathbb{Z}}$, i.e., $\mathcal{F}'^i=\mathcal{F}_i/\mathcal{F}_{i+1}$. However, this Levi might not be semistable yet, therefore we need to further take the leading term HN filtration of this ρ_{λ_1}-sheaf. In that case, we obtain a new reduction to a parabolic $P_{\lambda_2}\subset L_{\lambda_1}\subset G$ and repeat this process until we obtain a reduction whose associated Levi sheaf is semistable. The final associated filtration will be called the Gieseker HN filtration (see [**GHZ24**, Section 8.3].

4.6. Example of a Gieseker Harder-Narasimhan filtration. We illustrate the process of producing the Gieseker HN filtration with a concrete numerical example.

Let $G=\mathrm{Sp}(6)$ with $\rho:\mathrm{Sp}(6)\to\mathrm{GL}(6)$ the standard representation. Let $X=\mathbb{P}^3$. Let p and C be, respectively, a given fixed point and a fixed line in $\mathbb{P}^3$. Let $\mathcal{I}_p$ and $\mathcal{I}_C$ be their respective ideal sheaves. Set

$$\mathcal{F}=\mathcal{I}_C\oplus(\mathcal{I}_p\oplus\mathcal{I}_p)\oplus(\mathcal{O}_X\oplus\mathcal{O}_X)\oplus\mathcal{O}_X$$

[1]It is unique up to multiplication of all the indexes $m\in\mathbb{Z}$ by a scalar. This corresponds to reparametrization of the one-parameter subgroup by $z\mapsto z^a$

and let $\varphi : \mathcal{F} \otimes \mathcal{F} \to \mathcal{O}_X$ be the symplectic form defined by the matrix

$$\varphi = \begin{pmatrix} 0 & \cdots & & & & 1 \\ \vdots & & & & 1 & \\ & & & 1 & & \\ & & -1 & & & \\ & -1 & & & & \vdots \\ -1 & & & & \cdots & 0 \end{pmatrix}$$

The Hilbert polynomials are

$$P_{\mathcal{O}}(n) = \frac{1}{6}(n^3 + 6n^2 + 10n + 3)\,,$$

$$P_{\mathcal{I}_p}(n) = \frac{1}{6}(n^3 + 6n^2 + 10n - 3)\,,$$

$$P_{\mathcal{I}_C}(n) = \frac{1}{6}(n^3 + 6n^2 - 6n - 3)\,.$$

We will consider one-parameter subgroups of Sp(6) of the form

$$f(t) = \operatorname{diag}(t^a, t^b, t^c, t^{-c}, t^{-b}, t^{-a})$$

with $a \geq b \geq c \geq 0$. The associated filtration $\mathcal{F}_m$ satisfies $\mathcal{F}_m^{\perp} = \mathcal{F}_{-m+1}$, where the orthogonal of a subsheaf $\mathcal{F}_i \subset \mathcal{F}$ is given by

$$\mathcal{F}_i{}^{\perp} := \ker\big(\mathcal{F} \xrightarrow{\varphi} \mathcal{F}^{\vee} \longrightarrow \mathcal{F}_i{}^{\vee}\big).$$

The bilinear form gives a reduction of structure group of $\mathcal{F}|_{\mathbb{P}^3\setminus(C\cup p)}$ to Sp(6), so the data $(\mathcal{F}, \varphi)$ defines a principal ρ-sheaf. It can be checked that it is ν-unstable (Definition 4.8), and the one-parameter subgroup associated to the leading term HN filtration maximizing ν is $\lambda_1 = \operatorname{diag}(t, 1, 1, 1, 1, t^{-1})$. The filtration of sheaves $\big({}^{(1)}\mathcal{F}_m\big)_{m\in\mathbb{Z}}$ associated to λ_1 is

$$\overbrace{\mathcal{F}}^{\mathcal{F}_{-1}} \quad \supset \quad \overbrace{\mathcal{I}_p \oplus \mathcal{I}_p \oplus \mathcal{O} \oplus \mathcal{O}}^{\mathcal{F}_0} \quad \supset \quad \overbrace{\mathcal{O}}^{\mathcal{F}_1}$$

We have the associated Levi $L_{\lambda_1} = \mathrm{GL}(1) \times \mathrm{Sp}(4)$ and the representation ρ_{λ_1}, which is given by

$$\begin{array}{cccc} \rho_{\lambda_1}: & \mathrm{GL}(1) \times \mathrm{Sp}(4) & \longrightarrow & \mathrm{GL}(1) \times \mathrm{GL}(4) \times \mathrm{GL}(1) \\ & (t, g) & \mapsto & (t, g, t^{-1}) \end{array}.$$

The 1st associated Levi graded sheaf $\mathrm{gr}_1(\mathcal{F})$ is

$$\overbrace{\mathcal{I}_L}^{\mathcal{F}^{-1}:=\mathcal{F}_{-1}/\mathcal{F}_0} \quad \oplus \quad \overbrace{\mathcal{I}_p \oplus \mathcal{I}_p \oplus \mathcal{O} \oplus \mathcal{O}}^{\mathcal{F}^0:=\mathcal{F}_0/\mathcal{F}_{-1}} \quad \oplus \quad \overbrace{\mathcal{O}}^{\mathcal{F}^1:=\mathcal{F}_1}\,.$$

The principal ρ_1-sheaf defined by this is not ν-semistable, then we take its leading term HN filtration having associated one-parameter subgroup

$$\begin{array}{cccc} \lambda_2: & \mathbb{G}_m & \longrightarrow & \mathrm{GL}(1) \times \mathrm{Sp}(4) \\ & t & \mapsto & (1, (t, t, t^{-1}, t^{-1})) \end{array}$$

The corresponding 2nd filtration $\big({}^{(2)}\mathcal{F}_{(m_1,m_2)}\big)_{(m_1,m_2)\in\mathbb{Z}^2}$ is

$$\mathcal{F} \supset (\mathcal{I}_p \oplus \mathcal{I}_p) \oplus (\mathcal{O}_X \oplus \mathcal{O}_X) \oplus \mathcal{O}_X \supset (\mathcal{O}_X \oplus \mathcal{O}_X) \oplus \mathcal{O}_X \supset \mathcal{O}_X \tag{4.4}$$

whose 2nd associated Levi graded sheaf $\mathrm{gr}_2(\mathcal{F})$ is

$$\mathcal{I}_C \quad \oplus \quad (\mathcal{I}_p \oplus \mathcal{I}_p) \quad \oplus \quad (\mathcal{O}_X \oplus \mathcal{O}_X) \quad \oplus \quad \mathcal{O}_X$$

The second Levi factor L_{λ_2} is $\mathrm{GL}(1) \times \mathrm{GL}(2)$ (note that the embedding $\mathrm{GL}(2) \subset \mathrm{Sp}(4)$ is given by $A \mapsto (A, A^{-1})$) and the principal ρ_2-sheaf is ν-semistable. Since this is semistable, the process stops, and the Gieseker Harder-Narasimhan reduction has (4.4) as associated filtration.

Acknowledgments

We thank Andrés Ibáñez Núñez, Sveta Makarova and Nitin Nitsure, as well as the anonymous referee, for helpful comments on the manuscript. We also thank the VBAC community lead by Peter Newstead and the editors of this contributed volume for their continuous work.

This work is supported by grants CEX2019-000904-S and PID2019-108936GB-C21 (funded by MCIN/AEI/ 10.13039/501100011033).

References

[AAB02] B. Anchouche, H. Azad, and I. Biswas, *Harder-Narasimhan reduction for principal bundles over a compact Kähler manifold*, Math. Ann. **323** (2002), no. 4, 693–712, DOI 10.1007/s002080200322. MR1924276

[AB83] M. F. Atiyah and R. Bott, *The Yang-Mills equations over Riemann surfaces*, Philos. Trans. Roy. Soc. London Ser. A **308** (1983), no. 1505, 523–615, DOI 10.1098/rsta.1983.0017. MR702806

[ABB+22] J. Alper, P. Belmans, D. Bragg, J. Liang, and T. Tajakka, *Projectivity of the moduli space of vector bundles on a curve*, Stacks Project Expository Collection (SPEC), London Math. Soc. Lecture Note Ser., vol. 480, Cambridge Univ. Press, Cambridge, 2022, pp. 90–125. MR4480534

[ABHLX20] J. Alper, H. Blum, D. Halpern-Leistner, and C. Xu, *Reductivity of the automorphism group of K-polystable Fano varieties*, Invent. Math. **222** (2020), no. 3, 995–1032, DOI 10.1007/s00222-020-00987-2. MR4169054

[AHLH19] J. Alper, D. Halpern-Leistner, and J. Heinloth, *Existence of moduli spaces for algebraic stacks*, Preprint, `arXiv:1812.01128v4`, 2019.

[AHR19] J. Alper, J. Hall, and D. Rydh, *The étale local structure of algebraic stacks*, Preprint, `arXiv:1912.06162v2`, 2019.

[AHR20] J. Alper, J. Hall, and D. Rydh, *A Luna étale slice theorem for algebraic stacks*, Ann. of Math. (2) **191** (2020), no. 3, 675–738, DOI 10.4007/annals.2020.191.3.1. MR4088350

[Alp13] J. Alper, *Good moduli spaces for Artin stacks* (English, with English and French summaries), Ann. Inst. Fourier (Grenoble) **63** (2013), no. 6, 2349–2402. MR3237451

[Alp14] J. Alper, *Adequate moduli spaces and geometrically reductive group schemes*, Algebr. Geom. **1** (2014), no. 4, 489–531, DOI 10.14231/AG-2014-022. MR3272912

[Alp23] J. Alper, *Notes on stacks and moduli*, `https://sites.math.washington.edu/~jarod/moduli.pdf`, 2023, Unpublised lecture notes.

[Art69] M. Artin, *Algebraization of formal moduli. I*, Global Analysis (Papers in Honor of K. Kodaira), Univ. Tokyo Press, Tokyo, 1969, pp. 21–71. MR260746

[Art74] M. Artin, *Versal deformations and algebraic stacks*, Invent. Math. **27** (1974), 165–189, DOI 10.1007/BF01390174. MR399094

[Bal07] V. Balaji, *Principal bundles on projective varieties and the Donaldson-Uhlenbeck compactification*, J. Differential Geom. **76** (2007), no. 3, 351–398. MR2331525

[BDF+22] P. Belmans, C. Damiolini, H. Franzen, V. Hoskins, S. Makarova, and T. Tajakka, *Projectivity and effective global generation of determinantal line bundles on quiver moduli*, 2022.

[Beh91] K. A. Behrend, *The Lefschetz trace formula for the moduli stack of principal bundles*, ProQuest LLC, Ann Arbor, MI, 1991. Thesis (Ph.D.)–University of California, Berkeley. MR2686745

[BGLM21] L. Braun, D. Greb, K. Langlois, and J. Moraga, *Reductive quotients of klt singularities*, Preprint, arXiv:2111.02812v2, 2021.

[BH04] I. Biswas and Y. I. Holla, *Harder-Narasimhan reduction of a principal bundle*, Nagoya Math. J. **174** (2004), 201–223, DOI 10.1017/S0027763000008850. MR2066109

[BLX22] H. Blum, Y. Liu, and C. Xu, *Openness of K-semistability for Fano varieties*, Duke Math. J. **171** (2022), no. 13, 2753–2797, DOI 10.1215/00127094-2022-0054. MR4505846

[BP03] V. Balaji and A. J. Parameswaran, *Semistable principal bundles. II. Positive characteristics*, Transform. Groups **8** (2003), no. 1, 3–36, DOI 10.1007/s00031-003-0713-2. MR1959761

[BS02] V. Balaji and C. S. Seshadri, *Semistable principal bundles. I. Characteristic zero*, J. Algebra **258** (2002), no. 1, 321–347, DOI 10.1016/S0021-8693(02)00502-1. Special issue in celebration of Claudio Procesi's 60th birthday. MR1958909

[BT05] L. Bruasse and A. Teleman, *Harder-Narasimhan filtrations and optimal destabilizing vectors in complex geometry* (English, with English and French summaries), Ann. Inst. Fourier (Grenoble) **55** (2005), no. 3, 1017–1053. MR2149409

[BZ16] I. Biswas and A. Zamora, *On the Gieseker Harder-Narasimhan filtration for principal bundles*, Bull. Sci. Math. **140** (2016), no. 4, 58–69, DOI 10.1016/j.bulsci.2015.02.004. MR3498308

[CMW18] S. Casalaina-Martin and J. Wise, *An introduction to moduli stacks, with a view towards Higgs bundles on algebraic curves*, The geometry, topology and physics of moduli spaces of Higgs bundles, Lect. Notes Ser. Inst. Math. Sci. Natl. Univ. Singap., vol. 36, World Sci. Publ., Hackensack, NJ, 2018, pp. 199–399. MR3837871

[DM69] P. Deligne and D. Mumford, *The irreducibility of the space of curves of given genus*, Inst. Hautes Études Sci. Publ. Math. **36** (1969), 75–109. MR262240

[FGI+05] B. Fantechi, L. Göttsche, L. Illusie, S. L. Kleiman, N. Nitsure, and A. Vistoli, *Fundamental algebraic geometry*, Mathematical Surveys and Monographs, vol. 123, American Mathematical Society, Providence, RI, 2005. Grothendieck's FGA explained, DOI 10.1090/surv/123. MR2222646

[Gó01] T. L. Gómez, *Algebraic stacks*, Proc. Indian Acad. Sci. Math. Sci. **111** (2001), no. 1, 1–31, DOI 10.1007/BF02829538. MR1818418

[GHZ24] T. L. Gómez, A. F. Herrero, and A. Zamora, *The moduli stack of principal ρ-sheaves and Gieseker-Harder-Narasimhan filtrations*, Math. Z. **307**, 51 (2024). https://doi.org/10.1007/s00209-024-03497-6

[Gie77] D. Gieseker, *On the moduli of vector bundles on an algebraic surface*, Ann. of Math. (2) **106** (1977), no. 1, 45–60, DOI 10.2307/1971157. MR466475

[GLSS08] T. L. Gómez, A. Langer, A. H. W. Schmitt, and I. Sols, *Moduli spaces for principal bundles in arbitrary characteristic*, Adv. Math. **219** (2008), no. 4, 1177–1245, DOI 10.1016/j.aim.2008.05.015. MR2450609

[GLSS10] T. L. Gómez, A. Langer, A. H. W. Schmitt, and I. Sols, *Moduli spaces for principal bundles in large characteristic*, Teichmüller theory and moduli problem, Ramanujan Math. Soc. Lect. Notes Ser., vol. 10, Ramanujan Math. Soc., Mysore, 2010, pp. 281–371. MR2667560

[GN14] S. Gurjar and N. Nitsure, *Schematic Harder-Narasimhan stratification for families of principal bundles and Λ-modules*, Proc. Indian Acad. Sci. Math. Sci. **124** (2014), no. 3, 315–332, DOI 10.1007/s12044-014-0165-8. MR3258622

[GN16] S. Gurjar and N. Nitsure, *Schematic Harder-Narasimhan stratification for families of principal bundles in higher dimensions*, Math. Z. **289** (2018), no. 3-4, 1121–1142, DOI 10.1007/s00209-017-1990-0. MR3830241

[GN18] S. Gurjar and N. Nitsure, *Schematic Harder-Narasimhan stratification for families of principal bundles in higher dimensions*, Math. Z. **289** (2018), no. 3-4, 1121–1142, DOI 10.1007/s00209-017-1990-0. MR3830241

[GS05] T. Gómez and I. Sols, *Moduli space of principal sheaves over projective varieties*, Ann. of Math. (2) **161** (2005), no. 2, 1037–1092, DOI 10.4007/annals.2005.161.1037. MR2153406

[GSZ15] T. L. Gómez, I. Sols, and A. Zamora, *A GIT interpretation of the Harder-Narasimhan filtration*, Rev. Mat. Complut. **28** (2015), no. 1, 169–190, DOI 10.1007/s13163-014-0149-3. MR3296731

[Hei08a] J. Heinloth, *Bounds for Behrend's conjecture on the canonical reduction*, Int. Math. Res. Not. IMRN **14** (2008), Art. ID rnn045, 17, DOI 10.1093/imrn/rnn045. MR2440327

[Hei08b] J. Heinloth, *Semistable reduction for G-bundles on curves*, J. Algebraic Geom. **17** (2008), no. 1, 167–183, DOI 10.1090/S1056-3911-07-00476-6. MR2357683

[Hei10a] J. Heinloth, *Addendum to "Semistable reduction of G-bundles on curves" [MR2357683]*, J. Algebraic Geom. **19** (2010), no. 1, 193–197, DOI 10.1090/S1056-3911-09-00536-0. MR2551761

[Hei10b] J. Heinloth, *Lectures on the moduli stack of vector bundles on a curve*, Affine flag manifolds and principal bundles, Trends Math., Birkhäuser/Springer Basel AG, Basel, 2010, pp. 123–153, DOI 10.1007/978-3-0346-0288-4_4. MR3013029

[Hei17] J. Heinloth, *Hilbert-Mumford stability on algebraic stacks and applications to $\mathcal{G}$-bundles on curves*, Épijournal Géom. Algébrique **1** (2017), Art. 11, 37, DOI 10.46298/epiga.2018.volume1.2062. MR3758902

[Hes81] W. H. Hesselink, *Concentration under actions of algebraic groups*, Paul Dubreil and Marie-Paule Malliavin Algebra Seminar, 33rd Year (Paris, 1980), Lecture Notes in Math., vol. 867, Springer, Berlin, 1981, pp. 55–89. MR633514

[HL97] D. Huybrechts and M. Lehn, *The geometry of moduli spaces of sheaves*, Aspects of Mathematics, E31, Friedr. Vieweg & Sohn, Braunschweig, 1997, DOI 10.1007/978-3-663-11624-0. MR1450870

[HL21] D. Halpern-Leistner, *On the structure of instability in moduli theory*, `arXiv:1411.0627v5`, 2021.

[HLHJ21] D. Halpern-Leistner, A. F. Herrero, and T. Jones, *Moduli spaces of sheaves via affine Grassmannians*, `arXiv:2107.02172v1`, 2021.

[HN75] G. Harder and M. S. Narasimhan, *On the cohomology groups of moduli spaces of vector bundles on curves*, Math. Ann. **212** (1974/75), 215–248, DOI 10.1007/BF01357141. MR364254

[Kem78] G. R. Kempf, *Instability in invariant theory*, Ann. of Math. (2) **108** (1978), no. 2, 299–316, DOI 10.2307/1971168. MR506989

[Kir84] F. C. Kirwan, *Cohomology of quotients in symplectic and algebraic geometry*, Mathematical Notes, vol. 31, Princeton University Press, Princeton, NJ, 1984, DOI 10.2307/j.ctv10vm2m8. MR766741

[KM97] S. Keel and S. Mori, *Quotients by groupoids*, Ann. of Math. (2) **145** (1997), no. 1, 193–213, DOI 10.2307/2951828. MR1432041

[Knu71] D. Knutson, *Algebraic spaces*, Lecture Notes in Mathematics, Vol. 203, Springer-Verlag, Berlin-New York, 1971. MR302647

[Lan75] S. G. Langton, *Valuative criteria for families of vector bundles on algebraic varieties*, Ann. of Math. (2) **101** (1975), 88–110, DOI 10.2307/1970987. MR364255

[LMB00] G. Laumon and L. Moret-Bailly, *Champs algébriques* (French), Ergebnisse der Mathematik und ihrer Grenzgebiete. 3. Folge. A Series of Modern Surveys in Mathematics [Results in Mathematics and Related Areas. 3rd Series. A Series of Modern Surveys in Mathematics], vol. 39, Springer-Verlag, Berlin, 2000. MR1771927

[Mar78] M. Maruyama, *Moduli of stable sheaves. II*, J. Math. Kyoto Univ. **18** (1978), no. 3, 557–614, DOI 10.1215/kjm/1250522511. MR509499

[MFK94] D. Mumford, J. Fogarty, and F. Kirwan, *Geometric invariant theory*, 3rd ed., Ergebnisse der Mathematik und ihrer Grenzgebiete (2) [Results in Mathematics and Related Areas (2)], vol. 34, Springer-Verlag, Berlin, 1994. MR1304906

[Mum63a] D. Mumford, *Projective invariants of projective structures and applications*, Proc. Internat. Congr. Mathematicians (Stockholm, 1962), Inst. Mittag-Leffler, Djursholm, 1963, pp. 526–530. MR175899

[Mum63b] D. Mumford, *Projective invariants of projective structures and applications*, Proc. Internat. Congr. Mathematicians (Stockholm, 1962), Inst. Mittag-Leffler, Djursholm, 1963, pp. 526–530. MR0175899

[Nes84] L. Ness, *A stratification of the null cone via the moment map*, Amer. J. Math. **106** (1984), no. 6, 1281–1329, DOI 10.2307/2374395. With an appendix by David Mumford. MR765581

[Nit11] N. Nitsure, *Schematic Harder-Narasimhan stratification*, Internat. J. Math. **22** (2011), no. 10, 1365–1373, DOI 10.1142/S0129167X11007264. MR2851213

[Ols16] M. Olsson, *Algebraic spaces and stacks*, American Mathematical Society Colloquium Publications, vol. 62, American Mathematical Society, Providence, RI, 2016, DOI 10.1090/coll/062. MR3495343

[Ram75] A. Ramanathan, *Stable principal bundles on a compact Riemann surface*, Math. Ann. **213** (1975), 129–152, DOI 10.1007/BF01343949. MR369747

[Ram79] A. Ramanathan, *Moduli for principal bundles*, Algebraic geometry (Proc. Summer Meeting, Univ. Copenhagen, Copenhagen, 1978), Lecture Notes in Math., vol. 732, Springer, Berlin, 1979, pp. 527–533. MR555715

[Sch02] A. H. W. Schmitt, *Singular principal bundles over higher-dimensional manifolds and their moduli spaces*, Int. Math. Res. Not. **23** (2002), 1183–1209, DOI 10.1155/S1073792802107069. MR1903952

[Ses67] C. S. Seshadri, *Space of unitary vector bundles on a compact Riemann surface*, Ann. of Math. (2) **85** (1967), 303–336, DOI 10.2307/1970444. MR233371

[Sha77] S. S. Shatz, *The decomposition and specialization of algebraic families of vector bundles*, Compositio Math. **35** (1977), no. 2, 163–187. MR498573

[Sim94] C. T. Simpson, *Moduli of representations of the fundamental group of a smooth projective variety. I*, Inst. Hautes Études Sci. Publ. Math. **79** (1994), 47–129. MR1307297

[Sta23] The Stacks Project Authors, *Stacks Project*, `https://stacks.math.columbia.edu`, 2023.

[Taj21] T. Tajakka, *Projective Moduli Spaces of Complexes of Sheaves*, ProQuest LLC, Ann Arbor, MI, 2021. Thesis (Ph.D.)–University of Washington. MR4315165

[Zam16] A. Zamora, *Harder-Narasimhan filtration for rank 2 tensors and stable coverings*, Proc. Indian Acad. Sci. Math. Sci. **126** (2016), no. 3, 305–327, DOI 10.1007/s12044-016-0283-6. MR3531851

Instituto de Ciencias Matemáticas (CSIC-UAM-UC3M-UCM), Nicolás Cabrera 15, Campus Cantoblanco UAM, 28049 Madrid, Spain
Email address: `tomas.gomez@icmat.es`

Mathematics Hall MC 4404, 2990 Broadway, New York, New York 10027
Email address: `af3358@columbia.edu`

Departamento de Matemática Aplicada a las TIC, ETSI Informáticos, Universidad Politécnica de Madrid, Campus de Montegancedo, 28660 Madrid, Spain
Email address: `alfonso.zamora@upm.es`

Contemporary Mathematics
Volume **803**, 2024
https://doi.org/10.1090/conm/803/16101

Stratification of $\mathrm{SU}(r)$-character varieties of twisted Hopf links

Ángel González-Prieto, Marina Logares, Javier Martínez, and Vicente Muñoz

Dedicated to Prof. Peter E. Newstead on the occasion of his 80th birthday.

ABSTRACT. We describe the geometry of the character variety of representations of the fundamental group of the complement of a Hopf link with n twists, namely $\Gamma_n = \langle x, y \,|\, [x^n, y] = 1\rangle$ into the group $\mathrm{SU}(r)$. For arbitrary rank, we provide geometric descriptions of the loci of irreducible and totally reducible representations. In the case $r = 2$, we provide a complete geometric description of the character variety, proving that this $\mathrm{SU}(2)$-character variety is a deformation retract of the larger $\mathrm{SL}(2, \mathbb{C})$-character variety, as conjectured by Florentino and Lawton. In the case $r = 3$, we also describe different strata of the $\mathrm{SU}(3)$-character variety according to the semi-simple type of the representation.

1. Introduction

Let Γ be a finitely generated group and G a real or complex reductive algebraic group. A representation of Γ into G is a group homomorphism $\rho : \Gamma \to G$. Consider a presentation $\Gamma = \langle \gamma_1, \ldots, \gamma_k \,|\, \{r_\lambda\}\rangle$, where $\{r_\lambda\}$ is a finite generating set of relations of Γ. The map ρ is completely determined by the k-tuple $(A_1, \ldots, A_k) = (\rho(\gamma_1), \ldots, \rho(\gamma_k))$ subject to the relations $r_\lambda(A_1, \ldots, A_k) = \mathrm{id}$, for all λ. In this way, the set of representations of Γ into G is in bijection with the algebraic set

$$R(\Gamma, G) = \{(A_1, \ldots, A_k) \in G^k \,|\, r_\lambda(A_1, \ldots, A_k) = \mathrm{id}, \forall \lambda\} \subset G^k .$$

Two representations ρ and ρ' are said to be equivalent if there exists $g \in G$ such that $\rho'(\gamma) = g^{-1}\rho(\gamma)g$, for every $\gamma \in \Gamma$. When ρ is a faithful representation into $G \subset \mathrm{GL}(V)$, the equivalence of representations means that ρ and ρ' are the same representation up to a G-change of basis of V. The moduli space of representations (or character variety) can be thus obtained as the GIT quotient

$$X(\Gamma, G) = R(\Gamma, G) /\!\!/ G .$$

An important instance happens when $G = \mathrm{GL}(r, \mathbb{C})$, for which we recover the classical notion of a linear representation as a Γ-module structure on the vector

2020 *Mathematics Subject Classification.* Primary 14M35; Secondary 57K31.
Key words and phrases. Character variety, representation varieties, unitary group, knots, links.

space $\mathbb{C}^r$. It is worth noticing that, when Γ is a finite group, the vector space $\mathbb{C}^r$ can be equipped with a Γ-invariant hermitian metric. Hence, any representation $\rho : \Gamma \to \mathrm{GL}(r, \mathbb{C})$ descends to an $\mathrm{U}(r)$-representation

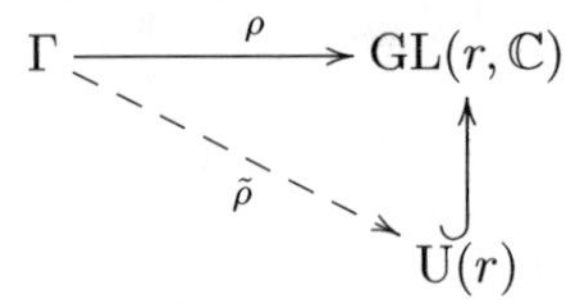

In other words, $X(\Gamma, \mathrm{GL}(r, \mathbb{C})) \cong X(\Gamma, \mathrm{U}(r))$. However, in the general case in which Γ is only finitely generated, such an invariant metric may not exist, so $X(\Gamma, \mathrm{U}(r))$ is only a semialgebraic subvariety of $X(\Gamma, \mathrm{GL}(r, \mathbb{C}))$. Similar considerations can be done in the case in which we fix the determinant of the representation, so we analyze the descending property of representations induced by the inclusion $\mathrm{SU}(r) \hookrightarrow \mathrm{SL}(r, \mathbb{C})$, which exhibits $X(\Gamma, \mathrm{SU}(r))$ as a semialgebraic subvariety of $X(\Gamma, \mathrm{SL}(r, \mathbb{C}))$ (see [**8**, Theorem 4.3]).

For non-finite groups Γ, the situation is different. In [**11**], it is proved that, when Γ is a free product of nilpotent groups or a star-shaped RAAG (Right Angled Artin Group), the inclusion $X(\Gamma, K) \hookrightarrow X(\Gamma, G)$ is a deformation retract for any reductive group G and its maximal compact subgroup $K \subset G$ (a property called 'flawed'). On the contrary, for $\Gamma = \pi_1(\Sigma_g)$, the fundamental group of a compact orientable surface Σ_g of genus $g \geq 2$, this inclusion is never a homotopy equivalence when G is reductive and non-abelian (it is said that Γ is a 'flawless' group in the language of [**11**]).

The study of the homotopy type of these character varieties in the case of knot groups was initiated in [**20**]. For a knot $K \subset S^3$, the finitely presented group $\Gamma = \pi_1(S^3 - K)$ is an important invariant of the knot, called the knot group. The first non-trivial family of knots studied in the literature has been the torus knot of type (m, n), for which the knot group is $\Gamma = \langle x, y \,|\, x^n = y^m \rangle$. For the torus knot group, [**18**] and [**20**] prove that such inclusion is a deformation retract in the case $\mathrm{SU}(2) \hookrightarrow \mathrm{SL}(2, \mathbb{C})$, and [**13**] proves the same result in the case $\mathrm{SU}(3) \hookrightarrow \mathrm{SL}(3, \mathbb{C})$.

The aim of this work is to extend this program to a family of 3-dimensional links. In analogy with the cases of knots, given a link $L \subset S^3$, the link group is the fundamental group $\Gamma = \pi_1(S^3 - L)$ of its complement. For an algebraic group G, we have the G-character variety of the link

$$X(L, G) = \mathrm{Hom}\,(\pi_1(S^3 - L), G) /\!/ G,$$

as before. Despite the advances for representation varieties of knots, much less is known in the case of links. An obvious case is the character variety of trivial links, i.e. representations of the free group for which a whole body of work has been developed in the literature (see for instance [**1, 4, 7, 10, 17**]). Very recently, more complicated links have been studied, such as the twisted Alexander polynomial for the Borromean link in [**5**]. In [**15**], the analysis of the geometry of character varieties is extended to the "twisted" Hopf link H_n, obtained by twisting a classical Hopf link with 2 crossings to get $2n$ crossings, as depicted in Figure 1.

The fundamental group of the link complement of H_n can be computed through a Wirtinger presentation [**15**], giving rise to the group $\Gamma_n = \langle a, b \,|\, [a^n, b] = 1 \rangle$. Therefore, the associated G-character variety is

$$X(H_n, G) = \left\{ (A, B) \in G^2 \,|\, [A^n, B] = 1 \right\} /\!/ G.$$

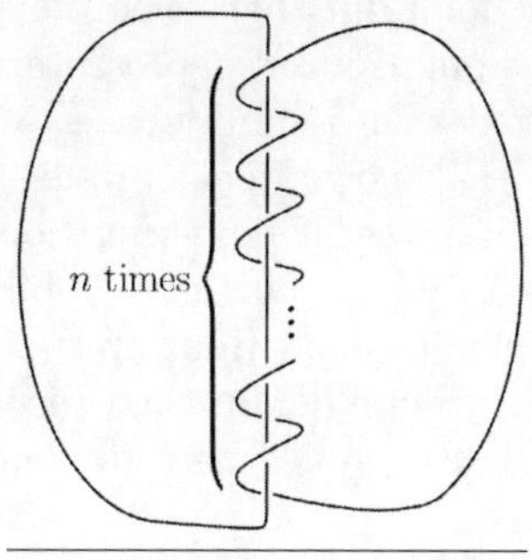

FIGURE 1. The twisted Hopf link of n twists.

In this sense, $X(H_n, G)$ should be understood as the variety counting "supercommuting" elements of G, generalizing the case $n = 1$ of the usual Hopf link that corresponds to commuting elements.

In this setting, we shall show in this paper how the character varieties of twisted Hopf links can be stratified according to the type of its semisimple decomposition in the cases $G = \mathrm{SU}(r)$ and $G = \mathrm{U}(r)$, an approach that was initiated in [**14**]. Moreover, we shall provide an explicit description of its irreducible locus (Section 2.3) and locus of representations that fully split into 1-dimensional representations (Section 2.4). This will lead us to explore other auxiliary spaces such as symmetric spaces of tori. In Section 3, we will apply these techniques to describe the global geometry of the SU(2)-character variety. Furthermore, in Section 4 we will compare its topology with the one of the $\mathrm{SL}(2, \mathbb{C})$-character variety, a variety that was previously studied in [**15**]. These analyses lead to the main result of this work.

THEOREM. *The* SU(2)*-character variety of a twisted Hopf link is a strong deformation retract of the corresponding* $\mathrm{SL}(2, \mathbb{C})$*-character variety. Moreover, they are homotopically equivalent to the wedge of n copies of* 2*-spheres.*

Notice that this result provides the first non-trivial extension of the results of [**11**] to link groups. In the cases of U(2) and SU(3), we also provide a thorough description of the semisimple stratification arising in the corresponding character varieties. Even though each piece is fully described, the geometry involved is very complicated and a complete description of the intersection pattern that arises is unavailable. This prevents further cohomological calculations that are needed to characterize the homotopy type of these varieties.

Despite this essential obstruction, our results suggest that an alternative approach can be considered to address the problem of comparing the homotopy type of SU(r) and $\mathrm{SL}(r, \mathbb{C})$ character varieties. Indeed, the methods of this work show that, even though the retraction $X(H_n, \mathrm{SL}(2, \mathbb{C})) \to X(H_n, \mathrm{SU}(2))$ does not respect the semisimple strata, it does keep their closures invariant. In this way, even though we do not expect the existence of any retraction that shrinks each stratum locally, such deformation retract may be possible if we allow the homotopy to use neighbouring strata in the closure. With this idea, we pose the following conjecture.

CONJECTURE. *For any* $r \geq 1$, *the* SU(r)*-character variety of a twisted Hopf link is a strong deformation retract of the corresponding* $\mathrm{SL}(r, \mathbb{C})$*-character variety.*

Finally, we would like to point out that, even though throughout this paper all these spaces will be studied from a purely representation-theoretic approach, they have been traditionally related with moduli spaces of parabolic bundles. Indeed, the work of Furuta and Steer [**12**] shows an algebro-geometric approach to the study of the topology of the moduli space of representations of the fundamental group of a Seifert fibered space into $\mathrm{SU}(r)$. This idea applies to the twisted Hopf links addressed in the present work. Twisted Hopf links are $(2, 2n)$-torus links, that is 2-bridge links which satisfy Burde and Murasugi theorem [**3**, Theorem 1], so that their complement is a Seifert fibered 3-sphere over an orbifold with one singular point of order n.

In [**2**, Section 6.7], Boden shows that, given such Seifert fibered homology 3-sphere M, there exists an orbifold surface Σ such that the character variety $X(\pi_1(M), \mathrm{SU}(2))$ is isomorphic to $X^*(\pi_1^{\mathrm{orb}}(\Sigma), \mathrm{SU}(2))$, the moduli space of irreducible representations of the orbifold fundamental group of Σ. This later space can be actually understood as an union of moduli spaces of stable parabolic bundles with appropriate parabolic data [**19**]. From this viewpoint, our work honours the work of Newstead in [**22**] in the study of the homotopy of moduli spaces of bundles.

The aforementioned relation with moduli spaces of parabolic bundles is completed with an analogous relation between $\mathrm{SL}(2, \mathbb{C})$-character varieties of Seifert fibered homology 3-spheres and moduli spaces of traceless parabolic Higgs bundles, thanks to the work of Nasatyr and Steer [**21**]. This would motivate us to seek a deformation retraction from the moduli space of traceless parabolic Higgs bundles to the moduli space of degree 0 stable parabolic bundles. We do not expect there is such a retraction, as it is in fact the case in the non-parabolic situation (see [**6**]). The reason may be the same: the moduli space of traceless parabolic Higgs bundles retracts to its well-known nilpotent cone, which contains the moduli space of degree 0 stable parabolic bundles as one of its irreducible components, but it is indeed reducible. Nevertheless, we shall address this question in future work.

2. Representation spaces and character varieties

We fix notation and outline here some of the properties that will be used throughout the paper. Let us consider a finitely generated group with a presentation $\Gamma = \langle \gamma_1, \ldots, \gamma_k \,|\, r_\lambda(\gamma_1, \ldots, \gamma_k) \rangle$, where $r_\lambda(\gamma_1, \ldots, \gamma_k)$ is the set of relations satisfied by the generators $\gamma_1, \ldots, \gamma_k$. Given an algebraic linear group G, the G-representation variety of Γ is the space

$$R(\Gamma, G) = \mathrm{Hom}\,(\Gamma, G) = \{(g_1, \ldots, g_k) \in G^k \mid r_\lambda(g_1, \ldots, g_k) = \mathrm{id}\}.$$

For our purposes, we shall focus on the classical matrix cases $G = \mathrm{GL}(r, \mathbb{C})$, $\mathrm{SL}(r, \mathbb{C})$, $\mathrm{U}(r)$ and $\mathrm{SU}(r)$. Given $\rho \in R(\Gamma, G)$, we will call ρ *reducible* if there exists some proper linear subspace $W \subset \mathbb{C}^r$ that is invariant for all $\gamma \in \Gamma$, that is, such that $\rho(\Gamma)(W) \subset W$; and we will say that ρ is *irreducible* otherwise. In the case that there exists a ρ-equivariant decomposition $\mathbb{C}^r = W_1 \oplus \ldots \oplus W_s$ into irreducible components, ρ is said to be *semisimple*.

There is a natural action of the group G on $R(\Gamma, G)$ given by $g \cdot \rho = g\rho g^{-1}$ for $g \in G$ and $\gamma \in \Gamma$, namely the conjugation action. Two representations are said to be equivalent if they belong to the same G-orbit under the conjugation action. The moduli space of G-representations of Γ is the GIT quotient (see [**23**] for an introduction to these algebraic quotients) with respect to this action,

$$X(\Gamma, G) = R(\Gamma, G) /\!/ G.$$

It is well known (see [**23**]) that $X(\Gamma, G)$ parametrizes representations up to an equivalence relation called S-equivalence: two representations ρ and ρ' are S-equivalent if the closures of their G-orbits intersect. It turns out that, for the classical groups $G = \mathrm{GL}(r, \mathbb{C}), \mathrm{SL}(r, \mathbb{C})$, every representation is S-equivalent to a semisimple one. Furthermore, in the compact case $G = \mathrm{U}(r)$ and $\mathrm{SU}(r)$, every representation is semisimple (see [**13**, Proposition 1]). Hence, in all these cases, $X(\Gamma, G)$ parametrizes semisimple representations up to (classical) equivalence. Moreover, in the compact case $X(\Gamma, G)$ is the usual topological quotient.

2.1. Representations of the twisted Hopf link group. We study here representations of the following group. For $n \geq 1$, let $H_n \subset S^3$ be the so-called twisted Hopf link, a generalization of the classical Hopf link with $2n$ crossings instead of 2, as represented in Figure 1.

Using its Wirtinger presentation, in [**15**, Proposition 3.1] it was shown that the fundamental group of the complement of the link H_n is

$$\Gamma_n = \pi_1(S^3 - H_n) = \langle a, b \mid [a^n, b] = 1 \rangle,$$

where $[x, y] = xyx^{-1}y^{-1}$ denotes the group commutator. In this way, the associated G-representation variety is

$$R(\Gamma_n, G) = \mathrm{Hom}\,(\Gamma_n, G) = \{(A, B) \in G^2 \mid [A^n, B] = \mathrm{id}\}.$$

In this paper we shall study the corresponding representation spaces and character varieties of Γ_n for $G = \mathrm{U}(r)$ and $G = \mathrm{SU}(r)$. We will denote them as

$$X_r = X(\Gamma_n, \mathrm{U}(r)), \qquad \mathrm{S}X_r = X(\Gamma_n, \mathrm{SU}(r)).$$

2.2. Stratification of representations. Let us fix $r \geq 1$. Recall that a partition π of r is a decomposition

$$r = a_1 r_1 + \cdots + a_s r_s\,,$$

where $r_1 > r_2 > \cdots > r_s > 0$ and $a_i \geq 0$. We shall denote this partition by $\pi = (r_1, \overset{(a_1)}{\ldots}, r_1, \ldots, r_s, \overset{(a_s)}{\ldots}, r_s)$. Associated to this partition, we can consider equivalence classes of semisimple representations $\rho : \Gamma_n \to \mathrm{U}(r)$ of type π, described as:

$$\rho = \bigoplus_{t=1}^{s} \bigoplus_{l=1}^{a_l} \rho_{tl}, \quad \rho_{tl} : \Gamma_n \to \mathrm{U}(r_t). \tag{2.1}$$

We denote the set of such representations by $X_r^\pi \subset X_r = X(\Gamma_n, \mathrm{U}(r))$. Analogously, for $\mathrm{SU}(r)$ we set $\mathrm{S}X_r^\pi = X_r^\pi \cap \mathrm{S}X_r$, which is made of those representations in (2.1) with $\prod_{t,l} \det(\rho_{tl}) = 1$.

As a consequence, the character varieties X_r and SX_r decompose into locally closed subvarieties indexed by the set Π_r of all partitions of r:

$$X_r = \bigsqcup_{\pi \in \Pi_r} X_r^\pi, \qquad \mathrm{S}X_r = \bigsqcup_{\pi \in \Pi_r} \mathrm{S}X_r^\pi.$$

Among all possible partitions, the following two cases will be important: $X_r^{\mathrm{TR}} = X_r^{\pi_1}$, the set of *totally reducible* representations, that corresponds to $\pi_1 = (1, \overset{(r)}{\ldots}, 1)$, and $X_r^* = X_r^{\pi_0}$, which will denote the set of *irreducible* representations, where $\pi_0 = (r)$. Analogous notation will be used for the $\mathrm{SU}(r)$ counterparts.

Finally, given a (semi) algebraic variety X, let us consider the symmetric product of r copies of X, $\mathrm{Sym}^r(X) = X^r/S_r$, where the symmetric group S_r acts by permutation of the factors. With this symmetric product, we have the following simple characterization of semisimple representations up to equivalence.

LEMMA 2.1. *For any partition* $\pi = (r_1, \overset{(a_1)}{\ldots}, r_1, \ldots, r_s, \overset{(a_s)}{\ldots}, r_s) \in \Pi_r$, *we have an isomorphism*

$$X_r^\pi \cong \prod_{i=1}^s \mathrm{Sym}^{a_i}(X_{r_i}^*).$$

2.3. Irreducible representations. First, let us observe the following useful property of irreducible representations of the twisted Hopf link group.

LEMMA 2.2. *Let* $\rho = (A, B) : \Gamma_n \to G$ *be an irreducible representation. Then* A^n *is a multiple of the identity.*

PROOF. The relation in Γ_n says that A^n is an equivariant linear map from $\mathbb{C}^r$ to $\mathbb{C}^r$, i.e. A^n commutes with A and B. Hence, by Schur's lemma, A^n is a multiple of the identity. □

Consider the natural projection map

$$\omega : X_r^* \to \mathrm{Sym}^r(S^1).$$

that assigns, to each irreducible representation $(A, B) \in X_r^*$, the collection of eigenvalues of the matrix A. To study this map, we stratify $\mathrm{Sym}^r(S^1)$ according to the number of repeated eigenvalues. To be precise, given a partition $\sigma = (r_1, \overset{(a_1)}{\ldots}, r_1, \ldots, r_s, \overset{(a_s)}{\ldots}, r_s) \in \Pi_r$, let us denote by $\mathrm{Sym}^r_\sigma(S^1)$ the collection of sets $\{\lambda_1, \ldots, \lambda_r\}$ such that there are a_i groups of r_i equal eigenvalues (i.e. a_i eigenvalues with multiplicity r_i). In particular, $\sigma_0 = (1, \overset{(r)}{\ldots}, 1)$ corresponds to the case of different eigenvalues. In this way, given $\sigma \in \Pi_r$, we set

$$X_\sigma^* = \omega^{-1}(\mathrm{Sym}^r_\sigma(S^1)).$$

These are equivalence classes of representations (A, B) such that A has repeated eigenvalues given by σ. Hence, we get a further stratification of the irreducible locus X_r^* according to the number of repeated eigenvalues

$$X_r^* = \bigsqcup_{\sigma \in \Pi_r} X_\sigma^*. \tag{2.2}$$

Notice that some of the strata X_σ^* may be empty in this decomposition. The maximal component of X_r^* corresponds to the partition $\sigma_0 = (1, \overset{(r)}{\ldots}, 1)$ of non-repeated eigenvalues. Provided that $n \geq r$, this stratum is non-empty: once we diagonalize A the irreducibility of the representation ensures that we can choose two non-coincident basis for the eigenvectors of A and B, which provides an element of the stratum.

Given $\sigma \in \Pi_r$, let us pick $\{\lambda_1, \ldots, \lambda_r\} \in \mathrm{Sym}^r_\sigma(S^1)$, and let us denote by $\mathrm{Stab}(\sigma)$ the $\mathrm{U}(r)$-stabilizer of the diagonal matrix $A_\sigma = \mathrm{diag}(\lambda_1, \ldots, \lambda_r)$. Notice that $\mathrm{Stab}(\sigma)$ does not depend on the particular choice of eigenvalues $\{\lambda_1, \ldots, \lambda_r\} \in \mathrm{Sym}^r_\sigma(S^1)$. Indeed, if $\sigma = (r_1, \overset{(a_1)}{\ldots}, r_1, \ldots, r_s, \overset{(a_s)}{\ldots}, r_s)$, we have the explicit description

$$\mathrm{Stab}(\sigma) = \prod_{i=1}^s \mathrm{SU}(r_i)^{a_i}.$$

2.3.1. SU(r)*-irreducible representations.* If we specialize to the case of SU(r), directly from Lemma 2.2 we get the following.

COROLLARY 2.3. *Let* $\rho = (A, B) : \Gamma_n \to \mathrm{SU}(r)$ *be an irreducible representation. Then* $A^n = \xi\,\mathrm{id}$, *where* $\xi \in \boldsymbol{\mu}_r$ *is an* r*-th root of unity. In particular,* A *admits finitely many eigenvalues.*

This result implies that the map

$$\omega : \mathrm{S}X_r^* \to \mathrm{SSym}^r(S^1)$$

has finite image, where $\mathrm{SSym}^r(S^1) \subset \mathrm{Sym}^r(S^1)$ is the collection of sets of the form $\{\lambda_1, \ldots, \lambda_r\} \in \mathrm{Sym}^r(S^1)$ such that $\lambda_1 \cdots \lambda_r = 1$. The image is precisely those eigenvalues additionally satisfying $\lambda_1^n = \lambda_2^n = \ldots = \lambda_r^n \in \boldsymbol{\mu}_r$. Analogously to the previous setting, we also have stratifications

$$\mathrm{SSym}^r(S^1) = \bigsqcup_{\sigma \in \Pi_r} \mathrm{SSym}^r_\sigma(S^1), \qquad \mathrm{S}X_r^* = \bigsqcup_{\sigma \in \Pi_r} \mathrm{S}X_\sigma^*, \tag{2.3}$$

according to the number of repeated eigenvalues.

Given $\sigma \in \Pi_r$, let us denote by N_σ the number of elements $\{\lambda_1, \ldots, \lambda_r\} \in \mathrm{SSym}^r_\sigma(S^1)$ such that $\lambda_1^n = \lambda_2^n = \cdots = \lambda_r^n = \xi \in \boldsymbol{\mu}_r$ (and $\lambda_1 \cdots \lambda_r = 1$). Fixed $\xi \in \boldsymbol{\mu}_r$, the number of possibilities was computed in [**14**, Corollary 6.7] for the case $\gcd(n, r) = 1$. Hence, accounting for the r choices of ξ, we get that

$$N_\sigma = \frac{r}{n}\binom{n}{a_1, a_2, \ldots, a_s},$$

at least for n and r coprime (conjecturally, for any n and r). Here, we have used the multinomial coefficients

$$\binom{n}{a_1, a_2, \ldots, a_s} = \frac{n!}{a_1! a_2! \cdots a_s! (n - a_1 - \cdots - a_s)!}.$$

LEMMA 2.4. *Let* $\sigma \in \Pi_r$. *We have that* $\mathrm{S}X_\sigma^*$ *is a disjoint union of* N_σ *copies of a subspace*

$$\mathbf{SF}_\sigma \subset \mathrm{SU}(r)/\,\mathrm{Stab}(\sigma),$$

where the action of $\mathrm{Stab}(\sigma)$ *on* SU(r) *is by conjugation.*

PROOF. Let $(A, B) \in \mathrm{S}X_\sigma^*$. Since A is diagonalizable, (A, B) has a representative with the matrix $A = \mathrm{diag}(\lambda_1, \ldots, \lambda_r)$ for some $\{\lambda_1, \ldots, \lambda_r\} \in \mathrm{SSym}^r_\sigma(S^1)$. This form is unique up to the action of $\mathrm{Stab}(A) = \mathrm{Stab}(\sigma)$ by conjugation on B, so it defines a point in $\mathrm{SU}(r)/\,\mathrm{Stab}(\sigma)$, as claimed. □

REMARK 2.5. The subspace $\mathbf{SF}_\sigma$ corresponding to $\mathrm{S}X_\sigma^*$ can be characterized explicitly. Let $\mathbf{SF}_\sigma^0$ by the preimage of $\mathbf{SF}_\sigma$ under the projection map $\mathrm{SU}(r) \to \mathrm{SU}(r)/\,\mathrm{Stab}(\sigma)$. Then $\mathbf{SF}_\sigma^0$ is given by the classes of matrices $B \in \mathrm{SU}(r)$ such that B has no proper invariant subspace in common with $A = \mathrm{diag}(\lambda_1, \ldots, \lambda_r)$. Notice that the invariant subspaces of A are certain subspaces generated by the canonical basis vectors of $\mathbb{C}^r$, depending on the partition σ.

REMARK 2.6. The situation of homogeneous spaces that strong deformation retract has been thoroughly studied in [**1**, Theorem 10].

2.3.2. $\mathrm{SU}(r)$-*irreducible representations with distinct eigenvalues.* In this section, we shall further study the case of the partition $\sigma_0 = (1, \overset{(r)}{\ldots}, 1)$ corresponding to non-repeated eigenvalues. In this setting, $\mathbf{SF}^0_{\sigma_0}$ is the collection of orthonormal bases $\{b_1, \ldots, b_r\}$ of $\mathbb{C}^r$ with volume 1 such that $\langle b_{i_1}, \ldots, b_{i_k}\rangle \neq \langle e_{i_1}, \ldots, e_{i_k}\rangle$ for any proper subset $\{i_1, \ldots, i_k\} \subset \{1, \ldots, r\}$. Moreover $\mathrm{Stab}(\sigma_0) = (S^1)^r$ with the action $(\lambda_1, \ldots, \lambda_r) \cdot (b_{ij}) = (\lambda_i \lambda_j^{-1} b_{ij})$ for $(\lambda_1, \ldots, \lambda_r) \in (S^1)^r$ and $B = (b_{ij}) \in \mathrm{SU}(r)$.

Let us first analyze the space $\mathrm{SU}(r)/(S^1)^r$ with the previous action. Recall that $\mathrm{SU}(r)$ can be understood as a fiber bundle over S^{2r-1} (the choice of the first basis vector) whose fiber is a fiber bundle over S^{2r-3} (the choice of the second basis vector), whose fiber is again a fiber bundle over S^{2r-5} and so on so forth until the last basis vector, that in principle belongs to S^1 but is actually fixed by the volume condition.

However, the action of $(S^1)^r$ makes things more involved. Given a matrix $B = (b_1 \,|\, \ldots \,|\, b_r)$, where $b_j = (b_{1j}, \ldots, b_{rj})$ are its column vectors, the action allows us to arrange $b_{i1} \in \mathbb{R}_{\geq 0}$ for all $i \geq 2$. Such representant is unique, so we get a projection map

$$\varphi : \mathrm{SU}(r)/(S^1)^r \longrightarrow B_r$$

onto the "coarse orthant" (c.f. [**13**, Theorem 10])

$$B_r = \{(z, x_2, \ldots, x_r) \in \mathbb{C} \times \mathbb{R}^{r-1}_{\geq 0} \,|\, |z|^2 + x_2^2 + \cdots + x_r^2 = 1\} \subset S^r \subset S^{2r-1}.$$

The fiber of φ over $b = b_1 = (z, x_2, \ldots, x_r)$ is determined by the number of ways in which b can be completed to an orthonormal basis of volume 1. On the interior of B_r the fiber $\varphi^{-1}(b)$ is an iterated bundle of spheres, but when b belongs to the boundary of B_r, there remains a residual action of $G_b = \{(\lambda_1, \ldots, \lambda_r) \in (S^1)^r \,|\, \lambda_1 = \lambda_i \text{ if } x_i = 0 \text{ for } 2 \leq i \leq r\}$ acting on $S^{2r-3} \times \ldots \times S^3$.

If we restrict our attention to $\mathbf{SF}_{\sigma_0} = \mathbf{SF}^0_{\sigma_0}/(S^1)^r$, then we must remove those bases that lead to a reducible representation. Such bases only occur at the boundary of B_r, so the previous description also holds for $\mathbf{SF}_{\sigma_0}$ on the interior of B_r.

2.3.3. $\mathrm{U}(r)$-*irreducible representations.* The case of $\mathrm{U}(r)$-representations is analogous to the previous one, and we get an eigenvalue map

$$\omega : X_r^* \to \mathrm{Sym}^r(S^1).$$

However, in this case the map no longer has finite image since, for a configuration of eigenvalues $\{\lambda_1, \ldots, \lambda_r\} \in \mathrm{Sym}^r(S^1)$ in the image, the element $\xi = \lambda_1^n = \lambda_2^n = \cdots = \lambda_r^n$ is an arbitrary point of S^1. We start the analysis of these representations with the simplest case.

LEMMA 2.7. *We have* $X_1^* \cong S^1 \times S^1$.

PROOF. Since $\mathrm{U}(1) = S^1$ is an abelian group, any pair $(A, B) \in \mathrm{U}(1) \times \mathrm{U}(1) = S^1 \times S^1$ leads to a representation for the twisted Hopf link. Moreover, since they are 1-dimensional representations, they are automatically irreducible. □

The higher rank case $r > 1$ is more involved. As a first step, we have an analogous result to Lemma 2.4.

PROPOSITION 2.8. *Fix* $\sigma \in \Pi_r$. *There is a Zariski locally trivial fiber bundle*

$$\omega_\sigma : X_\sigma^* \to \mathrm{Sym}^r_\sigma(S^1)$$

with fiber $\mathbf{F}_\sigma \subset \mathrm{U}(r)/\mathrm{Stab}(\sigma)$ *and base the collection of eigenvalues* $\{\lambda_1, \ldots, \lambda_r\} \in \mathrm{Sym}^r_\sigma(S^1)$ *such that* $\lambda_1^n = \lambda_2^n = \cdots = \lambda_r^n$.

PROOF. The spectrum map $\sigma : \mathrm{U}(r) \to \mathrm{Sym}^r_\sigma(S^1)$ is locally trivial in the Zariski topology. Indeed, on a trivializing open set $U \subset \mathrm{Sym}^r_\sigma(S^1)$ we may conjugate each representation (A, B) in $\omega_\sigma^{-1}(U)$ so that $A = \mathrm{diag}(\lambda_1, \ldots, \lambda_r)$. In this form, $B \in \mathrm{U}(r)$ is uniquely determined up to the conjugacy action of $\mathrm{Stab}(A) = \mathrm{Stab}(\sigma)$. $\square$

REMARK 2.9. Again, the subspace $\mathbf{F}_\sigma$ can be characterized explicitly in full analogy with $\mathbf{SF}_\sigma$ as in Remark 2.5, with the only difference that now $B \in \mathrm{U}(r)$ instead of $B \in \mathrm{SU}(r)$.

To look closer at this fibration, let us consider the variety $\hat{X}^*_r = X^*_r \times_{\mathrm{Sym}^r(S^1)} (S^1)^r$ given as the pullback

$$\begin{array}{ccc} \hat{X}^*_r & \longrightarrow & X^*_r \\ \Big\downarrow & & \Big\downarrow{\scriptstyle \omega_\sigma} \\ (S^1)^r & \longrightarrow & \mathrm{Sym}^r(S^1) \end{array}$$

where $(S^1)^r \to \mathrm{Sym}^r(S^1)$ is the quotient map. Analogously, if we want to restrict the number of coincident eigenvalues, we set $\hat{X}^*_\sigma = X^*_\sigma \times_{\mathrm{Sym}^r_\sigma(S^1)} (S^1)^r$ for $\sigma \in \Pi_r$, and we denote by $\Sigma_\sigma \subset (S^1)^r$ the image $\hat{X}^*_\sigma \to (S^1)^r$.

LEMMA 2.10. *The fibration $\hat{X}^*_\sigma \to \Sigma_\sigma$ is trivial, so we have an isomorphism*

$$\hat{X}^*_\sigma \cong \mathbf{F}_\sigma \times \Sigma_\sigma. \tag{2.4}$$

PROOF. The elements of $\hat{X}^*_\sigma$ are tuples $(A, B, \lambda_1, \ldots, \lambda_r) \in \mathrm{U}(r)^2 \times \Sigma_\sigma$ such that (A, B) is an irreducible representation and the spectrum of A is $\{\lambda_1, \ldots, \lambda_r\}$. Any such representation has a canonical representative with $A = \mathrm{diag}(\lambda_1, \ldots, \lambda_r)$ and thus $B \in \mathbf{F}_\sigma$, leading to the desired isomorphism. $\square$

In the case that there exists at least one eigenvalue with multiplicity one, we can describe the fiber of the fibration (2.4) in terms of the one for SU(r).

PROPOSITION 2.11. *If $\sigma = (r_1, \overset{(a_1)}{\ldots}, r_1, \ldots, r_s, \overset{(a_s)}{\ldots}, r_s) \in \Pi_r$ is a partition with $r_1 = 1$ (i.e., there exists at least one eigenvalue with multiplicity one), then we have an isomorphism*

$$\mathbf{F}_\sigma \cong \mathbf{SF}_\sigma \times S^1.$$

PROOF. Fix eigenvalues $\{\lambda_1, \ldots, \lambda_r\} \in \mathrm{Sym}^r_\sigma(S^1)$. To simplify the notation, we shall suppose that the first eigenvalue λ_1 is simple. Then, given an element $(A, B, \lambda_1, \ldots, \lambda_r) \in \mathbf{F}_\sigma$, the representation can be conjugated to one of the form

$$(A, B) = (\mathrm{diag}(\lambda_1, \ldots, \lambda_r), (b_1 \,|\, b_2 \,|\, \ldots \,|\, b_r)),$$

where b_i are the column vectors of B. Moreover, since λ_1 is a simple eigenvalue, the vector b_1 is well defined up to re-scaling by action of S^1, the stabilizer of the first eigenvector of A. In this way, if set $\mu = \det(B)$, we have that $B' = (\mu^{-1} b_1 \,|\, b_2 \,|\, \ldots \,|\, b_r) \in \mathrm{SU}(r)/\mathrm{Stab}(\sigma)$. Hence, the map $\mathbf{F}_\sigma \ni B \mapsto (B', \mu) \in \mathbf{SF}_\sigma \times S^1$ provides the desired isomorphism. $\square$

Regarding the base of the fibration in (2.4), we can easily describe it in combinatorial terms as follows.

LEMMA 2.12. *Given* $\sigma = (r_1, \overset{(a_1)}{\ldots}, r_1, \ldots, r_s, \overset{(a_s)}{\ldots}, r_s) \in \Pi_r$, *let* $N = a_1 + \cdots + a_s$ *be the number of different eigenvalues. Then, we have that*

$$\Sigma_\sigma \cong S^1 \times \Delta^{N-1}_{\boldsymbol{\mu}_n},$$

where $\Delta^{N-1}_{\boldsymbol{\mu}_n}$ *is the collection of tuples* $(\epsilon_2, \ldots, \epsilon_N)$ *with* $\epsilon_i \in \boldsymbol{\mu}_n^* = \boldsymbol{\mu}_n - \{1\}$ *such that* $\epsilon_i \neq \epsilon_j$ *for* $i \neq j$.

PROOF. Let $(\lambda_1, \ldots, \lambda_N)$ be the ordered different eigenvalues of an element of Σ_σ. Since $\lambda_1^n = \lambda_2^n = \cdots = \lambda_N^n$, for $i \geq 2$ we can uniquely write $\lambda_i = \lambda_1 \epsilon_i$ for some $\epsilon_i \in \boldsymbol{\mu}_n^*$ with all the roots $\epsilon_2, \ldots, \epsilon_N$ different. Hence, the map $(\lambda_1, \ldots, \lambda_N) \mapsto (\lambda_1, \epsilon_2, \ldots, \epsilon_N)$ yields the isomorphism. □

Putting together all these pieces, we finally get an explicit description of X_σ^*, at least in the case where there exists a simple eigenvalue.

COROLLARY 2.13. *Let* $\sigma = (r_1, \overset{(a_1)}{\ldots}, r_1, \ldots, r_s, \overset{(a_s)}{\ldots}, r_s) \in \Pi_r$ *be a partition with at least one eigenvalue with multiplicity one, and let* $N = a_1 + \ldots + a_s$ *be the number of different eigenvalues. Then we have an isomorphism*

$$X_\sigma^* = (\Sigma_\sigma \times \mathbf{F}_\sigma)/S_n = \left(S^1 \times \Delta^{N-1}_{\boldsymbol{\mu}_n} \times \mathbf{SF}_\sigma \times S^1\right)/S_n,$$

where the action of the symmetric group S_n *on* $S^1 \times \Delta^{N-1}_{\boldsymbol{\mu}_n}$ *is given by permutation of eigenvalues and on* $\mathbf{F}_\sigma = \mathbf{SF}_\sigma \times S^1$ *by permutation of columns.*

2.4. Totally reducible representations. In this section, we analyze the representations corresponding to the partition $\pi_1 = (1, \overset{(r)}{\ldots}, 1)$, that is, the spaces of totally reducible representations X_r^{TR} and $\mathrm{S}X_r^{\mathrm{TR}}$ for the $\mathrm{U}(r)$ and $\mathrm{SU}(r)$ cases, respectively. By Lemma 2.7 we have that for $G = \mathrm{U}(r)$

$$X_r^{\mathrm{TR}} = \mathrm{Sym}^r(X_1^*) = \mathrm{Sym}^r(S^1 \times S^1).$$

Analogously, for $G = \mathrm{SU}(r)$ we get

$$\mathrm{S}X_r^{\mathrm{TR}} = \mathrm{SSym}^r(S^1 \times S^1),$$

where $\mathrm{SSym}^r(S^1 \times S^1)$ is the subset of $\mathrm{Sym}^r(S^1 \times S^1)$ of sets $\{(\lambda_1, \mu_1), \ldots, (\lambda_r, \mu_r)\}$ such that $\lambda_1 \cdots \lambda_r = \mu_1 \cdots \mu_r = 1$.

Our first result in this direction is that the former space is a fibration with fiber the later one.

PROPOSITION 2.14. *We have a Zariski locally trivial fibration*

$$\varpi_\sigma : \mathrm{Sym}^r_\sigma(S^1 \times S^1) \to S^1 \times S^1, \quad \varpi_\sigma(\{(\lambda_1, \mu_1), \ldots, (\lambda_r, \mu_r)\}) = (\lambda_1 \cdots \lambda_r, \mu_1 \cdots \mu_r),$$

whose fiber is isomorphic to $\mathrm{SSym}^r(S^1 \times S^1)$.

PROOF. The base of the fibration is parametrized by coordinates $(t, s) = (\lambda_1 \cdots \lambda_r, \mu_1 \cdots \mu_r) \in S^1 \times S^1$, and it is clear that $\mathrm{SSym}^r(S^1 \times S^1) = \phi^{-1}(1, 1)$. The total space can be trivialized over any $(t, s) \in S^1 \times S^1$ via the map

$$\phi : \varpi_\sigma^{-1}(t, s) \times (S^1 - \{-t\}) \times (S^1 - \{-s\}) \to \varpi_\sigma^{-1}((S^1 - \{-t\}) \times (S^1 - \{-s\})),$$

given, for $\theta, \alpha \in (-\pi, \pi)$, by

$$\phi(\{(\lambda_1, \mu_1), \ldots, (\lambda_r, \mu_r)\}, te^{i\theta}, se^{i\alpha}) = \{(e^{i\theta/r}\lambda_1, e^{i\alpha/r}\mu_1), \ldots, (e^{i\theta/r}\lambda_r, e^{i\alpha/r}\mu_r)\}.$$

□

Now, let us consider the projection onto the first component

$$\varpi : \mathrm{Sym}^r(S^1 \times S^1) \to \mathrm{Sym}^r(S^1).$$

As in Section 2.3.1, given a partition $\sigma \in \Pi_r$, we shall focus on the subset $\mathrm{Sym}^r_\sigma(S^1) \subset \mathrm{Sym}^r(S^1)$ of configurations of points with coincident entries given by σ, and we set $\mathrm{Sym}^r_\sigma(S^1 \times S^1) = \varpi^{-1}(\mathrm{Sym}^r_\sigma(S^1))$. We analogously consider the determinant 1 case and denote by $\mathrm{SSym}^r_\sigma(S^1 \times S^1)$ the corresponding space.

Furthermore, given a partition $\sigma = (r_1, \overset{(a_1)}{\ldots}, r_1, \ldots, r_s, \overset{(a_s)}{\ldots}, r_s) \in \Pi_r$ of r, let $S_\sigma = S_{r_1}^{a_1} \times \cdots \times S_{r_s}^{a_s} < S_r$ be the subgroup of permutations of type σ. Then, we define

$$\mathrm{Sym}^\sigma(S^1) = (S^1)^r / S_\sigma,$$

and analogously $\mathrm{SSym}^\sigma(S^1) \subset \mathrm{Sym}^\sigma(S^1)$ for the configurations $\{\mu_1, \ldots, \mu_r\}$ such that $\mu_1 \cdots \mu_r = 1$. Observe in particular that we have identifications $\mathrm{Sym}^r(S^1) = \mathrm{Sym}^{(r)}(S^1)$ and $\mathrm{SSym}^r(S^1) = \mathrm{SSym}^{(r)}(S^1)$.

PROPOSITION 2.15. *Fixed* $\sigma = (r_1, \overset{(a_1)}{\ldots}, r_1, \ldots, r_s, \overset{(a_s)}{\ldots}, r_s) \in \Pi_r$, *the fibration*

$$\mathrm{SSym}^r_\sigma(S^1 \times S^1) \to \mathrm{SSym}^r_\sigma(S^1). \tag{2.5}$$

is Zariski locally trivial with fiber $\mathrm{SSym}^\sigma(S^1)$.

PROOF. Fix $\{\lambda_1, \ldots, \lambda_r\} \in \mathrm{SSym}^r_\sigma(S^1)$ corresponding to the first factor. The algorithm described in [**13**, Proposition 5] provides a unique way of choosing a logarithm cut α such that we have a natural ordering $\lambda_{i_1} = e^{2\pi i(\alpha+\theta_1)}, \lambda_{i_2} = e^{2\pi i(\alpha+\theta_2)}, \ldots, \lambda_{i_r} = e^{2\pi i(\alpha+\theta_r)}$ with $\theta_1 \leq \theta_2 \leq \cdots \leq \theta_r$. Hence, the r points in the second factor of the fiber $\varpi^{-1}(\{\lambda_1, \ldots, \lambda_r\})$ are uniquely ordered except in the cases where we have coincident values in the first factor. Hence, to remove this ordering, we need to quotient by the action of S_σ, as claimed.

Observe that, once a configuration of repeated values is fixed, the choice of the logarithm cut α varies algebraically with the points $\{\lambda_1, \ldots, \lambda_r\}$. Indeed, the cut α corresponds to the argument of one of them, and the chosen value only changes when it surpasses another one. Hence, the construction above is locally trivial in the Zariski topology, as stated. □

REMARK 2.16. Notice that the base space $\mathrm{SSym}^r(S^1)$ has already been studied in [**13**, Proposition 5] and turns out to be isomorphic to the $(r-1)$-simplex

$$\Delta_{r-1} = \left\{ (u_1, \ldots, u_r) \in \mathbb{R}^r_{\geq 0} \,\middle|\, \sum_{i=1}^r u_i = 0 \right\}.$$

In particular, all the base spaces $\mathrm{SSym}^r_\sigma(S^1)$ of Proposition 2.15 are simplices and thus contractible, so the fiber bundle (2.5) is topologically trivial.

REMARK 2.17. Using some basic algebraic geometry, it is possible to give an equivalent description of $\mathrm{Sym}^r(S^1 \times S^1)$. Let us fix an complex structure on $S^1 \times S^1$, which turns it into an elliptic curve Σ. In this setting, the elements of $\mathrm{Sym}^r(S^1 \times S^1)$ correspond exactly to effective divisors on Σ of degree r. Under this interpretation, if $\mathrm{Pic}_r(\Sigma)$ is the collection of (holomorphic) line bundles on Σ of degree r, we can consider the map

$$\mathrm{Sym}^r(S^1 \times S^1) \to \mathrm{Pic}_r(\Sigma), \qquad D \mapsto \mathcal{O}(D) \tag{2.6}$$

that sends each divisor D into its induced line bundle $\mathcal{O}(D)$. Note that, since Σ is an elliptic curve, $\mathrm{Pic}_r(\Sigma) \cong S^1 \times S^1$ topologically [**16**, p. 331]. The fiber of this map at $\mathcal{O}(D)$ is precisely the collection of effective line bundles linearly equivalent to D, which is the complex projective space $\mathbb{P}(H^0(\mathcal{O}(D)))$. Using that the canonical divisor K of Σ has degree zero, we get that $\mathcal{O}(K-D)$ has no global holomorphic sections and thus, by the Riemann-Roch theorem, we get $h^0(\mathcal{O}(D)) = r$. Hence, we get that (2.6) is a bundle over $S^1 \times S^1$ with fiber $\mathbb{CP}^{r-1}$.

2.4.1. *The spaces* $\mathrm{Sym}^2(S^1 \times S^1)$ *and* $\mathrm{SSym}^2(S^1 \times S^1)$. In this section, we shall study more closely configurations of two points in $S^1 \times S^1$. In the first place, $\mathrm{SSym}^2(S^1 \times S^1)$ corresponds to collections of sets $\{(\lambda, \lambda^{-1}), (\mu, \mu^{-1})\}$. We stratify this space according to the possible repetitions of partitions $\sigma \in \Pi_2$, as in Proposition 2.15. Recall from Remark 2.16 that $\mathrm{SSym}^2(S^1) = \bar{\mathbf{I}} = [-1, 1]$, the closed interval. Its strata are $\mathrm{SSym}^2_{(1,1)}(S^1) = \mathbf{I} = (-1, 1)$, the open interval (different eigenvalues), and $\mathrm{SSym}^2_{(2)}(S^1) = \{-1, 1\}$, the endpoints (repeated eigenvalues).

(1) For $\sigma = (1, 1)$ we have different values in the first factor. In this case,we get an open cylinder since
$$\mathrm{SSym}^2_{(1,1)}(S^1 \times S^1) = \mathrm{SSym}^2_{(1,1)}(S^1) \times \mathrm{SSym}^{(1,1)}(S^1) = \mathbf{I} \times S^1,$$
where we have used that $\mathrm{SSym}^{(1,1)}(S^1) = S^1$ and that the fiber bundle is topologically trivial since the base space $\mathbf{I}$ is contractible.

(2) For $\sigma = (2)$ we have repeated eigenvalues in the first factor. Hence, we get
$$\mathrm{SSym}^2_{(2)}(S^1 \times S^1) = \mathrm{SSym}^2_{(2)}(S^1) \times \mathrm{SSym}^2(S^1) = \{\pm 1\} \times \bar{\mathbf{I}}.$$

Hence, globally $\mathrm{SSym}^2(S^1 \times S^1)$ is the cylinder $\bar{\mathbf{I}} \times S^1$ with a gluing in the two boundaries $\{-1, 1\} \times S^1$ that collapses each circle into an interval. Hence, $\mathrm{SSym}^2(S^1 \times S^1)$ is topologically (but not smoothly) a 2-sphere: it is a pillowcase with four orbifold points (see Figure 2).

Remark 2.18. An alternative way of understanding $\mathrm{SSym}^2(S^1 \times S^1)$ is as the quotient $\mathrm{SSym}^2(S^1 \times S^1) = (S^1 \times S^1)/\mathbb{Z}_2$, where the $\mathbb{Z}_2$ action is given by $(\lambda, \mu) \mapsto (\lambda^{-1}, \mu^{-1})$. Regarding the torus as a quotient of $[0, 1]^2$, the action becomes $(s, t) \sim (1 - s, 1 - t)$, $s, t \in [0, 1]$, whose quotient space is homeomorphic to the 2-sphere.

Remark 2.19. We also would like to notice that the fact that $\mathrm{SSym}^2(S^1 \times S^1) = (S^1 \times S^1)/\mathbb{Z}_2$, was also proved in [**9**].

With respect to the space $\mathrm{Sym}^2(S^1 \times S^1)$, notice that by Proposition 2.14 we have a description as S^2-bundle
$$S^2 \to \mathrm{Sym}^2(S^1 \times S^1) \to S^1 \times S^1.$$

Observe that this perfectly fits with the description in Remark 2.17 since $\mathbb{CP}^1 \cong S^2$. The monodromy of this fiber bundle is given by $(\lambda, \mu) \mapsto (e^{i\pi}\lambda, e^{i\pi}\mu) = (-\lambda, -\mu)$. In coordinates (s, t), it is given by $(s, t) \mapsto (s + 1/2, t + 1/2)$, which is orientation preserving, and interchanges the four orbifold points of S^2 in pairs.

2.4.2. *The spaces* $\mathrm{Sym}^3(S^1 \times S^1)$ *and* $\mathrm{SSym}^3(S^1 \times S^1)$. Now, let us look at the 3-fold symmetric product $\mathrm{SSym}^3(S^1 \times S^1)$. We stratify this space according to the number of repeated eigenvalues in the first factor:

(1) $\sigma = (1,1,1)$ (three different eigenvalues). In this case, we have

$$\begin{aligned}\mathrm{SSym}^3_{(1,1,1)}(S^1 \times S^1) &= \mathrm{SSym}^3_{(1,1,1)}(S^1) \times \mathrm{SSym}^{(1,1,1)}(S^1)\\ &= \mathbf{T} \times S^1 \times S^1,\end{aligned}$$

where $\mathbf{T} = \mathrm{SSym}^3_{(1,1,1)}(S^1)$ is the open 2-dimensional triangle as described by Remark 2.16.

(2) $\sigma = (1,2)$ (two coincident eigenvalues). This corresponds to the interior of the edges of the triangle $\mathbf{T}$ above. In this case, we have

$$\mathrm{SSym}^3_{(1,2)}(S^1 \times S^1) = \mathrm{SSym}^3_{(1,2)}(S^1) \times \mathrm{SSym}^{(1,2)}(S^1).$$

Observe that, solving for the last eigenvalue, we have $\mathrm{SSym}^{(1,2)}(S^1) = \mathrm{Sym}^2(S^1)$, which is a Möbius band (see [**13**, Corollary 6]). Hence, over each point of the interior of the edges $\mathbf{T}$ we find a Möbius band.

(3) $\sigma = (1,3)$ (three equal eigenvalues). This corresponds to the vertices of the triangle $\mathbf{T}$. Now,

$$\mathrm{SSym}^3_{(3)}(S^1 \times S^1) = \mathrm{SSym}^3_{(3)}(S^1) \times \mathrm{SSym}^3(S^1),$$

where this later space is a closed triangle. Hence, over each vertex of $\mathbf{T}$, we attach a triangle. Each of these triangles glues with the Möbius bands of the incoming edges through their boundaries.

With respect to the total space $\mathrm{Sym}^2(S^1 \times S^1)$, notice that by Remark 2.17 we have a description as a fiber bundle

$$\mathbb{CP}^2 \to \mathrm{Sym}^3(S^1 \times S^1) \to S^1 \times S^1.$$

3. SU(2)-character variety

For $n = 2$, there are only two partitions, $\pi_0 = (2)$ and $\pi_1 = (1,1)$, that correspond to the set of irreducible representations and the set of totally reducible representations, respectively.

(1) The partition $\pi_0 = (2)$ yields the space of irreducible representations $\mathrm{S}X_2^*$. Notice that the only non-empty stratum for the stratification (2.3) of $\mathrm{S}X_2^*$ corresponds to $\sigma = (1,1)$ since, in the case $\sigma = (2)$, the matrix A of a representation $\rho = (A, B)$ is a multiple of the identity, so every representation is reducible.

Following the notation of Section 2.3, the partition $\sigma = (1,1)$ has $s = 1$, $r_1 = 1$ and $a_1 = 2$. Hence, according to Lemma 2.4, we have that $\mathrm{S}X^*_{(1,1)}$ has as many components $\mathbf{SF}_{(1,1)}$ as

$$N_{(1,1)} = \frac{2}{n}\binom{n}{2} = \frac{2}{n}\frac{n!}{2!\,(n-2)!} = n-1.$$

The stabilizer for this stratum is $\mathrm{Stab}((1,1)) = S^1 \times S^1$. To determine the space $\mathbf{SF}_{(1,1)} \subset \mathrm{SU}(2)/(S^1 \times S^1)$, we observe that, by the discussion of Section 2.3.2, we have a fibration

$$\varphi : \mathbf{SF}_{(1,1)} \to B_2 \subset S^2.$$

On the interior of B_2, this map is an isomorphism. On the boundary, we get no irreducible representations since the first vector of the basis must be colinear with e_1. Hence, we get that $\mathbf{SF}_{(1,1)}$ is homeomorphic to $\mathbf{D} = \mathrm{Int}(B_2)$, which is a 2-dimensional open disc.

REMARK 3.1. An equivalent way of describing $\mathbf{SF}_{(1,1)}$ is the following. With the notation of Remark 2.5, we have that $\mathbf{SF}^0_{(1,1)}$ is the collection of orthonormal bases of $\mathbb{C}^2$ whose vectors are not proportional to any of the vectors of the canonical basis. Since these bases are determined by the first vector, which can be seen as an element $(a,b) \in S^3 \subset \mathbb{C}^2$, we get that $\mathbf{SF}^0_{(1,1)} = S^3 - S^1$ is the 3-sphere minus the 1-sphere of equation $b=0$. Now, the action of S^1 on $(a,b) \in S^3 \subset \mathbb{C}^2$ is $\lambda \cdot (a,b) = (a, \lambda b)$, so the quotient is the weighted projective space

$$\mathbf{SF}^0_{(1,1)}/(S^1 \times S^1) = (S^3 - S^1)/S^1 = \mathbb{CP}^1_{(0,1)} - \{[1:0]\}.$$

This latter space can be understood as the set of $(a,b) \in S^3 \subset \mathbb{C}^2$ such that $b \in \mathbb{R}_{>0}$ so $\mathbb{CP}^1_{(0,1)} - \{[1:0]\}$ is the upper hemisphere of S^2 and thus homeomorphic to a disc.

(2) The partition $\pi_1 = (1,1)$ corresponds to the set of totally reducible representations $\mathrm{S}X_2^{\mathrm{TR}}$. By the results of Section 2.4, we have $\mathrm{S}X_2^{\mathrm{TR}} = \mathrm{SSym}^2(S^1 \times S^1)$, which is topologically (but not smoothly) a 2-sphere.

Summarising from the discussion above, we have that

$$\mathrm{S}X_2 = \mathrm{S}X_2^{(1,1)} \sqcup \mathrm{S}X_2^{(2)} = \mathrm{S}X_2^{\mathrm{TR}} \sqcup \mathrm{S}X^*_{(1,1)}\,.$$

The global picture of $\mathrm{S}X_2$ is as follows (see also Figure 2). First, the space $\mathrm{S}X_2^{\mathrm{TR}}$ is a 2-dimensional sphere. To it, we attach $n-1$ open discs $\mathbf{D}$ corresponding the the components of $\mathrm{S}X^*_{(1,1)}$, each of them attached to $\mathrm{S}X_2^{\mathrm{TR}}$ through their boundary $S^1 \subset \overline{\mathbf{D}}$. These boundaries correspond to reducible representations given by $A \sim \mathrm{diag}(\lambda, \lambda^{-1})$ and $B \sim \mathrm{diag}(\mu, \mu^{-1})$, where $\lambda \in \boldsymbol{\mu}^+_{2n}$, $\mu \in S^1$. They all inject into $\mathrm{S}X_2^{\mathrm{TR}}$ as $n-1$ disjoint circles. The space is homotopically equivalent to the wedge of n copies of 2-spheres.

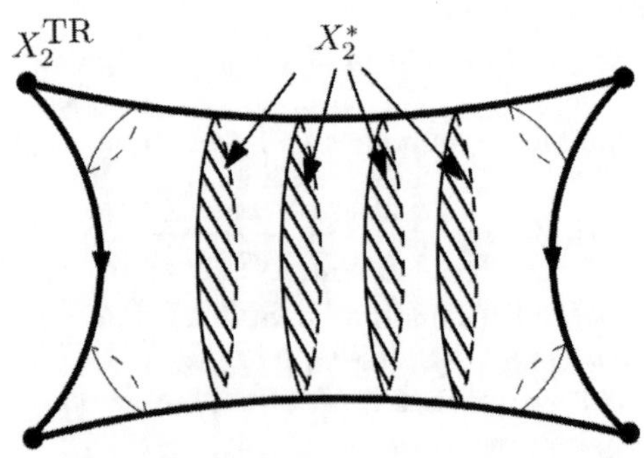

FIGURE 2. The global picture of $X_2 = X(\Gamma_n, \mathrm{SU}(2))$.

4. Homotopy type of the SL(2, $\mathbb{C}$)-character variety of twisted Hopf links

In this section, we study the homotopy type of the SL(2, $\mathbb{C}$)-character variety of the twisted Hopf link. Recall that this variety is

$$\begin{aligned} X(\Gamma_n, \mathrm{SL}(2,\mathbb{C})) &= \mathrm{Hom}\,(\Gamma_n, \mathrm{SL}(2,\mathbb{C}))/\!/\,\mathrm{SL}(2,\mathbb{C}) \\ &= \{(A,B) \in \mathrm{SL}(2,\mathbb{C}) \,|\, [A^n, B] = \mathrm{id}\}/\!/\,\mathrm{SL}(2,\mathbb{C}). \end{aligned}$$

This is a complex algebraic variety whose motive has been previously studied in [**15**]. The aim of this section is to prove that the natural inclusion map $SX_2 = X(\Gamma_n, \mathrm{SU}(2)) \hookrightarrow X(\Gamma_n, \mathrm{SL}(2,\mathbb{C}))$ is a deformation retract. Notice that $\mathrm{SU}(2) \hookrightarrow \mathrm{SL}(2,\mathbb{C})$ is the maximal compact subgroup and that $\mathrm{SL}(2,\mathbb{C})$ is the complexification $\mathrm{SL}(2,\mathbb{C}) = \mathrm{SU}(2)_{\mathbb{C}}$ of SU(2).

As in the SU(2)-case, we can decompose the SL(2, $\mathbb{C}$)-character variety into its reducible and irreducible locus

$$X(\Gamma_n, \mathrm{SL}(2,\mathbb{C})) = X^{\mathrm{TR}}(\Gamma_n, \mathrm{SL}(2,\mathbb{C})) \sqcup X^*(\Gamma_n, \mathrm{SL}(2,\mathbb{C})).$$

In sharp contrast with the SU(2)-case, not every reducible representation is semisimple. However, it turns out that, for the conjugacy action of SL(2, $\mathbb{C}$), every representation has a semisimple representation in the closure of its orbit. This implies that every reducible representation is identified in the GIT quotient with a semisimple one, so the reducible locus of the character variety $X^{\mathrm{TR}}(\Gamma_n, \mathrm{SL}(2,\mathbb{C}))$ parametrizes totally reducible representations up to equivalence, which justifies the assumed notation.

Regarding the irreducible locus $X^*(\Gamma_n, \mathrm{SL}(2,\mathbb{C}))$, recall from the results of [**15**] that it has $n-1$ components, corresponding to the possible eigenvalues of the matrix A of a representation $(A,B) \in X^*(\Gamma_n, \mathrm{SL}(2,\mathbb{C}))$. Fixing eigenvalues $\lambda, \lambda^{-1} \in \mathbb{C}^*$ for A, we can choose a representative of the form

$$(A,B) \sim \left(\begin{pmatrix} \lambda & 0 \\ 0 & \lambda^{-1} \end{pmatrix}, \begin{pmatrix} a & c \\ b & d \end{pmatrix} \right),$$

with $bc \neq 0$ since (A,B) is irreducible. This representative is not unique, since we get a residual action of $\mathbb{C}^*$ on $B \in \mathrm{SL}(2,\mathbb{C})$ given by

$$\lambda \cdot \begin{pmatrix} a & c \\ b & d \end{pmatrix} = \begin{pmatrix} a & \lambda^{-1}c \\ \lambda b & d \end{pmatrix}, \; \lambda \in \mathbb{C}^*.$$

As a consequence, we get that each of these components of $X^*(\Gamma_n, \mathrm{SL}(2,\mathbb{C}))$ is isomorphic to the space

$$\mathbf{SF}_{\mathbb{C}} := \left\{(a,b,c,d) \in \mathbb{C}^4 \,|\, ad - bc = 1, bc \neq 0\right\} /\!/\, \mathbb{C}^*.$$

The invariant coordinates for this $\mathbb{C}^*$-action are $p := bc$ and a, d, so we finally get the description

$$\mathbf{SF}_{\mathbb{C}} = \left\{(a,d,p) \in \mathbb{C}^3 \,|\, ad - p = 1, p \neq 0\right\} \cong \mathbb{C}^2 - H,$$

where H is the hyperbola $H = \{ad = 1\} \cong \mathbb{C}^*$ and the later isomorphism is $(a,d,p) \mapsto (a,d)$. Inside this variety, the corresponding component **SF** of the SU(2)-character variety (see Section 3) is the open disc

$$\mathbf{SF} = \mathbf{SF}_{(1,1)} = \left\{(a, \bar{a}, |a|^2 - 1) \in \mathbf{SF}_{\mathbb{C}} \;\middle|\; |a| < 1\right\} \cong \mathbf{D}.$$

Indeed, given $(a, \bar{a}, |a|^2 - 1) \in \mathbf{SF}$, the corresponding matrix B is

$$B = \begin{pmatrix} a & -\sqrt{1-|a|^2} \\ \sqrt{1-|a|^2} & \bar{a} \end{pmatrix}.$$

REMARK 4.1. The spaces $\mathbf{SF}$ and $\mathbf{SF}_{\mathbb{C}}$ are not homotopic. A straightforward computation with compactly supported cohomology shows that $H_c^k(\mathbf{SF}_{\mathbb{C}}) = H_c^k(\mathbb{C}^2 - \mathbb{C}^*) = \mathbb{Z}$ for $k = 2, 3, 4$ and $H_c^k(\mathbf{SF}_{\mathbb{C}}) = 0$ otherwise. Hence, the Betti numbers of $\mathbf{SF}_{\mathbb{C}}$ are $b_k(\mathbf{SF}_{\mathbb{C}}) = 1$ for $k = 0, 1, 2$ and $b_k(\mathbf{SF}_{\mathbb{C}}) = 0$ for $k \geq 3$, showing that $\mathbf{SF}_{\mathbb{C}}$ cannot be contractible, as it is $\mathbf{SF}$.

Now, in $\mathbf{SF}_{\mathbb{C}}$, the set $p = 0$ corresponds exactly to totally reducible representations in the boundary of the irreducible locus $X^*(\Gamma_n, \mathrm{SL}(2,\mathbb{C}))$. Hence, the closure

$$\overline{\mathbf{SF}}_{\mathbb{C}} := \left\{(a, d, p) \in \mathbb{C}^3 \,|\, ad - p = 1\right\} \cong \mathbb{C}^2,$$

with coordinates $(a, d) \in \mathbb{C}^2$, is precisely the collection of representations $(A, B) \in X^*(\Gamma_n, \mathrm{SL}(2,\mathbb{C}))$ where A has fixed eigenvalues λ, λ^{-1}. The corresponding closure of the component of the $\mathrm{SU}(2)$-character variety is thus the closed disc

$$\overline{\mathbf{SF}} = \left\{(a, \bar{a}, |a|^2 - 1) \in \overline{\mathbf{SF}}_{\mathbb{C}} \;\middle|\; |a| \leq 1\right\} \cong \overline{\mathbf{D}},$$

with coordinates $(a, \bar{a}) \in \mathbb{C}^2$.

LEMMA 4.2. *Let* $\overline{\mathbf{SF}}'_{\mathbb{C}} = \left\{(a, d) \in \overline{\mathbf{SF}}_{\mathbb{C}} \,|\, |a| = |d|\right\}$. *There exists a smooth homotopy*

$$H_t : \overline{\mathbf{SF}}_{\mathbb{C}} \to \overline{\mathbf{SF}}_{\mathbb{C}},$$

for $t \in [0, 1]$ *with* $H_0 = \mathrm{id}_{\overline{\mathbf{SF}}_{\mathbb{C}}}$, $H_1(\overline{\mathbf{SF}}_{\mathbb{C}}) \subseteq \overline{\mathbf{SF}}'_{\mathbb{C}}$, *and* $H_t|_{\overline{\mathbf{SF}}'_{\mathbb{C}}} = \mathrm{id}_{\overline{\mathbf{SF}}'_{\mathbb{C}}}$ *for all* t. *Under this homotopy, the space* $\overline{\mathbf{SF}}_{\mathbb{C}} - \mathbf{SF}_{\mathbb{C}}$ *of* $\mathrm{SL}(2,\mathbb{C})$*-reducible representations remains invariant and is rescaled into the space* $\overline{\mathbf{SF}} - \mathbf{SF}$ *of* $\mathrm{SU}(2)$*-reducible representations.*

PROOF. Given $(a, d) \in \overline{\mathbf{SF}}_{\mathbb{C}} = \mathbb{C}^2$, let us use polar coordinates $a = re^{i\alpha}$ and $d = se^{i\beta}$, with $r, s \in \mathbb{R}_{\geq 0}$ and $\alpha, \beta \in [0, 2\pi)$. Consider the auxiliary continuous homotopies $h^1, h^2 : \mathbb{R}^2_{\geq 0} \times [0, 1] \to \mathbb{R}_{\geq 0}$ given by

$$h_t^1(r, s) = \begin{cases} (1-t)r + t\sqrt{rs} & \text{for } r \geq s, \\ \frac{rs}{(1-t)s + t\sqrt{rs}} & \text{for } s > r, \end{cases} \quad h_t^2(r, s) = \begin{cases} \frac{rs}{(1-t)r + t\sqrt{rs}} & \text{for } r > s, \\ (1-t)s + t\sqrt{rs} & \text{for } s \geq r. \end{cases}$$

Observe that for all $r, s \geq 0$ we have $h_0^1(r, s) = r$, $h_0^2(r, s) = s$, $h_1^1(r, s) = h_1^2(r, s) = \sqrt{rs}$. Moreover, we have $h_t^1(r, r) = h_t^2(r, r) = r$ and $h_t^1(r, s) \cdot h_t^2(r, s) = rs$ for all $t \in [0, 1]$.

In this setting, we consider the homotopy $H : \overline{\mathbf{SF}}_{\mathbb{C}} \times [0, 1] \to \overline{\mathbf{SF}}_{\mathbb{C}}$ given by

$$H_t(a, d) = \left(h_t^1(r, s)e^{i\alpha}, h_t^2(r, s)e^{i\beta}\right).$$

Notice that this map makes sense even for $r = 0$ or $s = 0$ since, in these cases, $h_t^1(0, s) = 0$ and $h_t^2(r, 0) = 0$, respectively. This map satisfies the following properties:

(1) For $t = 0$ we get, for any $(a, d) \in \overline{\mathbf{SF}}_{\mathbb{C}}$,

$$H_0(a, d) = \left(re^{i\alpha}, se^{i\beta}\right) = (a, d).$$

Thus, $H_0 = \mathrm{id}_{\overline{\mathbf{SF}}_{\mathbb{C}}}$.

(2) For $t = 1$ we get, for any $(a, d) \in \overline{\mathbf{SF}}_{\mathbb{C}}$,
$$H_1(a, d) = \left(\sqrt{rs}\, e^{i\alpha}, \sqrt{rs}\, e^{-i\beta}\right).$$
In particular, $H_1(a, d) \in \overline{\mathbf{SF}}'_{\mathbb{C}}$ for all $a, d \in \overline{\mathbf{SF}}_{\mathbb{C}}$.

(3) For any point of the form $(a = re^{i\alpha}, d = re^{i\beta}) \in \overline{\mathbf{SF}}'_{\mathbb{C}}$, we get
$$H_t(a, d) = \left(re^{i\alpha}, re^{i\beta}\right) = (a, d)$$
for all $t \in [0, 1]$. In particular, $H_t|_{\overline{\mathbf{SF}}'_{\mathbb{C}}} = \mathrm{id}_{\overline{\mathbf{SF}}'_{\mathbb{C}}}$.

(4) For a point of the form $(a = re^{i\alpha}, a^{-1} = r^{-1}e^{-i\alpha}) \in \overline{\mathbf{SF}}_{\mathbb{C}} - \mathbf{SF}_{\mathbb{C}}$ (equivalently, for $p = 0$), we have
$$H_t(a, a^{-1}) = \left(h_t^1\left(r, r^{-1}\right) e^{i\alpha}, h_t^2\left(r, r^{-1}\right) e^{-i\alpha}\right)$$
for all $t \in [0, 1]$. Hence, since $h_t^1(r, r^{-1}) \cdot h_t^2(r, r^{-1}) = 1$ for all t, we have that $H_t(a, a^{-1}) \in \overline{\mathbf{SF}}_{\mathbb{C}} - \mathbf{SF}_{\mathbb{C}}$ for all t or, in other words, $\overline{\mathbf{SF}}_{\mathbb{C}} - \mathbf{SF}_{\mathbb{C}}$ is invariant. The homotopy there is a rescaling.

Therefore, property (1) shows that this map defines a homotopy equivalence between $H_0 = \mathrm{id}_{\overline{\mathbf{SF}}_{\mathbb{C}}}$ and H_1. Moreover, by properties (2) and (3), H defines a strong deformation retraction onto $\overline{\mathbf{SF}}'_{\mathbb{C}}$. By (4), this homotopy has the desired property on $\overline{\mathbf{SF}}_{\mathbb{C}} - \mathbf{SF}_{\mathbb{C}}$. □

REMARK 4.3. Since $\mathbf{SF} \subset \overline{\mathbf{SF}}'_{\mathbb{C}}$, this space remains fixed under the previous retraction.

Figure 3 shows the geometric interpretation of the homotopy of the proof of Lemma 4.2 in the quadrant (r, s): the hyperbolas $rs = k$, for constant k, are retracted onto the plane $\{s = r\}$ through the natural $\mathbb{R}_{\geq 0}$-action $\lambda \cdot (r, s) = (\lambda^{-1} r, \lambda s)$ for $\lambda \in \mathbb{R}_{\geq 0}$. In particular, the hyperbola $\overline{\mathbf{SF}}_{\mathbb{C}} - \mathbf{SF}_{\mathbb{C}} = \{rs = 1\}$ remains invariant, and the plane $\overline{\mathbf{SF}}'_{\mathbb{C}} = \{r = s\}$ is fixed.

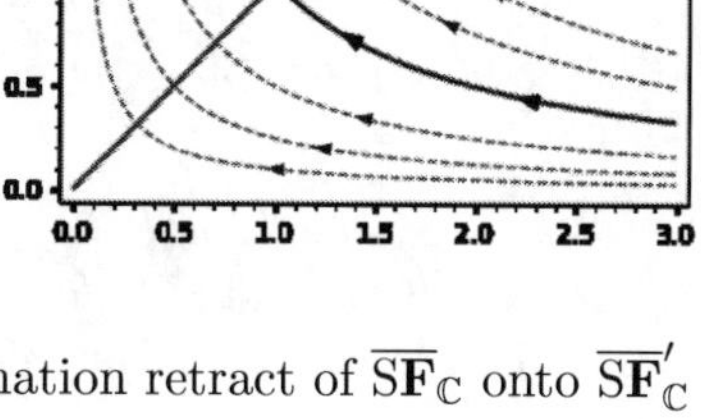

FIGURE 3. Deformation retract of $\overline{\mathbf{SF}}_{\mathbb{C}}$ onto $\overline{\mathbf{SF}}'_{\mathbb{C}}$ in the (r, s)-plane.

LEMMA 4.4. *Let* $\overline{\mathbf{SF}}''_{\mathbb{C}} = \left\{(a, \bar{a}) \in \overline{\mathbf{SF}}_{\mathbb{C}}\right\}$. *There exists a smooth homotopy*
$$H_t : \overline{\mathbf{SF}}'_{\mathbb{C}} \to \overline{\mathbf{SF}}_{\mathbb{C}},$$
for $t \in [0, 1]$ *with* $H_0 = \mathrm{id}_{\overline{\mathbf{SF}}'_{\mathbb{C}}}$, $H_1(\overline{\mathbf{SF}}'_{\mathbb{C}}) \subseteq \overline{\mathbf{SF}}''_{\mathbb{C}}$, *and* $H_t|_{\overline{\mathbf{SF}}''_{\mathbb{C}}} = \mathrm{id}_{\overline{\mathbf{SF}}''_{\mathbb{C}}}$ *for all* t.

PROOF. Let us write $a = re^{i\alpha}$ and $d = re^{i\beta}$ with $r, s \in \mathbb{R}_{\geq 0}$ and $\alpha, \beta \in [0, 2\pi)$. We consider the homotopy $H : \overline{\mathbf{SF}}'_{\mathbb{C}} \times [0,1] \to \overline{\mathbf{SF}}_{\mathbb{C}}$ given by

$$H_t(a,d) = \begin{cases} \left(re^{i\alpha}, r\left((1-t)e^{i\beta} + te^{-i\alpha}\right)\right) & \text{if } a, d \neq 0, \\ (0,0) & \text{if } a = d = 0. \end{cases}$$

Observe that, since $|a| = |d|$ in $\overline{\mathbf{SF}}'_{\mathbb{C}}$, then any of them vanishes if and only if both vanish.

We obviously have $H_0 = \mathrm{id}_{\overline{\mathbf{SF}}'_{\mathbb{C}}}$ and $H_1(a,d) = (a, \bar{a}) \in \overline{\mathbf{SF}}''_{\mathbb{C}}$. Moreover, the points of the form $(a = re^{i\alpha}, \bar{a} = re^{-i\alpha})$ remain fixed, so we directly get that H is the required homotopy. □

REMARK 4.5. Since $\overline{\mathbf{SF}} \subset \overline{\mathbf{SF}}''_{\mathbb{C}}$, this space remains fixed under the previous retraction.

REMARK 4.6. Notice that, throughout the homotopy of Lemma 4.4, the norm of the second component varies, so the target space of this homotopy is the whole $\overline{\mathbf{SF}}_{\mathbb{C}}$ and not the subspace $\overline{\mathbf{SF}}'_{\mathbb{C}}$. This is related to the following fact. Let $\Delta = \{(a,a)\} \subseteq \overline{\mathbf{SF}}'_{\mathbb{C}}$ be the diagonal. The space $\overline{\mathbf{SF}}'_{\mathbb{C}} - \Delta$ can be easily retracted to $\overline{\mathbf{SF}}''_{\mathbb{C}}$ by joining (a,d) with $(a,\bar{a})$ through the shortest arc joining d and $\bar{a}$. However, on $\Delta = \mathbb{C}$, retracting it into $\overline{\mathbf{SF}}''_{\mathbb{C}}$ is the same as homotopying the conjugation map $f : \mathbb{C} \to \mathbb{C}$, $f(a) = \bar{a}$, into the identity map. If we restrict to $f|_{S^1} : S^1 \to S^1$, this is impossible, but it is possible as maps on $\mathbb{C}$ if we allow the homotopy to pass through $0 \in \mathbb{C}$ with the simple linear homotopy.

Now, observe that $\mathbf{SF}''_{\mathbb{C}} = \{(a,\bar{a})\}$ radially retracts onto the disc $\overline{\mathbf{SF}} = \mathbf{SF}''_{\mathbb{C}} \cap \{|a| \leq 1\}$. Therefore, we have proven the following result.

PROPOSITION 4.7. *Each of the components $\overline{\mathbf{SF}}_{\mathbb{C}}$ of the closure of the irreducible locus of the* $\mathrm{SL}(2,\mathbb{C})$*-character variety strongly retracts onto the corresponding component $\overline{\mathbf{SF}}$ of the closure of the irreducible locus of the* $\mathrm{SU}(2)$*-character variety. On the set of totally reducible representations $\overline{\mathbf{SF}}_{\mathbb{C}} - \mathbf{SF}_{\mathbb{C}}$, the homotopy is just linear rescaling onto $\overline{\mathbf{SF}} - \mathbf{SF}$.*

REMARK 4.8. In light of Remark 4.1, Proposition 4.7 has a clear interpretation: the non-trivial elements of $H_1(\mathbf{SF}_{\mathbb{C}})$ and $H_2(\mathbf{SF}_{\mathbb{C}})$ are annihilated when we glue back the hyperbola $\{ad = 1\}$ of totally reducible representations. Hence, only after taking the closures, the inclusion $\overline{\mathbf{SF}} \hookrightarrow \overline{\mathbf{SF}}_{\mathbb{C}}$ becomes a homotopy equivalence.

On the other hand, we can easily prove that the natural inclusion $\mathrm{S}X_2^{\mathrm{TR}} \hookrightarrow X^{\mathrm{TR}}(\Gamma_n, \mathrm{SL}(2,\mathbb{C}))$ is a deformation retract.

PROPOSITION 4.9. *The totally reducible locus $\mathrm{S}X_2^{\mathrm{TR}}$ of the* $\mathrm{SU}(2)$*-character variety is a strong deformation retract of the reducible locus of the* $\mathrm{SL}(2,\mathbb{C})$*-character variety $X^{\mathrm{TR}}(\Gamma_n, \mathrm{SL}(2,\mathbb{C}))$ through a linear rescaling.*

PROOF. By [**15**], the reducible locus is given by pairs $(\lambda,\mu) \in (\mathbb{C}^*)^2$ quotiented by the $\mathbb{Z}_2$-action that identifies $(\lambda,\mu) \sim (\lambda^{-1},\mu^{-1})$. The radial deformation retract of $(\mathbb{C}^*)^2$ onto $(S^1)^2$ descends to a deformation retraction of $(\mathbb{C}^*)^2/\mathbb{Z}_2$ onto $\mathrm{S}X_2^{\mathrm{TR}} \cong \mathrm{SSym}^2(S^1 \times S^1)$ (c.f. Remark 2.18), since it commutes with the $\mathbb{Z}_2$-action. □

REMARK 4.10. Proposition 4.9 also follows directly from [**1**, Theorem 10].

Now, observe that on the common locus of totally reducible representations that lie in the closure of irreducible ones, the homotopies of Proposition 4.7 and 4.9 coincide. Hence, we can glue them together to give rise to a global homotopy $H : X(\Gamma_n, \mathrm{SL}(2,\mathbb{C})) \times [0,1] \to X(\Gamma_n, \mathrm{SL}(2,\mathbb{C}))$ such that $H_0 = \mathrm{id}_{X(\Gamma_n,\mathrm{SL}(2,\mathbb{C}))}$, $H_t|_{X(\Gamma_n,\mathrm{SU}(2))} = \mathrm{id}_{X(\Gamma_n,\mathrm{SU}(2))}$ and $H_1(X(\Gamma_n, \mathrm{SL}(2,\mathbb{C}))) \subseteq X(\Gamma_n, \mathrm{SU}(2))$. Therefore, we have proven the following result.

THEOREM 4.11. *The* $\mathrm{SU}(2)$*-character variety* $\mathrm{S}X_2 = X(\Gamma_n, \mathrm{SU}(2))$ *of a twisted Hopf link with n twists is a strong deformation retract of the* $\mathrm{SL}(2,\mathbb{C})$*-character variety* $X(\Gamma_n, \mathrm{SL}(2,\mathbb{C}))$.

5. U(2) and SU(3) character varieties

In this section, using the results of Section 2.3 and 2.4, we shall describe the stratification in the character varieties $X_2 = X(\Gamma_n, \mathrm{U}(2))$ and $\mathrm{S}X_3 = X(\Gamma_n, \mathrm{SU}(3))$. As we will see, a much more involved geometry arises in these cases, with many strata interacting with non-trivial intersection patterns.

5.1. U(2)-character variety. As in Section 3, we get again two possible cases arising from the partitions $\pi_0 = (2)$ and $\pi_1 = (1,1)$.

(1) $\pi_0 = (2)$ corresponds to irreducible representations X_2^* and, as for SU(2), the only configuration of eigenvalues that contributes is $\sigma = (1,1)$ (which has parameters $r_1 = 1$ and $a_1 = 2$ in the notation of Section 2.3). Hence, the number of different eigenvalues is $N = a_1 = 2$ and thus by Corollary 2.13, we have that

$$X_2^* = X_{(1,1)}^* = (S^1 \times \boldsymbol{\mu}_n^* \times \mathbf{F}_{(1,1)})/\mathbb{Z}_2,$$

where we have used that $\Delta^1_{\boldsymbol{\mu}_n} = \boldsymbol{\mu}_n^*$. The action of $\mathbb{Z}_2$ on $S^1 \times \boldsymbol{\mu}_n^* \times \mathbf{F}_{(1,1)}$ is $(\lambda, \epsilon, B) \mapsto (\lambda\epsilon, \epsilon^{-1}, P_0 B P_0^{-1})$, where P_0 is the permutation matrix that exchanges the columns of B. We have two options:

(a) If n is odd, then there exists a unique representative (λ, ϵ, B) with $\mathrm{Im}(\epsilon) > 0$. Hence, using that $\mathbf{F}_{(1,1)} = \mathbf{SF}_{(1,1)} \times S^1 = \mathbf{D} \times S^1$ with $\mathbf{D}$ an open 2-dimensional disc (c.f. Section 3), we get

$$X_2^* = S^1 \times \boldsymbol{\mu}_n^+ \times \mathbf{D} \times S^1,$$

where $\boldsymbol{\mu}_n^+ = \{\epsilon \in \boldsymbol{\mu}_n \,|\, \mathrm{Im}(\epsilon) > 0\}$. These are $(n-1)/2$ copies of $S^1 \times \mathbf{D} \times S^1$.

(b) If n is even, we still get $(n-2)/2$ copies of $S^1 \times \mathbf{D} \times S^1$, corresponding to the roots of unity with $\mathrm{Im}(\epsilon) > 0$. However, for $\epsilon = -1$ we get a residual action of $\mathbb{Z}_2$ on $S^1 \times \mathbf{D} \times S^1$. Since the action in the first copy of S^1 is $\lambda = e^{2\pi i\theta} \mapsto -\lambda = e^{2\pi i(\theta+1/2)}$, there is an unique representative with $0 \le \theta < 1/2$. Hence, in this case we get a contribution $\mathbf{I}_0 \times \mathbf{D} \times S^1$, where $\mathbf{I}_0 = [0, 1/2)$.

(2) $\pi_1 = (1,1)$ corresponds to the set of totally reducible representations $X_2^{\mathrm{TR}} = \mathrm{Sym}^2(S^1 \times S^1)$. As proven in Section 2.4.1, we have a S^2-fibration

$$S^2 \to \mathrm{Sym}^2(S^1 \times S^1) \to S^1 \times S^1.$$

The monodromy of this fibration is $(\lambda, \mu) \mapsto (e^{i\pi}\lambda, e^{i\pi}\mu) = (-\lambda, -\mu)$, that is an orientation preserving action interchanging the four orbifold points of S^2 in pairs.

5.2. SU(3)-character variety. In this case, we have three possible semisimple types corresponding to the partitions $\pi_0 = (3)$, $\pi_1 = (1,1,1)$ and $\pi_2 = (1,2)$.

- For $\pi_0 = (3)$ we get irreducible representations $\mathrm{S}X_3^*$. We stratify according to the repeated eigenvalues of the matrix A of a representation $(A,B) \in \mathrm{S}X_3^*$ as in (2.3) by

$$\mathrm{S}X_3^* = \mathrm{S}X_{(1,1,1)}^* \sqcup \mathrm{S}X_{(1,2)}^* \sqcup \mathrm{S}X_{(3)}^*.$$

(1) $\sigma = (1,1,1)$ (three different eigenvalues). By the results of Section 2.3.2, we have that the space $\mathrm{S}X_{(1,1,1)}^*$ is a collection of

$$N_{(1,1,1)} = \frac{3}{n}\binom{n}{3} = \frac{(n-1)(n-2)}{2}$$

copies of a certain subspace $\mathbf{SF}_{(1,1,1)} \subset \mathrm{SU}(3)/(S^1)^3$.
For the projection map onto the coarse orthant $\varphi : \mathrm{SU}(3)/(S^1)^3 \to B_3 = \{(z, x_2, x_3) \in \mathbb{C}\times\mathbb{R}_{\geq 0}^2 \,|\, |z|^2 + x_2^2 + x_3^2 = 1\} \subset S^3$, on the interior of B_3 the preimage is exactly $\mathbf{SF}_{(1,1,1)}$ and the fiber is S^3. On an edge of ∂B_3, let us say the one in which $x_2 = 0$, fixed a first column $(z, 0, x_3) \in \partial B_3$, the fiber is isomorphic to the quotient under the action of $S^1 \times S^1$ of the space

$$\left\{ (w_1, w_2, w_3) \in \mathbb{C}^3 \;\middle|\; \begin{array}{c} w_3 = -\frac{w_1\bar{z}}{x_3},\ |w_1|^2 + |w_2|^2 + |w_3|^2 = 1, \\ (A,B) \text{ is irreducible} \end{array} \right\}.$$

Since, (A,B) must be irreducible, we have that $w_1, w_3 \neq 0$ and hence, using the action of $S^1 \times S^1$ we find a unique point in the orbit with $w_1 \in \mathbb{R}_{>0}$. Hence, the fiber of $\mathbf{SF}_{(1,1,1)}$ on an edge is the orthant $B_2 \subset S^3$, which is a 2-dimensional disc. Finally, the vertices of the orthant are not included since otherwise the representation would be reducible.

(2) $\sigma = (1,2)$ (two coincident eigenvalues). In this case, the space $\mathrm{S}X_{(1,2)}^*$ is a collection of

$$N_{(1,2)} = \frac{3}{n}\binom{n}{2,1} = \frac{3(n-1)(n-2)}{2}$$

copies of a certain subspace $\mathbf{F}_{(1,2)} \subset \mathrm{SU}(3)/(\mathrm{U}(2)\times S^1)$, where $\mathrm{U}(2)\times S^1$ is the stabilizer of the type $\sigma = (2,1)$. Using the action of this stabilizer, any representation (A,B) can be put in the form $A = \mathrm{diag}(\lambda_1, \lambda_1, \lambda_2)$ and

$$B = \begin{pmatrix} x_1 & y_1 & z_1 \\ 0 & y_2 & z_2 \\ x_3 & y_3 & z_3 \end{pmatrix}.$$

The stabilizer of this shape of matrices is now $(S^1)^3$, with action as in Section 2.3.2. This corresponds exactly to an edge in case (1), so we get that $\mathbf{F}_{(1,2)} \cong \mathbf{I} \times B_2$.

(3) $\sigma = (3)$ (three equal eigenvalues). There are no elements in this stratum since they would be reducible.

- For $\pi_1 = (1,1,1)$ we get totally reducible representations, which are $\mathrm{S}X_3^{\mathrm{TR}} = \mathrm{SSym}^3(S^1 \times S^1)$. This space is described in Section 2.4.2.

- For $\pi_2 = (1, 2)$, we get that $\mathrm{S}X_3^{(1,2)} = X_2$. This space was described in Section 5.1.

Even though the description above provides an accurate geometric picture, the intersection pattern it describes is too involved to compute the homotopy types of these character varieties. A prospective future work is to seek alternative approaches that fully characterize these intersections, enabling effective homological and homotopical calculations in this higher rank case.

Acknowledgments. The first author has been supported by the Madrid Government (Comunidad de Madrid – Spain) under the Multiannual Agreement with the Universidad Complutense de Madrid in the line Research Incentive for Young PhDs, in the context of the V PRICIT (Regional Programme of Research and Technological Innovation) through the project PR27/21-029, COMPLEXFLUIDS research grant sponsored by the BBVA Foundation and by the Ministerio de Ciencia e Innovación Project PID2021-124440NB-I00 (Spain). The second author has been supported by the Ministerio de Ciencia e Innovación Project PID2021-124440NB-I00 (Spain), Santander-UCM PR44/21-29924 research grant and by the National Science Foundation under grant No. DMS-1928930 while she was in residence at the Simons Laufer Mathematical Sciences Institute (previously known as MSRI) Berkeley, California, during Fall 2022 Semester. The fourth author has been partially supported by Ministerio de Ciencia e Innovación Project PID2020-118452GB-I00 (Spain).

References

[1] I. Biswas, C. Florentino, S. Lawton, and M. Logares, *The topology of parabolic character varieties of free groups*, Geom. Dedicata **168** (2014), 143–159, DOI 10.1007/s10711-012-9822-1. MR3158035

[2] H. U. Boden, *Representations of orbifold groups and parabolic bundles*, Comment. Math. Helv. **66** (1991), no. 3, 389–447, DOI 10.1007/BF02566657. MR1120654

[3] G. Burde and K. Murasugi, *Links and Seifert fiber spaces*, Duke Math. J. **37** (1970), 89–93. MR253313

[4] A. Casimiro, C. Florentino, S. Lawton, and A. Oliveira, *On the homotopy type of free group character varieties*, Bol. Soc. Port. Mat. **Special issue** (2016), 53–57. MR3526614

[5] H. Chen and T. Yu, *The* $\mathrm{SL}(2,\mathbb{C})$*-character variety of the Borromean link*, arXiv:2202.07429.

[6] C. Florentino, P. B. Gothen, and A. Nozad, *Homotopy type of moduli spaces of G-Higgs bundles and reducibility of the nilpotent cone*, Bull. Sci. Math. **150** (2019), 84–101, DOI 10.1016/j.bulsci.2018.10.002. MR3891762

[7] C. Florentino and S. Lawton, *The topology of moduli spaces of free group representations*, Math. Ann. **345** (2009), no. 2, 453–489, DOI 10.1007/s00208-009-0362-4. MR2529483

[8] C. Florentino and S. Lawton, *Character varieties and moduli of quiver representations*, In the tradition of Ahlfors-Bers. VI, Contemp. Math., vol. 590, Amer. Math. Soc., Providence, RI, 2013, pp. 9–38, DOI 10.1090/conm/590/11720. MR3087924

[9] C. Florentino and S. Lawton, *Topology of character varieties of Abelian groups*, Topology Appl. **173** (2014), 32–58, DOI 10.1016/j.topol.2014.05.009. MR3227204

[10] C. Florentino, S. Lawton, and D. Ramras, *Homotopy groups of free group character varieties*, Ann. Sc. Norm. Super. Pisa Cl. Sci. (5) **17** (2017), no. 1, 143–185. MR3676044

[11] C. Florentino and S. Lawton, *Flawed groups and the topology of character varieties*, arXiv:2012.08481.

[12] M. Furuta and B. Steer, *Seifert fibred homology 3-spheres and the Yang-Mills equations on Riemann surfaces with marked points*, Adv. Math. **96** (1992), no. 1, 38–102, DOI 10.1016/0001-8708(92)90051-L. MR1185787

[13] Á. González-Prieto, J. Martínez, and V. Muñoz, *Geometry of SU(3)-character varieties of torus knots.* part A, Topology Appl. **339** (2023), no. part A, Paper No. 108586, DOI 10.1016/j.topol.2023.108586. MR4644371

[14] Á. González-Prieto and V. Muñoz, *Motive of the* SL_4*-character variety of torus knots*, J. Algebra **610** (2022), 852–895, DOI 10.1016/j.jalgebra.2022.06.008. MR4475981

[15] Á. González-Prieto and V. Muñoz, *Representation varieties of twisted Hopf links*, Mediterr. J. Math. **20** (2023), no. 2, Paper No. 89, 30, DOI 10.1007/s00009-023-02300-w. MR4541571

[16] P. Griffiths and J. Harris, *Principles of algebraic geometry*, Pure and Applied Mathematics, Wiley-Interscience [John Wiley & Sons], New York, 1978. MR507725

[17] S. Lawton and C. Manon, *Character varieties of free groups are Gorenstein but not always factorial*, J. Algebra **456** (2016), 278–293, DOI 10.1016/j.jalgebra.2016.01.044. MR3484145

[18] J. Martínez-Martínez and V. Muñoz, *The* $SU(2)$*-character varieties of torus knots*, Rocky Mountain J. Math. **45** (2015), no. 2, 583–602, DOI 10.1216/RMJ-2015-45-2-583. MR3356629

[19] V. B. Mehta and C. S. Seshadri, *Moduli of vector bundles on curves with parabolic structures*, Math. Ann. **248** (1980), no. 3, 205–239, DOI 10.1007/BF01420526. MR575939

[20] V. Muñoz, *The* $\mathrm{SL}(2,\mathbb{C})$*-character varieties of torus knots*, Rev. Mat. Complut. **22** (2009), no. 2, 489–497, DOI 10.5209/rev_REMA.2009.v22.n2.16290. MR2553945

[21] B. Nasatyr and B. Steer, *Orbifold Riemann surfaces and the Yang-Mills-Higgs equations*, Ann. Scuola Norm. Sup. Pisa Cl. Sci. (4) **22** (1995), no. 4, 595–643. MR1375314

[22] P. E. Newstead, *Topological properties of some spaces of stable bundles*, Topology **6** (1967), 241–262, DOI 10.1016/0040-9383(67)90037-7. MR232015

[23] P. E. Newstead, *Introduction to moduli problems and orbit spaces*, Tata Institute of Fundamental Research Lectures on Mathematics and Physics, vol. 51, Tata Institute of Fundamental Research, Bombay; Narosa Publishing House, New Delhi, 1978. MR546290

Departamento de Álgebra, Geometría y Topología, Facultad de Ciencias Matemáticas, Universidad Complutense de Madrid, Plaza Ciencias 3, 28040 Madrid Spain; and Instituto de Ciencias Matemáticas (CSIC-UAM-UCM-UC3M), C. Nicolás Cabrera 13-15, 28049 Madrid Spain.

Email address: `angelgonzalezprieto@ucm.es`

Departamento de Álgebra, Geometría y Topología, Facultad de Ciencias Matemáticas, Universidad Complutense de Madrid, Plaza Ciencias 3, 28040 Madrid Spain.

Email address: `mlogares@ucm.es`

Departamento de Matemática Aplicada, Ciencia e Ingeniería de los Materiales y Tecnología Electrónica. E.S. Ciencias Experimentales y Tecnología, Universidad Rey Juan Carlos, C. Tulipán 0, 28933 Móstoles, Madrid Spain.

Email address: `javier.martinezmar@urjc.es`

Instituto de Matemática Interdisciplinar (IMI) and Departamento de Álgebra, Geometría y Topología, Facultad de Ciencias Matemáticas, Universidad Complutense de Madrid, Plaza Ciencias 3, 28040 Madrid Spain.

Email address: `vicente.munoz@ucm.es`

Contemporary Mathematics
Volume **803**, 2024
https://doi.org/10.1090/conm/803/16102

Secant loci of scrolls over curves

George H. Hitching

Dedicated to Professor Peter E. Newstead on the occasion of his 80th birthday conference

Abstract. Given a curve C and a linear series ℓ on C, the secant locus $V_e^{e-f}(\ell)$ parametrises effective divisors of degree e which impose at most $e-f$ conditions on ℓ. For $E \to C$ a vector bundle of rank r, we define determinantal subschemes $H_e^{e-f}(\ell) \subseteq \mathrm{Hilb}^e(\mathbb{P}E)$ and $Q_e^{e-f}(V) \subseteq \mathrm{Quot}^{0,e}(E^*)$ which generalise $V_e^{e-f}(\ell)$, giving several examples. We describe the Zariski tangent spaces of $Q_e^{e-f}(V)$, and give examples showing that smoothness of $Q_e^{e-f}(V)$ is not necessarily controlled by injectiveness of a Petri map. We generalise the Abel–Jacobi map and the notion of linear series to the context of Quot schemes.

We give some sufficient conditions for nonemptiness of generalised secant loci, and a criterion in the complete case when $f = 1$ in terms of the Segre invariant $s_1(E)$. This leads to a geometric characterisation of semistability similar to that in [Quot schemes, Segre invariants, and inflectional loci of scrolls over curves, Geom. Dedicata 205 (2020), 1–19]. Using these ideas, we also give a partial answer to a question of Lange on very ampleness of $\mathcal{O}_{\mathbb{P}E}(1)$, and show that for any curve, $Q_e^{e-1}(V)$ is either empty or of the expected dimension for sufficiently general E and V. When $Q_e^{e-1}(V)$ has and attains expected dimension zero, we use formulas of Oprea–Pandharipande and Stark to enumerate $Q_e^{e-1}(V)$.

We mention several possible avenues of further investigation.

1. Introduction

The purpose of this article is to investigate secant loci of scrolls over curves, and to begin a study of generalised secant loci on Quot schemes.

Let C be a complex projective smooth curve of genus g. For $e \geq 1$, let C_e be the eth symmetric product of C. For $n \geq 1$ and $0 \leq f < e$, we consider the locus

$$C_e^f := \{D \in C_e : h^0(C, \mathcal{O}_C(D)) \geq f+1\} = \{D \in C_e : h^0(C, K_C(-D)) \geq g-e+f\}.$$

This is a determinantal subvariety of C_e, whose tangent spaces are defined by a Petri map. The construction and properties of C_e^f are explained in detail in [**ACGH85**, Chapter IV]. The locus C_e^f is the preimage of the Brill–Noether locus

$$W_e^f := \{L \in \mathrm{Pic}^e(C) : h^0(C, L) \geq f+1\}$$

by the Abel–Jacobi map $C_e \to \mathrm{Pic}^e(C)$.

2020 *Mathematics Subject Classification.* Primary 14H60, 14N07, 14M12.

An important characterisation of the divisors in C_e^f is given by the geometric Riemann–Roch theorem, which may be interpreted as saying that

$$C_e^f = \{D \in C_e : \dim \mathrm{Span}\,(\varphi_{K_C}(D)) \leq e - f - 1\}.$$

More generally, let L be a line bundle and $\ell = (L, V)$ a linear series of dimension n. For $0 \leq f \leq e$, we then have a *secant locus* associated to ℓ, defined as

$$V_e^{e-f}(\ell) := \{D \in C_e : \dim\left(V \cap H^0(C, L(-D))\right) \geq n + 1 - e + f\}. \tag{1.1}$$

Equivalently, writing φ for the natural map $C \to \mathbb{P}V^*$, we have

$$V_e^{e-f}(\ell) = \{D \in C_e : \dim \mathrm{Span}\,(\varphi(D)) \leq e - f - 1\}.$$

Note that $C_e^f = V_e^{e-f}\left(K_C, H^0(C, K_C)\right)$.

The loci C_e^f are important invariants of the curve, as illustrated by the theorems of Martens and Mumford [**ACGH85**, § IV.5], and their generalisations to $V_e^{e-f}(\ell)$ proven in [**Baj15**]. Various instances and properties of C_e^f have been investigated in [**Cop95**], [**CM91**], [**GP82**] and elsewhere. The tangent spaces of C_e^f and $V_e^{e-f}(\ell)$ have been studied in [**CJ91**], [**CJ96**], [**Baj15**] and [**Baj18**].

A major topic of interest is the enumerative geometry of $V_e^{e-f}(\ell)$. This is studied for C_e^f in [**ACGH85**, Chap. VII], and that of $V_e^{e-f}(\ell)$ is investigated in [**LeB06**], [**Far08**], [**Far23**], [**Cot11**], [**Ung21a**], [**Ung21b**], [**CHZ21**], and elsewhere. In contrast to the Brill–Noether loci W_e^f, even when $V_e^{e-f}(\ell)$ has nonnegative expected dimension it is not always defined by a map with suitable ampleness properties, so the question of nonemptiness may be nontrivial. In [**Ung19**], examples are given of empty secant loci with positive expected dimension, and in [**Ung21b**] the question is addressed of when intersection-theoretic computations have enumerative meaning.

The purpose of the present work is to study two closely related generalisations of $V_e^{e-f}(\ell)$, obtained by increasing $\dim(C)$ and $\mathrm{rk}(L)$ respectively. The first is very familiar. Let S be a smooth projective variety of dimension $r \geq 1$, and $\ell = (\mathcal{L}, V)$ a linear series of dimension n on S. Let $\mathrm{Hilb}^e_{\mathrm{sm}}(S) \subseteq \mathrm{Hilb}^e(S)$ be the component containing reduced subschemes. We define

$$H_e^{e-f}(\ell) := \{Z \in \mathrm{Hilb}^e_{\mathrm{sm}}(S) : \dim\left(V \cap H^0(S, \mathcal{L} \otimes \mathcal{I}_Z)\right) \geq n + 1 - e + f\}.$$

(The restriction to the smoothable component $\mathrm{Hilb}^e_{\mathrm{sm}}(S)$ is not necessary; see Remark 2.3.) Clearly, $H_e^{e-f}(\ell)$ reduces to $V_e^{e-f}(\ell)$ when S is the curve C. In case S is a surface, the enumerative geometry of $H_e^{e-f}(\ell)$ has been studied via integrals of Segre classes in [**EGL01**], [**Voi19**], [**MOP2019**] and elsewhere. Our focus is primarily on the case where S is a projective bundle $\pi\colon \mathbb{P}E \to C$ and $\mathcal{L} = \mathcal{O}_{\mathbb{P}E}(1) \otimes \pi^*M$, where techniques on vector bundles over curves can be brought to bear.

For the second generalisation of $V_e^{e-f}(\ell)$, let $E \to C$ be a vector bundle of rank r and degree d. The scheme $\mathrm{Quot}^{0,e}(E^*)$ parametrises torsion quotients of E^* of length e; equivalently, subsheaves $F^* \subset E^*$ of rank r and degree $-d-e$. Let M be a line bundle, and suppose that $V \subseteq H^0(C, E^* \otimes M)$ is a subspace of dimension $n+1$. We define

$$Q_e^{e-f}(E, M, V) := \{[F^* \to E^*] \in \mathrm{Quot}^{0,e}(E^*) : h^0(C, F^* \otimes M) \geq n + 1 - e + f\}.$$

If E and M are clear from the context, we write simply $Q_e^{e-f}(V)$; and we abbreviate $Q_e^{e-f}\left(H^0(C, E^* \otimes M)\right)$ to Q_e^{e-f}. Note that $\mathrm{Quot}^e(L^{-1}) = C_e$ for all line bundles

L. In this case, we have

$$Q_e^{e-f}(\mathcal{O}_C, L, V) \;=\; Q_e^{e-f}(L, \mathcal{O}_C, V) \;=\; V_e^{e-f}(L^*, V).$$

Note in particular that

$$C_e^f \;=\; Q_e^{e-f}\left(K_C, \mathcal{O}_C, H^0(C, K_C)\right). \tag{1.2}$$

It is straightforward to see (§ 2.3) that $H_e^{e-f}(\ell)$ and $Q_e^{e-f}(E, M, V)$, like $V_e^{e-f}(\ell)$, are determinantal varieties with expected codimension $f(n+1-e+f)$. Moreover, $Q_e^{e-f}(H^0(C, E^*))$ is an analogue of a Brill–Noether locus for Quot schemes; see Remark 2.8. The loci $Q_e^{e-f}(V)$ are better behaved than $H_e^{e-f}(\ell)$ from many points of view. However, as the geometric intuition associated with $H_e^{e-f}(\ell)$ is illuminating, we have as far as possible developed the two notions in parallel.

Let us now give a summary of the article. After some preliminaries on scrolls and secant defect, we construct the loci $H_e^{e-f}(\ell)$ and $Q_e^{e-f}(V)$ and give some examples. In § 2.5, using an approach from [**Sta21a**] and [**Hit20**], we describe a natural surjective rational map $\alpha\colon H_e^{e-f}(\mathcal{O}_{\mathbb{P}E}(1) \otimes \pi^* M, V) \dashrightarrow Q_e^{e-f}(E, M, V)$ which is birational for $f = 0$. This is a useful tool in what follows. The indeterminacy locus of α is discussed briefly in § 2.5.1. We show in § 2.6 that $Q_e^{e-f}(V)$ is not contained in $Q_e^{e-f-1}(V)$ for $n+1-e+f \geq 0$, with a partial analogue for $H_e^{e-f}(\ell)$. In § 2.7 we discuss the effect of a general projection on secant loci both of type $H_e^{e-f}(\ell)$ and $Q_e^{e-f}(V)$.

Next, suppose that $\mathrm{Quot}^{0,e}(E^*)$ contains a point $[F^* \to E^*]$ where F is stable. Then there are rational classifying maps from $\mathrm{Hilb}^e_{\mathrm{sm}}(S)$ and $\mathrm{Quot}^{0,e}(E^*)$ to the moduli space $U_C(r, d+e)$ of stable bundles of rank r and degree $d+e$. These generalise the Abel–Jacobi map $C_e \to \mathrm{Pic}^e(C)$, and map secant loci to Brill–Noether loci. In § 3, we study this situation and generalise the notion of linear series to the context of Quot schemes.

In § 4, we give some conditions for nonemptiness of the two types of generalised secant loci, extending various statements in the literature for $V_e^{e-f}(\ell)$. In § 5 the Zariski tangent spaces of $Q_e^{e-f}(E, M, V)$ are described. In the complete case, we show that smoothness is controlled by a Petri map; in general, however, the criterion for smoothness is not injectiveness of the Petri map. Indeed, in § 5.4 we show by example that when Q_e^{e-f} admits a map to a Brill–Noether locus, smoothness of the secant locus is neither necessary nor sufficient for smoothness of the Brill–Noether locus in general.

For the remainder of the article, we focus on the case $f = 1$, generalising the study of inflectional loci of linearly normal scrolls in [**Hit19**]; and using similar methods. In § 6, we construct a parameter space for Q_e^{e-1} using Quot schemes of invertible subsheaves of E. This has several applications. In Proposition 7.4, we give a criterion for nonemptiness of $\left(H_e^{e-1}\right)_{\mathrm{nd}}$ (defined in § 2.5.1) and Q_e^{e-1}, in terms of the Segre invariant $s_1(E)$. In Theorem 7.7, we offer a partial answer to a question of Lange [**Lan92**] on the relation between the Segre invariant and very ampleness of line bundles over ruled surfaces, which in fact is valid for projective bundles of any dimension over C. In § 7.4, we characterise semistable bundles in terms of secant loci. And in § 8, we show that for a general bundle E over any curve and a general $V \subseteq H^0(C, E^* \otimes M)$, the loci $Q_e^{e-f}(V)$ are either empty or of the expected dimension. The approach for the latter is similar to [**Hit19**, § 6], and relies on Kleiman's theorem on transversality of translates. We conclude by showing how results from [**EGL01**], [**OP21**] and [**Sta21b**] can be used to enumerate $Q_e^{e-1}(V)$ when it is has and attains expected dimension zero.

Topics of future inquiry. Many questions which can be asked about Brill–Noether loci on $U_C(r,d)$ or about moduli of coherent systems (see for example [**GT09**] and [**New22**]) can be formulated for $H_e^{e-f}(\ell)$ and $Q_e^{e-f}(V)$. For example, many questions of nonemptiness, components, dimension and smoothness of $H_e^{e-f}(\ell)$ and $Q_e^{e-f}(V)$ are open. The relation between smoothness of Q_e^{e-f} and the Brill–Noether locus to which it maps, and the role of semistability in the whole, would be interesting to investigate further.

The enumerative geometry of $Q_e^{e-f}(V)$ is of interest as a natural generalisation of that of $V_e^{e-f}(\ell)$, and a natural application for the recent works [**OP21**] and [**Sta21b**]. It appears a priori more tractable than that of $H_e^{e-f}(\ell)$: As a scroll contains many linear spaces, for $e \geq 3$ the loci H_e^{e-f} have components of excess dimension.

In another direction: Both $H_e^{e-f}(\ell)$ and $Q_e^{e-f}(V)$ appear naturally in the study of Mumford's notion of linear stability [**Mum77**] for scrolls in $\mathbb{P}^n$. This has relevance for questions of moduli and, as explained in [**MS12**] and [**CT-L18**], for Butler's well-known conjecture. This will be a topic of future investigation.

An object in some sense "sitting between" $H_e^{e-f}(\ell)$ and the inflectional loci studied in [**Hit19**] is the generalised Terracini locus defined in [**BC21**]. It would be interesting to see how the present techniques may be applicable to Terracini loci.

2. Construction and first properties

In this section we define generalised secant loci. After recalling or proving some background material on linear spans, secant defect and scrolls, we construct the loci $H_e^{e-f}(\ell)$ and $Q_e^{e-f}(V)$ and give some basic results. We give a number of examples.

2.1. Linear spans and secant defect. Let S be a variety and $\mathcal{L} \to S$ a line bundle with nonempty linear series, and write $\varphi_{\mathcal{L}} \colon S \dashrightarrow |\mathcal{L}|^*$ for the standard map to the complete linear series of $\mathcal{L}$. For any closed subscheme $Z \subseteq S$, let $r_Z \colon H^0(S,\mathcal{L}) \to H^0(Z,\mathcal{L}|_Z)$ be the restriction map. Then it is easy to see that

$$\text{(2.1)}\quad \operatorname{Span}\varphi_{\mathcal{L}}(Z) = \mathbb{P}\operatorname{Im}\left(H^0(Z,\mathcal{L}|_Z)^* \to H^0(S,\mathcal{L})^*\right) = \mathbb{P}\operatorname{Ker}\left(H^0(S,\mathcal{L})^* \to H^0(S,\mathcal{I}_Z\otimes\mathcal{L})^*\right).$$

Now let $V \subseteq H^0(S,\mathcal{L})$ be a nonzero subspace, and let $\psi =: S \dashrightarrow \mathbb{P}V^*$ be the natural map. We have an exact diagram

$$\begin{array}{ccccccc}
0 & \longrightarrow & H^0(S,\mathcal{I}_Z\otimes\mathcal{L}) & \longrightarrow & H^0(S,\mathcal{L}) & \xrightarrow{r_Z} & H^0(Z,\mathcal{L}|_Z) \\
 & & \uparrow & & \uparrow & & \uparrow \\
0 & \longrightarrow & V\cap H^0(S,\mathcal{I}_Z\otimes\mathcal{L}) & \longrightarrow & V & \xrightarrow{r_Z} & r_Z(V)
\end{array}$$

Dualising and projectivising, we obtain a diagram of varieties

$$\text{(2.2)}\quad \begin{array}{ccccc}
\operatorname{Span}\varphi_{\mathcal{L}}(Z) & \hookrightarrow & |\mathcal{L}|^* & \dashrightarrow\!\!\!\!\rightarrow & \mathbb{P}H^0(S,\mathcal{I}_Z\otimes\mathcal{L})^* \\
\downarrow & & \downarrow p_V & & \downarrow \\
\operatorname{Span}\psi(Z) & \hookrightarrow & \mathbb{P}V^* & \dashrightarrow\!\!\!\!\rightarrow & \mathbb{P}\left(V\cap H^0(S,\mathcal{I}_Z\otimes\mathcal{L})\right)^*.
\end{array}$$

This description of $\operatorname{Span}\psi(Z)$ will be useful in what follows.

We recall also the notion of secant defect of a zero-dimensional scheme.

DEFINITION 2.1. Let $Z \subset S$ be a subscheme of dimension zero. Let ℓ be a linear series on S and $\psi\colon S \dashrightarrow \mathbb{P}^n$ the associated map. The *defect* $\operatorname{def}\psi(Z)$ is the number

$$\operatorname{length} Z - \dim \operatorname{Span}\psi(Z) - 1.$$

(Here we take the empty set to have dimension -1.) We say that Z is *ℓ-nondefective* if $\operatorname{def}\psi(Z) = 0$, and *$\ell$-defective* (or simply *defective*) otherwise.

2.2. Scrolls. We briefly review scrolls over curves, which will be our primary objects of study. Let C be a complex projective smooth curve of genus g. Let $E \to C$ be a vector bundle of rank r and degree d, and $\pi\colon \mathbb{P}E \to C$ the associated projective bundle. For $M \in \operatorname{Pic}(C)$, we write $\mathcal{L}_M$ for the line bundle $\mathcal{O}_{\mathbb{P}E}(1) \otimes \pi^* M$ over $\mathbb{P}E$. By the projection formula and the definition of direct image, we have a natural identification $H^0(\mathbb{P}E, \mathcal{L}_M) \xrightarrow{\sim} H^0(C, E^* \otimes M)$. Fix a subspace $V \subseteq H^0(\mathbb{P}E, \mathcal{L}_M)$ of dimension $n+1$, and consider the evaluation map

$$V \otimes \mathcal{O}_C \to E^* \otimes M. \tag{2.3}$$

Dualising, projectivising and projecting to $\mathbb{P}V^*$, we obtain a map

$$\psi\colon \mathbb{P}E \dashrightarrow \mathbb{P}V^* \times C \to \mathbb{P}V^*. \tag{2.4}$$

This is naturally identified with the standard map $\mathbb{P}E \dashrightarrow |\mathcal{L}_M|^* \dashrightarrow \mathbb{P}V^*$.

2.3. Secant loci on Hilbert schemes. The first secant locus we will consider is a familiar and direct generalisation of the locus $C_e^f \subset C_e$. Let S be a smooth projective variety of dimension $r \geq 1$. For $e \geq 1$, we write $\operatorname{Hilb}^e_{\mathrm{sm}}(S)$ for the component of $\operatorname{Hilb}^e(S)$ containing smoothable subschemes; that is, containing an open subset of $\operatorname{Sym}^e S$.

DEFINITION 2.2. Let S be as above, and let $\ell = (\mathcal{L}, V)$ be a linear series of dimension n. Assume that $e \geq 1$ and $f \geq 0$ and $e \geq f$. We define

$$H_e^{e-f}(\ell) \ := \ \{Z \in \operatorname{Hilb}^e_{\mathrm{sm}}(S) : \dim(\operatorname{Span}\psi(Z)) \leq e - f - 1\}$$

or, equivalently,

$$\begin{aligned} H_e^{e-f}(\ell) \ &= \ \{Z \in \operatorname{Hilb}^e_{\mathrm{sm}}(S) : \operatorname{def}\psi(Z) \geq f\} = \\ &\qquad \{Z \in \operatorname{Hilb}^e_{\mathrm{sm}}(S) : \dim\left(V \cap H^0(S, \mathcal{L} \otimes \mathcal{I}_Z)\right) \geq n + 1 - e + f\}. \end{aligned}$$

Depending on the context, we may write $H_e^{e-f}(V)$ instead. We write H_e^{e-f} for $H_e^{e-f}\left(H^0(S, \mathcal{L})\right)$.

REMARK 2.3. If $r \geq 3$ then $\operatorname{Hilb}^e(S)$ has multiple irreducible components. It is of course not necessary to restrict to the distinguished component $\operatorname{Hilb}^e_{\mathrm{sm}}(S)$; the study of secant loci on the other components of $\operatorname{Hilb}^e(S)$ is no less interesting. For the present, we restrict to $\operatorname{Hilb}^e_{\mathrm{sm}}(S)$ simply for convenience, as the other components are less well understood at the present time.

The locus $H_e^{e-f}(\ell)$ can be constructed as a determinantal variety in a standard way. Let p and q be the projections of $\operatorname{Hilb}^e_{\mathrm{sm}}(S) \times S$ to the first and second factors respectively, and let $\mathcal{Z} \subset \mathrm{Hilb}^e_{\mathrm{sm}}(S) \times S$ be the universal subscheme. Over $\mathrm{Hilb}^e_{\mathrm{sm}}(S)$ we have the diagram

$$\begin{array}{ccccc} & & V \otimes \mathcal{O}_{\mathrm{Hilb}^e_{\mathrm{sm}}(S)} & & \\ & & \downarrow & \searrow^{\varepsilon} & \\ p_* \left(\mathcal{I}_{\mathcal{Z}} \otimes q^* \mathcal{L} \right) & \longrightarrow & p_* q^* \mathcal{L} & \longrightarrow & p_* \left(\mathcal{O}_{\mathcal{Z}} \otimes q^* \mathcal{L} \right) \end{array} \quad . \tag{2.5}$$

As $\mathcal{Z}$ is flat over $\mathrm{Hilb}^e(S)$, the sheaf $p_* \left(\mathcal{O}_{\mathcal{Z}} \otimes q^* \mathcal{L} \right)$ is locally free of rank e. Thus $H^{e-f}_e(\ell)$ is the determinantal variety

$$\{ Z \in \mathrm{Hilb}^e_{\mathrm{sm}}(S) : \mathrm{rk}(\varepsilon|_Z) \le e - f \}$$

defined by the $(e - f + 1)$st Fitting ideal of ε. The expected dimension of $H^{e-f}_e(\ell)$ is therefore $re - f(n + 1 - e + f)$.

EXAMPLE 2.4. Let $L_1, \ldots, L_r$ be line bundles on C. For $1 \le i \le r$, let $V_i \subseteq H^0(C, L_i)$ be a subspace of dimension $n_i + 1$. Set $E := \bigoplus_{i=1}^r L_i^{-1}$ and

$$V \ := \ \bigoplus_{i=1}^r V_i \ \subseteq \ H^0(C, E^*) \ = \ H^0(\mathbb{P}E, \mathcal{O}_{\mathbb{P}E}(1)),$$

and set $\ell = (\mathcal{O}_{\mathbb{P}E}(1), V)$. Now for each i, we have a natural map

$$\iota_i \colon C \ \to \ \mathbb{P}V_i^* \ \hookrightarrow \ \mathbb{P}\left(\bigoplus_{i=1}^r V_i^* \right).$$

For each i, let e_i, f_i be such that $V^{e_i - f_i}_{e_i}(L_i, V_i)$ (defined in (1.1)) has a nonempty component of dimension s_i. Write $e := \sum e_i$ and $f := \sum f_i$. Then

$$\left\{ \iota_1(D_1) \cup \cdots \cup \iota_r(D_r) : D_i \in V^{e_i - f_i}_{e_i}(L_i, V_i) \right\} \tag{2.6}$$

is a nonempty locus in $H^{e-f}_e(\ell)$ of dimension $\sum s_i$.

In § 2.5.1, Example 2.4 will motivate a conjectural dimension bound for certain secant loci.

EXAMPLE 2.5. Using the fact that $\mathbb{P}E$ is ruled, we can obtain loci in $H^{e-f}_e(\ell)$ of excess dimension. Suppose $e \ge r + 1$. If $Z \in \mathrm{Hilb}^e_{\mathrm{sm}}(S)$ is reduced and has support in a fibre $\mathbb{P}E|_x$ then $\mathrm{Span}\, \psi(Z) \subseteq \mathbb{P}E|_x = \mathbb{P}^{r-1}$. Hence any such Z imposes at most r conditions on sections of $\mathcal{O}_{\mathbb{P}E}(1)$. Thus the locus

$$\bigcup_{x \in C} \{ Z \text{ reduced of length } e : Z \subset \mathbb{P}E|_x \}$$

belongs to $H^r_e(V)$ and has dimension

$$\dim \left(\mathrm{Sym}^e \, \mathbb{P}^{r-1} \right) + \dim C \ = \ e(r-1) + 1 \ = \ re - (e-1).$$

This exceeds the expected dimension $re - (e - r)(n + 1 - r)$ of $H^r_e(V)$ if $2r \le n$.

We will return in § 2.5.1 to the study of subschemes with defect arising from points lining up inside a single fibre in this way.

REMARK 2.6. A linear series is base point free (resp., embedding) if and only if $H^0_1(\ell)$ (resp., $H^1_2(\ell)$) is empty. Most often we will consider embedded subvarieties of $\mathbb{P}^n$, but many of our results apply to arbitrary projective models.

2.4. Secant loci on Quot schemes. Let us now give another generalisation of $V_e^{e-f}(\ell)$. This time, instead of increasing $\dim(C)$, we increase $\mathrm{rk}(L)$.

DEFINITION 2.7. Let E be a vector bundle of rank r and degree d over C, and let $\pi\colon \mathbb{P}E \to C$ be the associated $\mathbb{P}^{r-1}$-bundle. Let $V \subseteq H^0(C, E^* \otimes M)$ be a subspace of dimension $n+1$. We define the generalised secant locus $Q_e^{e-f}(E, M, V)$ as

$$\left\{ [F^* \to E^*] \in \mathrm{Quot}^{0,e}(E^*) : \dim(V \cap H^0(C, F^* \otimes M)) \geq n+1-e+f \right\}.$$

This can be constructed as a determinantal variety in a similar way to $H_e^{e-f}(\ell)$. We may write simply $Q_e^{e-f}(V)$ if no confusion should arise. Also, we often abbreviate $Q_e^{e-f}\left(H^0(C, E^* \otimes M)\right)$ to Q_e^{e-f}.

REMARK 2.8. As in [**HHN21**, § 3.1], given a line bundle M and any family of vector bundles $\mathcal{V} \to B \times C$, one can define the *twisted Brill–Noether locus*

$$B^k(\mathcal{V}, M) \;=\; \{ b \in B : h^0(C, \mathcal{V}_b \otimes M) \geq k \} \;\subseteq\; B.$$

Let $\mathcal{F}^*$ be the universal subsheaf over $\mathrm{Quot}^{0,e}(E^*) \times C$. Then we have simply

$$Q_e^{e-f}\left(E, M, H^0(C, E^* \otimes M)\right) \;=\; Q_e^{e-f} \;=\; B^{h^0(C, E^* \otimes M)-e+f}(\mathcal{F}^*, M).$$

Other relations between secant loci and Brill–Noether loci will be studied in § 3.

2.5. The link between $H_e^{e-f}(V)$ and $Q_e^{e-f}(V)$. Following [**Hit20**] and [**Sta21a**], for $S = \mathbb{P}E$ we now describe the relation between the two generalised secant loci above. This will generalise the fact that if L is a line bundle, then

$$D \mapsto \left[L^{-1}(-D) \to L^{-1} \right] \text{ defines an isomorphism } C_e \xrightarrow{\sim} \mathrm{Quot}^{0,e}(L^{-1}).$$

Let $Z \subset \mathbb{P}E$ be a subscheme of length e. Taking direct images of $0 \to \mathcal{I}_Z(1) \to \mathcal{O}_{\mathbb{P}E}(1) \to \mathcal{O}_Z(1) \to 0$ on C, we have a sequence

$$0 \;\to\; \pi_* \mathcal{I}_Z(1) \;\to\; E^* \;\to\; \pi_* \mathcal{O}_Z(1) \;\to\; \cdots$$

As $\pi_* \mathcal{O}_Z(1)$ is a torsion sheaf on C, we see that $\pi_* \mathcal{I}_Z(1)$ is a full rank subsheaf of E^*. Following [**Hit20**], we denote this by E_Z^*. Dualising, for general $Z \in \mathrm{Hilb}^e_{\mathrm{sm}}(S)$, we have $\deg(E_Z) = \deg(E) + e$; for example if $\pi(Z)$ consists of e distinct points.

PROPOSITION 2.9. *Let $E \to C$ be a vector bundle.*

(a) *The association $Z \mapsto [E_Z^* \to E^*]$ defines a rational map $\alpha\colon \mathrm{Hilb}^e(S) \dashrightarrow \mathrm{Quot}^{0,e}(E^*)$.*

(b) *The restriction of α to the component $\mathrm{Hilb}^e_{\mathrm{sm}}(S)$ is surjective. Moreover, α is bijective on the open set*

$$\{ Z \in \mathrm{Hilb}^e_{\mathrm{sm}}(S) : \mathrm{Supp}\, \pi(Z) \text{ consists of } e \text{ distinct points} \}.$$

In particular, $\alpha|_{\mathrm{Hilb}^e_{\mathrm{sm}}(S)}$ is a birational equivalence.

PROOF. (a) The existence of a suitable family of length e quotients $E^* \to \pi_* \mathcal{O}_Z(1)$ parametrised by $\mathrm{Hilb}^e(\mathbb{P}E)$ is proven[1] in [**Sta21a**, Theorem 3 (i)] for $e = 2$ when $\mathbb{P}E$ is a projective bundle over a surface, and the same argument works for $e \geq 1$ when the base is a curve. In this case, over the open subset

$$\{ Z \in \mathrm{Hilb}^e(S) : E^* \to \pi_* \mathcal{O}_Z(1) \text{ is surjective} \}$$

[1] It was stated, unfortunately without an adequate justification, in [**Hit20**, Theorem 2.6 (a)].

we obtain a family of elementary transformations $E_Z^* \to E^*$ of degree $\deg(E^*) - e$. This locus is nonempty, as it contains all Z projecting to e distinct points of C, so we obtain the desired rational map α.

(b) This follows from the proof of [**Hit20**, Theorem 2.6]. □

REMARK 2.10. If $e \geq 2$ and $r \geq 2$, then α has fibres of positive dimension. For example, suppose $e = 2$ and let μ_1 and μ_2 be points of a fibre $\mathbb{P}E|_x$ spanning a $\mathbb{P}^1$. Then for any μ_1', μ_2' spanning the same $\mathbb{P}^1$, we have $E_{\{\mu_1',\mu_2'\}} = E_{\{\mu_1,\mu_2\}}$.

2.5.1. *Relatively defective subschemes.* In contrast to the situation studied in [**Sta21a**], the map α is not a morphism when $e \geq 3$. The indeterminacy locus is the set of $Z \in \mathrm{Hilb}^e(S)$ such that $E^* \to \pi_*(\mathcal{O}_Z \otimes \mathcal{O}_{\mathbb{P}E}(1))$ is not surjective; equivalently, $\deg(E_Z^*) > \deg E^* - e$. For example, if Z consists of three collinear points in a fibre of $\mathbb{P}E$, then $\deg(E_Z^*) = \deg E^* - 2$. This motivates the following, which is [**Hit20**, Definition 3.4]:

DEFINITION 2.11. Let $Z \subset \mathbb{P}E$ be a subscheme of dimension zero. Then Z is said to be *π-nondefective* if $\deg(E_Z^*) = \deg(E^*) - \mathrm{length}(Z)$, and *$\pi$-defective* otherwise.

Now we can formulate an important statement and a conjecture. We denote by $\mathrm{Hilb}^e_{\mathrm{sm}}(S)_{\mathrm{nd}}$ the locus of π-nondefective subschemes in $\mathrm{Hilb}^e_{\mathrm{sm}}(S)$, and we write

$$H_e^{e-f}(V)_{\mathrm{nd}} := H_e^{e-f}(V) \cap \mathrm{Hilb}^e_{\mathrm{sm}}(S)_{\mathrm{nd}}.$$

PROPOSITION 2.12. *Let V be an $n+1$-dimensional subspace of*

$$H^0(C, E^* \otimes M) \cong H^0(S, \mathcal{O}_{\mathbb{P}E}(1) \otimes \pi^* M).$$

Suppose $n+1 \geq e-f$. Let $Z \in \mathrm{Hilb}^e_{\mathrm{sm}}(S)$ be π-nondefective. Then $Z \in H_e^{e-f}(V)$ if and only if $[E_Z^* \to E^*]$ belongs to $Q_e^{e-f}(V)$.

PROOF. By construction of E_Z, we have

$$H^0(C, E_Z^* \otimes M) \cap V = H^0(S, \mathcal{I}_Z(1) \otimes \pi^* M) \cap V. \tag{2.7}$$

The statement follows. □

In view of the proposition, we will switch freely between $H_e^{e-f}(V)_{\mathrm{nd}}$ and $Q_e^{e-f}(V)$, depending on what is more convenient or illuminating. We refer to [**Hit20**, § 3.1] for discussion and some equivalent definitions of π-nondefectivity. Example 2.5 illustrates that π-defective subschemes can give rise to artificially large components of $H_e^{e-f}(\ell)$, which do not reflect the behaviour of $Q_e^{e-f}(V)$. For the most part, we will exclude π-defective subschemes from our consideration.

Next, we turn briefly to the question of dimension bounds for $H_e^{e-f}(V)$ and $Q_e^{e-f}(V)$. Returning to Example 2.4: If for each i the pair (C, L_i) is one of the four special types described in [**Baj15**, Theorem 4.6], and if $e_i \leq h^0(C, L_i) - 2$, then we may take s_i to be the maximal value $e_i - f_i - 1$ proven in loc. cit., and then the locus (2.6) above has dimension $e - f - r$. Note that in this situation we have

$$e = \sum_{i=1}^r e_i \leq \sum_{i=1}^r \left(h^0(C, L_i) - 2\right) = h^0(C, E) - 2r.$$

Based on this information, we make the following naive conjecture.

CONJECTURE 2.13. *Suppose C is nonhyperelliptic of genus $g \geq 9$. Let $E \to C$ be a bundle of rank r and slope at most $g-2$ such that $\mathcal{O}_{\mathbb{P}E}(1)$ is very ample. For $e \leq h^0(C,E) - 2r$, we have*

$$\dim Q_e^{e-f} \leq \dim \left(H_e^{e-f}\right)_{\mathrm{nd}} \leq e - f - r.$$

We also remark the following fact, which generalises [**Hit20**, Proposition 4.4 and Remark 4.5 (a)] in the complete case.

LEMMA 2.14. *Let $[F^* \to E^*]$ be a point of $\mathrm{Quot}^{0,e}(E^*)$. Let $Z \subset \mathbb{P}E$ be any subscheme such that $F = E_Z$. Then*

$$\mathrm{Span}\,\psi(Z) = \mathbb{P}\mathrm{Ker}\left(V^* \to (V \cap H^0(C, E_Z^* \otimes M)^*\right) = \mathbb{P}\left(V \cap H^0(C, E_Z^* \otimes M)\right)^{\perp}.$$

In particular, $\mathrm{Span}\,\psi(Z)$ depends only on F^*, and in fact does not require that Z be π-nondefective or even zero-dimensional.

PROOF. The statement is a consequence of (2.2) and (2.7). □

2.6. Distinctness of the strata. For fixed e, the determinantal varieties $H_e^{e-f}(\ell)$ and $Q_e^{e-f}(V)$ define stratifications of $\mathrm{Hilb}^e_{\mathrm{sm}}(S)$ and $\mathrm{Quot}^{0,e}(E^*)$ respectively as f varies. Generalising [**Baj15**, Lemma 4.1], [**AS15**, Lemma 2.2] and [**ACGH85**, Lemma IV.1.7], we will now show that $Q_e^{e-f}(V)$ is strictly bigger than $Q_e^{e-f-1}(V)$ whenever $n+1-e+f \geq 0$, and similarly for $H_e^{e-f}(\ell)$ under a certain condition. The result for higher dimensional varieties rests on the following statement for curves, which is a straightforward generalisation of [**AS15**, Lemma 2.2]; as the proof is essentially identical, we omit it.

LEMMA 2.15. *Let T be a smooth projective curve and $\ell = (\mathcal{L}, V)$ a very ample linear series on T of dimension $n+1$. Let e and f be integers satisfying $0 \leq f \leq e$ and $n+1-e+f \geq 0$. Then no component of $V_e^{e-f}(\ell)$ is fully contained in $V_e^{e-f-1}(\ell)$.*

PROPOSITION 2.16. *Let S be a smooth projective variety, and $\ell = (\mathcal{L}, V)$ a very ample linear series of dimension $n+1$. Let e and f be integers with $e \geq 1$ and $f \geq 0$ and $e \geq f$, and satisfying $n+1-e+f \geq 0$.*

(a) *Let $\mathcal{X}$ be a component of $H_e^{e-f}(\ell)$. Suppose $Z_0 \in \mathcal{X}$ is such that the subscheme $Z_0 \subset S$ is curvilinear. Then $\mathcal{X}$ is not fully contained in $H_e^{e-f-1}(\ell)$.*

(b) *Let $E \to C$ be a vector bundle, and $\mathcal{L}_M = \mathcal{O}_{\mathbb{P}E}(1) \otimes \pi^* M$. Let $V \subseteq H^0(C, E^* \otimes M)$ be a subspace of dimension $n+1$. Then no component of $Q_e^{e-f}(V)$ is fully contained in $Q_e^{e-f-1}(V)$.*

PROOF. (a) As curvilinearity is an open condition, clearly we may assume that Z_0 belongs to no component of $H_e^{e-f}(\ell)$ other than $\mathcal{X}$. By curvilinearity, $T_\nu Z_0$ has dimension one for all $\nu \in \mathrm{Supp}(Z_0)$. Hence, by considering the closure of a suitable affine curve contained in some affine open subset of S, we may assume that Z_0 belongs to $\mathrm{Hilb}^e(T) \cong T_e$ for a curve $T \subseteq S$ whose image in $\mathbb{P}V^*$ is nondegenerate. Then we obtain an identification of V with a subspace of $H^0(T, \mathcal{L}|_T)$. Setting $\ell_T := (\mathcal{L}|_T, V)$, we can consider the usual secant locus $V_e^{e-f}(\ell_T)$ as defined in (1.1). For all $f \geq 0$, there are natural inclusions $V_e^{e-f}(\ell_T) \hookrightarrow H_e^{e-f}(\ell)$. As Z_0 belongs only to the component $\mathcal{X}$ of $H_e^{e-f}(\ell)$, if $\mathcal{V}$ is any component of $V_e^{e-f}(\ell_T)$ containing Z_0 then we have $\mathcal{V} \subseteq \mathcal{X}$.

Now suppose that $Z_0 \in H_e^{e-f-1}(\ell)$. Then $Z_0 \in V_e^{e-f-1}(\ell_T) \cap \mathcal{V}$ also. Since $e \leq n+1+f$, by Lemma 2.15 we may deform Z_0 inside $\mathcal{V}$ to a point $Z_1 \in \mathcal{V} \backslash V_e^{e-f-1}(\ell)$. Then Z_1 has defect precisely f in $\mathbb{P}V^*$. But as $\mathcal{V} \subseteq \mathcal{X}$, we see that $\mathcal{X}$ contains the point $Z_1 \in H_e^{e-f}(\ell) \setminus H_e^{e-f-1}(\ell)$. This proves (a).

(b) Let $\mathcal{X}$ be a component of $Q_e^{e-f}(V)$. By Lemma 2.9 (a) and Proposition 2.12, we see that $\alpha^{-1}(\mathcal{X})$ is a union $\bigcup_i \widetilde{\mathcal{X}}_i$ of nonempty open subsets of components of $H_e^{e-f}(V)_{\mathrm{nd}}$. By the proof of [**Hit20**, Theorem 2.6 (a)], at least one $\widetilde{\mathcal{X}}_{i_0}$ contains a point Z_0 corresponding to a curvilinear subscheme. By part (a), therefore, $\widetilde{\mathcal{X}}_{i_0}$ is not contained in $H_e^{e-f-1}(V)$. Again by Proposition 2.12, the image of any point of $\widetilde{\mathcal{X}}_{i_0} \setminus H_e^{e-f-1}(V)$ is a point of $\mathcal{X} \setminus Q_e^{e-f-1}(V)$, and we obtain (b). □

2.7. Secant loci under general projections. We will now show that the secant loci $H_e^{e-f}(V)$ and $Q_e^{e-f}(V)$ behave predictably under projection from a general centre. The main ingredient is the theorem of Kleiman on transversality of a general translate, which will be used several times in what follows:

THEOREM 2.17 ([**Kle74**, Theorem 2 (i)]). *Let G be a connected algebraic group and Z an irreducible variety admitting a transitive action of G. Suppose that $a\colon X \to Z$ and $b\colon Y \to Z$ are maps of irreducible varieties. For any $\gamma \in G$, we write $\gamma \cdot Y$ for Y considered as a Z-variety via the map $y \mapsto \gamma \cdot b(y)$. Then there is a dense open subset $U \subseteq G$ such that $(\gamma \cdot Y) \times_Z X$ is either empty or equidimensional of dimension $\dim X + \dim Y - \dim Z$ for all $\gamma \in U$.*

Our results will follow from this theorem and linear algebra. To avoid repetition, **during this subsection, we fix integers n, m, e, and f such that**

$$0 \leq f \leq e \leq m+1 \leq n+1.$$

Let H be a variety and $\beta\colon \mathcal{E} \to V^* \otimes \mathcal{O}_H$ a map of vector bundles over H, where $\mathrm{rk}\, \mathcal{E} = e$ and $V^* = \mathbb{C}^{n+1}$. We consider the degeneracy loci

$$D_{e-f}(\beta) = \{h \in H : \mathrm{rk}\,(\beta|_h) \leq e-f\} \quad \text{and} \quad D_{e-f}^\circ(\beta) = \{h \in H : \mathrm{rk}\,(\beta|_h) = e-f\}.$$

For each f' with $0 \leq f' \leq f$, we write

$$d_m(f') := \dim H - f'(m+1-e+f).$$

Lastly, if $W \subseteq V$ is a subspace, we write p_W for the projection $V^* \to W^*$, and set $W^\perp := \mathrm{Ker}(p_W)$.

PROPOSITION 2.18. *Let H and $\beta\colon \mathcal{E} \to V^* \otimes \mathcal{O}_H$ be as above. Assume that for each $f' \in \{0, \ldots, f\}$ satisfying $f - f' \leq n - m$ we have*

$$\dim D_{e-f'}(\beta) \leq d_m(f'). \tag{2.8}$$

Then for a general subspace $W \subseteq V$ of dimension $m+1$, the degeneracy locus $D_{e-f}(p_W \circ \beta)$ is empty or of the expected dimension $\dim H - f(m+1-e+f)$ when this is nonnegative.

PROOF. Firstly, suppose $f' = f$. Clearly $D_{e-f}(\beta) \subseteq D_{e-f}(p_W \circ \beta)$. In this case, $d_m(f)$ is exactly the expected dimension of $D_{e-f}(p_W \circ \beta)$. Thus, by (2.8), any component of $D_{e-f}(p_W \circ \beta)$ of excess dimension must intersect $D_{e-f'}^\circ(\beta)$ for some $f' < f$.

Suppose that $h \in D_{e-f'}^\circ(\beta)$ for some $f' \in \{0, \ldots, f-1\}$. Then $\mathrm{rk}(p_W \circ \beta|_h) \leq e - f$ if and only if

$$\dim\left(W^\perp \cap \mathrm{Im}\, \beta|_h\right) \geq f - f'. \tag{2.9}$$

If $f - f' > n - m = \dim W^\perp$, then condition (2.9) is not satisfied for any $h \in D^\circ_{e-f'}(\beta)$, so $D_{e-f}(p_W \circ \beta) \cap D^\circ_{e-f'}(\beta)$ is empty for all W of dimension $m+1$.

On the other hand, suppose that $1 \le f - f' \le n - m$. Let $a\colon D^\circ_{e-f'}(\beta) \to \mathrm{Gr}(e - f', V^*)$ be the map $h \mapsto \operatorname{Im} \beta|_h$. Consider the Schubert cycle

$$\Sigma^{f-f'}_{W^\perp} := \{\Lambda \in \mathrm{Gr}(e-f', V^*) : \dim(\Lambda \cap W^\perp) \ge f - f'\}.$$

In the notation of [**GH78**, p. 194–196], we have $\Sigma^{f-f'}_{W^\perp} = \overline{W_{a_1,\ldots,a_{e-f'}}}$ where

$$a_1 = \cdots = a_{f-f'} = m+1-e+f \quad \text{and} \quad a_{f-f'+1} = \cdots = a_{e-f'} = 0.$$

By loc. cit.,

$$\dim \Sigma^{f-f'}_{W^\perp} = \dim \mathrm{Gr}(e-f', V^*) - (f-f')(m+1-e+f). \tag{2.10}$$

For each component Y of $D^\circ_{e-f'}(\beta)$, we consider the fibre product diagram

$$\begin{array}{ccc} Y_W & \longrightarrow & Y \\ \downarrow & & \downarrow a|_Y \\ \Sigma^{f-f'}_{W^\perp} & \hookrightarrow & \mathrm{Gr}(e-f', V^*). \end{array}$$

If $h \in Y$, then by (2.9) we see that $h \in D_{e-f}(p_W \circ \beta)$ if and only if $h \in Y_W$.

Now $\mathrm{GL}(V^*)$ acts transitively on $\mathrm{Gr}(\ell, V^*)$ for all ℓ. For $\gamma \in \mathrm{GL}(V^*)$, unwinding definitions, we see that $\gamma \cdot \Sigma^{f-f'}_{W^\perp} = \Sigma^{f-f'}_{\gamma\cdot W^\perp}$. Hence by Theorem 2.17 and by (2.10), for a general choice of W the locus Y_W is empty or equidimensional of dimension

$$\dim Y - (f-f')(m+1-e+f). \tag{2.11}$$

As by hypothesis (2.8) we have $\dim Y \le d_m(f')$, the number (2.11) is bounded above by the expected dimension of $D_{e-f}(p_W \circ \beta)$, as desired. (In fact, as the latter is determinantal, Y_W is nonempty only if $\dim Y = d_m(f')$.) □

REMARK 2.19. An easy computation shows that for $f' \ge 0$, the condition $f - f' \le n - m$ in the Proposition 2.18 is equivalent to

$$d_m(f') \ge \dim H - f'(n+1-e-f'),$$

the expected dimension of $D_{e-f'}(\beta)$.

We will apply the foregoing to the study of secant loci. The following generalises [**Hit19**, Theorem A.1]. It may be viewed as an elementary contribution to the line of inquiry in for example [**GP13**] and [**Ran15**]. For a projective model $\psi\colon S \dashrightarrow \mathbb{P}V^*$ and $W \subseteq V$, let ψ_W be the composition $S \dashrightarrow \mathbb{P}V^* \dashrightarrow \mathbb{P}W^*$.

COROLLARY 2.20. *Let S be a projective variety, and $\psi\colon S \dashrightarrow \mathbb{P}^n = \mathbb{P}V^*$ a projective model. Let $\mathcal{Z} \subset H \times S$ be a family of length e subschemes of S parametrised by a variety H. For each subspace $W \subseteq V$ and for each f', we define*

$$H^{e-f'}_e(H, W) := \{h \in H : \operatorname{def} \psi_W(\mathcal{Z}_h) \ge f'\}.$$

Assume that for each $f' \in \{0, \ldots, f\}$ such that $f - f' \le n - m$ we have

$$\dim H^{e-f'}_e(H, V) \le d_m(f').$$

Then for a general subspace $W \subseteq V$ of dimension $m+1$, the secant locus $H^{e-f}_e(H, W)$ is empty or of the expected dimension $\dim H - f(m+1-e+f)$.

PROOF. For any $W \subseteq V$, the locus $H_e^{e-f'}(H, W)$ can be constructed as a determinantal variety exactly as in § 2.3. Let $\mathcal{L}$ be the line bundle inducing the map to $\mathbb{P}^n$. Let p and q be the projections of $H \times S$ to the first and second factors respectively. We have a diagram of sheaves over $H \times S$ as follows:

$$\begin{array}{ccccc} & & W \otimes \mathcal{O}_H & & \\ & & \downarrow {\scriptstyle {}^t p_W} & \searrow & \\ & & V \otimes \mathcal{O}_H & \xrightarrow{\varepsilon} & \\ & & \downarrow & & \\ p_*\left(\mathcal{I}_{\mathcal{Z}} \otimes q^*\mathcal{L}\right) & \longrightarrow & p_*q^*\mathcal{L} & \longrightarrow & p_*\left(\mathcal{O}_{\mathcal{Z}} \otimes q^*\mathcal{L}\right) \end{array}$$

As $\mathcal{Z}$ is flat over H, the sheaf $p_*\left(\mathcal{O}_{\mathcal{Z}} \otimes q^*\mathcal{L}\right)$ is locally free of rank e. Now for $h \in H$, by (2.1) and (2.2) we have $\operatorname{Span} \psi_W(\mathcal{Z}_h) = \mathbb{P}\mathrm{Im}\,(p_W \circ {}^t\varepsilon|_h)$. Thus $H_e^{e-f'}(H, W)$ is the determinantal variety

$$\{Z \in H : \mathrm{rk}(p_W \circ {}^t\varepsilon|_Z) \leq e - f'\} \;=\; D_{e-f'}(p_W \circ {}^t\varepsilon).$$

Now by assumption, $D_{e-f'}({}^t\varepsilon)$ satisfies the hypothesis of Proposition 2.18 for $0 \leq f' \leq f$. Therefore, for W general, $H_e^{e-f}(H, W) = D_{e-f}(p_W \circ {}^t\varepsilon)$ is empty or of the expected dimension by Proposition 2.18 when this is nonnegative, and empty otherwise. □

We will now give a corresponding statement for secant loci of type $Q_e^{e-f}(V)$. This will be used in § 8.

COROLLARY 2.21. *Let C be a curve and $E \to C$ a vector bundle. Let $V \subseteq H^0(C, E^*)$ be a subspace of dimension $n+1$. Let $\mathcal{F}^\vee \to p_C^*E^*$ be a family of elementary transformations of degree* $\deg E^* - e$ *parametrised by a variety Q. For each f' and for each subspace $W \subseteq V$ of dimension $m+1$, set*

$$Q_e^{e-f'}(Q, W) \;:=\; \left\{q \in Q : \dim\left(H^0(C, \mathcal{F}_q^*) \cap W\right) \geq m + 1 - e + f'\right\}.$$

Assume that for each $f' \in \{0, \dots, f\}$ such that $f - f' \leq n - m$ we have

$$\dim Q_e^{e-f'}(Q, V) \;\leq\; d_m(f').$$

Then for a general subspace $W \subseteq V$ of dimension $m+1$, the locus $Q_e^{e-f}(Q, W)$ is empty or of the expected dimension $\dim Q - f(m+1-e+f)$ *when this is nonnegative, and empty otherwise.*

PROOF. This is similar to the proof of the previous corollary. □

This completes our overview of basic properties of scrolls and their secant loci. In the sections that follow, we will apply these to various questions on $H_e^{e-f}(\ell)$ and $Q_e^{e-f}(V)$.

3. Secant loci and Brill–Noether loci

We now further discuss the connection between secant loci and Brill–Noether loci, and generalise the notions of Abel–Jacobi map and linear series to the context of Quot schemes. This expands [**Hit20**, Remark 4.7].

It is well known that for $e \geq f \geq 0$ there is a diagram

(3.1)
$$\begin{array}{ccccc} C_e & \xrightarrow{\sim} & \mathrm{Quot}^{0,e}(\mathcal{O}_C) & \longrightarrow & \mathrm{Pic}^e(C) \\ \uparrow & & \uparrow & & \uparrow \\ C_e^f & \xrightarrow{\sim} & Q_e^{e-f}\left(\mathcal{O}_C, K_C, H^0(C, K_C)\right) & \longrightarrow & B_{1,e}^{1+f} \end{array}$$

The rows are given by $D \mapsto [\mathcal{O}_C(-D) \to \mathcal{O}_C] \mapsto \mathcal{O}_C(D)$. This composition is an Abel–Jacobi map, with fibre over L given by the linear series $|L|$.

Now let $U_C(r, d+e)$ be the moduli space of stable bundles of rank r and degree $d+e$ over C. For $k \geq 0$, we consider the higher rank Brill–Noether locus

$$B_{r,d+e}^k = \{F \in U_C(r, d+e) : h^0(F) \geq k\}.$$

Let E be a vector bundle of rank r and degree d, not necessarily semistable. Suppose that $0 \leq f \leq e$ and $Q_e^{e-f}(E, K_C, H^0(E^* \otimes K_C))$ contains a point $[F^* \to E^*]$ such that F is a stable bundle. Now by Serre duality, there is an exact sequence of vector spaces

$$0 \to H^0(C,E) \to H^0(C,F) \to H^0(C,F/E) \to \\ H^0(C, K_C \otimes E^*)^* \to H^0(C, K_C \otimes F^*)^* \to 0,$$

from which it follows that

(3.2)
$$h^0(C,F) = h^0(C,E) + f \text{ if and only if } h^0(C, F^* \otimes K_C) = h^0(C, E^* \otimes K_C) - e + f.$$

Thus, generalising (3.1), we have a Cartesian diagram

(3.3)
$$\begin{array}{ccccc} \mathrm{Hilb}^e_{\mathrm{sm}}(S)_{\mathrm{nd}} & \xrightarrow{\alpha} & \mathrm{Quot}^{0,e}(E^*) & \overset{a}{\dashrightarrow} & U_C(r, d+e)\,, \\ \uparrow & & \uparrow & & \uparrow \\ H_e^{e-f}(\mathcal{O}_{\mathbb{P}E}(1) \otimes \pi^* K_C)_{\mathrm{nd}} & \xrightarrow{\alpha} & Q_e^{e-f}\left(E, K_C, H^0(C, E^* \otimes K_C)\right) & \overset{a}{\dashrightarrow} & B_{r,d+e}^{h+f} \end{array}$$

where α is as defined in Proposition 2.9, and $h := h^0(C,E)$, and $a([F^* \to E^*]) = F$. The maps a and $a \circ \alpha$ are generalisations of the Abel–Jacobi map.

Now for any F of rank r, we write $H^0(C, \mathrm{Hom}(F^*, E^*))_{\mathrm{inj}}$ for the open subset of generically injective maps. Notice that there is a natural action of $\mathrm{Aut}\,(F^*)$ on $H^0(C, \mathrm{Hom}(F^*, E^*))$ that preserves $H^0(C, \mathrm{Hom}(F^*, E^*))_{\mathrm{inj}}$.

PROPOSITION 3.1. *Let $E \to C$ be a bundle of rank r and degree d.*

(a) *For a fixed bundle F of rank r and degree $d+e$, the locus*

$$\{[F_1^* \xrightarrow{j} E^*] \in \mathrm{Quot}^{0,e}(E^*) : F_1 \cong F \text{ as vector bundles}\} =: Q_E(F)$$

is in bijection with $H^0(C, \mathrm{Hom}(F^*, E^*))_{\mathrm{inj}}/\mathrm{Aut}\,(F^*)$.

(b) *Suppose that $\mathrm{Quot}^{0,e}(E^*)$ contains a point $[F^* \to E^*]$ such that F is a stable vector bundle. Let a be as defined in (3.3). Then*

$$a^{-1}(F) = Q_E(F) \cong \mathbb{P}H^0(C, \mathrm{Hom}(F^*, E^*))_{\mathrm{inj}}.$$

PROOF. (a) The association $j \mapsto [F^* \xrightarrow{j} E^*]$ defines a map

$$H^0(C, \mathrm{Hom}(F^*, E^*))_{\mathrm{inj}} \to Q_E(F)$$

which is clearly surjective. By definition of the Quot scheme and the universal property of kernels, two maps j' and j define the same element of $\mathrm{Quot}^{0,e}(E^*)$ if and only if $j' = j \circ \gamma$ for some $\gamma \in \mathrm{Aut}\,(F^*)$. This proves (a). Part (b) follows from (a) since $\mathrm{Aut}\,(F^*) = \mathbb{C}^*$ if F is stable. □

REMARK 3.2. If L is a line bundle of degree e, by Proposition 3.1 we obtain

$$a^{-1}(L) \cong Q_{\mathcal{O}_C}(L) \cong \mathbb{P}H^0(C, \mathrm{Hom}(L^{-1}, \mathcal{O}_C))_{\mathrm{inj}} \cong |L|.$$

Thus it makes sense to call $Q_E(F)$ a "generalised linear series of F with respect to E". If $\Omega \subseteq H^0(C, \mathrm{Hom}(F^*, E^*))$ is an $\mathrm{Aut}\,(F^*)$-invariant subspace, then the locus $\Omega_{\mathrm{inj}}/\mathrm{Aut}\,(F^*)$ generalises the notion of possibly incomplete linear subseries.

REMARK 3.3. The Riemann–Kempf singularity theorem [**ACGH85**, Chap. VI.2], gives a description of the tangent cones to the rank one Brill–Noether loci W_e^f. For $k \geq r$ and $W \in B_{r,d}^k$ a generated vector bundle, there is a generalisation of the Riemann–Kempf singularity theorem given as follows. For $\ell \geq r$, let $U^\ell \subseteq \mathrm{Gr}(\ell, H^0(C, W))$ be the open subset of subspaces which generically generate W. Assuming this is nonempty, by [**Hit20**, Theorem 5.6] the projectivised tangent cone to $B_{r,d}^k$ at W is given by

$$\mathbb{P}\mathcal{T}_W B_{r,d}^k = \overline{\bigcup_{\Lambda \in U^k} \left(\bigcap_{\Pi \in (U^r \cap \mathrm{Gr}(r,\Lambda))} \mathrm{Span}\,\psi\,(Z_\Pi \times_C \mathbb{P}W) \right)}. \tag{3.4}$$

Here $\psi\,(Z_\Pi \times_C \mathbb{P}W)$ is a certain generalisation of the canonical image of the divisor associated to a section of a line bundle, defined in [**Hit20**, § 5.2]. The point of interest for us here is that $U^r \cap \mathrm{Gr}(r, \Lambda)$ appears in (3.4) in place of the linear subseries g_d^{k-1} in the original Riemann–Kempf singularity theorem [**ACGH85**, p. 241]. Continuing from the previous remark, we will interpret $U^r \cap \mathrm{Gr}(r, \Lambda)$ as a generalised linear series.

Firstly, the map $(s_1, \ldots, s_r) \mapsto \mathrm{Span}\{s_1, \ldots, s_r\}$ defines a morphism

$$H^0(C, \mathrm{Hom}(\mathcal{O}_C{}^{\oplus r}, W))_{\mathrm{inj}} \to \mathrm{Gr}(r, H^0(C, W)),$$

whose image is exactly U^r, and which descends to an isomorphism

$$H^0(C, \mathrm{Hom}(\mathcal{O}_C{}^{\oplus r}, W))_{\mathrm{inj}}/\mathrm{GL}_r \xrightarrow{\sim} U^r. \tag{3.5}$$

By Proposition 3.1(a), this is exactly the generalised linear series $Q_{W^*}(\mathcal{O}_C{}^{\oplus r})$. Moreover, for $\Lambda \in U^k$ the isomorphism (3.5) restricts to an isomorphism

$$\Lambda_{\mathrm{inj}}^{\oplus r}/\mathrm{GL}_r \xrightarrow{\sim} U^r \cap \mathrm{Gr}(r, \Lambda).$$

Thus $U^r \cap \mathrm{Gr}(r, \Lambda)$ does indeed play the role of a linear series in the generalised Riemann–Kempf theorem.

4. Some nonemptiness results

In this section we give various sufficient or necessary conditions for nonemptiness of $H_e^{e-f}(V)_{\mathrm{nd}}$ and $Q_e^{e-f}(V)$, for some values of e and f. We begin with an easy "theoretical bound" valid for any smooth nondegenerate variety S of dimension at least 2 (it is trivial for curves).

LEMMA 4.1. *Let S be a smooth variety of dimension r and $\ell = (\mathcal{L}, V)$ a very ample linear series of dimension n on S. Set $e_0 := n - r + 2$. Then for any $e > e_0$ and $f \leq e - e_0$, the secant locus $H_e^{e-f}(\ell)$ is nonempty.*

PROOF. By Theorem 2.17, for a general $\Pi = \mathbb{P}^{n-r+1} \subset \mathbb{P}V^*$ the intersection $\Pi \cap \psi(S)$ is of dimension 1. Any reduced subscheme $Z \subset (\Pi \cap \psi(S))$ of length $e > e_0$ has linear span contained in Π, so has defect at least $e - e_0$. Thus Z defines a point of $H_e^{e-f}(\ell)$ for any $f \leq e - e_0$. □

REMARK 4.2. If $e = n$, then $e - e_0 = r - 2$. Thus for $r \geq 3$, Lemma 4.1 implies that $H_n^{n-f}(\ell)$ is nonempty for $1 \leq f \leq r - 2$. This generalises the nonemptiness statement of [**AS15**, Lemma 2.1] for $r \geq 3$.

Let us now focus on secant loci of scrolls over curves. Firstly, we have a technical statement.

LEMMA 4.3. *Let $\pi \colon \mathbb{P}E \to C$ be a projective bundle, and $\iota \colon C \to \mathbb{P}E$ a section. Let $x \in C$ be a point.*

(a) *There exists a local parameter z on C at x and a frame $\phi_1, \ldots, \phi_r$ for E^* near x such that for $k \geq 1$ the one-point subscheme $\iota(kx) \subset \mathbb{P}E$ is defined by the ideal*

$$\left(\pi^* z^k, \frac{\phi_2}{\phi_1}, \ldots, \frac{\phi_r}{\phi_1} \right). \tag{4.1}$$

(b) *Write $Z := \iota(kx)$ and set $E_Z^* = \pi_* \mathcal{I}_Z(1)$. Then $\{\pi^* z^k \cdot \phi_1, \phi_2, \ldots, \phi_r\}$ is a frame for E_Z^* near x.*

PROOF. Let $U := \operatorname{Spec} A$ be an affine neighbourhood of x in C upon which E is trivial. Let $\nu_1, \ldots, \nu_r$ be a frame such that $\iota \colon C \to \mathbb{P}E$ corresponds to the subbundle $\mathcal{O}_C \cdot \nu_1$. Let $\phi_1, \ldots, \phi_r$ be the dual frame of $E^*|_U$, and let z be a local parameter on C at x. We consider the affine variety

$$\Omega \ := \ \operatorname{Spec} A \left[\frac{\phi_2}{\phi_1}, \ldots, \frac{\phi_r}{\phi_1} \right] \ \cong \ U \times \mathbb{A}^{r-1} \ \subset \ \mathbb{P}E.$$

In these coordinates, the embedding

$$kx \ = \ \operatorname{Spec} \left(A/\mathfrak{m}_x^k \right) \ \hookrightarrow \ U \ \xrightarrow{\iota} \ \Omega$$

corresponds to the composed map of rings

$$A \left[\frac{\phi_2}{\phi_1}, \ldots, \frac{\phi_r}{\phi_1} \right] \ \xrightarrow{\sigma^*} \ A \ \to \ A/\mathfrak{m}_x^k.$$

where $\sigma^* z = z$ and $\sigma^*(\phi_i/\phi_1) = 0$ for $2 \leq i \leq r$. The last composed map has kernel exactly (4.1), and we obtain (a).

For the rest: Using (a), we see that $\mathcal{I}_{\iota(kx)}(1)$ is generated on $\pi^{-1}(U)$ by the elements $\pi^* z^k \cdot \phi_1, \phi_2, \ldots, \phi_r$. Statement (b) follows. □

COROLLARY 4.4. *Let $\mathbb{P}E$ and ι be as above. For any effective divisor D of degree e on C, the subscheme $\iota(D) \subset \mathbb{P}E$ defines a point of* $\operatorname{Hilb}^e(\mathbb{P}E)$ *which is π-nondefective and smoothable.*

PROOF. Firstly, as $\iota(D)$ is the image of a curvilinear scheme by an embedding, $\iota(D)$ is smoothable. For the rest: As π-nondefectivity is local on C, it suffices to consider the case $D = ex$. In this case, by Lemma 4.3 (b) we see that

$$E_Z^*|_U \ = \ \pi_* \mathcal{I}_Z(1)|_U \ \cong \ \mathcal{O}_U(-ex) \oplus \mathcal{O}_U^{\oplus(r-1)}.$$

Therefore, E^*/E_Z^* has length e; whence $\iota(D)$ is π-nondefective by Definition 2.11. □

Suppose now that $S = \mathbb{P}E$ is a scroll of dimension $r \geq 2$ over C. We will use existing results on secant loci of curves to prove nonemptiness of secant loci of S whose expected dimension is large.

PROPOSITION 4.5. *Let E be a bundle of rank $r \geq 2$ and degree d, and write $S := \mathbb{P}E$. Suppose that $V \subseteq H^0(S, \mathcal{L}_M)$ is a linear subspace of dimension $n+1$. Suppose that*

$$0 < f < e \quad \text{and} \quad re - f(n+1-e+f) \geq (r-1)e. \tag{4.2}$$

Then $H_e^{e-f}(V)_{\mathrm{nd}}$ and $Q_e^{e-f}(V)$ are nonempty.

PROOF. Write $d' := \deg(E \otimes M^{-1}) = d - r \cdot \deg(M)$. By [**PR03**, Proposition 6.1], for all $k \gg 0$, there exists a short exact sequence $0 \to L \to E \otimes M^{-1} \to F \to 0$ where F is a stable vector bundle of rank $r-1$ and degree k. We choose

$$k \geq \max\{0, 2g + d'\}. \tag{4.3}$$

Since $k \geq 0$, we have $h^0(C, F^*) = 0$ by stability. Taking global sections of $0 \to F^* \to E^* \otimes M \to L^{-1} \to 0$ and dualising, we obtain

$$\cdots \to H^0(L^{-1})^* \to H^0(E^* \otimes M)^* \to 0. \tag{4.4}$$

Projectivising and identifying C with $\mathbb{P}L$, we obtain a commutative diagram

$$\begin{array}{ccc} |L^{-1}|^* & \dashrightarrow\!\!\!\!\rightarrow & \mathbb{P}V^* \\ \uparrow{\scriptstyle \varphi_{L^{-1}}} & & \uparrow \\ C & \xhookrightarrow{\iota} & \mathbb{P}E. \end{array}$$

Now $\deg L^{-1} = k - d'$, which by (4.3) is at least $2g$. Hence $h^1(C, L^{-1}) = 0$ and $\varphi_{L^{-1}}$ is an embedding. As C is nondegenerate in $|L^{-1}|^*$, we obtain an identification of $\mathbb{P}V$ with a subseries $g^n_{k-d'}$ of $|L^{-1}|$.

We recall next from [**ACGH85**, p. 355] that the *relative deficiency* of this $g^n_{k-d'}$ is defined to be $\deg L^{-1} - g - n$. As $h^1(C, L^{-1}) = 0$, we have

$$\deg L^{-1} - g - n = \chi(C, L^{-1}) - (n+1) = h^0(C, L^{-1}) - \dim V,$$

which is nonnegative since we have identified $\mathbb{P}V$ with a subseries of $|L^{-1}|$ in (4.4). Furthermore, by (4.2) we have $e - f(n+1-e+f) \geq 0$. Hence by loc. cit. the usual secant locus $V_e^{e-f}(g^n_{d+k}) \subseteq C_e$ is nonempty. Thus there exists a divisor D on C of degree e such that $\operatorname{Span} \psi(\iota(D))$ has dimension $e - f - 1$. Then $\iota(D)$ is a length e subscheme of $\mathbb{P}E$ spanning at most a $\mathbb{P}^{e-f-1}$ in $|\mathcal{L}_M|^*$; and $\iota(D)$ is π-nondefective and smoothable by Corollary 4.4. Hence $H_e^{e-f}(V)_{\mathrm{nd}}$ is nonempty, and so therefore is $Q_e^{e-f}(V)$ by Proposition 2.12. □

REMARK 4.6. If $f = 1$, then (4.2) reduces to $e \geq 2$ and $2e - 1 \geq n + 1$. If $e = n$, then this is satisfied for $n \geq 2$. Thus for $r \geq 2$, the statement of Proposition 4.5 again generalises the nonemptiness part of [**AS15**, Lemma 2.1].

We conclude this section with a necessary condition for nonemptiness.

PROPOSITION 4.7. *Let E be a bundle of rank r and degree d. Set $\widehat{h} := h^0(C, K_C \otimes E)$. If Q_e^{e-f} is nonempty, then the twisted Brill–Noether locus*

$$B_{1,e}^{\widehat{h}+f}(K_C \otimes E) = \{L \in \mathrm{Pic}^e(C) : h^0(C, K_C L \otimes E) \geq f\}$$

contains an effective line bundle.

PROOF. Suppose $[F^* \to E^*]$ belongs to Q_e^{e-f}. Using (3.2), we obtain

$$h^0(C, K_C \otimes F) \geq h^0(K_C \otimes E) + f.$$

Now $F \subseteq E(D)$ for some $D \in C_e$. Then $\mathcal{O}_C(D)$ belongs to $B_{1,e}^{\widehat{h}+f}(K_C \otimes E)$. $\square$

This observation will be further developed in § 6 and thereafter.

5. Tangent spaces to secant loci

In this section, we will describe the Zariski tangent spaces of $Q_e^{e-f}(V)$, generalising [**ACGH85**, Lemma IV.1.5] and [**Cop95**, Theorem 0.3]. In preparation, we give a construction of the first-order infinitesimal deformations of an element of any Quot scheme, which will be convenient for calculations.

5.1. Tangent spaces to Quot schemes. Let $(X, \mathcal{O}_X(1))$ be a polarised variety and $\mathcal{V}$ a coherent sheaf over X. The scheme $\mathrm{Quot}^\Phi(\mathcal{V})$ parametrises equivalence classes of coherent quotients $\mathcal{V} \xrightarrow{q} \mathcal{Q}$ where $\mathcal{Q}$ has Hilbert polynomial Φ. The first-order infinitesimal deformations of a given $[\mathcal{V} \xrightarrow{q} \mathcal{Q}]$ are parametrised by the Zariski tangent space $T_q\mathrm{Quot}^\Phi(\mathcal{V})$. This is canonically isomorphic to $H^0(X, \mathrm{Hom}(\mathcal{W}, \mathcal{Q}))$, where $\mathcal{W} = \mathrm{Ker}(q)$; see for example [**HL10**, Chapter 2]. These deformations naturally induce deformations of the sheaf $\mathcal{W}$, which we will now construct. (A similar approach to deformations is given in [**Eis95**, Exercise 6.12 c].)

For each $u \in H^0(X, \mathrm{Hom}(\mathcal{W}, \mathcal{Q}))$, let $\mathbb{W}_u$ be the sheaf over $\mathrm{Spec}\,\mathbb{C}[\varepsilon] \times X$ given over each open set $U \subseteq X$ by

$$\mathbb{W}_u(U) \;=\; \{\varepsilon t + s \in H^0(U, \varepsilon \cdot \mathcal{V} \oplus \mathcal{W}) : q(t) = u(s)\} \tag{5.1}$$

(Here we use the fact that Spec $\mathbb{C}[\varepsilon] \times X$ has the same ambient topological space as X.) The sheaf $\mathbb{W}_u$ is naturally contained in $\varepsilon \cdot \mathcal{V} \oplus \mathcal{V}$, and $\mathbb{W}_u = \mathbb{W}$ is easily checked to fit into an exact diagram of the form

$$\begin{array}{ccccccccc}
 & & 0 & & 0 & & 0 & & \\
 & & \downarrow & & \downarrow & & \downarrow & & \\
0 & \longrightarrow & \varepsilon \cdot \mathcal{W} & \longrightarrow & \mathbb{W} & \longrightarrow & \mathcal{W} & \longrightarrow & 0 \\
 & & \downarrow{\scriptstyle j} & & \downarrow & & \downarrow{\scriptstyle j} & & \\
0 & \longrightarrow & \varepsilon \cdot \mathcal{V} & \longrightarrow & \varepsilon \cdot \mathcal{V} \oplus \mathcal{V} & \longrightarrow & \mathcal{V} & \longrightarrow & 0 \\
 & & \downarrow{\scriptstyle q} & & \downarrow{\scriptstyle q'} & & \downarrow{\scriptstyle q} & & \\
0 & \longrightarrow & \varepsilon \cdot \mathcal{Q} & \longrightarrow & \mathbb{Q} & \longrightarrow & \mathcal{Q} & \longrightarrow & 0 \\
 & & \downarrow & & \downarrow & & \downarrow & & \\
 & & 0 & & 0 & & 0. & &
\end{array} \tag{5.2}$$

Thus $\mathbb{W}_u$ defines a first-order infinitesimal deformation of $\mathcal{V} \xrightarrow{q} \mathcal{Q}$.

LEMMA 5.1. *The assignment*

$$u \;\mapsto\; \left[\varepsilon \cdot \mathcal{V} \oplus \mathcal{V} \to \frac{\varepsilon \cdot \mathcal{V} \oplus \mathcal{V}}{\mathbb{W}_u} \right]$$

defines a bijection between $H^0(C, \mathrm{Hom}(\mathcal{W}, \mathcal{Q}))$ and the set of isomorphism classes of first-order infinitesimal deformations of $[\mathcal{V} \xrightarrow{q} \mathcal{Q}]$.

PROOF. We indicate only the main steps. Recall firstly that the *graph* of a map of $\mathcal{O}_X$-modules $u\colon \mathcal{W} \to \mathcal{Q}$ is the $\mathcal{O}_X$-submodule $\Gamma_u \subseteq \mathcal{Q} \oplus \mathcal{W}$ given over open $U \subseteq X$ by

$$\Gamma_u(U) = \{(u(s), s) : s \in \mathcal{W}(U)\}.$$

By construction, $\mathbb{W}_u$ is exactly the inverse image of $\Gamma_{\varepsilon u}$ by the projection $\varepsilon \cdot \mathcal{V} \oplus \mathcal{W} \to \varepsilon \cdot \mathcal{Q} \oplus \mathcal{W}$.

Now by for example [**HL10**, § 2.2], each first-order infinitesimal deformation of $\mathcal{V} \xrightarrow{q} \mathcal{Q}$ corresponds to an exact diagram of the form (5.2). As mentioned above, $\mathbb{W} = \mathbb{W}_u$ fits into such a diagram. Conversely, given a diagram of the form (5.2), write $\widetilde{\mathbb{W}} := (q \oplus \mathrm{Id}_{\mathcal{V}})(\mathbb{W})$. Then

(i) $\widetilde{\mathbb{W}} \cap (\varepsilon \cdot \mathcal{Q} \oplus 0) = 0$ since $\mathbb{W} \cap (\varepsilon \cdot \mathcal{V} \oplus 0) = \varepsilon \cdot \mathcal{W}$; and

(ii) the image of $\widetilde{\mathbb{W}}$ in $\mathcal{V}$ is $\mathcal{W}$ since $\mathbb{W} \to \mathcal{W}$ is surjective.

Linear algebra arguments then show that $\widetilde{\mathbb{W}} = \Gamma_{\varepsilon u}$ for a uniquely determined $u\colon \mathcal{W} \to \mathcal{Q}$, and that that $\mathbb{W} = \mathbb{W}_u$. □

5.2. Zariski tangent spaces of $Q_e^{e-f}(V)$. We now consider a smooth curve C and a vector bundle $E \to C$ of rank r and degree d. Let M be a line bundle and $V \subseteq H^0(C, E^* \otimes M)$ a subspace of dimension $n+1$. We will study the tangent spaces to $Q_e^{e-f}(E, M, V) =: Q_e^{e-f}(V)$.

For the remainder of the paper, for any sheaf $\mathcal{W}$ over C we abbreviate $H^i(C, \mathcal{W})$ and $h^i(C, \mathcal{W})$ to $H^i(\mathcal{W})$ and $h^i(\mathcal{W})$ respectively.

Following the treatment in [**ACGH85**, Chap. IV] of the tangent spaces of C_e^f, which is exactly $Q_e^{e-f}(K_C, \mathcal{O}_C, H^0(K_C))$, we fix the following. Consider an elementary transformation $0 \to E \to F \to \mathcal{T} \to 0$ where τ has length e. Taking $\mathrm{Hom}(-, M)$, we obtain a sequence

$$0 \to F^* \otimes M \to E^* \otimes M \xrightarrow{q} \tau \to 0 \tag{5.3}$$

where $\tau := \mathcal{E}xt^1_{\mathcal{O}_C}(\mathcal{T}, M)$ is noncanonically isomorphic to $\mathcal{T}$. By the previous section, and since M is invertible,

$$T_{F^*}\mathrm{Quot}^{0,e}(E^*) \cong H^0(C, \mathrm{Hom}(F^*, \tau \otimes M^{-1})) \cong H^0(C, \mathrm{Hom}(F^* \otimes M, \tau)).$$

Also, recall that there is a natural map

$$c\colon H^0(\mathrm{Hom}(F^* \otimes M, \tau)) \to \mathrm{Hom}\left(H^0(F^* \otimes M), H^0(\tau)\right).$$

PROPOSITION 5.2. *Let E, M and V be as above. Suppose that $[F^* \to E^*]$ is a point of $Q_e^{e-f}(V) \setminus Q_e^{e-f-1}(V)$. Then*

$$T_{F^*}Q_e^{e-f}(V) = \{u \in H^0(\mathrm{Hom}(F^* \otimes M, \tau)) : c(u)(V \cap H^0(F^* \otimes M)) \subseteq q(V)\}.$$

Moreover, this is precisely the kernel of the map

$$H^0(\mathrm{Hom}(F^* \otimes M, \tau)) \to \mathrm{Hom}\left(V \cap H^0(F^* \otimes M), \frac{H^0(\tau)}{q(V)}\right)$$

given by composing c, restriction to $V \cap H^0(F^* \otimes M)$ and quotient by $q(V)$.

PROOF. A tangent vector $u \in H^0(\mathrm{Hom}(F^* \otimes M, \tau))$ belongs to $T_{F^*}Q_e^{e-f}(V)$ if and only if every $s \in V \cap H^0(F^* \otimes M)$ lifts to the deformation $\mathbb{W}_u$ given as in (5.1) by

$$\mathbb{W}_u(U) = \{\varepsilon t + s \in \varepsilon \cdot (E^* \otimes M)(U) \oplus (F^* \otimes M)(U) : q(t) = u(s)\}$$

over each open subset $U \subseteq C$. This is equivalent to saying that for each element $s \in V \cap H^0(F^* \otimes M)$, there exists $t \in V$ such that $q(t) = c(u)(s)$; in other words, that $c(u)\left(V \cap H^0(\widehat{F})\right) \subseteq q(V)$. This proves the first statement; and the second is then clear. $\square$

5.3. The complete case. During this subsection, we fix $M = K_C$ and assume that $V = H^0(E^* \otimes K_C)$. In this case, we can say more about $T_{F^*}Q_e^{e-f}$.

Firstly, we recall some familiar facts. To ease notation, for any locally free sheaf W over C we will denote $K_C \otimes W^*$ by $\widehat{W}$. Fix an elementary transformation $0 \to E \to F \to \mathcal{T} \to 0$. This canonically determines a sequence $0 \to \widehat{F} \to \widehat{E} \xrightarrow{q} \tau \to 0$ and an element of $\mathrm{Quot}^{0,e}(E^*)$ as above. Let $\partial \colon H^0(\tau) \to H^1(\widehat{F})$ denote the associated coboundary map. Then we obtain a commutative diagram with exact rows

(5.4)

$$\begin{array}{ccccccc}
H^0(\mathrm{Hom}(\widehat{F},\tau)) & \xrightarrow{\ \delta\ } & H^1(\mathrm{End}\,\widehat{F}) & \longrightarrow & H^1(\mathrm{Hom}(\widehat{F},\widehat{E})) & \longrightarrow & 0 \\
\downarrow{\scriptstyle c} & & \downarrow{\scriptstyle \cup} & & \downarrow & & \\
\mathrm{Hom}\left(H^0(\widehat{F}), H^0(\tau)\right) & \xrightarrow{\ \partial_*\ } & \mathrm{Hom}\left(H^0(\widehat{F}), H^1(\widehat{F})\right) & \longrightarrow & \mathrm{Hom}\left(H^0(\widehat{F}), H^1(\widehat{E})\right) & \longrightarrow & 0
\end{array}$$

where $\cup$ is the cup product map. We recall the well known fact that via Serre duality, $\cup$ is dual to the Petri map

$$\mu \colon H^0(\widehat{F}) \otimes H^0(F) \to H^0(\widehat{F} \otimes F) = H^0(K_C \otimes \mathrm{End}\,F).$$

PROPOSITION 5.3. *Let $0 \to E \to F \to \mathcal{T} \to 0$ be as above. Suppose that $[F^* \to E^*]$ belongs to $Q_e^{e-f} \setminus Q_e^{e-f-1}$.*

(a) *We have $T_{F^*}Q_e^{e-f} = \mathrm{Ker}(\cup \circ \delta) = \mathrm{Im}\left({}^t\delta \circ \mu\right)^\perp$.*
(b) *The locus Q_e^{e-f} is smooth and of the expected dimension at $[F^* \to E^*]$ if and only if $\mu^{-1}\left(H^0(\widehat{F} \otimes E)\right) = H^0(\widehat{F}) \otimes H^0(E)$.*

PROOF. (a) Since $V = H^0(\widehat{E})$, we have $\mathrm{Im}(c(u)) \subseteq q(V)$ if and only if $\partial \circ c(u)$ is zero in $\mathrm{Hom}\left(H^0(\widehat{F}), H^1(\widehat{F})\right)$. Hence by Proposition 5.2 and commutativity of (5.4), it follows that

$$T_{F^*}Q_e^{e-f} = \mathrm{Ker}(\partial_* \circ c) = \mathrm{Ker}(\cup \circ \delta).$$

As $\mathrm{Ker}(\varphi) = \mathrm{Im}({}^t\varphi)^\perp$ for any vector space map φ, and since ${}^t\cup = \mu$, this in turn coincides with $\mathrm{Im}\left({}^t\delta \circ \mu\right)^\perp$.

(b) By part (a), the codimension of $T_{F^*}Q_e^{e-f}$ in $T_{F^*}\mathrm{Quot}^{0,e}(E^*)$ is exactly $\dim \mathrm{Im}({}^t\delta \circ \mu)$. Let us compute the latter. Dualising (5.4), we obtain

(5.5)

$$\begin{array}{ccccccc}
0 & \longrightarrow & H^0(\widehat{F}) \otimes H^0(E) & \longrightarrow & H^0(\widehat{F}) \otimes H^0(F) & \longrightarrow & H^0(\widehat{F}) \otimes H^0(\tau)^* \\
 & & \downarrow & & \downarrow{\scriptstyle \mu} & & \downarrow \\
0 & \longrightarrow & H^0(\widehat{F} \otimes E) & \longrightarrow & H^0(\widehat{F} \otimes F) & \xrightarrow{\ {}^t\delta\ } & H^0(\mathrm{Hom}(\widehat{F},\widehat{\tau}))^*.
\end{array}$$

Thus $\mathrm{Ker}({}^t\delta \circ \mu) = \mu^{-1}\left(H^0(\widehat{F} \otimes E)\right)$, and so $\mathrm{Im}({}^t\delta \circ \mu)$ has dimension

$$h^0(\widehat{F}) \cdot h^0(F) - \dim \mu^{-1}\left(H^0(\widehat{F} \otimes E)\right) = h^0(\widehat{F}) \cdot \left(h^0(F) - h^0(E)\right) - \dim \frac{\mu^{-1}\left(H^0(\widehat{F} \otimes E)\right)}{H^0(\widehat{F}) \otimes H^0(E)}.$$

By hypothesis, $h^0(\widehat{F}) = h^0(\widehat{E}) - e + f$. Hence by (3.2) we have $h^0(F) - h^0(E) = f$. Thus $\mathrm{Im}({}^t\delta \circ \mu)$ has dimension

$$\left(h^0(\widehat{E}) - e + f\right) \cdot f - \dim \frac{\mu^{-1}\left(H^0(\widehat{F} \otimes E)\right)}{H^0(\widehat{F}) \otimes H^0(E)}.$$

As the expected codimension of Q_e^{e-f} in $\mathrm{Quot}^{0,e}(E^*)$ is $(h^0(\widehat{E}) - e + f) \cdot f$ (for in this case $n+1 = h^0(\widehat{E})$), we see that $T_{F^*} Q_e^{e-f}$ has the expected codimension if and only if $\mu^{-1}\left(H^0(\widehat{F} \otimes E)\right) = H^0(\widehat{F}) \otimes H^0(E)$. □

5.4. Some examples. Continuing to use the notation of the previous subsection, let us return to the study of the generalised Abel–Jacobi map

$$a \colon \mathrm{Quot}^{0,e}(E^*) \dashrightarrow U_C(r, d+e)$$

given by $[F^* \to E^*] \mapsto F$. As noted in § 3, we have

$$a^{-1}\left(B_{r,d+e}^{h^0(E)+f}\right) = Q_e^{e-f}(E, K_C, H^0(E^* \otimes K_C)).$$

It is well known that $B_{r,d+e}^{h^0(E)+f} \backslash B_{r,d+e}^{h^0(E)+f+1}$ is smooth and of the expected dimension at a stable bundle F if and only if μ is injective. In the special case $E = \mathcal{O}_C$ and $F = \mathcal{O}_C(D)$ with $D \in C_e^f \backslash C_e^{f+1}$, by [**ACGH85**, Lemma IV.1.6], injectiveness of μ is also equivalent to smoothness of C_e^f at D. One can check that in this situation, the condition in Proposition 5.3 (b) is equivalent to injectiveness of μ, so there is no contradiction. However, we will now give examples showing that in general, smoothness of $B_{r,d+e}^{h^0(E)+f}$ at F is neither necessary nor sufficient for smoothness of $Q_e^{e-f}(E, K_C, H^0(E^* \otimes K_C))$ at F^*.

EXAMPLE 5.4. Let D be an effective divisor of degree $e \leq g-1$, and $E := \mathcal{O}_C(-D)$. Set $F = E(D) = \mathcal{O}_C$, so that $\widehat{F} = K_C$. Here $\deg(E) = -e$, and $f = 1$, and

$$F \in Q_e^{e-1}(E, K_C, H^0(E^* \otimes K_C)) = Q_e^{e-1}(\mathcal{O}_C(-D), K_C, H^0(K_C(D))).$$

Trivially, $B_{1,d+e}^1 = B_{1,0}^1 = \{\mathcal{O}_C\}$ is smooth and of the expected dimension at F.

Now $H^0(\widehat{F}) \otimes H^0(F) = H^0(K_C) \otimes H^0(\mathcal{O}_C)$. Clearly μ is an isomorphism, and

$$\mu^{-1}\left(H^0(\widehat{F} \otimes E)\right) = \mu^{-1}\left(H^0(K_C(-D))\right) = H^0(\mathcal{O}_C) \otimes H^0(K_C(-D)),$$

which is nonzero since $\deg(D) \leq g-1$. On the other hand, $H^0(\mathcal{O}_C(-D)) \otimes H^0(K_C)$ is zero. By Proposition 5.3 (b), the secant locus Q_e^{e-1} is not smooth at F. (One can check that the expected dimension of Q_e^{e-1} is $e - g < 0$.)

The next example shows that Q_e^{e-f} can be smooth at F^* even though $B^k_{r,d}$ is not smooth at F. The idea is to produce an E and F such that $\operatorname{Ker}\mu$ is nonzero but contained in $H^0(\widehat{F})\otimes H^0(E)$ so that the condition $\mu^{-1}H^0(\widehat{F}\otimes E) = H^0(\widehat{F})\otimes H^0(E)$ can still obtain. Perhaps somewhat artificially, we will start with the bundle F and choose an appropriate E. The bundle F will have rank two and canonical determinant; the Brill–Noether loci of such bundles have been studied in [**TiB04**], [**TiB08**] and elsewhere, and the Petri map is known to have nonzero kernel in many cases.

EXAMPLE 5.5. Suppose $g \geq 6$, so that $B^2_{1,g-2}$ is nonempty. Let L_1 be a line bundle of degree $g-2$ with $h^0(L_1) = 2$. Let x_1 and x_2 be general points of C, and set $L_2 := K_C L_1^{-1}(-x_1 - x_2)$. Then $h^0(L_2) = 1$. Let

$$0 \to L_1 \oplus L_2 \to F \to \mathcal{O}_{x_1} \oplus \mathcal{O}_{x_2} \to 0 \tag{5.6}$$

be an elementary transformation such that neither L_1 nor L_2 is a quotient of F. By [**Mer99**, Théorème A.5], we may assume that F is stable.

We now claim that $H^0(F) = H^0(L_1)\oplus H^0(L_2)$. It will suffice to show that the composed map

$$H^0(\mathcal{O}_{x_1}\oplus\mathcal{O}_{x_2}) \to H^1(L_1)\oplus H^1(L_2) \to H^1(L_1) \cong H^0(K_C L_1^{-1})^* \tag{5.7}$$

is injective. It is not hard to see that the image of this composed map is exactly the cone over the secant spanned by the points x_1 and x_2 on the plane curve $\varphi_{K_C L_1^{-1}}(C)$. As already used, by generality of x_1 and x_2 we have $h^0(K_C L_1^{-1}(-x_1 - x_2))) = 1$, so this cone has dimension 2. Hence (5.7) is an isomorphism, and in particular injective as desired.

Thus F defines a point of $B^3_{2,2g-2}$. Moreover, since $\det(F) \cong K_C$, we have $F \cong K_C \otimes F^*$, and the Petri map can be identified with the natural map

$$H^0(F)\otimes H^0(F) \to H^0(F\otimes F).$$

Abusing notation, we also denote this map by μ. Let s_1, s_2 be a basis for $H^0(L_1)$. Then $s_1 \wedge s_2$ is a nonzero element of $\operatorname{Ker}(\mu)$, so $B^3_{2,2g-2}$ is *not* smooth at F.

We will now construct a bundle E such that $[F^* \to E^*]$ is a smooth point of a certain $Q_e^{e-f}(E^*, K_C, H^0(E^*\otimes K_C))$. Let E be an elementary transformation $0 \to E \to F \to \mathcal{O}_D \to 0$ where D is a general effective divisor of degree $e \geq 2$, and

$$\operatorname{Im}(E|_p \to F|_p) = L_1|_p \text{ for each } p \in D. \tag{5.8}$$

As $L_1 \to F \to \mathcal{O}_D$ is zero, $L_1 \subset E$. (Note that E is therefore not stable, as $\deg(L_1) = g-2 \geq \frac{2g-2-e}{2}$, but this does not affect the argument.) In particular $H^0(E)$ contains $H^0(L_1)$. Furthermore, let s_3 be a generator of $H^0(L_2)$. By generality of D, we may assume that $s_3(p) \notin L_1|_p$ for all $p \in D$. Thus s_3 is not a section of E, whence $h^0(E) = h^0(L_1) = 2$. By Riemann–Roch, $h^0(E^*\otimes K_C) = e+2$. As

$$h^0(F^*\otimes K_C) = h^0(F) = 3 = h^0(E^*\otimes K_C) - e + 1,$$

we have $[F^* \to E^*] \in Q_e^{e-1}\left(E, K_C, H^0(C, E^*\otimes K_C)\right)$. We will show that it is a smooth point.

By Proposition 5.3 (b), noting that $F \cong F^*\otimes K_C = \widehat{F}$, we must show that

$$\mu^{-1}\left(H^0(F\otimes E)\right) = H^0(F)\otimes H^0(E). \tag{5.9}$$

Let $s := \sum_{i,j=1}^3 \alpha_{ij} s_i \otimes s_j$ be an element of $H^0(F)\otimes H^0(F)$, and assume that $(\mu(s))(p)$ belongs to the image of $F\otimes E|_p$ for all $p \in C$. This is a nonempty condition only at $p \in D$, since $E|_p = F|_p$ for $p \notin D$. For $p \in D$, using (5.6), there is a canonical splitting of $(F \otimes F)|_p$ as

$$(L_1|_p \otimes L_1|_p) \oplus (L_2|_p \otimes L_1|_p) \oplus (L_1|_p \otimes L_2|_p) \oplus (L_2|_p \otimes L_2|_p).$$

We write out $(\mu(s))(p)$ in terms of this splitting:

$$\begin{aligned}(5.10)\quad &(\alpha_{11}s_1(p) \otimes s_1(p) + \alpha_{21}s_2(p) \otimes s_1(p) + \alpha_{12}s_1(p) \otimes s_2(p) + \alpha_{22}s_2(p) \otimes s_2(p),\\ &\alpha_{31}s_3(p) \otimes s_1(p) + \alpha_{32}s_3(p) \otimes s_2(p), \alpha_{13}s_1(p) \otimes s_3(p) + \alpha_{23}s_2(p) \otimes s_3(p),\\ &\alpha_{33}s_3(p) \otimes s_3(p)).\end{aligned}$$

Now in view of (5.8), by hypothesis $(\mu(s))(p)$ belongs to

$$\mathrm{Im}\,(E|_p \otimes F|_p \to F|_p \otimes F|_p) = (L_1|_p \oplus L_2|_p) \otimes L_1|_p$$

for $p \in D$. Thus the third and fourth components of (5.10) must be zero for all $p \in D$. Firstly, since $s_3(p)$ can be assumed to be nonzero for all $p \in D$ by generality of D, we obtain $\alpha_{33} = 0$. Furthermore, considering the third component of (5.10), we obtain

$$(\alpha_{13}s_1(p) + \alpha_{23}s_2(p)) \otimes s_3(p) = 0 \text{ for all } p \in D.$$

Using the nonvanishing of $s_3(p)$ above, we obtain $\alpha_{13}s_1(p) + \alpha_{23}s_2(p) = 0$ for all $p \in D$. Again using generality of D, we may assume that no two points of D are identified under the map $C \to |L_1|^* = \mathbb{P}^1$. Thus this condition places e independent linear conditions on $(\alpha_{13}, \alpha_{23})$. Since $e \geq 2$, the only solution is $\alpha_{13} = \alpha_{23} = 0$. It follows that $s \in H^0(F) \otimes H^0(E)$. Thus (5.9) is satisfied and Q_e^{e-1} is smooth at $[F^* \to E^*]$.

REMARK 5.6. In both of these examples, the singularity arises from a determinantal variety which does not have the expected dimension[2]. Moreover, the bundle E in Example 5.5 is not semistable. It would be interesting to have examples of other kinds, and to investigate more thoroughly the relationships between smoothness of $B^k_{r,d}$, smoothness of Q_e^{e-f} and semistability of E.

6. Parameterisation of Q_e^{e-1}

As before, let E be a vector bundle of rank r and degree d over C. In [**Hit19**, § 4.3], a parameterisation is given for the inflectional loci of a map $\mathbb{P}E \dashrightarrow |\mathcal{L}_M|^*$. We will now construct a similar parameterisation of $Q_e^{e-1}(E, M, H^0(E^* \otimes M))$. In the sections that follow, we will give some applications similar to those in [**Hit19**, § 5–6] for this parameterisation.

For $M \in \mathrm{Pic}^0(C)$, consider the Quot scheme

$$\mathrm{Quot}^{r-1, d+2r(g-1)+e}(K_C M^{-1} \otimes E)$$

parametrising equivalence classes of quotients $[K_C M^{-1} \otimes E \to Q]$ where Q is coherent of rank $r-1$ and degree $\deg(K_C M^{-1} \otimes E) + e$; equivalently, invertible subsheaves of $K_C M^{-1} \otimes E$ of degree $-e$. As we wish to focus on subsheaves, and to ease notation, we denote this Quot scheme by $Q_{1,-e}(K_C M^{-1} \otimes E)$.

[2]Note however that the fixed determinant Brill–Noether locus B^3_{2,K_C} may well have the expected dimension as a *symmetric* determinantal variety. See for example [**TiB08**].

Let $a\colon C_e \to \mathrm{Pic}^{-e}(C)$ be the Abel–Jacobi-type map $D \mapsto \mathcal{O}_C(-D)$, and let $b\colon Q_{1,-e}(K_C M^{-1} \otimes E) \to \mathrm{Pic}^{-e}(C)$ be the forgetful map $\left[L \xrightarrow{\sigma} K_C M^{-1} \otimes E\right] \mapsto L$. Consider the fibre product

$$
\begin{array}{ccc}
S_M^e & \longrightarrow & Q_{1,-e}(K_C M^{-1} \otimes E) \\
\big\downarrow & & \big\downarrow{\scriptstyle b} \\
C_e & \xrightarrow{\quad a \quad} & \mathrm{Pic}^{-e}(C).
\end{array}
\tag{6.1}
$$

Set-theoretically, S_M^e parametrises pairs (σ, D) where D is an effective divisor of degree e on C and $\sigma\colon \mathcal{O}_C(-D) \to K_C M^{-1} \otimes E$ is a sheaf injection. We will now construct a family of elementary transformations of E^* parametrised by S_M^e.

Consider the following diagram where all maps are projections.

$$
\begin{array}{ccccc}
 & & Q_{1,-e}(K_C M^{-1} \otimes E) \times C_e \times C & & \\
 & \swarrow^{p_1} & \big\downarrow{\scriptstyle q_0} & ^{p_2}\searrow & \\
Q_{1,-e}(K_C M^{-1} \otimes E) \times C & & & & C_e \times C \\
 & \searrow_{q_1} & & _{q_2}\swarrow & \\
 & & C & &
\end{array}
$$

Recall also the exact sequence $0 \to \mathcal{O}_{C_e \times C}(-\mathscr{D}) \to \mathcal{O}_{C_e \times C} \xrightarrow{\rho} \mathcal{O}_{\mathscr{D}} \to 0$ where $\mathscr{D}$ is the universal divisor.

Let $\Sigma\colon \mathcal{L} \to q_1^*(K_C M^{-1} \otimes E)$ be the universal subsheaf over $Q_{1,-e}(K_C M^{-1} \otimes E) \times C$. A key point is that $\mathcal{L}$ is locally free, and hence $p_1^*\mathcal{L}$ is too (this is the reason we work with $Q_{1,-e}(K_C M^{-1} \otimes E)$ instead of $\mathrm{Quot}^{1,e}({K_C}^{-1} M \otimes E^*)$ directly). Thus $p_1^*\Sigma$ can be considered as a map of vector bundles over $Q_{1,-e}(K_C M^{-1} \otimes E) \times C_e \times C$ (with a degeneracy locus). Hence we can take its transpose, twist by $q_0^*(K_C M^{-1})$ and form the composed map

$$
q_0^* E^* \;\to\; q_0^*(K_C M^{-1}) \otimes p_1^* \mathcal{L}^\vee \;\to\; q_0^*(K_C M^{-1}) \otimes p_1^* \mathcal{L}^\vee \otimes p_2^* \mathcal{O}_{\mathscr{D}}.
\tag{6.2}
$$

Now C_e is also a Quot scheme, parametrising degree e torsion quotients of any line bundle over C. Thus, by the universal property of Quot schemes, the restriction of this map to any fibre

$$
\left\{ [L \xrightarrow{\sigma} K_C M^{-1} \otimes E], D \right\} \times C
$$

is identified with the composed map

$$
E^* \xrightarrow{\widehat{\sigma}} K_C M^{-1} L^{-1} \to K_C M^{-1} L^{-1}|_D
$$

where $\widehat{\sigma} = {}^t\sigma \otimes \mathrm{Id}_{K_C M^{-1}}$. The kernel of this is an elementary transformation of E^*. As the restriction $K_C M^{-1} L^{-1} \to K_C M^{-1} L^{-1}|_D$ is always surjective, the elementary transformation has degree $\deg E^* - e$ if and only if σ is a vector bundle injection at each point of D. Pulling back to S_M^e as defined in (6.1), we obtain a family of elements of $\mathrm{Quot}^{0,e}(E^*)$ parametrised by the open subset

$$
\Big\{ \Big([L \xrightarrow{\sigma} K_C M^{-1} \otimes E], D \Big) : L \cong \mathcal{O}_C(-D) \text{ and } \sigma \text{ is a vector bundle injection along } D \Big\}
$$

of S^e_M. By the universal property of $\mathrm{Quot}^{0,e}(E^*)$, we obtain a map $\gamma^e_M : S^e_M \dashrightarrow \mathrm{Quot}^{0,e}(E^*)$ defined precisely on the above open set, given by sending

$$(6.3) \qquad (\sigma, D) \mapsto \mathrm{Ker}\left(E^* \xrightarrow{\widehat{\sigma}} K_C M^{-1}(D) \to K_C M^{-1}(D)|_D\right).$$

Now we characterise the locus $Q^{e-1}_e(E, M, H^0(E^* \otimes M)) = Q^{e-1}_e$ using the spaces S^e_M. The description is analogous to that of the osculating spaces to the scroll $\mathbb{P}E$ given in [**Hit19**, Proposition 4.1].

PROPOSITION 6.1. *Let E and M be as above, and γ^e_M as defined in* (6.3).

(a) *We have* $\mathrm{Im}(\gamma^e_M) \subseteq Q^{e-1}_e$.

(b) *Suppose that $[F^* \to E^*]$ belongs to Q^{e-1}_e. Then for some $e' \in \{1, \ldots, e\}$ there exists $[G^* \to E^*]$ belonging to $\mathrm{Im}\left(\gamma^{e'}_M\right) \subseteq \mathrm{Quot}^{e'}(E^*)$ such that $[F^* \to G^*]^* \in \mathrm{Quot}^{e-e'}(G^*)$.*

PROOF. (a) Suppose that $[F^* \to E^*]$ is of the form $\gamma^e_M(\sigma, D)$. By definition of γ^e_M, we have a exact diagram

$$\begin{array}{ccccc} F^* & \longrightarrow & E^* & \longrightarrow & E^*/F^* \\ \downarrow \widehat{\sigma}' & & \downarrow \widehat{\sigma} & & \downarrow \wr \\ K_C M^{-1} & \longrightarrow & K_C M^{-1}(D) & \longrightarrow & K_C M^{-1}(D)|_D \end{array}$$

Thus $\widehat{\sigma}'$ defines an element of $H^0(K_C M^{-1} \otimes F)$. Now by hypothesis, $\widehat{\sigma}$ is a vector bundle surjection along D. It follows that $\widehat{\sigma}'$ does not factorise via E^*. Hence $h^0(K_C M^{-1} \otimes F) \geq h^0(K_C M^{-1} \otimes E) + 1$. Using (3.2), we obtain $h^0(F^* \otimes M) \geq h^0(E^* \otimes M) - e + 1$. Hence $[F^* \to E^*]$ belongs to Q^{e-1}_e.

(b) Suppose that $h^0(F^* \otimes M) \geq h^0(E^* \otimes M) - e + 1$. Again using (3.2), we see that there exists $\sigma \in H^0(C, K_C M^{-1} \otimes F)$ whose image in $H^0(C, K_C M^{-1} \otimes (F/E))$ is nonzero.

Now we view sections of F as rational sections of E with at worst poles limited by F/E. (This generalises the definition of $\mathcal{O}_C(D)$ for an effective divisor D.) Let $x_1, \ldots, x_s \in \mathrm{Supp}(F/E)$ be the points where ($\sigma \mod K_C M^{-1} \otimes E$) is supported; as $\sigma \notin H^0(K_C M^{-1} \otimes E)$, there is at least one such point. For $1 \leq j \leq s$, we may choose a suitable neighbourhood U_j of x_j and a frame $\nu^{(j)}_1, \ldots, \nu^{(j)}_r$ for $K_C M^{-1} \otimes E|_{U_j}$ and a local parameter z_j on C at x_j, such that the image of $\sigma|_{U_j}$ is $\mathcal{O}_{U_j} \cdot \left(z_j^{-e_j} \cdot \nu^{(j)}_1\right)$. Let G be the elementary transformation of E such that $K_C M^{-1} \otimes G$ has frame

$$(6.4) \qquad z_j^{-e_j} \cdot \nu^{(j)}_1, \ \nu^{(j)}_2, \ \ldots, \ \nu^{(j)}_r$$

over each U_j, and is equal to $K_C M^{-1} \otimes E$ otherwise. Now $z_j^{-e_j} \cdot \nu^{(j)}_1 \subseteq \mathrm{Im}(\sigma)$ belongs to F, and the remaining frame elements $\nu^{(j)}_2, \ldots, \nu^{(j)}_r$ belong to $E \subset F$. Therefore, we have inclusions $E \subset G \subseteq F$. Writing $e' := \sum_{i=1}^s e_j$, we have

$$\deg E + e' = \deg G \leq \deg F = \deg E + e.$$

In particular $e' \leq e$, and

$$(6.5) \qquad [F^* \to G^*] \text{ is an element of } \mathrm{Quot}^{e-e'}(G^*).$$

Let D' be the effective divisor $\sum_{i=1}^{s} e_j x_j \in C_{e'}$. From the construction of G it follows that

$$\sigma \in H^0(K_C M^{-1} \otimes G) \subseteq H^0(K_C M^{-1}(D') \otimes E)$$

and moreover, that the image of $\sigma\colon \mathcal{O}_C \to K_C M^{-1} \otimes G$ is nonzero at each of the x_j. Twisting by $\mathcal{O}_C(-D')$, we see that σ defines a sheaf injection $\sigma'\colon \mathcal{O}_C(-D') \to K_C M^{-1} \otimes E$ which is a vector bundle injection along $\operatorname{Supp} D'$.

Now we claim that

$$[G^* \to E^*] = \gamma_M^{e'}\left([\sigma'\colon \mathcal{O}_C(-D') \to K_C M^{-1} \otimes E], D'\right).$$

To see this: For each U_j, let $\phi_1^{(j)}, \ldots, \phi_r^{(j)}$ be the frame for $E^*|_{U_j}$ dual to $\nu_1^{(j)}, \ldots, \nu_1^{(j)}$. Then by construction of G, we see that $G^*|_{U_j} \subset E^*|_{U_j}$ has the frame

$$z_j^{e_j} \cdot \phi_1^{(j)},\ \phi_2^{(j)},\ \ldots,\ \phi_r^{(j)} \tag{6.6}$$

dual to (6.4); and G^* coincides with E^* on $C \setminus D'$. On the other hand, the map

$$E^* \xrightarrow{{}^t\sigma' \otimes \mathrm{Id}_{K_C M^{-1}}} K_C M^{-1}(D') \to K_C M^{-1}(D')|_{D'}$$

is given in our chosen coordinates over U_j by

$$\sum_{i=1}^{r} f_i \phi_i^{(j)} \mapsto z_j^{-e_j} \cdot f_1 \mapsto z_j^{-e_j} \cdot f_1 \mod \mathcal{O}_{U_j}.$$

Comparing with (6.6), we see that this map has kernel exactly G^*, proving the claim. Statement (b) now follows from the claim and (6.5). □

REMARK 6.2. Suppose $[L \xrightarrow{\sigma} K_C \otimes E]$ is an element of $Q_{1,-e}(K_C M^{-1} \otimes E)$ where $|L^{-1}|$ has dimension $k \geq 1$. For simplicity, suppose $|L^{-1}|$ is base point free. Then $S^e_{\mathcal{O}_C}$ contains the locus

$$\{(\sigma, D) : D \in |L^{-1}|\} \cong \mathbb{P}^k.$$

By Proposition 6.1 (a), we obtain an element of Q_e^{e-1} for each $D \in |L^{-1}|^*$ along which σ is a bundle injection. By Propositions 2.9 and 2.12 the projective model $\mathbb{P}E \dashrightarrow |\mathcal{L}_M|^*$ has a family of 1-defective e-secants parametrised by an open subset of $|L^{-1}|$.

REMARK 6.3. One may ask whether there is a relation between S^e_M and H_e^{e-1} akin to that described in Proposition 6.1 for Q_e^{e-1}. There does exist a map $\beta\colon S^e_M \dashrightarrow \mathrm{Hilb}^e_{\mathrm{sm}}(S)_{\mathrm{nd}}$ by $(\sigma, D) \mapsto \mathbb{P}\sigma(D)$, globalising the construction of Lemma 4.3. This is again defined exactly when σ is a vector bundle injection at D, and in fact one can show that $\alpha \circ \beta = \gamma$, where $\alpha\colon \mathrm{Hilb}^e(S) \dashrightarrow \mathrm{Quot}^{0,e}(E^*)$ is the map studied in § 2.5. An analogue of Proposition 6.1 (a) then follows from Proposition 2.12. However, Proposition 6.1 (b) is harder to generalise, as one cannot always choose a subscheme playing the role of the sheaf G above; essentially due to issues of identifiability (defined for example in [**BC21**, Definition 6.1]) arising from the existence of linear spaces contained in $\mathbb{P}E$. Thus we content ourselves for now with the study of Q_e^{e-1}.

In the sections that follow, we use Proposition 6.1 to characterise semistability, to study questions of very ampleness of $\mathcal{O}_{\mathbb{P}E}(1)$, and to show that the secant loci are well behaved when certain parameters are chosen generally.

7. The Segre invariant s_1 and secant loci

In this section, we use the parameter spaces S^e_M to characterise nonempty secant loci for $f = 1$ in terms of the Segre invariant s_1. Firstly, we review the notion of Segre invariants.

7.1. Segre invariants.

DEFINITION 7.1. Let $E \to C$ be a vector bundle of rank r and degree d. For $1 \le t \le r-1$, the Segre invariant $s_t(E)$ is defined by

$$s_t(E) \ := \ \min\{td - r \cdot \deg(F) : F \subset E \text{ a subbundle of rank } t\}.$$

Segre invariants are "degrees of stability": E is stable if and only if $s_t(E) > 0$ for $1 \le t \le r-1$. It is clear that $s_1(E) = s_1(E \otimes L)$ for any line bundle L. Moreover, $s_t(E)$ is lower semicontinuous as E varies in families. Taking direct sums of line bundles, one can produce bundles E with arbitrarily low $s_t(E)$. However, Hirschowitz [**Hir86**, Théorème 4.4] (see also [**CH10**]) proved the following sharp upper bound on $s_t(E)$.

THEOREM 7.2 (Hirschowitz's bound). *Let $E \to C$ be a vector bundle of rank r and degree d. Then for $1 \le t \le r-1$, we have $s_t(E) \le t(r-t)(g-1) + \delta$, where δ is the unique integer such that $0 \le \delta \le r-1$ and $t(r-t)(g-1) + \delta \equiv td \mod r$. Moreover, a general stable E attains this upper bound.*

In particular, this implies that Quot schemes of subsheaves of E are always nonempty for low enough degree. Let us make this precise for line subbundles. The following is well known, but we include a proof for convenience.

LEMMA 7.3. *Let C be any curve. Let $E \to C$ be a vector bundle which is general in moduli. Then $Q_{1,-e}(K_C M^{-1} \otimes E)$ is nonempty and smooth of dimension $re+d+(r+1)(g-1)$ whenever this is nonnegative, and empty otherwise. Moreover, when $Q_{1,-e}(K_C M^{-1} \otimes E)$ is nonempty, a general element is a vector bundle injection.*

PROOF. One computes easily that $re + d + (r+1)(g-1) \ge 0$ if and only if

$$\deg(K_C \otimes E) - r \cdot (-e) \ \ge \ (r-1)(g-1) + \delta,$$

with δ as in Theorem 7.2, using the fact that left side is congruent to d modulo r. In this case, by the Hirschowitz bound, $K_C \otimes E$ has a line subbundle of degree at least $-e$, so $Q_{1,-e}(K_C M^{-1} \otimes E)$ is nonempty. Smoothness and dimension of $Q_{1,-e}(K_C M^{-1} \otimes E)$ for general E then follow from [**LN03**, Lemma 3.3] and the discussion before it. To see that a general element is saturated, we note that the locus of nonsaturated subsheaves in $Q_{1,-e}(K_C M^{-1} \otimes E)$ has dimension at most

$$\max\{\dim Q_{1,-e_1}(K_C \otimes E) + \dim C_{e-e_1} : e_1 < e\},$$

which one checks is strictly less than $re + d + (r+1)(g-1)$.

On the other hand, if if $re + d + (r+1)(g-1) < 0$ then $\deg(K_C \otimes E) + re < (r-1)(g-1)$. For such e, there is an invertible subsheaf of degree $-e$ in $K_C \otimes E$ if and only if $s_1(E)$ is smaller than the generic value, so E is not general. ☐

7.2. Nonemptiness criterion for $f = 1$. The following is a generalisation of [**Hit19**, Theorem 5.2], with a very similar proof.

THEOREM 7.4. *Let $E \to C$ be a vector bundle of rank r and degree d. For $e \ge 1$, the following are equivalent.*

(1) $s_1(E) > d + r(2g - 2 + e)$.
(2) *For all* $M \in \mathrm{Pic}^0(C)$ *and all* $Z \in \mathrm{Hilb}^e_{\mathrm{sm}}(S)_{\mathrm{nd}}$, *the span of* $\phi_{\mathcal{L}_M}(Z)$ *is of dimension* $e - 1$ *in* $|\mathcal{O}_{\mathbb{P}E}(1) \otimes \pi^* M|^*$.
(3) *For all* $M \in \mathrm{Pic}^0(C)$ *and all* $[F^* \subset E^*] \in \mathrm{Quot}^{0,e}(E^*)$, *we have*

$$h^0(F^* \otimes M) = h^0(E^* \otimes M) - e.$$

PROOF. The equivalence of (2) and (3) follows from Proposition 2.12 and surjectivity of $\alpha\colon \mathrm{Hilb}^e_{\mathrm{sm}}(S) \to \mathrm{Quot}^{0,e}(E^*)$. Let us show the equivalence of (1) and (3).

Assume (1). As $s_1(K_C M^{-1} \otimes E) = s_1(E) > d + r(2g-2+e)$, if $L \subset K_C M^{-1} \otimes E$ is an invertible subsheaf, then $\deg(L) < -e$. In particular, $Q_{1,-e'}(K_C M^{-1} \otimes E)$ and hence $S^{e'}_M$ are empty for all $M \in \mathrm{Pic}^0(C)$ and $e' \le e$. By Proposition 6.1(b), also Q^{e-1}_e is empty.

Conversely, suppose that $s_1(E) = s_1(K_C \otimes E) \le d + r(2g - 2 + e)$. Then there exists $L \in \mathrm{Pic}^{-e}(C)$ and a sheaf injection $\sigma'\colon L \to K_C \otimes E$ of degree $-e$. Let $D \in C_e$ be any divisor along which $L \to K_C \otimes E$ is a vector bundle injection. Set $M := L(D) \in \mathrm{Pic}^0(C)$ and

$$\sigma := \sigma' \otimes \mathrm{Id}_{L^{-1}(-D)}\colon \mathcal{O}_C(-D) \to K_C M^{-1} \otimes E.$$

Then (σ, D) defines an element of S^e_M at which γ^e_M is defined. Hence Q^{e-1}_e is nonempty by Proposition 6.1(a). □

REMARK 7.5. As in [**Hit19**, Remark 5.6], we note that Theorem 7.4 does not hold for incomplete linear series. For any $Z \in \mathrm{Hilb}^e_{\mathrm{sm}}(S)$, if we project from a point of $\mathrm{Span}\,\psi(Z)$ then, whatever the value of $s_1(E)$, the image of $\mathbb{P}E$ has a defective e-secant space. (However, as noted in § 2.7, under certain conditions the H^{e-1}_e and Q^{e-1}_e do behave as expected under projection from a general centre.)

We mention now some special cases, generalising those discussed in [**Hit19**, Corollary 5.4]. We omit the proofs, as they are practically identical to those in [**Hit19**]. Recall that the *slope* of a bundle $W \to C$ is the ratio $\mu(W) := \deg(W)/\mathrm{rk}(W)$.

COROLLARY 7.6. *Let* E *be a bundle of rank* r *and degree* d, *where* $d \le r(1-2g)$.
(a) *Suppose that* $s_1(E) > 0$ *(this is the case for example if* E *is stable). Then for* $e \le \mu(E^*) - (2g-2)$, *for all* $M \in \mathrm{Pic}^0(C)$ *the loci* $\left(H^{e-1}_e\right)_{\mathrm{nd}}$ *and* Q^{e-1}_e *are empty.*
(b) *If* $e \ge \mu(E^*) - \frac{1}{r}(r+1)(g-1)$, *then for some* $M \in \mathrm{Pic}^0(C)$, *the loci* $\left(H^{e-1}_e\right)_{\mathrm{nd}}$ *and* Q^{e-1}_e *are nonempty.*
(c) *If* $s_1(E)$ *is the generic value* $(r-1)(g-1) + \delta$, *then the converse to (b) also holds.*

7.3. A question of Lange. If $r = 2$, then $\mathbb{P}E$ is a ruled surface over C. For rational curves and elliptic curves, Lange [**Lan92**, Lecture 2 (a)] (see also [**Har77**, Exercise V.2.12]) gives criteria for very ampleness of $\mathcal{O}_{\mathbb{P}E}(1) \otimes \pi^* M$ in terms of $\deg(M)$ and $s_1(E)$. He then poses the problem of finding similar criteria for ruled surfaces when $g \ge 2$. We now offer a result in this direction, which in fact holds for projective bundles of any dimension $r \ge 2$ over C.

THEOREM 7.7. *Let* $E \to C$ *be a bundle of rank* r *and degree* d. *Then the line bundle* $\mathcal{O}_{\mathbb{P}E}(1) \otimes \pi^* L$ *is very ample for all* $L \in \mathrm{Pic}^\ell(C)$ *if and only if*

$$\ell > \mu(E) - \frac{s_1(E)}{r} + 2g. \tag{7.1}$$

PROOF. A map $S \to \mathbb{P}^n$ is an embedding if and only if all 2-secants are non-defective. Furthermore, when S is a projective bundle $\mathbb{P}E$ over C, it follows from the argument of [**Sta21a**, Theorem 3 (i)] that all 2-secants are π-nondefective (cf. Definition 2.11). Thus it suffices to show that (7.1) is equivalent to emptiness of the secant locus $H_2^1(\mathcal{O}_{\mathbb{P}E}(1) \otimes \pi^* L)_{\mathrm{nd}}$ for all $L \in \mathrm{Pic}^\ell(C)$.

Fix now a line bundle L_0 of degree ℓ. For any $M \in \mathrm{Pic}^0(C)$, we have

$$H_2^1(H^0(\mathbb{P}E, \mathcal{O}_{\mathbb{P}E}(1) \otimes \pi^* L_0 M)) \cong H_2^1\left(H^0(\mathbb{P}(E \otimes L_0^{-1}, \mathcal{O}_{\mathbb{P}(E \otimes L_0^{-1})}(1) \otimes \pi^* M))\right).$$

By Theorem 7.4, the right-hand side is empty for all $M \in \mathrm{Pic}^0(C)$ if and only if

$$s_1(E \otimes L_0^{-1}) > d - r\ell + 2rg.$$

As $s_1(E \otimes L_0^{-1}) = s_1(E)$, this is equivalent to $\ell > \mu(E) - \frac{s_1(E)}{r} + 2g$, as desired. □

Theorem 7.7 is a statement concerning $E \otimes L$ for all line bundles L of a given degree. We make one more observation showing that very ampleness may also depend on $L \in \mathrm{Pic}^\ell(C)$. Fix such an L, and write $h := h^0(K_C L^{-1} \otimes E)$. We consider the twisted Brill–Noether locus

$$B_{1,2}^{h+1}(K_C L^{-1} \otimes E) = \{N \in \mathrm{Pic}^2(C) : h^0(K_C L^{-1} N \otimes E) \geq h+1\}.$$

Also on $\mathrm{Pic}^2(C)$ we have the standard Brill–Noether locus

$$B_{1,2}^1 = \{\mathcal{O}_C(x+y) : x, y \in C\} \subseteq \mathrm{Pic}^2(C)$$

parametrising effective line bundles of degree two over C.

PROPOSITION 7.8. *Let E be a bundle of rank r and degree d. For $L \in \mathrm{Pic}^\ell(C)$, the line bundle $\mathcal{O}_{\mathbb{P}E}(1) \otimes \pi^* L$ is very ample if and only if*

$$B_{1,2}^{h+1}(K_C L^{-1} \otimes E) \cap B_{1,2}^1 = \emptyset.$$

PROOF. By [**Hit20**, Proposition 4.1] (substituting $K_C L^{-1} \otimes E$ for "V"), the line bundle $\mathcal{O}_{\mathbb{P}E}(1) \otimes \pi^* L$ is very ample on $\mathbb{P}E$ if and only if

$$h^0(C, K_C L^{-1} \otimes E(x+y)) = h^0(C, K_C L^{-1} \otimes E) \text{ for all } x+y \in C_2. \tag{7.2}$$

This is equivalent to saying that $h^0(C, K_C L^{-1} \otimes E(x+y)) = h$ for all $x + y \in C_2$; in other words, that no line bundle of the form $\mathcal{O}_C(x+y)$ belongs to the locus $B_{1,2}^{h+1}(K_C L^{-1} \otimes E)$. □

7.4. A criterion for semistability. Here we use Theorem 7.4 to give a criterion for semistability. This is entirely analogous to [**Hit19**, Theorem 5.7]; as the proof is practically identical, we only sketch it. For a given $E \to C$, for $t \geq 1$ we write $\pi_t \colon \mathbb{P}(\wedge^t E) \to C$ for the projection.

THEOREM 7.9. *Let E be a vector bundle of rank r with $\mu(E) < 1 - 2g$. Then the following are equivalent.*

(1) *The bundle E is semistable.*
(2) *For $1 \leq t \leq r-1$, for all $M \in \mathrm{Pic}^0(C)$, and for $1 \leq e < t \cdot \mu(E^*) - (2g-2)$, the secant locus*

$$H_e^{e-1}\left(H^0(\mathbb{P}(\wedge^t E), \mathcal{O}_{\mathbb{P}(\wedge^t E)}(1) \otimes \pi_t^* M\right)_{\mathrm{nd}}$$

is empty.

(3) *For $1 \le t \le r-1$, and for M and e as in (2), the secant locus*
$$Q_e^{e-1}\left(\wedge^t E, M, H^0(C, \wedge^t E \otimes \pi_t^* M)\right)$$
is empty; that is, $h^0(C, \mathcal{F}_n) = h^0(C, M \otimes \wedge^n E^*) - e$ for all elementary transformations $0 \to \mathcal{F}_n \to \wedge^n E^* \to T \to 0$ where T is torsion of length e.

PROOF. The equivalence of (2) and (3) is proven as in Theorem 7.4. If E is semistable then, since $\mathbb{C}$ has characteristic zero, also $\wedge^t E$ is semistable for $1 \le t \le r-1$. In particular, $s_1(\wedge^t E) \ge 0$. Theorem 7.4 then implies (3).

Conversely, assume (3). Then for each $t \in \{1, \ldots, r-1\}$, by Theorem 7.4 we have
$$s_1(\wedge^t E) \;>\; \deg(\wedge^t E) + \mathrm{rk}(\wedge^t E) \cdot (2g - 2 - e_t)$$
where $e_t \;=\; \max\{e : 1 \le e < t \cdot \mu(E^*) - (2g-1)\}$. By definition of s_1, the left and right sides of the last inequality are congruent modulo $\mathrm{rk}(\wedge^t E)$. Therefore
$$s_1(\wedge^t E) \;\ge\; \deg(\wedge^t E) + \mathrm{rk}(\wedge^t E) \cdot (2g - 2 + e_t + 1).$$
By definition of e_t, this becomes $s_1(\wedge^t E) \ge 0$. As in the proof of [**Hit19**, Theorem 5.7], we conclude that $\mu(F) \le \mu(E)$ for all rank t subbundles $F \subset E$. □

8. Secant loci of general scrolls for $f = 1$

In this final section we study questions of expected dimension, nonemptiness and enumeration for the secant loci $Q_e^{e-1}(E, M, V) =: Q_e^{e-1}(V)$ when the parameters E, M and V are chosen generally. The statements are valid for any curve C. We obtain also the emptiness of certain $H_e^{e-1}(\mathcal{L}_M, V)_{\mathrm{nd}}$; however, in the lack of a complete description of the fibres of $\alpha\colon \mathrm{Hilb}^e(\mathbb{P}E) \dashrightarrow \mathrm{Quot}^{0,e}(E^*)$, we do not prove further dimension bounds on $H_e^{e-1}(\mathcal{L}_M, V)$ at present.

8.1. Expected dimension of $Q_e^{e-1}(V)$. Here we show that if the parameters are chosen generally, $\left(H_e^{e-1}(V)\right)_{\mathrm{nd}}$ and $Q_e^{e-1}(V)$ are empty if the expected dimension is zero, and $Q_e^{e-1}(V)$, if nonempty, is of the expected dimension otherwise. This generalises [**Hit19**, Theorem 6.4], and uses a similar method. We begin by treating the complete case. Fix $r \ge 2$ and $d < r(1-g) - 1$, so that $n := -d - r(g-1) - 1 \ge 0$. Recall that for $f = 1$, the expected dimension of H_e^{e-1} and Q_e^{e-1} is $re - (n + 1 - e + 1) = (r+1)e - n - 2$.

THEOREM 8.1. *Let $E \to C$ be a general vector bundle of rank r and degree d. Then there is a nonempty open subset $U \in \mathrm{Pic}^0(C)$ such that for $M \in U$, the following hold.*

(a) *$h^0(C, E^* \otimes M) = n + 1$, so $|\mathcal{L}_M|$ has dimension n.*
(b) *If $(r+1)e - n - 2 < 0$, then Q_e^{e-1} and $\left(H_e^{e-1}\right)_{\mathrm{nd}}$ are empty.*
(c) *If $(r+1)e - n - 2 \ge 0$, then Q_e^{e-1} is empty or of the expected dimension.*

PROOF. Part (a) is exactly [**Hit19**, Theorem 6.4 (a)]. For (b): We recall the construction of S_M^e as a fibre product in (6.1), and the forgetful map
$$b\colon Q_{1,-e}(K_C M^{-1} \otimes E) \;\to\; \mathrm{Pic}^{-e}(C).$$
Unwinding definitions, we see for any $M \in \mathrm{Pic}^0(C)$ that
$$M^{-1} \cdot b\left(Q_{1,-e}(K_C \otimes E)\right) \;=\; b\left(Q_{1,-e}(K_C M^{-1} \otimes E)\right),$$

where the action on the left hand side is that of $\mathrm{Pic}^0(C)$ on itself by translation. Thus, by Theorem 2.17, for general $M \in \mathrm{Pic}^0(C)$ the space S_M^e is empty or of dimension

$$\dim Q_{1,-e}(K_C M^{-1} \otimes E) + \dim C_e - \dim \mathrm{Pic}^{-e}(C)$$

when this is nonnegative. By Lemma 7.3 and the assumption of generality of E, this number is

$$re + d + (r+1)(g-1) + e - g \;=\; (r+1)e - n - 2,$$

the last equality by the definition of n above. This is exactly the expected dimension of Q_e^{e-1} and H_e^{e-1}.

Now by Proposition 6.1 (b), the locus Q_e^{e-1} is nonempty only if $S_M^{e'}$ is nonempty for some $e' \leq e$. The above computation shows that $\dim S_M^{e'} < \dim S_M^e$ for $e' < e$. Therefore, if $(r+1)e - n - 2 < 0$ then Q_e^{e-1} is empty. As $\alpha\left(H_e^{e-1}\right)_{\mathrm{nd}} = Q_e^{e-1}$ by Proposition 2.12, also $\left(H_e^{e-1}\right)_{\mathrm{nd}}$ is empty. This proves (b).

For the rest: If $[F^* \to E^*]$ is a point of Q_e^{e-1} then, by Proposition 6.1 (b), we have $F^* \subseteq G^*$, where $[G^* \to E^*]$ belongs to the image of

$$\gamma_M^{e'} \colon S_M^{e'} \dashrightarrow \mathrm{Quot}^{0,e'}(E^*),$$

for some $e' \leq e$. Thus $\dim Q_e^{e-1}$ is at most

$$\dim S_M^{e'} + \dim \mathrm{Quot}^{e-e'}(G^*) \;=\; (r+1)e' - n - 2 + r(e-e') \;=\; (r+1)e - n - 2 - (e-e').$$

As $e \geq e'$, part (c) follows. (Note that we have equality for $e' = e$; compare with Proposition 2.16 (b).) □

Using Corollary 2.21, we can extend Theorem 8.1 to the case of incomplete linear series.

THEOREM 8.2. *Let E and M be general in the sense of Theorem 8.1. Let $W \subseteq H^0(C, E^* \otimes M)$ be a general subspace of dimension $m+1 < n+1$.*

(a) *If $(r+1)e - m - 2 < 0$, then $Q_e^{e-1}(W)$ is empty.*
(b) *If $(r+1)e - m - 2 < 0$, then $\left(H_e^{e-1}(W)\right)_{\mathrm{nd}}$ is empty.*
(c) *If $(r+1)e - m - 2 \geq 0$, then $Q_e^{e-1}(W)$ is empty or of the expected dimension.*

PROOF. For (a) and (c), we will apply Corollary 2.21. As $f = 1$, we have $0 \leq f' \leq 1$, so $f - f' \leq 1 \leq n - m$. Thus, by Remark 2.19, to satisfy the hypothesis of Corollary 2.21 it will suffice to show that $Q_e^{e-1}(H^0(E^* \otimes M))$ is empty or has the expected dimension (trivially, Q_e^e is always of the expected dimension). This follows from Theorem 8.1. Then (b) follows as before from Propositions 2.9 (b) and 2.12. □

8.2. Nonemptiness of $Q_e^{e-1}(V)$. Theorems 8.1 and 8.2 show that $Q_e^{e-1}(V)$, if nonempty, is of the expected dimension for a general choice of the parameters. We now describe two situations in which $Q_e^{e-1}(V)$ is in fact nonempty.

PROPOSITION 8.3. *Let $E \to C$ be a bundle which is general in moduli. Suppose $e \geq \max\{g, -d/r - (g-1)(r+1)/r\}$. Then $Q_e^{e-1}(V)$ is nonempty for generic $M \in \mathrm{Pic}^0(C)$.*

PROOF. As $e \geq -d/r - (g-1)(r+1)/r$, we have $re + d + (r+1)(g-1) \geq 0$. Since E is general, by Lemma 7.3 the scheme $Q_{1,-e}(K_C M^{-1} \otimes E)$ is nonempty and a general element is of the form $[L \xrightarrow{\sigma} K_C M^{-1} \otimes E]$ where σ is a vector bundle injection. But since also $e \geq g$, for any L so occurring we have $L = \mathcal{O}_C(-D)$ for some $D \in C_e$. Therefore, (σ, D) is a point of S^e_M where γ^e_M is defined. Hence Q^{e-1}_e is nonempty by Proposition 6.1 (a). □

PROPOSITION 8.4. *Let E be any bundle, and suppose that $Q_{1,-e}(K_C \otimes E)$ is nonempty for some $e \geq 1$. Then $Q^{e-1}_e(E, M, H^0(E^* \otimes M))$ is nonempty for some $M \in \mathrm{Pic}^0(C)$.*

PROOF. Suppose $[L \xrightarrow{\sigma} K_C \otimes E]$ is a point of $Q_{1,-e}(K_C \otimes E)$. Now σ may not be a bundle injection at all points, but we can choose a sufficiently general $D \in C_e$ along which σ is a bundle injection. Then $L^{-1}(-D)$ has degree zero, and

$$\left(\left[\mathcal{O}_C(-D) \xrightarrow{\sigma \otimes L^{-1}(-D)} K_C L^{-1}(-D)) \otimes E\right], D\right)$$

is a point of $S^e_{L^{-1}(-D)}$ at which $\gamma^e_{L^{-1}(-D)}$ is defined. Then

$$Q^{e-1}_e(E, L^{-1}(-D), H^0(C, E^* \otimes L^{-1}(-D)))$$

is nonempty by Proposition 6.1 (a). □

REMARK 8.5. Proposition 8.4 applies whenever $Q_{1,-e}(K_C \otimes E)$ has nonnegative expected dimension. If E is generic, then this follows from Lemma 7.3. For arbitrary E, one can find possibly nonsaturated elements of $Q_{1,-e}(K_C M^{-1} \otimes E)$ by taking subsheaves of elements of $Q_{1,h}(K_C \otimes E)$ for $h > -e$.

8.3. Enumeration of $Q^{e-1}_e(V)$. By the previous two subsections, one knows that $Q^{e-1}_e(E, M, V) = Q^{e-1}_e(V)$, if nonempty, is of the expected dimension for any C and for a general choice of (E, M, V); and furthermore that $Q^{e-1}_e(V)$ is indeed nonempty under certain conditions. We now use results of [**EGL01**], [**OP21**] and [**Sta21b**] to enumerate $Q^{e-1}_e(V)$ when it has and attains expected dimension zero.

Following the notation of [**OP21**] and [**Sta21b**], we denote by $M^{[e]}$ the vector bundle over $\mathrm{Quot}^{0,e}(E^*)$ with fibre $H^0(C, (E^*/F^*) \otimes M)$ at the point $[F^* \to E^*]$. For $V \subseteq H^0(E^* \otimes M)$, let $\varepsilon\colon V \otimes \mathcal{O}_{\mathrm{Quot}^{0,e}(E^*)} \to M^{[e]}$ be the evaluation map. Then $Q^{e-1}_e(V)$ is the determinantal locus $\{F^* \in \mathrm{Quot}^{0,e}(E^*) : \mathrm{rk}\, \varepsilon|_{F^*} \leq e-1\}$. When $re = n + 2 - e$, so that $Q^{e-1}_e(V)$ has expected dimension zero, using the Porteous formula we obtain

$$[Q^{e-1}_e(V)] \;=\; s_{n+2-e}\left(M^{[e]}\right) \;=\; \int_{[\mathrm{Quot}^{0,e}(E^*)]} s(M^{[e]}). \tag{8.1}$$

The main goal of this subsection will be to sketch the proof of the following.

THEOREM 8.6. *Let C be a smooth curve, $E \to C$ a vector bundle of rank r and $M \to C$ a line bundle. Then there is an equality of formal power series*

$$\sum_{e \geq 0} \int_{[\mathrm{Quot}^{0,e}(E^*)]} s(M^{[e]})\, q^e \;=\; A_1(q)^{\deg(E^* \otimes M)} \cdot B(q)^{1-g} \tag{8.2}$$

where

$$q = (-1)^r t(1+t)^r, \quad A_1(q) = 1 + t \quad \textit{and} \quad B(q) = \frac{(1+t)^{r+1}}{1 + t(r+1)}.$$

SKETCH OF PROOF. All the ingredients for proving this theorem are present in [**EGL01**], [**OP21**] and [**Sta21b**], so we give only an outline. The formula (8.2) is proven in [**OP21**, Theorem 6 and Theorem 8] when E is the trivial bundle of rank $r \geq 1$. The generalisation of the corresponding formula [**OP21**, Corollary 16] for surfaces to arbitrary E is [**Sta21b**, Proposition 4.3], and we use the same strategy.

To show the existence of the power series A_1 and B, we follow essentially word for word the proof of [**EGL01**, Theorem 4.2], replacing $\mathcal{K}_r$ with the set

$$\mathcal{T}_r := \{(C, E, M) : C \text{ a smooth curve}, E \to C \text{ a bundle of rank } r \text{ and } M \in \mathrm{Pic}(C)\},$$

and $H_{\Psi,\Phi}(S, x)$ with the map $\mathcal{T}_r \to \mathbb{Q}[[q]]$ given by

$$(C, E, M) \mapsto \sum_{e \geq 0} q^e \int_{[\mathrm{Quot}_C^{0,e}(E^*)]} s(M^{[e]});$$

and γ with the map $(C, E, M) \mapsto (\deg(E^* \otimes M), \chi(\mathcal{O}_C))$. The role of [**EGL01**, Theorem 4.1] is played in our situation by the following statement which emerges in the proof of [**Sta21b**, Theorem 3.3]:

PROPOSITION 8.7. *For each $e \geq 0$, there exists a polynomial $u_e(x, y)$ which is universal in the sense that for any $(C, E, M) \in \mathcal{T}_r$ we have*

$$\int_{[\mathrm{Quot}^{0,e}(E^*)]} s(M^{[e]}) = u_e\left(\deg(E^* \otimes M), \chi(\mathcal{O}_C)\right).$$

Once the existence of A_1 and B is established, they can be computed by comparing with [**OP21**, Theorem 6 and Theorem 8] for suitable triples (C, E, M) with E trivial and C rational or elliptic. □

The following enumeration result is immediate from Theorem 8.6 and (8.1).

COROLLARY 8.8. *Suppose E, M and $V \subseteq H^0(E^* \otimes M)$ are such that $Q_e^{e-1}(V)$ has and attains expected dimension zero. Then the number of points of $Q_e^{e-1}(V)$, counted with multiplicity, is the coefficient of q^e in the expression* (8.2).

Acknowledgments

I thank the organisers of the VBAC 2022 meeting at the University of Warwick for financial support and for a very rewarding and enjoyable conference. I thank Ali Bajravani, Abel Castorena, Gavril Farkas and Peter Newstead for enjoyable and helpful communication. I thank Dragos Oprea for generous advice on enumeration questions. I am grateful to Samuel Stark for suggesting the strategy for Theorem 8.6 and for detailed explanation of the papers [**EGL01**], [**OP21**] and [**Sta21b**]. I thank the referee for a careful reading and for useful comments.

References

[AS15] M. Aprodu and E. Sernesi, *Secant spaces and syzygies of special line bundles on curves*, Algebra Number Theory **9** (2015), no. 3, 585–600, DOI 10.2140/ant.2015.9.585. MR3340545

[ACGH85] E. Arbarello, M. Cornalba, P. A. Griffiths, and J. Harris, *Geometry of algebraic curves. Vol. I*, Grundlehren der mathematischen Wissenschaften [Fundamental Principles of Mathematical Sciences], vol. 267, Springer-Verlag, New York, 1985, DOI 10.1007/978-1-4757-5323-3. MR770932

[Baj15] A. Bajravani, *Martens-Mumford theorems for Brill-Noether schemes arising from very ample line bundles*, Arch. Math. (Basel) **105** (2015), no. 3, 229–237, DOI 10.1007/s00013-015-0805-y. MR3391482

[Baj17] A. Bajravani, *Remarks on the geometry of secant loci*, Arch. Math. (Basel) **108** (2017), no. 4, 373–381, DOI 10.1007/s00013-016-1004-1. MR3627396

[Baj18] A. Bajravani, *A note on the tangent cones of the scheme of secant loci*, Rend. Circ. Mat. Palermo (2) **67** (2018), no. 3, 599–608, DOI 10.1007/s12215-018-0399-9. MR3912014

[BC21] E. Ballico and L. Chiantini, *On the Terracini locus of projective varieties*, Milan J. Math. **89** (2021), no. 1, 1–17, DOI 10.1007/s00032-020-00324-5. MR4277359

[CMTiB11] S. Casalaina-Martin and M. Teixidor i Bigas, *Singularities of Brill-Noether loci for vector bundles on a curve*, Math. Nachr. **284** (2011), no. 14-15, 1846–1871, DOI 10.1002/mana.200910093. MR2838286

[CT-L18] A. Castorena and H. Torres-López, *Linear stability and stability of syzygy bundles*, Internat. J. Math. **29** (2018), no. 11, 1850080, 14, DOI 10.1142/S0129167X18500805. MR3871724

[CH10] I. Choe and G. H. Hitching, *Secant varieties and Hirschowitz bound on vector bundles over a curve*, Manuscripta Math. **133** (2010), no. 3-4, 465–477, DOI 10.1007/s00229-010-0381-1. MR2729263

[Cop95] M. Coppens, *Brill-Noether theory for non-special linear systems*, Compositio Math. **97** (1995), no. 1-2, 17–27. Special issue in honour of Frans Oort. MR1355115

[CJ96] M. Coppens, *An infinitesimal study of secant space divisors*, J. Pure Appl. Algebra **113** (1996), no. 2, 121–144, DOI 10.1016/0022-4049(95)00098-4. MR1415555

[CJ91] M. Coppens and T. Johnsen, *Secant lines of smooth projective curves; an infinitesimal study of the symmetric products*, Enumerative algebraic geometry (Copenhagen, 1989), Contemp. Math., vol. 123, Amer. Math. Soc., Providence, RI, 1991, pp. 61–87, DOI 10.1090/conm/123/1143547. MR1143547

[CM91] M. Coppens and G. Martens, *Secant spaces and Clifford's theorem*, Compositio Math. **78** (1991), no. 2, 193–212. MR1104787

[Cot11] E. Cotterill, *Geometry of curves with exceptional secant planes: linear series along the general curve*, Math. Z. **267** (2011), no. 3-4, 549–582, DOI 10.1007/s00209-009-0635-3. MR2776048

[CHZ21] E. Cotterill, X. He, and N. Zhang, *Secant planes of a general curve via degenerations*, Geom. Dedicata **211** (2021), 165–201, DOI 10.1007/s10711-020-00545-3. MR4228498

[EGL01] G. Ellingsrud, L. Göttsche, and M. Lehn, *On the cobordism class of the Hilbert scheme of a surface*, J. Algebraic Geom. **10** (2001), no. 1, 81–100. MR1795551

[Eis95] D. Eisenbud, *Commutative algebra: With a view toward algebraic geometry*, Graduate Texts in Mathematics, vol. 150, Springer-Verlag, New York, 1995, DOI 10.1007/978-1-4612-5350-1. MR1322960

[Far08] G. Farkas, *Higher ramification and varieties of secant divisors on the generic curve*, J. Lond. Math. Soc. (2) **78** (2008), no. 2, 418–440, DOI 10.1112/jlms/jdn038. MR2439633

[Far23] G. Farkas, *Generalized de Jonquières divisors on generic curves*,Studia Univ. Babes-Bolyai Math. **68** (2023), 13–27 (special volume dedicated to the memory of Cs. Varga).

[GH78] P. Griffiths and J. Harris, *Principles of algebraic geometry*, Wiley Classics Library, John Wiley & Sons, Inc., New York, 1994. Reprint of the 1978 original, DOI 10.1002/9781118032527. MR1288523

[GP82] L. Gruson and C. Peskine, *Courbes de l'espace projectif: variétés de sécantes* (French), Enumerative geometry and classical algebraic geometry (Nice, 1981), Progr. Math., vol. 24, Birkhäuser, Boston, MA, 1982, pp. 1–31. MR685761

[GP13] L. Gruson and C. Peskine, *On the smooth locus of aligned Hilbert schemes, the k-secant lemma and the general projection theorem*, Duke Math. J. **162** (2013), no. 3, 553–578, DOI 10.1215/00127094-2019817. MR3024093

[GT09] I. Grzegorczyk and M. Teixidor i Bigas, *Brill-Noether theory for stable vector bundles*, Moduli spaces and vector bundles, London Math. Soc. Lecture Note Ser., vol. 359, Cambridge Univ. Press, Cambridge, 2009, pp. 29–50. MR2537065

[Har77] R. Hartshorne, *Algebraic geometry*, Graduate Texts in Mathematics, No. 52, Springer-Verlag, New York-Heidelberg, 1977. MR463157

[Hir86] A. Hirschowitz: *Problèmes de Brill–Noether en rang supérieur.* Prépublications Mathématiques n. 91, Nice (1986).

[Hir86] A. Hirschowitz, *Problèmes de Brill-Noether en rang supérieur* (French, with English summary), C. R. Acad. Sci. Paris Sér. I Math. **307** (1988), no. 4, 153–156. MR956606

[Hit19] G. H. Hitching, *Quot schemes, Segre invariants, and inflectional loci of scrolls over curves*, Geom. Dedicata **205** (2020), 1–19, DOI 10.1007/s10711-019-00463-z. MR4076817

[Hit20] G. H. Hitching, *A Riemann-Kempf singularity theorem for higher rank Brill-Noether loci*, Bull. Lond. Math. Soc. **52** (2020), no. 4, 620–640, DOI 10.1112/blms.12354. MR4171391

[HHN21] G. H. Hitching, M. Hoff, and P. E. Newstead, *Nonemptiness and smoothness of twisted Brill-Noether loci*, Ann. Mat. Pura Appl. (4) **200** (2021), no. 2, 685–709, DOI 10.1007/s10231-020-01009-x. MR4229545

[HL10] D. Huybrechts and M. Lehn, *The geometry of moduli spaces of sheaves*, 2nd ed., Cambridge Mathematical Library, Cambridge University Press, Cambridge, 2010, DOI 10.1017/CBO9780511711985. MR2665168

[Kle74] S. L. Kleiman, *The transversality of a general translate*, Compositio Math. **28** (1974), 287–297. MR360616

[Lan92] H. Lange, *Some geometrical aspects of vector bundles on curves*, Topics in algebraic geometry (Guanajuato, 1989), Aportaciones Mat. Notas Investigación, vol. 5, Soc. Mat. Mexicana, México, 1992, pp. 53–74. MR1308330

[LN03] H. Lange and P. E. Newstead, *Maximal subbundles and Gromov-Witten invariants*, A tribute to C. S. Seshadri (Chennai, 2002), Trends Math., Birkhäuser, Basel, 2003, pp. 310–322. MR2017590

[LeB06] P. Le Barz, *Sur les espaces multisécants aux courbes algébriques* (French, with English summary), Manuscripta Math. **119** (2006), no. 4, 433–452, DOI 10.1007/s00229-005-0621-y. MR2223626

[Mer99] V. Mercat, *Le problème de Brill-Noether et le théorème de Teixidor* (French, with English summary), Manuscripta Math. **98** (1999), no. 1, 75–85, DOI 10.1007/s002290050126. MR1669595

[MS12] E. C. Mistretta and L. Stoppino, *Linear series on curves: stability and Clifford index*, Internat. J. Math. **23** (2012), no. 12, 1250121, 25, DOI 10.1142/S0129167X12501212. MR3019424

[Mum77] D. Mumford, *Stability of projective varieties*, Enseign. Math. (2) **23** (1977), no. 1-2, 39–110. MR450272

[MOP2019] A. Marian, D. Oprea, and R. Pandharipande, *The combinatorics of Lehn's conjecture*, J. Math. Soc. Japan **71** (2019), no. 1, 299–308, DOI 10.2969/jmsj/78747874. MR3909922

[New22] P. E. Newstead, *Higher rank Brill-Noether theory and coherent systems open questions*, Proyecciones **41** (2022), no. 2, 449–480. MR4432600

[OP21] D. Oprea and R. Pandharipande, *Quot schemes of curves and surfaces: virtual classes, integrals, Euler characteristics*, Geom. Topol. **25** (2021), no. 7, 3425–3505, DOI 10.2140/gt.2021.25.3425. MR4372634

[PR03] M. Popa and M. Roth, *Stable maps and Quot schemes*, Invent. Math. **152** (2003), no. 3, 625–663, DOI 10.1007/s00222-002-0279-y. MR1988296

[Ran15] Z. Ran, *Unobstructedness of filling secants and the Gruson-Peskine general projection theorem*, Duke Math. J. **164** (2015), no. 4, 697–722, DOI 10.1215/00127094-2881483. MR3322308

[Sta21a] S. Stark, *On the Quot scheme* $\mathrm{Quot}^{\ell}(\mathscr{E})$, `arXiv:2107.03991`.

[Sta21b] S. Stark, *Cosection localization and the Quot scheme* $\mathrm{Quot}^{l}_{S}(\mathcal{E})$, Proc. A. **478** (2022), no. 2268, Paper No. 20220419, 16. MR4529797

[TiB04] M. Teixidor i Bigas, *Rank two vector bundles with canonical determinant*, Math. Nachr. **265** (2004), 100–106, DOI 10.1002/mana.200310138. MR2033069

[TiB08] M. Teixidor i Bigas, *Petri map for rank two bundles with canonical determinant*, Compos. Math. **144** (2008), no. 3, 705–720, DOI 10.1112/S0010437X07003442. MR2422346

[Ung19] M. Ungureanu, *Refined de Jonquières divisors and secant varieties on algebraic curves*, `arXiv:1911.09457`.

[Ung21a] M. Ungureanu, *Geometry of intersections of some secant varieties to algebraic curves*, J. Lond. Math. Soc. (2) **103** (2021), no. 1, 288–313, DOI 10.1112/jlms.12374. MR4203050

[Ung21b] M. Ungureanu, *Dimension theory and degenerations of de Jonquières divisors.* Int. Math. Res. Not. 2021, no. 20 (2021), 15911–15958.

[Voi19] C. Voisin, *Segre classes of tautological bundles on Hilbert schemes of surfaces*, Algebr. Geom. **6** (2019), no. 2, 186–195, DOI 10.14231/AG-2019-010. MR3914750

Oslo Metropolitan University, Postboks 4, St. Olavs plass, 0130 Oslo, Norway
Email address: gehahi@oslomet.no

Contemporary Mathematics
Volume **803**, 2024
https://doi.org/10.1090/conm/803/16103

Moduli spaces and geometric invariant theory: old and new perspectives

Victoria Hoskins

For Peter Newstead on his 80th Birthday

ABSTRACT. Many moduli spaces are constructed as quotients of group actions; this paper surveys the classical theory, as well as recent progress and applications. We review geometric invariant theory for reductive groups and how it is used to construct moduli spaces, and explain two new developments extending this theory to non-reductive groups and to stacks, which enable the construction of new moduli spaces.

Introduction

Mumford's geometric invariant theory (GIT) [**75**] for reductive groups provides a method for constructing quotients of reductive group actions in algebraic geometry. For a reductive group acting on a projective scheme, GIT provides an open *semistable* set admitting a categorical quotient which is projective and constructed from the invariant ring; furthermore, the semistable set can be explicitly described using (torus) weights via the Hilbert–Mumford criterion, rather than in terms of non-vanishing invariants. An excellent guide to GIT is Peter Newstead's lecture notes [**80**], which introduced the theory to several generations of mathematicians, including the author. The aim of this note is to give an updated treatment which highlights recent progress.

Whilst reductive GIT has been successfully employed to construct numerous moduli spaces, it has some limitations. First, it only provides moduli spaces of *semistable* objects. Second, it only applies to *reductive* group actions. Third, it only applies in the situation where the moduli problem is presented in terms of a *group action.* Two recent developments aim to overcome some of these issues: GIT for *non-reductive groups* and *stacky generalisations* of GIT.

One of the first challenges for non-reductive group actions is the possibility of non-finitely generated invariant rings; the best-known example is Nagata's counterexample [**77**] to Hilbert's 14th Problem. However, even when non-reductive invariant rings are finitely generated, the corresponding 'GIT quotient' is not well-behaved (for example, the quotient map might not even be surjective and its image

2020 *Mathematics Subject Classification.* Primary 14L24, 14D20; Secondary 14D23.
Key words and phrases. Geometric invariant theory, non-reductive group, moduli, good moduli space.

may only be constructible, see §5.1). Although it is possible to construct geometric quotients of open subsets [**34, 85, 98**], these open subsets are typically hard to describe explicitly. However, recent work on GIT [**14, 15**] for non-reductive groups with *graded unipotent radical* (e.g. $\mathbb{G}_a \rtimes \mathbb{G}_m$ or parabolic subgroups) has enabled the construction of projective quotients of certain stable sets, which admit explicit Hilbert–Mumford type descriptions; the price to pay for obtaining these explicit projective non-reductive GIT quotients is that one must impose certain stabiliser assumptions. One of the goals of this survey is to explain the origins, assumptions and results of non-reductive GIT in as simple a context as possible to make them accessible to a broad audience, as well as to highlight some exciting applications.

We also outline another significant development that extends ideas of (reductive) GIT to stacks, as pioneered by Alper, Halpern–Leistner and Heinloth [**2, 7, 48, 52**]. Alper's notion of *good and adequate moduli spaces* of stacks enables GIT-free constructions of moduli spaces. This is even more tangible following the recent existence criteria of Alper–Halpern-Leistner–Heinloth [**7**], which equates the existence of moduli spaces to two simple valuative criteria and has been applied to various moduli problems [**4, 5, 12, 13**].

However, adequate and good moduli spaces are locally modelled on reductive GIT and require closed points to have reductive stabiliser groups. Thus, in some senses these two recent developments are orthogonal to each other and ideally there should eventually be an extension of non-reductive GIT to stacks.

Structure. In §1, we describe moduli problems and their relation to group actions; then we survey Mumford's reductive GIT in §2 and describe (in)stability in §3. In §4, we briefly outline stacky generalisations of reductive GIT as a new approach to constructing moduli spaces. In §5, we provide an introduction to non-reductive GIT and explain some interesting applications in §6.

Conventions. Throughout we will assume k is an algebraically closed field and all schemes are assumed to be finite type k-schemes, unless otherwise stated. By a point, we will mean a k-point (or equivalently, a closed point).

1. Moduli problems and group actions

We start with an example-driven introduction to moduli problems in §1.1 and describe the relation to group actions in §1.2. Finally in §1.3, we give some basic definitions on algebraic groups, actions and quotients, which lays the foundations for GIT in §2.

1.1. Moduli functors and spaces. Naively, a moduli problem is a collection $\mathcal{A}$ of objects with an equivalence relation $\sim$ on $\mathcal{A}$ and we would like to give the set of equivalence classes $\mathcal{A}/\sim$ the structure of a scheme that encodes how objects vary continuously in 'families'.

EXAMPLE 1.1.

(1) Let $\mathcal{A}$ be the set of r-dimensional linear subspaces of an n-dimensional k-vector space and $\sim$ be equality.
(2) Let $\mathcal{A}$ be the set of finite-dimensional k-vector spaces with an endomorphism and $\sim$ be isomorphisms commuting with the endomorphism.
(3) Let $\mathcal{A}$ be the set of $n \times n$ matrices over k and $\sim$ be the equivalence relation given by similarity of matrices.

(4) Let $\mathcal{A}$ to be the set of hypersurfaces of degree d in $\mathbb{P}^n$ and $\sim$ be the relation given by projective change of coordinates.
(5) Let $\mathcal{A}$ be the collection of smooth projective curves of fixed genus and $\sim$ be the relation given by isomorphism.
(6) Let $\mathcal{A}$ be the collection of vector bundles on a fixed scheme X and $\sim$ be the relation given by isomorphisms of vector bundles.

Often there is a natural notion of families of objects over a scheme S and an extension of $\sim$ to families over S, such that we can pullback families by morphisms $T \to S$ compatibly with the notion of equivalence.

EXAMPLE 1.2.
(1) A family over S of r-dimensional linear subspaces of an n-dimensional vector space is a rank r vector subbundle $\mathcal{V} \subset \mathcal{O}_S^{\oplus n}$.
(2) A family over S of vector spaces with endomorphisms is a vector bundle $\mathcal{V}$ over S with an endomorphism $\Phi : \mathcal{V} \to \mathcal{V}$.

The next example shows there may be several ways to extend $(\mathcal{A}, \sim)$ to families.

EXAMPLE 1.3. For vector bundles on a fixed scheme X up to isomorphism, the natural notion for a family over S is a vector bundle $\mathcal{F}$ over $X \times S$ over S, but there are at least two natural equivalence relations:

$$\begin{aligned} \mathcal{F} \sim'_S \mathcal{G} &\iff \mathcal{F} \cong \mathcal{G} \\ \mathcal{F} \sim_S \mathcal{G} &\iff \mathcal{F} \cong \mathcal{G} \otimes \pi_S^* \mathcal{L} \text{ for a line bundle } \mathcal{L} \to S, \end{aligned}$$

where $\pi_S : X \times S \to S$. Since $\mathcal{L} \to S$ is locally trivial, there is a cover S_i of S such that $\mathcal{F}|_{X \times S_i} \cong \mathcal{G}|_{X \times S_i}$. Hence $\sim_S$ can be thought of a Zariski local version of $\sim'_S$.

We can now give a more precise definition of a moduli problem using families.

DEFINITION 1.4 (Moduli problem and moduli functor). A *moduli problem* consists of
(1) for each scheme S, a collection $\mathcal{A}_S$ of families over S with an equivalence relation $\sim_S$,
(2) for each morphism $T \to S$ of schemes, a pullback map $f^* : \mathcal{A}_S \to \mathcal{A}_T$

such that
(i) for $f : T \to S$ and equivalent families $\mathcal{F} \sim_S \mathcal{G}$ over S, we have $f^*\mathcal{F} \sim_T f^*\mathcal{G}$;
(ii) for any family $\mathcal{F}$ over S, we have $\mathrm{Id}_S^*\mathcal{F} = \mathcal{F}$;
(iii) for any morphisms $f : T \to S$ and $g : S \to R$, and a family $\mathcal{F}$ over R, we have an equivalence $(g \circ f)^*\mathcal{F} \sim_T f^*g^*\mathcal{F}$.

This gives rise to a *moduli functor* $\mathcal{M} : \mathrm{Sch}^{\mathrm{op}} \to \mathrm{Set}$ where

$$\mathcal{M}(S) := \{\text{families over } S\}/\sim_S \quad \text{and} \quad \mathcal{M}(f : T \to S) = f^* : \mathcal{M}(S) \to \mathcal{M}(T).$$

Notation: For a family $\mathcal{F}$ over S and a point $s : \operatorname{Spec} k \to S$, we write $\mathcal{F}_s := s^*\mathcal{F}$ to denote the corresponding family over $\operatorname{Spec} k$. We write $(\mathcal{A}, \sim) := (\mathcal{A}_{\operatorname{Spec} k}, \sim_{\operatorname{Spec} k})$.

In particular, a moduli functor is a presheaf on the category Sch of schemes. Recall that the Yoneda Lemma gives an embedding of the category of schemes into the category of presheaves; more precisely there is a fully faithful functor $h : \mathrm{Sch} \to \mathrm{PSh}(\mathrm{Sch})$ which on objects sends a scheme X to its functor of points

$\mathrm{Hom}(-, X) : \mathrm{Sch}^{\mathrm{op}} \to \mathrm{Set}$. A presheaf is called *representable* if it is in the essential image of the Yoneda embedding.

A moduli functor being representable is the ideal situation and leads to the notion of a fine moduli space, but if that fails, one can instead ask for a universal natural transformation from $\mathcal{M}$ to the functor of points of a scheme, which leads to the notion of a coarse moduli space.

DEFINITION 1.5 (Fine and coarse moduli spaces). Let $\mathcal{M} : \mathrm{Sch} \to \mathrm{Set}$ be a moduli functor.

i) A scheme M is a *fine moduli space* for $\mathcal{M}$ if it represents $\mathcal{M}$; that is, there is a natural isomorphism $\mathcal{M} \to \mathrm{Hom}(-, M)$. In this case, $\mathrm{Id}_M \in \mathrm{Hom}(M, M)$ corresponds to an element of $\mathcal{M}(M)$ called the *universal family* $\mathcal{U}$, which is a family over M up to the notion of equivalence.
ii) A *coarse moduli space* for $\mathcal{M}$ is a scheme M with a natural transformation of functors $\eta : \mathcal{M} \to h_M$ which is universal (for any natural transformation $\nu : \mathcal{M} \to \mathrm{Hom}(-, N)$ to the functor of points of a scheme, there exists a unique morphism $f : M \to N$ such that $\nu = f_* \circ \eta$) such that $\eta_{\mathrm{Spec}\, k} : \mathcal{M}(\mathrm{Spec}\, k) \to h_M(\mathrm{Spec}\, k)$ is bijective.

EXAMPLE 1.6. For 1-dimensional subspaces of k^n, a fine moduli space is given by $\mathbb{P}^n$, with the tautological line bundle $\mathcal{O}_{\mathbb{P}^n}(-1) \subset \mathcal{O}_{\mathbb{P}^n}^{\oplus n}$ giving a universal family (see [**51**, II Theorem 7.1]).

REMARK 1.7.

(1) If a fine or coarse moduli space exists, then it is unique up to unique isomorphism by its universal property.
(2) If a fine moduli space exists, then the universal family $\mathcal{U}$ over M describes all other families in the following sense: for any scheme S, we have that a family $\mathcal{F} \in \mathcal{M}(S)$ is equivalent to a morphism $f : S \to M$ with $f^*\mathcal{U} \sim_S \mathcal{F}$.
(3) Since $\mathrm{Hom}(-, M)$ is a sheaf in the Zariski topology (that is, for every scheme S and Zariski cover $\{S_i\}$ of S, the natural map
$$\{f \in F(S)\} \longrightarrow \{(f_i \in F(S_i))_i : f_i|_{S_i \cap S_j} = f_j|_{S_j \cap S_i} \text{ for all } i, j\}$$
is a bijection), for a moduli functor $\mathcal{M}$ to admit a fine moduli space it must be a sheaf in the Zariski topology.

EXERCISE 1.8. Show that the equivalence relation $\sim_S'$ on families over S of vector bundles on a fixed scheme defines a moduli functor that is not representable. (Hint: Show it fails to be a Zariski sheaf by considering the other equivalence relation $\sim_S$ for families of vector bundles).

Unfortunately, there are moduli problems which do not admit even a coarse moduli space.An excellent guide to GIT is Peter Newstead's lecture notes [**80**], which introduced the theory to several generations of mathematicians, including the author. The aim of this note is to give an updated treatment which highlights recent progress.

EXERCISE 1.9. Let $\mathcal{M}$ be a moduli functor with the *jump phenomenon*; that is, there is a family $\mathcal{F}$ over $\mathbb{A}^1$ such that $\mathcal{F}_s \sim \mathcal{F}_1$ for all $s \neq 0$ and $\mathcal{F}_0 \nsim \mathcal{F}_1$. Show that there is no coarse moduli space for $\mathcal{M}$ by showing for any natural transformation $\eta : \mathcal{M} \to \mathrm{Hom}(-, M)$, the morphism $\eta_{\mathbb{A}^1}(\mathcal{F}) : \mathbb{A}^1 \to M$ is constant.

EXAMPLE 1.10. Moduli of rank 2 degree 0 vector bundles on $\mathbb{P}^1$ exhibit the jump phenomenon: there is a family $\mathcal{F}$ of rank 2 degree 0 vector bundles over $\mathbb{A}^1$ such that

$$\mathcal{F}_s = \begin{cases} \mathcal{O}_{\mathbb{P}^1}^{\oplus 2} & s \neq 0 \\ \mathcal{O}_{\mathbb{P}^1}(1) \oplus \mathcal{O}_{\mathbb{P}^1}(-1) & s = 0. \end{cases}$$

Indeed this family is constructed using the isomorphisms

$$\mathrm{Ext}^1(\mathcal{O}_{\mathbb{P}^1}(1), \mathcal{O}_{\mathbb{P}^1}(-1)) \cong H^1(\mathbb{P}^1, \mathcal{O}_{\mathbb{P}^1}(-2)) \cong H^0(\mathbb{P}^1, \mathcal{O}_{\mathbb{P}^1})^* \cong k.$$

Another reason for a coarse moduli space to fail to exist is if the moduli problem is *unbounded*: there does not exist a family $\mathcal{F}$ over a scheme S (of finite type over k) such that for any object E (i.e. family over k), we have $E \sim \mathcal{F}_s$ for some (possibly non-unique) $s \in S$.

EXERCISE 1.11. Show the moduli problem of rank 2 degree 0 vector bundles on $\mathbb{P}^1$ is unbounded: suppose there exists a family $\mathcal{F}$ over S such that for any such vector bundle E we have $\mathcal{F}_s \sim E$ for some $s \in S$, then show that S cannot be Noetherian by considering the subschemes

$$S_n := \{s \in S : \dim H^0(\mathbb{P}^1, \mathcal{F}_s) \geq n\},$$

which are closed by the semi-continuity theorem.

A further obstruction to the existence of a coarse moduli space is that there may be non-trivial families which are fibrewise trivial; this may happen when there are non-trivial automorphisms.

One possible solution for dealing with some of these issues is to instead work with a moduli stack and try to find a coarse or good moduli space for this stack as in §4. However, often by imposing a notion of stability on objects, which can be moduli-theoretic or arising from the GIT construction, one obtains a moduli problem for which one can construct moduli spaces.

1.2. Construction of moduli spaces using group actions. Many moduli spaces can be constructed as quotients of group actions via the following strategy (after fixing any discrete invariants and restricting to a bounded class of objects):

(1) Find an overparametrisation: find a parameter scheme X with a family $\mathcal{F}$ such that any other family can be locally obtained by pullback from $\mathcal{F}$ (possibly non-uniquely).
(2) Find a group action describing the symmetries: find a group G acting on X such that the orbits correspond to the equivalence classes.
(3) Take a quotient (in the category of schemes if possible).

The third step is typically performed using Geometric Invariant Theory (GIT).

Let us give some examples, before describing algebraic group actions in §1.3.

EXAMPLE 1.12.

(1) 1-dimensional vector subspaces in $V = k^n$ can be parametrised by fixing a basis vector in $X = V \setminus \{0\}$. Since two basis vectors are related by scalar multiplication, $\mathbb{G}_m$ acting on X describes the symmetries and $\mathbb{P}^{n-1} = X/\mathbb{G}_m$ is the fine moduli space.
(2) An r-dimensional subspace in $V = k^n$ can be parametrised by choosing a basis, which gives an element in $\mathrm{Mat}_{r\times n}$ of rank r and the choice of basis is controlled by the action of GL_r by left multiplication. In Exercise 3.14,

we will see that the open locus of rank r matrices is a GIT semistable locus that admits a quotient, namely a Grassmannian.

(3) A projective hypersurface of degree d in $\mathbb{P}^n$ is given by the vanishing locus $\{F = 0\}$ of a degree d homogeneous polynomial in $n+1$ variables. Since λF defines the same hypersurface, one can consider $X = \mathbb{P}(k[x_0, \dots, x_n]_d)$ and the action of $G = \mathrm{PGL}_n = \mathrm{Aut}(\mathbb{P}^n)$ describes these hypersurfaces up to change of coordinates.

(4) For vector bundles of rank n and degree d on a smooth projective curve C, provided d is sufficiently large, any *semistable*[1] vector bundle E can be parametrised as a quotient of a fixed vector bundle (see [**80**, Lemma 5.2]): for E semistable of sufficiently large degree, the evaluation map is surjective and E has vanishing higher cohomology, so by choosing a basis of global sections we obtain a quotient

$$\mathcal{O}_C^{\oplus \chi} \cong H^0(E) \otimes \mathcal{O}_C \overset{\mathrm{ev}}{\twoheadrightarrow} E$$

where $\chi = d + n(1-g)$ is the Euler characteristic of E. Consequently, a Quot scheme parametrising quotients of $\mathcal{O}_C^{\oplus \chi}$ with fixed invariants gives an overparametrisation and the action of GL_χ describes the symmetries.

1.3. Algebraic groups, actions and quotients. To realise the above strategy, we first need to formally introduce algebraic groups and their actions (for detailed expositions, see [**20, 22, 72**]).

DEFINITION 1.13. An *algebraic group* over k is a a group object in the category of k-schemes (that is, a k-scheme G with identity element $e : \operatorname{Spec} k \to G$, group operation $m : G \times G \to G$ and inversion $i : G \to G$ given by morphisms of schemes such that the usual group axioms are stated as commutativity of certain diagrams). We say G is an *affine algebraic group* if the underlying scheme G is affine.

REMARK 1.14. The k-algebra $\mathcal{O}(G)$ of regular functions on G is a commutative *Hopf algebra* with comultiplication $m^* : \mathcal{O}(G) \to \mathcal{O}(G) \otimes \mathcal{O}(G)$, coinversion $i^* : \mathcal{O}(G) \to \mathcal{O}(G)$ and counit $e^* : \mathcal{O}(G) \to k$ and dualised commutative diagrams. In fact, there is a one-one correspondence between finitely generated Hopf algebras over k and affine algebraic groups over k [**72**, II Theorem 5.1].

As the next example shows, many familiar groups are affine algebraic groups.

EXAMPLE 1.15.

(1) The *additive group* $\mathbb{G}_a = \operatorname{Spec} k[t]$ over k is the algebraic group whose underlying scheme is the affine line $\mathbb{A}^1$ over k and whose group operation is given by addition:

$$m^*(t) = t \otimes 1 + 1 \otimes t \quad \text{and} \quad i^*(t) = -t.$$

For a k-algebra R, we have $\mathbb{G}_a(R) = (R, +)$.

(2) The *multiplicative group* $\mathbb{G}_m = \operatorname{Spec} k[t, t^{-1}]$ over k is the algebraic group whose underlying variety is the $\mathbb{A}^1 - \{0\}$ and whose group operation is given by multiplication:

$$m^*(t) = t \otimes t \quad \text{and} \quad i^*(t) = t^{-1}.$$

For a k-algebra R, we have $\mathbb{G}_m(R) = (R^\times, \cdot)$.

[1]There is a natural notion of semistability involving verifying an inequality of slopes for all subbundles, which turns out to be related to a corresponding GIT notion of semistability.

(3) The *general linear group* GL_n over k is an open subvariety of $\mathbb{A}^{n^2}$ cut out by the non-vanishing of the determinant. It is an affine variety with coordinate ring $k[x_{ij} : 1 \leq i, j \leq n]_{\det(x_{ij})}$. The co-group operations are defined by:

$$m^*(x_{ij}) = \sum_{k=1}^{n} x_{ik} \otimes x_{kj} \quad \text{and} \quad i^*(x_{ij}) = (x_{ij})_{ij}^{-1}$$

where, using Cramer's rule, inversion is given by $1/det(x_{ij})$ multiplied by the adjugate matrix.

(4) For a finite group G, the group algebra $k[G]$ is a Hopf algebra and determines an affine algebraic group $\underline{G}_k := \mathrm{Spec}(k[G])$, whose k-points are identified with elements of G.

(5) For $n \geq 1$, the group of nth roots of unity is $\mu_n := \mathrm{Spec}\, k[t, t^{-1}]/(t^n - 1) \subset \mathbb{G}_m$. Write I for the ideal $(t^n - 1)$ of $R := k[t, t^{-1}]$. Then

$$m^*(t^n - 1) = t^n \otimes t^n - 1 \otimes 1 = (t^n - 1) \otimes t^n + 1 \otimes (t^n - 1) \in I \otimes R + R \otimes I$$

which implies that μ_n is an algebraic subgroup of $\mathbb{G}_m$. If n is different from $\mathrm{char}(k)$, the polynomial $X^n - 1$ is separable and there are n distinct roots in k. Then the choice of a primitive nth root of unity in k determines an isomorphism $\mu_n \simeq \underline{\mathbb{Z}/n\mathbb{Z}}_k$. However, if $n = \mathrm{char}(k)$, then $X^n - 1 = (X-1)^n$ in $k[X]$, which implies that the scheme μ_n is non-reduced (with 1 as the only closed point).

A *linear algebraic group* is a closed subgroup of GL_n; hence, any linear algebraic group is an affine algebraic group. The converse statement is also true: any affine algebraic group is a linear algebraic group (see Remark 1.19).

DEFINITION 1.16. An (algebraic) *action* of an affine algebraic group G on a scheme X is a morphism of schemes $\sigma : G \times X \to X$ such that the following diagrams commute

$$\begin{array}{ccc} \mathrm{Spec}\, k \times X & \xrightarrow{e \times \mathrm{id}_X} & G \times X \\ & \searrow^{\cong} & \downarrow{\scriptstyle \sigma} \\ & & X \end{array} \qquad \begin{array}{ccc} G \times G \times X & \xrightarrow{\mathrm{id}_G \times \sigma} & G \times X \\ {\scriptstyle m_G \times \mathrm{id}_X}\downarrow & & \downarrow{\scriptstyle \sigma} \\ G \times X & \xrightarrow[\sigma]{} & X. \end{array}$$

A subscheme $Z \subset X$ is called *G-invariant* if it is preserved by the action; that is, $\sigma(G \times Z) \subset Z$.

A morphism $f : X \to Y$ between schemes with actions $\sigma_X : G \times X \to X$ and $\sigma_Y : G \times Y \to Y$ is *G-equivariant* if the following diagram commutes

$$\begin{array}{ccc} G \times X & \xrightarrow{\mathrm{id}_G \times f} & G \times Y \\ {\scriptstyle \sigma_X}\downarrow & & \downarrow{\scriptstyle \sigma_Y} \\ X & \xrightarrow[f]{} & Y. \end{array}$$

If Y is given the trivial action $\sigma_Y = \pi_Y : G \times Y \to Y$, then we say $f : X \to Y$ is *G-invariant.*

REMARK 1.17. For an action $\sigma : G \times X \to X$, there is an induced action of G on the ring of regular functions $\mathcal{O}(X)$: for $g \in G$, $f \in \mathcal{O}(X)$ and $x \in X$, we have

$$(g \cdot f)(x) = f(g^{-1} \cdot x).$$

By the following lemma, any action of an affine algebraic group G on an affine scheme X gives a G-action on the k-algebra $\mathcal{O}(X)$ which is *rational*; that is, every $f \in \mathcal{O}(X)$ is contained in a finite dimensional G-invariant linear subspace of $\mathcal{O}(X)$.

LEMMA 1.18. *For an affine algebraic group G acting on an affine scheme X, any finite dimensional vector subspace of $\mathcal{O}(X)$ is contained in a finite dimensional G-invariant vector subspace.*

PROOF. Let $\sigma^* : \mathcal{O}(X) \to \mathcal{O}(G) \otimes \mathcal{O}(X)$ denote the coaction. Let $W = \mathrm{Span}_k(f_1, \dots, f_n) \subset \mathcal{O}(X)$ and write $\sigma^*(f_i) = \sum_{j=1}^{n_i} h_{ij} \otimes f_{ij}$ with $h_{ij} \in \mathcal{O}(G)$ and $g_{ij} \in \mathcal{O}(X)$. The vector space spanned by f_{ij} is a G-invariant finite-dimensional subspace containing W, as $g \cdot f_i = \sum_j h_{ij}(g) f_{ij}$. □

REMARK 1.19. By applying this to the action of G on itself by left multiplication, if we let W be a vector space spanned by a finite choice of algebra generators for $\mathcal{O}(G)$, then W is contained in a finite dimensional G-invariant vector subspace $V \subset \mathcal{O}(X)$. One can prove that there is an embedding $G \to \mathrm{GL}(V)$ to show any affine algebraic group over k is a linear algebraic group.

We can define orbits and stabilisers in this setting; the latter has a scheme structure by definition, and we shall soon see the former can also be equipped with a scheme structure.

DEFINITION 1.20 (Orbits and stabilisers). For an action $\sigma : G \times X \to X$ of an affine algebraic group G on a scheme X and for a k-point $x \in X$, we define

(1) the *orbit* $G \cdot x$ of x to be the (set-theoretic) image of the action morphism $\sigma_x = \sigma(-, x) : G(k) \to X(k)$ given by $g \mapsto g \cdot x$;
(2) the *stabiliser* G_x of x to be the fibre product of the morphisms $\sigma_x : G \to X$ and $x : \mathrm{Spec}\, k \to X$.

The stabiliser G_x of x is a closed subscheme of G (as it is the preimage of a closed subscheme of X under $\sigma_x : G \to X$) and a subgroup of G. By Chevalley's Theorem [**51**, II Exercise 3.19], as the image of a morphism of schemes, the orbit is a priori only a constructible subset of X. However, we claim it is a locally closed subset and so can be equipped with the structure of a reduced locally closed subscheme of X. Indeed, $G \cdot x$ is open in its closure: since the orbit is constructible, there is a dense open subset U of $\overline{G \cdot x}$ with $U \subset G \cdot x$ and, as G acts transitively on the orbit, every point in the orbit is contained in a G-translate of U.

The boundary of an orbit is a union of orbits of strictly smaller dimension, and so in particular every orbit closure contains a closed orbit (of minimal dimension). There is also an orbit stabiliser theorem:

$$\dim G = \dim G_x + \dim G \cdot x$$

as $\sigma_x : G \to G \cdot x$ is flat (by transitivity of the G-action, we can deduce this from generic flatness) and so we can apply the dimension formula for fibres of a flat morphism [**51**, Proposition III.9.5].

In general, for an action $\sigma : G \times X \to X$, the set of orbits X/G is not a scheme. Instead, we ask for a universal quotient in the category of schemes.

DEFINITION 1.21 (Categorical quotient). For an action of an affine algebraic group G on a scheme X, a *categorical quotient* is a G-invariant morphism $\varphi : X \to Y$ of schemes which is universal (that is, every other G-invariant morphism $f : X \to Z$ factors uniquely through φ so that there exists a unique morphism $h : Y \to Z$ such that $f = h \circ \varphi$). If the preimage of each k-point in Y is a single orbit, then we say φ is an *orbit space.*

If a categorical quotient exists, then it is unique up to unique isomorphism. In cases where a categorical quotient does not exist, one may want to enlarge the category of schemes to algebraic spaces or even algebraic stacks.

A categorical quotient is constant on orbits and orbit closures. Hence, a categorical quotient is an orbit space only if the action of G on X is closed; that is, all the orbits $G \cdot x$ are closed.

EXAMPLE 1.22.

(1) For $\mathbb{G}_m$ acting on $\mathbb{A}^n$ by scalar multiplication $t \cdot (a_1, \ldots, a_n) = (ta_1, \ldots, ta_n)$, there are two types of orbits:
 - punctured lines through the origin,
 - the origin (closed of dimension 0)

 Every orbit contains the origin in its closure. As any $\mathbb{G}_m$-invariant function on $\mathbb{A}^n$ is constant on orbits and their closures, it must be constant and so factors via the structure map $\pi : \mathbb{A}^n \to \operatorname{Spec} k$. Hence, the structure map is a categorical quotient.

(2) For the action of $\mathbb{G}_m$ on $\mathbb{A}^2$ by $t \cdot (x, y) = (tx, t^{-1}y)$, the orbits are
 - conics $\{(x, y) : xy = \alpha\}$ for $\alpha \in \mathbb{A}^1 \setminus \{0\}$ (closed of dimension 1),
 - the punctured x-axis,
 - the punctured y-axis,
 - the origin (closed of dimension 0).

 The punctured axes both contain the origin in their orbit closures. We will see that $\mathbb{A}^2 \to \mathbb{A}^1$ given by $(x, y) \mapsto xy$ is a categorical quotient.

We see the sort of problems that may occur when we have non-closed orbits. In the first example, our geometric intuition tells us that we would ideally like to remove the origin and then take the quotient of $\mathbb{G}_m$ acting on $\mathbb{A}^n \setminus \{0\}$ to obtain the projective space $\mathbb{P}^{n-1} = (\mathbb{A}^n \setminus \{0\})/\mathbb{G}_m$, which is an orbit space for this action. We will return to this example and see that by introducing a non-trivial notion of GIT semistability, we can remove the origin (see Example 2.20).

The following stronger notion of quotient arises in GIT [**89**, Definition 1.5].

DEFINITION 1.23 (Good quotient). A morphism $\varphi : X \to Y$ is a *good quotient* for an action of G on X if

i) φ is G-invariant, surjective and affine;
ii) The map $\mathcal{O}_Y \to \varphi_* \mathcal{O}_X^G$ is an isomorphism.
iii) If W_1 and W_2 are disjoint G-invariant closed subschemes, then $\varphi(W_1)$ and $\varphi(W_2)$ are disjoint closed subschemes.

If moreover the preimage of each point is a single orbit, then we say φ is a *geometric quotient.*

REMARK 1.24.

(1) The definition of a good quotient is local in the target, which enables the construction of good quotients via gluing.

(2) The last two conditions imply that φ is surjective: the second property shows that φ is dominant (i.e. the image of φ is dense in Y) and the third condition shows that the image of φ is closed. Furthermore it implies that for each $y \in Y$, the preimage $\varphi^{-1}(y)$ contains a unique closed orbit, whose stabiliser is reductive (see Definition 2.3 and [**69**], as the quotient of the reductive group G by the subgroup G_x is affine if and only if G_x is reductive). In particular, if all orbits are closed, then φ is a geometric quotient.

(3) The third condition also enables us to determine when two orbit closures meet: we have $\overline{G \cdot x_1} \cap \overline{G \cdot x_2} \neq \phi$ if and only if $\varphi(x_1) = \varphi(x_2)$.

(4) Any good quotient is a categorical quotient; see [**80**, Proposition 3.11].

Let us relate the construction of moduli spaces with categorical quotients. For a moduli problem $\mathcal{M}$, a family $\mathcal{F}$ over a scheme S has the *local universal property* if for any other family $\mathcal{G}$ over a scheme T and for any k-point $t \in T$, there exists a neighbourhood U of t in T and a morphism $f : U \to S$ such that $\mathcal{G}|_U \sim_U f^*\mathcal{F}$.

PROPOSITION 1.25 ([**80**, Proposition 2.13]). *Let $\mathcal{M}$ be a moduli problem for which there exists a family $\mathcal{F}$ over S with the local universal property. Suppose that there is an algebraic group G acting on S such that two k-points s, t lie in the same G-orbit if and only if $\mathcal{F}_t \sim \mathcal{F}_s$. Then*

i) any coarse moduli space is a categorical quotient of the G-action on S;

ii) a categorical quotient of the G-action on S is a coarse moduli space if and only if it is an orbit space.

PROOF. For any scheme M, we claim that there is a bijective correspondence

$$\left\{ \begin{array}{c} \text{natural transformations} \\ \eta : \mathcal{M} \to \operatorname{Hom}(-, M) \end{array} \right\} \longleftrightarrow \left\{ \begin{array}{c} G\text{-invariant morphisms} \\ f : S \to M \end{array} \right\}$$

given by $\eta \mapsto \eta_S(\mathcal{F})$, which is G-invariant by our assumptions about the G-action on S. Conversely, given a G-invariant morphism $f : S \to M$, we define $\eta : \mathcal{M} \to \operatorname{Hom}(-, M)$ to associate to a family $\mathcal{G}$ over T, a morphism $\eta_T(\mathcal{G}) : T \to M$ glued together locally by using the local universal property of $\mathcal{F}$ over S. More precisely, we can cover T by open subsets U_i such that there is a morphism $h_i : U_i \to S$ and $h_i^*\mathcal{F} \sim_{U_i} \mathcal{G}|_{U_i}$. For $u \in U_i \cap U_j$, we have

$$\mathcal{F}_{h_i(u)} \sim (h_i^*\mathcal{F})_u \sim \mathcal{G}_u \sim (h_j^*\mathcal{F})_u \sim \mathcal{F}_{h_j(u)}$$

and so by assumption $h_i(u)$ and $h_j(u)$ lie in the same G-orbit. Since f is G-invariant, the compositions $f \circ h_i : U_i \to M$ glue to a morphism $\eta_T(\mathcal{G}) : T \to M$.

Hence, if $(M, \eta : \mathcal{M} \to h_M)$ is a coarse moduli space, then $\eta_S(\mathcal{F}) : S \to M$ is G-invariant and the universal G-invariant morphism from S, which proves statement *i)*. Furthermore, the G-invariant morphism $\eta_S(\mathcal{F}) : S \to M$ is an orbit space if and only if $\eta_{\operatorname{Spec} k}$ is bijective, which proves statement *ii)*. □

2. Mumford's reductive geometric invariant theory

The origins of GIT go back to 19th century invariant theory and a question of Hilbert on the finite generation of invariant rings. We begin with Hilbert's 14th problem in §2.1 and describe other techniques for constructing quotients in §2.2. We give various definitions and examples related to reductive and unipotent groups in §2.3, and prove finite generation results for invariant rings in §2.4. We describe

Mumford's GIT [**75**] for affine schemes in §2.5, for projective schemes in §2.6 and in general in §2.7, and state a twisted affine version in §2.8. Beyond Mumford's book [**75**], the notes of Newstead [**80**] and Thomas [**94**] provide excellent introductions to GIT.

2.1. Hilbert's 14th Problem. Consider an action $\sigma : G \times X \to X$ of an affine algebraic group on an affine scheme. The coaction determines a linear representation $G \to \mathrm{GL}(\mathcal{O}(X))$. Any G-invariant morphism $\phi : X \to Z$ induces a homomorphism $\phi^* : \mathcal{O}(Z) \to \mathcal{O}(X)$ whose image is contained in the ring of G-invariant functions

$$\mathcal{O}(X)^G := \{f \in \mathcal{O}(X) : g \cdot f = f \text{ for all } g \in G\}.$$

Consequently one can ask if the inclusion of the ring of G-invariant functions corresponds to a morphism of affine schemes.

QUESTION 2.1 (Hilbert's 14th Problem). Is $\mathcal{O}(X)^G$ a finitely generated k-algebra?

Hilbert showed that for the general linear group over the complex numbers, the answer was yes. However, in general $\mathcal{O}(X)^G$ is not finitely generated, due to a counterexample of Nagata constructed using an action of a product of additive groups [**76, 77**]; see [**35**] for a survey of counterexamples to Hilbert's 14th problem. Fortunately, Nagata also showed that the answer is yes for a large number of groups, namely reductive groups and in this case Mumford showed that taking the spectrum of the inclusion $\mathcal{O}(X)^G \subset \mathcal{O}(X)$ gives a categorical quotient of the action. For non-reductive groups, we will see in §5.1 that even when $\mathcal{O}(X)^G$ is finitely generated, taking the spectrum of the inclusion of invariants does not always yield a categorical quotient.

2.2. Constructions of quotients by affine algebraic groups. Before turning to Mumford's GIT for reductive groups, let us give a brief summary of some results about the construction of quotients of affine algebraic groups in general.

For a free action of an affine algebraic group on a scheme, there is a geometric quotient in the category of *algebraic spaces* by results of Artin [**8**] and Kollár [**65**]. For a finite group, it is easy to see that a free action induces an étale equivalence relation and any quotient of a scheme by an étale equivalence relation is an algebraic space. There are examples of free actions whose geometric quotient is not a scheme: an example of Hironaka gives an action of a finite group whose geometric quotient is not a scheme (see [**51**, Appendix B, Example 3.4.1]) and an example of Derksen gives an action of the additive group $\mathbb{G}_a$ whose geometric quotient is not a scheme (see [**29**, Example 18]).

Rosenlicht [**85**] showed that for a connected affine algebraic group G acting on an irreducible variety X, there is a dense open subset which admits a geometric quotient. Unfortunately, this set is non-explicit, as his proof involves showing that the field $k(X)^G$ of invariant *rational* functions is finitely generated (see [**43**, §19 Appendix Lemma 1]).

REMARK 2.2 (Transfer Principle). Let $H < G$ be a closed subgroup of an affine algebraic group G, then H acts on G by left multiplication and this action has a geometric quotient G/H and the quotient map $G \to G/H$ is étale locally trivial (see [**20**, p181] and [**22**, Theorem 1.16]).

For $H < G$ as above, suppose there is an action of H on X; then for the diagonal action of H on $G \times X$ by $h \cdot (g, x) = (gh^{-1}, hz)$, there is a geometric quotient $G \times^H X$ (as H acts set-theoretically freely, this quotient always exists as an algebraic space, but is a scheme if for example H is special or H is reductive and X is quasi-projective with an ample linearisation); for details on this construction, see [**28**, III §4]. The action of G on itself by left multiplication induces a G-action on $G \times^H X$ such that there is a bijective correspondence between H-orbits in X and G-orbits in $G \times^H X$. A scheme is a geometric H-quotient of X if and only if it is a geometric G-quotient of $G \times^H X$. Furthermore, if X is affine and the H-action on X extends to G, then we have the following transfer principle (due to Roberts, see [**43**, §9])

$$\mathcal{O}(X)^H \cong (\mathcal{O}(X) \otimes \mathcal{O}(G/H))^G.$$

Note G/H may not be affine, so $\mathcal{O}(G/H)$ is not necessarily finitely generated (see Remark 2.10).

2.3. Reductive groups. Let us fix some definitions for unipotent and reductive groups; see also [**26, 72**] for alternative equivalent formulations.

DEFINITION 2.3. An affine algebraic k-group G is

i) *unipotent* if it is isomorphic to a subgroup of a standard unipotent group $\mathbb{U}_n \subset \mathrm{GL}_n$ consisting of upper triangular matrices with diagonal entries equal to 1.
ii) *reductive* if it is smooth and every connected unipotent normal subgroup is trivial (this second condition is often phrased as asking for the unipotent radical to be trivial).
iii) *geometrically reductive* if for every finite dimensional linear representation $\rho : G \to \mathrm{GL}(V)$ and every non-zero G-invariant point $v \in V$, there is a non-constant G-invariant homogeneous polynomial $f \in \mathcal{O}(V)$ such that $f(v) \neq 0$.
iv) *linearly reductive* if for every finite dimensional linear representation $\rho : G \to \mathrm{GL}(V)$ and every non-zero G-invariant point $v \in V$, there is a non-constant G-invariant linear polynomial $f \in \mathcal{O}(V)$ such that $f(v) \neq 0$.

REMARK 2.4.

(1) G is unipotent if and only if every finite dimensional linear representation $\rho : G \to \mathrm{GL}(V)$ has a non-zero fixed point.
(2) G is linearly reductive if and only if every finite dimensional linear representation $\rho : G \to \mathrm{GL}(V)$ is completely reducible (that is, ρ decomposes as a direct sum of irreducible representations) or equivalenty if taking G-invariants on finite dimensional linear G-representations is exact.

EXAMPLE 2.5.

(1) The additive group $\mathbb{G}_a$ is unipotent, as we have an embedding $\mathbb{G}_a \hookrightarrow \mathbb{U}_2$ given by

$$c \mapsto \begin{pmatrix} 1 & c \\ 0 & 1 \end{pmatrix}.$$

(2) In characteristic p, there is a finite subgroup $\alpha_p \subset \mathbb{G}_a$ where we define the functor of points of α_p by associating to a k-algebra R,

$$\alpha_p(R) := \{c \in \mathbb{G}_a(R) : c^p = 0\}.$$

This is represented by the scheme $\operatorname{Spec} k[t]/(t^p)$ and so α_p is a unipotent group which is not smooth.

(3) The multiplicative group $\mathbb{G}_m$ or any algebraic torus $T = \mathbb{G}_m^r$ is linearly reductive: as any linear representation $\rho : T \to \mathrm{GL}(V)$ admits a weight decomposition

$$V = \bigoplus_{\chi \in X^*(T)} V_\chi \quad \text{where } V_\chi = \{v \in V : t \cdot v = \chi(t)v \ \text{ for all } \ t \in T\}.$$

EXERCISE 2.6. Prove that any finite group of order not divisible by the characteristic of k is linearly reductive. (Hint: consider averaging over the group.)

For smooth affine algebraic group schemes over k, we have

$$\text{linearly reductive} \implies \text{geometrically reductive} \iff \text{reductive}$$

and all three notions coincide in characteristic zero. The first implication is immediate from the definitions and, in characteristic zero, the opposite implication goes back to Weyl and uses the representation theory of compact Lie groups (this is known as Weyl's unitary trick). The equivalence between reductive and geometrically reductive for smooth affine group schemes was conjectured by Mumford after Nagata proved that every geometrically reductive group is reductive [**77**]; the opposite implication was proved by Haboush [**45**].

Let us state an important property of geometrically reductive group actions.

LEMMA 2.7 (Geometrically reductive group actions separate closed orbits, [**80**, Lemma 3.3]). *Let G be a geometrically reductive group acting on an affine scheme X. If W_1 and W_2 are disjoint G-invariant closed subsets of X, then there is an invariant function $f \in \mathcal{O}(X)^G$ which separates these sets, i.e.*

$$f(W_1) = 0 \quad \text{and} \quad f(W_2) = 1.$$

2.4. Finitely generated rings of invariants. Recall that for an action of an affine algebraic group G on an affine scheme X, the associated action on the coordinate ring $\mathcal{O}(X)$ is rational.

THEOREM 2.8 (Nagata, [**77**]). *Let G be a geometrically reductive group acting rationally on a finitely generated k-algebra A. Then the G-invariant subalgebra A^G is finitely generated.*

We will outline the proof of this theorem in the significantly easier case when G is linearly reductive; see [**80**, Theorem 3.4] for the full proof. In this case, one can construct a *Reynolds operator*, which is a projection $R : A \twoheadrightarrow A^G$ onto the G-invariants that satisfies $R(ab) = aR(b)$ for all $a \in A^G$ and $b \in A$. If G is finite, R can be viewed as averaging over the group. Using the Reynolds operator, one can show that A^G is Noetherian and then prove it is finitely generated. This approach is similar to Hilbert's proof that over the complex numbers the ring of invariants for GL_n is finitely generated.

PROOF OF THEOREM 2.8 (FOR LINEARLY REDUCTIVE GROUPS). Since A is a finitely generated k-algebra, it has a countable basis as a k-vector space; thus A can be written as an increasing union of finite dimensional vector spaces. By applying Lemma 1.18 to these vector spaces, we can write A as an increasing union of finite dimensional G-invariant vector spaces W_n over $n \in \mathbb{N}$.

Our assumption that G is linearly reductive implies that the finite dimensional G-representation W_n is completely reducible. In particular, we can write W_n as a sum of G-representations

$$W_n = W_n^G \oplus W_n'$$

and obtain a projection $R_n : W_n \twoheadrightarrow W_n^G$, which together induce a projection $R : A \to A^G$.

To show that R is a Reynolds operator, we need to show $R(ab) = aR(b)$ for all $a \in A^G$ and $b \in A$. For this take n, so $a, b \in W_n$ and pick $m \geq n$ such that left multiplication $l_a : A \to A$ restricts to a homomorphism of G-representations

$$l_a : W_n \to W_m.$$

As above, we write $W_n = W_n^G \oplus W_n'$. Since $a \in A^G$, we have $l_a(W_n^G) \subset W_m^G$ and by Schur's Lemma, the image of each irreducible representation appearing in W_n' is either zero or isomorphic to that irreducible representation, thus $l_a(W_n') \subset W_m'$. If we write $b = b^G + b' \in W_n^G \oplus W_n'$, then

$$ab = l_a(b) = l_a(b^G) + l_a(b') = ab^G + ab' \in W_m^G \oplus W_m'.$$

Hence, $R(ab) = ab^G = aR(b)$ as required.

For any ideal $I \subset A^G$, we have $I \subset IA \cap A^G$, and using the Reynolds operator, one can show the opposite inclusion. Hence $I = IA \cap A^G$ and from this we deduce A^G is Noetherian: any increasing chain of ideals I_n in A^G must stabilise, as the corresponding chain of ideals $I_n A$ stabilises due to A being Noetherian.

By choosing generators for the k-algebra A, we can realise it as a quotient of a polynomial ring with linear G-action $\mathrm{Sym}^*(V) \twoheadrightarrow A$. Since any G-equivariant homomorphism of algebras commutes with their Reynolds operators, we obtain a surjection $\mathrm{Sym}^*(V)^G \twoheadrightarrow A^G$ and so to show A^G is finitely generated, it suffices to show $\mathrm{Sym}^*(V)^G$ is finitely generated. Thus we may assume $A = \mathrm{Sym}^*(V)$ is a polynomial ring with linear action $G \to \mathrm{GL}(V)$. Since A^G is Noetherian, the ideal $A_+^G := \oplus_{n>0} \mathrm{Sym}^n(V)^G$ is finitely generated and the generators of this ideal are generators of A^G as a k-algebra. □

Popov [**82**] proved a converse to Nagata's theorem: for any non-reductive group G there is an affine scheme X such that $\mathcal{O}(X)^G$ is not finitely generated.

In some simple situations, the ring of invariants for a non-reductive group is finitely generated; however, the corresponding morphism of schemes may fail to be a good quotient (see §5.1).

THEOREM 2.9 (Weitzenböck [**97**]). *Assume that the characteristic of k is zero, then any linear $\mathbb{G}_a$-action on $\mathbb{A}^n$ extends to SL_2. In this case, the invariant ring $\mathcal{O}(\mathbb{A}^n)^{\mathbb{G}_a}$ is finitely generated.*

PROOF (AFTER SESHADRI [**87**]). The first statement follows by putting the associated locally nilpotent derivation (see §5.3) in Jordan normal form, so that each length n Jordan block corresponds to the standard SL_2-representation $\mathrm{Sym}^{n-1}(k^2)$; for details, see [**43**, Lemma 10.2].

Assuming that the linear $\mathbb{G}_a$-action extends to SL_2, let us prove that the ring of invariants is finitely generated. By the transfer principle (Remark 2.2), the ring of $\mathbb{G}_a$-invariants on $\mathbb{A}^n$ is isomorphic to the ring of SL_2-invariants on $\mathbb{A}^n \times^{\mathbb{G}_a} \mathrm{SL}_2$.

For $\mathbb{G}_a$-acting on SL_2 by left multiplication, the bottom row is invariant and thus the map $\mathrm{SL}_2 \to \mathbb{A}^2 \setminus \{0\}$ given by sending a matrix $A \in \mathrm{SL}_2$ to its bottom row

(a_{21}, a_{22}) is $\mathbb{G}_a$-invariant. In fact, this map is an orbit space and $\mathrm{SL}_2/\mathbb{G}_a \cong \mathbb{A}^2 \setminus \{0\}$; one way to see this is to note that $\mathbb{G}_a$ is the SL_2-stabiliser of $(1,0) \in \mathbb{A}^2$ and its orbit $\mathrm{SL}_2 \cdot (1,0) = \mathbb{A}^2 \setminus \{0\}$ is isomorphic to $\mathrm{SL}_2/\mathbb{G}_a$. Since the $\mathbb{G}_a$-action on $\mathbb{A}^n$ extends to SL_2, we have an SL_2-equivariant isomorphism

$$\mathbb{A}^n \times^{\mathbb{G}_a} \mathrm{SL}_2 \cong \mathbb{A}^n \times \mathrm{SL}_2/\mathbb{G}_a.$$

This is only quasi-affine (so its coordinate ring may not be finitely generated), but as $\{0\} \subset \mathbb{A}^2$ has codimension 2, any regular function extends from $\mathbb{A}^2 \setminus \{0\}$ to $\mathbb{A}^2$ by Hartogs' lemma. Hence,

$$\mathcal{O}(\mathbb{A}^n)^{\mathbb{G}_a} \cong \mathcal{O}(\mathbb{A}^n \times^{\mathbb{G}_a} \mathrm{SL}_2)^{\mathrm{SL}_2} \cong \mathcal{O}(\mathbb{A}^n \times \mathrm{SL}_2/\mathbb{G}_a)^{\mathrm{SL}_2} \cong \mathcal{O}(\mathbb{A}^{n+2})^{\mathrm{SL}_2}$$

is finitely generated. □

REMARK 2.10. The second part of this proof for $H = \mathbb{G}_a < G = \mathrm{SL}_2$ can be extended to any *Grosshans subgroup*, which is a closed subgroup $H < G$ of a reductive group such that G/H is quasi-affine and $\mathcal{O}(G/H) = \mathcal{O}(G)^H$ is finitely generated (see [**43**] for a detailed treatment). Grosshans [**41**] shows $\mathcal{O}(G)^H$ is finitely generated if and only if G/H can be emdedded in an affine variety with complement of codimension 2, and proves that unipotent radicals of parabolic subgroups in a reductive group are Grosshans subgroups [**42**]. For a Grosshans subgroup $H < G$, the same proof shows that *if* a H-action an an affine scheme X *extends to* G (which is not immediate as in the case of Weitzenböck's Theorem), then $\mathcal{O}(X)^H$ is finitely generated.

The proof shows that for a non-reductive group, even if the ring of invariants is finitely generated, taking its spectrum does not necessarily provide a categorical quotient: for $\mathbb{G}_a$-acting on SL_2, we have $\mathcal{O}(\mathrm{SL}_2)^{\mathbb{G}_a} = k[x_{21}, x_{22}]$, but the induced map $\mathrm{SL}_2 \to \mathbb{A}^2$ is not surjective, so $\mathbb{A}^2$ is not the categorical quotient. In fact, even worse, the image may only be a constructible subset (see §5.1). In the next subsection we will see that when G is reductive, taking the spectrum of the ring of invariants does give a categorical quotient.

2.5. Affine geometric invariant theory for reductive groups. Let G be a reductive group acting on an affine scheme X. There is an induced action of G on the coordinate ring $\mathcal{O}(X)$ and the ring of invariants $\mathcal{O}(X)^G$ is a finitely generated k-algebra by Nagata's Theorem.

DEFINITION 2.11 (Affine GIT quotient). For an action of a reductive group G on an affine scheme X, the *affine GIT quotient* is the morphism $\varphi : X \to X//G := \operatorname{Spec} \mathcal{O}(X)^G$ of affine schemes associated to the inclusion $\varphi^* : \mathcal{O}(X)^G \hookrightarrow \mathcal{O}(X)$.

The double slash notation $X//G$ used for the GIT quotient is a reminder that this quotient is not necessarily an orbit space and so it may identify some orbits. In nice cases, the GIT quotient is an orbit space and in this case we shall write X/G.

THEOREM 2.12 (Mumford, [**75**, Theorem 1.1]). *For a reductive group G acting on an affine scheme X, the affine GIT quotient $X \to X//G$ is a good quotient and thus categorical quotient.*

We will not include the proof of this result, but we note that the proof that the affine GIT quotient is good uses several properties of reductive group actions beyond simply the finite generation of the invariant ring: for a geometrically reductive group, invariant functions can be used to separate closed orbits (see Lemma 2.7) and, for linearly reductive groups, the proof can be simplified by using the fact that taking invariants is exact.

The affine GIT quotient restricts to a geometric quotient on an open stable subset $X^s \subset X$.

DEFINITION 2.13. A point $x \in X$ is *stable* if its orbit is closed in X and $\dim G_x = 0$ (or equivalently, $\dim G \cdot x = \dim G$). We let X^s denote the set of stable points.

For $x \in X$, we note that x is stable if and only if $\sigma_x : G \to X$ is proper. Indeed if σ_x is proper, then its image $G \cdot x$ is closed and the fibres, being both affine and proper, must be finite. Conversely if x is stable, then $\sigma_x : G \to G \cdot x$ has finite fibres and one can show it is finite.

EXAMPLE 2.14. For the $\mathbb{G}_m$-action on $\mathbb{A}^2$ by $t \cdot (x,y) = (tx, t^{-1}y)$, we have $\mathcal{O}(\mathbb{A}^2)^{\mathbb{G}_m} = k[xy]$ with affine GIT quotient $\varphi : \mathbb{A}^2 \to \mathbb{A}^1$ is given by $(x,y) \mapsto xy$. It is not a geometric quotient, as the three orbits consisting of the punctured axes and the origin are all identified. The stable locus is the complement of $xy = 0$, which admits a geometric quotient $\mathbb{A}^1 \setminus \{0\}$.

If we remove the origin, the affine line with a double origin is a geometric quotient of $\mathbb{A}^2 \setminus \{0\}$. In this case, we obtain a non-separated quotient as a categorical quotient of a separated scheme.

.

EXAMPLE 2.15. Consider $G = \mathrm{GL}_2$ acting by conjugation on the space $\mathrm{Mat}_{2\times 2}$ of 2×2 matrices with k-coefficients. The trace and determinant (which are the coefficients of the characteristic polynomial) are invariant functions, and so

$$k[\mathrm{tr}, \det] \subset \mathcal{O}(M_{2\times 2})^{\mathrm{GL}_2}.$$

We will soon see this is in fact an equality.

First, we describe the orbits using the theory of Jordan normal forms. As any orbit contains a matrix in Jordan normal form, there are three types of orbits:

- Matrices with distinct eigenvalues α, β and Jordan normal form
$$\begin{pmatrix} \alpha & 0 \\ 0 & \beta \end{pmatrix}.$$
These are closed 2 dimensional orbits, with 2 dimensional stabiliser (diagonal matrices).
- Matrices with repeated eigenvalue and Jordan normal form with one block
$$\begin{pmatrix} \alpha & 1 \\ 0 & \alpha \end{pmatrix}.$$
These orbits are also 2 dimensional but are not closed: for example
$$\lim_{t\to 0} \begin{pmatrix} t & 0 \\ 0 & t^{-1} \end{pmatrix} \begin{pmatrix} \alpha & 1 \\ 0 & \alpha \end{pmatrix} \begin{pmatrix} t^{-1} & 0 \\ 0 & t \end{pmatrix} = \begin{pmatrix} \alpha & 0 \\ 0 & \alpha \end{pmatrix}.$$
- Matrices with repeated eigenvalue and Jordan normal form with two blocks
$$\begin{pmatrix} \alpha & 0 \\ 0 & \alpha \end{pmatrix}.$$
The stabiliser of such a matrix is GL_2 and its orbit is a point, which is closed.

Every orbit closure of the second type contains an orbit of the third type.

Let us show that $\mathcal{O}(\mathrm{Mat}_{2\times 2})^{\mathrm{GL}_2} = k[\mathrm{tr}, \det]$. Since any orbit closure contains a diagonal matrix, any invariant function is completely determined by its values on the diagonal matrices and is invariant under permuting the diagonal entries. Hence

$$\mathcal{O}(\mathrm{Mat}_{2\times 2})^{\mathrm{GL}_2} \subset k[x_{11}, x_{22}]^{S_2} = k[x_{11} + x_{22}, x_{11}x_{22}] = k[\mathrm{tr}, \det]$$

by the theory of (elementary) symmetric polynomials.

The affine GIT quotient is $\varphi = (\mathrm{tr}, \det) : \mathrm{Mat}_{2\times 2} \to \mathrm{Mat}_{2\times 2} /\!/ \mathrm{GL}_2 = \mathbb{A}^2$. Since scalar multiples of the identity fix every point, there are no stable points for this action; however, the restriction to the locus of matrices with distinct eigenvalues is a geometric quotient.

EXERCISE 2.16. Show the GIT quotient of GL_n acting by conjugation on $\mathrm{Mat}_{n\times n}$ is $\mathbb{A}^n$.

We note that Newstead resolves the issue of having no stable points for the conjugation action on endomorphisms by constructing moduli spaces of *cyclic endomorphisms* of a vector space [**80**, Chapter 2].

2.6. Projective geometric invariant theory. Suppose that a reductive group G acts on a projective scheme $X \subset \mathbb{P}^n$ linearly (i.e. by a representation $G \to \mathrm{GL}_{n+1}$). The homogeneous coordinate ring of X is the graded ring

$$R(X) = k[x_0, \dots, x_n]/I_X = \bigoplus_{r\geq 0} H^0(X, \mathcal{O}(r))$$

where $\mathcal{O}(1)$ denotes the pullback of $\mathcal{O}_{\mathbb{P}^n}(1)$ to X. By Nagata's theorem, $R(X)^G$ is finitely generated. The inclusion $R(X)^G \hookrightarrow R(X)$ determines a rational map of projective schemes

$$X \dashrightarrow \mathrm{Proj}\, R(X)^G \tag{2.1}$$

whose indeterminacy locus is the closed subscheme of X defined by the homogeneous ideal $R(X)^G_+ := \oplus_{r>0} R(X)^G_r$. The domain of definition of this map is the GIT semistable locus.

DEFINITION 2.17. Let G be a reductive group acting linearly on a projective scheme $X \subset \mathbb{P}^n$.

i) We say $x \in X$ is *semistable* if there exists a G-invariant homogeneous function $f \in R(X)^G_r$ for some $r > 0$ such that $f(x) \neq 0$. We write X^{ss} for the open set in X of semistable points; this is the domain of definition of (2.1).
ii) We say $x \in X$ is *stable* if $\dim G_x = 0$ and there exists a G-invariant homogeneous function $f \in R(X)^G_r$ for some $r > 0$, which is non-vanishing at x and such that the G-action on X_f is closed. We write X^s for the open set in X of stable points.
iii) The restriction of (2.1) to the semistable locus $X^{ss} \to X/\!/G := \mathrm{Proj}\, R(X)^G$ is called the *projective GIT quotient*, which is projective over k.

Rather confusingly, a point is called *unstable* if it is not semistable; this terminology is now standard and there is not much we can do to change it! We refer to points which are semistable but not stable, as *strictly semistable*. If there are several groups acting on X, we clarify which group we mean by talking about *G-(semi)stability*.

By definition, X^{ss} is open as it is the domain of definition of the rational map (2.1). To see that X^s is open, we note that it is the intersection of two opens: the set of points with zero dimensional stabiliser is open as $x \mapsto \dim G_x$ is upper semi-continuous and the union of X_f for $f \in R(X)^G_+$ on which the action on X_f is closed is open.

THEOREM 2.18 (Mumford, see [**80**, Theorem 3.14]). *For a reductive group G acting linearly on a projective scheme $X \subset \mathbb{P}^n$, the projective GIT quotient $\varphi : X^{ss} \to X//G$ is a projective and good quotient, which restricts to a quasi-projective and geometric quotient of X^s.*

This result can be proved by gluing together affine GIT quotients: for $f \in R(X)^G_+$, the non-vanishing locus X_f is affine with affine GIT quotient $X_f \to X_f//G$, and we can write X^{ss} as the union of these open affines X_f, so $X//G$ is covered by the open affines $X_f//G$.

We have $\varphi(x) = \varphi(y)$ if and only if the orbit closures of x and y meet in X^{ss}. Furthermore, the preimage of any point in $X//G$ contains a unique closed orbit (of minimal dimension in this preimage), whose stabiliser is reductive (see Remark 1.24).

REMARK 2.19. It is important to note that the semistable set and the GIT quotient both depend on the G-equivariant embedding $X \hookrightarrow \mathbb{P}^n$, as the homogeneous coordinate ring depends on this embedding (or equivalently on the line bundle $\mathcal{O}(1)$ pulled back from $\mathbb{P}^n$).

Alternatively, rather than fixing a linear G-equivariant projective embedding of X, one can instead fix an ample *G-equivariant* line bundle $\mathcal{L}$ on X, which is often called an ample *G-linearisation*: $\mathcal{L} = (L, \Phi)$ is an ample invertible sheaf L on X together with a G-equivariant structure given by an isomorphism $\Phi : \sigma^* L \to \pi_2^* L$, where $\sigma, \pi_2 : G \times X \to X$ denote the action and second projection, which satisfies a cocycle condition $\pi_{23}^* \Phi \circ (\mathrm{Id}_G \times \sigma)^* \Phi = (m \times \mathrm{Id}_G)^* \Phi$ on $G \times G \times X$. In terms of the associated geometric line bundle, which by abuse of notation we shall also call L, this is equivalent to a G-action on L commuting with the projection $L \to X$ such that the action on the fibres $L_{g \cdot x} \to L_x$ is linear.

Given an ample *G-equivariant* line bundle $\mathcal{L}$ on X, we obtain a graded ring with a G-action

$$R(X, \mathcal{L}) = \bigoplus_{r \geq 0} H^0(X, L^{\otimes r})$$

such that the inclusion of invariants induces a rational map whose domain of definition is the semistable set and whose codomain is the GIT quotient (both with respect to $\mathcal{L}$)

$$X^{ss}(\mathcal{L}) \to X//_{\mathcal{L}} G := \operatorname{Proj} R(X, \mathcal{L})^G.$$

Since replacing $\mathcal{L}$ with a positive power just has the effect of changing the grading on this ring[2], we can assume $\mathcal{L}$ is very ample and then we obtain a linear G-equivariant embedding $X \hookrightarrow \mathbb{P}(V)$ where $V := H^0(X, L)^*$, which recovers the above setting of a linear action.

The effect of changing $\mathcal{L}$ is called *variation of GIT* and can be described in terms of certain birational transformations known as VGIT flips [**31, 93**]. Furthermore, the space of G-linearisations admits a wall and chamber decomposition

[2]By construction, the projective GIT quotient comes with a line bundle and this regrading does not change the GIT quotient but does change this line bundle.

describing how semistability varies: in chambers, semistability coincides with stability, but semistability changes on crossing a wall.

2.7. General GIT quotients. More generally, given a scheme X with a G-linearisation $\mathcal{L}$, Mumford defines a GIT quotient using invariant sections of positive powers of L whose non-vanishing locus is affine (so that one can take affine GIT quotients and glue them). This produces a good quotient of a 'semistable locus' ([**75**, Definition 1.7]), which in this situation is defined to be the set of points $x \in X$ such that there exists $\sigma \in H^0(X, L^{\otimes r})^G$ for $r > 0$ with $\sigma(x) \neq 0$ and such that X_σ is affine[3]. The semistable set and quotient obtained in this way are both quasi-projective (see [**80**, Theorem 3.21]).

Let us remark that in this survey we have assumed that we are working over an algebraically closed field k. The assumption that k is algebraically closed can be dropped, but one has to be careful about rationality questions and work with geometric points for certain statements (for example, the Hilbert–Mumford criterion). Moreover, Seshadri [**90**] extended GIT to work relative to a base scheme S with mild assumptions on S.

2.8. Affine GIT linearised by a character. As a special case of §2.7, consider a linear action of G on an affine scheme $X \subset \mathbb{A}^n$; then the structure sheaf $\mathcal{O}_X$ is naturally equipped with a G-equivariant structure, where if we view this as a geometric line bundle $X \times \mathbb{A}^1$, the G-action on $\mathbb{A}^1$ is trivial. In this case, the GIT quotient with respect to this ample G-linearisation $\mathcal{O}_X$ is just the affine GIT quotient, as

$$R(X, \mathcal{L}) = \mathcal{O}(X)[z]$$

with trivial G-action on z and this ring is graded by the degree of z, thus

$$X/\!/_{\mathcal{O}_X} G := \operatorname{Proj} R(X, \mathcal{L})^G = \operatorname{Proj} \mathcal{O}(X)^G[z] = \operatorname{Spec} \mathcal{O}(X)^G = X/\!/G.$$

This linearisation can be modified by using a character $\rho : G \to \mathbb{G}_m$ to obtain a linearisation $\mathcal{O}_\rho$ which is given by G acting linearly on the geometric line bundle $X \times \mathbb{A}^1$ by the given action on X and acting via multiplication with ρ on $\mathbb{A}^1$. The outcome of applying GIT in this situation of twisting the linearisation by a character was described by King [**62**] and results in an open subset $X^{\rho-ss}$ of ρ-semistable points and a GIT quotient

$$X^{\rho-ss} \to X/\!/_\rho G := \operatorname{Proj} \bigoplus_{r \geq 0} H^0(X, (\mathcal{O}_\rho)^{\otimes r})^G.$$

In this case, the G-invariant sections of $(\mathcal{O}_\rho)^{\otimes r} \cong \mathcal{O}_{\rho^r}$ are identified with $f \in \mathcal{O}(X)$ such that $f(g \cdot x) = \rho^r(g) f(x)$ for all $g \in G$ and $x \in X$, which we refer to as *ρ-semi-invariant functions of weight r*. By definition, x is ρ-semistable if there exists a ρ-semi-invariant function of weight $r > 0$ which is non-vanishing at x. Furthermore, $X/\!/_\rho G$ is projective over the spectrum of the 0th-graded piece which is just the affine GIT quotient $X/\!/G = \operatorname{Spec} \mathcal{O}(X)^G$.

EXAMPLE 2.20. For $\mathbb{G}_m$ acting on $\mathbb{A}^n$ by scalar multiplication linearised by $\mathcal{O}_\rho$ for $\rho : \mathbb{G}_m \to \mathbb{G}_m$ given by $t \to t$, the coordinate functions are ρ-semi-invariant functions of weight 1 and these generate the ring of invariants. Consequently, we obtain the GIT quotient

$$(\mathbb{A}^n)^{\rho-ss} = \mathbb{A}^n \setminus \{0\} \to \mathbb{A}^n/\!/_\rho \mathbb{G}_m = \operatorname{Proj} k[x_1, \ldots, x_n] = \mathbb{P}^{n-1}.$$

[3]If X is projective and $\mathcal{L}$ is ample, then this non-vanishing locus is always affine.

3. Semistability and instability in reductive GIT

Since the reductive GIT quotient only provides a quotient of an open semistable locus, this naturally leads to two questions: can we describe the semistable points and what can we say about unstable (i.e. not semistable) points? For actions on projective (over affine) schemes, the first question is tackled by the Hilbert–Mumford criterion for semistability described in §3.1. In moduli problems with a natural notion of subobjects, the Hilbert–Mumford criterion often leads to a clean moduli-theoretic interpretation of GIT semistability. We state some application of reductive GIT to moduli in §3.2. We then turn to the second question in §3.3 and describe how work of Kempf [**61**], Hesselink [**53**], Kirwan [**63**] and Ness [**79**] gives a stratification of the unstable locus, with a largely combinatorial flavour; we survey some applications of these stratifications and discuss the question of construction quotients of unstable strata, where naturally non-reductive groups (namely, parabolic subgroups, representing an instability flag) appear.

3.1. Semistability and the Hilbert–Mumford criterion. By definition, semistability in reductive GIT is given in terms of the existence of a non-vanishing invariant section. From this definition, it is extremely challenging to determine semistability, as it is essentially equivalent to computing invariant rings, which is a notoriously challenging problem. Fortunately, in certain situations (projective GIT, affine GIT linearised by a character or more generally a projective over affine set-up), the Hilbert–Mumford criterion reduces semistability to checking semistability for $\mathbb{G}_m$-actions, which in turn can be combinatorially described using the weights of the action. More precisely, a G-semistable point is semistable for any subgroup, and thus in particular, for any $\mathbb{G}_m$ contained in G; the Hilbert–Mumford criterion gives a converse to this statement.

For simplicity, throughout this section, we assume we have a linear representation $G \to \mathrm{GL}(V)$ of a reductive group G and consider the associated linear action on $X = \mathbb{P}(V)$. We will describe the semistable points in this setting. For a closed subscheme $Y \subset X$ with a linear G-action, we have $Y^{ss} = Y \times_X X^{ss}$ and so it suffices to understand semistability on the ambient projective space. For an ample G-linearisation $\mathcal{L}$ on X, using a power of $\mathcal{L}$ puts us in this linear setting.

We will see several different versions of the Hilbert–Mumford criterion, which make it possible to determine semistability in practice. The first, and weakest, version is a topological criterion.

PROPOSITION 3.1 (Topological Hilbert–Mumford criterion, [**75**, Proposition 2.2]). *For a linear action of a reductive group G on $\mathbb{P}(V)$, the following statements hold for $x = [v] \in \mathbb{P}(V)$.*

i) x is semistable if and only if $0 \notin \overline{G \cdot v}$;
ii) x is stable if and only if $\dim G_v = 0$ and $G \cdot v$ is closed in V.

PROOF. We will just give the proof of the first statement. By definition $x = [v]$ is semistable if and only if there is a G-invariant homogeneous polynomial $f \in R(X)^G$ which is non-zero at x. Since f is G-invariant it is constant on orbit closures, and so f separates the closed schemes $\overline{G \cdot v}$ and 0, which shows these closed subschemes are disjoint. Conversely, if the closed G-invariant schemes $\overline{G \cdot v}$ and 0 in V are disjoint, then as G is geometrically reductive, there exists a

G-invariant polynomial $f \in \mathcal{O}(V)^G$ separating these subsets

$$f(\overline{G \cdot v}) = 1 \quad \text{and} \quad f(0) = 0$$

by Lemma 2.7. By considering the decomposition of $f = \sum_i f_i$ into (G-invariant) homogeneous pieces, we see there is a G-invariant homogeneous piece f_i which is non-vanishing at x. $\square$

DEFINITION 3.2. For a linear action of a torus $T = \mathbb{G}_m^n$ on $\mathbb{P}(V)$, consider the associated weight decomposition $V = \oplus_{\chi \in X^*(T)} V_\chi$. We refer to the support of this decomposition as the T-weights on $\mathbb{P}(V)$. For $x = [v] \in \mathbb{P}(V)$, we write $v = \sum v_\chi$ and define the *T-weight set* of this point to be

$$\mathrm{wt}_T(x) = \mathrm{wt}_T(v) = \{\chi : v_\chi \neq 0\} \subset X^*(T) \cong \mathbb{Z}^n.$$

We write $\mathrm{conv}(\mathrm{wt}_T(x))$ for the convex hull of the T-weight set and $\mathrm{Int}(\mathrm{conv}(\mathrm{wt}_T(x)))$ for its interior.

For a $\mathbb{G}_m$-action on a separated scheme, we will often use the following notation.

NOTATION 3.3. If a morphism $f : \mathbb{G}_m \to S$, with S separated, extends to $\tilde{f} : \mathbb{A}^1 \to S$, then this extension is unique and we write $\lim_{t\to 0} f(t) := \tilde{f}(0)$. Similarly if f extends to $\mathbb{P}^1$, we write $\lim_{t\to\infty} f(t) := \tilde{f}(\infty)$.

We can now give a combinatorial description of (semi)stability for a $\mathbb{G}_m$-action in terms of whether or not the origin lies in (the interior of) the convex hull of $\mathbb{G}_m$-weights.

PROPOSITION 3.4 (Hilbert–Mumford for $\mathbb{G}_m$-actions). *For a linear action of $\mathbb{G}_m$ on $\mathbb{P}(V)$ and $x \in \mathbb{P}(V)$, the following statements hold:*

i) x is $\mathbb{G}_m$-semistable if and only if $0 \in \mathrm{conv}(\mathrm{wt}_{\mathbb{G}_m}(x))$.
ii) x is $\mathbb{G}_m$-stable if and only if $0 \in \mathrm{Int}(\mathrm{conv}(\mathrm{wt}_{\mathbb{G}_m}(x)))$.

PROOF. We again just prove the statement for semistability. By the topological Hilbert–Mumford criterion, we have that $x = [v] \in \mathbb{P}(V)$ is $\mathbb{G}_m$-semistable if and only if $0 \notin \overline{\mathbb{G}_m \cdot v}$. Any point in the boundary of this orbit closure is either

$$\lim_{t\to 0} t \cdot v \quad \text{or} \quad \lim_{t\to\infty} t \cdot v = \lim_{t\to 0} t^{-1} \cdot v.$$

Moreover, we have $\lim_{t\to 0} t \cdot v = 0$ if and only if $\mathrm{wt}_{\mathbb{G}_m}(v) \subset \mathbb{Z}_{>0}$ (and similarly $\lim_{t\to\infty} t \cdot v = 0$ if and only if $\mathrm{wt}_{\mathbb{G}_m}(v) \subset \mathbb{Z}_{<0}$). Hence $x = [v] \in \mathbb{P}(V)$ is $\mathbb{G}_m$-semistable if and only if there exists $r_0 \leq 0$ and $r_\infty \geq 0$ in $\mathrm{wt}_{\mathbb{G}_m}(v)$, or equivalently $0 \in \mathrm{conv}(\mathrm{wt}_{\mathbb{G}_m}(x))$. $\square$

EXAMPLE 3.5. The linear action of $\mathbb{G}_m$ on $X = \mathbb{P}^n$ by

$$t \cdot [x_0 : x_1 : \cdots : x_n] = [t^{-1}x_0 : tx_1 : \cdots : tx_n]$$

has weights ± 1. Hence, for a point x to be (semi)stable it needs both these weights, which means its first coordinate x_0 must be non-zero and at least one of the other coordinates x_i for $i > 0$ must be non-zero. One can also see this by directly proving that

$$R(\mathbb{P}^n, \mathcal{O}(1))^{\mathbb{G}_m} = k[x_0x_1, \ldots, x_0x_n].$$

In particular, $X^{ss} \cong \mathbb{A}^n \setminus \{0\}$ and $\mathbb{P}^n /\!/ \mathbb{G}_m = \mathbb{P}^{n-1}$ is a geometric $\mathbb{G}_m$-quotient.

The Hilbert–Mumford criterion will ultimately be a numerical criterion that phrases semistability in terms of the weights of *1-parameter subgroups* (1-PS), which are non-trivial group homomorphisms $\lambda : \mathbb{G}_m \to G$.

DEFINITION 3.6 (Hilbert–Mumford weight). For a linear action of a reductive group G on $\mathbb{P}(V)$, we define the Hilbert-Mumford weight of $x = [v]$ at a 1-parameter subgroup $\lambda : \mathbb{G}_m \to G$ to be

$$\mu(x, \lambda) := -\min \mathrm{wt}_{\lambda(\mathbb{G}_m)}(x).$$

Let us note some useful properties of the Hilbert–Mumford weight.

EXERCISE 3.7. Show that the Hilbert–Mumford weight of $x = [v]$ has the following properties.

(1) $\mu(x, \lambda)$ is the unique integer μ such that $\lim_{t\to 0} t^{\mu}\lambda(t) \cdot v$ exists and is non-zero.
(2) $\mu(x, \lambda) = \mu(x_0, \lambda)$ where $x_0 = \lim_{t\to 0} \lambda(t) \cdot x$ (and this limit exists as X is projective).
(3) $\mu(x, \lambda) \leq 0 \iff \lim_{t\to 0} \lambda(t) \cdot v$ exists, with equality if and only if $\lim_{t\to 0} \lambda(t) \cdot v \neq 0$.
(4) $\mu(g \cdot x, g\lambda g^{-1}) = \mu(x, \lambda)$ for all $g \in G$.
(5) $\mu(x, \lambda^n) = n\mu(x, \lambda)$ for a positive integer n.

For a linear $\mathbb{G}_m$-action on $\mathbb{P}(V)$, we see that for the 1-PS given by $\lambda(t) = t$, we have

$$\mu(x, \lambda) \geq 0 \iff \lim_{t\to 0} t \cdot v \neq 0$$

and

$$\mu(x, \lambda^{-1}) \geq 0 \iff \lim_{t\to \infty} t \cdot v \neq 0$$

Hence x is semistable if $\mu(x, -) \geq 0$ for λ and λ^{-1}. Furthermore, x is stable if and only if this inequality is strict for both 1-PSs. This is precisely the numerical version of the Hilbert–Mumford criterion that we now can state.

THEOREM 3.8 (Hilbert–Mumford criterion, [**75**, Theorem 2.1]). *For a reductive group G acting linearly on a projective scheme $X \subset \mathbb{P}^n$, the following statements hold for $x \in X$.*

i) x is semistable if and only if $\mu(x, \lambda) \geq 0$ for all 1-PS $\lambda : \mathbb{G}_m \to G$,
ii) x is stable if and only if $\mu(x, \lambda) > 0$ for all 1-PS $\lambda : \mathbb{G}_m \to G$.

Note that it suffices to check these inequalities for *primitive* 1-PSs (i.e. 1-PSs which are not positive powers of another 1-PS); see Exercise 3.7.

The full proof of the Hilbert–Mumford criterion is slightly too technical to include in this survey, but following the topological version (Proposition 3.1), it suffices to show that the reductive group G has enough 1-PSs to detect if the origin is contained in the closure of orbits of linear actions $G \to \mathrm{GL}(V)$, which is precisely the following result (see [**75**, p53] and [**61**, Theorem 1.4]), whose proof involves the Cartan-Iwahori decomposition for the reductive group G.

THEOREM 3.9 (Fundamental Theorem of GIT). *Let G be a reductive group acting on an affine space V. If $v \in V$ and $0 \in \overline{G \cdot v}$, then there is a 1-PS λ of G such that $\lim_{t\to 0} \lambda(t) \cdot v = 0$.*

REMARK 3.10 (Hilbert–Mumford weight for a linearised action). In the case of a linear G-action on a projective scheme $X \subset \mathbb{P}^n$, the Hilbert–Mumford weight for $x \in X$ defined above depends on the choice of G-representation $G \to \mathrm{GL}_{n+1}$ (as the weights depend on this representation).

In general, for a G-linearisation $\mathcal{L} = (L, \Phi)$ on a projective G-scheme X, we consider the $\lambda(\mathbb{G}_m)$-fixed point $x_0 = \lim_{t \to 0} \lambda(t) \cdot x$. The linearisation Φ induces a $\mathbb{G}_m$-representation on the fibre of L over x_0

$$L_{x_0} = L_{\lambda(t) \cdot x_0} \xrightarrow{\cdot \lambda(t)^{-1}} L_{x_0}$$

of weight r (that is $\lambda(t)^{-1}$ acts on this fibre by $t \mapsto t^r$). Then the Hilbert–Mumford weight (with respect to $\mathcal{L}$) is defined to be minus the weight on this fibre

$$\mu^{\mathcal{L}}(x, \lambda) = -r.$$

In this linearised situation, the Hilbert–Mumford criterion says $x \in X$ is semistable (with respect to $\mathcal{L}$) if and only if $\mu^{\mathcal{L}}(x, \lambda) \geq 0$ for all 1-PS $\lambda : \mathbb{G}_m \to G$.

If $X \subset \mathbb{P}^n$ and $\mathcal{L} = \mathcal{O}(1)$ is the pullback of $\mathcal{O}_{\mathbb{P}^n}(1)$, then these two definitions coincide: we have $\mu^{\mathcal{O}(1)}(x, \lambda) = \mu(x, \lambda)$ by [**75**, Proposition 2.3].

As any 1-PS can be conjugated to lie in a fixed maximal torus $T < G$, one can phrase G-semistability of a point in terms of T-semistability of all G-translates of that point by Exercise 3.7 above. Then T-semistability can be stated combinatorially using the torus weights analogous to Proposition 3.4 above. This gives a combinatorial Hilbert–Mumford criterion.

PROPOSITION 3.11 (Torus weights version of Hilbert–Mumford criterion, [**30**, §9.4]). *For a reductive group G acting on a projective scheme $X \subset \mathbb{P}^n$ linearly, fix a maximal torus $T < G$. For $x \in X$, the following statements hold.*

i) x is G-(semi)stable if and only if $g \cdot x$ is T-(semi)stable for all $g \in G$.
ii) x is T-semistable if and only if $0 \in \mathrm{conv}(\mathrm{wt}_T(x))$.
iii) x is T-stable if and only if $0 \in \mathrm{Int}(\mathrm{conv}(\mathrm{wt}_T(x)))$.

By the first statement, the G-semistable set is the G-sweep of the T-semistable set:

$$X^{G-ss} = \bigcap_{g \in G} g \cdot X^{T-ss}.$$

EXERCISE 3.12 (Semistability for binary forms). Consider the action of SL_2 on the space of degree d binary forms $\mathbb{P}^d = \mathbb{P}(k[x,y]_d)$. For $p_F \in \mathbb{P}^d$ corresponding to $F(x,y) \in k[x,y]_d$, show

i) F is semistable if and only if all roots of F have multiplicity less than or equal to $d/2$;
ii) F is stable if and only if all roots of F have multiplicity strictly less than $d/2$.

REMARK 3.13 (Hilbert–Mumford criterion for action on affine scheme twisted by a character). For a linear action of a reductive group G on an affine scheme $X \subset \mathbb{A}^n$ linearised via $\mathcal{O}_\rho$ for a character $\rho : G \to \mathbb{G}_m$ (see §2.8), King proved a topological Hilbert–Mumford criterion [**62**, Lemma 2.2], by using the total space of the dual linearisation to replace the affine cone, and obtained the following numerical Hilbert–Mumford criterion [**62**, Proposition 2.5]:

(1) x is ρ-semistable if and only if $\langle \rho, \lambda \rangle \geq 0$ for all 1-PS $\lambda : \mathbb{G}_m \to G$ such that $\lim_{t \to 0} \lambda(t) \cdot x$ exists,

(2) x is ρ-stable if and only if $\langle \rho, \lambda \rangle > 0$ for all 1-PS $\lambda : \mathbb{G}_m \to G$ such that $\lim_{t\to 0} \lambda(t) \cdot x$ exists,

where $\langle \rho, \lambda \rangle = r$ if $\rho \circ \lambda(t) = t^r$, i.e. this is the natural pairing between characters and cocharacters. Using the abstract definition of the Hilbert–Mumford weight in terms of the weight of the action on the fibre over the limit point of the $\mathbb{G}_m$-action (see Remark 3.10), we see that if $x_0 = \lim_{t\to 0} \lambda(t) \cdot x$ exists, then $\mu^{\mathcal{O}_\rho}(x, \lambda) = \langle \rho, \lambda \rangle$.

EXERCISE 3.14. Using Remark 3.13, show that for GL_2 acting on $\mathrm{Mat}_{2\times n}$ by left multiplication the (semi)stable locus for the character $\rho = \det$ is the matrices of maximal rank (namely rank 2), and so the GIT quotient is the Grassmannian $\mathrm{Gr}(2, n)$.

We note that there is a Hilbert–Mumford criterion in the more general setting of a projective over affine variety with an action of a linearly reductive group [**44**].

3.2. A brief survey of applications of reductive GIT to moduli. The notion of moduli functor is heavily influenced by Grothendieck's approach to algebraic geometry. Furthermore, Grothendieck proved that the Hilbert and Quot functors are representable by projective schemes; these are fine moduli spaces and provide parameter spaces in the GIT constructions of moduli of smooth projective curves and moduli of vector bundles on curves.

The first truly interesting application of GIT was Mumford's construction of moduli spaces of curves [**75**, Chapter 5]. Mumford constructed a coarse moduli space M_g for smooth projective curves of genus $g \geq 2$ by using a power of the canonical bundle to give a projective embedding $C \hookrightarrow \mathbb{P}^N$ and constructing M_g as a quotient of a suitable Chow variety parametrising pluricanonical curves. Gieseker [**38**] provided an alternative GIT construction of M_g and its Deligne–Mumford compactification $\overline{M_g}$ via stable curves as a quotient of the PGL_{N+1}-action on a suitable Hilbert scheme with a linearisation given by embedding in a Grassmannian associated to a sufficiently large choice of m. Although there are direct proofs that smooth curves [**38, 75**] are (asymptotically) GIT stable, the proof that stable curves are (asymptotically) GIT stable is indirect (see [**73**, §3.1]). Gieseker's Hilbert scheme construction is now the prevalent perspective, which has been generalised to give GIT constructions of moduli spaces of pointed stable curves and stable maps and their (birational) geometry is studied using VGIT (see [**68, 73**]).

The other influential and successful application of GIT was the construction of moduli spaces of vector bundles (of fixed rank and degree) on a fixed smooth projective curve C. One of the first ideas to construct vector bundle moduli spaces over $k = \mathbb{C}$ was to use unitary representations of the fundamental group $\pi_1(C)$ and led to the Narasimhan–Seshadri Theorem [**78**] relating irreducible representations with *stable* vector bundles considered by Mumford [**74**], where Mumford's notion of stability came from the Hilbert–Mumford criterion in GIT and involves verifying an inequality of slopes (the ratio of the degree and the rank) for all subbundles. For moduli problems with a natural notion of subobjects, the study of 1-PSs in GIT often corresponds to filtrations by subobjects and stability can be phrased as an inequality for all subobjects. The GIT construction of moduli spaces of (semi)stable vector bundles was given by Seshadri [**88**] (see [**80**, Chapter 5]), and was later generalised by Simpson [**91**] to construct moduli spaces of sheaves (and Higgs sheaves) on higher dimensional schemes as GIT quotients of Quot schemes. Quot schemes appear as semistable vector bundles can be parametrised as quotients of a fixed vector bundle as mentioned in Example 1.12(4). This construction has been generalised to construct various bundle moduli spaces [**86**].

Mumford also applied GIT to construct moduli spaces of projective hypersurfaces $X \subset \mathbb{P}^n$ of degree d by taking a quotient of PGL_{n+1} acting on $\mathbb{P}(k[x_1, \dots, x_n]_d)$ as in Example 1.12 (3). He showed smooth hypersurfaces are GIT stable if $n \geq 2$ and $d \geq 3$ (see [**75**, Chapter 4.2]).

King [**62**] developed GIT for a linear action on an affine space with respect to a character (see §2.8) to construct reasonable moduli spaces of *semistable* representations of a quiver, where semistability depends on a stability parameter; the Hilbert–Mumford criterion gives a moduli-theoretic interpretation of semistability as an inequality holding for all subrepresentations.

3.3. Instability. In this section, we continue to suppose that G is a reductive group acting on a projective scheme $X \subset \mathbb{P}^n$ linearly. Since the GIT quotient provides a categorical quotient of the semistable locus X^{ss}, it is natural to ask what can be said about the unstable points (i.e. not semistable points)

$$X^{us} := X \setminus X^{ss}.$$

By the Hilbert–Mumford criterion, if a point is unstable, then it has a negative Hilbert–Mumford weight for some 1-PS. Starting from this observation, Kempf [**61**] associated to an unstable orbit a conjugacy class of 1-PSs which are 'most responsible' for its instability, in the sense that they minimise a 'normalised Hilbert–Mumford weight'. Hesselink then used Kempf's work to statify the unstable locus [**53**]. This stratification was described more explicitly and, when $k = \mathbb{C}$, compared with a Morse stratification associated to the norm square of the moment map for the action of a maximal compact subgroup by Kirwan [**63**] and Ness [**79**].

Let us describe how to fix a conjugation invariant norm on 1-PSs of G.

DEFINITION 3.15. A conjugation invariant norm on 1-PSs of a reductive group G is given by fixing a maximal torus $T < G$ and a Weyl-invariant integral-valued bilinear form on the 1-PSs $X_*(T)$ of T with associated norm $\| - \|$. For any 1-PS $\lambda : \mathbb{G}_m \to G$, there exists $g \in G$ such that $g\lambda g^{-1} \in X_*(T)$ and we define

$$\|\lambda\| := \|g\lambda g^{-1}\|,$$

which is independent of the choice of g due to the Weyl invariance.

Over the complex numbers, such a norm can be constructed by fixing a Weyl invariant inner product on the Lie algebra $\mathfrak{t}$ of T, which gives an identification $\mathfrak{t} \cong \mathfrak{t}^*$.

EXAMPLE 3.16. If $G = \mathrm{GL}_n$ and T is the diagonal maximal torus, then the Euclidean norm on $\mathbb{R}^n \cong X_*(T)_{\mathbb{R}}$ is invariant under the Weyl group S_n.

Using the norm $\| - \|$, we define a normalised Hilbert–Mumford weight and state Kempf's notion [**61**] of an *adapted* 1-PS for an unstable point.

DEFINITION 3.17 (Normalised Hilbert–Mumford weight and adapted 1-PS). For a reductive group G acting on a projective scheme $X \subset \mathbb{P}^n$ linearly and a fixed conjugation invariant norm on 1-PSs of G, we define the *normalised Hilbert–Mumford weight* of $x \in X$ at a 1-PS λ to be $\mu(x, \lambda)/\|\lambda\|$ and we define the *minimum normalised Hilbert–Mumford weight of* x to be

$$M(x) := \inf_{\lambda \in X_*(G)} \frac{\mu(x, \lambda)}{\|\lambda\|}.$$

If x is unstable, a primitive 1-PS is said to be *adapted* to x if it achieves this minimum and we write Λ_x for the set of primitive 1-PSs adapted to x.

Although there is not a unique 1-PS adapted to x, Kempf proves there is a unique parabolic subgroup associated to x (see Theorem 3.19 below), playing the role of an instability flag similar to Harder–Narasimhan filtrations for vector bundles. First, we recall how a 1-PS of G naturally determines a parabolic subgroup.

DEFINITION 3.18 (Parabolic and Levi group associated to a 1-PS). For a 1-PS $\lambda : \mathbb{G}_m \to G$ of a reductive group G, we define

$$P_\lambda := \left\{ g \in G : \lim_{t\to 0} \lambda(t) g \lambda(t)^{-1} \text{ exists in } G \right\} \overset{q_\lambda}{\twoheadrightarrow} L_\lambda := \left\{ \lim_{t \to 0} \lambda(t) g \lambda(t)^{-1} : g \in P_\lambda \right\};$$

then $P_\lambda = U_\lambda \rtimes L_\lambda$ is a parabolic subgroup with Levi subgroup L_λ and unipotent radical U_λ and $q_\lambda : P_\lambda \to L_\lambda$ is a retraction onto the Levi.

Let us collect Kempf's results on adapted 1-PSs in the following theorem.

THEOREM 3.19 (Kempf, [**61**]). *Let G be a reductive group acting on a projective scheme $X \subset \mathbb{P}^n$ linearly and fix a conjugation invariant norm on 1-PSs of G. Then for an unstable point $x \in X$, we have $\Lambda_x \neq \emptyset$ and there is a parabolic subgroup $P_x < G$ with the following properties.*

i) For any $\lambda \in \Lambda_x$, we have $P_x = P_\lambda$.
ii) Any two 1-PSs in Λ_x are conjugate by an element of P_x.
iii) If $T < G$ is a maximal torus with $T < P_x$, then $\Lambda_x \cap X_*(T)$ is a single Weyl orbit.
iv) We have $g \Lambda_x g^{-1} = \Lambda_{g\cdot x}$ for all $g \in G$.
v) If $\lambda \in \Lambda_x$, then $\lambda \in \Lambda_{x_0}$ and also $M(x) = M(x_0)$ where $x_0 := \lim_{t\to 0} \lambda(t) \cdot x$.

Although we will not give the details on the proof of this theorem, the existence of an adapted 1-PS boils down to the fact that one can work in a maximal torus (by translating using the G-action) and then the normalised Hilbert–Mumford weight for 1-PSs in a given maximal torus can be determined from subsets of the torus weights of the action, which is a finite set (see also Remark 3.26). The last two properties follow from Exercise 3.7 and will correspond to the unstable strata being G-invariant and closed under these types of limits.

Hesselink [**53**] stratified the unstable locus by pairs $\beta = ([\lambda], m)$ consisting of a conjugacy class of an adapted 1-PS $[\lambda]$ and a minimum normalised Hilbert–Mumford weight m. Before we give the concrete construction of this stratification, we provide a summary of its properties and give an overview of various applications. As is customary, we include the semistable set as the lowest stratum in this stratification.

THEOREM 3.20 (Kempf [**61**], Hesselink [**53**], Kirwan [**63**], Ness [**79**]). *For a reductive group G acting linearly on a projective scheme $X \subset \mathbb{P}(V)$ and a conjugation invariant norm on 1-PSs of G, there is a finite* instability stratification

$$X = \bigsqcup_{\beta \in \mathcal{B}} S_\beta \tag{3.1}$$

into locally closed subschemes with a partial ordered index set $\mathcal{B}$ with the following properties.

i) *The lowest stratum is indexed by* $\beta = 0$ *and we have* $S_0 = X^{ss}$.
ii) *The closure of any stratum is contained in the union of higher strata:* $\overline{S_\beta} \subset \bigsqcup_{\gamma \geq \beta} S_\gamma$.
iii) $\mathcal{B}$ *is determined combinatorially from the weights on* V *of a maximal torus* $T < G$.
iv) *The strata* S_β *can be determined from simpler limit sets, which are GIT semistable loci for smaller reductive group actions with a twisted linearisation.*

We will soon make the last two statements more precise. First, let us state some applications of these instability (or Hesselink–Kempf–Kirwan–Ness) stratifications.

REMARK 3.21. When X is smooth, these stratifications have been used in the following ways:

(1) By Kirwan [**63**] over $k = \mathbb{C}$, to compute the G-equivariant rational Betti numbers of X^{ss} (which coincides with the rational Betti numbers of $X//G$ when $X^s = X^{ss}$) by showing the Gysin long exact sequences in equivariant cohomology for this stratification split. Hence, there is a surjection known as the *Kirwan map*

$$H_G^*(X, \mathbb{Q}) \twoheadrightarrow H_G^*(X^{ss}, \mathbb{Q})$$

with explicit kernel.
(2) By Dolgachev–Hu [**31**] and Thaddeus [**93**], to describe the birational transformations between GIT quotients given by varying the linearisation.
(3) By Halpern-Leistner [**47**] and Ballard–Favero–Katzarkov [**10**], to construct semi-orthogonal decompositions of the derived category of the quotient stack $[X^{ss}/G]$.
(4) By Halpern-Leistner [**48**], as inspiration to formulate an abstract notion of a Θ-stratification on a stack.

REMARK 3.22. Over $k = \mathbb{C}$, there is a close relationship between GIT quotients and symplectic reductions, which are quotients in symplectic geometry. A smooth projective variety $X \subset \mathbb{P}^n_{\mathbb{C}}$ is Kähler and thus has a symplectic form (inherited from the Fubini-Study form on $\mathbb{P}^n_{\mathbb{C}}$). For a representation $G \to \mathrm{GL}_{n+1}$, there is a maximal compact subgroup $K < G$ which acts by unitary transformations, and K preserves the Kähler form. Moreover, there is a moment map $\mu : X \to \mathfrak{k}^*$ to the co-Lie algebra of K such that the GIT quotient is homeomorphic to the symplectic reduction (the quotient of the zero level set of the moment map by K, see [**70**]):

$$X//G \simeq \mu^{-1}(0)/K$$

via the Kempf–Ness Theorem [**60**] (see also [**94**]). More precisely, $x \in X$ is semistable if and only if its G-orbit closure meets $\mu^{-1}(0)$ (and if this intersection is non-empty, it consists of a unique K-orbit). Furthermore, by work of Kirwan [**63**] and Ness [**79**], the GIT instability stratification with respect to a conjugation invariant norm coincides with the Morse stratification associated to the norm square of the moment map $\|\mu\|^2 : X \to \mathbb{R}$ (see also [**75**, Chapter 8]).

We now turn to the (first set-theoretic) construction of the unstable strata in the situation of a reductive group G acting linearly on a projective scheme $X \subset \mathbb{P}^n$ with a fixed conjugation invariant norm $\| - \|$ on 1-PSs on G.

Definition 3.23 (Unstable strata, attracting sets and limit sets). For a pair $\beta = ([\lambda], m) \in (X_*(G)/G) \times \mathbb{R}_{<0}$, we define the associated unstable stratum

$$S_\beta = \{x \in X : \Lambda_x \cap [\lambda] \neq \emptyset \text{ and } M(x) = m\}.$$

For a representative λ of $[\lambda]$, we define the associated *limit set* Z_β^{ss} and *attracting set* Y_β^{ss} as follows

$$Z_\beta^{ss} = X^{\lambda(\mathbb{G}_m)} \cap Y_\beta^{ss} \xleftarrow{p_\beta} Y_\beta^{ss} = \{x \in X : \lambda \in \Lambda_x \text{ and } M(x) = m\},$$

where $p_\lambda(x) := \lim_{t\to 0} \lambda(t) \cdot x$. We define the index set $\mathcal{B} = \{0\} \sqcup \{\beta : S_\beta \neq \emptyset\}$, which turns out to be finite (see Remark 3.26 below).

Let us also introduce a closed subscheme Z_β of the λ-fixed locus on which the normalised Hilbert–Mumford weight of β is m and its attracting set Y_β under the flow by $\lambda(\mathbb{G}_m)$ as $t \to 0$:

$$Z_\beta := \left\{ x \in X^{\lambda(\mathbb{G}_m)} : \frac{\mu(x,\lambda)}{\|\lambda\|} = m \right\} \xleftarrow{p_\beta} Y_\beta := \{x \in X : \lim_{t\to 0} \lambda(t) \cdot x \in Z_\beta\}.$$

Note that $Z_\beta^{ss} \subset Z_\beta$ and $Y_\beta^{ss} \subset Y_\beta$; we will soon see these are open subsets and thus we can give these sets a scheme structure. Furthermore, these schemes all depend on the chosen representative λ of $[\lambda]$. Recall that $Y_\beta^{(ss)}$ and $Z_\beta^{(ss)}$ depend on a choice of 1-PS $\lambda \in [\lambda]$ as well as m, whereas P_λ only depends on λ (and not m). Note that in [**63**], P_λ is denoted by P_β.

Since the notion of adapted 1-PS depends on the choice of norm, the unstable strata also depend on this choice of norm, but $S_0 = X^{ss}$ does not.

Proposition 3.24 (Kirwan, [**63**, §12]). *Assume that $X \subset \mathbb{P}^n$ is smooth. For an unstable index $\beta = ([\lambda], m) \neq 0 \in \mathcal{B}$, the stratum S_β can be described as follows.*

$$S_\beta = GY_\beta^{ss} \cong G \times^{P_\lambda} Y_\beta^{ss}. \tag{3.2}$$

Furthermore, the attracting and limit sets can be described as follows.

- *i) The retraction $p_\beta : Y_\beta \to Z_\beta$ is a Zariski locally trivial affine space fibration.*
- *ii) The scheme Y_β is preserved by the P_λ-action and Z_β is preserved by the L_λ-action. Moreover, p_β is equivariant with respect to the retraction $q_\lambda : P_\lambda \to L_\lambda$.*
- *iii) For $p_\beta : Y_\beta \to Z_\beta$, we have $p_\beta^{-1}(Z_\beta^{ss}) = Y_\beta^{ss}$.*
- *iv) Z_β^{ss} is the GIT semistable set for the action of the reductive Levi subgroup L_λ on Z_β with respect to a canonical linearisation $\mathcal{L}_\beta$ obtained by twisting by a rational multiple of a character corresponding to λ.*

Remark 3.25. The fact that the retraction $p_\beta : Y_\beta \to Z_\beta$ is a Zariski locally trivial affine space fibration follows by work of Białynicki-Birula [**19**] describing the decomposition of a smooth projective variety with a $\mathbb{G}_m$-action by taking the flow as $t \to 0$. Here it is crucial that X is *smooth* for the fibres to be affine *spaces*.

In iv) of Proposition 3.24, recall that via the norm $\|-\|$, we can identify characters and co-characters, thus we can twist the linearisation by the rational character χ corresponding to the rational 1-PS $\frac{-m}{\|\lambda\|}\lambda$, so that $\mu^{\mathcal{L}_\beta}(x,\lambda) = \mu(x,\lambda) + \langle \chi, \lambda \rangle = 0$ for $x \in X^\lambda$ to 'cancel' the effect of λ.

Remark 3.26. For G acting linearly on $X = \mathbb{P}(V)$ with a fixed choice of norm $\|-\|$, we can compute the index set $\mathcal{B}$ of the instability stratification in terms of the finitely many weights of the action of a maximal torus $T < G$. For any subset of the T-weights whose convex hull does not contain the origin (that is, this is a weight set of an unstable point), we let λ be the primitive 1-PS of T corresponding under $\|-\|$ to the ray in the $X_*(T)$ through the closest point to 0 in the convex hull of this weight set and define m to be the minimum normalised Hilbert–Mumford weight of any point with this weight set, then $\beta = ([\lambda], m) \in \mathcal{B}$ (see [**63**, Lemma 12.6]). In particular, $\mathcal{B}$ is finite as there are only finitely many T-weights.

EXERCISE 3.27 (Instability stratification for binary forms). Consider the action of SL_2 on $\mathbb{P}^d = \mathbb{P}(k[x,y]_d)$ as in Exercise 3.12 and show for each integer $\frac{d}{2} < r \leq d$, there is an unstable stratum corresponding to binary forms $F(x,y)$ with a root of exactly multiplicity r.

In the situation of a reductive group acting linearly on an affine space linearised by a character, using King's Hilbert–Mumford criterion [**62**, Proposition 2.5] (see also Remark 3.13), one can construct an instability stratification [**54**]. For the action of GL_r on $\mathrm{Mat}_{r\times n}$ by left multiplication generalising Exercise 3.14, the (semi)stable locus for the character $\rho = \det$ is the matrices of maximal rank and the instability stratification is given by the rank (see [**55**, Example 2.14]).

Given an instability stratification (3.1), we can ask if there is a categorical G-quotient of an unstable strata S_β, or equivalently via the isomorphism (3.2), a categorical P_λ-quotient of Y_β^{ss}.

PROPOSITION 3.28 (Categorical quotients of unstable strata, [**57**, Lemma 3.1]). *The composition* $Y_\beta^{ss} \xrightarrow{p_\beta} Z_\beta^{ss} \xrightarrow{\pi} Z_\beta /\!/_{\mathcal{L}_\beta} L_\lambda$ *of* p_β *with the reductive GIT quotient* π *is a categorical* P_λ*-quotient.*

PROOF. This composition is P_λ-invariant, as p_β is equivariant with respect to $q_\lambda : P_\lambda \twoheadrightarrow L_\lambda$ and π is L_λ-invariant. Given a P_λ-invariant morphism $f : Y_\beta^{ss} \to S$, its restriction $f|$ to Z_β^{ss} is L_λ-invariant and since f is constant on orbit closures, we have $f = f| \circ \pi$. Then by the universal property of π, we see that $f|$ (and thus also f) factors uniquely via $Z_\beta /\!/_{\mathcal{L}_\beta} L_\lambda$. □

By [**57**, Lemma 3.1] we have

$$R(\overline{Y_\beta}, \mathcal{L}_\beta)^{P_\lambda} \cong R(Z_\beta, \mathcal{L}_\beta)^{L_\lambda},$$

which is finitely generated, and the categorical quotient coincides with the projective spectrum of the invariants. However, this categorical quotient factors via the retraction p_β and so identifies every x with $p_\beta(x) = \lim_{t\to 0} \lambda(t) \cdot x$. Thus this categorical quotient is far from being an orbit space. Since the closed subscheme Z_β^{ss} causes these identifications, we would like to further twist the linearisation to make Z_β^{ss} become unstable and prevent these unwanted identifications. It is at this point where we see that the non-reductive action of P_λ on Y_β^{ss} is preferable to the reductive action of G on S_β: the parabolic subgroup has more characters which can be used to twist the given ample linearisation; in §6.3, we explain how to apply non-reductive GIT.

In fact, since linearisations give line bundles on quotient stacks, we have a line bundle

$$\mathcal{L}_\beta \to [Y_\beta^{ss}/P_\lambda] \cong [S_\beta/G],$$

but whilst the corresponding P_λ-equivariant line bundle on Y_β^{ss} is ample, the corresponding G-equivariant line bundle on S_β is not ample (see [**57**, Remark 9]) and so is not suitable for working with from the perspective of GIT.

4. Generalisations of reductive GIT to stacks

In §4.1 we describe different types of moduli spaces for stacks and in §4.2, we state a recent existence criterion [**7**] for stacks to admit a good moduli space.

Throughout this section, for simplicity, we will assume that our algebraically closed field k is of characteristic 0 to avoid the distinction between linearly reductive and geometrically reductive groups in positive characteristic, which in turn leads to a distinction between *good moduli spaces* and *adequate moduli spaces* for stacks. We will assume all stacks are noetherian algebraic stacks over k; however, everything in this section also extends to a relative setting. For a detailed introduction to algebraic spaces and stacks with a focus on moduli, see [**1**, §3].

4.1. Moduli spaces for stacks. Associated to an action $G \times X \to X$, there is a quotient stack $[X/G]$ whose points (and residual gerbes) describe the orbits (and stabilisers) of the action. A categorical quotient of this action is equivalent to a universal map from $[X/G]$ to a scheme. One can naturally ask if an arbitrary stack has a universal map to a scheme (or possibly an algebraic space, as was necessary in §2.2). If additionally, as in the definition of a coarse moduli space for a moduli functor, we ask for a bijection on k-points, this leads to the following notion.

DEFINITION 4.1. A *coarse moduli space* (CMS) for a stack $\mathfrak{X}$ is a map $\mathfrak{X} \to X$ to an algebraic space which is initial for maps from $\mathfrak{X}$ to algebraic spaces and such that the induced map $|\mathfrak{X}(k)| \to X(k)$ is bijective.

Keel and Mori [**59**] studied quotients of groupoids in the category of algebraic spaces and proved the existence of a categorical quotient under the assumption of finite stabilisers; let us state a stacky reformulation as in [**25**].

THEOREM 4.2 (Keel–Mori Theorem). *An algebraic stack $\mathfrak{X}$ over k with finite inertia stack admits a coarse moduli space $\pi : \mathfrak{X} \to X$ with the following properties*

i) $\mathcal{O}_X \to \pi_* \mathcal{O}_\mathfrak{X}$ is an isomorphism;
ii) If $\mathfrak{X}$ is separated (resp. of finite type) over k, then X is separated (resp. of finite type) over k;
iii) π is a proper universal homeomorphism;
iv) Any flat base change of π is also a coarse moduli space.

In particular, the Keel–Mori Theorem applies to separated Deligne–Mumford stacks. Requiring $\mathfrak{X} \to X$ to induce a bijection on closed points is very strong: as soon as there are non-closed orbits this condition fails. Fortunately Alper [**2**] adapted the reductive GIT notion of good quotient to the setting of stacks as follows.

DEFINITION 4.3. A *good moduli space* (GMS) for a stack $\mathfrak{X}$ is a quasi-compact quasi-separated morphism $f : \mathfrak{X} \to X$ to an algebraic space such that

(1) the pushforward map $f_* : \mathcal{Q}Coh(\mathfrak{X}) \to \mathcal{Q}Coh(X)$ on quasi-coherent sheaves is exact,
(2) the natural map $\mathcal{O}_X \to f_* \mathcal{O}_\mathfrak{X}$ is an isomorphism.

The following example shows this theory applies to quotients of linearly reductive groups.

EXAMPLE 4.4 (Alper, [**2**, Example 12.9]). For the classifying stack $BG = [\operatorname{Spec} k/G]$ of an affine algebraic group, the map $\pi : BG \to \operatorname{Spec} k$ satisfies the second property in the definition of a good moduli space. Then $\pi_* : \mathcal{V}ect_k^G \to \mathcal{V}ect_k$ is given by taking G-invariants of a linear representation of G. Hence, the first condition holds if and only if G is linearly reductive.

Subsequently, Alper later adapted his theory to include geometrically reductive groups in positive and mixed characteristic by giving a notion of an *adequate moduli space* [**3**], which weakens the first property in the definition of good moduli spaces.

EXAMPLE 4.5 (Alper, [**2**, Theorem 13.6]). For a linearly reductive group G acting on an affine scheme X, the quotient stack admits a good moduli space $[X/G] \to X/\!/G$ given by the affine GIT quotient. More generally, for a linearly reductive group acting on a scheme Y with respect to an ample linearisation $\mathcal{L}$, the morphism $[Y^{ss}(\mathcal{L})/G] \to Y/\!/_{\mathcal{L}}G$ is a good moduli space.

Let us note some important properties of good moduli spaces.

REMARK 4.6. If $f : \mathfrak{X} \to X$ is a good moduli space, then it has the following properties.

(1) The morphism f is initial among maps from $\mathfrak{X}$ to algebraic spaces [**2**, Theorem 6.6] and is surjective and universally closed [**2**, Theorem 4.16 (i) and (ii)].
(2) For every point $x \in X$, there is a unique closed point x_0 in $f^{-1}(x)$ and the automorphism group of x_0 is linearly reductive [**2**, Theorem 9.1 and Proposition 12.14].
(3) The morphism f induces a bijection between closed points in $\mathfrak{X}$ and closed points in X [**2**, Theorem 4.16 (iv)].
(4) If $\mathfrak{X}$ is of finite type over k, then so is X [**2**, Theorem 4.16 (xi)].
(5) Any base change of f along a morphism of quasi-separated algebraic spaces is also a good moduli space [**2**, Proposition 4.7].

4.2. Stability and existence criteria. Halpern-Leistner [**46**] and Heinloth [**52**] studied how ideas in reductive GIT, such as the Hilbert–Mumford criterion, can be applied to stacks. The role of 1-PSs and their limits can be replaced by the stack *Theta*:

$$\Theta := [\mathbb{A}^1/\mathbb{G}_m]$$

over $\operatorname{Spec}\mathbb{Z}$, for the $\mathbb{G}_m$-action on $\mathbb{A}^1 = \operatorname{Spec}\mathbb{Z}[x]$ by scalar multiplication. This stack plays a prominent role in the work of Halpern-Leistner [**46**] and led to a notion of Θ-*stability* for stacks and a generalisation of GIT instability stratifications to Θ-*stratifications* of stacks [**48**].

In this section, we will state a recent existence theorem of Alper, Halpern-Leistner and Heinloth [**7**] which gives necessary and sufficient conditions for a stack to admit a good moduli space. These conditions are valuative criteria known as Θ-reductivity and S-completeness.

DEFINITION 4.7 (Valuative criteria for stacks). A noetherian algebraic stack $\mathfrak{X}$ is said to be

i) *Θ-reductive* if for any DVR R, any morphism $\Theta_R \setminus \{0\} \to \mathfrak{X}$ extends uniquely to Θ_R, where $\Theta_R = \Theta \times_{\mathbb{Z}} \operatorname{Spec} R$ and $0 \in \Theta_R$ denotes the unique closed point.
ii) *S-complete* if for any DVR R, any morphism $\overline{\mathrm{ST}}_R \setminus \{0\} \to \mathfrak{X}$ extends uniquely to

$$\overline{\mathrm{ST}}_R := [\operatorname{Spec}(R[s,t]/(st-\pi))/\mathbb{G}_m]$$

for a uniformiser π and $\mathbb{G}_m$-action with weights $+1, -1$ on s, t.

The stack $\overline{\mathrm{ST}}_R$ originates from work of Heinloth [**52**] and naturally generalises Example 2.14 where $\mathbb{G}_m$ acts on $\mathbb{A}^2$ with weights $+1, -1$. Recall that after removing the origin, $\mathbb{A}^2 \setminus \{0\}$ has non-separated geometric quotient given by the affine line with two origins. S-completeness should be thought of as a stacky valuative criterion for separatedness.

Let us explain how we can think about morphisms from Θ_R and $\overline{\mathrm{ST}}_R$ to moduli stacks of objects in an abelian category.

REMARK 4.8. Assume that $\mathfrak{M}$ is a moduli stack for objects in an abelian category as in [**7**, §7]. Then a morphism $\Theta_k \to \mathfrak{M}$ is a (weighted) filtration on a family of $\mathfrak{M}$ over k such that the associated graded lies in $\mathfrak{M}$. Let R be a DVR with fraction field K and residue field $k = R/(\pi)$, where π denotes a uniformiser. In this case, we can interpret the above conditions as follows.

(1) A morphism $\Theta_R \setminus \{0\} \to \mathfrak{M}$ is given by a family of $\mathfrak{M}$ over the DVR R together with a filtration on the generic fibre K (whose associated graded object lies in $\mathfrak{M}$). This extends uniquely to Θ_R if and only if the filtration on the generic fibre extends uniquely to the special fibre (again with the associated graded object lying in $\mathfrak{M}$).
(2) A morphism $\overline{\mathrm{ST}}_R \setminus \{0\} \to \mathfrak{M}$ is equivalent to two families of $\mathfrak{M}$ over R whose generic fibres over K are equivalent. The condition that this extends uniquely to $\overline{\mathrm{ST}}_R$ implies that the special fibres have filtrations whose associated graded objects are isomorphic[4] in $\mathfrak{M}$ with oppositely ordered graded pieces; this condition often characterises S-completeness in practice.

The above claims can be seen from looking at the following diagrams appearing in [**1**, §6.7.2]

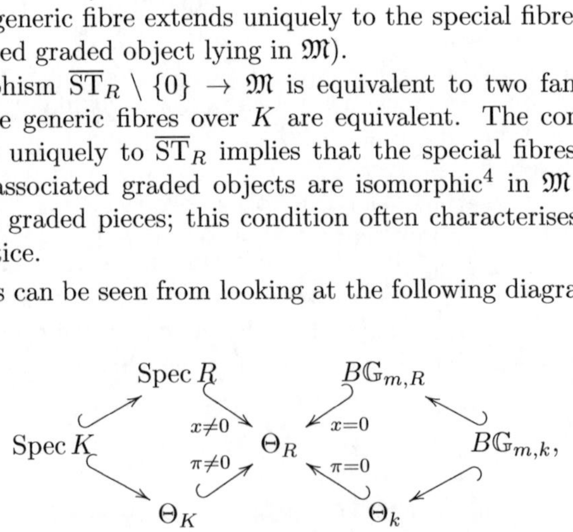

where the left side of the diagram corresponds to the open immersion $\Theta_R \setminus \{0\} \hookrightarrow \Theta_R$ and all morphisms on the left are open immersions, and the right side represents the closed immersion $\{0\} \hookrightarrow \Theta_R$ and all morphisms on the right are closed immersions.

[4]For the stack of semistable vector bundles on a curve, this asks for the bundles on the special fibre to be S-equivalent.

There is a similar diagram for $\overline{\mathrm{ST}}_R$:

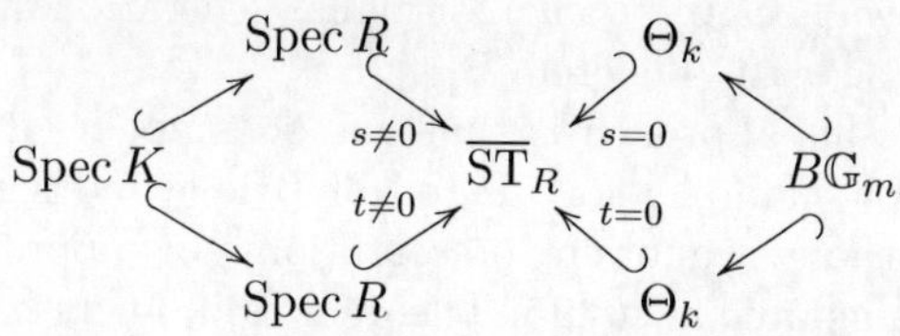

We can now state the existence theorem of Alper–Halpern-Leistner–Heinloth [**7**].

THEOREM 4.9 (Existence criterion for GMS, [**7**, Theorem A]). *Let $\mathfrak{X}$ be an algebraic stack of finite type over k of characteristic zero, with affine diagonal. Then $\mathfrak{X}$ admits a separated good moduli space X if and only if $\mathfrak{X}$ is Θ-reductive and S-complete. Moreover, X is proper if and only if $\mathfrak{X}$ satisfies the existence part of the valuative criterion for properness.*

In characteristic zero, the existence criterion is much nicer due to the existence of *étale local quotient presentations* around closed points with reductive stabiliser due to Alper–Hall–Rydh [**6**], which is a stacky generalisation of Luna's étale slice theorem [**69**]. Showing these conditions are necessary for $\mathfrak{X}$ to have a good moduli space is relatively formal. To show these conditions suffice to construct a good moduli space, one first uses S-completeness to show closed points have reductive stabilisers and then one glues the affine GIT quotients associated to the étale local quotient presentations, which uses both Θ-reductivity and S-completeness.

In positive characteristic there is also an existence theorem for adequate moduli spaces, but as an input it requires the existence of étale local quotient presentations, which is part of the *local reductivity* assumption in [**7**, Theorem A].

With this existence criterion in hand, one can construct moduli spaces without GIT as follows.

(1) Interpret the moduli problem as an algebraic stack $\mathfrak{M}$.

(2) Apply an existence theorem to give a (proper) good moduli space $\mathfrak{M} \to M$.

However, this only yields a proper good moduli space in cases where GIT would provide a projective moduli space. Consequently, one further step is required for this strategy:

(3) Find an ample line bundle on M to show it is projective.

This approach has been implemented for moduli of smooth projective curves in [**24**], where the second step uses the Keel–Mori Theorem and the third step follows Kollar's proof of projectivity using the determinant of a relative pluricanonical sheaf for the universal family; and for moduli of vector bundles on a smooth projective curve in characteristic zero in [**4**], where the second step uses Theorem 4.9 with Langton's proof of the valuative criteria [**67**] and the third step follows Faltings' proof of projectivity using a determinantal line bundle constructed from the universal family. In fact, Weissmann and Zhang [**96**] prove that the semistable locus is the unique maximal open substack admitting a schematic good moduli space. For moduli of representations of an acyclic quiver (in arbitrary characteristic), a new moduli-theoretic proof of projectivity was given in [**13**], where again a determinantal line bundle is used. Of course, in all these cases, reductive GIT could be applied to produce the moduli space without this extra work! However, this intrinsic moduli-theoretic approach to projectivity does give new insights: the line bundle obtained in the final step is of inherent interest to the moduli problem and the techniques give new effective bounds for global generation of determinantal line bundles (for example, see [**13**, Theorem B]).

The advantage of this approach is that one does not need a quotient presentation of the stack; for example, for moduli of Bridgeland semistable objects in a derived category without a quotient presentation, this approach can be applied provided the stack is algebraic as in [**95**, Theorem 1.3]. There have been impressive applications to moduli of K-polystable Fano varieties [**5**] and moduli of torsors for a Bruhat–Tits group scheme over a curve [**7**, Theorem 8.1].

Since the theory of good (and adequate) moduli spaces is based on GIT for reductive groups, there are still many interesting moduli problems which do not admit good (or adequate) moduli spaces, such as moduli stacks of weighted projective hypersurfaces, where non-reductive groups naturally appear (see §6.2).

5. Non-reductive geometric invariant theory

In this section, we will describe recent progress on non-reductive GIT (in characteristic zero) due to Bérczi, Doran, Hawes and Kirwan [**14, 15**] concerning non-reductive groups with *graded unipotent radicals*. Before we describe their approach, we start by illustrating the issues in constructing non-reductive quotients in §5.1 and give a survey of some previous contributions in §5.2. We then proceed to explain the relationship between $\mathbb{G}_a$-actions and locally nilpotent derivations in §5.3, as well as their interaction with $\mathbb{G}_m$-actions in §5.4. In §5.5 we introduce graded unipotent groups and state the main results of [**14, 15**] in §5.6, before giving details on some of the proofs in §5.7. Throughout this section, we assume k is of characteristic zero.

5.1. Examples of bad behaviour of additive actions. The ring of invariants being non-finitely generated for a non-reductive group is not the only issue that arises when trying to construct quotients of non-reductive group actions. Even when the ring of invariants is finitely generated (for example, see Theorem 2.9), there can be further issues:

(1) The quotient morphism given by the inclusion of invariants may fail to be surjective (and in general, its image is only a constructible subset), so will not be a good quotient.
(2) There may not be enough invariants to separate disjoint closed orbits in contrast to the case for geometrically reductive groups (Lemma 2.7).
(3) Invariants may not extend to the ambient space.

Let us give some concrete examples of linear $\mathbb{G}_a$-actions that demonstrate these issues.

EXAMPLE 5.1. The following examples are all built from powers (or symmetric powers) of the standard representation of the additive group on $\mathbb{A}^2$ given by $\mathbb{G}_a < \mathrm{SL}_2$ (as the upper triangular unipotent radical) acting on $V = \mathbb{A}^2$ by left multiplication; thus Theorem 2.9 applies.

(1) [**81**, Section 4.3] For $\mathbb{G}_a$ acting on $V \times V = \mathbb{A}^4$ via $u \cdot (x_1, x_2, x_3, x_4) = (x_1 + ux_2, x_2, x_3 + ux_4, x_4)$,

$$\mathcal{O}(V \times V)^{\mathbb{G}_a} = k[x_2, x_4, x_1x_4 - x_2x_3]$$

and the (invariant-theoretic) quotient $\pi : \mathbb{A}^4 \to \mathbb{A}^3 \cong \operatorname{Spec} \mathcal{O}(V \times V)^{\mathbb{G}_a}$ only has constructible image, as it misses the punctured line $\{(0,0,\eta) : \eta \neq 0\}$.

(2) [**32**, Example 3.1.3] For $\mathbb{G}_a$ acting on $\operatorname{Sym}^2(V) = \mathbb{A}^3$ via $u \cdot (x_1, x_2, x_3) = (x_1 + 2ux_2 + u^2x_3, x_2 + ux_3, x_3)$,

$$\mathcal{O}(\operatorname{Sym}^2(V))^{\mathbb{G}_a} = k[x_3, x_2^2 - x_1x_3]$$

and these invariants do not separate the closed orbits $\mathbb{G}_a \cdot (1,1,0) = \{(\eta,1,0)\}$ and $\mathbb{G}_a \cdot (1,-1,0) = \{(\eta,-1,0)\}$.

(3) For $\mathbb{G}_a$ acting on $\operatorname{Sym}^2(V) = \mathbb{A}^3$ as in (2), the $\mathbb{G}_a$-invariant closed subscheme $X = \{x_3 = 0\}$ has an invariant function $x_2 \in \mathcal{O}(X)^{\mathbb{G}_a}$ which does not extend to $\mathcal{O}(\mathbb{A}^3)^{\mathbb{G}_a}$.

In particular, just trying to get rings of invariants to be finitely generated will not suffice to generalise the nice properties of reductive GIT.

Even in the case of free $\mathbb{G}_a$-actions, there are examples which do not admit a geometric quotient (for example, see [**29**, Example 18] which is proved via a geometrical argument).

5.2. Short historical note on non-reductive group actions. Let us summarise some of the contributions towards the development of non-reductive GIT, before turning to [**14, 15**].

As mentioned in Remarks 2.2 and 2.10, the transfer principle was used by Grosshans [**41**] to prove finite generation in certain cases. Grosshans proved the unipotent radical U of a parabolic subgroup in a reductive group G is a Grosshans group and thus if a U-action on an affine variety extends to G, then the ring of U-invariant functions is finitely generated [**42**]; his proof uses a grading by weights of a maximal torus (see [**22**, Theorem 2.7]), which will also play an important role in non-reductive GIT.

Fauntleroy defines global intrinsic notions of semistability for a connected unipotent group action on a quasi-affine normal variety in terms of properties of invariant sections and showed if the stabilisers are finite that the open semistable set admits a categorical quotient [**33**, Theorem 5]. In subsequent work [**34**], he more generally defined a notion of properly stable points for a linearised action of a connected linear algebraic group on a normal projective variety and shows this locus has a quasi-projective geometric quotient. Fauntleroy combined ideas from reductive GIT with Seshadri covers, which Seshadri used to show actions of connected linear algebraic groups on a normal variety with finite stabiliser groups admit geometric quotients up to a finite equivariant replacement [**89**].

For non-reductive actions on normal varieties, Winkelmann [**98**, Theorem 3] showed there is a rational quotient map to a quasi-affine variety, whose ring of regular functions is the ring of invariants (and is not necessarily finitely generated). In fact, he showed the study of rings of rational functions on quasi-affine varieties corresponds to the study of rings of invariants for $\mathbb{G}_a$-actions on affine varieties [**98**, Theorem 1].

Alternatively, one can ignore the issue of whether or not rings of invariants are finitely generated and work in the category of all schemes (not necessarily of finite type) over k; Greuel and Pfister take this approach in [**40**] to define a notion of stability for unipotent group actions and show there is a geometric quotient in the category of varieties. Their motivation came from the study of singularities, where non-reductive translation actions appear.

We note that additive group actions also arise as translation actions in affine geometry and have led to progress on classical questions on affine spaces such as the cancellation problem, the existence of exotic affine spaces and the Jacobian conjecture (for example, see [**66**]).

Doran and Kirwan [**32**] give various notions of (semi)stability for non-reductive GIT using properties of invariants and by transferring the problem to a reductive GIT setting using a notion of 'fine reductive envelope' and they obtain various types of quotients of these (semi)stable sets.

The recent progress on non-reductive GIT [**15**], which we explain below, uses a multiplicative group to grade the unipotent radical; this enables the construction of geometric unipotent quotients, as well as providing a natural projective completion and an explicit Hilbert–Mumford type description of stability. To trace back the origin of this grading multiplicative group, we begin with the correspondence between $\mathbb{G}_a$-actions and locally nilpotent derivations, and will see these multiplicative group actions naturally appear when there is a slice of the $\mathbb{G}_a$-action.

5.3. Actions of the additive group. Since k is of characteristic zero, we can utilise the dictionary between additive group actions and locally nilpotent derivations (e.g. see [**36**]).

For an action $\sigma : \mathbb{G}_a \times X \to X$ on an affine scheme, the coaction

$$\sigma^* : \mathcal{O}(X) \to \mathcal{O}(\mathbb{G}_a \times X) \cong \mathcal{O}(X) \otimes_k k[t]$$

can be used to define a derivation $D_\sigma : \mathcal{O}(X) \to \mathcal{O}(X)$ given by $D_\sigma(f) := \frac{\partial}{\partial t}(\sigma^*(f))|_{t=0}$ satisfying the Leibniz rule $D_\sigma(fg) = fD_\sigma(g) + D_\sigma(f)g$. This derivation is *locally nilpotent* (i.e. for any $f \in \mathcal{O}(X)$, there is $n \in \mathbb{N}$ such that $D_\sigma^n(f) = 0$), as one can inductively show

$$\sigma^*(f) = \sum_{n \geq 0} \frac{D_\sigma^n(f) t^n}{n!}$$

and since the left side is a polynomial in t, we must have $D_\sigma^n(f) = 0$ for all n sufficiently large.

Conversely, given a locally nilpotent derivation $D : A \to A$ on the k-algebra $A = \mathcal{O}(X)$ of functions on an affine scheme, we can exponentiate to construct a coaction

$$\sigma_D^* := \exp(tD) = \sum_{n \geq 0} \frac{D^n t^n}{n!} : A \to A[t]$$

which is well-defined, as D is locally nilpotent.

These constructions are inverse to each other and we collect some other results relating these geometric and algebraic points of view.

PROPOSITION 5.2. *For an affine variety X with coordinate ring $A = \mathcal{O}(X)$, there is a bijective correspondence*

$$\begin{array}{ccc} \{\text{actions } \mathbb{G}_a \times X \to X\} & \longleftrightarrow & \{\text{locally nilpotent derivations } D : A \to A\} \\ \sigma : \mathbb{G}_a \times X \to X & \mapsto & D_\sigma := \frac{\partial}{\partial t}(\sigma^*(-))|_{t=0} \\ \sigma_D^* = \exp(tD) : A \to A[t] & \leftarrow\!\shortmid & D : A \to A. \end{array}$$

For an action σ corresponding to a locally nilpotent derivation D, the following statements hold.

i) The ring of invariants is the kernel of the derivation: $\mathcal{O}(X)^{\mathbb{G}_a} = \ker(D)$*,*
ii) $x \in X$ *is* $\mathbb{G}_a$*-fixed if and only if* $D(A) \subset \mathfrak{m}_x$*.*
iii) If D *has a slice (i.e. there exists* $s \in A = \mathcal{O}(X)$ *such that* $D(s) = 1$*), then* $\mathcal{O}(X)^{\mathbb{G}_a}$ *is a finitely generated* k*-algebra, the subscheme* $S := \{s = 0\} \subset X$ *is a geometric slice of the* $\mathbb{G}_a$*-action (i.e. there is a* $\mathbb{G}_a$*-equivariant isomorphism* $\mathbb{G}_a \times S \simeq X$*) and* $X \to \operatorname{Spec} \mathcal{O}(X)^{\mathbb{G}_a}$ *is a trivial principal* $\mathbb{G}_a$*-bundle.*

PROOF. Let us just give some details on the third statement, as this will be important in what follows. Suppose that $s \in \mathcal{O}(X)$ is a slice for D. Then define a k-algebra homomorphism

$$\Phi : A \to A, \quad f \mapsto \exp(tD(f))|_{t=-s}$$

such that $\operatorname{Im}(\Phi) = \ker(D)$. In particular, $\mathcal{O}(X)^{\mathbb{G}_a} = \ker(D) = \operatorname{Im}(\Phi)$ is a finitely generated k-algebra: the images under Φ of generators for the algebra $A = \mathcal{O}(X)$ are generators of $\mathcal{O}(X)^{\mathbb{G}_a}$.

By induction on the nilpotency index of elements in $\mathcal{O}(X)$, we can write any $f \in \mathcal{O}(X)$ as a polynomial in s with coefficient in $\operatorname{Im}(\Phi) = \mathcal{O}(X)^{\mathbb{G}_a}$:

$$f = \sum_{n \geq 0} \frac{\Phi(D^n(f))s^n}{n!}$$

and so $\mathcal{O}(X) \simeq \mathcal{O}(X)^{\mathbb{G}_a}[s]$ and $\mathcal{O}(X)^{\mathbb{G}_a} \simeq \mathcal{O}(X)/(s)$. Hence $S = \{s = 0\} \subset X$ has coordinate ring $\mathcal{O}(X)/(s) \simeq \mathcal{O}(X)^{\mathbb{G}_a}$ and there is an isomorphism $X \simeq \operatorname{Spec} \mathcal{O}(S)[s] \simeq \mathbb{G}_a \times S$. Furthermore, the isomorphism $X \simeq \mathbb{G}_a \times S$ is $\mathbb{G}_a$-equivariant, as under the isomorphism $\mathcal{O}(X) \simeq \mathcal{O}(X)^{\mathbb{G}_a}[s]$, the derivation D corresponds to $\partial/\partial s$, since D is trivial on $\mathcal{O}(X)^{\mathbb{G}_a}$. This shows that S is a geometric slice and $\pi : X \to S$ is a trivial principal $\mathbb{G}_a$-bundle. Note that the k-algebra homomorphism Φ induces a morphism $\Phi^* : X \to X$ which factors as the composition of the projection $\pi : X \twoheadrightarrow S$ with the inclusion $S \hookrightarrow X$. □

Note that if there is a slice, then the coordinate ring is a polynomial ring in a single variable with coefficients in the ring of invariants. In particular, the coordinate ring is graded by $\mathbb{N}$ (the degree of this polynomial) and thus this gives a $\mathbb{G}_m$-action. This naturally leads to considering actions of semi-direct products of $\mathbb{G}_a$ and $\mathbb{G}_m$.

5.4. Semi-direct products of additive and multiplicative groups. For an affine scheme X, a $\mathbb{G}_m$-action on X is equivalent to a $\mathbb{Z}$-grading of the k-algebra $\mathcal{O}(X)$. The limit under the $\mathbb{G}_m$-action as $t \to 0$ exists for all $x \in X$ if and only if $\mathbb{G}_m$ acts with non-negative weights on X, which is if and only if the grading on $\mathcal{O}(X)$ is supported in non-positive degrees.

DEFINITION 5.3. A semi-direct product of $\mathbb{G}_a$ and $\mathbb{G}_m$ is given by specifying a group homomorphism $\varphi : \mathbb{G}_m \to \operatorname{Aut}(\mathbb{G}_a)$ and defining the semi-direct product $\mathbb{G}_a \rtimes_\varphi \mathbb{G}_m$ to have underlying set $\mathbb{G}_a \times \mathbb{G}_m$ with group operation

$$(u, t) \cdot (v, s) = (v + \varphi_s(u), ts).$$

For $n \in \mathbb{Z}$, let us define φ by $t \mapsto t^{-n}$ and write $\mathbb{G}_a \rtimes_n \mathbb{G}_m$ for the associated semi-direct product, where $tut^{-1} = t^n u$.

EXAMPLE 5.4. The upper triangular Borel $B < \mathrm{SL}_2$ is isomorphic to a semi-direct product $\mathbb{G}_a \rtimes_2 \mathbb{G}_m$, via $(u, t) \mapsto \left(\begin{smallmatrix} t & ut \\ 0 & t^{-1} \end{smallmatrix}\right)$.

REMARK 5.5. For an action of a semi-direct product $\mathbb{G}_a \rtimes_n \mathbb{G}_m$ on an affine scheme, the locally nilpotent derivation D associated to the $\mathbb{G}_a$-action is homogeneous of degree n with respect to the grading $\mathcal{O}(X) = \bigoplus_{r\in\mathbb{Z}} \mathcal{O}(X)_r$ determined by the $\mathbb{G}_m$-action; that is, $D(\mathcal{O}(X)_r) \subset \mathcal{O}(X)_{r+n}$.

The following result is the key proposition which enables the inductive construction of quotients of unipotent group actions in the presence of an appropriate $\mathbb{G}_m$-action grading the unipotent action. This is a modification of [**14**, Lemma 7.3].

PROPOSITION 5.6 (Key Proposition). *For an affine $\mathbb{G}_a$-scheme X with locally nilpotent derivation $D : \mathcal{O}(X) \to \mathcal{O}(X)$, the following statements are equivalent:*

(1) D has a slice (i.e. there exists $s \in \mathcal{O}(X)$ with $D(s) = 1$),

(2) The $\mathbb{G}_a$-action extends to $\mathbb{G}_a \rtimes_n \mathbb{G}_m$ for some $n > 0$ such that

  i) $\lim_{t\to 0} t \cdot x$ exists for all $x \in X$ and

  ii) $\mathrm{Stab}_{\mathbb{G}_a}(z) = \{e\}$ for all $z \in Z := \{\lim_{t\to 0} t \cdot x\}$.

In particular, if (2) holds, then there is a trivial $\mathbb{G}_a$-quotient $X \mapsto \mathcal{O}(X)^{\mathbb{G}_a}$ (i.e. this map is a trivial principal $\mathbb{G}_a$-bundle).

PROOF. Suppose D has a slice s; then $\mathcal{O}(X) = \mathcal{O}(X)^{\mathbb{G}_a}[s]$ as in the proof of Proposition 5.2. Since $\mathcal{O}(X)$ is a polynomial ring in s, it is naturally graded. However, we choose a $\mathbb{Z}_{\leq 0}$-grading given by $\mathcal{O}(X)_{-r} = \mathcal{O}(X)^{\mathbb{G}_a} s^r$ so that $\lim_{t\to 0} t\cdot x$ exists for all $x \in X$. Since $D(s) = 1$, we see that D is homogeneous of degree 1 and thus there is an action of $\mathbb{G}_a \rtimes_1 \mathbb{G}_m$. Since $X \to \operatorname{Spec} \mathcal{O}(X)^{\mathbb{G}_a}$ is a trivial $\mathbb{G}_a$-bundle (see Proposition 5.2), all $\mathbb{G}_a$-stabilisers are trivial.

Let us outline the converse direction (see [**14**, Lemma 7.3] for the full proof). Given $n > 0$ and a $\mathbb{G}_a \rtimes_n \mathbb{G}_m$-action on X such that $\lim_{t\to 0} t \cdot x$ exists for all $x \in X$ and $\mathrm{Stab}_{\mathbb{G}_a}(z) = \{e\}$ for all $z \in Z$, we note that D has homogeneous degree n with respect to the grading $\mathcal{O}(X) = \oplus_r \mathcal{O}(X)_r$ given by the $\mathbb{G}_m$-action. By the assumption that $\lim_{t\to 0} t \cdot x$ exists for all $x \in X$, this grading is supported in non-positive degrees. Since $D(\mathcal{O}(X)_r) \subset \mathcal{O}(X)_{r+n}$ and $\mathcal{O}(X)_{>0} = 0$, we have

$$D(\mathcal{O}(X)_r) \subset \begin{cases} 0 & \text{if } r > -n \\ \mathcal{O}(X)_0 & \text{if } r = -n \\ \mathcal{O}(X)_{<0} & \text{if } r < -n \end{cases} \tag{5.1}$$

and we will show that $D(\mathcal{O}(X)_{-n}) = \mathcal{O}(X)_0$, so there exists $s \in \mathcal{O}(X)_{-n}$ with $D(s) = 1$. Let

$$I := D(\mathcal{O}(X)_{-n}) \oplus \mathcal{O}(X)_{<0}, \tag{5.2}$$

which is $\mathbb{G}_a$-stable by (5.1) and also $\mathbb{G}_m$-stable, as $\mathcal{O}(X)_{<0}$ is a sum of $\mathbb{G}_m$-weight spaces and the $\mathbb{G}_m$-action on $D(\mathcal{O}(X)_{-n}) \subset \mathcal{O}(X)_0$ is trivial. Then it suffices to show that I is an ideal (Claim 1) and moreover $I = \mathcal{O}(X)$ (Claim 2). Indeed, as $I = \mathcal{O}(X)$ we must have $D(\mathcal{O}(X)_{-n}) = \mathcal{O}(X)_0$ and so $D(s) = 1$ for some $s \in \mathcal{O}(X)_{-n}$.

To prove Claim 1, we need to show for $f \in I$ and $h \in \mathcal{O}(X)$ that $hf \in I$; for this we can write $h = \sum h_r$ with respect to the $\mathbb{G}_m$-grading and it suffices to show $h_r f \in I$ for all r. We can also write $f = D(p_{-n}) + \sum_{r<0} f_r$ by (5.2). The only non-trivial case is for $h = h_0$ and $f = D(p_{-n})$; however, by the Leibniz rule $D(h_0 p_{-n}) = h_0 D(p_{-n})$ and so $h_0 f = D(h_0 p_{-n}) \in I$.

To prove Claim 2, we argue by contradiction. If $I \subsetneq \mathcal{O}(X)$, then it is contained in a maximal ideal $\mathfrak{m}_x$. Since I is $\mathbb{G}_a \rtimes_n \mathbb{G}_m$-stable, so is $\mathfrak{m}_x$ and so it corresponds to a $\mathbb{G}_a \rtimes_n \mathbb{G}_m$-fixed point $x \in Z$. However, this contradicts the assumption that $\mathrm{Stab}_{\mathbb{G}_a}(z) = \{e\}$ for all $z \in Z$. □

Note that there is a choice of sign here. One could consider actions of $\mathbb{G}_a \rtimes_n \mathbb{G}_m$ for $n < 0$ such that $\lim_{t\to\infty} t \cdot x$ exists for all $x \in X$ and obtain an analogous result.

REMARK 5.7. For a linear representation $\mathbb{G}_a \rtimes_n \mathbb{G}_m \to \mathrm{GL}(V)$, the derivation sends a $\mathbb{G}_m$-weight space V_i to V_{n+i}. Thus $V_{\max} \subset V^{\mathbb{G}_a}$, where $V_{\max}$ is the maximal $\mathbb{G}_m$-weight space.

EXAMPLE 5.8. Consider the upper triangular Borel $\mathbb{G}_a \rtimes_2 \mathbb{G}_m \cong B < \mathrm{SL}_2$ acting on $V = \mathbb{A}^2$ by left multiplication. We have

$$V = V_{+1} \oplus V_{-1} = \{(*, 0)\} \oplus \{(0, *)\}$$

and $V_{\max} = V_{+1} \subset V^{\mathbb{G}_a}$.

5.5. Graded unipotent groups. Generalising Proposition 5.6, we will consider actions of unipotent groups which are graded by a $\mathbb{G}_m$-action as follows.

DEFINITION 5.9. A *graded unipotent group* $\widehat{U} := U \rtimes \mathbb{G}_m$ is a semi-direct product of a unipotent group with a multiplicative group such that the conjugation of $\mathbb{G}_m$ on the Lie algebra of U has strictly positive weights.

EXAMPLE 5.10. For $n > 0$, the group $\mathbb{G}_a \rtimes_n \mathbb{G}_m$ is a graded unipotent group. In particular, the upper triangular Borel $B < \mathrm{SL}_2$ is a graded unipotent group.

PROPOSITION 5.11. *Let X be an affine scheme with an action of a graded unipotent group $\widehat{U} := U \rtimes \mathbb{G}_m$ such that $\lim_{t\to 0} t \cdot x$ exists for all $x \in X$ and $\mathrm{Stab}_U(z) = \{e\}$ for all $z \in Z := \{\lim_{t\to 0} t \cdot x\}$, then $\mathcal{O}(X)^U$ is finitely generated and $X \to \mathrm{Spec}\, \mathcal{O}(X)^U$ is a trivial U-quotient.*

PROOF. The idea is to iteratively apply Proposition 5.6 as in the proof of [**14**, Proposition 7.4]. By lifting a filtration on the Lie algebra via the exponential map, we obtain normal subgroups

$$\{e\} = U_0 < U_1 < \cdots < U_r = U$$

whose successive quotients are copies of $\mathbb{G}_a$ on which $\mathbb{G}_m$ acts by conjugation with strictly positive weights. Assume that we have constructed a quotient $q_j : X \to X_j := \mathrm{Spec}\, \mathcal{O}(X)^{U_j}$ which is a trivial U_j-quotient. The base case for $j = 1$ is given by Proposition 5.6. For the inductive step, we claim that we can apply Proposition 5.6 to the $\mathbb{G}_a \cong U_{j+1}/U_j$ action on X_j to show there is a trivial U_{j+1}/U_j-quotient $X_j \to \mathcal{O}(X_j)^{U_{j+1}/U_j}$. Then the composition

$$X \to X_j := \mathrm{Spec}\, \mathcal{O}(X)^{U_j} \to \mathcal{O}(X_j)^{U_{j+1}/U_j} = \mathcal{O}(X)^{U_{j+1}}$$

is a principal U_j-bundle (see [**17**, Proposition 4.7]), which is trivial as the base is affine by [**9**, Theorem 3.12].

To complete the proof, we must show for the $\mathbb{G}_a \cong U_{j+1}/U_j$ action on X_j graded by $\mathbb{G}_m < \widehat{U}$ that all limits $\lim_{t\to 0} t \cdot x_j$ exists for $x_j \in X_j$, and $\mathrm{Stab}_{U_{j+1}/U_j}(x_j) = \{e\}$ for all $x_j \in X_j^{\mathbb{G}_m}$ in order to be able to apply Proposition 5.6. Since q_j is $\mathbb{G}_m$-equivariant, we have

$$\lim_{t\to 0} t \cdot q_j(x) = \lim_{t\to 0} q_j(t \cdot x) = q_j\left(\lim_{t\to 0} t \cdot x\right).$$

Thus the set Z_j of such limits in X_j is contained in $q_j(Z)$. For a point $q_j(z) \in Z_j$, suppose that $uU_j \in \mathrm{Stab}_{U_{j+1}/U_j}(q_j(z))$; that is, there exists $u' \in U_j$ such that $u'z = uz$. However, by assumption $\mathrm{Stab}_U(z) = \{e\}$ and so we have $u = u' \in U_j$, which means $\mathrm{Stab}_{U_{j+1}/U_j}(q_j(z))$ is trivial and we can apply Proposition 5.6 to X_j as claimed. □

REMARK 5.12. For a representation $\rho : \widehat{U} \to \mathrm{GL}(V)$, note that $V_{\max} \subset V^U$.

In particular, rather than trying to find U-invariant sections, we can use $\mathbb{G}_m$-maximal sections to construct a non-reductive GIT quotient as in the next subsection.

5.6. Statement of the key results in non-reductive GIT. Instead of working with a linearised action on a projective scheme, for simplicity we will assume that we have a linear action on $X = \mathbb{P}(V)$ as in §2.6. As before, in the case of a (very ample) linearisation $\mathcal{L}$ on X, we get a projective embedding $X \hookrightarrow \mathbb{P}(V)$ where $V = H^0(X, \mathcal{L})^*$ has a linear action.

This concerns actions of affine algebraic groups, whose unipotent radical is graded by a $\mathbb{G}_m$.

DEFINITION 5.13. Let $G = U \rtimes R$ be an affine algebraic group with unipotent radical U and reductive Levi factor R. A 1-PS $\lambda : \mathbb{G}_m \to Z(R)$ is said to *grade* U if its conjugation action on $\mathrm{Lie}\, U$ has strictly positive weights. In this case, we say *G has graded unipotent radical (by λ).*

This is what is called an *internal grading* in [**14**], where there is also a more general notion of an *external grading*; however, we will stick to the simpler internal point of view, as this suffices in all examples and applications we consider.

There could be several central 1-PSs in $Z(R)$ that grade U, but we will assume we have fixed a grading $\mathbb{G}_m$ and write $\widehat{U} = U \times \mathbb{G}_m$. Different grading 1-PSs gives rise to different quotients, so can be thought of as an additional choice to the linearisation (see Remark 5.17 below).

EXAMPLE 5.14. A parabolic subgroup P of GL_n (or more generally any reductive group) has graded unipotent radical: write $P = P_\lambda$ for a 1-PS λ, then $P_\lambda = U_\lambda \rtimes L_\lambda$, with λ being central in L_λ and grading U_λ.

DEFINITION 5.15 (Minimal weight space and attracting set). For a multiplicative group $\mathbb{G}_m$ acting linearly on $X = \mathbb{P}(V)$, we define the minimal weight space $Z_{\min}$ and minimal attracting set $X_{\min}$ as follows:

$$\begin{array}{rlcrl} Z_{\min} & := \mathbb{P}(V_{\min}) & \stackrel{p}{\longleftarrow} & X_{\min} & := \{x \in X : \lim_{t\to 0} t \cdot x \in Z_{\min}\} \\ & = \{x : \mathrm{wt}_{\mathbb{G}_m}(x) = \{\omega_{\min}\}\} & & & = \{x : \omega_{\min} \in \mathrm{wt}_{\mathbb{G}_m}(x)\} \end{array}$$

where $\omega_{\min} = \omega_0 < \omega_1 < \cdots < \omega_n$ are the $\mathbb{G}_m$-weights on V and $V_{\min} = V_{\omega_{\min}}$.

In [**15**], the minimal attracting set $X_{\min}$ is denoted $X^0_{\min}$. For the associated Białynicki-Birula stratification [**19**] (flowing as $t \to 0$), the variety $X_{\min}$ is the open stratum and $p : X_{\min} \to Z_{\min}$ is a Zariski locally trivial affine space fibration.

In [**14**], there are two types of assumptions needed for the construction of non-reductive GIT quotients: the first (Definition 5.16) concerns the positioning of the weights for the linearised action of the grading $\mathbb{G}_m$, which selects a particular VGIT chamber for $\mathbb{G}_m$ and can be achieved by twisting the linearisation by a (rational) character $\chi : \widehat{U} \to \mathbb{G}_m$, and the second (Assumption $[\widehat{U}]_0$ in Definition 5.18) requires certain unipotent stabiliser groups to be trivial, which is referred to as *semistability coincides with stability* in [**14**].

DEFINITION 5.16 (Adapted linearisation). Let $G = U \rtimes R$ be a group with unipotent radical graded by $\mathbb{G}_m < Z(R)$ acting linearly on $X = \mathbb{P}(V)$. We say this linearised action is *adapted* if the $\mathbb{G}_m$-weights on V satisfy

$$\omega_{\min} = \omega_0 < 0 < \omega_1 < \cdots < \omega_n.$$

REMARK 5.17. The assumption that the linearised action is adapted fixes a particular VGIT chamber for the $\mathbb{G}_m$-action, where (semi)stability is given by

$$X^{\mathbb{G}_m-(s)s} = X_{\min} \setminus Z_{\min}.$$

Since twisting the linearisation by a character $\chi : \widehat{U} \to \mathbb{G}_m$ shifts the weights by $-\chi$ (where we identify characters of $\widehat{U}$ with integers $\mathbb{Z}$ such that $1 \in \mathbb{Z}$ corresponds to a character $\widehat{U} \to \mathbb{G}_m$ with kernel U), if we are prepared to modify the linearisation we can always arrange for this condition to hold. Note that this shift of weights does not change the minimal weight space $Z_{\min}$ and attracting set $X_{\min}$.

DEFINITION 5.18 (Stabiliser assumptions). Let $G = U \rtimes R$ be a group with unipotent radical graded by $\mathbb{G}_m < Z(R)$ acting linearly on $X = \mathbb{P}(V)$.

(1) We say *the unipotent stabiliser assumption holds* if

$$[\widehat{U}]_0 \qquad \dim \mathrm{Stab}_U(z) = 0 \;\text{ for all }\; z \in Z_{\min},$$

(2) We say *the reductive stabiliser assumption holds* if

$$[\overline{R}]_0 \qquad \dim \mathrm{Stab}_{\overline{R}}(z) = 0 \;\text{ for all }\; z \in Z_{\min}^{\overline{R}-ss},$$

where as $\mathbb{G}_m$ is central in R, there is an induced R-action on the $\mathbb{G}_m$-fixed variety $Z_{\min}$ and we let $Z_{\min}^{\overline{R}-ss}$ denote the GIT semistable locus for $\overline{R} := R/\mathbb{G}_m$.

REMARK 5.19. The unipotent stabiliser assumptions are crucial to apply Proposition 5.11, whereas the reductive stabiliser assumptions are used to more readily obtain an explicit Hilbert–Mumford type description of the locus we obtain a quotient of. The reductive stabiliser assumption implies semistability coincides with stability for the $\overline{R}$-action on $Z_{\min}$.

If these stabiliser assumptions fail, then one would like to perform a sequence of equivariant blow-ups to arrange for this to hold on the blow-up (similar to Kirwan's partial desingularisation procedure [**64**]) and then construct a quotient of the original scheme using the quotient of the blow-up; however, this procedure is much more complicated in the non-reductive setting described in [**14**, §9]. Furthermore, if there are generically positive dimensional unipotent stabilisers, blowing up would only result in constant dimensional stabilisers (rather than trivial stabilisers), and so instead one must filter U by normal subgroups whose stabilisers are constant (assuming this can be done via blow-ups) and then proceed as in [**14**, Remark 7.1] and [**83**].

The strategy is to first use the grading $\mathbb{G}_m$ to obtain a projective quotient by $\widehat{U} = U \rtimes \mathbb{G}_m$ and then take a reductive GIT quotient by the residual group $\overline{R} = R/\mathbb{G}_m$. Therefore, we first state the result for quotients by graded unipotent groups $\widehat{U} = U \rtimes \mathbb{G}_m$.

THEOREM 5.20 (The $\widehat{U}$-Theorem, [**15**, Theorem 2.16]). *Let $\widehat{U} = U \rtimes \mathbb{G}_m$ be a graded unipotent group acting linearly on $X = \mathbb{P}(V)$. If the linearised action is adapted and the unipotent stabiliser assumption $[\widehat{U}]_0$ hold, then we have the following statements.*

i) There is a geometric U-quotient $q_U : X_{\min} \to X_{\min}/U$ such that $X_{\min}/U$ is a quasi-projective variety.
ii) There is a geometric $\widehat{U}$-quotient $q_{\widehat{U}} : X_{\min} \backslash UZ_{\min} \to (X_{\min} \backslash UZ_{\min})/\widehat{U} =: X/\!/\widehat{U}$ such that $X/\!/\widehat{U}$ is a projective variety.
iii) If the linearised action is well-adapted[5], then the ring of $\widehat{U}$-invariant sections (for an appropriate power) of the linearisation is finitely generated and taking the Proj construction gives the above geometric $\widehat{U}$-quotient $q_{\widehat{U}}$.

Let us outline the structure of the proof.

(1) Since $X_{\min} = \bigcup_{\sigma \in H^0(X,\mathcal{O}(1))_{\max}} X_\sigma$, the U-quotient is constructed by applying Proposition 5.11 to each affine $\widehat{U}$-variety X_σ for $\sigma \in H^0(X, \mathcal{O}(1))_{\max}$ (see Proposition 5.26).
(2) Construct a $\mathbb{G}_m$-equivariant embedding $X_{\min}/U \hookrightarrow \mathbb{P}(W)$ with $W := (H^0(X, \mathcal{O}(r))^U)^*$ for some $r > 0$, to show $X_{\min}/U$ is quasi-projective (see Proposition 5.27).
(3) By appropriately twisting the original linearisation (so it's *well-adapted*), the induced $\mathbb{G}_m$-action on $\mathbb{P}(W)$ is adapted and so the $\mathbb{G}_m$-(semi)stable locus is $\mathbb{P}(W)^{\mathbb{G}_m-(s)s} = \mathbb{P}(W)_{\min} \setminus \mathbb{P}(W_{\min})$; thus

$$q_U^{-1}(\overline{X_{\min}/U}^{\mathbb{G}_m-(s)s}) = X_{\min} \setminus UZ_{\min},$$

has a geometric $\widehat{U}$-quotient, as a closed subvariety of $\mathbb{P}(W)/\!/\mathbb{G}_m$ (see Proposition 5.29).
(4) To prove that the ring of $\widehat{U}$-invariant sections is finitely generated, one shows that $q_{\widehat{U}}$ coincides with an *enveloping quotient* (see [**16**, Definition 3.1.6]) and as the enveloping quotient is projective, the ring of $\widehat{U}$-invariant sections for an appropriately divisible power of the well-adapted linearisation is finitely generated by [**16**, Corollary 3.1.21].

The first two steps give the proof of Theorem 5.20 i), whereas (3) and (4) give statements ii) and iii) respectively; more details on the proofs of Steps (1) - (3) are given in §5.7 below.

In (1), it is crucial that the U-action is graded by $\mathbb{G}_m$, that the unipotent stabiliser assumption holds and that the action is adapted in order to apply Proposition 5.11. The grading $\mathbb{G}_m$-action is also used in (2) and (3) to firstly show $X_{\min}/U$ is quasi-projective and then obtain a projective quotient of $X_{\min} \setminus UZ_{\min}$ as a closed subvariety of $\mathbb{P}(W)/\!/\mathbb{G}_m$. In (3), in order for the induced $\mathbb{G}_m$-linearisation on $\mathbb{P}(W)$ to be adapted, we need the minimal weight $\omega_{\min}$ for the original linearised action to be negative but very small (see the proof of Proposition 5.29); this leads to the following notion.

DEFINITION 5.21 (Well-adapted linearisation). Let $G = U \rtimes R$ be a group with unipotent radical graded by $\mathbb{G}_m < Z(R)$ acting linearly on $X = \mathbb{P}(V)$. We say this

[5]See Definition 5.21 below, which can be achieved by further twisting by a rational character.

linearised action is *well-adapted* if there is $0 < \epsilon << 1$ such that the $\mathbb{G}_m$-weights on V satisfy

$$-\epsilon < \omega_{\min} = \omega_0 < 0 < \omega_1 < \cdots < \omega_n.$$

Let us present a simple example for matrices up to conjugation extending Example 2.15.

EXAMPLE 5.22 (To appear in upcoming joint work with E. Hamilton and J. Jackson). Consider the upper triangular Borel subgroup $\widehat{U} = \mathbb{G}_a \rtimes_2 \mathbb{G}_m < \mathrm{SL}_2$ acting by conjugation on $\mathrm{Mat}_{2\times 2}$. To apply Theorem 5.20, we consider the projective embedding $\mathrm{Mat}_{2\times 2} \hookrightarrow X := \mathbb{P}(\mathrm{Mat}_{2\times 2} \oplus k)$, with trivial action on k. The minimal weight space is 1-dimensional and spanned by the elementary matrix E_{21}, so $Z_{\min} = \{*\} = \mathbb{P}(kE_{21})$ is contained at infinity (i.e. in $\mathbb{P}(\mathrm{Mat}_{2\times 2})$), and $X_{\min} = \{[A : z] : a_{21} \neq 0\}$. The unique point in $Z_{\min}$ has trivial $\mathbb{G}_a$-stabiliser; thus $[\widehat{U}]_0$ holds. By twisting the linearisation to make it adapted, we obtain a geometric $\widehat{U}$-quotient

$$X^{\widehat{U}-s} = X_{\min} \setminus UZ_{\min} \to \mathbb{P}(1,1,2), \quad [A : z] \mapsto [z : \operatorname{tr} A : \det A].$$

As $UZ_{\min}$ is contained at infinity, $\mathrm{Mat}_{2\times 2} \cap X^{\widehat{U}-s} = \mathrm{Mat}_{2\times 2} \cap X_{\min} = (\mathrm{Mat}_{2\times 2})_{a_{21}}$ and we obtain a *geometric* $\widehat{U}$-quotient of matrices whose bottom left entry is non-zero

$$(\mathrm{Mat}_{2\times 2})_{a_{21}} \to \mathbb{A}^2, \quad A \mapsto (\operatorname{tr} A, \det A).$$

Suppose that $G = U \rtimes R$ acts on X; then the projective geometric $\widehat{U}$-quotient $X/\!/\widehat{U}$ of Theorem 5.20 has a residual action of the reductive group $\overline{R} = R/\mathbb{G}_m$. There is a quotient

$$X_{\min} \setminus UZ_{\min} \xrightarrow{q_{\widehat{U}}} X/\!/\widehat{U} \dashrightarrow X/\!/G := (X/\!/\widehat{U})/\!/\overline{R},$$

where the second morphism is the reductive GIT quotient; this gives a projective and good G-quotient of the open set $q_{\widehat{U}}^{-1}((X/\!/\widehat{U})^{\overline{R}-ss})$. The key challenge is to determine this preimage in terms of the original action on X. This is easiest when $[\overline{R}]_0$ holds (i.e. semistability coincides with stability for the $\overline{R}$-action on $Z_{\min}$); in this case, we define the following semistable sets.

DEFINITION 5.23 (Non-reductive stable set). Let $G = U \rtimes R$ be a group with unipotent radical graded by $\mathbb{G}_m < Z(R)$ acting linearly on $X = \mathbb{P}(V)$. If the linearisation is well-adapted and both $[\widehat{U}]_0$ and $[\overline{R}]_0$ hold, then we define the *G-stable set* by

$$X^{G-s} := X_{\min}^{\overline{R}-ss} \setminus UZ_{\min}^{\overline{R}-ss} = p^{-1}(Z_{\min}^{\overline{R}-ss}) \setminus UZ_{\min}^{\overline{R}-ss}.$$

REMARK 5.24. Since $[\widehat{U}]_0$ holds for the linear action of the graded unipotent group $\widehat{U}$ on $X = \mathbb{P}(V)$, the U-sweep of $Z_{\min}$ is a closed subvariety of $X_{\min}$ by [**14**, Lemma 5.4]. Consequently, the $\widehat{U}$-stable locus $X^{\widehat{U}-s} = X_{\min} \setminus UZ_{\min}$ is open in X, and so is the G-stable locus.

The next result is a special case of [**14**, Theorem 2.20] as stated in [**56**, Theorem 2.28], where the reductive stabiliser assumption $[\overline{R}]_0$ is used to describe the preimage of the reductive (semi)stable locus by using the reductive Hilbert–Mumford criterion and comparing the torus weight sets of points $x \in X_{\min}$ and their images under q_U. Although some torus weights are lost on applying q_U, all torus weights corresponding to maximal grading $\mathbb{G}_m$-weights survive, and if $x \in UZ_{\min}$ then it has at least one non-minimal weight; these observations together with the assumption that semistability coincides with stability for $\overline{R}$ on $Z_{\min}$ are central in the proof.

THEOREM 5.25 (Construction of non-reductive GIT quotients, [**17**, Theorem 4.28]). *Let $G = U \rtimes R$ be a group with unipotent radical graded by $\mathbb{G}_m < Z(R)$ acting linearly on $X = \mathbb{P}(V)$. If the linearisation is well-adapted and both $[\widehat{U}]_0$ and $[\overline{R}]_0$ hold, then there is a projective and geometric G-quotient*

$$X^{G-s} = X_{\min}^{\overline{R}-ss} \setminus UZ_{\min}^{\overline{R}-ss} \longrightarrow X/\!/G,$$

which coincides with the Proj construction associated to the (finitely generated) invariant ring.

5.7. Overview of the proof. We provide some details on the proof of Theorem 5.20 i) and ii). The structure of the proof is as follows.

(1) Using Proposition 5.11, construct a geometric U-quotient of $X_{\min}$ (see Proposition 5.26).
(2) Construct $X_{\min}/U \hookrightarrow \mathbb{P}(W)$ to show $X_{\min}/U$ is quasi-projective (see Proposition 5.27).
(3) Inside $\mathbb{P}(W)/\!/\mathbb{G}_m$ (taken with respect to an appropriate linearisation), construct a geometric $\widehat{U}$-quotient of $X^{\widehat{U}-s}$ (see Proposition 5.29).

The first step relies on the Key Proposition (Proposition 5.6) about $\mathbb{G}_a$-slices via $\mathbb{G}_m$-gradings.

PROPOSITION 5.26. *Let $\widehat{U} = U \rtimes \mathbb{G}_m$ be a graded unipotent group acting linearly on $X = \mathbb{P}(V)$. If the linearised action is adapted and $[\widehat{U}]_0$ holds, then there is a geometric U-quotient of $X_{\min}$.*

PROOF. Since $X_{\min}$ is the union of the open affine varieties X_σ over the sections $\sigma \in H^0(X, \mathcal{O}(1))_{\max}$, we will construct a geometric U-quotient by gluing trivial quotients $X_\sigma \to X_\sigma/U = \operatorname{Spec} \mathcal{O}(X_\sigma)^U$ which are constructed by applying Proposition 5.11 to each affine $\widehat{U}$-variety X_σ for $\sigma \in H^0(X, \mathcal{O}(1))_{\max}$.

In fact, by choosing a basis of V consisting of $\mathbb{G}_m$-weight vectors, which gives an identification $X \cong \mathbb{P}^n$, it suffices to construct these trivial quotients in the case where $\sigma = x_i \in H^0(X, \mathcal{O}(1))_{\max}$ is a coordinate function. Then $X_\sigma = \mathbb{P}(V)_{x_i} \cong \mathbb{A}^n$ has coordinates x_j/x_i. Since $x_i \in H^0(X, \mathcal{O}(1))_{\max}$, its $\mathbb{G}_m$-weight is $-\omega_{\min}$ (recall that $\omega_{\min}$ is the minimal weight in V and $V = H^0(X, \mathcal{O}(1))^*$). Hence the weights of the $\mathbb{G}_m$-action on $X_\sigma = \mathbb{P}(V)_{x_i} \cong \mathbb{A}^n$ are of the form $\omega_j - \omega_{\min} \geq 0$, where this inequality holds due to the linearised action being adapted. In particular, the flow under $\mathbb{G}_m$ as $t \to 0$ exists for all points in X_σ. Since $Z_\sigma \subset Z_{\min}$, the unipotent stabiliser assumption $[\widehat{U}]_0$ implies that the corresponding stabiliser assumption in Proposition 5.11 holds; hence we obtain the claimed trivial U-quotient $X_\sigma \to X_\sigma/U = \operatorname{Spec} \mathcal{O}(X_\sigma)^U$. □

The quotient obtained from this gluing construction is a priori just an abstract scheme, but the next result shows it is in fact quasi-projective. This result is [**14**, Lemma 7.6].

PROPOSITION 5.27. *Let $\widehat{U}$ be a graded unipotent group acting linearly on $X = \mathbb{P}(V)$ such that the linearised action is adapted and $[\widehat{U}]_0$ holds. There exists a positive integer r and an embedding*

$$X_{\min}/U \hookrightarrow \mathbb{P}(W)_{\min} \hookrightarrow \mathbb{P}(W)$$

where $W := (H^0(X, \mathcal{O}(r))^U)^*$, the first morphism is a closed immersion and the second morphism is the open inclusion of the minimal attracting set for the induced $\mathbb{G}_m$-action on W.

PROOF. Fix a basis $\sigma_1, \dots, \sigma_l$ of $H^0(X, \mathcal{O}(1))_{\max}$ such that $\mathcal{O}(X_{\sigma_i})$ is finitely generated (see the proof of Proposition 5.26 above). Then there exists a positive integer r, such that for $1 \leq i \leq l$, we have that $R(X, \mathcal{O}(1))^U_{(\sigma_i^r)}$ is generated by $\{\frac{f}{\sigma_i^r} : f \in H^0(X, \mathcal{O}(r))^U\}$. Let $\Sigma_i \in H^0(\mathbb{P}(W), \mathcal{O}(1)) \cong H^0(X, \mathcal{O}(r))$ correspond to σ_i^r; then $\operatorname{Sym}(W^*)_{\Sigma_i} \to \mathcal{O}(X)_{\sigma_i}$ is surjective.

The inclusion $H^0(X, \mathcal{O}(r))^U \hookrightarrow H^0(X, \mathcal{O}(r))$ induces a rational map $\phi : X \dashrightarrow \mathbb{P}(W)$, which is well-defined on $X_{\min}$, as maximal sections are U-invariant, and $\phi|_{X_{\min}} = \overline{\phi} \circ q_U$ for

$$\overline{\phi} : X_{\min}/U \to \mathbb{P}(W).$$

Since $\operatorname{Spec} \mathcal{O}(X_{\sigma_i})^U$ cover $X_{\min}/U$, we see that $\overline{\phi}$ factors via

$$\mathbb{P}(W)_{\min} = \bigcup_{\Sigma \in H^0(\mathbb{P}(W), \mathcal{O}(1))_{\max}} \mathbb{P}(W)_{\Sigma}.$$

For a partition $\underline{k} = (k_1, \dots, k_l)$ of r, the section $\sigma^{\underline{k}} := \prod_{i=1}^l \sigma_i^{k_i} \in H^0(X, \mathcal{O}(r))_{\max}$ corresponds to $\Sigma_{\underline{k}} \in H^0(\mathbb{P}(W), \mathcal{O}(1))_{\max}$.

Since being a closed immersion is a local property on the target, the morphism $\overline{\phi} : X_{\min}/U \to \mathbb{P}(W)_{\min}$ is a closed immersion if $\overline{\phi}_{\underline{k}} : \operatorname{Spec} \mathcal{O}(X_{\sigma^{\underline{k}}})^U \to \mathbb{P}(W)_{\Sigma_{\underline{k}}}$ is a closed immersion for each partition $\underline{k}$, or equivalently $\operatorname{Sym}(W^*)_{\Sigma_{\underline{k}}} \to \mathcal{O}(X)_{\sigma^{\underline{k}}}$ is surjective. This last statement is deduced from the fact that $\operatorname{Sym}(W^*)_{\Sigma_i} \to \mathcal{O}(X)_{\sigma_i}$ is surjective for each i by the choice of r. □

The next two results are described in the discussion after Lemma 7.7 in [**14**].

LEMMA 5.28. *For an adapted linear action of a graded unipotent group $\widehat{U}$ on $X = \mathbb{P}(V)$, assume that Assumption $[\widehat{U}]_0$ holds, so that there is a geometric U-quotient $q_U : X_{\min} \to X_{\min}/U$, which is locally closed in $\mathbb{P}(W)$ by Proposition 5.29. Let $x \in X_{\min}$; then $q_U \in \mathbb{P}(W_{\min})$ if and only if $x \in UZ_{\min}$.*

PROOF. For $q_U(x) \in X_{\min}/U \hookrightarrow \mathbb{P}(W)_{\min}$, we have that $q_U \in \mathbb{P}(W_{\min})$ if and only if

$$q_U(x) = \lim_{t \to 0} t \cdot q_u(x) = q_U(\lim_{t \to 0} t \cdot x) = q_U(p(x)),$$

or, as q_U is a geometric quotient, equivalently $U \cdot x = U \cdot p(x)$, i.e. $x \in UZ_{\min}$. □

PROPOSITION 5.29. *For an adapted linear action of a graded unipotent group $\widehat{U}$ on $X = \mathbb{P}(V)$, assume $[\widehat{U}]_0$ holds. There is a well-adapted rational twist of the $\widehat{U}$-linearisation on X such that the induced $\mathbb{G}_m$-linearisation on $\mathbb{P}(W)$ is adapted and*

$$\mathbb{P}(W)^{\mathbb{G}_m-(s)s} = \mathbb{P}(W)_{\min} \setminus \mathbb{P}(W_{\min}).$$

Furthermore, the preimage under q_U of the $\mathbb{G}_m$-stable locus of the closure of $X_{\min}/U$ in $\mathbb{P}(W)$

$$q_U^{-1}(\overline{X_{\min}/U}^{\mathbb{G}_m-(s)s}) = X_{\min} \setminus UZ_{\min}$$

admits a projective geometric $\widehat{U}$-quotient $X_{\min} \setminus UZ_{\min} \to (\overline{X_{\min}/U})//\mathbb{G}_m$.

PROOF. Recall that $\omega_{\min}$ is the minimal weight on $V = H^0(X, \mathcal{O}(1))^*$ and so $r\omega_{\min}$ is the minimal weight on $W := (H^0(X, \mathcal{O}(r))^U)^*$. Pick $\epsilon > 0$ so that $r\epsilon < 1$. Let $\chi : \widehat{U} \to \mathbb{G}_m$ be the rational character corresponding to $\omega_{\min} + \epsilon \in \mathbb{Q}$. Then the $\mathbb{G}_m$-weights α_j on $\mathcal{O}_{\mathbb{P}(W)}(1)^{r\chi}$ satisfy

$$\alpha_{\min} = \alpha_0 = r\omega_{\min} - r\chi = -r\epsilon < 0 < 1 - r\epsilon = r\omega_{\min} + 1 - r\chi \leq \alpha_1 < \cdots < \alpha_{\max}$$

so the induced $\mathbb{G}_m$-linearisation $\mathcal{O}_{\mathbb{P}(W)}(1)$ is adapted and

$$\mathbb{P}(W)^{\mathbb{G}_m-(s)s} = \mathbb{P}(W)_{\min} \setminus \mathbb{P}(W_{\min}).$$

This linearisation on $\mathbb{P}(W)$ is induced from the well-adapted twisted linearisation $\mathcal{O}_X(1)^\chi$. Since $X_{\min}/U \hookrightarrow \mathbb{P}(W)_{\min}$ is a closed subvariety, we have

$$\overline{X_{\min}/U}^{\mathbb{G}_m-(s)s} = \overline{X_{\min}/U} \cap \mathbb{P}(W)^{\mathbb{G}_m-(s)s} = (X_{\min}/U) \setminus ((X_{\min}/U) \cap \mathbb{P}(W_{\min}))$$

and so the final claim follows from Lemma 5.28. □

The proof of the finite generation of an appropriate power of the linearisation is given in the discussion proceeding Corollary 7.10 in [**14**].

6. Recent applications of non-reductive GIT

In this section we will give an overview of some recent applications of non-reductive GIT.

6.1. Moduli of jets of map germs and hyperbolicity. Bérczi and Kirwan [**18**] used non-reductive GIT to construct and study compactifications of spaces of invariant jet differentials in order to prove polynomial versions of the Green–Griffiths–Lang conjecture and Kobayashi conjecture concerning hyperbolicity properties of generic smooth projective hypersurfaces. Let us outline these conjectures and the approach using non-reductive GIT.

A complex projective manifold X is *Brody hyperbolic* if every holomorphic map $f : \mathbb{C} \to X$ is constant. For example, in dimension 1, a smooth projective curve is hyperbolic if and only if $g \geq 2$. Hyperbolic varieties are interesting from the point of view of complex geometry and also for their conjectural Diophantine properties (Lang conjectured that if a projective variety defined over $\mathbb{Q}$ is hyperbolic, then $X(\mathbb{Q})$ is finite).

The Kobayashi conjecture predicts that a very general hypersurface $X \subset \mathbb{P}^{n+1}$ of sufficiently large degree d_n is Brody hyperbolic. Green, Griffiths and Lang conjectured that every projective algebraic variety X of general type is *weakly (or pseudo Brody) hyperbolic*; that is, there exists a proper subvariety $Y \subsetneq X$ such that the image of every non-constant holomorphic map $f : \mathbb{C} \to X$ is contained in Y. These conjectures are related by a recent result of Riedl and Yang [**84**]: if the Green–Griffiths–Lang conjecture holds for projective hypersurfaces of dimension n and degree at least d_n, then the Kobayashi conjecture is true for projective hypersurfaces of dimension n with degree at least d_{2n-1}. The strategy for approaching these conjectures goes back to work of Demailly [**27**] and Siu [**92**], which involves studying *invariant jet differentials*; here non-reductive group actions naturally arise as reparametrisation groups.

For a smooth projective complex variety X of dimension n, the bundle of $J_kX \to X$ of *k-jet germs in* X has fibre over $p \in X$ is given by germs of holomorphic maps $f : (\mathbb{C}, 0) \to (X, p)$ for fixed local coordinates at p up to the equivalence relation given by equality of the first k-derivatives at 0; thus the fibres can be represented by truncated Taylor expansions or equivalently k-tuples of vectors in $\mathbb{C}^n$ given by the first k-derivatives. The transition functions are polynomial, but not linear, so $J_kX \to X$ is not a vector bundle. The group Diff_k of k-jets of maps $(\mathbb{C}, 0) \to (\mathbb{C}, 0)$ which are regular (i.e. non-zero first derivative) acts fibrewise on $J_kX \to X$ by reparametrisations; Diff_k is naturally an upper triangular subgroup of GL_k and fortunately is a graded unipotent group

$$\mathrm{Diff}_k = \left\{ \begin{pmatrix} \alpha_1 & \alpha_2 & \cdots & \alpha_k \\ 0 & \alpha_1^2 & \cdots & \vdots \\ \vdots & \ddots & \ddots & \vdots \\ 0 & \cdots & 0 & \alpha_1^k \end{pmatrix} : \alpha_1 \in \mathbb{C}^*, \alpha_2, \ldots, \alpha_k \in \mathbb{C}^+ \right\} \cong U_k \rtimes \mathbb{C}^*.$$

Green and Griffiths studied algebraic differential operators, which are polynomial functions on J_kX, and constructed a sheaf of algebraic differential operators of order k of fixed weighted degree (with respect to the $\mathbb{C}^*$-weights). Demailly considered a subbundle of jet differentials invariant under reparametrisations from U_k. A key approach to finding invariant jet differentials is to produce a projective completion of the fibrewise quotient of Diff_k acting on the jet bundle $J_kX \to X$ and take sections of an ample line bundle on this projective completion. The projective completion given by Bérczi and Kirwan [**18**] uses non-reductive GIT, where a blow-up at $Z_{\min}$, which is just a point, is needed for the unipotent stabiliser assumption to hold. They then use intersection theory for non-reductive GIT quotients (see [**17**] and §6.4 below) to prove a polynomial version of the Green–Griffiths–Lang conjecture.

THEOREM 6.1 (Polynomial Green–Griffiths–Lang Theorem, [**18**]). *A generic smooth projective hypersurface of dimension n and degree $d \geq 32n^4$ is weakly hyperbolic.*

By work of Riedl–Yang [**84**], this gives a polynomial Kobayashi theorem [**18**, Theorem 1.4] for generic smooth projective hypersurfaces of dimension n and degree $d \geq 32(2n-1)^4$.

6.2. Moduli spaces of hypersurfaces in weighted projective orbifolds. One classical application of reductive GIT is to construct moduli spaces of projective hypersurfaces as the GIT quotient of the PGL_{n+1}-action on $\mathbb{P}(k[x_0, \ldots, x_n]_d)$; this gives compactifications of moduli spaces of smooth hypersurfaces, as Mumford proved that any smooth hypersurface $X \subset \mathbb{P}^n$ of degree $d \geq 3$ is GIT stable when $n > 1$. In general determining precisely which other hypersurfaces are (semi)stable is challenging, even with the Hilbert–Mumford criterion in hand.

The advent of non-reductive GIT enables this to be extended to hypersurfaces in weighted projective spaces and more general projective toric varieties, whose automorphism groups are non-reductive affine algebraic groups and are explicitly described by the work of Cox as quotients of the graded automorphism group of the Cox ring.

EXAMPLE 6.2. The weighted projective plane $\mathbb{P}(1,1,2)$ has automorphism group given by

$$1 \longrightarrow \mathbb{G}_m \overset{(tI_2,t^2,0,0,0)}{\longrightarrow} (\mathrm{GL}_2 \times \mathbb{G}_m) \rtimes \mathbb{G}_a^3 \longrightarrow \mathrm{Aut}(\mathbb{P}(1,1,2)) \longrightarrow 1$$

where the unipotent group appears from automorphisms of the form $z \mapsto z + ax^2 + bxy + cy^2$ where $(a,b,c) \in \mathbb{G}_a^3$,

Fortunately the automorphism groups of weighted projective spaces have graded unipotent radicals; see [**15**, Lemma 4.1] and also [**23**].

Bunnett [**23**] studied the application of non-reductive GIT to moduli of weighted projective hypersurfaces and more generally hypersurfaces in toric orbifolds. In a well-formed weighted projective space $\mathbb{P}(a_0,\dots,a_n)$ he proves [**23**, Theorem 5.18] that any *quasi-smooth* hypersurface (see [**11**, §3]) of degree $d \geq 2+\max\{a_0,\dots,a_n\}$ is stable (in the sense of non-reductive GIT) provided the unipotent stabiliser assumption holds, so that no blow-ups are required.

In the case of a well-formed weighted projective space $\mathbb{P}(a_0,\dots,a_n)$, any quasi-smooth hypersurface of degree $d \geq 2+\max\{a_0,\dots,a_n\}$ has finitely many automorphisms coming from the ambient automorphisms of $\mathbb{P}(a_0,\dots,a_n)$ by [**23**, Theorem 3.13]. Consequently, the Keel–Mori Theorem gives the existence of a coarse moduli spaces as an algebraic space. However, non-reductive GIT gives the construction of a quasi-projective moduli space.

For a toric variety X, there is an A-discriminant (see [**37**]) for hypersurfaces of class α, but its non-vanishing locus does not coincide with the locus of quasi-smooth hypersurfaces (whereas the classical discriminant vanishes precisely on singular projective hypersurfaces). The A-discriminant vanishes on non-quasi-smooth hypersurfaces, but may be non-zero for some non-quasi-smooth surfaces (as the A-discriminant only checks for singularities in the sweep under $G = \mathrm{Aut}_\alpha(X)$ of the torus $T \subset X$, see [**23**, Remark 4.11]). The A-discriminant of [**37**] is interpreted as an invariant section of a twisted linearisation in [**23**, Corollary 4.12].

Let us explain the non-reductive GIT set-up for hypersurfaces in $\mathbb{P}(a_0,\dots,a_n)$ of degree d. Assume that hypersurfaces of degree d are Cartier divisors (i.e. the lowest common multiple of the weights divides d). Consider the non-reductive group $G = \mathrm{Aut}(\mathbb{P}(a_0,\dots,a_n))$ acting on the space $X = \mathbb{P}(k[x_0,\dots,x_n]_d)$ of weighted degree d homogeneous polynomials. Bunnett proves that quasi-smooth hypersurfaces are contained in the $\widehat{U}$-stable set $X_{\min} \setminus UZ_{\min}$ (if $[\widehat{U}]_0$ holds). Using the non-reductive GIT Hilbert–Mumford criterion of [**14**], he shows that quasi-smooth hypersurfaces are stable for the action of G assuming that d is a Cartier degree with $d \geq 2+\max\{a_0,\dots,a_n\}$ and the unipotent stabiliser assumption holds. Furthermore, if the weighted projective space has only two weights, then the unipotent stabiliser assumption holds (see [**23**, Proposition 5.9]).

Bunnett obtains the best results for Cartier hypersurfaces in a rational cone $\mathbb{P}(1,\dots,1,r)$ of degree $d \geq r+2$ (see [**23**, Theorem 5.20]): he explicitly describes the quasi-smooth locus as the non-vanishing locus of a section (which is obtained by multiplying the A-discriminant with a variable) and constructs a U-quotient of this open affine variety using Proposition 5.11, where the necessary unipotent stabiliser assumption is easily verified. He then directly obtains a geometric quotient of the locus of quasi-smooth projective hypersurfaces, which is a projective over affine variety, because it is constructed as a reductive GIT quotient of the affine U-quotient twisted by a character as in §2.8 rather than using the more complicated methods of [**14**].

6.3. Moduli of unstable objects. Recall from §3.3 that associated to a linear action of a reductive group G on $\mathbb{P}(V)$ and a choice of norm, there is an instability stratification

$$\mathbb{P}(V) = \bigsqcup_{\beta \in \mathcal{B}} S_\beta \tag{6.1}$$

where $S_\beta \cong G \times^{P_\lambda} Y_\beta^{ss}$ for a parabolic subgroup $P_\lambda < G$. A categorical P_λ-quotient of Y_β^{ss}, or equivalently a categorical G-quotient of S_β, is given by Proposition 3.28; however, as explained after this proposition, this is far from being an orbit space as it factors via the retraction $p_\beta : Y_\beta^{ss} \to Z_\beta^{ss}$ sending a point to its flow under λ as $t \to 0$. Instead, we would like to apply non-reductive GIT to the action of P_λ on the closure $\overline{Y_\beta}$ of $Y_\beta \subset X$, where we can twist the linearisation by (a rational multiple of) a character corresponding to λ to make it well-adapted. Fortunately, the non-reductive notion of stability precisely picks out the locus we would like and removes (the P_λ-sweep) of the limit set Z_β^{ss}; see Theorem 6.3 below.

For the parabolic group $P_\lambda = U_\lambda \rtimes L_\lambda$ acting on the blade closure $X = \overline{Y_\beta}$ of an unstable stratum S_β as in (6.1) above, there is a twisted rational linearisation $\mathcal{L}_{\beta(1+\epsilon)}$ which is well-adapted. Furthermore, we have that in the non-reductive GIT notation the map $p : X_{\min} \to Z_{\min}$ coincides with the retraction $p_\beta : Y_\beta \to Z_\beta$ appearing in the description of the unstable strata. Furthermore, Z_β^{ss} is defined to be the semistable locus for L_λ with respect to $\mathcal{L}_\beta$, or equivalently for $\overline{L_\lambda} := L_\lambda/\lambda(\mathbb{G}_m)$ as $\lambda(\mathbb{G}_m)$ acts trivially, which coincides with the semistable locus in $Z_{\min}$ appearing in the definition of the non-reductive stable locus.

THEOREM 6.3 (Non-reductive GIT quotients of unstable strata, [**56**, Theorem 1.1]). *For the parabolic group $P_\lambda = U_\lambda \rtimes L_\lambda$ graded by λ acting on the blade closure $X := \overline{Y_\beta}$ of an unstable stratum S_β as in* (6.1) *with the well-adapted linearisation $\mathcal{L}_{\beta(1+\epsilon)}$, the following statements hold.*

i) If $[\widehat{U}]_0$ holds, then there is a projective geometric $\widehat{U}_\lambda$-quotient

$$q_{\widehat{U}_\lambda} : \overline{Y_\beta}^{\widehat{U}_\lambda - s} = Y_\beta \setminus UZ_\beta \longrightarrow \overline{Y_\beta}/\!/\widehat{U}_\lambda$$

and by taking a reductive GIT quotient by $\overline{L_\lambda}$ one obtains a projective categorical P_λ-quotient of an open subset of the $\widehat{U}$-stable locus.

ii) If both $[\widehat{U}]_0$ and $[\overline{R}]_0$ hold, then there is a projective geometric P_λ-quotient

$$q_{P_\lambda} : \overline{Y_\beta}^{P_\lambda - s} = Y_\beta^{ss} \setminus UZ_\beta^{ss} \longrightarrow \overline{Y_\beta}/\!/P_\lambda.$$

Moreover, the ring of invariant sections if finitely generated and q_{P_λ} coincides with the Proj construction for this invariant ring.

We would like to apply this theorem to moduli of objects in an abelian category, where there are moduli-theoretic instability filtrations, such as the Harder–Narasimhan (HN) filtrations for vector bundles [**50**]; for example, moduli of sheaves on projective schemes or moduli of quiver representations. In these examples, the

GIT instability stratification has been compared with the moduli-theoretic Harder–Narasimhan stratification [**39**, **54**, **55**, **57**, **99**] and this suggests moduli of objects of fixed HN type should be constructed as non-reductive GIT quotients.

Unfortunately the stabiliser assumptions in Theorem 6.3 are quite restrictive and so it is only possible in quite limited situations. For example, for vector bundles (or Higgs bundles) on a smooth projective curve of fixed HN type, the reductive stabiliser assumption $[\overline{R}]_0$ only holds for coprime HN types of length 2 (i.e. the HN filtration has two terms and the invariants for the successive quotients are coprime, so that semistability coincides with stability) and even in this case, the unipotent stabiliser assumption rarely holds and so blow-ups are needed (see [**56**, §3.2.1] for a detailed discussion). In this length 2 coprime case, the non-reductive GIT quotient picks out non-split HN filtrations of length 2 whose automorphism groups have a fixed dimension; see [**21**, **58**] for the case of vector bundles and [**49**] for the case of Higgs bundles. To rectify the failure of the reductive stabiliser assumption $[\overline{R}]_0$, one can alternatively perform a quotient in stages, using different 1-PSs in the centre of L_λ to grade different subgroups of the unipotent radical as in [**56**]; this results in a natural notion of stability for sheaves of a fixed HN type, but again the unipotent stabiliser assumption is rarely satisfied, so a blow-up procedure would be required.

6.4. Interaction with symplectic geometry and cohomology. GIT quotients for complex reductive groups are closely related to symplectic quotients for a maximal compact group (see Remark 3.22); the close relationship between the reductive GIT instability stratification and a Morse-theoretic stratification for the norm square of the moment map was used in [**63**] to describe the rational Betti numbers of reductive GIT quotients.

Fortunately, for non-reductive groups with internally graded unipotent radicals, this close relationship with symplectic geometry has been extended Bérczi and Kirwan [**17**], and applied to compute cohomology of non-reductive GIT quotients.

For a reductive group G acting on a smooth complex projective variety Y, to construct a moment map one fixes a maximal compact subgroup $K < G$ and a symplectic form ω invariant under the K-action. The moment map for K and ω is a K-invariant map $\mu_{K,\omega} : Y \to \mathfrak{K}^* := \mathrm{Lie}(K)^*$ with the moment map property (that it lifts the infinitesimal action via the correspondence between vector fields and forms given by ω). However, any other maximal compact subgroup is of the form $g^{-1}Kg$ and $g^*\omega$ is invariant under the $g^{-1}Kg$-action with moment map $\mu_{g^{-1}Kg,g^*\omega} = \mathrm{Ad}^*_{g^{-1}} \circ \mu_{K,\omega} \circ g$. Therefore rather than defining a moment map $\mu_{K,\omega} : Y \to \mathfrak{K}^*$, Bérczi and Kirwan instead fix a *G-equivariant Kähler structure* Ω (namely a G-orbit in the space of pairs (K,ω) of maximal compact subgroups of G and Kähler forms on Y which are invariant under this maximal compact) and define an *Ω-moment map* to be a smooth G-equivariant map

$$m_{G,Y,\Omega} : \Omega \times Y \to \mathfrak{g}^*$$

such that $m_{G,Y,\Omega}(K,\omega,-) = \iota_K \circ \mu_{K,\omega} : Y \to \mathfrak{K}^* \hookrightarrow \mathfrak{g}^*$ is a moment map for the K-action on (Y,ω), where as $\mathfrak{g} = \mathfrak{K} \otimes \mathbb{C}$, we have a canonical embedding $\iota_K : \mathfrak{K}^* \hookrightarrow \mathfrak{g}^*$.

Let us explain how Bérczi and Kirwan define moment maps for a smooth complex projective variety with an action of a graded unipotent group $\widehat{U} = U \rtimes \mathbb{C}^*$. Assume that $\widehat{U} < G$ is a subgroup of a reductive group and that $X \subset Y$ is a submanifold of a compact Kähler manifold Y with a G-action on Y that restricts to the given $\widehat{U}$-action on X (note that X is not required to be invariant under the G-action). As above, fix a G-equivariant Kähler structure Ω on Y and let $m_{G,Y,\Omega} : \Omega \times Y \to \mathfrak{g}^*$ be an Ω-moment map, they define $m_{\widehat{U},X,\Omega} : X \times \Omega \to \hat{\mathfrak{u}}^*$ by restricting the Ω-moment map to X and composing with the restriction $\mathfrak{g}^* \to \hat{\mathfrak{u}}^*$

$$\begin{array}{ccc} \Omega \times X & \longrightarrow & \mathfrak{g}^* = \mathfrak{K}^* \oplus i\mathfrak{K}^* \\ & \searrow^{m_{\widehat{U},X,\Omega}} & \downarrow \\ & & \hat{\mathfrak{u}}^* = \mathbb{R} \oplus i\mathbb{R} \oplus \mathfrak{u}^*. \end{array}$$

Assuming $[\widehat{U}]_0$ holds, Bérczi and Kirwan provide a moment map description of the non-reductive GIT quotient: for any $(K, \omega) \in \Omega$, they show

$$X^{\widehat{U}-s} = \widehat{U}(\mu^{\widehat{U}}_{(K,\omega)})^{-1}(0), \quad \text{where} \quad \mu^{\widehat{U}}_{(K,\omega)} := m_{\widehat{U},X,\Omega}((K,\omega), -) : X \to \hat{\mathfrak{u}}^*$$

and that 0 is a regular value of $\mu^{\widehat{U}}_{(K,\omega)}$ and the inclusion of the zero level set of the moment map in the stable locus induces a diffeomorphism of orbifolds

$$(\mu^{\widehat{U}}_{(K,\omega)})^{-1}(0)/(K \cap \widehat{U}) = (\mu^{\widehat{U}}_{(K,\omega)})^{-1}(0)/S^1 \simeq X^{\widehat{U}-s}/U = X/\!/\widehat{U},$$

which can be viewed as a non-reductive Kempf-Ness Theorem. They extend this result to groups with internally graded unipotent radical [**17**, Theorem 1.1].

They apply this to compute Betti numbers of non-reductive GIT quotients. Assuming $[\widehat{U}]_0$ holds so that $X^{\widehat{U}-s} = X_{\min} \backslash U Z_{\min}$, they show that the stratification $X_{\min} = X^{\widehat{U}-s} \sqcup U Z_{\min}$ is $\widehat{U}$-equivariantly perfect and so the Poincaré series of $X/\!/\widehat{U}$ can be computed from that of $Z_{\min}$. Similarly for $H = U \rtimes R$ as above, they show the Poincaré series of $X/\!/H$ can be computed from that of the reductive GIT quotient of $Z_{\min}$ by $\overline{R} := R/\mathbb{G}_m$, whose Poincaré series can in turn be computed as in [**63**] using the reductive GIT instability stratification of §3.3.

Furthermore, they adapt methods of Martin [**71**] relating the rational cohomology of GIT quotients by reductive groups to that of GIT quotients for a maximal torus. Martin shows the intersection pairing for the reductive GIT quotient can be computed from that of the GIT quotient for the maximal torus via an integration formula, which can then be combined with torus localisation techniques. In the non-reductive case, this enables a description of the rational cohomology ring and a non-reductive integration formula [**17**, Theorems 1.4 and 1.5], which leads to a residue formula for the intersection pairing on the non-reductive GIT quotient.

These methods are used to prove the polynomial versions of the Green–Griffiths–Lang conjecture and Kobayashi conjecture described in §6.1. They can also be applied in the future to describe the cohomology of new moduli spaces constructed via non-reductive GIT.

Acknowledgments

I am indebted to both Peter Newstead and Frances Kirwan, as I learned the basics of reductive GIT from Peter Newstead's Tata lecture notes [**80**] and discussions with Frances Kirwan. I am very grateful to Greg Bérczi, Dominic Bunnett, Eloise Hamilton, Josh Jackson and Frances Kirwan for numerous conversations on non-reductive GIT. I would also like to thank the organisers of VBAC 2022 for soliciting this paper in honour of Peter Newstead. I am grateful to the referee for their careful reading of this text, as well as many helpful and thoughtful comments.

References

[1] J. Alper, *Stacks and moduli*, https://sites.math.washington.edu/~jarod/moduli.pdf.

[2] J. Alper, *Good moduli spaces for Artin stacks* (English, with English and French summaries), Ann. Inst. Fourier (Grenoble) **63** (2013), no. 6, 2349–2402. MR3237451

[3] J. Alper, *Adequate moduli spaces and geometrically reductive group schemes*, Algebr. Geom. **1** (2014), no. 4, 489–531, DOI 10.14231/AG-2014-022. MR3272912

[4] J. Alper, P. Belmans, D. Bragg, J. Liang, and T. Tajakka, *Projectivity of the moduli space of vector bundles on a curve*, Stacks Project Expository Collection (SPEC), London Math. Soc. Lecture Note Ser., vol. 480, Cambridge Univ. Press, Cambridge, 2022, pp. 90–125. MR4480534

[5] J. Alper, H. Blum, D. Halpern-Leistner, and C. Xu, *Reductivity of the automorphism group of K-polystable Fano varieties*, Invent. Math. **222** (2020), no. 3, 995–1032, DOI 10.1007/s00222-020-00987-2. MR4169054

[6] J. Alper, J. Hall, and D. Rydh, *The étale local structure of algebraic stacks*, arXiv:1912.06162.

[7] J. Alper, D. Halpern-Leistner, and J. Heinloth, *Existence of moduli spaces for algebraic stacks*, Invent. Math. **234** (2023), no. 3, 949–1038, DOI 10.1007/s00222-023-01214-4. MR4665776

[8] M. Artin, *Versal deformations and algebraic stacks*, Invent. Math. **27** (1974), 165–189, DOI 10.1007/BF01390174. MR399094

[9] A. Asok and B. Doran, *On unipotent quotients and some $\mathbb{A}^1$-contractible smooth schemes*, Int. Math. Res. Pap. IMRP **2** (2007), Art. ID rpm005, 51 pp. MR2335246

[10] M. Ballard, D. Favero, and L. Katzarkov, *Variation of geometric invariant theory quotients and derived categories*, J. Reine Angew. Math. **746** (2019), 235–303, DOI 10.1515/crelle-2015-0096. MR3895631

[11] V. V. Batyrev and D. A. Cox, *On the Hodge structure of projective hypersurfaces in toric varieties*, Duke Math. J. **75** (1994), no. 2, 293–338, DOI 10.1215/S0012-7094-94-07509-1. MR1290195

[12] A. Bayer, M. Lahoz, E. Macrì, H. Nuer, A. Perry, and P. Stellari, *Stability conditions in families*, Publ. Math. Inst. Hautes Études Sci. **133** (2021), 157–325, DOI 10.1007/s10240-021-00124-6. MR4292740

[13] P. Belmans, C. Damiolini, H. Franzen, V. Hoskins, S. Makarova, and T. Tajakka, *Projectivity and effective global generation of determinantal line bundles on quiver moduli*, arXiv:2210.00033.

[14] G. Bérczi, B. Doran, T. Hawes, and F. Kirwan, *Projective commpletions of graded unipotent quotients (with an appendix by E. Hamilton)*, arXiv:1607.04181.

[15] G. Bérczi, B. Doran, T. Hawes, and F. Kirwan, *Geometric invariant theory for graded unipotent groups and applications*, J. Topol. **11** (2018), no. 3, 826–855, DOI 10.1112/topo.12075. MR3989432

[16] G. Bérczi, T. Hawes, F. Kirwan, and B. Doran, *Constructing quotients of algebraic varieties by linear algebraic group actions*, Handbook of group actions. Vol. IV, Adv. Lect. Math. (ALM), vol. 41, Int. Press, Somerville, MA, 2018, pp. 341–446. MR3888691

[17] G. Bérczi and F. Kirwan, *Moment maps and cohomology of non-reductive quotients*, Invent. Math. **235** (2024), no. 1, 1–79, DOI 10.1007/s00222-023-01218-0. MR4688700

[18] G. Bérczi and F. Kirwan, *Non-reductive geometric invariant theory and hyperbolicity*, Invent. Math. **235** (2024), no. 1, 81–127, DOI 10.1007/s00222-023-01219-z. MR4688701

[19] A. Białynicki-Birula, *Some theorems on actions of algebraic groups*, Ann. of Math. (2) **98** (1973), 480–497, DOI 10.2307/1970915. MR366940

[20] A. Borel, *Linear algebraic groups*, 2nd ed., Graduate Texts in Mathematics, vol. 126, Springer-Verlag, New York, 1991, DOI 10.1007/978-1-4612-0941-6. MR1102012

[21] L. Brambila-Paz and R. Rios Sierra, *Moduli of unstable bundles of HN-length two with fixed algebra of endomorphisms*, arXiv:2203.08986.

[22] M. Brion, *Introduction to actions of algebraic groups*, Les cours du CIRM **1** (2010), no. 1, 1–22.

[23] D. Bunnett, *Moduli of Hypersurfaces in Toric Orbifolds*, ProQuest LLC, Ann Arbor, MI, 2019. Thesis (Ph.D.)–Freie Universitaet Berlin (Germany). MR4144635

[24] R. Cheng, C. Lian, and T. Murayama, *Projectivity of the moduli of curves*, Stacks Project Expository Collection (SPEC), London Math. Soc. Lecture Note Ser., vol. 480, Cambridge Univ. Press, Cambridge, 2022, pp. 1–43. MR4480532

[25] B. Conrad, *Keel-Mori theorem via stacks*, `http://math.stanford.edu/~conrad/papers/coarsespace.pdf`.

[26] B. Conrad, *Reductive group schemes* (English, with English and French summaries), Autour des schémas en groupes. Vol. I, Panor. Synthèses, vol. 42/43, Soc. Math. France, Paris, 2014, pp. 93–444. MR3362641

[27] J.-P. Demailly, *Algebraic criteria for Kobayashi hyperbolic projective varieties and jet differentials*, Algebraic geometry—Santa Cruz 1995, Proc. Sympos. Pure Math., vol. 62, Amer. Math. Soc., Providence, RI, 1997, pp. 285–360, DOI 10.1090/pspum/062.2/1492539. MR1492539

[28] M. Demazure and P. Gabriel, *Groupes algébriques. Tome I: Géométrie algébrique, généralités, groupes commutatifs* (French), Masson & Cie, Éditeurs, Paris; North-Holland Publishing Co., Amsterdam, 1970. MR302656

[29] H. Derksen, *Quotients of algebraic group actions*, Automorphisms of affine spaces (Curaçao, 1994), Kluwer Acad. Publ., Dordrecht, 1995, pp. 191–200. MR1352700

[30] I. Dolgachev, *Lectures on invariant theory*, London Mathematical Society Lecture Note Series, vol. 296, Cambridge University Press, Cambridge, 2003, DOI 10.1017/CBO9780511615436. MR2004511

[31] I. V. Dolgachev and Y. Hu, *Variation of geometric invariant theory quotients*, Inst. Hautes Études Sci. Publ. Math. **87** (1998), 5–56. With an appendix by Nicolas Ressayre. MR1659282

[32] B. Doran and F. Kirwan, *Towards non-reductive geometric invariant theory*, Pure Appl. Math. Q. **3** (2007), no. 1, Special Issue: In honor of Robert D. MacPherson., 61–105, DOI 10.4310/PAMQ.2007.v3.n1.a3. MR2330155

[33] A. Fauntleroy, *Categorical quotients of certain algebraic group actions*, Illinois J. Math. **27** (1983), no. 1, 115–124. MR684546

[34] A. Fauntleroy, *Geometric invariant theory for general algebraic groups*, Compositio Math. **55** (1985), no. 1, 63–87. MR791647

[35] G. Freudenburg, *A survey of counterexamples to Hilbert's fourteenth problem*, Serdica Math. J. **27** (2001), no. 3, 171–192. MR1917641

[36] G. Freudenburg, *Algebraic theory of locally nilpotent derivations*, Encyclopaedia of Mathematical Sciences, vol. 136, Springer-Verlag, Berlin, 2006. Invariant Theory and Algebraic Transformation Groups, VII. MR2259515

[37] I. M. Gel′fand, M. M. Kapranov, and A. V. Zelevinsky, *Discriminants, resultants, and multidimensional determinants*, Mathematics: Theory & Applications, Birkhäuser Boston, Inc., Boston, MA, 1994, DOI 10.1007/978-0-8176-4771-1. MR1264417

[38] D. Gieseker, *Lectures on moduli of curves*, Tata Institute of Fundamental Research Lectures on Mathematics and Physics, vol. 69, Published for the Tata Institute of Fundamental Research, Bombay by Springer-Verlag, Berlin-New York, 1982. MR691308

[39] T. L. Gómez, I. Sols, and A. Zamora, *A GIT interpretation of the Harder-Narasimhan filtration*, Rev. Mat. Complut. **28** (2015), no. 1, 169–190, DOI 10.1007/s13163-014-0149-3. MR3296731

[40] G.-M. Greuel and G. Pfister, *Geometric quotients of unipotent group actions*, Proc. London Math. Soc. (3) **67** (1993), no. 1, 75–105, DOI 10.1112/plms/s3-67.1.75. MR1218121

[41] F. Grosshans, *Observable groups and Hilbert's fourteenth problem*, Amer. J. Math. **95** (1973), 229–253, DOI 10.2307/2373655. MR325628

[42] F. D. Grosshans, *The invariants of unipotent radicals of parabolic subgroups*, Invent. Math. **73** (1983), no. 1, 1–9, DOI 10.1007/BF01393822. MR707345

[43] F. D. Grosshans, *Algebraic homogeneous spaces and invariant theory*, Lecture Notes in Mathematics, vol. 1673, Springer-Verlag, Berlin, 1997, DOI 10.1007/BFb0093525. MR1489234

[44] M. G. Gulbrandsen, L. H. Halle, and K. Hulek, *A relative Hilbert-Mumford criterion*, Manuscripta Math. **148** (2015), no. 3-4, 283–301, DOI 10.1007/s00229-015-0744-8. MR3414477

[45] W. J. Haboush, *Reductive groups are geometrically reductive*, Ann. of Math. (2) **102** (1975), no. 1, 67–83, DOI 10.2307/1970974. MR382294

[46] D. Halpern-Leistner, *On the structure of instability in moduli theory*, `arXiv:1411.0627`.

[47] D. Halpern-Leistner, *The derived category of a GIT quotient*, J. Amer. Math. Soc. **28** (2015), no. 3, 871–912, DOI 10.1090/S0894-0347-2014-00815-8. MR3327537

[48] A. Bayer, *Wall-crossing implies Brill-Noether: applications of stability conditions on surfaces*, Algebraic geometry: Salt Lake City 2015, Proc. Sympos. Pure Math., vol. 97, Amer. Math. Soc., Providence, RI, 2018, pp. 3–27, DOI 10.1090/pspum/097.1/01668. MR3821144

[49] E. Hamilton, *Stratifications and quasi-projective coarse moduli spaces for the stack of Higgs bundles*, arXiv:1911.13194.

[50] G. Harder and M. S. Narasimhan, *On the cohomology groups of moduli spaces of vector bundles on curves*, Math. Ann. **212** (1974/75), 215–248, DOI 10.1007/BF01357141. MR364254

[51] R. Hartshorne, *Algebraic geometry*, Graduate Texts in Mathematics, No. 52, Springer-Verlag, New York-Heidelberg, 1977. MR463157

[52] J. Heinloth, *Hilbert-Mumford stability on algebraic stacks and applications to $\mathcal{G}$-bundles on curves*, Épijournal Géom. Algébrique **1** (2017), Art. 11, 37 pp., DOI 10.46298/epiga.2018.volume1.2062. MR3758902

[53] W. H. Hesselink, *Desingularizations of varieties of nullforms*, Invent. Math. **55** (1979), no. 2, 141–163, DOI 10.1007/BF01390087. MR553706

[54] V. Hoskins, *Stratifications associated to reductive group actions on affine spaces*, Q. J. Math. **65** (2014), no. 3, 1011–1047, DOI 10.1093/qmath/hat046. MR3261979

[55] V. Hoskins, *Stratifications for moduli of sheaves and moduli of quiver representations*, Algebr. Geom. **5** (2018), no. 6, 650–685, DOI 10.14231/AG-2018-017. MR3871820

[56] V. Hoskins and J. Jackson, *Quotients by parabolic groups and moduli spaces of unstable objects*, arXiv:2111.07429.

[57] V. Hoskins and F. Kirwan, *Quotients of unstable subvarieties and moduli spaces of sheaves of fixed Harder-Narasimhan type*, Proc. Lond. Math. Soc. (3) **105** (2012), no. 4, 852–890, DOI 10.1112/plms/pds022. MR2989806

[58] J. Jackson, *Moduli spaces of unstable objects: Sheaves of Harder-Narasimhan length 2*, arXiv:2111.07428.

[59] S. Keel and S. Mori, *Quotients by groupoids*, Ann. of Math. (2) **145** (1997), no. 1, 193–213, DOI 10.2307/2951828. MR1432041

[60] G. Kempf and L. Ness, *The length of vectors in representation spaces*, Algebraic geometry (Proc. Summer Meeting, Univ. Copenhagen, Copenhagen, 1978), Lecture Notes in Math., vol. 732, Springer, Berlin, 1979, pp. 233–243. MR555701

[61] G. R. Kempf, *Instability in invariant theory*, Ann. of Math. (2) **108** (1978), no. 2, 299–316, DOI 10.2307/1971168. MR506989

[62] A. D. King, *Moduli of representations of finite-dimensional algebras*, Quart. J. Math. Oxford Ser. (2) **45** (1994), no. 180, 515–530, DOI 10.1093/qmath/45.4.515. MR1315461

[63] F. C. Kirwan, *Cohomology of quotients in symplectic and algebraic geometry*, Mathematical Notes, vol. 31, Princeton University Press, Princeton, NJ, 1984, DOI 10.2307/j.ctv10vm2m8. MR766741

[64] F. C. Kirwan, *Partial desingularisations of quotients of nonsingular varieties and their Betti numbers*, Ann. of Math. (2) **122** (1985), no. 1, 41–85, DOI 10.2307/1971369. MR799252

[65] J. Kollár, *Quotient spaces modulo algebraic groups*, Ann. of Math. (2) **145** (1997), no. 1, 33–79, DOI 10.2307/2951823. MR1432036

[66] H. Kraft, *Challenging problems on affine n-space*, Astérisque **237** (1996), Exp. No. 802, 5, 295–317. Séminaire Bourbaki, Vol. 1994/95. MR1423629

[67] S. G. Langton, *Valuative criteria for families of vector bundles on algebraic varieties*, Ann. of Math. (2) **101** (1975), 88–110, DOI 10.2307/1970987. MR364255

[68] R. Laza, *GIT and moduli with a twist*, Handbook of moduli. Vol. II, Adv. Lect. Math. (ALM), vol. 25, Int. Press, Somerville, MA, 2013, pp. 259–297. MR3184179

[69] D. Luna, *Slices étales* (French), Sur les groupes algébriques, Supplément au Bull. Soc. Math. France, Tome 101, Soc. Math. France, Paris, 1973, pp. 81–105, DOI 10.24033/msmf.110. MR342523

[70] J. Marsden and A. Weinstein, *Reduction of symplectic manifolds with symmetry*, Rep. Mathematical Phys. **5** (1974), no. 1, 121–130, DOI 10.1016/0034-4877(74)90021-4. MR402819

[71] S. Martin, *Symplectic quotients by a nonabelian group and by its maximal torus*, arXiv:0001002.

[72] J. S. Milne, *The basic theory of affine group schemes*, https://www.jmilne.org/math/CourseNotes/AGS.pdf.

[73] I. Morrison, *GIT constructions of moduli spaces of stable curves and maps*, Surveys in differential geometry. Vol. XIV. Geometry of Riemann surfaces and their moduli spaces, Surv. Differ. Geom., vol. 14, Int. Press, Somerville, MA, 2009, pp. 315–369, DOI 10.4310/SDG.2009.v14.n1.a12. MR2655332

[74] D. Mumford, *Projective invariants of projective structures and applications*, Proc. Internat. Congr. Mathematicians (Stockholm, 1962), Inst. Mittag-Leffler, Djursholm, 1963, pp. 526–530. MR175899

[75] D. Mumford, J. Fogarty, and F. Kirwan, *Geometric invariant theory*, 3rd ed., Ergebnisse der Mathematik und ihrer Grenzgebiete (2) [Results in Mathematics and Related Areas (2)], vol. 34, Springer-Verlag, Berlin, 1994. MR1304906

[76] M. Nagata, *On the 14-th problem of Hilbert*, Amer. J. Math. **81** (1959), 766–772, DOI 10.2307/2372927. MR105409

[77] M. Nagata, *Invariants of a group in an affine ring*, J. Math. Kyoto Univ. **3** (1963/64), 369–377, DOI 10.1215/kjm/1250524787. MR179268

[78] M. S. Narasimhan and C. S. Seshadri, *Stable and unitary vector bundles on a compact Riemann surface*, Ann. of Math. (2) **82** (1965), 540–567, DOI 10.2307/1970710. MR184252

[79] L. Ness, *A stratification of the null cone via the moment map*, Amer. J. Math. **106** (1984), no. 6, 1281–1329, DOI 10.2307/2374395. With an appendix by David Mumford. MR765581

[80] P. E. Newstead, *Introduction to moduli problems and orbit spaces*, Tata Institute of Fundamental Research Lectures on Mathematics and Physics, vol. 51, Tata Institute of Fundamental Research, Bombay; Narosa Publishing House, New Delhi, 1978. MR546290

[81] A. N. Parshin and I. R. Shafarevich (eds.), *Algebraic geometry. IV*, Encyclopaedia of Mathematical Sciences, vol. 55, Springer-Verlag, Berlin, 1994, Linear algebraic groups. Invariant theory, A translation of *Algebraic geometry. 4* (Russian), Akad. Nauk SSSR Vsesoyuz. Inst. Nauchn. i Tekhn. Inform., Moscow, 1989 [MR1100483 (91k:14001)].

[82] V. L. Popov, *On Hilbert's theorem on invariants* (Russian), Dokl. Akad. Nauk SSSR **249** (1979), no. 3, 551–555. MR553208

[83] Y. Qiao, *GIT for $\hat{U}$-actions on algebraic $\mathbb{C}$-schemes*, arXiv:2204.13884.

[84] E. Riedl and D. Yang, *Applications of a Grassmannian technique to hyperbolicity, Chow equivalency, and Seshadri constants*, J. Algebraic Geom. **31** (2022), no. 1, 1–12, DOI 10.1090/jag/786. MR4372405

[85] M. Rosenlicht, *A remark on quotient spaces*, An. Acad. Brasil. Ci. **35** (1963), 487–489. MR171782

[86] A. H. W. Schmitt, *Geometric invariant theory and decorated principal bundles*, Zurich Lectures in Advanced Mathematics, European Mathematical Society (EMS), Zürich, 2008, DOI 10.4171/065. MR2437660

[87] C. S. Seshadri, *On a theorem of Weitzenböck in invariant theory*, J. Math. Kyoto Univ. **1** (1961/62), 403–409, DOI 10.1215/kjm/1250525012. MR144914

[88] C. S. Seshadri, *Space of unitary vector bundles on a compact Riemann surface*, Ann. of Math. (2) **85** (1967), 303–336, DOI 10.2307/1970444. MR233371

[89] C. S. Seshadri, *Quotient spaces modulo reductive algebraic groups*, Ann. of Math. (2) **95** (1972), 511–556; errata, ibid. (2) 96 (1972), 599, DOI 10.2307/1970870. MR309940

[90] C. S. Seshadri, *Geometric reductivity over arbitrary base*, Advances in Math. **26** (1977), no. 3, 225–274, DOI 10.1016/0001-8708(77)90041-X. MR466154

[91] C. T. Simpson, *Moduli of representations of the fundamental group of a smooth projective variety. I*, Inst. Hautes Études Sci. Publ. Math. **79** (1994), 47–129. MR1307297

[92] Y.-T. Siu, *Hyperbolicity in complex geometry*, The legacy of Niels Henrik Abel, Springer, Berlin, 2004, pp. 543–566. MR2077584

[93] M. Thaddeus, *Geometric invariant theory and flips*, J. Amer. Math. Soc. **9** (1996), no. 3, 691–723, DOI 10.1090/S0894-0347-96-00204-4. MR1333296

[94] R. P. Thomas, *Notes on GIT and symplectic reduction for bundles and varieties*, Surveys in differential geometry. Vol. X, Surv. Differ. Geom., vol. 10, Int. Press, Somerville, MA, 2006, pp. 221–273, DOI 10.4310/SDG.2005.v10.n1.a7. MR2408226

[95] Y. Toda, *Moduli stacks and invariants of semistable objects on $K3$ surfaces*, Adv. Math. **217** (2008), no. 6, 2736–2781, DOI 10.1016/j.aim.2007.11.010. MR2397465

[96] D. Weissmann and X. Zhang, *A stacky approach to identify the semi-stable locus of vector bundles*, arXiv:2302.09245.

[97] R. Weitzenböck, *Über die Invarianten von linearen Gruppen* (German), Acta Math. **58** (1932), no. 1, 231–293, DOI 10.1007/BF02547779. MR1555349

[98] J. Winkelmann, *Invariant rings and quasiaffine quotients*, Math. Z. **244** (2003), no. 1, 163–174, DOI 10.1007/s00209-002-0484-9. MR1981881

[99] A. Zamora, *On the Harder-Narasimhan filtration for finite dimensional representations of quivers*, Geom. Dedicata **170** (2014), 185–194, DOI 10.1007/s10711-013-9876-8. MR3199484

Radboud University, IMAPP, PO Box 9010, 6525 AJ Nijmegen, Netherlands
Email address: v.hoskins@math.ru.nl

Selected Published Titles in This Series

803 **Peter Gothen, Margarida Melo, and Montserrat Teixidor i Bigas, Editors,** Moduli Spaces and Vector Bundles—New Trends, 2024

801 **K. A. Brown, T. J. Hodges, M. Vancliff, and J. J. Zhang, Editors,** Recent Advances in Noncommutative Algebra and Geometry, 2024

800 **Avinoam Mann, Louis H. Rowen, David J. Saltman, Aner Shalev, Lance W. Small, and Uzi Vishne, Editors,** Amitsur Centennial Symposium, 2024

799 **Alberto A. Condori, Elodie Pozzi, William T. Ross, and Alan A. Sola, Editors,** Recent Progress in Function Theory and Operator Theory, 2024

798 **Marat V. Markin, Igor V. Nikolaev, and Carsten Trunk, Editors,** Advances in Functional Analysis and Operator Theory, 2024

797 **Sangita Jha, Mrinal Kanti Roychowdhury, and Saurabh Verma, Editors,** Recent Developments in Fractal Geometry and Dynamical Systems, 2024

796 **John Cremona, John Jones, Jennifer Paulhus, Andrew V. Sutherland, and John Voight, Editors,** LuCaNT: LMFDB, Computation, and Number Theory, 2024

795 **Michael A. Jones, David McCune, and Jennifer M. Wilson, Editors,** Mathematical Analyses of Decisions, Voting and Games, 2024

794 **Jean-Pierre Magnot, Editor,** Recent Advances in Diffeologies and Their Applications, 2024

793 **Abba Gumel, Editor,** Mathematical and Computational Modeling of Phenomena Arising in Population Biology and Nonlinear Oscillations, 2024

792 **Shaoming Guo, Zane Kun Li, and Brian Street, Editors,** Recent Developments in Harmonic Analysis and its Applications, 2024

791 **Mee Seong Im, Bach Nguyen, and Arik Wilbert, Editors,** Algebraic and Topological Aspects of Representation Theory, 2024

790 **Peter Crooks and Alexandru I. Suciu, Editors,** Compactifications, Configurations, and Cohomology, 2023

789 **I. S. Krasil′shchik, A. B. Sossinsky, and A. M. Verbovetsky, Editors,** The Diverse World of PDEs, 2023

788 **I. S. Krasil′shchik, A. B. Sossinsky, and A. M. Verbovetsky, Editors,** The Diverse World of PDEs, 2023

787 **Hemen Dutta, Editor,** Mathematical Modelling, 2023

786 **Hemen Dutta, Editor,** Mathematical Modelling, 2023

785 **A. Leroy and S. K. Jain, Editors,** Algebra and Coding Theory, 2023

784 **Dinh-Liem Nguyen, Loc Hoang Nguyen, and Thi-Phong Nguyen, Editors,** Advances in Inverse Problems for Partial Differential Equations, 2023

783 **Alla Detinko, Michael Kapovich, Alex Kontorovich, Peter Sarnak, and Richard Schwartz, Editors,** Computational Aspects of Discrete Subgroups of Lie Groups, 2023

782 **Galina Filipuk, Alberto Lastra, and Sławomir Michalik, Editors,** Recent Trends in Formal and Analytic Solutions of Diff. Equations, 2023

781 **Emanuel Indrei, Diego Marcon, and Levon Nurbekyan, Editors,** Geometric and Functional Inequalities and Recent Topics in Nonlinear PDEs, 2023

780 **Erik Koelink, Stefan Kolb, Nicolai Reshetikhin, and Bart Vlaar, Editors,** Hypergeometry, Integrability and Lie Theory, 2022

779 **Samuele Anni, Valentijn Karemaker, and Elisa Lorenzo García, Editors,** Arithmetic, Geometry, Cryptography, and Coding Theory 2021, 2022

778 **Carlos Galindo, Alejandro Melle Hernández, Julio José Moyano-Fernández, and Wilson A. Zúñiga-Galindo, Editors,** p-Adic Analysis, Arithmetic and Singularities, 2022

For a complete list of titles in this series, visit the
AMS Bookstore at **www.ams.org/bookstore/conmseries/**.